Cell Signaling

This book provides a comprehensive understanding of cell signaling, molecular interactions, and their implications for human health and diseases. It introduces fundamental principles underlying cell communication through signaling molecules and their diverse transmission and reception mechanisms, highlighting their role in intercellular communication through voltage- and ion-gated channels, immunological and neuron synapses, and rhinovirus–receptor interaction involved in pathogenesis and disease development. Toward the end, the book highlights the profound implications of altered cell signaling pathways in the inflammation and immune response followed by the progression of various disorders, including cancer, endocrine disorders, and neurological illnesses. It explores the diagnostic and therapeutic implications of cell signaling in targeted therapies, highlighting advanced techniques for detecting signaling molecules and innovative therapeutic approaches to inspire new developments in precision medicine. It serves as an important resource for academics, students, and professionals in the fields of cell biology and biomedical sciences.

Key Features:

- Provides in-depth understanding of cell signaling, exploring its complexities and impact on human health and disease
- Introduces fundamental principles of cell communication, emphasizing the different signaling molecules and their various transmission pathways
- Focuses on complex structures and functions of receptors, highlighting their essential role in intercellular communication and regulating cellular behavior
- Examines the molecular aspects of cell surface adhesion receptors, elucidating protein–protein interactions, signaling cascades, and enzyme–substrate interactions
- Discusses the impact of cell signaling on inflammation, cancer, and endocrine and neurological disorders

Cell Signaling

Interplay, Mechanisms, and Therapeutic Implications

Edited by
Adeeb Shehzad

CRC Press
Taylor & Francis Group
Boca Raton London New York

CRC Press is an imprint of the
Taylor & Francis Group, an **informa** business

First edition published 2025
by CRC Press
2385 NW Executive Center Drive, Suite 320, Boca Raton FL 33431

and by CRC Press
4 Park Square, Milton Park, Abingdon, Oxon, OX14 4RN

CRC Press is an imprint of Taylor & Francis Group, LLC

ISBN: 9781032707440 (hbk)
ISBN: 9781032976914 (pbk)
ISBN: 9781003594901 (ebk)

DOI: 10.1201/9781003594901

Typeset in Times
by Deanta Global Publishing Services, Chennai, India

Contents

Preface....vii
Editor....ix
Contributors....x

SECTION I Cell Signaling: Interplay

Chapter 1 Introduction to Cell Signaling....1

Ishtiaq Ali, Muhammad Junaid Ahmad Tariq, Fazli Wahid, and Muhammad Imran Khan

Chapter 2 Regulation of Cell Signaling Cascades....27

Habab Ali Ahmad, Muhammad Ibrahim Khan, Fazli Wahid, and Muhammad Imran Khan

Chapter 3 Cellular Receptors and Cell Signaling....46

Maleeha Azhar, Aroob Hasan Khan, Aroosa Younis Nadeem, Areej Fatima Ebtesam A Al-Suaimi, and Adeeb Shehzad

Chapter 4 Signal Transduction and Integral Membrane Proteins....64

Zain Saeed, Zohra Jabeen, Sehrish Kakakhel, Saira Justin, and Aneela Javed

Chapter 5 Antigen-Antibody Interaction in Cell Signaling....92

Saman Noor, Anam Fatima, Iman Fatima, and Aneela Javed

Chapter 6 Immunological and Neuron Synapses....126

Muhammad Salman Khalid, Somia Mazhar, Ebtesam A Al-Suhaimi, and Adeeb Shehzad

Chapter 7 Rhinoviruses Receptors and Their Interactions....147

Ariba Qaiser, Areej A. Sattar, and Sobia Manzoor

Chapter 8 Role of Ligand- and Voltage-Gated Ion Channels in Cell Signaling....154

Maham Mehrangaze Khonza, Hira Ali, Sadaf Mumtaz, Rumeza Hanif, and Aneela Javed

Chapter 9 Hormone and Receptor Interactions in Cell Signaling....187

Tehmina Bibi, Shahar Bano, Eun Kyoung Seo, and Salman Khan

SECTION II Cell Signaling: Mechanisms in Health and Disease

Chapter 10 Signaling Networks in Development and Differentiation 202

Aroosa Younis Nadeem, Areej Fatima, Maleeha Azhar, and Adeeb Shehzad

Chapter 11 Cell Signaling and Metabolism 218

Saadia Zahid, Sanila Amber, Laila tul Qadar, and Akasha Khurshid

Chapter 12 Inflammatory Signaling and Immune Response 231

Naseem Zehra, Zia Uddin, Muhammad Imran Amir Zada, and Muhammad Ikram

Chapter 13 Cell Signaling in Cancer 268

Khoula Sharif Mughal, Nabi Shah, Muhammad Naeem, and Muhammad Ikram

Chapter 14 Cell Signaling Pathways in Endocrine Diseases 302

Gauhar Rehman and Adil Sher

Chapter 15 Cell Signaling in Neurological Disorders 319

Muhammad Ibrar Khan, Muhammad Junaid, Sana Zafar, and Salman Khan

SECTION III Cell Signaling: Diagnostics and Therapeutic Implications

Chapter 16 Cell Signaling and Diagnosis 334

Areej Fatima, Muhammad Ammad Jamil, Maleeha Azhar, Aroosa Younis Nadeem, and Adeeb Shehzad

Chapter 17 Signal Transduction and Targeted Therapies 353

Ayesha Khan, Ayesha Riaz, Asima Saeed, and Muhammad Imran Khan

Chapter 18 Role of Cell Signaling in Drug Development 383

Taimoor Musharaf, Anoosha Ahmed, Muhammad Asim Siddiqui, Wajahat Sultan Abbasi, and Saadia Andleeb

Index 409

Preface

Cell signaling is a complex regulatory communication process where cellular molecules transmit information between the cells of multicellular organisms through the binding of ligands with receptors expressed on target cells and orchestrating a multitude of biological processes that are essential for homeostasis in humans. Cell signaling is mediated in autocrine and paracrine passion regulating cell growth and differentiation in a complex network and modulating immune responses and endocrine and neural functions. However, abnormal cell signaling contributes to many diseases, including chronic inflammation, infection, cancers, neurodegeneration, and metabolic disorders. Consequently, research into the mechanisms underlying the control of cell signaling processes in health and disease has accelerated the discovery of disease biomarkers and new drug targets. In the last few decades, remarkable advances in molecular biology techniques and genome-wide high throughput experiments have unwrapped the complexities of cell signaling pathways. These discoveries have not only enhanced our comprehension of fundamental biological processes but also paved the way for novel diagnostic and therapeutic strategies. The integration of these findings into clinical practice holds immense promise for improving patient outcomes across a spectrum of diseases, including cancer, cardiovascular disorders, and neurodegenerative conditions. Therefore, the study of cell signaling is pivotal to our understanding of both health and disease. This book, *Cell Signaling: Interplay, Mechanisms, and Therapeutic Implications*, aims to explore the multifaceted network of cell signaling, providing insights into its dynamic interplay, underlying mechanisms, and potential therapeutic applications.

This book is structured into three comprehensive sections, each focusing on a critical aspect of cell signaling.

SECTION I—CELL SIGNALING: INTERPLAY

This section is composed of nine chapters that explore the intricate interplay between various cell signaling pathways. Starting with the introduction and regulation of cell signaling, this section underpins major types of cellular receptors and integral membrane proteins involved in signal transduction that regulate cell behavior. Disease development is a complex process requiring coordinated interactions between numerous proteins, signaling pathways, and cell types. Subsequent discussion focuses on the mechanisms of antibody–antigen interactions in cell signaling through immunological and neuron synapses and rhinovirus–receptor interactions involved in pathogenesis and disease development. It highlights how cells communicate and coordinate responses through voltage- and ion-gated channels, hormones, and receptors interaction in cell signaling. By examining the synergistic and antagonistic interactions among signaling pathways, readers will gain a deeper appreciation of the complexity and precision of cellular communication. The steps in these pathways have delineated a number of unique events in cells, marked by morphological and histological changes and altered expression of genes and proteins.

SECTION II—CELL SIGNALING: MECHANISMS IN HEALTH AND DISEASE

There is a total of six chapters in this second section, which delve into the molecular mechanisms that govern cell signaling in both healthy and diseased states. This section provides a detailed analysis of key signaling pathways in development and differentiation, metabolism, inflammatory signaling, and immune responses in cancers, endocrine diseases, and neurological disorders. By elucidating the role of these pathways in normal cellular functions and their dysregulation in various diseases, we aim to bridge the gap between basic science and clinical implications. Readers will discover how aberrant signaling contributes to the pathogenesis of diseases and how targeted interventions can restore normal cellular functions.

SECTION III—CELL SIGNALING: DIAGNOSTICS AND THERAPEUTIC IMPLICATIONS

The final section is composed of three chapters that focus on the diagnostic and therapeutic implications of cell signaling in targeted therapies and drug development. It explores cutting-edge techniques and tools used to detect and quantify signaling molecules, providing valuable tools for early disease diagnosis and monitoring. Additionally, this section discusses innovative therapeutic approaches, including small molecule inhibitors, monoclonal antibodies, and gene therapy, that target specific signaling pathways. By showcasing recent advancements and ongoing research, the aim is to inspire new avenues for the development of targeted therapies and precision medicine.

Through this book, the contributors and I aspire to offer a comprehensive resource for students, researchers, and healthcare professionals interested in the fascinating world of cell signaling. By presenting a blend of foundational knowledge and cutting-edge discoveries, we hope to foster a deeper understanding of cellular communication and its implications for human health. We believe that the insights gained from this book will contribute to the ongoing efforts to translate basic research into clinical innovations, ultimately improving the quality of life for numerous individuals.

I extend my gratitude to the authors and contributors who have dedicated their expertise and time to this project. Their invaluable input has enriched the content and ensured the accuracy and relevance of the information presented. I also thank the editors and reviewers for their interest and engagement, and hope that this book will serve as a valuable companion in their journey of exploration and discovery in the realm of cell signaling.

Adeeb Shehzad

Editor

Adeeb Shehzad is Associate Professor at the Biodiversity Unit, Research Center, Dhofar University, Oman. He obtained his BPharm from the University of Peshawar, Pakistan, in 2003. He completed his MS and PhD at Kyungpook National University, South Korea (2008–2014). He has published over 100 scientific publications and book chapters in refereed journals. He worked as a production pharmacist at GlaxoSmithKline Pvt Ltd, Karachi, Pakistan, and Medicraft Pharmaceutical Pvt Ltd, Peshawar, Pakistan, from 2003 to 2007. His research interest is mainly oriented to the screening and development of plant-isolated compounds as anticancer agents, as well as investigating the modulation of different signaling pathways in different types of cancers causing drug resistance and poor prognosis. He is also interested in the development of kits for diagnosis and drug release patches with validation in animal models. He is an editorial board member for several journals and has authored two books with international publishers. He has received a number of national and international awards in academics and research.

Contributors

Abbasi, Wajahat Sultan
National University of Sciences & Technology (NUST)
Atta ur Rahman School of Applied Biosciences
Islamabad, Pakistan

Ahmed, Anoosha
National University of Sciences & Technology (NUST)
Atta ur Rahman School of Applied Biosciences
Islamabad, Pakistan

Ahmad Tariq, Muhammad Junaid
Pak Austria Fachhochschule: Institute of Applied Sciences and Technology
Department of Biomedical Sciences
Haripur, Pakistan

Ahmad, Habab Ali
Pak Austria Fachhochschule: Institute of Applied Sciences and Technology
Department of Biomedical Sciences
Haripur, Pakistan

Al-Suhaimi, Ebtesam A.
Imam Abdulrahman bin Faisal University (IAU)
Vice President for Scientific Research and Innovation
Dammam, Saudi Arabia

Ali, Hira
National University of Sciences & Technology (NUST)
Department of Biomedicine
Islamabad, Pakistan

Ali, Ishtiaq
Pak Austria Fachhochschule: Institute of Applied Sciences and Technology
Department of Biomedical Sciences
Haripur, Pakistan

Andleeb, Saadia
National University of Sciences & Technology (NUST)
Atta ur Rahman School of Applied Biosciences
Islamabad, Pakistan

Amber, Sanila
University of Mianwali
Department of Biotechnology
Mianwali, Pakistan

Azhar, Maleeha
National University of Sciences & Technology (NUST)
Department of Biomedical Sciences
Islamabad, Pakistan

Bano, Shahar
Quaid-i-Azam University
Department of Pharmacy
Islamabad, Pakistan

Bibi, Tehmina
Quaid-i-Azam University
Department of Pharmacy
Islamabad, Pakistan

Fatima, Areej
National University of Sciences & Technology (NUST)
Department of Biomedical Sciences
Islamabad, Pakistan

Fatima, Anam
National University of Sciences & Technology (NUST)
Department of Biomedicine
Islamabad, Pakistan

Fatima, Iman
National University of Sciences & Technology (NUST)
Department of Biomedicine
Islamabad, Pakistan

Hanif, Rumeza
National University of Sciences & Technology (NUST)
Department of Biomedicine
Islamabad, Pakistan

Ibrahim Khan, Muhammad
Pak Austria Fachhochschule: Institute of Applied Sciences and Technology
Department of Biomedical Sciences
Haripur, Pakistan

Ikram, Muhammad
COMSATS University Islamabad
Department of Pharmacy
Abbottabad, Pakistan

Jabeen, Zohra
National University of Sciences & Technology (NUST)
Department of Biomedicine
Islamabad, Pakistan

Javed, Aneela
National University of Sciences & Technology (NUST)
Department of Biomedicine
Islamabad, Pakistan

Justin, Saira
National University of Sciences & Technology (NUST)
Department of Biomedicine
Islamabad, Pakistan

Junaid, Muhammad
Quaid-i-Azam University
Department of Pharmacy
Islamabad, Pakistan

Kakahel, Sehrish
Department of Biochemistry
Abdul Wali Khan University Mardan
Mardan, Pakistan

Khan, Aroob Hasan
National University of Sciences & Technology (NUST)
Department of Biomedical Sciences
Islamabad, Pakistan

Khan, Muhammad Imran
Pak Austria Fachhochschule: Institute of Applied Sciences and Technology
Department of Biomedical Sciences
Haripur, Pakistan

Khan, Muhammad Ibrar
Quaid-i-Azam University
Department of Pharmacy
Islamabad, Pakistan

Khalid, Muhammad Salman
National University of Sciences & Technology (NUST)
Department of Biomedical Sciences
Islamabad, Pakistan

Khan, Ayesha
Pak Austria Fachhochschule: Institute of Applied Sciences and Technology
Department of Biomedical Sciences
Haripur, Pakistan

Khan, Salman
Quaid-i-Azam University
Department of Pharmacy
Islamabad, Pakistan

Khurshid, Akasha
National University of Sciences & Technology (NUST)
Department of Healthcare Biotechnology
Islamabad, Pakistan

Mazhar, Somia
National University of Sciences & Technology (NUST)
Department of Biomedical Sciences
Islamabad, Pakistan

Manzoor, Sobia
National University of Sciences & Technology (NUST)
Atta ur Rahman School of Applied Biosciences
Islamabad, Pakistan

Mehrangaze Khonza, Maham
National University of Sciences & Technology (NUST)
Department of Biomedicine
Islamabad, Pakistan

Mughal, Khoula Sharif
COMSATS University Islamabad
Department of Pharmacy
Abbottabad, Pakistan

Mumtaz, Sadaf
National University of Sciences & Technology (NUST)
Department of Biomedicine
Islamabad, Pakistan
and
Department of Physiology
Rawalpindi Medical University
Rawalpindi, Pakistan

Musharaf, Taimoor
National University of Sciences & Technology (NUST)
Atta ur Rahman School of Applied Biosciences
Islamabad, Pakistan

Naeem, Muhammad
COMSATS University Islamabad
Department of Pharmacy
Abbottabad, Pakistan

Nadeem, Aroosa Younis
National University of Sciences & Technology (NUST)
Department of Biomedical Sciences
Islamabad, Pakistan

Noor, Saman
National University of Sciences & Technology (NUST)
Department of Biomedicine
Islamabad, Pakistan

Qadar, Laila tul
National University of Sciences & Technology (NUST)
Department of Healthcare Biotechnology
Islamabad, Pakistan

Qaiser, Ariba
National University of Sciences & Technology (NUST)
Atta ur Rahman School of Applied Biosciences
Islamabad, Pakistan

Rehman, Gauhar
Abdul Wali khan University
Department of Zoology
Mardan, Pakistan

Riaz, Ayesha
Pak Austria Fachhochschule: Institute of Applied Sciences and Technology
Department of Biomedical Sciences
Haripur, Pakistan

Saeed, Zain
National University of Sciences & Technology (NUST)
Department of Biomedicine
Islamabad, Pakistan

Saeed, Asima
Pak Austria Fachhochschule: Institute of Applied Sciences and Technology
Department of Biomedical Sciences
Haripur, Pakistan

Sattar, Areej A.
National University of Sciences & Technology (NUST)
Atta ur Rahman School of Applied Biosciences
Islamabad, Pakistan

Seo, Eun Kyoung
Ehwa Woman University
College of Pharmacy
Seoul, South Korea

Shehzad, Adeeb
Dhofar University
Biodiversity Research Center
Salalah, Oman

Sher, Adil
Abdul Wali khan University
Department of Zoology
Mardan, Pakistan

Siddiquie, Muhammad Asim
National University of Sciences & Technology (NUST)
Atta ur Rahman School of Applied Biosciences
Islamabad, Pakistan

Shah, Nabi
COMSATS University Islamabad
Department of Pharmacy
Abbottabad, Pakistan

Uddin, Zia
COMSATS University Islamabad
Department of Pharmacy
Abbottabad, Pakistan

Wahid, Fazli
Pak Austria Fachhochschule: Institute of Applied Sciences and Technology
Department of Biomedical Sciences
Haripur, Pakistan

Zada, Muhammad Imran Amir
COMSATS University Islamabad
Department of Pharmacy
Abbottabad, Pakistan

Zahid, Saadia
National University of Sciences & Technology (NUST)
Department of Healthcare Biotechnology
Islamabad, Pakistan

Zehra, Naseem
COMSATS University Islamabad
Department of Pharmacy
Abbottabad, Pakistan

1 Introduction to Cell Signaling

Ishtiaq Ali, Muhammad Junaid Ahmad Tariq, Fazli Wahid, and Muhammad Imran Khan
Department of Biomedical Sciences, Pak Austria Fachhochschule: Institute of Applied Sciences and Technology, Mang, Haripur, Khyber Pakhtunkhwa, Pakistan
*Department of Biomedical Sciences, Pak Austria Fachhochschule: Institute of Applied Sciences and Technology, Mang, Haripur, Khyber Pakhtunkhwa, Pakistan

1.1 OVERVIEW OF CELL SIGNALING

1.1.1 Defining Cell Signaling

Cell signaling is the complex communication system that leads to cellular activity. It encompasses all the processes through which cells receive and respond to signals from their environment, as well as the mechanisms by which they communicate with one another (1). This constant exchange of information is fundamental to the proper functioning of cells, tissues, and, ultimately, entire organisms (2).

Cell signaling involves the production, release, and reception of signaling molecules known as ligands or signaling factors. These molecules may originate from various sources, including other cells within the same organism (endocrine or paracrine signaling), the extracellular environment, or even the cell itself (autocrine signaling). When these signaling molecules bind to specific receptor proteins on the cell surface or within the cell, they initiate a cascade of biochemical reactions that alter the cell's behavior, gene expression, or physiological state (Figure 1.1).

1.1.2 Historical Perspective

The study of cell signaling has a rich and fascinating history dating back to the early 20th century. One of the pioneering discoveries in this field was made by Earl W. Sutherland in the 1950s, who explained the role of cyclic adenosine monophosphate (cAMP) as a second messenger in hormone-mediated signaling pathways. This groundbreaking work laid the foundation for our understanding of signal transduction mechanisms and earned Sutherland the Nobel Prize in Physiology or Medicine in 1971.

The latter half of the 20th century saw numerous other significant breakthroughs in cell signaling research (3), including the identification of various receptors, signaling molecules, and signaling pathways. Modern molecular biology techniques, such as gene cloning, sequencing, and protein analysis, greatly accelerated the pace of discovery in this field.

Today, cell signaling remains an active and rapidly evolving area of research, with implications spanning diverse fields, including cell biology, developmental biology, immunology, neuroscience, and cancer research, among others (4).

DOI: 10.1201/9781003594901-1

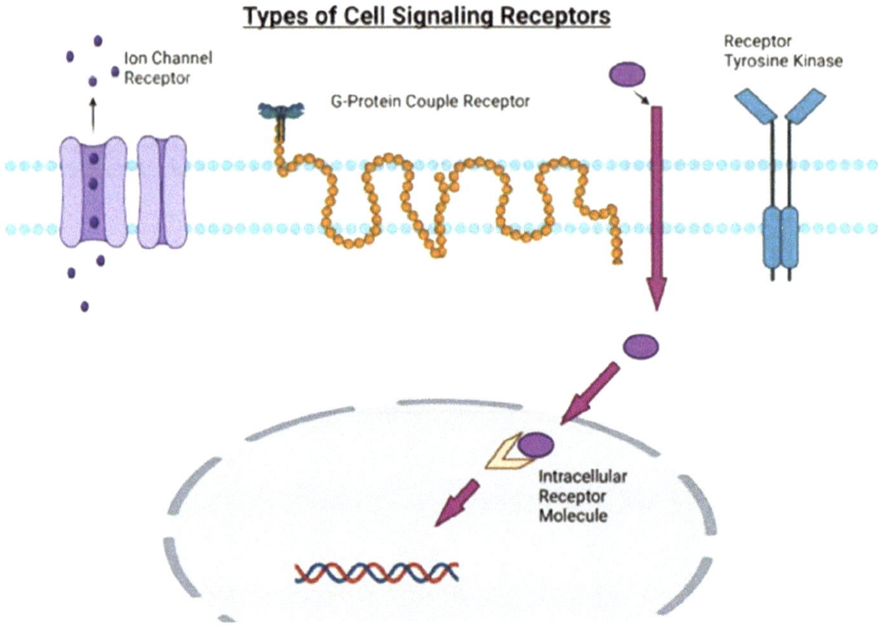

FIGURE 1.1 A generalized cell signaling mechanism.

1.1.3 Importance in Biology and Medicine

Cell signaling is of paramount importance in virtually all aspects of biology and medicine. It orchestrates the complex processes that govern cellular growth, differentiation, metabolism, and responses to environmental stimuli. Disruptions or dysregulations in cell signaling pathways can lead to a wide range of diseases, including cancer, autoimmune disorders, neurological disorders, and metabolic diseases (5).

An in-depth understanding of cell signaling mechanisms is vital for understanding the underlying causes of many diseases and developing effective therapeutic interventions. For example, many modern cancer therapies target specific signaling pathways that are dysregulated in tumor cells, aiming to selectively inhibit or modulate these pathways to achieve therapeutic effects (6). Also, cell signaling research has significant implications for regenerative medicine, stem cell biology, and tissue engineering, as these fields rely heavily on our ability to control and manipulate the signaling processes that govern cellular differentiation and tissue development (7).

Therefore, cell signaling represents a fundamental aspect of cellular function and communication, with far-reaching implications across biology and medicine. Ongoing research in this area promises to advance our understanding of physiological processes, disease mechanisms, and the development of novel therapeutic strategies.

1.2 HISTORY AND EVOLUTION IN CELL SIGNALING

1.2.1 Evolutionary Perspectives

Cell signaling pathways have evolved over billions of years, reflecting the complex adaptations of organisms to diverse environments. Studying the evolutionary origins and conservation of these pathways provides valuable insights into their fundamental importance and the selective pressures that have shaped them.

One of the earliest forms of cell signaling can be traced back to prokaryotic organisms, such as bacteria, which utilize simple signaling mechanisms to coordinate their behaviors, including quorum sensing and chemotaxis (8). These primitive signaling systems likely laid the foundation for the more complex signaling networks that evolved in multicellular organisms.

The emergence of multicellular life forms necessitated coordinated communication between cells. This evolutionary pressure drove the development of specialized signaling pathways that facilitated cell-to-cell interactions, tissue organization, and regulation of various physiological processes.

Many vital signaling pathways, such as those involving G protein-coupled receptors (GPCRs), receptor tyrosine kinases (RTKs), and intracellular signaling cascades, are highly conserved across diverse species, suggesting their ancient origins and fundamental importance in cellular function (9).

1.2.2 Landmarks in Cell Signaling Discovery

Several landmark discoveries have shaped our understanding of cellular communication. Some of the most significant milestones include:

The discovery of hormones and their role in signaling (late 19th century) (10)

The identification of second messengers, such as cyclic AMP (cAMP) by Earl W. Sutherland (1950s)

The elucidation of GPCR signaling pathways (1970s–1980s)

The discovery of RTKs and their role in signal transduction (1980s)

The characterization of intracellular signaling cascades, such as the mitogen-activated protein kinase (MAPK) pathways (1990s)

The development of advanced techniques, including gene cloning, protein analysis, and imaging technologies, which revolutionized the study of cell signaling (late 20th century) (11)

These discoveries, along with countless others, have continually reshaped our understanding of cell signaling and paved the way for new avenues of research and therapeutic development.

1.2.3 Comparative Signaling in Different Organisms

While cell signaling pathways share many common features across diverse organisms, notable differences exist in the specific mechanisms and components involved. Comparative studies of signaling pathways in various species have provided valuable insights into their evolutionary relationships, adaptations, and functional specializations (12).

For example, plants have evolved unique signaling pathways to respond to environmental cues such as light, temperature, and hormones that are essential for their growth, development, and survival. These signaling pathways often involve specialized receptors and signaling molecules that differ from those in animal systems (13). Conversely, signaling pathways in mammals and other vertebrates have evolved to regulate complex physiological processes, including embryonic development, immune responses, and neuronal functions. These signaling pathways often involve complex interaction and integration with other signaling networks, reflecting the increasing complexity of these organisms.

Comparative studies have also revealed surprising similarities in signaling mechanisms across distantly related species. The highly conserved Wnt signaling pathway, for example, plays vital roles in embryonic development and stem cell biology and is found in organisms ranging from sponges to humans, underscoring its ancient origins and fundamental importance (14).

1.3 FUNDAMENTALS OF CELL COMMUNICATION

1.3.1 Types of Cell Signaling

Cell signaling can be categorized based on the origin of the signaling molecule, the distance it travels, and the target cells it affects. Figure 1.2 illustrates the major types of cell signaling, which include (see also Table 1.1) (15):

1. **Autocrine signaling:** A cell secretes signaling molecules that bind to its own, creating a self-regulating feedback loop.
2. **Paracrine signaling:** Signaling molecules are released by a cell and diffuse locally to affect the behavior or function of nearby cells within the same tissue or organ.
3. **Endocrine signaling:** In this long-range signaling mechanism, signaling molecules (hormones) are secreted into the bloodstream by specialized endocrine glands and travel to target cells throughout the body.
4. **Juxtacrine signaling:** This form of signaling involves direct cell-to-cell contact where signaling molecules on the surface of one cell interact with receptors on an adjacent cell.

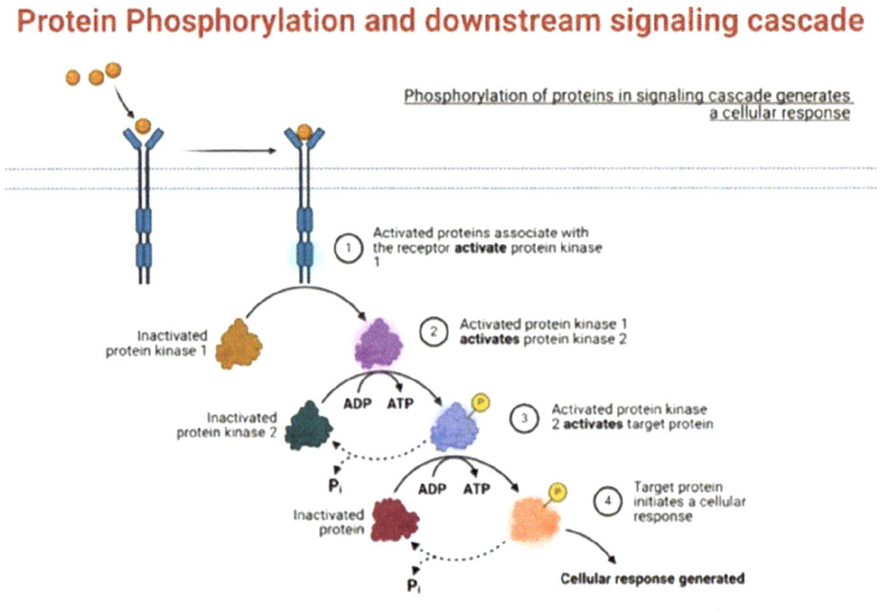

FIGURE 1.2 Types of cell signaling that occur between different cells of the body.

TABLE 1.1
Summary of the Different Types of Cell Signaling

Type of Cell Signaling	Signaling Molecule Origin	Target Cell(s)	Examples
Autocrine	Same cell	Same cell	Growth factors, cytokines
Paracrine	Nearby cells	Neighboring cells	Neurotransmitters, cytokines
Endocrine	Endocrine glands	Distant cells	Hormones (e.g., insulin, growth hormone)
Juxtacrine	Cell surface	Adjacent cell	Notch/Delta signaling
Synaptic	Presynaptic neuron	Postsynaptic neuron	Neurotransmitters (e.g., glutamate, GABA)

5. **Synaptic signaling:** Occurring primarily in the nervous system, this type of signaling involves the release of neurotransmitters from a presynaptic neuron, which diffuse across the synaptic cleft and bind to receptors on the postsynaptic neuron.

1.3.2 Components of Signaling Pathways

Cell signaling pathways consist of several essential components that work together to transmit and process signals within and between cells (16). These components include:

a. **Signaling molecules (Ligands):** These are the molecules that initiate the signaling process by binding to specific receptors. Examples include hormones, growth factors, neurotransmitters, and cytokines.
b. **Receptors:** Proteins located on the cell surface or within the cell that recognize and bind to specific signaling molecules. They are the primary targets for signaling molecules and initiate the downstream signaling cascade.
c. **Signal transducers:** Proteins or molecules that relay and amplify the signal received by the receptor, often through a series of biochemical reactions or conformational changes.
d. **Second messengers:** Many signaling pathways involve the production of small, diffusible molecules known as second messengers, which act as intracellular signals and propagate the signal within the cell.
e. **Effector proteins:** The proteins that execute the cellular response to the signal, such as enzymes, transcription factors, or ion channels.
f. **Feedback mechanisms:** Signaling pathways often involve negative or positive feedback loops that regulate the intensity and duration of the signal, ensuring proper cellular responses (Figure 1.3; Table 1.2) (17).

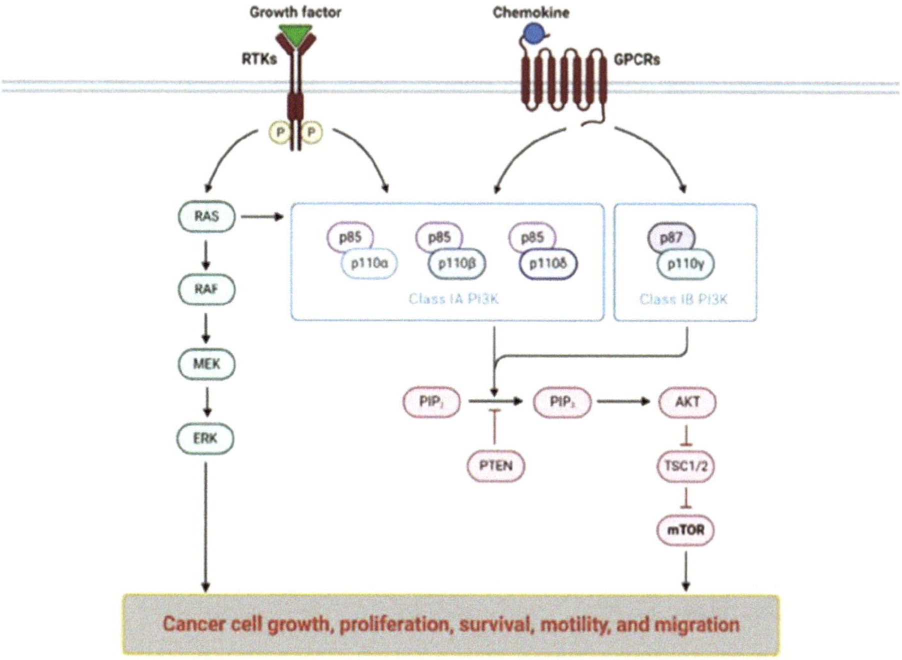

FIGURE 1.3 Components involved in cell signaling.

TABLE 1.2
Key Components of Cell Signaling Pathways

Component	Description	Examples
Signaling Molecules (Ligands)	Initiate the signaling process	Hormones, growth factors, neurotransmitters
Receptors	Recognize and bind to signaling molecules	G protein-coupled receptors, receptor tyrosine kinases
Signal Transducers	Relay and amplify the signal	G proteins, kinases, adaptor proteins
Second Messengers	Intracellular signals that propagate the signal	cAMP, Ca^{2+}, IP_3
Effector Proteins	Carry out the cellular response	Enzymes, transcription factors, ion channels
Feedback Mechanisms	Regulate the signal intensity and duration	Negative and positive feedback loops

1.3.3 Signal Reception and Transduction

Signal reception and transduction are critical processes in cell signaling, wherein the signaling molecule is recognized and the signal is propagated within the cell. This process typically involves the following steps:

1. **Signal recognition:** The signaling molecule (ligand) binds to a specific receptor on the cell surface or within the cell. This binding event triggers a conformational change in the receptor, activating it.
2. **Signal initiation:** The activated receptor initiates a series of biochemical reactions or interactions with other proteins, known as signal transducers. These signal transducers relay and amplify the signal within the cell.
3. **Signal transduction:** The signal propagates through a cascade or pathway, often involving the activation or inhibition of various enzymes, the production of second messengers, and the modulation of effector proteins.
4. **Cellular response:** The activated effector proteins, such as transcription factors, enzymes, or ion channels, carry out the cellular response to the signal. This can include changes in gene expression, metabolic processes, cell growth, differentiation, or other physiological responses.
5. **Signal termination:** The signaling pathway is terminated or attenuated through various mechanisms, such as the degradation or internalization of the signaling molecule or receptor, the inactivation of signal transducers, or the action of negative feedback loops.

Throughout the signal transduction process, regulatory mechanisms such as feedback loops, cross-talk between pathways, and signal integration ensure precise control over the cellular response (18). These mechanisms allow cells to respond appropriately to diverse environmental cues and maintain homeostasis.

1.4 TYPES OF CELL SIGNALING

1.4.1 Autocrine Signaling

Autocrine signaling is a type of cell signaling in which a cell secretes signaling molecules that bind to its own receptors, creating a self-regulating feedback loop (19). This signaling type is crucial in various cellular processes, including growth, differentiation, and survival.

1.4.1.1 Key Features of Autocrine Signaling

- The same cell that expresses the corresponding receptor produces and releases the signaling molecule (ligand).
- The signaling molecule binds to receptors on the surface of the same cell, initiating a signaling cascade within the cell.
- Autocrine signaling can regulate cell behavior autonomously, independent of external signals.
- It is often involved in the regulation of cell proliferation, apoptosis, and other cellular processes.

1.4.1.2 Examples of Autocrine Signaling

1. Growth factors (e.g., epidermal growth factor (EGF), platelet-derived growth factor (PDGF)) stimulating cell proliferation and survival (20)
2. Cytokines (e.g., interleukins) regulating immune cell function and inflammation
3. Neurotrophins (e.g., nerve growth factor (NGF)) promoting neuronal survival and differentiation

1.4.2 Paracrine Signaling

Paracrine signaling occurs when a cell releases a signaling molecule that diffuses locally to nearby cells within the same tissue or organ, affecting their behavior or function (21).

1.4.2.1 Key Features of Paracrine Signaling

- The signaling molecule (ligand) is released by a cell and diffuses over a short distance to target cells in the local environment.
- The target cells express receptors specific to the signaling molecule, allowing them to respond to the signal.
- Paracrine signaling plays a vital role in coordinating the activities of different cell types within a tissue or organ.
- It is often involved in regulating processes such as tissue development, repair, and homeostasis.

1.4.2.2 Examples of Paracrine Signaling

1. Neurotransmitters (e.g., acetylcholine, dopamine) released by neurons to act on nearby cells
2. Growth factors (e.g., vascular endothelial growth factor (VEGF)) regulating angiogenesis and vascular development (22)
3. Cytokines (e.g., interleukins, tumor necrosis factor (TNF)) mediating local inflammatory responses (Table 1.3)

TABLE 1.3
Key Differences between Autocrine and Paracrine Signaling

Characteristic	Autocrine Signaling	Paracrine Signaling
Signal Origin	Same cell	Nearby cell
Target Cell	Same cell	Neighboring cells
Distance	Local	Short-range
Function	Self-regulation	Coordination of tissue/organ activities

1.4.3 Endocrine Signaling

Endocrine signaling is a long-range form of cell signaling in which specialized endocrine glands secrete signaling molecules (hormones) into the bloodstream, affecting target cells throughout the body (23).

1.4.3.1 Key Features of Endocrine Signaling

- Signaling molecules (hormones) are produced and released by specialized endocrine glands or tissues.
- Hormones are transported through the circulatory system (bloodstream or lymphatic system) to reach distant target cells.
- Target cells express receptors specific to the hormones, allowing them to respond to the signal.
- Endocrine signaling regulates various physiological processes, including growth, metabolism, reproduction, and stress responses.

1.4.3.2 Examples of Endocrine Signaling

1. Insulin secreted by the pancreas regulating glucose metabolism
2. Growth hormones released by the pituitary gland promoting growth and development
3. Thyroid hormones (e.g., thyroxine) regulating metabolic rate and development
4. Adrenal hormones (e.g., cortisol) involved in stress responses and immune function (24)

1.4.4 Direct Cell-to-Cell Signaling

Direct cell-to-cell signaling, also known as juxtracrine signaling, involves the direct interaction between signaling molecules on the surface of one cell and receptors on an adjacent cell without the diffusion of soluble signaling molecules (25).

1.4.4.1 Key Features of Direct Cell-to-Cell Signaling

- Signaling molecules (ligands) are membrane-bound proteins or lipids on the surface of one cell.
- The receptor on the adjacent cell directly binds to the membrane-bound signaling molecule, initiating a signaling cascade.
- This form of signaling requires close physical contact between the cells.
- It plays a vital role in processes such as embryonic development, tissue patterning, and immune cell interactions.

1.4.4.2 Examples of Direct Cell-to-Cell Signaling

1. Notch/Delta signaling pathway involved in cell fate determination and tissue patterning during development
2. Ephrin/Eph signaling, regulating cell migration, axon guidance, and vascular development
3. Immune cell interactions (e.g., T-cell activation by antigen-presenting cells) mediated by direct cell-to-cell contact (Table 1.4)

These various types of cell signaling work together to orchestrate complex cellular processes and maintain homeostasis within tissues, organs, and the entire organism.

1.5 SIGNAL TRANSDUCTION PATHWAYS

1.5.1 G Protein-Coupled Receptors (GPCRs)

GPCRs are a large family of receptors that play a vital role in signal transduction across cell membranes (26). They are involved in physiological processes such as sensory perception, neurotransmission, and hormonal regulation (Figure 1.4).

TABLE 1.4
Key Characteristics of Cell Signaling Types

Type of Signaling	Signal Origin	Signal Molecule	Target Cell	Distance	Examples
Autocrine	Same cell	Growth factors, cytokines	Same cell	Local	EGF, PDGF, interleukins
Paracrine	Nearby cell	Neurotransmitters, growth factors, cytokines	Neighboring cells	Short-range	Acetylcholine, VEGF, interleukins
Endocrine	Endocrine glands	Hormones	Distant cells	Long-range	Insulin, growth hormone, cortisol
Direct Cell-to-Cell	Cell surface	Membrane-bound proteins/lipids	Adjacent cell	Close contact	Notch/Delta, Ephrin/Eph

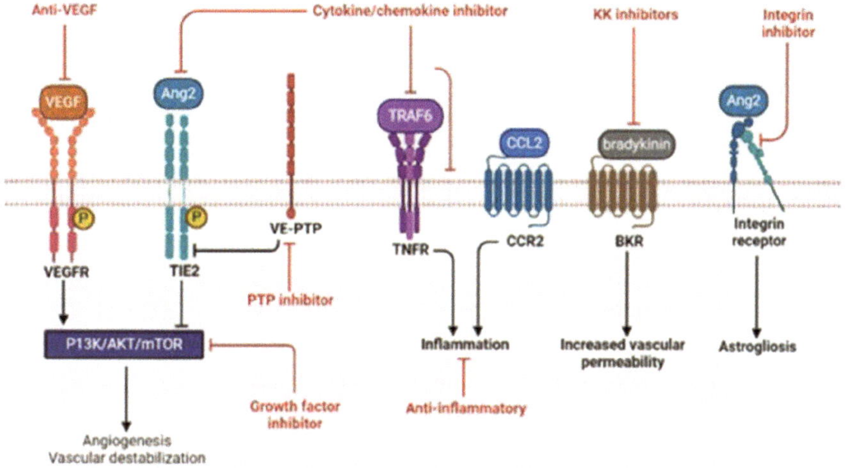

FIGURE 1.4 Types of cell signaling receptors.

1.5.1.1 Characteristics of GPCRs

- Consist of a single polypeptide chain with seven transmembrane domains
- Coupled to heterotrimeric G proteins (α, β, and γ subunits)
- Ligands include hormones, neurotransmitters, chemokines, and sensory signals
- Ligand binding activates G proteins through conformational changes to GPCRs

1.5.1.2 Signaling Mechanisms

- Ligand binding causes the dissociation of the G protein α subunit from the β γ complex (27).
- The activated α subunit can modulate the activity of various effector proteins (e.g., adenylyl cyclase, phospholipase C).
- These effectors generate second messengers (e.g., cAMP, IP3, Ca2+), which trigger downstream signaling cascades.
- The β γ complex can also activate specific signaling pathways.

1.5.1.3 Examples of GPCR Signaling

1. β-adrenergic receptors in the cardiovascular system, activated by epinephrine and norepinephrine (28)
2. Rhodopsin in the visual system, activated by photons of light
3. Chemokine receptors in the immune system, involved in leukocyte migration

1.5.2 Receptor Tyrosine Kinases (RTKs)

RTKs are a class of transmembrane receptors that play a vital role in cellular signaling pathways regulating cell growth, differentiation, and survival (29).

1.5.2.1 Characteristics of RTKs

- RTKs consist of an extracellular ligand-binding domain, a single transmembrane domain, and an intracellular tyrosine kinase domain.
- Ligands include growth factors, hormones, and cytokines.
- Upon ligand binding, RTKs undergo dimerization and autophosphorylation of specific tyrosine residues.

1.5.2.2 Signaling Mechanisms

1. Ligand binding induces RTK dimerization and autophosphorylation (30).
2. Phosphorylated tyrosine residues serve as docking sites for adaptor proteins and enzymes.
3. Recruitment and activation of signaling molecules (e.g., Ras, PI3K, PLCγ) initiate downstream signaling cascades.
4. Major signaling pathways activated include MAPK, PI3K/Akt, and PLCγ.

These pathways regulate gene expression, metabolism, cell cycle progression, and survival.

1.5.2.3 Examples of RTK Signaling

1. Epidermal growth factor receptor (EGFR), involved in cell proliferation and survival
2. Insulin receptor, regulating glucose metabolism
3. Vascular endothelial growth factor receptor (VEGFR), important for angiogenesis (31)

1.5.3 Ion Channel Receptors

Ion channel receptors regulate the flow of ions (e.g., Na+, K+, Ca2+) across cell membranes, leading to changes in membrane potential and cellular excitability (32).

1.5.3.1 Characteristics of Ion Channel Receptors

- Ion channel receptors consist of transmembrane protein complexes that form ion-selective pores.
- Ligand binding causes conformational changes, opening or closing the channel.
- Can be either ionotropic (ion-gated) or metabotropic (G protein-coupled).

1.5.3.2 Signaling Mechanisms

1. Ligand binding induces conformational changes in the receptor, allowing the flow of specific ions (33).
2. Ion flux alters the membrane potential and cellular excitability.
3. Downstream signaling pathways can be activated by changes in intracellular ion concentrations (e.g., Ca2+ signaling).

1.5.3.3 Examples of Ion Channel Receptor Signaling

1. Nicotinic acetylcholine receptors in the neuromuscular junction, mediating muscle contraction
2. Glutamate receptors (AMPA, NMDA) in the central nervous system, involved in synaptic transmission
3. GABA receptors in the brain, mediating inhibitory neurotransmission

1.5.4 Intracellular Receptors

Intracellular receptors, also known as nuclear receptors, reside within the cytoplasm or nucleus and regulate gene transcription in response to specific ligands (34).

1.5.4.1 Characteristics of Intracellular Receptors

- Soluble proteins that can translocate to the nucleus upon ligand binding.
- Ligands are typically hydrophobic molecules (e.g., steroid hormones, thyroid hormones, retinoids).
- Act as transcription factors, regulating the expression of target genes.

1.5.4.2 Signaling Mechanisms

1. In the absence of a ligand, intracellular receptors are often bound to chaperone proteins or corepressors (35).
2. Ligand binding causes a conformational change, leading to the dissociation of corepressors and binding of coactivators.
1. The ligand-receptor complex translocates to the nucleus and binds to specific DNA sequences (response elements).

Recruitment of coactivators and the transcriptional machinery regulates gene expression.

1.5.4.3 Examples of Intracellular Receptor Signaling

1. Glucocorticoid receptor, involved in stress response and immune regulation
2. Estrogen receptor, important in female reproductive function and bone metabolism
3. Thyroid hormone receptor, regulating metabolic processes and development

These major signal transduction pathways play vital roles in various physiological processes and are often targeted in the development of therapeutic interventions for various diseases.

1.6 MOLECULAR MECHANISMS OF SIGNAL TRANSDUCTION

1.6.1 Second Messengers

Second messengers are small, diffusible molecules or ions that relay and amplify the signal initiated by the binding of a signaling molecule (ligand) to its receptor (36). They play a vital role in propagating signals within the cell and activating downstream signaling pathways (37).

1.6.1.1 Common Second Messengers Include

1. **Cyclic adenosine monophosphate (cAMP):** Generated by the enzyme adenylyl cyclase, cAMP activates protein kinase A (PKA) and regulates various cellular processes such as metabolism, gene expression, and ion channel activity (38).

2. **Inositol trisphosphate (IP3) and diacylglycerol (DAG):** Produced by the cleavage of phosphatidylinositol 4,5-bisphosphate (PIP2) by phospholipase C (PLC), IP3 triggers the release of calcium from intracellular stores, while DAG activates protein kinase C (PKC).
3. **Calcium ions (Ca2+):** Changes in intracellular calcium levels serve as a versatile second messenger, regulating various processes such as muscle contraction, neurotransmitter release, and gene expression (39).
4. **Nitric oxide (NO):** A gaseous second messenger involved in vasodilation, immune response, and neurotransmission.

Second messengers can amplify the initial signal by activating multiple downstream effector proteins, allowing for signal amplification and the coordination of complex cellular responses.

1.6.2 Protein Phosphorylation

Protein phosphorylation is a key molecular mechanism in signal transduction, where kinases (enzymes that add phosphate groups) and phosphatases (enzymes that remove phosphate groups) regulate the activity, localization, and interactions of target proteins (40).

- Kinases catalyze the transfer of phosphate groups from ATP to specific amino acid residues (e.g., serine, threonine, or tyrosine) on target proteins.
- Phosphorylation can alter the conformation, activity, and binding properties of the target protein, thereby modulating its function.
- It plays a vital role in various cellular processes, including signal transduction cascades, gene expression, cell cycle regulation, and metabolism.
- Kinases and phosphatases are often organized into complex signaling networks, allowing for precise regulation and integration of multiple signaling pathways.
- Dysregulation of protein phosphorylation is implicated in various diseases, such as cancer, diabetes, and neurological disorders, making kinases and phosphatases attractive therapeutic targets (Figure 1.5).

1.6.3 Signaling Cascades and Networks

Signal transduction often occurs through cascades or networks of interacting molecules, allowing for signal amplification, integration, and the coordination of complex cellular responses (41).

- Signaling cascades involve sequential activation of multiple components, typically involving kinases that phosphorylate and activate downstream components.
- Examples include the MAPK cascades (e.g., ERK, JNK, and p38 pathways), which regulate various cellular processes such as proliferation, differentiation, and stress responses (42).
- Signaling networks feature extensive crosstalk and interconnections between different pathways, allowing for signal integration and modulation.
- Positive and negative feedback loops are common features of signaling networks, providing mechanisms for signal amplification, attenuation, and oscillations.
- Scaffolding proteins and signaling complexes facilitate the spatial organization and efficient transmission of signals by bringing together multiple components of signaling pathways.
- Computational modeling and systems biology approaches are increasingly used to study the complex dynamics and emergent properties of signaling networks.

The complex interplay between second messengers, protein phosphorylation, and signaling cascades and networks enables cells to precisely coordinate and integrate various signals, ultimately governing diverse cellular processes and maintaining homeostasis.

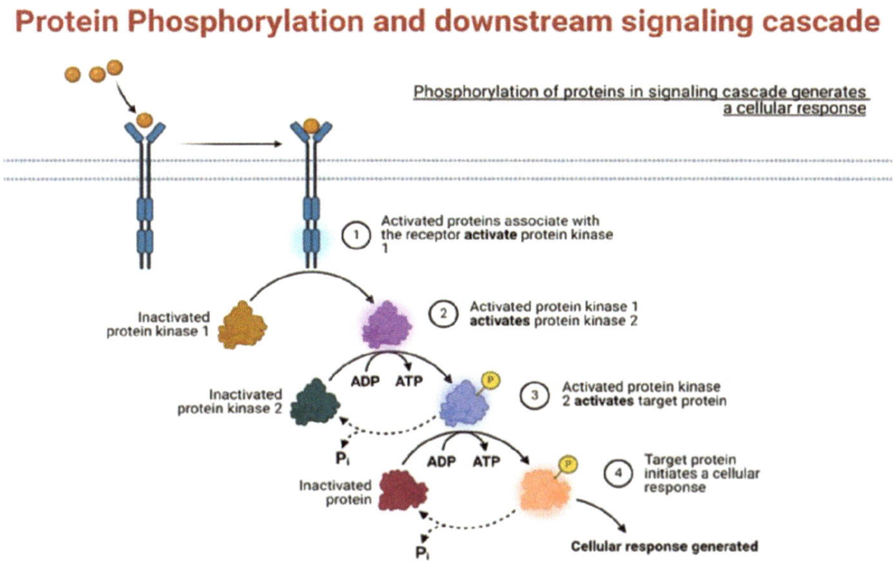

FIGURE 1.5 Initiation of downstream signaling due to the phosphorylation and hence activation of proteins involved in the cell signaling cascade.

1.7 REGULATION OF CELL SIGNALING

1.7.1 Feedback Mechanisms

Feedback mechanisms play a vital role in regulating the intensity, duration, and specificity of cell signaling pathways (43). They ensure that cellular responses are appropriately controlled and maintained within physiological limits.

1.7.1.1 Negative Feedback Loops

Negative feedback loops attenuate or terminate signaling pathways to prevent excessive or prolonged cellular responses (44).

Mechanisms include:

a. Induction of inhibitory proteins or phosphatases that counteract the signaling cascade
b. Receptor desensitization or internalization, reducing the cell's responsiveness to the signal
c. Degradation or inactivation of signaling molecules

Example: The MAPK pathway is negatively regulated by MAPK phosphatases and inhibitor proteins like DUSP and Sprouty.

1.7.1.2 Positive Feedback Loops

Positive feedback loops amplify and reinforce signaling pathways, enabling sustained or switch-like responses (45).

Mechanisms include:

a. Activation of signaling components that further stimulate the pathway (46)
b. Induction of gene expression to produce more signaling molecules
c. Cooperative binding of signaling proteins

Example: Wnt signaling is positively regulated through the inhibition of the negative regulator GSK-3β.

Feedback mechanisms are vital for maintaining homeostasis, preventing excessive or uncontrolled cellular responses, and ensuring the proper timing and coordination of signaling events.

1.7.2 Crosstalk between Signaling Pathways

Cell signaling pathways do not operate in isolation; instead, they exhibit extensive crosstalk and interconnections (47), allowing for signal integration, modulation, and the generation of complex cellular responses.

1.7.2.1 Crosstalk Can Occur at Various Levels

1. Shared signaling components (e.g., kinases, adaptor proteins)
2. Regulation of signaling components by other pathways
3. Convergence of multiple pathways on common effector molecules or transcription factors

1.7.2.2 Mechanisms of Crosstalk

a. Phosphorylation or dephosphorylation of signaling components
b. Protein-protein interactions and formation of signaling complexes
c. Regulation of gene expression and protein synthesis

Crosstalk enables cells to integrate multiple environmental and developmental cues, allowing for context-specific responses. It also provides robustness and redundancy in signaling networks, ensuring that critical cellular processes are maintained even if one pathway is disrupted (41).

Example: The MAPK and PI3K/Akt pathways exhibit extensive crosstalk, with components from one pathway regulating the other, leading to coordinated regulation of cell growth, survival, and metabolism (48).

1.7.3 Dysregulation and Disease

Dysregulation of cell signaling pathways can have severe consequences and is implicated in various diseases, including cancer (49), autoimmune disorders, metabolic disorders, and neurological diseases.

1.7.3.1 Mechanisms of Dysregulation

a. Mutations in signaling components (e.g., receptors, kinases, transcription factors)
b. Aberrant expression or activity of signaling molecules (50)
c. Disruption of feedback mechanisms or crosstalk
d. Environmental factors that perturb signaling pathways

1.7.3.2 Consequences of Dysregulation

1. Uncontrolled cell proliferation (cancer)
2. Inappropriate immune responses (autoimmune diseases)
3. Metabolic imbalances (diabetes, obesity)
4. Neuronal dysfunction (neurodegenerative diseases) (Figure)

1.7.3.3 Therapeutic Strategies

- Targeted therapies that modulate specific signaling components (e.g., kinase inhibitors, monoclonal antibodies) (51)
- Restoration of feedback mechanisms or crosstalk
- Manipulation of environmental factors that influence signaling pathways (52)

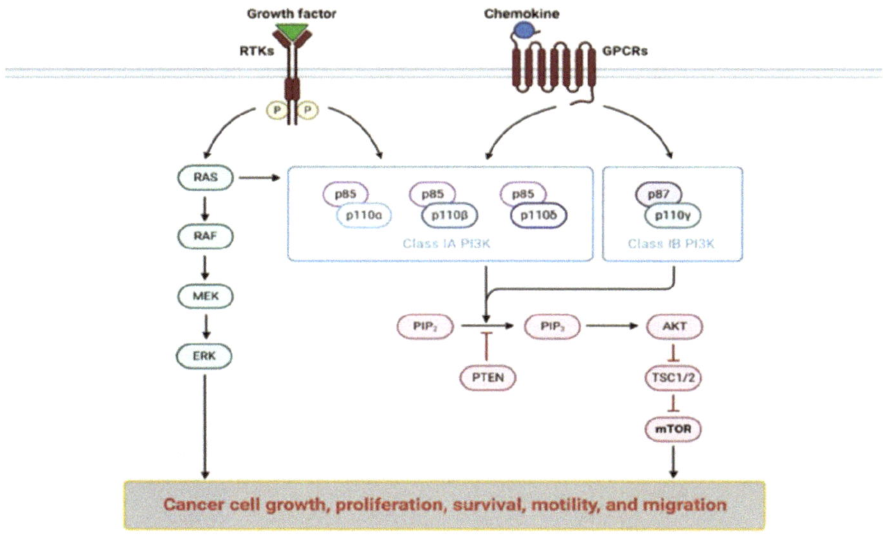

FIGURE 1.6 Disease onset due to dysregulated cellular signaling pathways.

1.8 CELL-TO-CELL AND EXTRACELLULAR MATRIX SIGNALING

1.8.1 Cell-to-cell Signaling Dynamics

Cell-to-cell signaling plays a vital role in coordinating the behavior and activities of cells within tissues and organs (53). This form of signaling involves direct communication between adjacent cells through various mechanisms, including:

1. **Gap junctions:** These are specialized intercellular channels that directly connect the cytoplasm of neighboring cells, allowing for the exchange of small molecules, ions, and second messengers (54). Gap junctions are vital for the coordinated electrical and metabolic coupling of cells in tissues such as the heart, brain, and liver.
2. **Adherens junctions and tight junctions:** These cell-to-cell junctions not only provide structural adhesion between cells but also participate in signaling pathways that regulate cell proliferation, differentiation, and tissue organization (55). Adherens junctions are composed of cadherin proteins and associated catenins, which can modulate signaling cascades like the Wnt and Hippo pathways.
3. **Notch signaling:** The Notch signaling pathway is a highly conserved juxtacrine signaling mechanism that plays a critical role in cell fate determination, tissue patterning, and stem cell maintenance (56). It involves direct cell-to-cell interactions between the Notch receptor on one cell and its ligands (e.g., Delta, Jagged) on adjacent cells (57).
4. **Synaptic signaling:** In the nervous system, synaptic signaling mediates communication between neurons and their target cells (58). Neurotransmitters released from the presynaptic neuron diffuse across the synaptic cleft and bind to receptors on the postsynaptic cell, triggering electrical or biochemical signals.

The dynamics of cell-to-cell signaling are tightly regulated, involving feedback loops, gradients of signaling molecules, and modulation of receptor expression and localization. Disruptions in these signaling mechanisms can lead to developmental abnormalities, tissue dysfunction, and various diseases, such as cancer and neurological disorders.

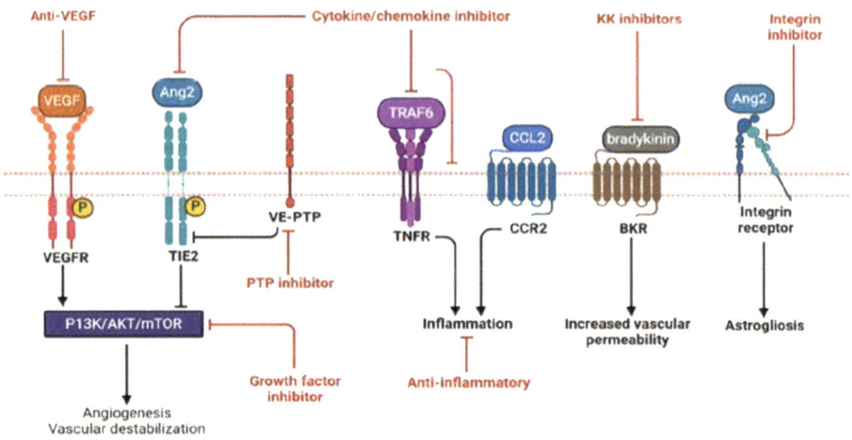

FIGURE 1.7 Disease onset due to dysregulated cellular signaling pathways.

1.8.2 Signaling through the Extracellular Matrix

The extracellular matrix (ECM) is a complex network of macromolecules, including collagen, laminin, and proteoglycans, surrounding and supporting cells within tissues (59). Beyond its structural role, the ECM plays a vital role in cell signaling by interacting with cell surface receptors and modulating various signaling pathways.

1. **Integrin signaling:** Integrins are transmembrane receptors that bind to specific ECM components, such as fibronectin and laminin (60). Upon binding, integrins activate signaling cascades that regulate cell survival, proliferation, migration, and differentiation. Integrin signaling is critical for processes like wound healing, angiogenesis, and tissue homeostasis.
2. **Growth factor sequestration and release:** The ECM can bind and sequester various growth factors, cytokines, and morphogens, creating localized reservoirs of these signaling molecules. The controlled release of these factors from the ECM can modulate cellular responses, including proliferation, migration, and differentiation.
3. **Mechanical signaling:** The ECM can transmit mechanical forces to cells, which can influence signaling pathways and cellular behavior (61). For example, the stiffness of the ECM can modulate integrin signaling, affecting processes like stem cell differentiation and cancer cell invasion.
4. **Matrix remodeling:** Cells can secrete enzymes like matrix metalloproteinases (MMPs) that degrade and remodel the ECM. This remodeling can release sequestered growth factors and generate new signaling molecules (matrikines) that can bind to cell surface receptors and initiate signaling cascades.

The ECM acts as a dynamic signaling hub, integrating biochemical and biomechanical cues to regulate cellular processes. Dysregulation of ECM signaling has been implicated in various pathological conditions, including fibrosis, cancer metastasis, and tissue degeneration.

1.8.3 Integrins and Cell-Matrix Interactions

Integrins are a family of transmembrane receptors that mediate cell-matrix interactions and play a central role in signaling pathways triggered by the ECM (62). Integrin signaling is bidirectional, with signals propagating both from the ECM to the cell (outside-in signaling) and from the cell to the ECM (inside-out signaling).

1. **Outside-in signaling:** Integrins binding to ECM ligands triggers undergo conformational changes, exposing binding sites for cytoplasmic adaptor proteins and kinases. This initiates a cascade of signaling events that can regulate diverse cellular processes, such as cell migration, proliferation, survival, and differentiation (63).
2. **Inside-out signaling:** Intracellular signals, such as those triggered by growth factors or cytokines, can modulate integrin affinity and avidity for their ECM ligands. This inside-out signaling regulates the strength of cell-matrix adhesion and can influence processes like cell spreading, migration, and ECM remodeling.
3. **Focal adhesions:** Integrins cluster with cytoskeletal and signaling proteins to form focal adhesion complexes at the cell-matrix interface (64). These complexes act as signaling hubs, integrating ECM and cytoskeleton signals and regulating downstream signaling pathways.
4. **Crosstalk with other signaling pathways:** Integrin signaling does not occur in isolation but exhibits extensive crosstalk with other signaling pathways, such as those mediated by growth factor receptors, G protein-coupled receptors, and ion channels (65). This crosstalk enables integration of multiple signals and the coordination of complex cellular responses.

Integrins and cell-matrix interactions are vital for tissue homeostasis, wound healing, and embryonic development. Dysregulated integrin signaling has been implicated in various pathological conditions, including cancer metastasis, fibrosis, and autoimmune diseases, making integrins attractive therapeutic targets.

1.9 METHODS AND TECHNOLOGY IN CELL SIGNALING

1.9.1 Experimental Techniques

Studying cell signaling pathways and their complex mechanisms requires diverse experimental techniques, from classical biochemical methods to advanced molecular and cellular approaches (66). Key experimental techniques include:

1. **Biochemical techniques:**
   a. Protein purification and characterization (e.g., chromatography, electrophoresis, mass spectrometry) (67)
   b. Enzyme activity assays (e.g., kinase assays, phosphatase assays)
   c. Protein-protein interaction studies (e.g., co-immunoprecipitation, pull-down assays, yeast two-hybrid system) (68)
   d. Post-translational modification analysis (e.g., phosphorylation, ubiquitination)

2. **Molecular biology techniques:**
   a. Gene expression analysis (e.g., RT-PCR, microarrays, RNA-seq) (69)
   b. Gene silencing and knockdown (e.g., siRNA, shRNA, CRISPR/Cas9)
   c. Mutagenesis and protein engineering (e.g., site-directed mutagenesis, domain swapping)
   d. Reporter assays (e.g., luciferase reporters, GFP reporters)

3. **Cell biology techniques:**
   a. Cell culture and transfection (e.g., primary cells, immortalized cell lines)
   b. Functional assays (e.g., proliferation, migration, apoptosis assays)
   c. Receptor activation and inhibition studies (e.g., ligand stimulation, pharmacological inhibitors) (70)

d. Subcellular fractionation and organelle isolation (e.g., membrane preparations, nuclear extracts)

4. **In vivo studies:**
   a. Animal models (e.g., transgenic mice, knockout mice, disease models)
   b. Tissue-specific gene manipulation (e.g., conditional knockouts, tissue-specific promoters) (71)
   c. In vivo imaging and biosensors (e.g., bioluminescence, fluorescence)

These experimental techniques are often combined to dissect cell signaling mechanisms from molecular to organismal levels, providing a comprehensive understanding of their functions and regulatory mechanisms.

1.9.2 Computational Modeling in Signaling Networks

The complexity of cell signaling networks, with their numerous components, interactions, and feedback loops, has necessitated computational modeling approaches to gain insights into their dynamics and emergent properties (72). Computational modeling contributes to cell signaling research by:

1. Integrating diverse experimental data from various sources, such as biochemical measurements, gene expression data, and imaging data, into a unified framework, enabling a systems-level understanding of signaling networks.
2. Simulating and predicting network behavior under different conditions, allowing researchers to predict cellular responses, identify potential therapeutic targets, and test hypotheses.
3. Revealing network properties that are difficult to discern from experimental data alone, such as robustness, sensitivity, and redundancy.
4. Generating testable hypotheses by identifying knowledge gaps or unexpected behaviors, guiding experimental design and drive further research.

Several modeling approaches are employed in cell signaling research, including:

Ordinary differential equation (ODE) models: These describe the temporal dynamics of signaling components using coupled differential equations (73), accounting for various processes such as synthesis, degradation, and interactions.

Boolean network models: These qualitative models represent signaling components as binary states (on/off) and capture the logical relationships between them (74), providing insights into the overall network topology and dynamics.

Petri net models: Petri nets are graphical modeling formalisms representing concurrent processes and resource sharing, suitable for modeling signaling pathways with complex interactions and feedback loops (75).

Agent-based models: These simulate the behavior of individual signaling components (agents) and their interactions, enabling investigation of emergent properties and spatial organization within cells.

Constraint-based models: These models, such as flux balance analysis, use stoichiometric constraints and optimization principles to analyze signal flow through metabolic and signaling networks.

1.9.3 Imaging Techniques and Their Applications

Imaging techniques have become indispensable tools in cell signaling research, providing valuable insights into the spatial and temporal dynamics of signaling processes within living cells and tissues (76). Key imaging techniques and their applications include:

1. **Fluorescence microscopy:**

   a. **Confocal microscopy:** Allows for high-resolution imaging of specific signaling components by using fluorescently labeled antibodies or genetically encoded fluorescent proteins.
   b. **Fluorescence resonance energy transfer (FRET):** Enables the detection of protein-protein interactions and conformational changes in signaling proteins by measuring energy transfer between fluorophores.
   c. **Fluorescence recovery after photobleaching (FRAP):** Monitors the mobility and dynamics of fluorescently labeled signaling components by measuring the recovery of fluorescence after photobleaching.
2. **Live-cell imaging:**
   a. **Biosensors and reporters:** Genetically encoded fluorescent biosensors and reporters (e.g., FRET-based, luminescence-based) can visualize the spatiotemporal dynamics of second messengers, enzyme activities, and protein-protein interactions in living cells.
   b. **Time-lapse microscopy:** Allows for the monitoring of signaling events over time, enabling the study of processes such as cell migration, differentiation, and morphogenesis.
3. **Super-resolution microscopy:**
   Techniques like stimulated emission depletion (STED) microscopy, structured illumination microscopy (SIM), and single-molecule localization microscopy (SMLM) overcome the diffraction limit of conventional microscopy, providing ultra-high-resolution imaging of signaling complexes and subcellular structures.
4. **In vivo imaging:**
   Intravital microscopy: Enables the visualization of signaling processes in living tissues and organisms, using techniques such as multiphoton microscopy and fluorescence molecular tomography (77).
   Bioluminescence and fluorescence imaging: Allows for non-invasive monitoring of signaling events in animal models by detecting the emission of light from genetically encoded or exogenously administered reporters (78).
5. **Correlative imaging:**
   Correlative light and electron microscopy (CLEM): Combines the strengths of fluorescence microscopy and electron microscopy, providing both functional and ultrastructural information about signaling complexes and organelles.
   These imaging techniques, coupled with advanced image analysis software and quantitative approaches, have revolutionized our understanding of cell signaling dynamics (79), enabling researchers to visualize and quantify signaling events in unprecedented detail, from the molecular level to the organismal level.

1.10 THE IMPORTANCE OF CELL SIGNALING IN BIOMEDICAL RESEARCH

1.10.1 Signaling Pathways in Disease

Dysregulation of cell signaling pathways is implicated in a wide range of diseases, making their study vital for understanding disease mechanisms and developing effective therapeutic interventions (80). Examples include:

1. **Cancer:** Aberrant signaling in pathways regulating cell proliferation, survival, and migration is a hallmark of cancer. Mutations in oncogenes (e.g., Ras, EGFR, PI3K) and tumor suppressor genes (e.g., PTEN, p53) lead to uncontrolled cell growth and metastasis.
2. **Autoimmune and inflammatory diseases:** Dysregulated signaling in immune cells and inflammatory pathways (e.g., cytokine signaling, NF-κB pathway) contributes to the development of autoimmune disorders like rheumatoid arthritis, lupus, and inflammatory bowel diseases (81).

3. **Metabolic disorders:** Defects in signaling pathways regulating metabolism, such as insulin signaling (in diabetes) and leptin signaling (in obesity), lead to metabolic imbalances and associated complications.
4. **Neurodegenerative diseases:** Aberrant signaling in neurotrophic factor pathways, oxidative stress signaling, and protein misfolding pathways contribute to the pathogenesis of diseases like Alzheimer's, Parkinson's, and Huntington's disease.
5. **Cardiovascular diseases:** Dysregulation of signaling pathways involved in vascular remodeling, angiogenesis, and myocardial function (e.g., VEGF, Notch, angiotensin II signaling) contributes to the development of hypertension, atherosclerosis, and heart failure.
6. **Developmental disorders:** Congenital malformations and developmental abnormalities can arise from defects in critical signaling pathways (e.g., Hedgehog, Wnt, TGF-β) that regulate embryonic patterning, tissue morphogenesis, and stem cell behavior (Table).

1.10.2 Therapeutic Targets in Cell Signaling

Cell signaling pathways offer a plethora of potential therapeutic targets for various diseases, including:

1. **Kinase inhibitors:** Many signaling pathways involve protein kinases that regulate various cellular processes. Kinase inhibitors, such as imatinib (targeting BCR-ABL in chronic myeloid leukemia) and trastuzumab (targeting HER2 in breast cancer) (82), have shown remarkable success in cancer treatment (Figure
2. **Monoclonal antibodies:** These can target cell surface receptors or soluble signaling molecules, preventing their interaction with downstream signaling components. Examples include rituximab (targeting CD20 in B-cell lymphomas) and bevacizumab (targeting VEGF in various cancers).
3. **Receptor antagonists and modulators:** Small molecules or biologics that block or modulate the activity of cell surface receptors can target specific signaling pathways. Examples include tamoxifen (targeting the estrogen receptor in breast cancer) and canakinumab (targeting IL-1β in inflammatory diseases).
4. **Enzyme inhibitors:** Inhibitors targeting enzymes involved in signaling cascades, such as matrix metalloproteinases (MMPs), cyclooxygenases (COXs), and histone deacetylases (HDACs), have therapeutic potential in various diseases, including cancer, inflammatory disorders, and neurological diseases.
5. **Transcription factor modulators:** Targeting transcription factors that regulate gene expression in response to signaling pathways can modulate cellular responses in diseases.

TABLE 1.5
Diseases and the Dysregulated Pathways Involved in Their Progression

Disease	Dysregulated Signaling Pathways
Cancer	MAPK (Ras, RAF, MEK, ERK), PI3K/Akt, NF-κB, Wnt/β-catenin, Notch, JAK/STAT, HER2/EGFR, VEGF
Autoimmune/Inflammatory Diseases	NF-κB, JAK/STAT, TNF, IL-1β, IL-6, TGF-β, T-cell receptor signaling
Metabolic Disorders (Diabetes, Obesity)	Insulin/IGF-1, Leptin, AMPK, mTOR, FGF, PPAR
Neurodegenerative Diseases	Neurotrophic factor signaling (NGF, BDNF), oxidative stress, protein misfolding (Tau, α-synuclein)
Cardiovascular Diseases	VEGF, Notch, Angiotensin II, TGF-β, Endothelin, MAPK, Wnt
Developmental Disorders	Hedgehog, Wnt, BMP, FGF, TGF-β, Notch

Examples include thiazolidinediones (targeting PPARγ in diabetes) and proteasome inhibitors (targeting NF-κB in cancer and inflammatory diseases).

6. **Gene Therapy and RNA Interference:** Strategies like gene therapy and RNA interference (RNAi) can be used to modulate the expression of signaling components or introduce therapeutic genes, offering potential treatments for genetic disorders and certain cancers (Table 1.6).

1.10.3 Cell Signaling in Drug Development

Cell signaling research is crucial in various stages of drug development, from target identification (83) and validation to preclinical testing and clinical trials:

1. **Target identification and validation:** Understanding the role of specific signaling pathways and components in disease mechanisms is vital for identifying potential therapeutic targets. Techniques like genomics, proteomics, and functional screens are used to identify dysregulated signaling components in disease models (84).
2. **Lead compound screening and optimization:** High-throughput screening (HTS) and structure-based drug design approaches are employed to identify and optimize lead compounds that modulate the activity of target signaling components (e.g., kinase inhibitors and receptor antagonists).
3. **Preclinical testing:** Cell-based assays and animal models are used to evaluate the efficacy, specificity, and toxicity of lead compounds targeting signaling pathways. Signaling biomarkers and imaging techniques are employed to monitor the modulation of signaling pathways and assess therapeutic responses.
4. **Clinical trials:** In clinical trials, signaling biomarkers (e.g., phosphoprotein levels and gene expression profiles) are used to stratify patient populations, monitor treatment responses, and predict drug resistance or toxicity. Companion diagnostics based on signaling pathway alterations can guide personalized treatment strategies.
5. **Drug resistance and combination therapies:** Understanding the mechanisms of drug resistance, often involving compensatory signaling pathways or mutations in target proteins, can inform the development of combination therapies targeting multiple signaling nodes.
6. **Precision medicine:** Genomic and proteomic profiling of signaling pathway alterations in individual patients can guide the selection of personalized, targeted therapies, leading to the development of precision medicine approaches in various disease areas.

TABLE 1.6
Cell Signaling Therapeutic Targets

Target Class	Examples	Therapeutic Applications
Kinase Inhibitors	Imatinib (BCR-ABL), Trastuzumab (HER2), Erlotinib (EGFR)	Cancer, inflammatory diseases
Monoclonal Antibodies	Rituximab (CD20), Bevacizumab (VEGF), Tocilizumab (IL-6R)	Cancer, autoimmune diseases
Receptor Antagonists	Tamoxifen (ER), Canakinumab (IL-1β), Ibrutinib (BTK)	Cancer, inflammatory diseases
Enzyme Inhibitors	MMP inhibitors, COX-2 inhibitors, HDAC inhibitors	Cancer, inflammatory diseases, neurological diseases
Transcription Factor Modulators	Thiazolidinediones (PPARγ), proteasome inhibitors (NF-κB)	Metabolic disorders, cancer, inflammatory diseases
Gene Therapy/RNAi	siRNA, shRNA, CRISPR/Cas9	Genetic disorders, cancer

1.11 FUTURE PERSPECTIVES IN CELL SIGNALING RESEARCH

The field of cell signaling research is poised for significant advancements, driven by rapid technological developments and a growing understanding of the intricacies of signaling pathways. One of the key areas of focus will be the integration of multi-omics data, including genomics, transcriptomics, proteomics, and metabolomics, to gain a systems-level view of signaling networks and their dysregulation in disease states. Advanced computational approaches, such as machine learning and artificial intelligence, will play a vital role in analyzing these vast datasets and generating predictive models of signaling dynamics. Additionally, novel imaging techniques, including super-resolution microscopy and in vivo imaging modalities, will enable the visualization of signaling events with unprecedented spatial and temporal resolution, providing insights into the subcellular organization and dynamics of signaling complexes. Furthermore, the emergence of novel gene editing tools, such as CRISPR/Cas systems, will facilitate the precise manipulation of signaling components and the generation of advanced disease models, accelerating the discovery of therapeutic targets and the development of personalized treatment strategies.

1.12 SUMMARY AND CONCLUSIONS

Cell signaling is fundamental to cellular function and communication, orchestrating physiological processes and governing the complex interplay between cells, tissues, and organisms. From the initial detection of signaling molecules to the propagation of signals through complex networks and the elicitation of appropriate cellular responses, cell signaling pathways exhibit remarkable complexity and sophistication. The study of cell signaling has not only deepened our understanding of biological processes but has also shed light on the molecular mechanisms underlying numerous diseases, paving the way for the development of targeted therapeutics and personalized medicine approaches. As we continue to unravel the intricacies of cell signaling, we will gain invaluable insights into life's fundamental principles, enabling innovative strategies for combating disease, promoting regeneration, and ultimately improving human health and well-being.

REFERENCES

1. Sims NA, Civitelli R. Cell–cell signaling: broadening our view of the basic multicellular unit. *Calcif Tissue Int*. 2014;94:2–3.
2. García-Mata C, Lamattina L. Gasotransmitters are emerging as new guard cell signaling molecules and regulators of leaf gas exchange. *Plant Sci*. 2013;201:66–73.
3. Lim W, Mayer B, Pawson T. *Cell signaling*. Garland Science; 2014.
4. Nichols SA, Dirks W, Pearse JS, King N. Early evolution of animal cell signaling and adhesion genes. *Proc Natl Acad Sci*. 2006;103(33):12451–6.
5. Smith RP, Barraza I, Quinn RJ, Fortoul MC. The mechanisms and cell signaling pathways of programmed cell death in the bacterial world. *Int Rev Cell Mol Biol*. 2020;352:1–53.
6. Sanchez-Vega F, Mina M, Armenia J, Chatila WK, Luna A, La KC, et al. Oncogenic signaling pathways in the cancer genome atlas. *Cell*. 2018;173(2):321-337. e10.
7. Johnson HE, Toettcher JE. Signaling dynamics control cell fate in the early Drosophila embryo. *Dev Cell*. 2019;48(3):361–370. e3.
8. Holm A, Vikström E. Quorum sensing communication between bacteria and human cells: signals, targets, and functions. *Front Plant Sci*. 2014;5:98757.
9. Nair A, Chauhan P, Saha B, Kubatzky KF. Conceptual evolution of cell signaling. *Int J Mol Sci*. 2019;20(13):3292.
10. Lee LT, Siu FK, Tam JK, Lau IT, Wong AO, Lin MC, et al. Discovery of growth hormone-releasing hormones and receptors in nonmammalian vertebrates. *Proc Natl Acad Sci*. 2007;104(7):2133–8.
11. Wen R, Zhou L, Peng Z, Fan H, Zhang T, Jia H, et al. Single-cell sequencing technology in colorectal cancer: a new technology to disclose the tumor heterogeneity and target precise treatment. *Front Immunol*. 2023;14:1175343.

12. Soncin F, Natale D, Parast MM. Signaling pathways in mouse and human trophoblast differentiation: a comparative review. *Cell Mol Life Sci.* 2015;72:1291–302.
13. Steinhorst L, Kudla J. Signaling in cells and organisms—calcium holds the line. *Curr Opin Plant Biol.* 2014;22:14–21.
14. Sokol SY. *Wnt Signaling in Embryonic Development.* Elsevier; 2011.
15. Hotamisligil GS, Davis RJ. Cell signaling and stress responses. *Cold Spring Harb Perspect Biol.* 2016;8(10):a006072.
16. Leivar P, Quail PH. PIFs: pivotal components in a cellular signaling hub. *Trends Plant Sci.* 2011;16(1):19–28.
17. Barone V, Lang M, Krens SG, Pradhan SJ, Shamipour S, Sako K, et al. An effective feedback loop between cell-to-cell contact duration and morphogen signaling determines cell fate. *Dev Cell.* 2017;43(2):198–211. e12.
18. Qi X, Li X. Mechanistic insights into the generation and transduction of hedgehog signaling. *Trends Biochem Sci.* 2020;45(5):397–410.
19. Mercurio AM, Bachelder RE, Bates RC, Chung J. Autocrine signaling in carcinoma: VEGF and the α6β4 integrin. In: Theresa Vincent, PhD editor, *Seminars in cancer biology.* Elsevier; 2004. p. 115–22.
20. Endocrine, paracrine, and autocrine signaling pathways that regulate ovulation: trends in endocrinology & metabolism. [Internet] [cited 2024 Mar 29]. Available from: https://www.cell.com/trends/endocrinology-metabolism/abstract/S1043-2760(18)30043-2
21. Paracrine mechanisms in adult stem cell signaling and therapy | Circulation Research. [Internet] [cited 2024 Mar 29]. Available from: https://www.ahajournals.org/doi/full/10.1161/CIRCRESAHA.108.176826
22. The fibroblast growth factor signaling pathway - Ornitz - 2015 - WIREs Developmental Biology - Wiley Online Library. [Internet] [cited 2024 Mar 29]. Available from: https://wires.onlinelibrary.wiley.com/doi/full/10.1002/wdev.176
23. Yang T, Fu M, Pestell R, Sauve AA. SIRT1 and endocrine signaling. *Trends Endocrinol Metab.* 2006 Jul 1;17(5):186–91. [Internet] [cited 2024 Mar 29]. Available from: https://www.cell.com/trends/endocrinology-metabolism/abstract/S1043-2760(06)00077-4
24. Frontiers | Signaling interactions in the adrenal cortex. [Internet] [cited 2024 Mar 29]. Available from: https://www.frontiersin.org/journals/endocrinology/articles/10.3389/fendo.2016.00017/full
25. Direct cadherin-activated cell signaling – PMC. [Internet] [cited 2024 Mar 29]. Available from: https://www.ncbi.nlm.nih.gov/pmc/articles/PMC2172751/
26. Brogi S, Tafi A, Désaubry L, Nebigil CG. Discovery of GPCR ligands for probing signal transduction pathways. *Front Pharmacol.* 2014 Nov 28;5. [Internet] [cited 2024 Mar 29]. Available from: https://www.frontiersin.org/journals/pharmacology/articles/10.3389/fphar.2014.00255/full
27. Gutkind JS. Cell growth control by G protein-coupled receptors: From signal transduction to signal integration. *Oncogene.* 1998 Oct 1;17:1331–42.
28. Benovic JL. Novel β2-adrenergic receptor signaling pathways. *J Allergy Clin Immunol.* 2002 Dec 1;110(6, Supplement):S229–35. [Internet] [cited 2024 Mar 29]. Available from: https://www.sciencedirect.com/science/article/pii/S0091674902014975
29. Pawson T. Regulation and targets of receptor tyrosine kinases. *Eur J Cancer.* 2002 Sep 1;38:S3–10. [Internet] [cited 2024 Mar 29]. Available from: https://www.sciencedirect.com/science/article/pii/S0959804902805974
30. Receptor tyrosine kinase activation: from the ligand perspective – ScienceDirect. [Internet] [cited 2024 Mar 29]. Available from: https://www.sciencedirect.com/science/article/abs/pii/S0955067420300235
31. Álvarez-Aznar A, Muhl L, Gaengel K. Chapter thirteen - VEGF receptor tyrosine kinases: key regulators of vascular function. In: Jenny A, editor. *Current Topics in Developmental Biology.* Academic Press; 2017. p. 433–82. (Protein Kinases in Development and Disease; vol. 123). [Internet] [cited 2024 Mar 29]. Available from: https://www.sciencedirect.com/science/article/pii/S0070215316301806
32. Glutamate receptor ion channels – ScienceDirect. [Internet] [cited 2024 Mar 29]. Available from: https://www.sciencedirect.com/science/article/abs/pii/S0959438805000693
33. Hattori M, Gouaux E. Molecular mechanism of ATP binding and ion channel activation in P2X receptors. *Nature.* 2012 May;485(7397):207–12. [Internet] [cited 2024 Mar 29]. Available from: https://www.nature.com/articles/nature11010
34. Gutjahr A, Tiraby G, Perouzel E, Verrier B, Paul S. Triggering intracellular receptors for vaccine adjuvantation. *Trends Immunol.* 2016 Sep 1;37(9):573–87. [Internet] [cited 2024 Mar 29]. Available from: https://www.cell.com/trends/immunology/abstract/S1471-4906(16)30074-6

35. Zacarías NVO, Lenselink EB, IJzerman AP, Handel TM, Heitman LH. Intracellular receptor modulation: novel approach to target GPCRs. *Trends Pharmacol Sci.* 2018 Jun 1;39(6):547–59. [Internet] [cited 2024 Mar 29]. Available from: https://www.cell.com/trends/pharmacological-sciences/abstract/S0165-6147(18)30053-1
36. Newton AC, Bootman MD, Scott JD. Second messengers. *Cold Spring Harb Perspect Biol.* 2016 Aug 1;8(8):a005926. [Internet] [cited 2024 Mar 29]. Available from: http://cshperspectives.cshlp.org/content/8/8/a005926
37. Schreck R, Baeuerle PA. A role for oxygen radicals as second messengers. *Trends Cell Biol.* 1991 Aug 1;1(2):39–42. [Internet] [cited 2024 Mar 29]. Available from: https://www.cell.com/trends/cell-biology/abstract/0962-8924(91)90072-H
38. Beyond receptors: Multiple second-messenger systems in brain - Worley - 1987 - Annals of Neurology - Wiley Online Library. [Internet] [cited 2024 Mar 29]. Available from: https://onlinelibrary.wiley.com/doi/abs/10.1002/ana.410210302?casa_token=A3eS6JsrC58AAAAA:gpy2FRKQuOlpecoB-7-q3vd2AGehiKTBRSFVY62ZzGIviiKh5VGfzNndwoZ3KC1XBLWXp98a1S6AKKzv
39. Second messenger systems and adaptation. – PMC. [Internet] [cited 2024 Mar 29]. Available from: https://www.ncbi.nlm.nih.gov/pmc/articles/PMC1434555/
40. Hormones, second messengers and the reversible phosphorylation of proteins: An overview - Cohen - 1985 - BioEssays - Wiley Online Library. [Internet] [cited 2024 Mar 29]. Available from: https://onlinelibrary.wiley.com/doi/abs/10.1002/bies.950020206?casa_token=7tw_YeARLuoAAAAA:0bDp34W5gOPs1aLtl12trjHxORWlbj1NO3_dlqa3ZKAp-kI3wNxfsyRSP5EDUGMzr0wW_b4BU_ktoJsO
41. Bouvier M. Cross-talk between second messengers. *Ann N Y Acad Sci.* 1990 Jan 1;594:120–9. [Internet] [cited 2024 Mar 29]. Available from: https://doi.org/10.1111/j.1749-6632.1990.tb40473.x
42. Crosstalk between secondary messengers, hormones and MAPK modules during abiotic stress signalling in plants – ScienceDirect. [Internet] [cited 2024 Mar 29]. Available from: https://www.sciencedirect.com/science/article/abs/pii/S0734975013001237
43. Lemmon MA, Freed DM, Schlessinger J, Kiyatkin A. The dark side of cell signaling: positive roles for negative regulators. *Cell.* 2016 Mar 10;164(6):1172–84. [Internet] [cited 2024 Mar 29]. Available from: https://www.sciencedirect.com/science/article/pii/S0092867416301970
44. Negative feedback mechanisms and their roles during pattern formation: cell. [Internet] [cited 2024 Mar 29]. Available from: https://www.cell.com/fulltext/S0092-8674(00)80710-2
45. Autocrine loops with positive feedback enable context-dependent cell signaling | American Journal of Physiology-Cell Physiology. [Internet] [cited 2024 Mar 29]. Available from: https://journals.physiology.org/doi/full/10.1152/ajpcell.00260.2001
46. Eungdamrong NJ, Iyengar R. Modeling cell signaling networks. *Biol Cell.* 2004 Jun 1;96(5):355–62. [Internet] [cited 2024 Mar 29]. Available from: https://www.sciencedirect.com/science/article/pii/S0248490004000772
47. Luo K. Signaling Cross Talk between TGF-β/Smad and other signaling pathways. *Cold Spring Harb Perspect Biol.* 2017 Jan 1;9(1):a022137. [Internet] [cited 2024 Mar 29]. Available from: http://cshperspectives.cshlp.org/content/9/1/a022137
48. Kunkel BN, Brooks DM. Cross talk between signaling pathways in pathogen defense. *Curr Opin Plant Biol.* 2002 Aug 1;5(4):325–31. [Internet] [cited 2024 Mar 29]. Available from: https://www.sciencedirect.com/science/article/pii/S1369526602002753
49. Wu WKK, Cho CH, Lee CW, Fan D, Wu K, Yu J, et al. Dysregulation of cellular signaling in gastric cancer. *Cancer Lett.* 2010 Sep 28;295(2):144–53. [Internet] [cited 2024 Mar 29]. Available from: https://www.sciencedirect.com/science/article/pii/S0304383510002405
50. Wu WKK, Wang XJ, Cheng ASL, Luo MXM, Ng SSM, To KF, et al. Dysregulation and crosstalk of cellular signaling pathways in colon carcinogenesis. *Crit Rev Oncol Hematol.* 2013 Jun 1;86(3):251–77. [Internet] [cited 2024 Mar 29]. Available from: https://www.sciencedirect.com/science/article/pii/S1040842812002363
51. Novel treatments targeting the dysregulated cell signaling pathway during sepsis – PMC. [Internet] [cited 2024 Mar 29]. Available from: https://www.ncbi.nlm.nih.gov/pmc/articles/PMC8725530/
52. Zhao WL. Targeted therapy in T-cell malignancies: dysregulation of the cellular signaling pathways. *Leukemia.* 2010 Jan;24(1):13–21. [Internet] [cited 2024 Mar 29]. Available from: https://www.nature.com/articles/leu2009223
53. Signaling pathways in anchoring junctions of epithelial cells: cell-to-cell and cell-to-extracellular matrix interactions: Journal of Receptors and Signal Transduction;35(1). [Internet] [cited 2024 Mar 29]. Available from: https://www.tandfonline.com/doi/abs/10.3109/10799893.2014.931426

54. Werle MJ. Cell-to-cell signaling at the neuromuscular junction. *Ann N Y Acad Sci.* 2008;1132(1):13–8. [Internet] [cited 2024 Mar 29]. Available from: https://onlinelibrary.wiley.com/doi/abs/10.1196/annals.1405.035
55. Extracellular matrix, junctional integrity and matrix metalloproteinase interactions in endothelial permeability regulation* - Alexander - 2002 - Journal of Anatomy - Wiley Online Library. [Internet] [cited 2024 Mar 29]. Available from: https://onlinelibrary.wiley.com/doi/full/10.1046/j.1469-7580.2002.00057.x
56. Notch Signaling Pathway | Science's STKE. [Internet] [cited 2024 Mar 29]. Available from: https://www.science.org/doi/abs/10.1126/stke.3642006cm7
57. Batchuluun K, Azuma M, Yashiro T, Kikuchi M. Notch signaling-mediated cell-to-cell interaction is dependent on E-cadherin adhesion in adult rat anterior pituitary. *Cell Tissue Res.* 2017 Apr 1;368(1):125–33. [Internet] [cited 2024 Mar 29]. Available from: https://doi.org/10.1007/s00441-016-2540-5
58. Dityatev A, Schachner M. The extracellular matrix and synapses. *Cell Tissue Res.* 2006 Nov 1;326(2):647–54. [Internet] [cited 2024 Mar 29]. Available from: https://doi.org/10.1007/s00441-006-0217-1
59. Biology of the extracellular matrix: an overview : journal of glaucoma. [Internet] [cited 2024 Mar 29]. Available from: https://journals.lww.com/glaucomajournal/fulltext/2014/10001/Biology_of_the_Extracellular_Matrix__An_Overview.7.aspx
60. The matrix reorganized: extracellular matrix remodeling and integrin signaling – ScienceDirect. [Internet] [cited 2024 Mar 29]. Available from: https://www.sciencedirect.com/science/article/abs/pii/S0955067406001207
61. Mechanical signals regulating extracellular matrix gene expression in fibroblasts - Sarasa-Renedo - 2005 - Scandinavian Journal of Medicine & Science in Sports - Wiley Online Library. [Internet] [cited 2024 Mar 29]. Available from: https://onlinelibrary.wiley.com/doi/full/10.1111/j.1600-0838.2005.00461.x?casa_token=L40h-IDnOOkAAAAA%3AtXOfShx2sHCVvExV_eA_XHL8eGe-CAtjQ-Mfa_iBefGjEDpce8Z1eycwYEX0RPtE3-Wc_NVVzSUWXZYC
62. Labat-Robert J. Cell–Matrix interactions, the role of fibronectin and integrins. A survey. *Pathol Biol.* 2012 Feb 1;60(1):15–19. [Internet] [cited 2024 Mar 29]. Available from: https://www.sciencedirect.com/science/article/pii/S0369811411001416
63. Gilchrist CL, Chen J, Richardson WJ, Loeser RF, Setton LA. Functional integrin subunits regulating cell–matrix interactions in the intervertebral disc. *J Orthop Res.* 2007;25(6):829–40. [Internet] [cited 2024 Mar 29]. Available from: https://onlinelibrary.wiley.com/doi/abs/10.1002/jor.20343
64. Journal of Cellular Physiology | Cell Biology Journal | Wiley Online Library. [Internet] [cited 2024 Mar 29]. Available from: https://onlinelibrary.wiley.com/doi/full/10.1002/jcp.21237
65. Tissue repair and fibrosis. [Internet] [cited 2024 Mar 29]. Available from: https://link.springer.com/book/10.1007/978-3-642-58456-5
66. Elucidating cell signaling mechanisms using antisense technology: trends in pharmacological sciences. [Internet] [cited 2024 Mar 29]. Available from: https://www.cell.com/trends/pharmacological-sciences/abstract/S0165-6147(00)01448-6
67. Analysis of protein phosphorylation and cellular signaling events by flow cytometry: techniques and clinical applications – ScienceDirect. [Internet] [cited 2024 Mar 29]. Available from: https://www.sciencedirect.com/science/article/abs/pii/S1521661603003097
68. Asthagiri AR, Lauffenburger DA. Bioengineering models of cell signaling. *Annu Rev Biomed Eng.* 2000 Aug 1;2:31–53. [Internet] [cited 2024 Mar 29]. Available from: https://www.annualreviews.org/content/journals/10.1146/annurev.bioeng.2.1.31
69. Yao Z, Petschnigg J, Ketteler R, Stagljar I. Application guide for omics approaches to cell signaling. *Nat Chem Biol.* 2015 Jun;11(6):387–97. [Internet] [cited 2024 Mar 29]. Available from: https://www.nature.com/articles/nchembio.1809
70. Sachs K, Gifford D, Jaakkola T, Sorger P, Lauffenburger DA. Bayesian network approach to cell signaling pathway modeling. *Sci STKE.* 2002 Sep 3;2002(148):pe38–pe38. [Internet] [cited 2024 Mar 29]. Available from: https://www.science.org/doi/abs/10.1126/stke.2002.148.pe38
71. Modulating cell signalling in vivo with magnetic nanotransducers | Nature Reviews Methods Primers. [Internet] [cited 2024 Mar 29]. Available from: https://www.nature.com/articles/s43586-022-00170-2
72. Computational modeling of mammalian signaling networks - Hughey - 2010 - WIREs Systems Biology and Medicine - Wiley Online Library. [Internet] [cited 2024 Mar 29]. Available from: https://wires.onlinelibrary.wiley.com/doi/full/10.1002/wsbm.52?casa_token=xfWQ3D-5PHQAAAAA%3ABWKKHwFnsRRHT5jqQOGgONfhunmKtw6SIui1XoRqKCIZHkMdqIXxcL9r5Te7Ui3qTn1v2-ufy6BG4qbz

73. Spatial modeling of cell signaling networks – ScienceDirect. [Internet] [cited 2024 Mar 29]. Available from: https://www.sciencedirect.com/science/article/abs/pii/B9780123884039000084
74. Boolean modeling: a logic-based dynamic approach for understanding signaling and regulatory networks and for making useful predictions - Albert - 2014 - WIREs Systems Biology and Medicine - Wiley Online Library. [Internet] [cited 2024 Mar 29]. Available from: https://wires.onlinelibrary.wiley.com/doi/full/10.1002/wsbm.1273?casa_token=HvRgfeQfvkoAAAAA%3Alx62x8azRhJuiQ8EPMzLpF5Rui1KR_BlMkoWwuknb7QGs324J1ir0zMYC-Y7OKYUrZAmDosXzN7kuJoN
75. Computational modeling of signal transduction networks without kinetic parameters: petri net approaches | American Journal of Physiology-Cell Physiology. [Internet] [cited 2024 Mar 29]. Available from: https://journals.physiology.org/doi/abs/10.1152/ajpcell.00487.2022
76. Full article: intravital imaging of cell signaling in mice. [Internet] [cited 2024 Mar 29]. Available from: https://www.tandfonline.com/doi/full/10.4161/intv.20802
77. Pharmaceutics | Free Full-Text | Live cell in vitro and in vivo imaging applications: accelerating drug discovery. [Internet] [cited 2024 Mar 29]. Available from: https://www.mdpi.com/1999-4923/3/2/141
78. Aoki K, Kamioka Y, Matsuda M. Fluorescence resonance energy transfer imaging of cell signaling from in vitro to in vivo: basis of biosensor construction, live imaging, and image processing. *Dev Growth Differ.* 2013;55(4):515–22. [Internet] [cited 2024 Mar 29]. Available from: https://onlinelibrary.wiley.com/doi/abs/10.1111/dgd.12039
79. Bagnall J, Boddington C, Boyd J, Brignall R, Rowe W, Jones NA, et al. Quantitative dynamic imaging of immune cell signalling using lentiviral gene transfer. *Integr Biol.* 2015 Jun 8;7(6):713–25. [Internet] [cited 2024 Mar 29]. Available from: https://doi.org/10.1039/c5ib00067j
80. Hancock JT. Cell signalling is the music of life. *Br J Biomed Sci.* 2008 Jan 1;65(4):205–8. [Internet] [cited 2024 Mar 29]. Available from: https://doi.org/10.1080/09674845.2008.11732830
81. Berridge MJ. Unlocking the secrets of cell signaling. *Annu Rev Physiol.* 2005 Feb 1;67:1–21. [Internet] [cited 2024 Mar 29]. Available from: https://www.annualreviews.org/content/journals/10.1146/annurev.physiol.67.040103.152647
82. Yen CF, Wang HS, Lee CL, Liao SK. Roles of integrin-linked kinase in cell signaling and its perspectives as a therapeutic target. *Gynecol Minim Invasive Ther.* 2014 Aug 1;3(3):67–72. [Internet] [cited 2024 Mar 29]. Available from: https://www.sciencedirect.com/science/article/pii/S2213307014000562
83. Journal of Cellular Biochemistry | Wiley Online Library. [Internet] [cited 2024 Mar 29]. Available from: https://onlinelibrary.wiley.com/doi/abs/10.1002/jcb.21500
84. Panda M, Biswal BK. Cell signaling and cancer: a mechanistic insight into drug resistance. *Mol Biol Rep.* 2019 Oct 1;46(5):5645–59. [Internet] [cited 2024 Mar 29]. Available from: https://doi.org/10.1007/s11033-019-04958-6

2 Regulation of Cell Signaling Cascades

Habab Ali Ahmad, Muhammad Ibrahim Khan, Fazli Wahid, and Muhammad Imran Khan**
Department of Biomedical Sciences, Pak Austria Fachhochschule: Institute of Applied Sciences and Technology, Mang, KPK, Pakistan
*Department of Biomedical Sciences, Pak Austria Fachhochschule: Institute of Applied Sciences and Technology, Mang, KPK, Pakistan

2.1 INTRODUCTION

2.1.1 Overview of Cell Signaling and Its Importance in Biological Processes

Cell signaling encompasses intricate communication networks essential for orchestrating the foundational activities of cells and complex physiological responses in multicellular organisms. These mechanisms regulate a wide array of biological processes, including cell growth, proliferation, differentiation, survival, apoptosis, metabolism, homeostasis, immune responses, inflammation, embryonic development, tissue patterning, neuronal signaling, synaptic plasticity, and hormonal regulation alongside endocrine functions. The capacity of cells to accurately detect and respond to their surroundings, external stimuli, and internal conditions is vital for the health and adaptability of all living beings. Disruptions in these signaling pathways can lead to a variety of diseases, including cancer, autoimmune diseases, neurological disorders, and metabolic syndromes, underscoring the essential role of well-regulated signaling pathways in maintaining biological integrity and function (1).

2.1.2 Types of Signaling Molecules (e.g., Proteins, Lipids, Ions, Gases)

Cell signaling involves the action of diverse signaling molecules called ligands that can be classified into several categories:

a) **Proteins:**
   Various proteins, such as growth factors (e.g., EGF, FGF, PDGF, TGF-β), cytokines (e.g., interleukins, interferons, chemokines), hormones (e.g., insulin, glucagon, estrogen, testosterone), and neurotransmitters (e.g., acetylcholine, dopamine, serotonin), function as ligands to initiate signaling pathways
b) **Lipids:**
   Various lipids, such as steroid hormones (e.g., cortisol, aldosterone), eicosanoids (e.g., prostaglandins, leukotrienes), and phospholipids (e.g., phosphatidylinositol's, sphingolipids), can induce signaling cascades.
c) **Ions and gases**
   Certain ions, such as Calcium (Ca^{2+}), Sodium (Na^{+}), and Potassium (K^{+}), and gaseous molecules such as nitric oxide (NO) and carbon monoxide (CO), can induce biosignaling (Figure 2.1).
   These signaling molecules can act as extracellular signals (e.g., growth factors, hormones) or intracellular second messengers (e.g., cyclic nucleotides, calcium ions) and can

DOI: 10.1201/9781003594901-2

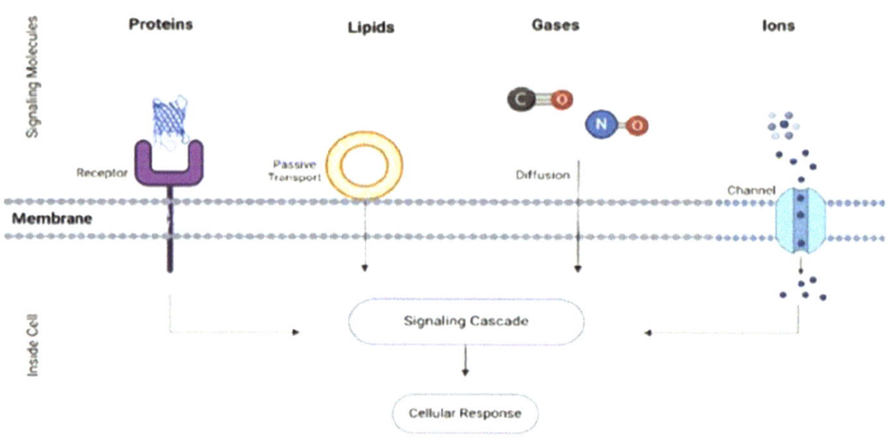

FIGURE 2.1 Different signaling molecules and their mode of entry into the cell.

elicit specific cellular responses through their interactions with corresponding receptors and downstream signaling cascades (2).

2.1.3 Extracellular and Intracellular Signaling Pathways

Cell signaling pathways can be broadly categorized into extracellular and intracellular signaling pathways:

a) **Extracellular signaling pathways:**
   - These involve the binding of extracellular signaling molecules (e.g., growth factors, hormones) to cell surface receptors.
   - These receptors can be classified into various types, such as G protein-coupled receptors (GPCRs), receptor tyrosine kinases (RTKs), and ion channel receptors.
   - The ligand binding to the receptor initiates a cascade of intracellular signaling events, often involving second messengers and activation of kinase cascades.

b) **Intracellular signaling pathways:**
   - These pathways regulate cellular processes in response to intracellular cues or stimuli.
   - Examples include signaling pathways involved in cell cycle regulation, apoptosis, stress responses, and metabolic regulation.
   - These pathways involve the action of signaling molecules, such as kinases, phosphatases, transcription factors, and other regulatory proteins.
   - These pathways can be triggered by extracellular signals or by intracellular events (e.g., DNA damage and oxidative stress) (Figure 2.2).

Both extracellular and intracellular signaling pathways are interconnected and often converge, forming complex signaling networks that enable cells to integrate multiple signals and respond appropriately to different stimuli and environmental conditions (3).

2.2 SIGNALING CASCADES AND THEIR COMPONENTS

2.2.1 Signaling Cascades and Their Characteristics

Signal transduction pathways, commonly referred to as signaling cascades, facilitate internal cellular communication, carrying and intensifying signals through a series of biochemical reactions

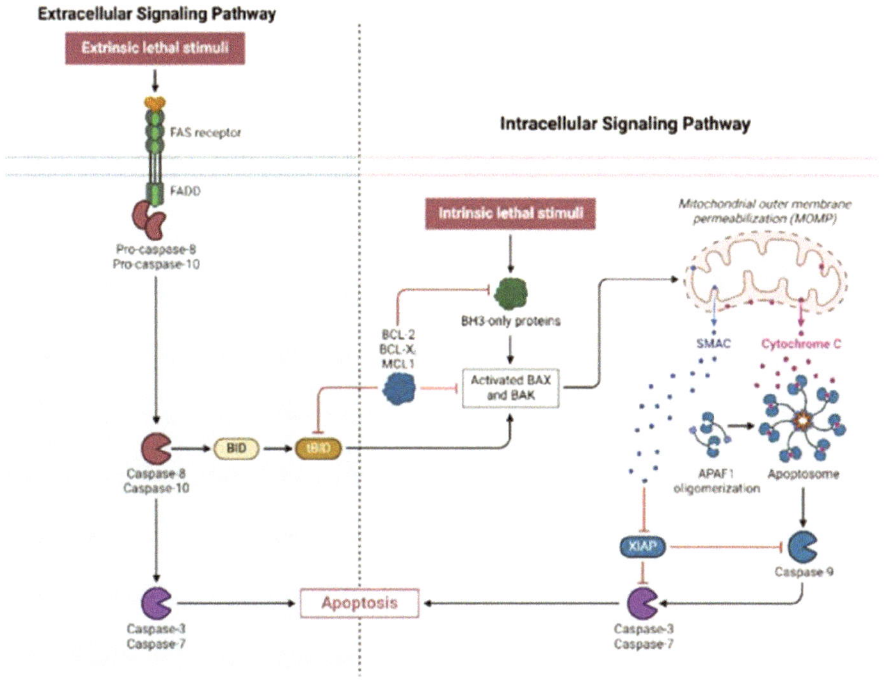

FIGURE 2.2 Examples of extracellular and intracellular signaling pathways.

(4). These pathways are distinguished by several key characteristics. Firstly, they are inherently multistep processes, wherein each phase entails the activation or modification of distinct proteins or molecules. This setup ensures that even a minimal initial stimulus can be amplified at each step, resulting in substantial cellular reactions. Importantly, these cascades are reversible, enabling cells to fine-tune and terminate the signaling process as needed. Moreover, they demonstrate remarkable specificity, guaranteeing that cells respond correctly to precise signals. Additionally, these pathways are not isolated; they possess the ability to interact with and integrate signals from other cascades, facilitating a complex web of communication within the cell. This intricate system enables cells to process and respond to diverse signals, ensuring coordinated responses to various stimuli.

2.2.2 Key Components

Signaling cascades involve the coordinated action of various molecular components (5), including:

a) **Receptors:** These proteins, located on the cell surface or within the cell, bind to specific signaling molecules (e.g., growth factors, hormones, cytokines) to initiate the signaling cascade (Example of FAS receptor in Figure 2.2).
b) **Adaptor proteins:** These molecules serve as bridges, facilitating interactions between different components of the signaling cascade and enabling signal propagation.
c) **Kinases:** These enzymes catalyze the transfer of phosphate groups from high-energy molecules (e.g., ATP) to specific target proteins (for example, see ERK and JNK in Figure 2.3). Phosphorylation can activate or deactivate proteins, thereby regulating their function and the flow of the signaling cascade.
d) **Phosphatases:** These enzymes remove phosphate groups from proteins, counteracting kinase action and providing a mechanism for signal termination or modulation.
e) **Transcription factors:** These proteins regulate gene expression by binding to specific DNA sequences and controlling the transcription of target genes (for example, see

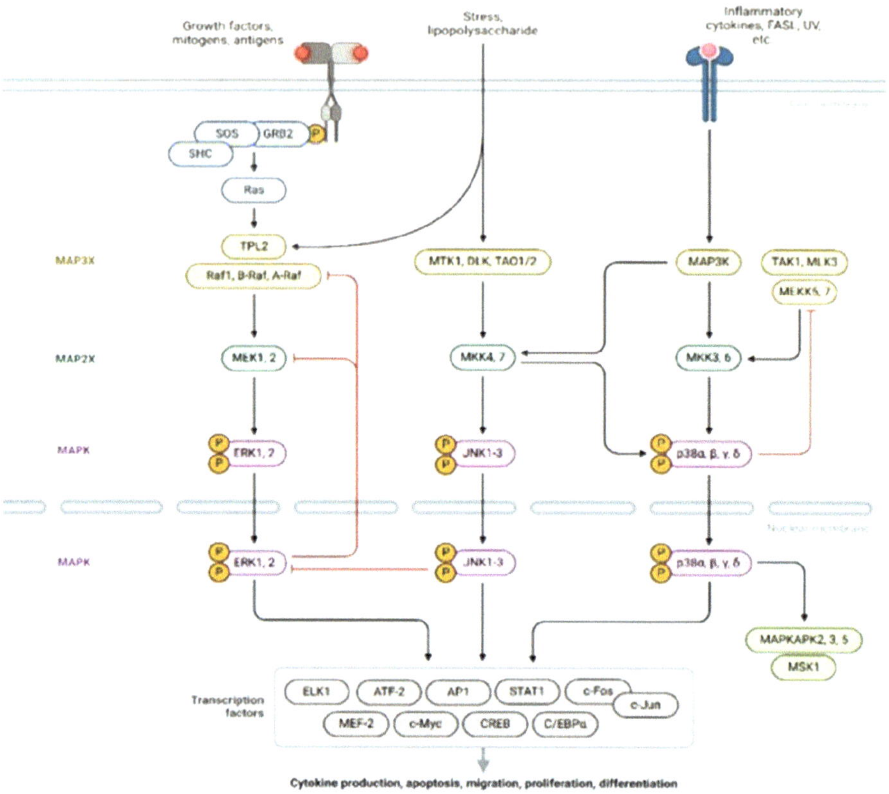

FIGURE 2.3 Role of MAPK in regulating different pathways for different cellular responses.

transcription factor box in Figure 2.3). Many signaling cascades culminate in the activation or repression of transcription factors, resulting in changes in gene expression patterns and cellular behavior.

2.2.3 Examples of Major Signaling Cascades

Several well-studied signaling cascades play crucial roles in various cellular processes:

a) **Mitogen-activated protein kinase (MAPK) cascade:**

The MAPK cascades are evolutionarily conserved signaling pathways that regulate diverse cellular processes, including proliferation, differentiation, survival, apoptosis, and stress responses (6). These cascades are organized into three-tiered modules, comprising a MAPK kinase kinase (MAP3K), a MAPK kinase (MAPKK), and a MAPK, which sequentially activate each other through phosphorylation events (7).

The extracellular signal-regulated kinase (ERK) cascade is the best-characterized MAPK pathway, typically activated by growth factors, cytokines, and mitogenic stimuli (8). Ligand binding to RTKs or GPCRs recruits and activates the small GTPase Ras, leading to sequential activation of the MAP3K Raf, the MAPKK MEK, and the MAPK ERK (9). Activated ERK can phosphorylate and regulate various cytoplasmic and nuclear substrates, including transcription factors (e.g., Elk-1, c-Myc), kinases, and cytoskeletal proteins, thereby controlling gene expression, cell cycle progression, and cellular morphology (10) (Figure 2.3).

Other major MAPK cascades include the c-Jun N-terminal kinase (JNK) and p38 pathways, which are primarily activated by environmental stresses, inflammatory cytokines, and genotoxic agents (11). These stress-activated MAPK pathways regulate diverse cellular responses, such as apoptosis, inflammation, and cellular differentiation.

b) **Phosphoinositide 3-kinase (PI3K)/Akt cascade:**

The PI3K/Akt signaling cascade is a critical regulator of cellular metabolism, growth, proliferation, and survival. Various extracellular signals, including growth factors, hormones, and cytokines, activate this pathway by binding to their respective receptors on the cell surface (12).

Receptor activation recruits the lipid kinase PI3K to the membrane, where it phosphorylates the lipid phosphatidylinositol (4,5)-bisphosphate (PIP2) to generate the second messenger phosphatidylinositol (3–5)-trisphosphate (PIP3). PIP3 serves as a docking site for the serine/threonine kinase Akt (also known as protein kinase B, PKB) and its upstream activator, the kinase PDK1 (phosphoinositide-dependent kinase 1). PDK1 phosphorylates and partially activates Akt, which is further activated by mTORC2 (mammalian target of rapamycin complex 2) through additional phosphorylation events (13) (Figure 2.4).

Activated Akt regulates a diverse array of downstream targets, including transcription factors (e.g., FOXO, NF-κB), kinases (e.g., mTORC1, GSK3β), and other signaling molecules, thereby controlling cellular processes such as metabolism, cell cycle progression, angiogenesis, and apoptosis, leading to different health conditions (14–16). The PI3K/Akt pathway is frequently dysregulated in various cancers, making it an attractive target for therapeutic intervention.

c) **Janus kinase/signal transducers and activators of transcription (Jak/STAT) cascade:**

The Jak/STAT signaling cascade is a critical pathway for transducing signals from extracellular cytokines and growth factors, regulating diverse biological processes, including immunity, inflammation, cell proliferation, and differentiation (17).

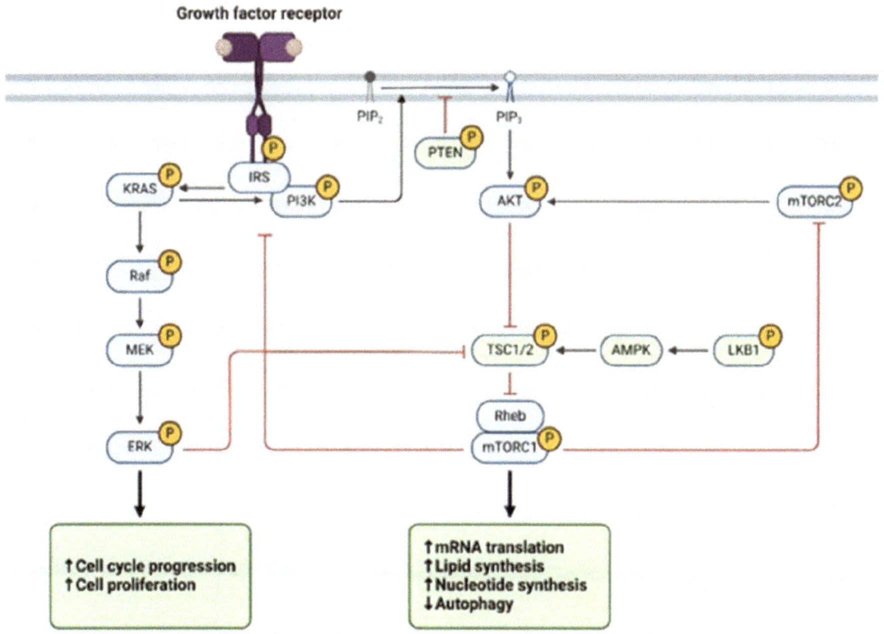

FIGURE 2.4 Illustration of PI3K/AKT mediated signaling pathway influencing different cellular functions.

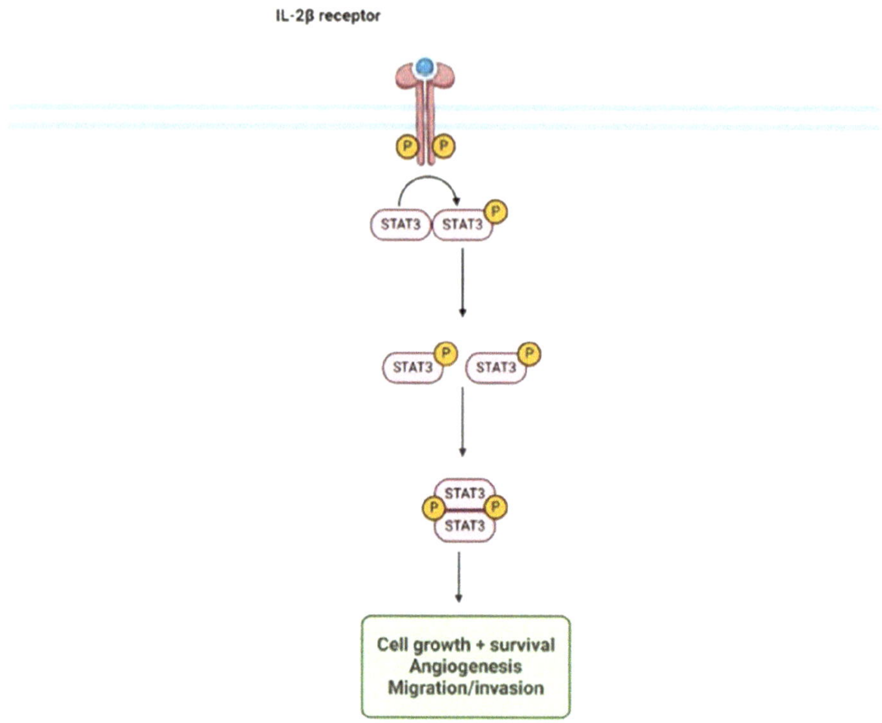

FIGURE 2.5 IL-2B binding with receptor activates the downstream activation and phosphorylation of STAT3, leading to cellular functions.

Cytokine binding to their cognate receptors activates the associated Janus kinases (Jaks) through trans-phosphorylation events. Activated Jaks then phosphorylate specific tyrosine residues on the receptor cytoplasmic tails, creating docking sites for the recruitment of STAT transcription factors. Jaks phosphorylate the recruited STATs, leading to their dimerization and subsequent translocation to the nucleus, where they bind to specific DNA sequences and regulate the transcription of target genes (18) (Figure 2.5).

There are seven mammalian STAT proteins (STAT1–6, including STAT5A and STAT5B), each with distinct functions and tissue distributions. For example, STAT1 and STAT2 are primarily involved in antiviral and anti-inflammatory responses, while STAT3 and STAT5 play crucial roles in regulating cell proliferation, survival, and immune responses (19). The specific STAT protein activation mainly depends on the ligand type, which leads to the downstream regulation of STAT proteins. Dysregulation of the Jak/STAT pathway has been implicated in various diseases, including cancers (20), autoimmune disorders, and immunodeficiencies (21), making it an attractive therapeutic target (22).

d) **Nuclear factor-kappa B (NF-κB) cascade:**

The NF-κB signaling pathway is a critical regulator of immune responses, inflammation, cell survival, and proliferation. NF-κB is a family of transcription factors that exist in an inactive state in the cytoplasm, bound to inhibitory proteins called IκBs (inhibitors of NF-κB) (23).

Upon stimulation by various signals, such as pro-inflammatory cytokines (e.g., tumor necrosis factor-alpha, TNF-α), pathogen-associated molecular patterns (PAMPs), or oxidative stress, the IκB kinase (IKK) complex becomes activated. Activated IKK complex phosphorylates IκB proteins, marking them for ubiquitination and subsequent proteasomal degradation. This releases the NF-κB dimers (typically composed of p65 and p50 subunits), allowing them to translocate to the nucleus and bind to specific DNA sequences, inducing the expression of target genes involved in immune responses, inflammation, cell survival, and proliferation (24, 25).

The NF-κB pathway is tightly regulated by various negative feedback mechanisms and crosstalk with other signaling pathways, such as the MAPK and PI3K/Akt (26). Dysregulation of the NF-κB pathway has been implicated in various chronic inflammatory diseases, autoimmune disorders, and certain cancers, making it an attractive target for therapeutic intervention (27).

2.3 MECHANISMS OF REGULATION OF CELL SIGNALING CASCADES

2.3.1 Spatial and Temporal Regulation of Signaling Cascades

Signaling cascades are precisely regulated in both space and time to ensure appropriate cellular responses. Spatial regulation involves the compartmentalization and localization of signaling components within specific subcellular regions, enabling efficient signal transduction and preventing undesired crosstalk between pathways (28).

a) **Subcellular localization:** Signaling molecules can be selectively recruited to specific cellular compartments, such as the plasma membrane, cytosol, organelles, or the nucleus, through interactions with anchoring proteins or lipid modifications.
b) **Scaffold proteins:** Scaffold proteins act as molecular platforms, co-localizing multiple components of a signaling cascade and facilitating their interaction, thereby enhancing the efficiency and specificity of signal propagation.
c) **Endocytosis and trafficking:** Internalization and trafficking of activated receptors and signaling complexes through endocytic pathways can regulate the duration and intensity of signaling, as well as facilitate signal termination or propagation to different cellular locations.

   Temporal regulation ensures that signaling cascades are activated or inactivated at the appropriate times, allowing for dynamic and transient cellular responses (28).
d) **Rapid activation and inactivation:** Signaling cascades often involve rapid and reversible post-translational modifications (e.g., phosphorylation, ubiquitination) that can quickly switch signaling components between active and inactive states.
e) **Oscillatory behavior:** Some signaling cascades exhibit oscillatory dynamics, with periodic cycles of activation and inactivation, enabling specific gene expression patterns and cellular responses.
f) **Feedback loops:** Positive and negative feedback mechanisms (discussed below) can modulate the temporal dynamics of signaling cascades, leading to signal amplification, attenuation, or termination.

2.3.2 Regulation at the Receptor Level

Signaling cascades are often regulated at the receptor level, which serves as the initial point of signal detection and transduction (29).

a) **Ligand binding:** The binding of extracellular signaling molecules (e.g., growth factors, cytokines, hormones) to their respective receptors can induce conformational changes or oligomerization, initiating receptor activation and downstream signaling events.
b) **Receptor dimerization/oligomerization:** Some receptors, such as RTKs, require dimerization or oligomerization upon ligand binding to become fully activated and initiate signaling cascades.

c) **Receptor internalization and trafficking:** The internalization and trafficking of activated receptors through endocytic pathways can regulate signaling duration and intensity, as well as facilitate signal termination or propagation to different cellular locations.
d) **Receptor desensitization:** Mechanisms such as phosphorylation, ubiquitination, or association with inhibitory proteins can lead to receptor desensitization, preventing further signaling and promoting receptor internalization or degradation.

2.3.3 Regulation by Post-Translational Modifications

Post-translational modifications (PTMs) are crucial in regulating the activity, localization, and interactions of signaling proteins within cascades (30).

a) **Phosphorylation:** This reversible modification, catalyzed by kinases, can activate or inactivate signaling proteins, modulate their interactions, or alter their subcellular localization.
b) **Ubiquitination:** The covalent attachment of ubiquitin molecules can target signaling proteins for proteasomal degradation or alter their activity, localization, or interactions.
c) **Acetylation:** Acetylation of specific lysine residues can modulate the activity, stability, or interactions of signaling proteins, particularly transcription factors.
d) **Other PTMs:** Additional modifications, such as methylation, SUMOylation, and lipidation, can also regulate various aspects of signaling cascades (Figure 2.6).

2.3.4 Regulation by Protein-Protein Interactions

Protein-protein interactions (PPIs) are essential for the assembly, regulation, and propagation of signaling cascades. These interactions facilitate the formation of multi-protein complexes, enable

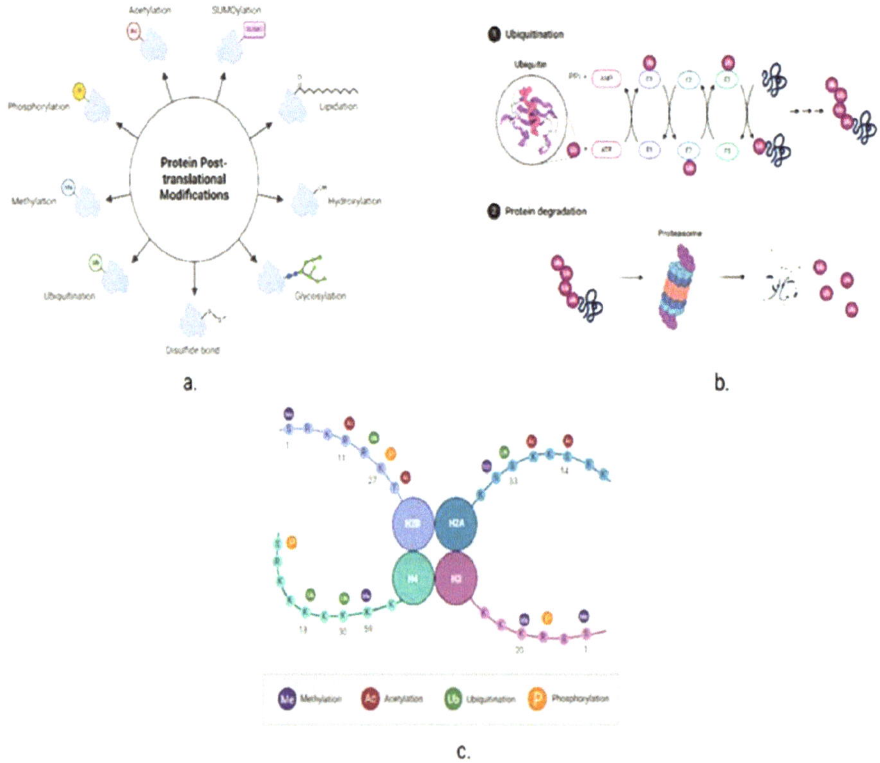

FIGURE 2.6 Receptor regulation through different processes.

the transfer of signals between pathway components, and modulate the activity, localization, and stability of signaling proteins (31). Table 2.1 outlines the various roles of PPIs in regulating signaling cascades.

Importantly, many signaling proteins contain modular protein-binding domains (e.g., SH2, SH3, PDZ, WW domains) that mediate specific PPIs, enabling the assembly of signaling complexes and the propagation of signals within the cascade (43, 44) (Figure 2.7).

Furthermore, PPIs can be regulated by various mechanisms, such as PTMs, allosteric regulation, or the availability of binding partners, providing additional layers of control over signaling pathways.

TABLE 2.1
Represents Mechanistic Description of Pathways

Various molecules and aspects of pathway	Mechanistic description of pathway
Scaffold proteins	Scaffold proteins act as molecular platforms that coordinate the spatial and temporal organization of signaling pathways (32). They provide a structural framework for the co-localization and interaction of multiple signaling components, thereby enhancing the efficiency, specificity, and fidelity of signal transduction. Example: The scaffold proteins KSR and JIP (JNK-interacting protein) facilitate the assembly and activation of the MAPK signaling cascades by bringing together upstream kinases, MAPKs, and their substrates (33).
Adaptor proteins	Adaptor proteins serve as molecular bridges, linking different components of signaling cascades and facilitating their interactions (34). They typically lack enzymatic activity but possess modular protein-binding domains that enable them to recruit and assemble signaling complexes. Example: The adaptor protein Grb2 (growth factor receptor-bound protein 2) links activated RTKs to the Ras signaling pathway by binding to phosphorylated tyrosine residues on the receptor and to the Ras activator protein Sos (35).
Inhibitory proteins	Inhibitory proteins can bind to and inhibit the activity of signaling proteins, providing a mechanism for signal termination or modulation (36). These interactions can prevent the activation, reduce the activity, or alter the localization of signaling components, thereby attenuating or terminating the signaling cascade. Example: The inhibitory proteins IκBs (inhibitors of NF-κB) bind to and sequester the transcription factor NF-κB in the cytoplasm (37), preventing its nuclear translocation and transcriptional activity (see Figure 2.6).
Conformational changes	PPIs can induce conformational changes in signaling proteins, altering their activity, localization, or interaction partners. These conformational changes can expose or mask functional domains, catalytic sites, or binding surfaces, thereby regulating the protein's function within the signaling cascade. Example: The binding of the Ras protein to its effector kinase Raf induces a conformational change in Raf, exposing its catalytic domain and enabling its activation and subsequent phosphorylation of downstream kinases (38).
Allosteric regulation	PPIs can also regulate signaling proteins through allosteric mechanisms, where the binding of a regulatory protein induces long-range conformational changes that modulate the activity or affinity of the target protein. Example: The binding of cyclin proteins to cyclin-dependent kinases (CDKs) induces conformational changes that activate the CDKs, enabling them to phosphorylate and regulate their substrates involved in cell cycle progression (39).
Post-translational modifications	PPIs can facilitate or promote PTMs of signaling proteins, such as phosphorylation, ubiquitination, or acetylation (40, 41). These PTMs can alter the activity, stability, or localization of the modified proteins, thereby regulating the signaling cascade. Example: The binding of E3 ubiquitin ligases to their target proteins can promote the ubiquitination and subsequent proteasomal degradation of the target, providing a mechanism for signal termination (42).

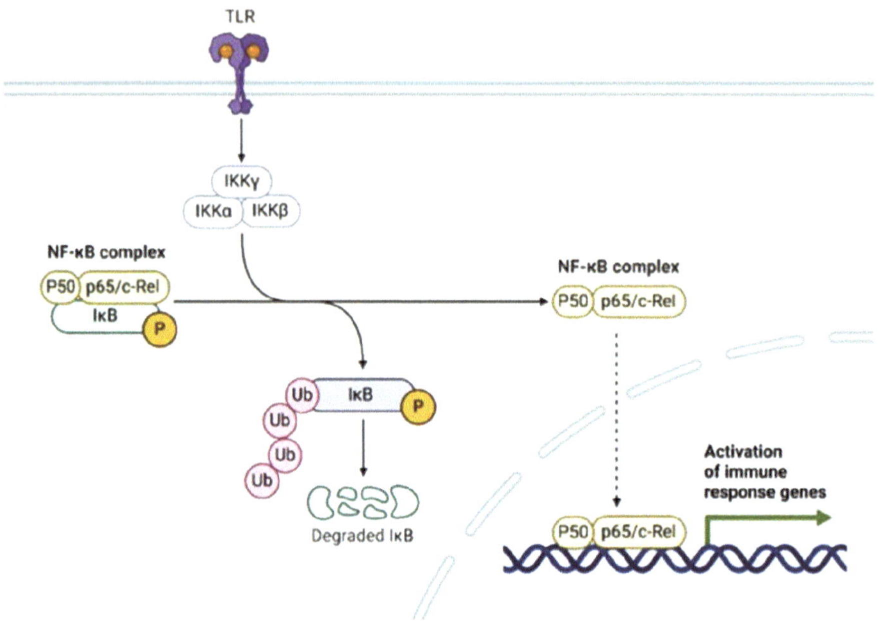

FIGURE 2.7 Intracellular signaling of NF-κB and IκBs in regulating cellular response to stress.

2.3.5 Regulation by Feedback Loops

Feedback loops are crucial mechanisms for maintaining homeostasis and ensuring appropriate cellular responses to signaling cascades. These loops provide a means of self-regulation, enabling cells to fine-tune and modulate signaling pathways in response to various stimuli or environmental conditions.

2.3.5.1 Negative Feedback Loops

Negative feedback loops are inhibitory mechanisms that counteract or attenuate the initial signaling event, preventing excessive or prolonged signaling. These loops act as a brake on the signaling cascade, ensuring that the cellular response is transient and proportional to the initial stimulus. Negative feedback loops can operate through various mechanisms:

a) **Transcriptional induction of inhibitory proteins:**
   Activation of signaling pathways can lead to the transcriptional induction of inhibitory proteins, such as phosphatases or inhibitors of upstream components. For example, in the MAPK pathway, the transcription factor AP-1, activated downstream of ERK, can induce the expression of MAPK phosphatases (e.g., MKP-1), which dephosphorylate and inactivate MAPKs, thereby terminating the signal (45).
b) **Activation of phosphatases:**
   Signaling cascades often involve the activation of phosphatases that can dephosphorylate and inactivate upstream components of the pathway. For example, in the PI3K/Akt pathway, the lipid phosphatase PTEN can dephosphorylate PIP3, the product of PI3K, thereby inhibiting Akt activation (Figure 2.4).
c) **Receptor internalization and degradation:**
   Prolonged stimulation can lead to activated receptor internalization and subsequent degradation, resulting in signal termination. For example, in the EGF receptor (EGFR)

signaling pathway, ligand-induced receptor internalization and lysosomal degradation contribute to negative feedback regulation (46).

d) **Induction of inhibitory factors:**

Signaling pathways can induce the expression or activation of inhibitory factors that directly bind and inhibit upstream components of the cascade. For instance, in the NF-κB pathway, the transcriptional target IκBα can bind and sequester NF-κB in the cytoplasm, preventing its nuclear translocation and transcriptional activity (Figure 2.7).

2.3.5.2 Positive Feedback Loops

In contrast to negative feedback loops, positive feedback loops amplify and reinforce the initial signaling event, leading to signal propagation and sustained cellular responses. These loops can promote switch-like or bistable behaviors, where the signaling pathway becomes locked in an active or inactive state. Positive feedback loops can operate through various mechanisms:

a) **Activation of downstream components that further activate upstream components:**

In some signaling cascades, activated downstream components can directly or indirectly activate upstream components, creating a self-amplifying loop. For example, in the MAPK pathway, activated ERK can phosphorylate and inhibit the MAPK phosphatase MKP-3, leading to sustained ERK activation (Figure 2.3).

b) **Transcriptional induction of signaling components:**

Activation of signaling pathways can induce the transcription of genes encoding components of the same or interacting pathways, creating a positive feedback loop. For instance, in the Jak/STAT pathway, activated STAT transcription factors can induce the expression of cytokine receptors or Jak kinases, amplifying the signal (Figures 2.5 and 2.8).

2.3.5.3 Integrated Feedback Circuits

Complex signaling networks often involve intricate combinations of positive and negative feedback loops, creating integrated feedback circuits that allow for precise temporal control and dynamic regulation of cellular responses. These integrated circuits can exhibit various behaviors, such as

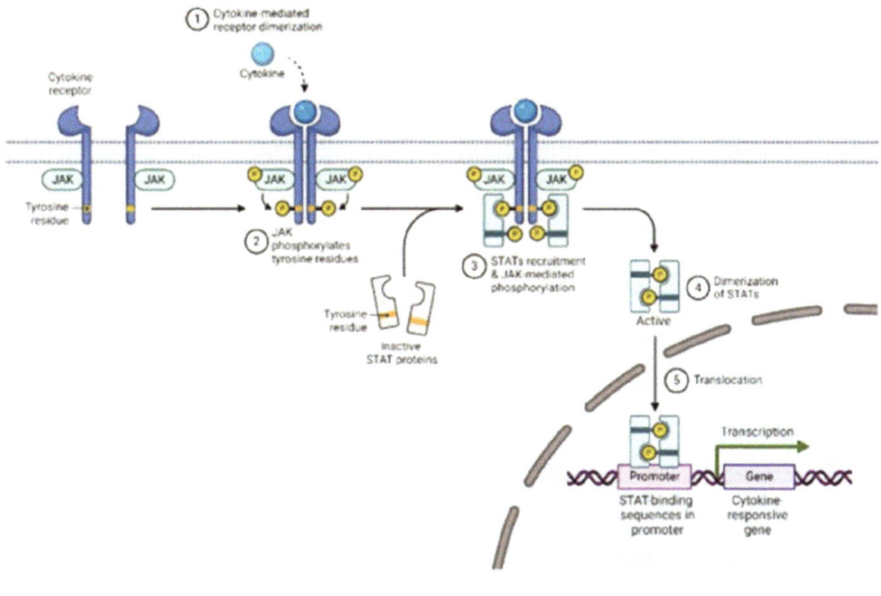

FIGURE 2.8 Cytokine-mediated JAK/STAT pathway activation to regulate cellular responses.

oscillations, bistability, or transient pulses, depending on the specific interactions and kinetics of the feedback loops (47).

a) **Oscillatory behavior:**
Certain signaling pathways, such as the NF-κB and p53 pathways, exhibit oscillatory dynamics due to the interplay between positive and negative feedback loops (48). These oscillations can encode specific gene expression patterns and cellular responses, such as cell cycle regulation or apoptosis induction.
b) **Bistability and switch-like behavior:**
Positive feedback loops can create bistable systems, where the signaling pathway can exist in two stable states (on or off) based on the strength and duration of the initial stimulus. This switch-like behavior can be important for irreversible cellular decisions, such as cell differentiation or apoptosis (47).
c) **Transient pulse responses:**
Combinations of positive and negative feedback loops can generate transient pulse-like responses, where the signaling pathway is activated for a specific duration before returning to its basal state. Such transient responses can be crucial for processes like cell cycle progression or stress responses, where sustained signaling may be detrimental (49).

2.4 CROSSTALK AND INTEGRATION OF SIGNALING PATHWAYS

2.4.1 Convergence and Divergence of Signaling Cascades

Signaling cascades exhibit complex patterns of convergence and divergence, allowing for the integration and coordination of multiple signals and cellular responses.

a) **Convergence:**
Convergence occurs when different signaling pathways converge on shared downstream components or nodes. This convergence enables different upstream signals to ultimately regulate the same cellular process or response. For example, the MAPK, PI3K/Akt, and NF-κB pathways can converge on the regulation of gene expression by modulating the activity of shared transcription factors or co-regulators (see Pathway Figure 2.3 for reference).
b) **Divergence:**
Divergence occurs when a single signaling pathway bifurcates or diverges to regulate multiple downstream effectors or cellular processes. This divergence allows for the propagation of a signal to elicit diverse cellular responses. For instance, the Jak/STAT pathway can diverge to regulate gene expression (via STAT transcription factors), cell survival (via PI3K/Akt), and cytoskeletal reorganization (via Rho GTPases) (18).

2.4.2 Interaction between Different Signaling Pathways

Signaling pathways do not operate in isolation but instead interact with each other through various mechanisms, enabling crosstalk and integration of signals. These interactions can occur at multiple levels:

a) **Shared components:**
Different signaling pathways can share common components, such as kinases, adaptor proteins, or transcription factors, facilitating crosstalk and signal integration. For example, the MAPK and PI3K/Akt pathways share common upstream activators (e.g., Ras) and downstream effectors (e.g., mTOR), enabling crosstalk between these pathways.

b) **Feedback loops:**

Signaling pathways can regulate each other through feedback loops, either positive or negative, enabling crosstalk and modulation of pathway activities. For example, the PI3K/Akt pathway can activate the NF-κB pathway through the phosphorylation and degradation of IκB, while NF-κB can transcriptionally induce the expression of PI3K inhibitors, creating a negative feedback loop.

c) PTMs:

Components of one signaling pathway can directly modify and regulate components of another pathway through PTMs, such as phosphorylation, acetylation, or ubiquitination. For example, the MAPK pathway can phosphorylate and modulate the activity of components of the Wnt signaling pathway, enabling crosstalk between these pathways.

2.4.3 Mechanisms of Crosstalk

Cell signaling pathways exhibit complex crosstalk and integration through various mechanisms, enabling cells to respond appropriately to complex environmental cues and coordinate diverse cellular processes. One prominent mechanism is the sharing of common components, such as kinases, adaptor proteins, or transcription factors, between different signaling cascades. For instance, the MAPK and PI3K pathways share the small GTPase Ras as a common upstream activator, while the transcription factor CREB serves as a convergence point for multiple pathways, including MAPK, PI3K/Akt, and Ca2+/calmodulin-dependent kinase pathways (50). Another mechanism of crosstalk involves transcriptional regulation, where signaling pathways modulate the expression of components of other pathways. The NF-κB pathway, for example, can induce the expression of genes encoding components of the Jak/STAT pathway, such as cytokine receptors and Jak kinases, thereby affecting its activity. Similarly, the Hypoxia-inducible factor 1 (HIF-1) transcription factor, activated by the PI3K/Akt pathway, can induce the expression of various components of the MAPK pathway, facilitating crosstalk (51) (Figure 2.9).

PTMs also play a crucial role in crosstalk between signaling pathways. Components of one pathway can directly modify and regulate components of another pathway through mechanisms such as phosphorylation, acetylation, or ubiquitination. For instance, the PI3K/Akt pathway can phosphorylate and inhibit the activity of the GSK-3β kinase, a key regulator of the Wnt/β-catenin signaling pathway (52), while the MAPK pathway can phosphorylate and activate the IKK complex, leading to the activation of the NF-κB pathway (53).

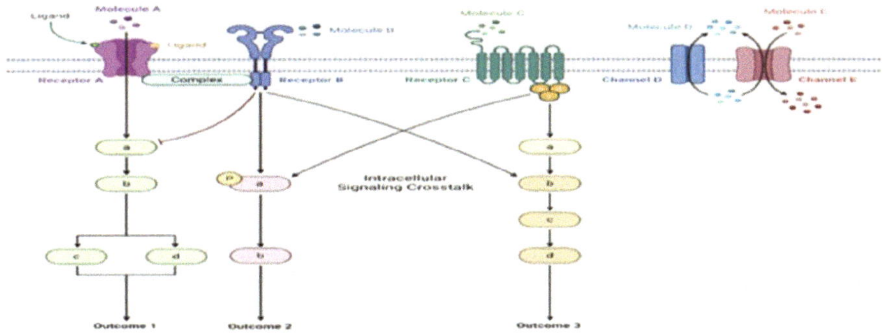

FIGURE 2.9 General illustration of intracellular crosstalk between different signaling molecules to regulate cellular outcome.

Scaffolding proteins facilitate crosstalk by assembling and co-localizing components from different signaling pathways. The scaffolding protein KSR (Kinase Suppressor of Ras) can coordinate the interaction between the MAPK and PI3K/Akt pathways by binding to components of both pathways, such as Raf and PDK1. Similarly, the scaffolding protein IQGAP1 can integrate signals from the MAPK, PI3K/Akt, and Wnt/β-catenin pathways by interacting with key components of these pathways (54).

Finally, the spatial and temporal regulation of signaling components through subcellular localization or compartmentalization can modulate crosstalk between different pathways. The localization of signaling components to specific subcellular compartments, such as endosomes or the nucleus, can facilitate or inhibit their interactions with components of other pathways. Additionally, the nucleocytoplasmic shuttling of signaling proteins, such as the MAPK ERK, can enable crosstalk with transcriptional regulators and modulate gene expression in response to different stimuli.

2.4.4 Biological Implications of Pathway Integration

The crosstalk and integration of signaling pathways have profound biological implications, enabling cells to respond appropriately to complex environmental cues and coordinate various cellular processes:

a) **Amplification or attenuation of signals:**
   Pathway integration can lead to the amplification or attenuation of specific signals, depending on the nature of the crosstalk (positive or negative feedback loops).
b) **Specificity and context-dependent responses:**
   The integration of multiple signaling pathways allows cells to generate specific and context-dependent responses based on the combination and intensity of the input signals.
c) **Coordination of complex cellular processes:**
   Signaling pathway integration enables the coordination of complex cellular processes, such as cell growth, differentiation, survival, and apoptosis, by integrating multiple inputs and regulatory mechanisms.
d) **Robustness and redundancy:**
   Crosstalk between signaling pathways can provide robustness and redundancy, ensuring that critical cellular processes are maintained even if one pathway is disrupted or perturbed.
e) **Pathological implications:**
   Dysregulation of signaling pathway crosstalk and integration can contribute to the development and progression of various diseases, such as cancer, autoimmune disorders, and metabolic disorders.

   Crosstalk and integration of signaling pathways represent a fundamental aspect of cellular signaling networks, enabling cells to integrate and process multiple inputs, generate context-dependent responses, and coordinate complex biological processes.

2.5 DYSREGULATION OF SIGNALING CASCADES IN DISEASE

Precise regulation of cell signaling cascades is crucial for maintaining cellular homeostasis and coordinating various physiological processes. However, dysregulation or aberrant signaling can contribute to the development and progression of numerous diseases, underscoring the importance of these pathways in disease pathogenesis.

2.5.1 Role of Aberrant Signaling in Disease Pathogenesis

Dysregulated cell signaling can disrupt various cellular processes, leading to pathological consequences. Several mechanisms contribute to aberrant signaling, including:

a) **Genetic alterations:** Mutations in genes encoding signaling components (e.g., receptors, kinases, transcription factors) can result in constitutive activation, inactivation, or altered regulation of signaling cascades.
b) **Epigenetic modifications:** Epigenetic changes, such as DNA methylation or histone modifications, can affect the expression or activity of signaling components, leading to dysregulated signaling.
c) **Altered PPIs:** Disruptions in the interactions between signaling proteins, such as scaffolding proteins or adaptor proteins, can impair the assembly and propagation of signaling cascades.
d) **Dysregulated PTMs**: Abnormalities in PTMs (e.g., phosphorylation, ubiquitination) can affect the activity, localization, or stability of signaling proteins, contributing to aberrant signaling.
e) **Imbalanced feedback loops:** Dysregulation of positive or negative feedback loops can lead to excessive or insufficient signaling, disrupting the homeostatic control of cellular processes.

2.5.2 Examples of Diseases Linked to Dysregulated Signaling

a) **Cancer:**
Aberrant signaling in pathways regulating cell proliferation, survival, and growth (e.g., MAPK, PI3K/Akt, Wnt/β-catenin) is a hallmark of many cancers. Examples include Wnt, Hippo, and Notch pathways. Similarly, mutations in oncogenes (e.g., Ras, Raf, PI3K) or tumor suppressor genes (e.g., PTEN, p53) can lead to constitutive activation or inactivation of signaling cascades, promoting uncontrolled cell growth and survival.
b) **Autoimmune disorders:**
Dysregulation of signaling pathways involved in immune cell development, activation, and inflammation (e.g., NF-κB, Jak/STAT) can contribute to the pathogenesis of autoimmune diseases, such as rheumatoid arthritis, lupus, and multiple sclerosis. Aberrant signaling can lead to excessive or inappropriate immune responses, causing tissue damage and inflammation.
c) **Neurological diseases:**
Dysregulated signaling in pathways regulating neuronal survival, plasticity, and neurotransmission (e.g., MAPK, PI3K/Akt, Wnt) has been implicated in various neurological disorders, including Alzheimer's disease, Parkinson's disease, and amyotrophic lateral sclerosis (ALS) (14). Aberrant signaling can contribute to neuronal death, synaptic dysfunction, and protein misfolding or aggregation, leading to neurodegenerative processes.
d) **Metabolic disorders:**
Dysregulation of signaling pathways involved in metabolic processes, such as insulin signaling (e.g., PI3K/Akt, MAPK), can contribute to the development of metabolic disorders, including obesity, type 2 diabetes, and insulin resistance (55). Aberrant signaling can impair glucose and lipid metabolism, leading to metabolic dysregulation and associated complications.

2.5.3 Therapeutic Strategies Targeting Dysregulated Signaling Pathways

Given the crucial role of dysregulated signaling in disease pathogenesis, targeting aberrant signaling cascades has emerged as a promising therapeutic approach. Several strategies have been explored:

1. **Small molecule inhibitors:** These compounds target specific components of dysregulated signaling pathways (e.g., kinase inhibitors, inhibitors of PPIs) and have shown therapeutic potential in various diseases, including cancer and inflammatory disorders.

2. **Monoclonal antibodies:** Monoclonal antibodies targeting dysregulated receptors or signaling proteins have been developed for various cancers and autoimmune diseases, aiming to block aberrant signaling and restore homeostasis.
3. **Gene therapy:** Gene therapy approaches, such as gene editing or RNA interference, can be used to correct genetic alterations or modulate the expression of dysregulated signaling components.
4. **Epigenetic modulators:** Drugs targeting epigenetic mechanisms, such as DNA methyltransferase inhibitors or histone deacetylase inhibitors, can modulate the expression and activity of signaling components affected by epigenetic dysregulation.
5. **Combination therapies:** Combining multiple therapeutic agents targeting different components of dysregulated signaling cascades or complementary pathways can enhance therapeutic efficacy and overcome resistance mechanisms.

It is important to note that while targeting dysregulated signaling cascades holds promise, these pathways often exhibit complex crosstalk and redundancy, posing challenges in developing specific and effective therapeutic interventions. Additionally, the pleiotropic nature of many signaling cascades necessitates careful consideration of potential off-target effects and toxicities.

2.6 FUTURE PERSPECTIVES AND CHALLENGES IN CELL SIGNALING RESEARCH

The field of cell signaling has witnessed remarkable progress in recent years, driven by advances in various disciplines, including genomics, proteomics, systems biology, and computational approaches. However, despite these advancements, several challenges and unanswered questions remain, presenting exciting opportunities for future research. One emerging research area in cell signaling is the study of signal integration and crosstalk between different pathways. As our understanding of individual signaling cascades deepens, it has become increasingly evident that these pathways do not operate in isolation but rather form complex signaling networks with extensive crosstalk and feedback loops. Unraveling the complex interplay between these pathways and elucidating the mechanisms governing signal integration will be crucial for understanding cellular decision-making processes and developing more effective therapeutic strategies. Another active area of investigation is the role of cell signaling in the regulation of non-coding RNAs, such as microRNAs (miRNAs) and long non-coding RNAs (lncRNAs). These non-coding RNAs have emerged as important regulators of gene expression and are involved in various cellular processes, including cell signaling. Discovering the complex crosstalk between cell signaling pathways and non-coding RNAs may provide new insights into the complexity of gene regulation and open up novel therapeutic avenues (56).

The study of cell signaling in the context of multicellular systems and tissue microenvironments is also gaining increasing attention. While most signaling studies have traditionally focused on isolated cell lines or primary cultures, it is becoming increasingly clear that the tissue microenvironment and cell-cell interactions play a crucial role in modulating signaling pathways and cellular responses (57). Integrating these contextual factors into the study of cell signaling will be essential for understanding the physiological relevance of these pathways and their dysregulation in disease settings.

A major challenge in studying complex signaling networks is the development of robust experimental and computational approaches to capture the complex dynamics and spatiotemporal regulation of these pathways. Advanced imaging techniques, such as live-cell imaging and super-resolution microscopy, coupled with biosensors and reporters, are providing unprecedented insights into the spatial and temporal dynamics of signaling events. Furthermore, the integration of these experimental approaches with computational modeling and simulation techniques is enabling the generation of predictive models and the identification of emergent properties of signaling networks. Another significant challenge in the field of cell signaling is the translation of basic research findings into clinical applications and therapeutic interventions. While significant progress has been made in targeting dysregulated signaling pathways in diseases such as cancer and autoimmune disorders, the complexity and redundancy of these pathways often limit the efficacy of single-target therapies. Developing combination therapies that target multiple nodes within signaling networks or leveraging synthetic lethality approaches may hold promise for overcoming these limitations.

Additionally, the integration of cell signaling research with emerging technologies, such as gene editing, RNA therapeutics, and precision medicine approaches, presents exciting opportunities for developing personalized and targeted therapies. For example, the CRISPR/Cas9 system has revolutionized gene editing capabilities, enabling the precise modulation of signaling components and the dissection of their roles in disease pathogenesis (58).

REFERENCES

1. Knapp S. Cell signalling pathways. In: Stefan Knapp, editor, *Oxford Textbook of Oncology*. Oxford Academic;2016.p.23–30.
2. Cooper G. Signaling molecules and their receptors. In: *The Cell: A Molecular Approach*, 2nd edition. Sinauer Associates;2014.
3. Hastings JF, Skhinas JN, Fey D, Croucher DR, Cox TR. The extracellular matrix as a key regulator of intracellular signalling networks. *Br J Pharmacol.* 2019;176.
4. Klingmüller U. Biological foundations of signal transduction and the systems biology perspective. *Comput Systems Biol.* 2005 Jan 1;149–68.
5. Hancock JT. The principles of cell signalling. *On Growth, Form and Computers.* 2003 Jan 1;64–81.
6. Maik-Rachline G, Wexler S, Seger R. The MAPK signaling cascades. In: Ralph A. Bradshaw, Gerald W. Hart, Philip D. Stahl, editor, *Encyclopedia of Cell Biology*, Vol. 1–6, 2nd edition. Elsevier;2022.
7. Braicu C, Buse M, Busuioc C, Drula R, Gulei D, Raduly L, et al. A comprehensive review on MAPK: A promising therapeutic target in cancer. *Cancers.* 2019;11.
8. Guo Y, Pan W, Liu S, Shen Z, Xu Y, Hu L. ERK/MAPK signalling pathway and tumorigenesis (Review). *Exp Ther Med.* 2020;19:1997–2007.
9. Ullah R, Yin Q, Snell AH, Wan L. RAF-MEK-ERK pathway in cancer evolution and treatment. *Semin Cancer Biol.* 2022;85.
10. Gao Y, Pu J, Ye L. The extracellular signal-regulated kinase signaling pathway in biology of pluripotent stem cells. In: Massimo Loda, editor, *Molecular Players in iPSC Technology.* AACR;2021.
11. Pua LJW, Mai CW, Chung FFL, Khoo ASB, Leong CO, Lim WM, et al. Functional roles of JNK and p38 MAPK signaling in nasopharyngeal carcinoma. *Int J Mol Sci.* 2022;23.
12. Bairstow SF, Bunce MW, Anderson RA. Kinases/phosphatases | Phosphoinositide 4- and 5-kinases and phosphatases. In: Joseph Jez, editor, *Encyclopedia of Biological Chemistry: Third Edition.* Elsevier;2021.
13. Xue G, Hemmings BA. PKB/akt-dependent regulation of cell motility. *J Natl Cancer Inst.* 2013;105(6).
14. Kumar M, Bansal N. Implications of phosphoinositide 3-Kinase-Akt (PI3K-Akt) pathway in the pathogenesis of Alzheimer's disease. *Mol Neurobiol.* 2022;59.
15. Lee JJX, Loh K, Yap YS. PI3K/Akt/mTOR inhibitors in breast cancer. *Cancer Biol Med.* 2015;12.
16. Mayer IA, Arteaga CL. The PI3K/AKT pathway as a target for cancer treatment. *Annu Rev Med.* 2016;67.
17. Hu Q, Bian Q, Rong D, Wang L, Song J, Huang HS, et al. JAK/STAT pathway: Extracellular signals, diseases, immunity, and therapeutic regimens. *Front Bioeng Biotechnol.* 2023 Feb 23;11:1110765.
18. Hu X, li J, Fu M, Zhao X, Wang W. The JAK/STAT signaling pathway: From bench to clinic. *Signal Transduct Target Ther.* 2021 Dec 1;6(1). [Internet] [cited 2024 Apr 4]. Available from: /pmc/articles/PMC8617206/
19. Loh CY, Arya A, Naema AF, Wong WF, Sethi G, Looi CY. Signal transducer and activator of transcription (STATs) proteins in cancer and inflammation: Functions and therapeutic implication. *Front Oncol.* 2019;9.
20. Shao F, Pang X, Baeg GH. Targeting the JAK/STAT signaling pathway for breast cancer. *Curr Med Chem.* 2020 Dec 8;28(25):5137–51.
21. Seif F, Khoshmirsafa M, Aazami H, Mohsenzadegan M, Sedighi G, Bahar M. The role of JAK-STAT signaling pathway and its regulators in the fate of T helper cells. *Cell Commun Signal.* 2017 Jun 21;15(1):1–13. [Internet] [cited 2024 Apr 4]. Available from: https://biosignaling.biomedcentral.com/articles/10.1186/s12964-017-0177-y

22. Garg M, Shanmugam MK, Bhardwaj V, Goel A, Gupta R, Sharma A, et al. The pleiotropic role of transcription factor STAT3 in oncogenesis and its targeting through natural products for cancer prevention and therapy. *Med Res Rev.* 2021 May 1;41(3):1291–336.
23. Solt LA, May MJ. The IκB kinase complex: Master regulator of NF-κB signaling. Immunol Res. 2008;42.
24. Liu T, Zhang L, Joo D, Sun SC. NF-κB signaling in inflammation. *Signal Transduct Target Ther.* 2017;2.
25. H. Z, L. W, G. Y, Y. C, M. Z, Y. W. Inflammation and tumor progression: Signaling pathways and targeted intervention. *Signal Transduct Target Ther.* 2021;6(1).
26. Xiao K, Liu C, Tu Z, Xu Q, Chen S, Zhang Y, et al. Activation of the NF-κ B and MAPK signaling pathways contributes to the inflammatory responses, but not cell injury, in IPEC-1 cells challenged with hydrogen peroxide. *Oxid Med Cell Longev.* 2020;2020.
27. Serasanambati M, Chilakapati SR. Function of Nuclear Factor Kappa B (NF-kB) in human diseases-a review. *South Indian J Biol Sci.* 2016;2(4).
28. Housden BE, Perrimon N. Spatial and temporal organization of signaling pathways. *Trends Biochem Sci.* 2014;39.
29. Bhatla SC. Signal perception and transduction. In: Satish C Bhatla, Manju A. Lal, editor, *Plant Physiology, Development and Metabolism.* Springer Singapore;2018.
30. Suskiewicz MJ. The logic of protein post-translational modifications (PTMs): Chemistry, mechanisms and evolution of protein regulation through covalent attachments. *BioEssays.* 2024 Mar 1;46(3):2300178. [Internet] [cited 2024 Apr 4]. Available from: https://onlinelibrary.wiley.com/doi/full/10.1002/bies.202300178
31. Lu H, Zhou Q, He J, Jiang Z, Peng C, Tong R, et al. Recent advances in the development of protein–protein interactions modulators: Mechanisms and clinical trials. *Signal Transduct Target Ther.* 2020;5.
32. Su Z, Dhusia K, Wu Y. Understand the functions of scaffold proteins in cell signaling by a mesoscopic simulation method. *Biophys J.* 2020;119(10).
33. Roy F, Laberge G, Douziech M, Ferland-McCollough D, Therrien M. KSR is a scaffold required for activation of the ERK/MAPK module. *Genes Dev.* 2002;16(4).
34. Borowicz P, Chan H, Hauge A, Spurkland A. Adaptor proteins: Flexible and dynamic modulators of immune cell signalling. *Scand J Immunol.* 2020;92(5).
35. Belov AA, Mohammadi M. Grb2, a double-edged sword of receptor tyrosine kinase signaling. Sci Signal. 2012;5.
36. Margiotta A. All good things must end: Termination of receptor tyrosine kinase signal. *Int J Mol Sci.* 2021;22.
37. Oeckinghaus A, Ghosh S. The NF-κB Family of transcription factors and its regulation. *Cold Spring Harb Perspect Biol.* 2009;1(4). [Internet] [cited 2024 Apr 4]. Available from: /pmc/articles/PMC2773619/
38. Terrell EM, Morrison DK. Ras-mediated activation of the Raf family kinases. *Cold Spring Harb Perspect Med.* 2019;9(1).
39. Ding L, Cao J, Lin W, Chen H, Xiong X, Ao H, et al. The roles of cyclin-dependent kinases in cell-cycle progression and therapeutic strategies in human breast cancer. *Int J Mol Sci.* 2020;21.
40. Kadam A, Jadiya P, Tomar D. Post-translational modifications and protein quality control of mitochondrial channels and transporters. *Front Cell Dev Biol.* 2023;11.
41. Du Toit A. Post-translational modification: Sweetening protein quality control. *Nat Rev Mol Cell Biol.* 2014;15.
42. De Bie P, Ciechanover A. Ubiquitination of E3 ligases: Self-regulation of the ubiquitin system via proteolytic and non-proteolytic mechanisms. *Cell Death and Differ.* 2011;18.
43. Mayer BJ. Protein-protein interactions in signaling cascades. *Methods Mol Biol (Clifton, N.J.).* 2006;332.
44. Schlessinger J. SH2/SH3 signaling proteins. *Curr Opin Genet Dev.* 1994;4(1).
45. Yue J, López JM. Understanding MAPK signaling pathways in apoptosis. *Int J Mol Sci.* 2020;21.
46. Burke P, Schooler K, Wiley HS. Regulation of epidermal growth factor receptor signaling by endocytosis and intracellular trafficking. *Mol Biol Cell.* 2001;12(6).
47. Tyson JJ, Chen KC, Novak B. Sniffers, buzzers, toggles and blinkers: Dynamics of regulatory and signaling pathways in the cell. *Curr Opin Cell Biol.* 2003;15.
48. Mengel B, Hunziker A, Pedersen L, Trusina A, Jensen MH, Krishna S. Modeling oscillatory control in NF-κB, p53 and Wnt signaling. *Curr Opin Genet Dev.* 2010;20.
49. Brandman O, Meyer T. Feedback loops shape cellular signals in space and time. *Science.* 2008;322.
50. Xing J, Ginty DD, Greenberg ME. Coupling of the RAS-MAPK pathway to gene activation by RSK2, a growth factor-regulated CREB kinase. *Science (1979).* 1996;273(5277).

51. Zhang Z, Yao L, Yang J, Wang Z, Du G. PI3K/Akt and HIF-1 signaling pathway in hypoxia-ischemia (Review). *Mol Med Rep.* 2018;18.
52. McCubrey JA, Steelman LS, Bertrand FE, Davis NM, Abrams SL, Montalto G, et al. Multifaceted roles of GSK-3 and Wnt/β-catenin in hematopoiesis and leukemogenesis: Opportunities for therapeutic intervention. *Leukemia.* 2014;28.
53. Israël A. The IKK complex, a central regulator of NF-κB activation. *Cold Spring Harb Perspect Biol.* 2010;2(3). [Internet] [cited 2024 Apr 4]. Available from: /pmc/articles/PMC2829958/
54. Thines L, Roushar FJ, Hedman AC, Sacks DB. The IQGAP scaffolds: Critical nodes bridging receptor activation to cellular signaling. *J Cell Biol.* 2023;222.
55. Huang X, Liu G, Guo J, Su ZQ. The PI3K/AKT pathway in obesity and type 2 diabetes. *Int J Biol Sci.* 2018;14.
56. Zarkou V, Galaras A, Giakountis A, Hatzis P. Crosstalk mechanisms between the WNT signaling pathway and long non-coding RNAs. *Non-coding RNA Res.* 2018;3.
57. Radhakrishnan K, Halász Á, Vlachos D, Edwards JS. Quantitative understanding of cell signaling: The importance of membrane organization. *Curr Opin Biotechnol.* 2010;21.
58. Ma X, Ashnest JR. Genome editing: Small-size CasΦ is shining. *Mol Plant.* 2020;13(9).

3 Cellular Receptors and Cell Signaling

Maleeha Azhar[1], Aroob Hasan Khan[1], Aroosa Younis Nadeem[1], Areej Fatima Ebtesam A Al-Suaimi[3,4], and Adeeb Shehzad[2]

[1]Department of Biomedical Sciences, School of Mechanical and Manufacturing Engineering, National University of Sciences & Technology, H-12 Islamabad, Pakistan

[2]Biodiversity Unit, Research Center, Dhofar University, Salalah 211, Oman

[3]Vice presidency for Scientific Research and Innovation, Imam Abdulrahman bin Faisal University (IAU), P.O. Box 1982, Dammam, 31441, Saudi Arabia

[4]King Abdulaziz and his Companions Foundation for Giftedness and Creativity "Mawhiba" Riyad, 11372, Saudi Arabia

3.1 INTRODUCTION

The emergence of multicellular organisms from unicellular ancestors is a remarkable milestone in our planet's evolutionary timeline. This transition required cells to develop distinct ways of communicating with each other to coordinate their functions. Cell-to-cell communication is crucial in the early stages of embryonic development, determining cell fate and specialization through interactions with nearby cells. In well-developed organisms, coordinating activities among different cell types is essential for complex processes like growth, immune response, and internal balance [1]. Defects in cell-to-cell communication, such as autocrine signaling dysregulation, play a crucial role in various medical conditions, including cancer, autoimmune diseases, and metabolic disorders. Cellular responsiveness and adaptability to the surrounding environment are fundamental in cell biology. Rapid response is primarily achieved through specifically tailored receptors on the cell surface that detect and respond to environmental changes [2].

Cellular signaling pathways, with their intricate and interconnected nature, are a source of communication within cells. They integrate information from various sources, including growth factor receptors, cell-matrix, and cell-cell contacts. This assimilation of information is pivotal in regulating a wide array of cellular processes, from protein synthesis to cell growth, motility, polarity, architecture, differentiation, and programmed cell death. The same signaling molecules can control different processes within various signaling complexes or at different intracellular locations. These pathways are dynamic and subject to developmental regulation, generating different outcomes depending on the cell type and context. Unraveling the functioning of these complex signaling networks in vivo and their alterations in cancer cells presents a significant intellectual challenge [3]. Studying cellular anatomy, physiology, and signaling molecule interactions with associated receptors provides valuable insights into how these systems work and how they might be manipulated for desired outcomes. This knowledge is essential for developing new therapies and treatments for various diseases and conditions [4].

Signaling proteins in intracellular networks operate as nodes, responding to multiple inputs and regulating multiple outputs. Modular protein domains facilitate protein-protein interactions, directing signaling proteins to specific substrates, locations, and multiprotein signaling pathways. These modules respond to post-translational modifications, including phosphorylation,

DOI: 10.1201/9781003594901-3

acetylation, and methylation, and can regulate signaling functions through intramolecular interactions. Understanding the signaling network's architecture is key to identifying all inter- and intramolecular interactions, thus unlocking the mysteries of cellular signaling [2].

3.2 RECEPTOR TYPES

Cell-surface receptors, localized in various cell types, including immune cells (B and T cells) and sensory organs, are crucial for cell communication and intracellular processes such as cell growth, immune response, and animal development [5]. They typically comprise an extracellular ligand-binding domain, a hydrophobic transmembrane domain, and a cytoplasmic domain. When a ligand binds to the receptor, it triggers a change in the extracellular domain, leading to the activation of enzymatic machinery attached to the cytoplasmic domain. This enzymatic machinery produces second messengers, such as kinases and phosphatases, for signal transduction. The three main categories of cell-surface receptors are G-protein-coupled receptors (GPCRs), ionotropic receptors, and receptor tyrosine kinases [6].

3.2.1.1 Cytoplasmic Receptors

Cytoplasmic receptors, located within the cell, possess a unique ability in cellular signaling. They can directly interact with membrane-permeable molecules like steroid hormones. When these signaling molecules, also known as ligands, bind to cytoplasmic receptors, a significant change occurs. The ligand-receptor complex transforms into an active transcription factor, which then integrates into the nucleus. Inside the nucleus, the complex interacts with DNA and regulates gene expression, controlling which genes are turned on or off. This unique ability, exclusive to cytoplasmic receptors, allows them to influence gene activity without relying on additional cellular messengers [7]. Cytoplasmic receptors are further categorized into Intracellular receptors and transmembrane receptors.

3.2.1.2 Intracellular Receptors

Intracellular receptors, such as the estrogen receptor and the glucocorticoid receptor, are a class of proteins found within the cytoplasm or nucleus of cells. They mediate cellular responses to various signaling molecules, including hormones, steroids, and neurotransmitters. These responses include modifications in gene expression, alterations in cell metabolism, and changes in cell growth and differentiation patterns. Unlike cell-surface receptors, which transduce extracellular signals into intracellular responses through second messenger systems, intracellular receptors directly regulate gene expression upon ligand binding. This direct modulation of gene expression allows for precise and long-lasting cellular responses to changes in the extracellular environment. Intracellular receptors have a modular domain structure that typically consists of three main domains: the ligand-binding domain (LBD), the DNA-binding domain (DBD), and the transactivation domain (TAD) [8].

The ligand-binding domain is the primary site for ligand recognition and binding and plays a pivotal role in the activation of intracellular receptors. Ligands, such as hormones, steroids, and neurotransmitters, bind to the LBD, inducing conformational changes in the receptor that facilitate its activation. This activation is a crucial step for gene regulation and highlights the significance of ligand recognition and binding in the functioning of intracellular receptors. The DNA-binding domain binds to specific DNA sequences, known as hormone response elements (HREs), within the promoter regions of target genes. This binding influences the transcriptional activity of these genes, impacting gene expression. The pivotal role of the DBD in this process underscores its importance in the regulation of gene expression and, consequently, in cellular responses. In some intracellular receptors, the TAD facilitates interactions with co-regulatory proteins, such as co-activators or co-repressors, which play a crucial role in modulating the receptor's transcriptional activity. Co-activators enhance the receptor's ability to activate gene expression by promoting the assembly of the transcriptional machinery, while co-repressors inhibit this activity by preventing it. These interactions are essential for precise regulation of target gene expression [9, 10].

Steroid hormones known as glucocorticoids can bind to intracellular receptors. The glucocorticoid receptor (GR) is expressed in various tissues and regulates gene expression and signaling pathways. It can modulate immune and inflammatory responses by inhibiting transcription factors such as NF-κB and AP-1. Additionally, it can suppress kinases such as mitogen-activated protein kinases (MAPKs), which play a role in downstream signaling pathways and gene expression. GR signaling pathway dysregulation has been linked to various diseases, including autoimmune disorders, inflammatory bowel disease, and cancer [11, 12].

Intracellular receptors' actions can be broadly categorized as genomic and non-genomic based on their mechanism of action. When a ligand binds to intracellular receptors, it induces a conformational change, which helps it integrate into the nucleus. Inside the nucleus, the activated receptor selectively attaches to certain hormone response elements (HREs) in the promoter regions of target genes. This interaction attracts co-regulatory proteins, which play a crucial role in adjusting the receptor's transcriptional function. This process ultimately controls the expression of specific genes. The transcriptional response initiated by intracellular receptors is known for its specificity, duration, and ability to coordinate intricate cellular processes [13]. Importantly, certain intracellular receptors also enable rapid non-genetic reactions upon ligand binding. These responses, occurring within seconds or minutes, differ from alterations in gene expression. They often involve kinase cascade activation or other signaling pathways, regulating a wide range of cellular functions, including ion transport, cell proliferation, and apoptosis, thereby maintaining the homeostasis of the cell [14].

3.2.2 Transmembrane Receptors

These are proteins located at the cell surface that span the membrane. Some types of transmembrane receptors are discussed below.

3.2.2.1 G-Protein-Coupled Receptors

G-protein-coupled receptors are a part of the largest group of membrane receptors, comprising over 800 proteins in subfamilies involved in different physiological responses in eukaryotes, from synaptic transmission to basic metabolism. GPCRs are complex structures consisting of seven transmembrane bundles that interact with ligands and facilitate activation of intracellular signal transduction pathways [15, 16]. GPCRs are classified into subfamilies: class A (rhodopsin), class B (secretin and adhesion), class C (glutamate), and some frizzled and seven transmembrane protein receptors [17]. A diverse array of ligands bind with the GPCRs, including proteins, peptides, lipids, and sugars from energy metabolism. This ligand binding causes conformational changes that expose and activate intracellular proteins like kinases (GRKs), β-arrestins, and heterotrimeric G-proteins. These pathways regulate normal physiological processes like synaptic transmission, neurological functions, cell differentiation, proliferation, inflammation, and immunity [18]. In heterotrimeric G-proteins, part of the GTPase family, guanosine diphosphate (GDP) dissociates from its Gα subunit and is replaced by guanosine triphosphate (GTP), leading to dissociation from Gβ/Gγ. This frees the α, β, and γ subunits for downstream signaling. The Gα subunit has four families based on function and morphology: Gαs, Gαi/o, Gαq, and Gα12/13. Gαs/i regulate cyclic adenosine monophosphate (cAMP) activity by activating or inhibiting adenyl cyclase, Gα13 activates the Ras family of GTPases, and Gαq activates inositol 1,4,5-phosphate (IP3) signaling by activating phospholipase-Cβ and intracellular mobilization of Ca^{2+}. GRKs play a role in phosphorylating GCPRs in the intracellular C-terminal and facilitate β-arrestin activation [19]. β-arrestins bind to different signaling proteins that activate pathways distinct from the G-proteins [15]. Most of these pathways are mutated and play influential roles in cancer development and progression.

3.2.2.2 Enzyme-Based Receptors

Enzyme-based receptors are transmembrane proteins with an extracellular domain that interacts with the ligands, a membrane-bound domain, and an intracellular cytoplasmic domain that interacts with proteins that function as enzymes. These receptors form a large, primarily consisting of tyrosine kinases or proteins associated with kinases [20]. They classified into receptor tyrosine kinases that phosphorylate intracellular proteins, tyrosine kinase-associated receptors, tyrosine phosphatases that detach phosphate groups, receptor serine/threonine kinases, receptor guanylyl cyclases that regulate cGMP, and histidine kinase-associated receptors [21].

Cytosolic proteins have intracellular binding sites for tyrosine residues. Ligands, usually growth factors, bind to the extracellular domain, resulting in the phosphorylation of the tyrosine residues on these proteins, which initiates intracellular signal transduction by interacting with different proteins [22]. When a ligand binds to the receptor, the intracellular domain forms dimer by the linking of two receptors. The tails of the two receptors of the dimer interact and phosphorylate each other through the process known as autophosphorylation, which results in receptor activation. One is the tyrosine kinase activation of Ras. Ras, a group of GTP-binding proteins bound to the cytoplasmic side of the plasma membrane, is activated by the tyrosine kinase receptor through GDP to GTP phosphorylation. Activated Ras then phosphorylates intracellular kinase enzymes, triggering a series of protein phosphorylation reactions from the membrane to the cell's nuclear region. This reaction cascade, called the mitogen-activated protein kinase (MAPK) pathway, further phosphorylates serine and threonine residues on the transcription-binding domains. The Ras/MAPK pathway is crucial for cell proliferation and differentiation, and is often mutated in the cancer cells [23].

3.2.2.3 Chemical Receptors

Chemical receptors are categorized into two types based on function and structure:

3.2.2.3.1 Ligand-Gated Ion Channels

Ligand-gated ion channels are specific ion channels in the plasma membrane that allow the movement of Na^{1+}, Ca^{+2}, Cl^{-1}, or K^{+1} across the membrane, maintaining membrane potential by hyperpolarizing or depolarizing it, according to the flow of ions. These large receptors have four to five subunits forming pores, each with hydrophobic domains interacting with hydrophilic groups [22]. Ion flow across the membrane is determined by the electrochemical gradient of the permeable ions and is usually passive. The working of these channels is simple. The channels are either open or gated, with gated channels opened by the binding of a specific ligand (usually a neurotransmitter), causing conformational changes in the channel subunit that allow conduction. Regulation of these channels is done by certain exogenous and endogenous modulators. Ion-gated channels allow for rapid transmission of synaptic information from pre-synaptic to post-synaptic neurons within milliseconds.

Ligand-gated channels are classified as excitatory or inhibitory, and fall into five receptor families:

1) The pentameric cys-loop family, including nicotinic acetylcholine (nACh) receptors, 5-hydroxytryptamine3 (5HT3) receptors, γ-amino-butyric acid (GABA) receptors, glycine (Gly) receptors, and the zinc-activated channel receptor
2) The tetrameric glutamate-gated receptors
3) The ATP-gated receptors
4) The transient receptor potential (TRP) channels
5) The acid-sensitive ion receptors (ASIC)

Traditionally, nACh, 5-HT3, and glutamate receptors are excitatory, while GABA and glycine receptors are inhibitory. TRP and ASIC receptors have sensory functions but are also found in the central nervous system. Most ion-gated receptors are expressed in neural pathways, muscles, lymphocytes, endothelial cells, keratinocytes, etc. [24, 25].

3.2.2.3.2 Voltage-Gated Ion Channels

Voltage-gated ion channels are plasma membrane proteins responsible for generating action potentials in excitable cells like neurons and muscles. These pore-forming ion channels are selectively permeable to Na^+, K^+, Ca^{2+}, and Cl^- ions across the membrane. The three major classes are Na^+, K^+, and Ca^{2+} channels, encoded by several genes with varying physiological functions and localization [26]. The general structure includes four subunits (in K^+ channels) or one protein with four subunits (in Na^+ and Ca^+ channels). Each subunit comprises six transmembrane units and a pore loop, with specific functions: the fourth unit is a voltage sensor, the fifth and sixth units including the pore loop are involved in ion conduction, and the second and third units have aspartate and glutamate residues that modify receptor functions of [27].

Na^+ channels activate through outward charge movement resulting from transmembrane potential changes, causing conformational changes in the receptor that opens the pore. The channel is inactivated within milliseconds after activation [26]. Ca^{2+} channels activate similarly and are involved in crucial functions inside the cell beyond generation action potentials, such as controlling neurotransmitter release at synapses and major biochemical pathways. K^+ channels form the widest-ranging class of voltage-gated channels, playing crucial roles in generating resting membrane potential and entering the hyperpolarized state. Their inactivation varies by subtype, with some activating instantly and others taking a few minutes [28].

3.3 DYNAMICS OF LIGANDS

In the realm of cell biology, cells exhibit remarkable adaptability, responding to a myriad of signals, be they electrical, mechanical, or chemical. Chemical signals, including growth factors, hormones, cytokines, neurotransmitters, and extracellular matrix components, reign supreme. Prokaryotic cells, for instance, can detect nutrients and facilitate mechanotransduction towards richer nutrient gradients. Eukaryotic cells, on the other hand, showcase a staggering level of complexity in their response to signals, enabling them to adapt to environmental changes and uphold cellular homeostasis dynamically. Ligands, chemical signals produced by signaling cells and transmitted to target cells, play a pivotal role in this process. These molecules range from small proteins to small ions and are classified as small hydrophobic ligands (which can traverse plasma membranes) and water-soluble ligands (which cannot) [4, 29].

3.3.1 Small Hydrophobic Ligands

Lipid-soluble ligands, also known as hydrophobic ligands, can penetrate the plasma membrane and directly bind directly to interior receptors. Among these ligands, steroid hormones constitute a significant class. Steroids are a type of lipid characterized by a hydrocarbon skeleton composed of four fused rings. The functional groups attached to this carbon skeleton differentiate the various types of steroids. For instance, estradiol, a type of estrogen, serves as a female sex hormone, while testosterone acts as a male sex hormone. Cholesterol, another steroid hormone, is an essential structural element in biological membranes and a precursor to other steroid hormones. Other hydrophobic hormones include thyroid hormones and vitamin D. These ligands require carrier proteins for transportation in the bloodstream [30].

3.3.2 Water-Soluble Ligands

Water-soluble ligands, which are polar and often too large to pass through the plasma membrane unaided, typically bind to the extracellular domain of cell-surface receptors. This diverse group includes small molecules, peptides, and proteins, all of which are essential in cell signaling processes. Examples of water-soluble ligands include protein hormones like growth hormone (GH), insulin, and insulin-like growth factors (IGFs). These ligands, released by endocrine glands, bind to cell-surface receptors such as GPCRs and receptor tyrosine kinases. Upon binding, these ligands induce conformational changes that activate intracellular signaling pathways, regulating key cellular functions like metabolism, growth, and differentiation [31]. Other water-soluble ligands include cytokines, chemokines, and neurotransmitters.

3.4 TYPES OF CELLULAR SIGNALING

3.4.1 Paracrine

Cells commonly communicate through the release of chemicals, which spread and are detected by another cell, as observed in endocrine signaling. Paracrine signaling is characterized by the localized action of released chemicals near the signaling cell. The diffusion of signaling molecules is constrained due to their rapid degradation by extracellular enzymes or their quick immobilization on or uptake by adjacent cells. An example of such signaling is the discharge of neurotransmitters from neurons, affecting adjacent neurons or nearby cells like muscle cells. In immune responses, various types of white blood cells secrete cytokines and chemokines to regulate the other white blood cells and nearby cells in reaction to organism damage, such as pathogen invasion [32, 33].

Epidermal growth factor (EGF) is a classic paracrine signaling molecule that controls cell growth, survival, and differentiation. EGF binds to its receptor, EGFR, located exterior to target cells, initiating signaling cascades such as the Ras-MAPK and PI3K-Akt pathways. EGF signaling dysregulation is associated with the advancement of cancer and other medical conditions.

Transforming growth factor-β (TGF-β) is a versatile cytokine that plays a role in cell proliferation, differentiation, and immune system regulation. TGF-β binding to its receptors activates Smad proteins and other downstream proteins. Abnormal TGF-β signaling is linked to the development of fibrosis, autoimmune diseases, and metastasis [34].

Acetylcholine (ACh) functions as another paracrine neurotransmitter in both the central and peripheral nervous systems, as well as in non-neuronal tissues. ACh produced by neurons binds to either nicotinic or muscarinic receptors on neighboring cells, leading to depolarization or modified intracellular signaling pathways. Abnormal modulation of ACh signaling is associated with neurological disorders like Alzheimer's disease and myasthenia gravis.

Glutamate, the main neurotransmitter that causes excitement in the central nervous system, is involved in transmitting signals between synapses, facilitating changes in neural connections, and promoting the survival of neurons. Pre-synaptic neurons release glutamate, which then activates ionotropic (such as NMDA, AMPA, and kainate) and metabotropic glutamate receptors in nearby cells, triggering various physiological responses. Dysregulated glutamate signaling is associated with neurodegenerative diseases, epilepsy, and psychiatric disorders [35].

3.4.2 Autocrine

Autocrine effect, a term used to describe the localized action of diffusible signaling molecules on the parent cell, is a fascinating mechanism in cellular signaling. This form of signaling, comparable to paracrine signaling, is prevalent among growth hormones and occasionally among chemicals classified as eicosanoids. It is commonly observed in cells during development or differentiation, prolonging required responses. After a cell has received instructions to undergo a specific developmental change, it releases an autocrine signal to reinforce the intended pathway and ensure that it is followed. Notably, tumor cells frequently exhibit autocrine signaling, producing growth hormones that stimulate unregulated growth and proliferation, a phenomenon that warrants further investigation [36, 37].

Interleukin-2 (IL-2) is a cytokine generated by activated T cells that plays a role in controlling immunological responses by stimulating the growth and activation of T cells. T cells release IL-2, which binds to IL-2 receptors comprising at least three subunits: alpha (IL-2R alpha), beta (IL-2R beta), and gamma (IL-2R gamma) chains. While the precise role of IL-2R gamma in IL-2 signaling remains uncertain, it is known that IL-2R beta is essential for receptor-mediated signaling. This self-stimulating cycle boosts the reproduction of T cells and maintains the immune system's effectiveness during inflammation and infection. Abnormal modulation of IL-2 signaling is associated with autoimmune disorders and resistance to cancer therapy [38].

Insulin-like growth factor-1 (IGF-1) regulates cellular development, viability, and metabolism. Autocrine IGF-1 signaling generates and releases IGF-1, which binds to transmembrane IGF-1 receptors. This activates PI3K-Akt and Ras-MAPK pathways, stimulating cell proliferation and enhancing cell survival. Abnormal IGF-1 signaling is associated with cancer, metabolic disorders, and age-related diseases [37, 39].

Vascular endothelial growth factor (VEGF) is essential for developmental and pathological angiogenesis. Many types of tumors commonly produce VEGF, which is responsible for promoting angiogenesis. Autocrine VEGF signaling takes place in endothelial cells, where VEGF is localized to VEGF receptors located in the interior of the cells. VEGF receptor activation initiates intracellular signaling pathways that enhance the proliferation, migration, and permeability of endothelial cells. VEGF binds to specific tyrosine kinase receptors, including VEGFR-1 (Flt-1) and VEGFR-2 (KDR), as well as non-tyrosine kinase receptors like Neuropilin-1 (NP1) and Neuropilin-2 (NP2) [40, 41]. Abnormal VEGF signaling contributes to disorders associated with the growth of new blood vessels, such as cancer and retinopathy [41].

3.4.3 Endocrine

Endocrine cells produce hormones that communicate with target cells located in remote organs. These hormones travel throughout the body via the circulatory system, exposing the entire organism to them. However, only cells that express specific hormone receptors or target cells are affected by them. This results in slow responses from the target cells, but the effects also last longer. Endocrine signaling involves two types of receptors: cell surface and intracellular.

Cell-surface receptors function similarly to other membrane-bound receptors. When hormones, also known as ligands, bind to hormone-specific G-protein-coupled receptors, they cause conformational changes in the receptor and release a G-protein subunit. This activates second messengers, which internalize the message by triggering signaling cascades and transcription factors [42]. Numerous hormones, including epinephrine, norepinephrine, insulin, prostaglandins, prolactin, and growth hormones, function through cell-surface receptors.

In contrast, steroid hormones, such as testosterone, estrogen, and progesterone, use intracellular receptors to transmit signals. Hormone binding causes a conformational change in the receptor, activating the receptor's potential as a transcription factor and initiating gene expression. These intracellular hormone receptors belong to a large superfamily of receptors, all sharing a similar single polypeptide chain structure with three distinct domains. The N-terminus contains the active transcription factor domain responsible for initiating gene expression. The middle region comprises a DNA-binding domain specific to the gene of interest, allowing the receptor to bind to specific DNA sequences. The C-terminus contains the hormone-binding domain, which triggers the conformational change in the receptor [43].

3.4.4 Direct

Direct signaling enables coordinated responses among multiple cells, even if only one receives the initial signal. Plasmodesmata, present in all plants, create a vast communication network throughout the organism. This primary method of cellular communication involves signaling molecules directly interacting with receptors on neighboring or surrounding cells, facilitating swift and accurate coordination of cellular reactions within nearby microenvironments.

Unlike other signaling mechanisms that rely on diffusible chemicals or long-range transmission pathways, direct signaling requires physical contact or proximity between signaling and target cells. Gap junctions in animals and plasmodesmata in plants link the plasma membranes of adjacent cells. These aqueous channels facilitate the diffusion of small signaling molecules between the two cells. Small molecules, like calcium ions (Ca^{2+}), are capable of intercellular movement, while larger molecules, such as proteins and DNA, cannot pass through the channels due to their size. This specificity guarantees cellular independence while facilitating rapid and efficient signal transmission [44].

Signal transduction occurs through either cell-surface receptor interactions or gap junction-mediated communication. In receptor interactions, signaling molecules on one cell's surface bind to neighboring cells' receptors, triggering intracellular signaling cascades and cellular responses. Hap junctions enable swift and synchronized communication between neighboring cells through specialized structures called gap junctions, crucial for functions like neuronal electrical coupling and heart muscle contraction synchronization [45].

The Notch signaling system, a highly preserved mechanism, plays a role in determining cell fate, organizing tissues, and regulating developmental processes. When Notch receptors on one cell interact with ligands on neighboring cells, it triggers proteolytic cleavage and release of the Notch intracellular domain (NICD). The NICD protein then translocates to the nucleus and exerts control over the expression of specific genes, thus impacting the determination of cell fate.

Cadherins, a group of calcium-dependent membrane proteins, facilitate cell-cell attachment. These connections between adjacent cells are essential for tissue development, epithelial structure maintenance, and cell movement. Disruption of cadherin-mediated adhesion can result in developmental abnormalities and the spread of cancer to other parts of the body [46].

3.5 SIGNAL TRANSDUCTION

Signal generation involves a series of biochemical events from the receptor on the membrane to the cytoplasmic effector that travels to the nucleus and induces physical or biochemical changes in the cell's transcriptional machinery. Signal transduction occurs via both the transmembrane receptors themselves and the protein complexes on the membrane that link intracellular components with the extracellular region [47]. Some receptors undergo conformational changes, which can open ion channels and generate a membrane potential, while others activate their intracellular parts, triggering intracellular proteins and enzymes [48]. In this section, we will highlight the steps involved in the activation of a receptor and the generation of a signal.

3.5.1 Receptor Binding

Transmembrane receptors comprise hydrophilic extracellular and intracellular domains and a hydrophobic transmembrane domain. As the ligand binds to the specific receptor, its subunits undergo a conformational change, initiating a signal in the receptor's intracellular part. This process of transforming an inactive receptor to an active, signal-initiating one is called receptor activation [49]. Ligands carry extracellular signals, both physical and chemical, from the environment. Different types of stimuli, such as biochemical (growth factors, hormones, and proteins), mechanical changes (shear, stretch), antigens, and stress signals, can bind to the receptor's extracellular portion [50]. In response to the ligand binding, the adjacent receptors cluster together on the membrane surface, a process called dimerization, and start crosslinking autophosphorylation. The intensity of the crosslinking determines the strength of the signal [47].

3.5.2 Receptor Dimerization and Autophosphorylation

Most cell-surface receptors are linked to protein tyrosine kinase tails inside the cell, known as receptor protein tyrosine kinases, which phosphorylate other proteins' tyrosine residues. During receptor dimerization, these tails can phosphorylate each other through autophosphorylation. Phosphorylation activates the receptors, enabling them to activate other proteins inside the cell. Some receptors don't directly possess tyrosine kinases but interact with intracellular enzymes, called receptor-associated tyrosine kinases, or non-receptor protein tyrosine kinases. These kinases can either be cytosolic or nuclear. Receptor clustering brings these kinases together, leading to mutual autophosphorylation [47, 50].

3.5.3 Activation of Intracellular Signaling Molecules

These receptor conformational changes initiate a cascade of protein-protein interactions inside the cells, facilitated by kinases, which add phosphate groups (phosphorylation), and phosphatases, which remove them (dephosphorylation). These events trigger downstream signaling activity in the nucleus [47, 51, 52]. Intracellular protein activation or inactivation through phosphorylation or dephosphorylation is a common mechanism regulating the biochemical processes inside the cell, controlling the rate of cell proliferation. Phosphorylation can also create additional binding sites on the protein without affecting its intrinsic activity. As a result, the phosphate acts as a tag and recruits auxiliary proteins from the cytosol. The recruited proteins interact with the receptors' phosphorylated sites and are activated by phosphates, initiating the transduction process.

Phosphorylation's advantages include its rapidity and reversibility, allowing easy regulation of the addition and removal of phosphate groups by the kinases and phosphatases. This enables quick inactivation of the signal transduction pathways, keeping cellular responses under control [47].

3.5.4 Signal Transduction and Amplification

After the signal has been transferred through the membrane, many cytosolic effectors are activated with the help of activated membrane enzymes and proteins. The cell cytosol consists of various effectors that influence signal propagation through calcium ion mobilization, reversible phosphorylation by kinases and phosphatases, cAMP or cGMP, and lipid mediators. Several parallel or serial cytosolic proteins help in signal transduction and amplification. [50] A common example is the phospholipase C-γ (PLC-γ) enzyme. Receptor-associated tyrosine kinases recruit PLC-γ and phosphorylate its tyrosine residue, activating it. PLC-γ then cleaves multiple phosphatidylinositol biphosphate (PIP2) membrane phospholipids into their components: inositol triphosphate (IP3) and diacylglycerol (DAG). This process amplifies the signal as one PLC-γ molecule cleaves numerous PIP2 molecules. IP3 binds to its receptors on the endoplasmic reticulum, which allows the release of Ca^{2+} ions in the cytosol. Reduced Ca^{2+} levels in the endoplasmic reticulum prompt the opening of calcium channels inside the plasma membrane. The resulting increased Ca^{2+} levels inside the cytosol interact with enzymes, transmitting the signal to the nucleus. Meanwhile, DAG activates protein kinase C (PKC), which also induces signal transmission to the nucleus [47].

3.5.5 Activation of Small G-proteins

Some small G-proteins, or GTP-binding proteins, also interact with tyrosine kinase-associated receptors. Ras is a common example. Inactivated Ras is bound to GDP, which converts to GTP in order to activate Ras. This process is also reversible: Ras-GTPase activity removes the phosphate from GTP to form GDP and inactivate Ras. Ras is only activated upon receiving signals from activated receptors. Activated Ras then triggers a series of protein kinases known as the mitogen-activated protein kinase cascade. The MAPK cascade is a common pathway in multicellular organisms that directly phosphorylates and activates transcription factors to regulate gene expression. MAPK mutations are frequent in cancer cells [47].

3.5.6 Regulation of Gene Expression

Ultimately, the extracellular signal reaches the nucleus, where it regulates de novo gene expression, influencing protein transcription and translation and producing cellular responses. Numerous enzymes in the nucleus phosphorylate transcriptional factors [50].

3.6 CROSSTALK BETWEEN SIGNALING PATHWAYS

Several signaling pathways have been studied to elucidate the biochemical processes underlying cellular responses. This section discusses some of the most common intracellular signaling pathways:

3.6.1 Ras/Raf/MAPK

The small G-protein Ras pathway is triggered by various growth factor receptors, cytokines, and hormones. In its active form, Ras is bound by GDP, which converts to GTP upon activation. Activated GTP-bound Ras then triggers several downstream cascades regulating proliferation and apoptosis under normal conditions [53–55]. Raf kinase proteins work with Ras to mediate the Ras/Raf/MAPK pathway [53]. There are three types of Raf kinases: A-Raf, B-Raf, and Raf-1, all involved in MAPK signaling. MAPK kinases activate MEK1 and MEK2, which in turn phosphorylate and activate the final effectors ERK1 and ERK2. ERK activation in the cytoplasm leads to the mediation of cell adhesion, cytoskeletal organization, cell movement, etc., while ERK activation in the nucleus leads to the phosphorylation of transcriptional factors involved in DNA synthesis and cell cycle regulation [56, 57].

3.6.2 mTOR

The mTOR (mammalian target of rapamycin) pathway is regulated by Akt, also known as protein kinase B. Akt, activated by growth factors binding to their receptors, mediates cell survival and proliferation in eukaryotes. Akt interacts both with PIP3 and mTOR, directly phosphorylating the latter. Akt also has an inhibitor that removes the phosphate attached to PIP3. The Akt/mTOR pathway promotes cell survival, allowing cells to evade apoptotic stimuli by inactivating the pro-apoptotic proteins [53, 58].

3.6.3 cAMP

The cyclic adenosine monophosphate (cAMP) pathway, also regulated by G-proteins, exemplifies signal amplification. Ligand (hormone) binding to the receptor activates hundreds of G-proteins, which trigger adenylyl cyclase and amplify the signal. Each adenylyl cyclase catalyzes the significant cAMP production in the cytosol. This activates protein kinase A, which phosphorylates phosphorylase enzyme, in turn phosphorylating and activating glycogen phosphorylase. The cAMP pathway also activates transcription of the cAMP response element (CRE) by phosphorylating the CRE-binding element (CREB), thereby regulating cell proliferation, differentiation, and survival [59].

3.6.4 Phospholipids

Membrane phospholipids, derived from the plasma membrane, form common signal transduction pathways. Phosphatidylinositol 4,5 bisphosphate (PIP2), a plasma membrane derivative, is cleaved by PLC to form IP3 and DAG protein (as discussed above, in section 3.5.4, "Signal Transduction and Amplification"). The PIP2 pathway is triggered by the activation of both protein tyrosine kinases and the G-protein-coupled receptors. GPCRs activate phospholipase C-β (PLC-β), while protein tyrosine kinases activate PLC-γ. Protein kinase C also activates the MAPK pathway, influencing gene expression and proliferation [55, 59].

3.6.5 JAK/STAT

While the MAPK pathway indirectly links extracellular signals to the nucleus through a cascade of reactions, the JAK/STAT pathway provides a direct link between membrane protein tyrosine kinases and transcription factors, influencing gene expression. The STAT proteins that act as transducers are localized in the cytoplasm. Ligand binding causes recruitment and binding of these proteins to the receptor's intercellular domain tyrosine residues. The non-receptor tyrosine kinase enzymes from the JAK family phosphorylate the STAT proteins, which dimerize and move to the nucleus, regulating cell survival, apoptosis, and hypertrophy [58–60].

The biochemical pathways occurring inside the cells are interconnected, with signaling molecules interacting and sharing pathways. This crosstalk can result in inhibition of one pathway affecting others, impacting the versatility of cell signaling [61]. For example, the cAMP pathway interacts with the Ras/Raf/MAPK pathway. cAMP regulates CREB transcription factors in the presence of intracellular Ca^{2+}. Upon ligand binding by growth factors, Ras is activated, which activates its effector, Raf. cAMP inhibits C-Raf via protein kinase A (PKA), while the same kinase activates B-Raf.

Pathway crosstalk also works as a feedback mechanism, influencing the dynamics of the signal transduction inside the cell. The feedback mechanism in MAPK requires positive feedback by ERK1 and ERK2 to activate Raf by inactivating the Raf-inhibiting protein. In the cyclic GMP pathway, the activation of phosphodiesterase reduces Ca^{2+} ions inside the cell by closing the Ca^{2+} gates, thereby reducing cGMP levels. Decreased Ca^{2+} levels reduce adenylyl cyclase activity, which decreases the level of cAMP in the cell [62]. Additionally, degradation of PLC-β alters the Ras/MAPK pathways, affecting proliferation and promoting apoptosis [63]. Significant interactions have also been reported between the PI3K, MEK, and Raf/MEK/ERK pathways [57] (Figure 3.1).

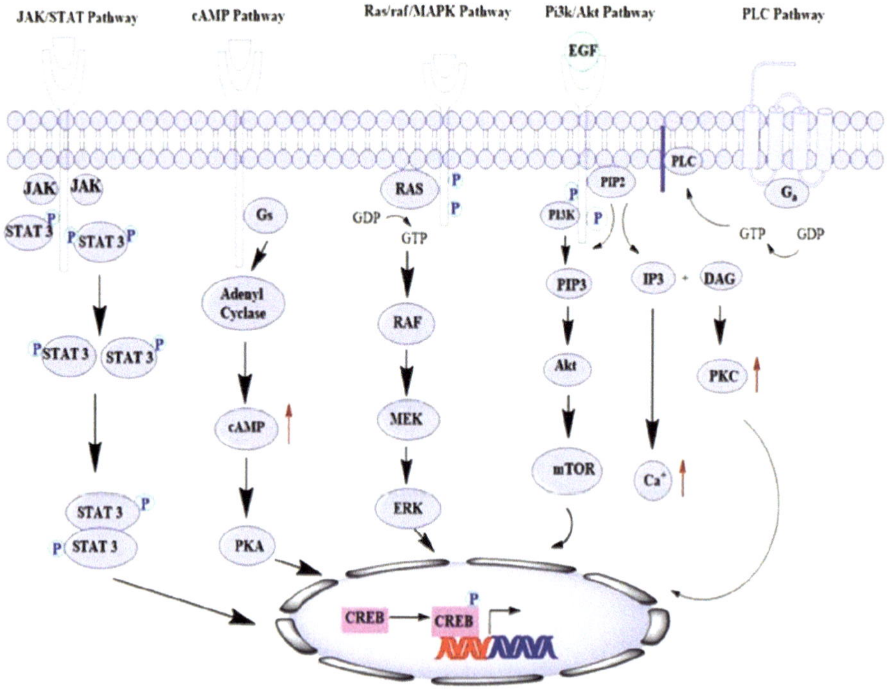

FIGURE 3.1 Crosstalk between different cellular pathways.

3.7 INHIBITION OF AN ABERRANT CELLULAR SIGNAL

Multicellular organisms are tightly controlled by cellular homeostasis during the cell cycle to maintain the integrity of the cells [64]. Apoptosis regulates the cell cycle at the molecular level during normal physiological processes and in response to toxicities, counterbalancing cell proliferation. Normal homeostatic processes induce apoptosis when a cell encounters an antigen or death signal. However, abnormal interference with normal cell signaling can lead to aberrant cell accumulation and disease [65]. Research has identified four types of molecules that trigger apoptosis in response to aberrant signaling: caspases, caspases activating adapter proteins, tumor necrosis factor (TNF) and its receptors, and BCL-2 proteins [66]. Additionally, elevated levels of secondary messengers like cAMP and cGMP can induce apoptosis in certain cell types by increasing intracellular nitric oxide following cAMP- and cGMP-induced Ca^{2+} release [67].

One apoptotic pathway involves the binding of a death ligand to a death receptor, which recruits the caspase-activating adaptor protein Fas-associated death domain (FADD). FADD then recruits the pro-apoptotic protein pro-caspase-8, which is subsequently hydrolyzed and activated. In another pathway, an apoptotic stimulus results in the release of cytochrome c from the mitochondria. Cytochrome c interacts with a cell death protein, apoptosis peptidase activating factor 1 (Apaf1), which activates pro-caspase-9. BCL-2 proteins, which are anti-apoptotic, inhibit apoptosis by interfering with cytochrome c release and its interaction with Apaf1 [68].

3.8 THERAPEUTIC APPROACHES TO ABERRANT SIGNALING

3.8.1 Cancer

The Ras/Raf/MAPK pathway is the most common unregulated pathway found in almost 30% of cancer types. Small molecules have been an attractive target for the inhibition of this pathway in cancer therapy [69]. In some cancers, cAMP plays a suppressive role; therefore, increasing the activity of cAMP inhibits these cancers. The cAMP signaling pathway is modulated by the phosphodiesterase (PDE) protein, which hydrolyzes cAMP into AMP. PDE inhibitors work by increasing cAMP levels [70]. On the other hand, forskolin, CREBtide, and KEMPtide work by modulating cAMP and PKA activity inside the cell [71]. The JAK/STAT pathway is also common in several solid tumors and cancers like lymphoma, lung cancer, and gastric cancer. Several novel treatments are being used in cancer therapy. Tofacitinib, a JAK inhibitor, inactivates the downstream STAT proteins. Oclactinib, another JAK inhibitor, targets the cytokines involved in JAK/STAT signaling. Ruxolitinib and peficitinib are also used as JAK inhibitors in cancer therapeutics [72]. In the mTOR pathway, rapamycin, everolimus, and deforolimus are used as mTOR inhibitors, blocking the phosphorylation of mTOR kinases [73].

3.8.2 Syndromes

The Costello, CFC, and Noonan syndromes are characterized by mutations in the regulation of the Ras/Raf/MAPK pathway. People with these syndromes have distinctive cranial and facial features, down-slanting palpebral fissures, hypertelorism, high-arched palate, heart defects, and musculoskeletal abnormalities, among many other common features. Many cancer therapeutic strategies are being applied to Costello, CFC, and Noonan syndromes. Farnesyltransferase inhibitors and Raf/MEK inhibitors are being investigated for targeting Ras in the treatment of Costello syndrome. Sorafenib, a kinase inhibitor, and PLX-4720, a Raf inhibitor, show promising results for CFC treatment [69]. Additionally, endocrine features are affected due to mutations in the Ras, SOS, and Raf genes, resulting in shorter stature. Growth hormone treatment is also being investigated for treating the nervous system-related phenotypic symptoms of these syndromes [74]. Mutations in PTEN, which is a PI3K/Akt negative downstream regulator, result in the development of tissue overgrowth syndromes. The PI3K/Akt pathway is activated by many receptors such as tyrosine kinases, small

G-proteins, and GCPR heterodimeric proteins, whose p110 domains also interact with the Ras pathway. P110 inhibitors have been developed for these tissue overgrowth syndromes, which result in symptoms like localized tissue overgrowth, and malformations in the organs, skin, nervous system, and skeleton. Rapamycin is used to block mTOR-induced vascular anomalies [75].

3.8.3 Nervous Disorders

Nervous system cells express PDE under normal conditions. PDE can be overexpressed under stress conditions with elevated levels of cytokines. Glucogenolysis, an important process of glycogen breakdown in the nervous system, is blocked by PKA and adenylyl cyclase inhibition, affecting neuronal functions like memory and learning, and contributing various nervous system disorders. Dibutyryl-cAMP (dbcAMP), which is an analog of cAMP, and forskolin, which activates adenylyl cyclase, both trigger glycogenolysis [76]. The JAK/STAT pathway plays a crucial role in neuronal inflammation and degeneration. Failure of this pathway leads to the pathogenesis of Parkinson's disease. Reduced inflammation of neuronal cells is achieved by AZD-1480, a JAK inhibitor, which also blocks STAT, inhibiting differentiation of neuronal cells [72, 77].

3.8.4 Other Disorders

The Ras/MAPK and PI3K/mTOR pathways are important players in cell growth, metabolism, and metabolic disorders. They are also involved in cardiac hypertrophy through the activation of $G_{\alpha q/11}$ receptors. $G_{\alpha q/11}$ activates the PLC pathway. Ras/MAPK and PI3K/Akt induction are enhanced by Ca^{2+}/calcineurine, and its overload leads to hypertrophy and ultimately heart failure. The mTOR inhibitor, rapamycin, blocks the activation of phenylephrine (PE) in the Ras pathway. This also shows the involvement of the Ras pathway in mTOR activation. Statins, which inhibit farnesyl synthesis, block the Ras pathway, reducing cardiac hypertrophy [78]. Forskolin, an adenyl cyclase-activating molecule, is being used to treat obesity, glaucoma, and cardiac and liver fibrosis by targeting the cAMP pathway. Several PDE inhibitor isoforms have been developed to treat asthma, neurological disorders, and tissue fibrosis, which are caused by dysregulated cAMP signaling [79]. dbcAMP and PKA, which help cAMP escape PDE hydrolysis, have been used to treat heart disease and inflammation [80]. In addition to cancer treatment, several cancer drugs are being tested for their therapeutic potential in other diseases linked with similar pathways. Tofacitinib and ruxolitinib, both JAK inhibitors, have also been found to treat arthritis, alopecia, and autoimmune disorders by reversibly binding to the ATP site of JAK proteins, thereby blocking STAT activation [72, 81].

Some of the therapeutic perspectives of signaling pathways are illustrated in Table 3.1.

3.9 FUTURE PROSPECTS

The cell signaling pathways discussed in this section play critical roles in the regulation of normal physiological processes like cell growth, proliferation, cell death, and metabolism. These pathways have been reported to have an essential influence on the development and progression of cancers. The crosstalk mechanisms are still poorly understood and promise improved therapeutic approaches by targeting several pathways together. Studying these cell signaling pathways will also lead to the development of novel biological markers and molecular and genetic targets for more efficient diagnosis and treatment strategies. Additionally, there exists an interesting similarity between these signaling pathways and electronic circuits pertaining to their feedback systems, hierarchies, and aberrant signaling. As a result, the study of these pathways combined with concepts from engineering can lead to the development of computational models incorporating experimental data for simulation-based studies. Such simulations can be used to predict and manipulate signaling behaviors with respect to therapeutic approaches.

TABLE 3.1
Receptors/Ligands Involved with Specific Pathways, with Their Associated Diseases and Treatment Strategies

Signaling Pathway	Associated Receptor/Ligands	Associated Diseases	Treatment Strategies	Ref
Ras/Raf/MAPK	EFGR, VEGFR, PDGFR	Mutations found in 30% of somatic cancers; pancreatic cancer, colorectal cancer, lung cancer, small and large intestine cancer, skin cancer, melanoma, etc.	MEK inhibitors (CI-1040, AZD-6244, RDEA-119, etc.), MEK, Ras, Raf inhibitors (GSK1120212), B-Raf inhibitors (PLX-4720, PLX-4032)	[82, 83]
	GFR, Cytokines, Extracellular	Noonan syndrome, Costello syndrome, Cardio-facio-cutaneous syndrome (CFC)	Growth hormone therapy for Noonan syndrome, potential Ras/MAPK inhibitors (farnesyltransferase) for Costello, potential PLX-4720 and kinase inhibitor (sorafenib) treatment for CFC	[69, 74, 84]
	GPCRs, Tyrosine Kinase Receptors	Cardiac hypertrophy	Inhibitors of angiotensin-II (candesartan, captopril), farnesyltransferase inhibitors (statins, FPT-III, siRNAs)	[78]
cAMP Pathway	GPCRs (Gαs)	Cardiac hypertrophy	Calcineurin inhibitor (cyclosporin A)	[78]
	GPCRs (Gαβγ)/ Ethanol Binding, Cytokines	Liver disease (steatosis, fibrosis)	Dibutyryl-cAMP (non-degradable cAMP), PDE-4 (phosphodiesterase-4) inhibitor, adenosine receptor inhibitors, PKA activators	[79, 80]
	GPCR, 5HT Receptors/ Neurotransmitters, Hormone Binding	Alzheimer's disease, multiple sclerosis, and major degenerative diseases of the nervous system	PDE-4 inhibitors (RO-204721), Small G receptor agonists	[76, 85]
	GPCRs, Small-G Proteins, Specific Androgen Receptors	Adenoma, Somatotrophinoma, germ-line tumors, hepatocellular carcinoma, brain tumor, etc.	PDE inhibitor (rolipram), cAMP analogue (8-Cl-cAMP), PKA inhibitors, peptides (forskolin, CREBtide, KEMPtide)	[70, 71, 86]
PI3K/mTOR	Tyrosine Kinases, Small-Gs, GPCRs	Breast cancer, colon cancer, endometrial cancer, squamous cell lung cancer, glioma, etc.	PI3K/mTOR inhibitor (Dactolisib, etc.), mTOR inhibitor (rapamycin, everolimus, etc.)	[75, 83]
	Nuclear Receptors (PPARγ)	Cardiac hypertrophy	mTOR inhibitor (rapamycin)	[78, 87]
	EGFR	Tissue overgrowth syndromes	Rapalogs, p110α inhibitors, everolimus	[75]
	EGFR, IGFR	Metabolic disorders (hyperglycemia, type 2 diabetes, obesity)	PI3K activators (resveratrol) Anti-diabetic (dapagliflozin, etc.)	[75, 88]
JAK/STAT	Tyrosine Kinase Receptors/Cytokines, Hormones	Chronic myeloid leukemia, lymphoma, solid tumors, and various cancers	JAK inhibitors (tofacitinib, ruxolitinib, INCB047986, pacritinib, etc.), STAT inhibitors (OPB-31121, AZD-9150, etc.)	[72, 81]
	Tyrosine Kinase Receptor/ Cytokines (IFN-γ, IL-6)	Parkinson's disease	JAK/STAT inhibitor (AZD-1480), JAK inhibitor (ruxolitinib, baricitinib, oclatinib, etc.)	[72, 89]
	Tyrosine Kinase	Skin disorders (dermatitis, eczema, psoriasis, etc.)	JAK inhibitor (ruxolitinib, baricitinib, oclatinib, etc.)	[89]

Abbreviations: cAMP: cyclic adenosine monophosphate; EFGR: epidermal growth factor receptor; GFR: growth factor receptor; GPCR: G-protein-coupled receptor; JAK: Janus kinase; MAPK: mitogen-activated protein kinase; MEK: mitogen-activated extracellular signal-regulated kinase; mTOR: mammalian target of rapamycin; PDE: phosphodiesterase; PDGFR: platelet-derived growth factor receptor; PI3K: phosphatidylinositol 3-kinase; STAT: signal transducers and activators of transcription; VEGFR: vascular endothelial growth factor receptor; 5HT: 5-hydroxytryptamine.

3.10 CONCLUSION

The intricate process of maintaining balance and stability within the human body, orchestrated by a vast network of cellular receptors—either cytoplasmic or transmembrane—and signaling pathways, is a testament to the complexity of our biological systems. Signal transduction occurs when a signaling molecule, or ligand, binds to a specific receptor on the surface of a cell. This activates the receptor and triggers a series of intracellular events that result in a cellular response. The response depends on the ligand, the receptor, and the downstream signaling pathways involved. A variety of receptors, such as G-protein-coupled receptors, enzyme-linked receptors, and ligand-gated ion channels, play a crucial role in initiating diverse signaling pathways that allow cells to react to their environment and maintain their physiological processes. However, when these receptors experience mutations or dysregulation, abnormal signaling can result and contribute to the development of various disorders, such as cancer and neurodegenerative and heart diseases.

In the battle against diseases, researchers have harnessed a promising approach involving targeting specific receptors and enzymes involved in cellular signaling. For instance, monoclonal antibodies, with their ability to inhibit tumor growth by targeting specific receptors involved in cancer cell proliferation, hold immense potential. Similarly, small molecule inhibitors, capable of blocking abnormal cellular signaling by targeting specific enzymes like kinases, offer a ray of hope in disease treatment.

REFERENCES

1. Combarnous, Y. and T.M.D. Nguyen, Cell communications among microorganisms, plants, and animals: Origin, evolution, and interplays. *International Journal of Molecular Sciences*, 2020. **21**(21): p. 8052.
2. Ramilowski, J.A., et al., A draft network of ligand–receptor-mediated multicellular signalling in human. *Nature Communications*, 2015. **6**(1): p. 7866.
3. Rozbesky, D. and E.Y. Jones, Cell guidance ligands, receptors and complexes–orchestrating signalling in time and space. *Current Opinion in Structural Biology*, 2020. **61**: p. 79–85.
4. Purvis, J.E. and G. Lahav, Encoding and decoding cellular information through signaling dynamics. *Cell*, 2013. **152**(5): p. 945–956.
5. Iwasaki, A. and R. Medzhitov, Control of adaptive immunity by the innate immune system. *Nature Immunology*, 2015. **16**(4): p. 343–353.
6. Uings, I. and S. Farrow, Cell receptors and cell signalling. *Molecular Pathology*, 2000. **53**(6): p. 295.
7. Brumbaugh, P.F. and M.R. Haussler, Nuclear and cytoplasmic receptors for 1, 25-dihydroxycholecalciferol in intestinal mucosa. *Biochemical and Biophysical Research Communications*, 1973. **51**(1): p. 74–80.
8. Rosen, J., et al., The intracellular receptor and signal transducers and activators of transcription factor superfamilies: Novel targets for small-molecule drug discovery. *Journal of Medicinal Chemistry*, 1995. **38**(25): p. 4855–4874.
9. Jones, J.D., R.E. Vance, and J.L. Dangl, Intracellular innate immune surveillance devices in plants and animals. *Science*, 2016. **354**(6316): p. aaf6395.
10. Bonardi, V. and J.L. Dangl, How complex are intracellular immune receptor signaling complexes? *Frontiers in Plant Science*, 2012. **3**: p. 35164.
11. Sacta, M.A., Y. Chinenov, and I. Rogatsky, Glucocorticoid signaling: An update from a genomic perspective. *Annual Review of Physiology*, 2016. **78**: p. 155–180.
12. Vitellius, G., et al. Pathophysiology of glucocorticoid signaling. In *Annales d'endocrinologie*, Frédéric Castinetti et al., editor. 2018, Elsevier: New York.
13. Giraldi, T., et al., Steroid signaling activation and intracellular localization of sex steroid receptors. *Journal of Cell Communication and Signaling*, 2010. **4**: p. 161–172.
14. Coleman, K.M. and C.L. Smith, Intracellular signaling pathways: Nongenomic actions of estrogens and ligand-independent activation of estrogen receptors. *Frontiers in Bioscience*, 2001. **6**(1): p. D1379–D1391.
15. Wingler, L.M. and R.J.J.T.I.C.B. Lefkowitz, Conformational basis of G protein-coupled receptor signaling versatility. *Trends in Cell Biology*, 2020. **30**(9): p. 736–747.

16. Patwardhan, A., N. Cheng, and J.J.P.r. Trejo, Post-translational modifications of G Protein–Coupled receptors control cellular signaling dynamics in space and time. *Pharmacological Reviews*, 2021. **73**(1): p. 120–151.
17. Ahmad, R., S. Wojciech, and R. Jockers, Hunting for the function of orphan GPCRs – beyond the search for the endogenous ligand. *British Journal of Pharmacology*, 2015. **172**(13): p. 3212–3228.
18. Ahmad, R. and J.E.J.F.I.P. Dalziel, G protein-coupled receptors in taste physiology and pharmacology. *Frontiers Pharmacology*, 2020. **11**: p. 587664.
19. Lynch, J.R. and J.Y. Wang, G protein-coupled receptor signaling in stem cells and cancer. *International Journal of Molecular Sciences*, 2016. **17**(5): p. 707.
20. Goodman, S.R., *Medical Cell Biology*. 2007, Academic Press: New York. p. 227–248.
21. Alberts, B., Signaling through enzyme-linked cell-surface receptors. In *Molecular Biology of the Cell*. 2002, Garland Science: New York.
22. Pacheco, M.A., Receptor types and subtypes. In *xPharm: The Comprehensive Pharmacology Reference*, S.J. Enna and D.B. Bylund, Editors. 2007, Elsevier: New York. p. 1–5.
23. Misra, A., The cell. In Challenges in Delivery of Therapeutic Genomics and Proteomics. Ambikanandan Misra, editor. 2010, Elsevier: New York. p. 1–43.
24. Alexander, S., A. Mathie, and J. Peters, Ligand-gated ion channels. *British Journal of Pharmacology*, 2011. **164**(s1): p. S115–S135.
25. Gopalakrishnan, M.J.M.R.W., Ion channels—Ligand gated. Comprehensive medicinal chemistry II, Edited by Triggle DJ et al. *Comprehensive Medicinal Chemistry*, 2006. **2**: p. 877–918.
26. Mantegazza, M. and W.A. Catterall, Voltage-Gated Na+ Channels: Structure, Function, and Pathophysiology. In *Jasper's Basic Mechanisms of the Epilepsies* [Internet], 4th edition. Noebels JL, Avoli M, Rogawski MA, et al., editors. Bethesda (MD): National Center for Biotechnology Information (US); 2012.
27. Bezanilla, F., Voltage-gated ion channels. In *Biological Membrane Ion Channels: Dynamics, Structure, and Applications*, S.-H. Chung, O.S. Andersen, and V. Krishnamurthy, Editors. 2007, Springer: New York, NY. p. 81–118.
28. Bezanilla, F., Voltage-gated ion channels. In *Neuroscience*, D. Purves, et al., Editors. 2001, Sinauer Associates: Sunderland, MA.
29. Pollard, T.D., et al., *Cell Biology E-Book: Cell Biology E-Book*. 2022, Elsevier Health Sciences: Philadelphia.
30. de Freitas, R.F. and M. Schapira, A systematic analysis of atomic protein–ligand interactions in the PDB. *Medchemcomm*, 2017. **8**(10): p. 1970–1981.
31. Weis, W.I. and B.K. Kobilka, The molecular basis of G protein–coupled receptor activation. *Annual Review of Biochemistry*, 2018. **87**: p. 897–919.
32. Roy, S. and T.B. Kornberg, Paracrine signaling mediated at cell–cell contacts. *Bioessays*, 2015. **37**(1): p. 25–33.
33. Gould, S.T., et al., The role of valvular endothelial cell paracrine signaling and matrix elasticity on valvular interstitial cell activation. *Biomaterials*, 2014. **35**(11): p. 3596–3606.
34. Wee, P. and Z. Wang, Epidermal growth factor receptor cell proliferation signaling pathways. *Cancers*, 2017. **9**(5): p. 52.
35. Lauder, J.M., Neurotransmitters as growth regulatory signals: Role of receptors and second messengers. *Trends in Neurosciences*, 1993. **16**(6): p. 233–240.
36. Caicedo, A. Paracrine and autocrine interactions in the human islet: More than meets the eye. In *Seminars in Cell & Developmental Biology*. Yanlan Mao et al. editor. 2013, Elsevier: New York.
37. Shehzad, A., J. Lee, and Y.S. Lee, Autocrine prostaglandin E2 signaling promotes promonocytic leukemia cell survival via COX-2 expression and MAPK pathway. *BMB Reports*, 2015. **48**(2): p. 109.
38. Ross, S.H. and D.A. Cantrell, Signaling and function of interleukin-2 in T lymphocytes. *Annual Review of Immunology*, 2018. **36**: p. 411–433.
39. Kasprzak, A., Insulin-like growth factor 1 (IGF-1) signaling in glucose metabolism in colorectal cancer. *International Journal of Molecular Sciences*, 2021. **22**(12): p. 6434.
40. Lee, S., et al., Autocrine VEGF signaling is required for vascular homeostasis. *Cell*, 2007. **130**(4): p. 691–703.
41. Kliche, S. and J. Waltenberger, VEGF receptor signaling and endothelial function. *IUBMB Life*, 2001. **52**(1): p. 61–66.
42. Sinha, R.A., B.K. Singh, and P.M. Yen, Reciprocal crosstalk between autophagic and endocrine signaling in metabolic homeostasis. *Endocrine Reviews*, 2017. **38**(1): p. 69–102.

43. Bianco, A.C., et al., Paradigms of dynamic control of thyroid hormone signaling. *Endocrine Reviews*, 2019. **40**(4): p. 1000–1047.
44. Yap, A.S. and E.M. Kovacs, Direct cadherin-activated cell signaling: A view from the plasma membrane. *The Journal of Cell Biology*, 2003. **160**(1): p. 11.
45. Vinayagam, A., et al., A directed protein interaction network for investigating intracellular signal transduction. *Science Signaling*, 2011. **4**(189): p. rs8–rs8.
46. Lai, E.C., *Notch Signaling: Control of Cell Communication and Cell Fate*. 2004. **131**(5): p. 965–973.
47. Janeway, C.A. Jr, T.P., Walport, M., et al., *Immunobiology: The Immune System in Health and Disease*. 5th ed. General principles of transmembrane signaling. 2001, Garland Science: New York.
48. Shehzad, A. and Y.S. Lee, Molecular mechanisms of curcumin action: Signal transduction. *Biofactors*, 2013. **39**(1): p. 27–36.
49. Ji, T.H., et al., Activation of and signal generation by membrane receptors. *Molecules and Cells*, 1995. **5**(1): p. 1–8.
50. Cornell, T.T. and T.P.J.C.C.M. Shanley, Signal transduction overview. *Critical Care Medicine*, 2005. **33**(12): p. S410–S413.
51. Sakkaff, Z., A. Immaneni, and M. Pierobon. Estimating the molecular information through cell signal transduction pathways. 2018 IEEE 19th International Workshop on Signal Processing Advances in Wireless Communications (SPAWC). 2018. IEEE.
52. Gonçalves, E., et al., Bridging the layers: Towards integration of signal transduction, regulation and metabolism into mathematical models. *Molecular Biosystems*, 2013. **9**(7): p. 1576–1583.
53. Bianco, R., et al., Key cancer cell signal transduction pathways as therapeutic targets. *European Journal of Cancer*, 2006. **42**(3): p. 290–294.
54. Shields, J.M., et al., Understanding Ras:'It ain't over'til it's over'. *Trends in Cell Biology*, 2000. **10**(4): p. 147–154.
55. Henson, E.S. and S.B. Gibson, Surviving cell death through epidermal growth factor (EGF) signal transduction pathways: Implications for cancer therapy. *Cellular Signalling*, 2006. **18**(12): p. 2089–2097.
56. Asl, E.R., et al., Interplay between MAPK/ERK signaling pathway and MicroRNAs: A crucial mechanism regulating cancer cell metabolism and tumor progression. *Life Sciences*, 2021. **278**: p. 119499.
57. Blalock, W., et al., Signal transduction, cell cycle regulatory, and anti-apoptotic pathways regulated by IL-3 in hematopoietic cells: Possible sites for intervention with anti-neoplastic drugs. *Leukemia*, 1999. **13**(8): p. 1109–1166.
58. Fang, Z., et al., Signaling pathways in cancer-associated fibroblasts: Recent advances and future perspectives. *Cancer Commun (Lond)*, 2023. **43**(1): p. 3–41.
59. Cooper, G.M.J.S. and Pathways of intracellular signal transduction. In The Cell: A Molecular Approach, 2nd ed. 2000, Sinauer Associates: Sunderland, MA.
60. Mascareno, E., M.A.Q. Siddiqui, and M. Wagner, Cell signaling events. In *Medical Cell Biology*, S.R. Goodman, Editor. 2007, Academic Press: London. p. 249–272.
61. Yang, Z.R. Crosstalk and signalling pathway complexity - A case study on synthetic models. In *Hybrid Artificial Intelligence Systems*. 2008, Springer: Berlin, Heidelberg.
62. Nair, A., et al., Conceptual evolution of cell signaling. *International Journal of Molecular Sciences*, 2019. **20**(13): p. 3292.
63. Fu, D., et al., Key signal transduction pathways and crosstalk in cancer: Biological and therapeutic opportunities. *Translational Oncology*, 2022. **26**: p. 101510.
64. Jin, Z., W.S.J.C.b. El-Deiry, and therapy, overview of cell death signaling pathways. *Cancer Biology and Therapy*, 2005. **4**(2): p. 147–171.
65. Corcoran, G.B., et al., Apoptosis: Molecular control point in toxicity. *Toxicology and Applied Pharmacology*, 1994. **128**(2): p. 169–181.
66. Strasser, A., L. O'Connor, and V.M. Dixit, Apoptosis signaling. *Annual Review of Biochemistry*, 2000. **69**(1): p. 217–245.
67. Payne, C.M., et al., Apoptosis overview emphasizing the role of oxidative stress, DNA damage and signal-transduction pathways. *Leukemia and Lymphoma*, 1995. **19**(1–2): p. 43–93.
68. Los, M., S. Wesselborg, and K.J.I. Schulze-Osthoff, The role of caspases in development, immunity, and apoptotic signal transduction: Lessons from knockout mice. *Immunity*, 1999. **10**(6): p. 629–639.
69. Tidyman, W.E. and K.A. Rauen, Noonan, Costello and cardio–facio–cutaneous syndromes: Dysregulation of the Ras–MAPK pathway. *Expert Reviews in Molecular Medicine*, 2008. **10**: p. e37.
70. Zhang, H., et al., Complex roles of cAMP–PKA–CREB signaling in cancer. *Experimental Hematology & Oncology*, 2020. **9**(1): p. 32.

71. Ahmed, M.B., et al., cAMP signaling in cancer: A PKA-CREB and EPAC-centric approach. *Cells*, 2022. **11**(13): p. 2020.
72. Xin, P., et al., The role of JAK/STAT signaling pathway and its inhibitors in diseases. *International Immunopharmacology*, 2020. **80**: p. 106210.
73. He, Y., et al., Targeting PI3K/Akt signal transduction for cancer therapy. *Signal Transduction and Targeted Therapy*, 2021. **6**(1): p. 425.
74. Dahlgren, J. and C. Noordam, Growth, endocrine features, and growth hormone treatment in noonan syndrome. *Journal of Clinical Medicine*, 2022. **11**(7): p. 2034.
75. Fruman, D.A., et al., The PI3K pathway in human disease. *Cell*, 2017. **170**(4): p. 605–635.
76. Zhou, Z., Y. Ikegaya, and R. Koyama, The astrocytic cAMP pathway in health and disease. *International Journal of Molecular Sciences*, 2019. **20**(3): p. 779.
77. Al-Suhaimi, E.A. and A. Shehzad, Leptin, resistin and visfatin: The missing link between endocrine metabolic disorders and immunity. *European Journal of Medical Research*, 2013. **18**: p. 1–13.
78. Ramos-Kuri, M., et al., Molecules linked to Ras signaling as therapeutic targets in cardiac pathologies. *Biological Research*, 2021. **54**(1): p. 23.
79. Elnagdy, M., et al., cAMP signaling in pathobiology of alcohol associated liver disease. *Biomolecules*, 2020. **10**(10): p. 1433.
80. Wahlang, B., et al., Role of cAMP and phosphodiesterase signaling in liver health and disease. *Cellular Signaling*, 2018. **49**: p. 105–115.
81. O'Shea, J.J., et al., The JAK-STAT pathway: Impact on human disease and therapeutic intervention. *The Annual Review of Medicine*, 2015. **66**(1): p. 311–328.
82. Santarpia, L., S.M. Lippman, and A.K.J.E.o.o.t.t. El-Naggar, Targeting the MAPK–RAS–RAF signaling pathway in cancer therapy. *Expert Opinion on Therapeutic Targets*, 2012. **16**(1): p. 103–119.
83. De Luca, A., et al., The RAS/RAF/MEK/ERK and the PI3K/AKT signalling pathways: Role in cancer pathogenesis and implications for therapeutic approaches. *Expert Opinion on Therapeutic Targets*, 2012. **16**(sup2): p. S17–S27.
84. Simanshu, D.K., D.V. Nissley, and F.J.C. McCormick, RAS proteins and their regulators in human disease. *Cell*, 2017. **170**(1): p. 17–33.
85. Christiansen, S.H., et al., Combined anti-inflammatory effects of β2-adrenergic agonists and PDE4 inhibitors on astrocytes by upregulation of intracellular cAMP. *Neurochemistry International*, 2011. **59**(6): p. 837–846.
86. Bolger, G.B.J.F.i.E., The cAMP-signaling cancers: Clinically-divergent disorders with a common central pathway. *Frontiers in Endocrinology*, 2022. **13**: p. 1024423.
87. Peng, X., et al., PPARγ–PI3K/AKT–NO signal pathway is involved in cardiomyocyte hypertrophy induced by high glucose and insulin. *Journal of Diabetes and its Complications*, 2015. **29**(6): p. 755–760.
88. Savova, M.S., et al., Targeting PI3K/AKT signaling pathway in obesity. *Biomedicine & Pharmacotherapy*, 2023. **159**: p. 114244.
89. Lashgari, N.-A., et al., The involvement of JAK/STAT signaling pathway in the treatment of Parkinson's disease. *Journal of Neuroimmunology*, 2021. **361**: p. 577758.

4 Signal Transduction and Integral Membrane Proteins

Zain Saeed[1], Zohra Jabeen[1], Sehrish Kakakhel[2], Saira Justin[1], and Aneela Javed[1]
[1]Department of Biomedicine, Atta ur Rahman School of Applied Biosciences (ASAB), National University of Sciences and technology (NUST), H12 Islamabad, Pakistan
[2]Department of Biochemistry, Abdul Wali Khan University Mardan, Mardan, Pakistan

4.1 INTRODUCTION

The cell membrane, also known as the plasma membrane, is the outer structural boundary of the cell. It divides the internal and external environments and is semipermeable in nature. This means that it allows only certain molecules to pass through while prohibiting the entry or exit of unnecessary contents. This unique property allows the cell membrane to not only provide structural support but also perform a variety of functions. For example, it plays a vital role in homeostasis by regulating the concentration and the type of molecules entering and exiting the cell. Additionally, it serves as a structural boundary, offering protection and rigidity to the cell, and supports the many biological molecules the cell uses to communicate with the external environment (Cell Membrane (Plasma Membrane), 2024).

The cell membrane's structure is most accurately represented by the fluid mosaic model. This model depicts the membrane as a fluid composed of a bilayer of phospholipid membranes. The lipid bilayers arrange themselves according to their interaction with neighboring water molecules; hence, the hydrophobic elements of the phospholipids are on the inside of the cell and the hydrophilic elements on the outside (Cooper, 2000a). Proteins embedded into the structure of the plasma membrane, known as membrane proteins, are the focus of this chapter.

4.1.1 History of Cell Membrane Discovery

The earliest research on the cell membrane focused on providing evidence of its existence or proposing that the cell was a colloidal particle without any definitive structural boundary. Osmotic studies proved crucial in supporting its existence. Osmosis, the movement of water from a high concentration to a low concentration across a semipermeable membrane, could not be understood without a semipermeable membrane. While Nollet's initial osmotic experiments received little recognition, Dutrochet's work on the principles of osmosis highlighted the importance of such studies. This fueled discussions about cell membranes, and by the early 20th century, the work of notable scientists such as Overton, Chambers, and Höber had strengthened the case for their existence. In 1925, Gorter and Grendel's experiment provided evidence of a bilayer, rather than a lipid monolayer, by quantifying lipids extracted from a specific area of erythrocytes (Lombard, 2014). The debate finally concluded when Singer and Nicolson presented their fluid mosaic model, describing the lipid membrane as a "two-dimensional oriented solution of integral proteins (or lipoproteins) in the viscous phospholipid bilayer solvent" (Singer & Nicolson, 1972). They also introduced the concept

 DOI: 10.1201/9781003594901-4

of integral membrane proteins, describing them as embedded in the lipid bilayer with polar groups that extend out of the membrane. This model not only confirmed the existence of membrane proteins but also explained their localization. Subsequent research has revealed the immense diversity of membrane proteins and their essential role in numerous cellular functions. Perhaps one of their most important functions is their ability to transduce signals from the external environment, which allows us to interact with and understand the world around us.

4.1.2 Classification of Membrane Proteins

Membrane proteins are classified into two types: integral membrane proteins, which will be the focus of this chapter, and peripheral membrane proteins (Figure 4.1).

1. Integral membrane proteins (IMPs):
   These proteins are predominantly attached to the cell membrane. They are sometimes used synonymously with transmembrane proteins; however, this category also includes monotopic membrane proteins (Vit & Petrak, 2017).
2. Peripheral membrane proteins: Unlike IMPs, these proteins are not embedded or permanently attached to the cell membrane but interact with it indirectly, usually modulated by integral membrane proteins (Cooper, 2000b). They are attached to the cell membrane based on electrostatic forces or via lipid interactions (Whited & Johs, 2015). These proteins are depicted in Figure 4.2.

4.1.2.1 Topological Classification of IMP

IMPs can be further classified based on their topology or their association with the cell membrane, as shown in Figure 4.3.

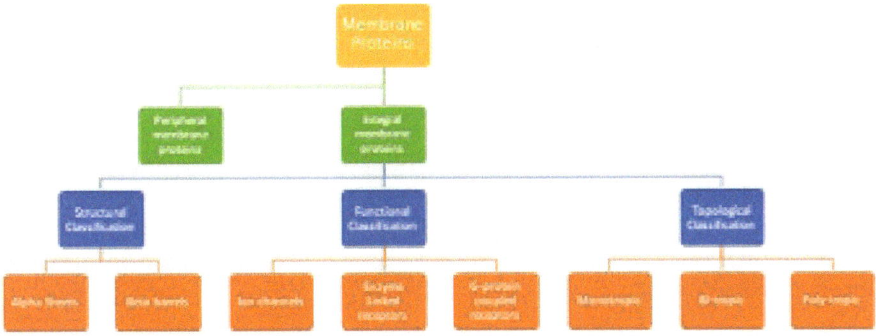

FIGURE 4.1 Classification of membrane proteins.

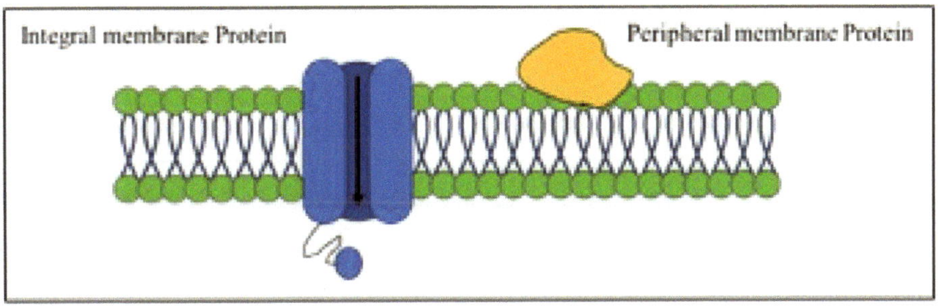

FIGURE 4.2 Diagrammatic representation of an integral membrane protein and a peripheral membrane protein. Integral membrane proteins are permanently attached to the cell membrane (e.g., an ion channel), whereas peripheral membrane proteins are not.

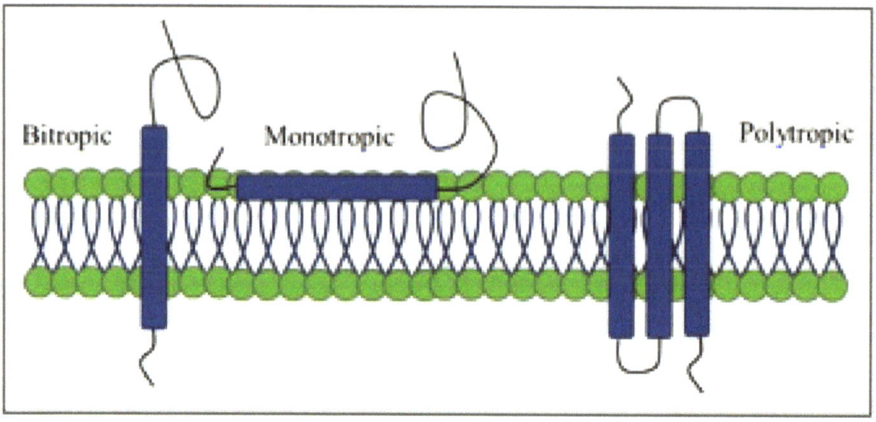

FIGURE 4.3 Topological classification of integral membrane proteins. Monotopic proteins are attached to the plasma membrane on only one side; bitopic proteins are attached on two sides; polytopic proteins are attached on multiple sides.

A) **Polytopic:** These proteins attach to the cell membrane from multiple sides; examples include multipass proteins.
B) **Bitopic:** These attach to the cell membrane on both sides.
C) **Monotopic:** These are irreversibly attached to one side of the cell membrane.

However, the single-sided attachment of monotopic proteins can lead to their misclassification as peripheral proteins, and vice versa. Furthermore, our structural and functional understanding of such proteins is limited (Allen et al., 2019). Consequently, this chapter will focus on bitopic and polytopic proteins and their functional role in signaling.

4.1.2.2 Structural Classification of IMP

Transmembrane proteins can be further divided based on their 3D structure; they can form either an alpha chain or a beta barrel (Alberts et al., 2002). Both structures are important for different types of membrane proteins and receptors; beta barrels can act as channels (e.g., ion channels), and alpha chains can act as receptors (e.g., enzyme-linked receptors) (Whitelegge, 2013). Transmembrane proteins can be single-pass or multipass, meaning that the protein can cross the membrane either once or multiple times.

4.1.2.3 Functional Classification of IMP

Regarding their functional use and involvement in signaling, transmembrane proteins can be broadly categorized into three different classes, which will be the focus of this chapter: (1) GPCRs, (2) channel-associated proteins, and (3) enzyme-linked receptors (Miller & Lappin, 2022; Purves et al., 2001).

4.2 ION CHANNELS

Ion channels are crucial integral proteins characterized by their ability to transport ions across the semipermeable plasma membrane. These channels usually contain a pore that allows only specific ions to pass through them. For example, a sodium channel will, under normal conditions, only allow sodium to pass through it. This selectivity is essential for maintaining the ionic gradient across the plasma membrane and producing action potentials in excitable cells (Roux, 2017). Furthermore, these channels can exist in multiple discrete conformations, such as being open or closed. This conformation is usually determined by the channel's interaction with an interacting molecule or an environmental stimulus. This process, known as gating, has the potential to block or open the ion channel's pore. In an open conformation, the ion channel allows the selective ion to pass through; however, upon receiving an adequate stimulus, it can change to a closed conformation and bar any other molecules from entering (Catacuzzeno et al., 2020). It should also be noted that these channels are extremely efficient and can discern between two positively charged ions with a selectivity of up to a thousand times. For example, potassium channels can block the entry of sodium ions but allow the flow of potassium ions at a near-maximum diffusion rate, meaning that such channels can allow their selected ions to pass through the cell membrane with little to no resistance (Tóth, 2021).

In a signaling cascade, ion channels are used for their selective permeability, allowing the transduction of specific signals from the environment into the cell. When in an open conformation, ions enter the cell by passive transport and are involved in downstream signaling cascades (Catterall, 2010). However, because it is crucial for the cell to return to its resting state, passive diffusion is counteracted by active transport carried out by pumps in the plasma membrane. These pumps maintain the established concentration gradient and, after excitation caused by the influx of ions, they regulate the internal conditions of the cell by pumping out the excess ions that have entered it (Gadsby, 2009).

When an ion channel is activated and induces a change in the electrical potential across the cell membrane, numerous downstream signaling events can occur, depending on the type of ion channel (Bezanilla, 2005). For example, in calcium signaling, calcium ion channels allow an influx of Ca^{2+} ions into the cell. This drastically elevates the concentration of ions inside the cell, causing deviation from the basal concentration of about 100 nM (Bootman et al., 2020). These calcium ions can directly interact with numerous proteins and enzymes to initiate downstream signaling events (Dolphin & Lee, 2020). For example, in neurons, calcium ions can bind to synaptotagmins on the surface of intracellular vesicles containing neurotransmitters, prompting their release from the presynaptic membrane (Südhof, 2012). Importantly, a reduction in ionic concentrations or a lack of signaling can also be considered a signaling event; for example, contrasting studies have shown that either an increase or a decrease in calcium signaling can cause the cell to undergo autophagy, with the direction of events being dictated by the cell's external environment and/or internal cell conditions (Bootman et al., 2020; Kondratskyi et al., 2018)

Ion channels can be classified into three distinct types based on their interaction with their environment: (1) voltage-gated ion channels, (2) ligand-gated ion channels, and (3) mechanosensitive ion channels (See Figure 4.4). These ion channels are briefly described below. For details, please refer to Chapter 8.

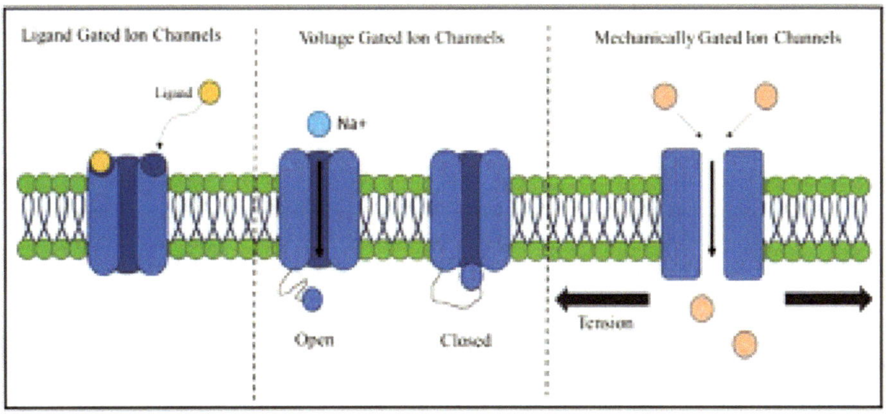

FIGURE 4.4 Overview of ion channel types. Ligand-gated ion channels depend on ligands for activation; voltage-gated ion channels are activated by electrochemical gradient changes; mechanically gated or mechanosensitive ion channels open in response to mechanical stress.

4.2.1 Voltage-Gated Ion Channels

Voltage-gated ion channels are essential IMPs that operate in the presence of an electrochemical gradient (Kariev & Green, 2012). They are highly selective for the ions they permit to pass through them; this selectivity is established via a diverse set of exclusion and inclusion mechanisms. For example, calcium ions operate in an environment with a very high relative concentration of sodium, an ion that has roughly the same ionic diameter as calcium (Catterall et al., 2017). Hence, calcium and sodium channels must adopt a chemical affinity-based exclusion system instead of a size-based exclusion system. These ion channels, once activated, can produce a result several times greater than the initial signal that activated them, resulting in signal amplification (Zheng & Trudeau, 2023).

Structurally, voltage-gated ion channels (VGICs) are highly conserved units, generally containing the same structural format throughout the superfamily (Oelstrom et al., 2014). A generalized overview of the structure consists of four transmembrane helical segments surrounding the ion pore. However, not all members of the VGICs contain the pore domain. The sensitivity of these channels stems from their helices containing positively charged ions, amino acids, at regular intervals within their 3D structure, thus allowing for the microscale detection of ionic charges (De Lera Ruiz & Kraus, 2015).

4.2.2 Ligand-Gated Channels

Ligand-gated channels transduce signals by interacting with specific ligands and undergoing a conformational change. For example, in neurons, these channels interact with neurotransmitters to transmit action potentials (Alexander et al., 2011a). They are classified based on the type of ligand that binds to these channels and their structure. Broadly, they can be classified into three different categories: the pentameric cys-loop channels, the tetrameric glutamate receptors, and the trimeric ligated-gated ion channels (P2X) (Alexander et al., 2011b; Collingridge et al., 2009). These channels bind to different ligands, as listed in Table 4.1. Their activation can lead to various outcomes, such as the generation of an action potential or hormonal release. Thus, like other ion channels, they are involved in regulating and maintaining the concentration of a certain chemical in the extracellular/intracellular matrix of the cell.

TABLE 4.1
Classification of Extracellular Ligand-Gated Ion Channels (LGICs)

Extracellular LGIC	Families	Ligand
Pentameric (Cys-loop superfamily)	5-HT3	Ca^{2+}, 5-hydroxytryptamine, Mg^{2+}, Zn^{2+}
	Nicotinic ACh	Acetylcholine
	GABAA	GABA, 5α-pregnan-3α-ol-20-one tetrahydrodeoxycorticosterone, Zn^{2+}
	Glycine	Cu^{2+}, glycine, H^+, Zn^{2+}
	Zinc-activated	Cu^{2+}, H^+, Zn^{2+}
Tetrameric (Ionotropic glutamate family)	AMPA	AMPA
	NMDA	NMDA
	Kainate	L-glutamic acid
Trimeric ligand-gated channels	P2X	ATP
	ASICs	H^+
	ENaC	Na^+

4.2.3 Mechanoreceptors

Cells constantly undergo mechanical stress due to internal conditions, such as cytokinesis, or external conditions, such as an external pressure applied to the cell. Mechanoreceptors and mechanosensitive ion channels respond to force applied to the cell membrane; these receptors then transduce this signal into an electrical potential, driving downstream signaling events (Kefauver et al., 2020).

Mechanoreceptors are structurally diverse and do not contain the conserved structural similarities found in voltage-gated channels. However, they are functionally conserved at a certain level. In general, they utilize the energy that is exerted onto the surface of the lipid bilayer or the cytoskeleton to undergo a conformational change, shifting from a closed to an open position and initiating the signal transduction cascade (Uray et al., 2021).

4.2.4 Ion Channels in Disease

Ion channels exert pleiotropic effects downstream; they can both activate and deactivate signaling events. Hence, mutations in these ion channels can also have various side effects, causing aberrant changes in their trafficking, pore structure, and gating potential.

Protein trafficking is an essential step and is constantly ongoing in several cell populations, such as cardiomyocytes. Trafficking refers to the process of transporting membrane proteins to and from the cell, renewing and recycling them. It is estimated that trafficking problems are more prevalent than other mutations in ion channels (Pitt, 2016). Mutations can occur during several steps of protein trafficking; for example, a mutation in the potassium channel Kv11.1 results in decreased expression in the sarcolemma (van der Heyden et al., 2018). Similarly, a mutation in the sodium channel NaV1.5 impairs its binding efficiency to ankyrin molecules, reducing this channel's presence across the membrane (Blandin et al., 2021).

Mutations in the genes responsible for pore-forming can result in disorders such as epilepsy. A missense in the SCN1A gene, which encodes the NaV1.1 ion channel, causes a failure of the pore-forming structure, leading to generalized anxiety with febrile seizures plus (GEFS+) (Pavone et al., 2022). Similarly, genetic defects in ligand-gated ion channels can also result in epileptic seizures and hyperekplexia (Absalom et al., 2020).

Defects in ion channels have also been associated with different types of cancers; for example, mutations in the potassium channel Kv11.1 correlate strongly with colon tumors (Iorio et al., 2020). Voltage-gated sodium channels have also been associated with leukemia, lymphoma, and melanoma (Angus et al., 2019). Furthermore, defects in chloride and potassium channels have been implicated in various gastrointestinal cancers, lung cancer, and head and neck carcinomas (Anderson et al., 2019; Bulk et al., 2021; Del-Río-Ibisate et al., 2021).

Abnormalities in the gating potential can also lead to neurological disorders; for example, alterations in the gating pore current of voltage-gated ion channels may contribute to the development of autism (Gamal El-Din et al., 2021).

4.3 G-PROTEIN-COUPLED RECEPTORS

G-protein-coupled receptors (GPCRs) are multiprotein receptors that comprise as much as 5% of the human genome. They display immense diversity as transducers, but feature characteristics that are conserved and shared by all eukaryotic organisms, from yeast to humans. Their signaling mechanism involves a series of interactions or information flow. These transmembrane proteins, found in the cell membrane, have an affinity for extracellular ligands, such as neurotransmitters, hormones, or other signaling chemicals. Ligand binding induces a substantial structural alteration, activating the corresponding G-protein (Wensel, 1999). G proteins are heterotrimeric, consisting of α, β, and γ subunits. Activation converts GDP to GTP, separating Gα and G$\beta\gamma$ subunits, controlling downstream effector molecules, and initiating intricate signaling cascades (Rosenbaum et al., 2009).

GPCRs regulate various physiological signaling pathways, including those involved in hormone and neurotransmitter responses, vision, olfaction, and taste. Studying the structure and function of GPCRs significantly enhances our understanding of fundamental biological processes and may facilitate the development of novel therapeutics for certain diseases or disorders.

4.3.1 General Structure of G-Protein-Coupled Receptors

GPCRs can be characterized by their distinct seven-transmembrane (7TM) helical structure, also known as the heptahelical serpentine. The helix loops around the cell membrane seven times, creating three extracellular loops (ECLs) and three intracellular loops with the N-terminus present in the extracellular space and the C-terminus in the intracellular space. These transmembrane helices are known as TM1, TM2, TM3, and so on (Figure 4.5).

The distinctive ligand-binding regions of GPCRs, which facilitate interactions with signaling molecules, can be classified into two groups: orthosteric and allosteric. Orthosteric binding sites directly induce conformational changes in the receptor, whereas allosteric binding sites indirectly affect the receptor. This can be done by altering the GPCR's affinity for orthosteric ligands (Hauser et al., 2017)

4.3.2 Diversity of GPCRs

GPCRs are categorized into distinct families according to their sequence homology and phylogenetic origin. These families exhibit unique structural features related to their ability to bind to ligands (Kamato et al., 2015a). While sequence homology classification applies to both vertebrates and invertebrates, phylogenetic classification applies only to vertebrates. This chapter focuses on human GPCRs, using the phylogenetic classification known as the GRAFS system (glutamate, rhodopsin, adhesion, Frizzled/taste 2, secretin) (Figure 4.6).

4.3.2.1 Glutamate Family

As the name suggests, glutamate receptors use glutamate as an extracellular ligand. These receptors are predominantly found in the brain, where they mediate neuron excitation and synaptic

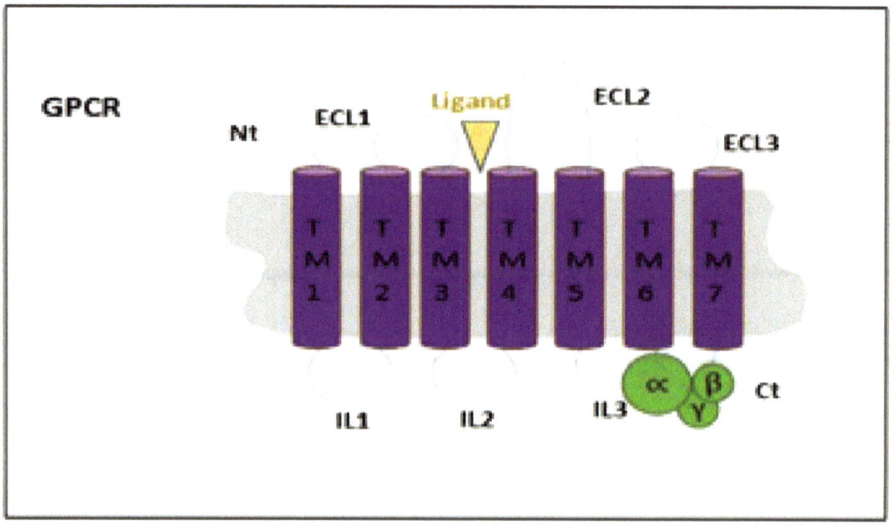

FIGURE 4.5 GPCR heptahelical structure with extracellular loops (ECL1, ECL2, ECL3) and intracellular loops (IL1, IL2, IL3).

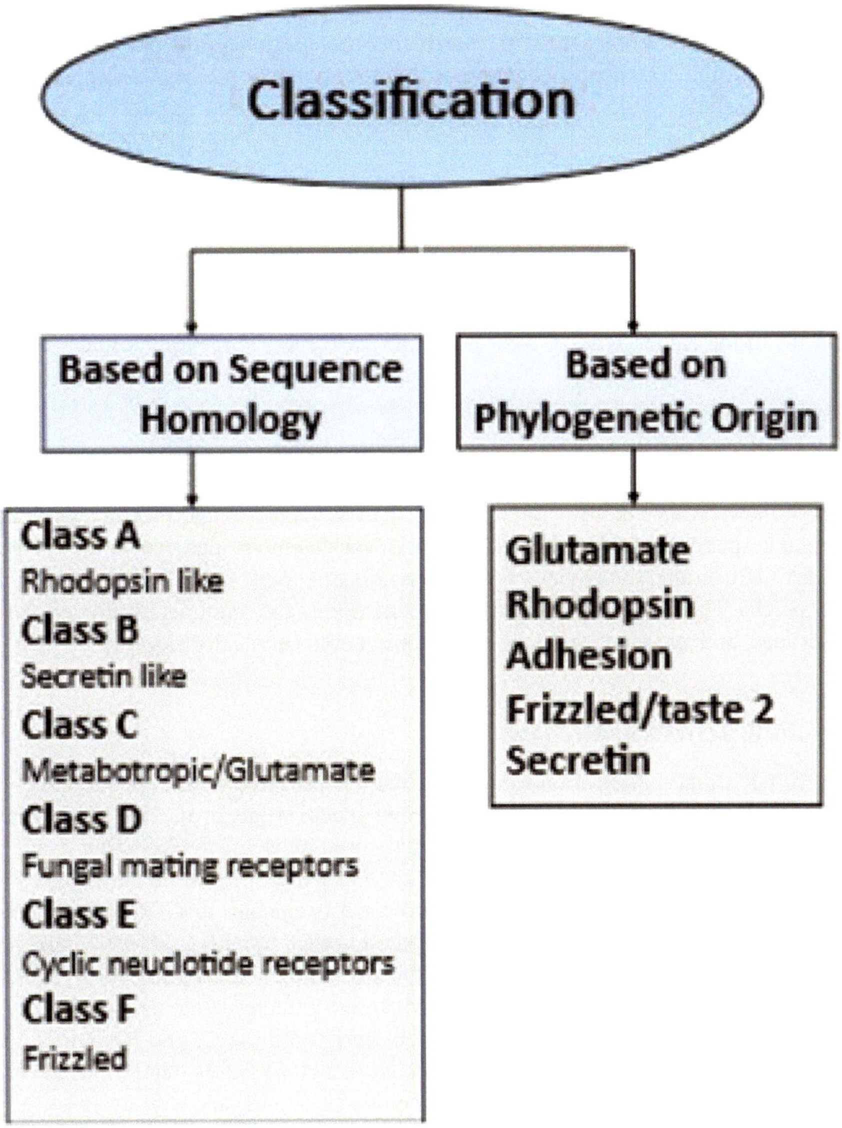

FIGURE 4.6 Classification of GPCRs based on sequence homology and phylogenetic origin (GRAFS system).

transmission. They feature a distinctive N-terminal domain known as the Venus flytrap, which serves as the glutamate binding site (Niswender & Conn, 2010).

4.3.2.2 Rhodopsin Family

A large group of cell surface signaling receptors belong to the rhodopsin family. They are primarily responsible for color vision (Zhou et al., 2012) and are classified as class A of the GPCR superfamily. These receptors transduce extracellular signals to intracellular ones by activating heterotrimeric protein complexes. Their structure includes a transmembrane region responsible for binding ligands and eight additional helices (Yang et al., 2021).

4.3.2.3 Adhesion Family

The adhesion family consists of a total of 33 members divided into nine subfamilies. These receptors contain extracellular domains with motifs that are specialized in facilitating cell attachment and movement. Another distinctive feature of this group is their ability to autophosphorylate their extracellular domains. They are involved in development, tissue repair, and immune function (Vizurraga et al., 2020).

4.3.2.4 Frizzled/Taste2 Families

The Frizzled GPCR family contains a cysteine-rich domain (CRD) in the extracellular region and consists of ten Frizzled FZD receptors. It plays an essential role in embryonic development, body structure formation, and adult cell maintenance and regeneration (Kamato et al., 2015). Furthermore, as the name suggests, these receptors are present in the tongue and are responsible for taste (Bachmanov & Beauchamp, 2007).

4.3.2.5 Secretin Family

The secretin family consists of 15 members, with secretin receptors being the first discovered. Members of the secretin subfamily are distinguished by their expansive extracellular domains and exhibit notable specificity in their interactions with various molecules, including vasoactive intestinal peptide (VIP), pituitary adenylate cyclase-activating peptide (PACAP), and corticotropin-releasing factor (CRF). Their diversity allows them to function in the immune, cardiovascular, endocrine, central nervous, and gastrointestinal systems (Karageorgos et al., 2018).

4.3.3 GPCR Activation Mechanisms

Classical GPCR signaling relies on signal propagation via three G-protein subunits: Gα, Gβ, and Gγ. Human Gα isoforms are encoded by 16 different genes (Hilger et al., 2018). G proteins bind with a variety of receptors, suggesting a preserved activation mechanism (Komolov & Benovic, 2018). When Gα is coupled to GDP, it combines with the Gβ γ dimer to create an inactive heterotrimer. Upon a ligand binding to the receptor, the heterotrimer exchanges its GDP with GTP, enabling the Gα and Gβ γ subunits to dissociate and participate in other signaling events (Figure 4.7) (Weis & Kobilka, 2018). The G-protein activation cycle ends when the Gα subunit hydrolyzes GTP to GDP, reuniting with Gβ γ. Gα subunits regulate downstream effector proteins, specifically interacting with adenylyl cyclases, cGMP phosphodiesterase, phospholipase C, and RhoGEFs (Hilger et al., 2018). The Gβ γ subunit binds to G-protein-coupled receptor kinases (GRKs), which regulate various channels, including potassium, voltage-dependent calcium, adenylyl cyclases, phospholipase C, phosphoinositide 3-kinase, and mitogen-activated protein kinases (Khan et al., 2013). GTPase-activating proteins, such as RGS proteins, interact with activated Gα subunits, accelerating GTP hydrolysis and regulating GPCR signaling (Kimple et al., 2011).

The two principal signal transmission pathways in GPCR signaling are the cAMP and phosphatidylinositol pathways (Gilman, 1987) (Figure 4.7).

4.3.4 GPCR Signaling Pathways

4.3.4.1 cAMP Pathway

cAMP (cyclic adenosine 3′,5′-monophosphate) was the first secondary messenger to be identified (Sassone-Corsi, 2012). Intracellular cAMP levels are regulated by two key enzymes: adenylyl cyclase (AC) and cyclic nucleotide phosphodiesterase (PDE). These enzymes can modulate cAMP to produce a cell-specific or a stimulus-specific response. This specificity is achieved through their immense diversity, differential expression patterns, and regulatory mechanism of the enzymes (Li et al., 2024).

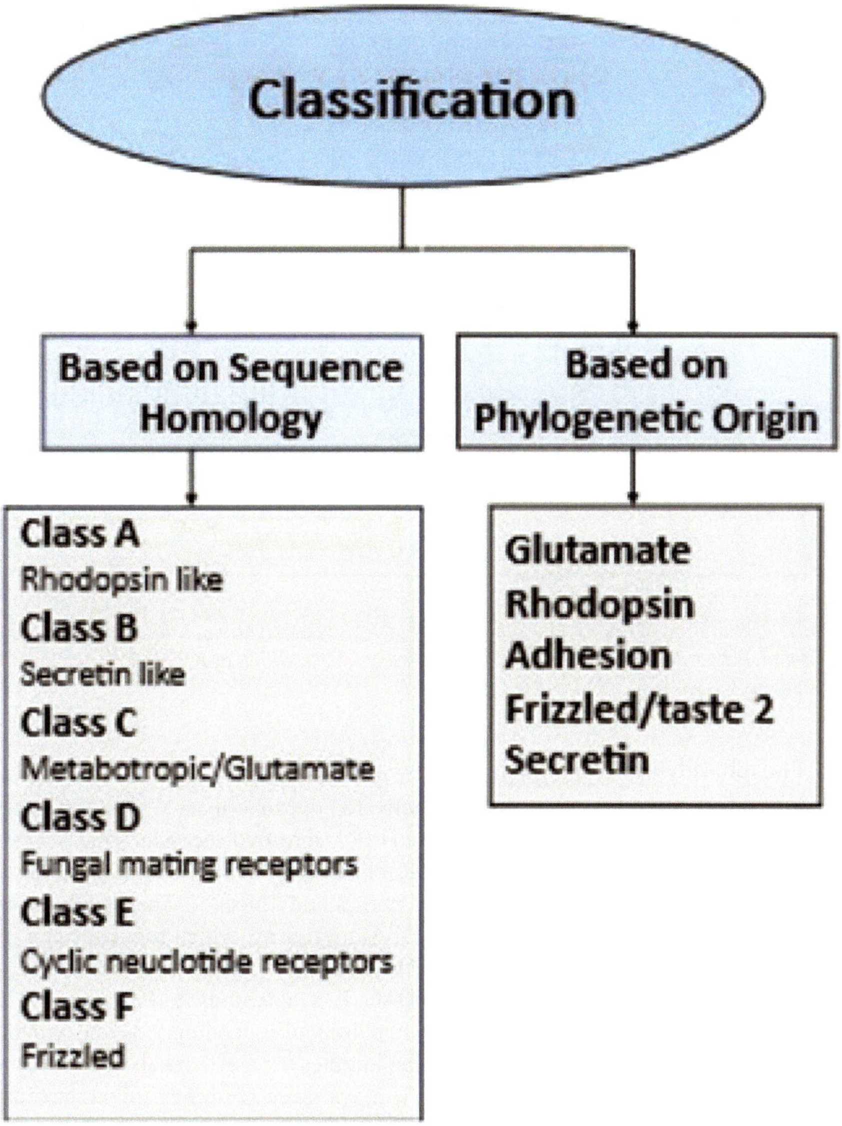

FIGURE 4.7 (A) Inactive GPCR with GDP attached to the alpha subunit of the G-protein trimer. (B) Ligand binding and dissociation of the trimers; the individual subunits can then activate other signaling pathways.

Adenyl cyclase is normally activated by GPCRs, such as the adrenoceptor, through interactions with a subunit of the Gs protein (Breitwieser, 2004). Activation of adenyl cyclase leads to cAMP production. cAMP then activates various effectors, with cAMP-dependent protein kinase (PKA) being the most prominent. PKA is a tetramer comprising two regulatory (R) subunits and two catalytic (C) subunits, with various isoforms of both subunits (Bos, 2006). The binding of cAMP to two sites on each R subunit activates their dissociation from the C subunits, freeing them. C subunits are catalytic kinases that can function by phosphorylating various transcription factors, including CREB, CREM, and ATF1, thereby promoting target gene transcription (Figure 4.8). Cyclic nucleotide phosphodiesterase degrades cAMP, serving as a regulatory mechanism to maintain equilibrium.

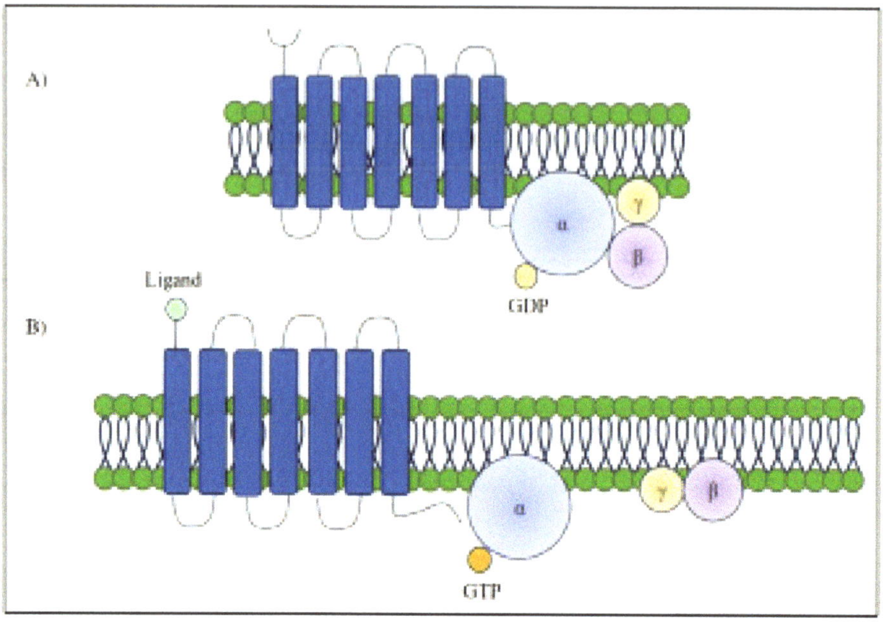

FIGURE 4.8 Generation of cAMP and activation of adenyl cyclase (AC). cAMP binds to the inactive heterotrimer of protein kinase A (PKA), causing subunit disassociation, which produces downstream effects through phosphorylation.

4.3.4.2 Phosphatidylinositol Signaling Pathway

In the phosphatidylinositol signaling pathway, activated phospholipase C (PLC) triggers the breakdown of phosphatidylinositol-(4,5)-bisphosphate (PIP2) into two secondary messengers: IP3 (inositol triphosphate) and DAG (diacylglycerol) (Rubio et al., 2018). Phosphoinositides PI3P, PIP2, and PIP3 are formed by the phosphorylation of phosphatidylinositol. These phosphoinositides are involved in signaling pathways such as the PI3K-Akt pathway, which regulates cell proliferation, survival, and metabolism (Zhang et al., 2013). Activation of GPCRs can initiate the activation of PLC (PLC-β), which cleaves PIP2 into IP3 and DAG. IP3 then binds to IP3-sensitive Ca^{2+} channels in the ER, releasing stored Ca^{2+}. Calcium binds to calmodulin, leading to a conformational change and activating calmodulin-dependent kinase. This initiates the Ca^{2+} signaling pathway, which plays diverse roles in inflammatory response, metabolism, apoptosis, cellular motion, fertilization, short-term and long-term memory, and immune response (Rubio et al., 2018). At the same time, calcium ions released from the endoplasmic reticulum raise the intracellular calcium concentration, translocating protein kinase C (PKC) from the cytosol to the inner plasma membrane, where it is activated by DAG. Activated PKC leads to the phosphorylation of multiple protein substrates, triggering diverse cellular responses (such as cell secretion, muscle contraction, or malignant cell proliferation and differentiation) (L. Zhang et al., 2013) (Figure 4.9).

3.5 DISEASES CAUSED BY MUTATIONS IN GPCRS

GPCRs constitute the most extensive group of receptors found in cell membranes. As mentioned previously, ligands such as hormones, neurotransmitters, and even photons can bind to GPCRs. Mutations affecting various stages of receptor maturation can lead to diseases with various causes.

Rhodopsin, the receptor responsible for visual perception, requires the formation of a chromophore apoprotein complex (CAM). This complex comprises 11-cis-retinol (the chromophore) and an apoprotein—an enzyme that lacks a cofactor (Hofmann & Palczewski, 2015). Mutations

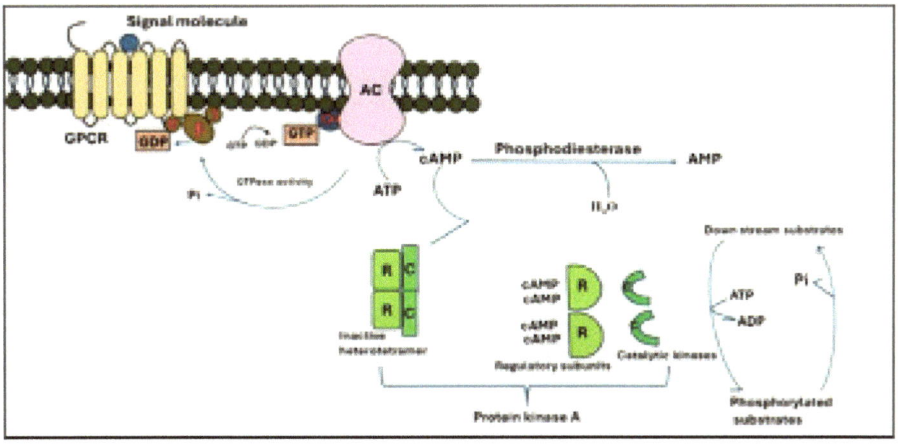

FIGURE 4.9 *PLC-β on the plasma membrane is activated by the interaction of a signal molecule with GPCR's α subunit, which hydrolyzes PIP2 to IP3 and DAG. IP3 binds to IP3-gated Ca^{2+} channels on the endoplasmic reticulum, opening them. Ca^{2+} binds to calmodulin (CaM) to form a Ca^{2+}-CaM complex, which interacts with target enzymes such as AC, PDEs, and CaMK. DAG activates PKC, which binds to the plasma membrane. PKC can phosphorylate serine or threonine residues of proteins, generating different cellular reactions.*

in rhodopsin-encoding genes can disrupt CAM formation, resulting in conditions such as retinitis pigmentosa (RP), which causes programmed cell death of rod cells (Zalewska et al., 2014).

Mutations in adrenergic receptors, such as single nucleotide polymorphisms (SNP), can impact protein expression, ligand interaction, G-protein binding, and signal transduction. These mutations can increase the risk of or trigger diseases such as hypertension, asthma, and cardiac failure.

Additionally, genetic mutations in dopamine, serotonin, and opioid receptors are implicated in mental disorders. Understanding the composition and operation of genetically altered GPCRs may help treat previously untreatable illnesses (Heng et al., 2013).

4.4 ENZYME-LINKED RECEPTORS

Enzyme-linked receptors are generally classified based on their cytosolic enzymatic domain. As such, they can possess tyrosine kinase, tyrosine phosphatase, serine/threonine kinase, or guanylyl cyclase activity (Dowd & Abel, 2017). The major classes are described below.

4.4.1 Guanylyl Cyclase Receptors (GC-Rs)

Guanylyl cyclase receptors are essential for different physiological functions, such as vision, bone development, blood pressure, and intestinal fluid regulation. There are seven subtypes, each with a specific ligand and a conserved structure featuring a homodimer and transmembrane segment. Their name derives from their ability to convert GTP (guanine triphosphate) to cGMP (cyclic guanine monophosphate), which serves as a downstream secondary messenger (Kuhn, 2016).

4.4.2 Receptor-Type Protein Tyrosine Phosphatase

The interplay between activating kinases and deactivating phosphatases is crucial for cellular homeostasis. While protein phosphatases have long been overshadowed by protein kinases, the enzymatic activity of both is required to maintain cellular homeostasis (Köhn, 2020). Receptor-type protein tyrosine phosphatases (RPTPs) counteract protein kinases by inactivating various signaling pathways, serving as a counterbalance. They comprise three distinct domains: an extracellular domain, a transmembrane region, and an intracellular enzymatic domain. RPTPs possess inherent phosphatase activity and thus play an essential role in cell proliferation and differentiation, implicating them in diseases such as cancer (Xu & Fisher, 2012). However, cancer development typically results from mutations in the regulation of these receptors, such as changes in gene copy numbers or incorrect promoter methylation. In normal cell physiology, these receptors are tightly regulated, but such errors can occur.

4.4.3 Receptor Serine/Threonine Kinases

Receptor serine/threonine kinases (RSKs) are enzyme-linked receptors with inherent kinase activity specific to serines and threonines. They fall into two subgroups, type I and type II RSKs, based on their amino acid sequences. Upon activation, these receptors form heterodimers, with type II RSKs activating type I RSKs, which then initiate downstream signaling cascades. Like other transmembrane receptors, RSKs have three characteristic structural features: an extracellular ligand-binding domain, a single-pass transmembrane region, and an intracellular enzymatic domain. As members of the transforming growth factor beta (TGF-β) superfamily, they contribute to tissue fibrosis, tissue homeostasis, embryogenesis, and even cancer (Moustakas et al., 2005).

4.4.4 Receptor Tyrosine Kinases (RTKs)

RTKs are classified as single-pass type I transmembrane receptors with an extracellular domain for ligand binding and an intracellular domain for kinase activity (Li & Hristova, 2006). These receptors are key components in signal transduction pathways that modulate different biological activities, such as cell growth, differentiation, homeostasis, migration, and motility (Li & Hristova, 2010). RTKs promote cell communication by phosphorylating tyrosine residues on intracellular substrate proteins, which activates downstream signaling cascades and elicits various cellular responses (Wheeler & Yarden, 2015).

This diverse group of membrane proteins comprises 20 subfamilies with 58 different RTKs, classified according to their structural features and ligand affinity. Key receptors include nerve growth factor receptors (NGFRs), hepatocyte growth factor receptors (HGFRs), erythropoietin-producing (Eph) receptors, platelet-derived growth factor receptors (PDGFRs), epidermal growth factor receptors (ErbBs), insulin and insulin-like growth factor receptors (IRs and IGFRs), vascular endothelial growth factor receptors (VEGFRs), and fibroblast growth factor receptors (FGFRs) (Hubbard, 1999). RTKs share evolutionarily conserved structural features across diverse organisms, including a cytoplasmic region, a single transmembrane helix, and an N-terminal ligand-binding domain. The cytoplasmic region consists of a protein tyrosine kinase (TK) domain, a juxtamembrane regulatory region of 40 to 80 amino acids, and a carboxy-terminal domain (Lemmon & Schlessinger, 2010). The TK module is made up of 12 subdomains in two lobes connected by the kinase insert motif (subdomain V) (Hubbard & Till, 2000). The kinase domain facilitates the transfer of the gamma phosphate from ATP to the hydroxyl group present on the side chain of the amino acid tyrosine. Additionally, autophosphorylation sites are frequently found in areas like the juxtamembrane and the carboxy-terminal regions, which serve as platforms for protein-protein interactions (Figure 4.10).

4.4.4.1 Activation Mechanism of RTKs under Normal Physiological Conditions

Generally, there are four modes of RTK activation: receptor-induced dimerization, ligand-induced dimerization, ligand-induced dimerization with receptor contacts, and ligand-induced dimerization with receptor contacts and accessory molecules (Sudhesh Dev et al., 2021). Receptor-induced dimerization is mediated through the direct receptor interaction. Ligand-induced dimerization is mediated through ligand binding (Wehrman et al., 2007). Ligand-induced dimerization with receptor

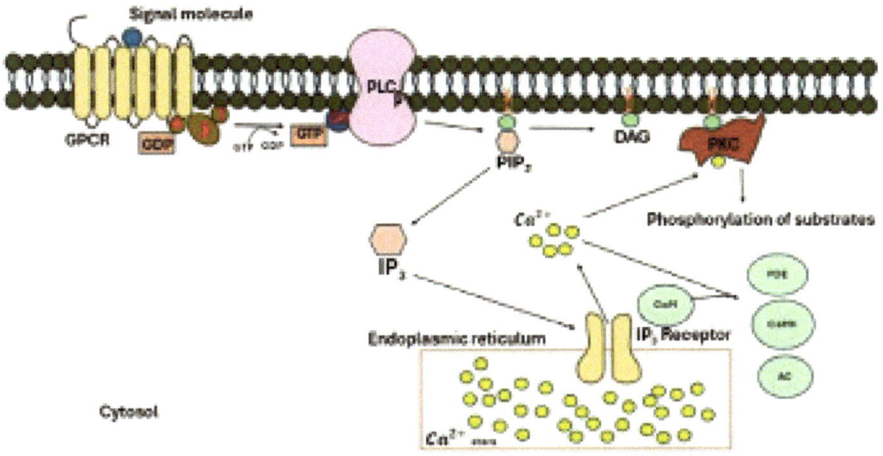

FIGURE 4.10 Schematic representation of RTK structure and dimerization process: (Left) RTKs are generally characterized by an N-terminal ligand-binding domain, single transmembrane domain, and intracellular region (juxtamembrane region, tyrosine kinase domain, and carboxy-terminal tail). (Right) Two RTKs dimerize upon ligand binding, leading to autophosphorylation, activation of the TK domain, and, finally, autoinhibition.

contacts involves the two receptor molecules binding to ligand homodimers. The fourth mode, ligand-induced dimerization with receptor contacts and accessory molecules, involves bivalent ligand binding, receptor-receptor interactions, and accessory molecule recruitment. For instance, the FGFR RTK family uses heparin or heparin sulfate as accessory molecules.

The activation mechanism of RTKs comprises a series of events triggered by the adhering of extracellular ligands, which promotes receptor dimerization (Schlessinger, 2000). This brings the intracellular tyrosine kinase domains (TKDs) into proximity, allowing for trans-autophosphorylation and cis-autoinhibition release (Lemmon & Schlessinger, 2010). These processes trigger a cascade of events that cause the recruitment and assembly of various downstream signaling proteins. These proteins contain Src homology-2 (SH2) or phosphotyrosine-binding (PTB) domains that interact with phosphotyrosine residues, either by direct or indirect binding (Pawson et al., 2001). Key docking proteins comprise Grb-associated binder (Gab1), FRS2, and insulin receptor substrate-1 (IRS1). They are responsible for activating various signaling pathways, such as mitogen-activated protein kinase (MAPK), PI3K/Akt, Janus kinase (JAK)/signal transducer and activator of transcription (STAT), and PKC, which regulate numerous cellular processes such as proliferation, survival, differentiation, and metabolism. The c-Jun N-terminal kinase (JNK) and p38 MAPK pathways primarily mediate cellular stress response and apoptosis, while the extracellular signal-regulated kinase (ERK)/MAPK signaling pathway regulates various processes, such as cell proliferation and differentiation, within the complex cellular signal transduction network (Kolch, 2005).

4.4.5 Signaling Pathways

4.4.5.1 Mitogen-Activated Protein (MAP) Kinase Pathway

MAP kinases are key components in cellular biology, playing a crucial role in cellular signaling. They comprise three kinase types: MAP kinase kinase kinase (MAPKKK), MAP kinase kinase (MAPKK), and lastly, the MAP kinase (MAPK). These cascades are activated by various external signals, including growth factors, stress, cytokines, and ceramides (Zhang & Liu, 2002). These signals trigger specific MAPK cascades, starting with MAPKKK activation (Cuevas et al., 2007). Activated MAPKKK phosphorylates and activates MAPKK, which then activates MAPK.

Activated MAPK plays a pivotal role in regulating cellular activities by phosphorylating a range of substrates in the nucleus and cytoplasm, leading to altered gene expression and protein function. These changes can promote cell growth, differentiation, or programmed cell death. MAPKs also have docking sites for MAPKKs and their substrates, facilitating high-affinity protein-protein interactions. This specificity ensures that MAPKs are activated by the appropriate upstream MAPKKs to effectively identify and act upon their downstream targets (Morrison, 2012).

The MAPK signaling cascade primarily comprises three main pathways: p38/SAPKs (stress-activated protein kinases), JNKs (Jun amino-terminal kinases), and ERKs. These pathways are initiated by different members of the evolutionarily conserved upstream MAPKK family, including MEK1/2, MKK3, MKK4, MKK6, and MKK7. MKK4/7 activates the JNK pathway, and MKK3/6 activates the p38 pathway (Caliz et al., 2022). Some studies suggest MKK4 is also involved in the activation of the p38 pathway, while MEK1/2 activates the ERK signaling pathway (Wang & Xia, 2012) (Figure 4.11).

4.4.5.2 ERK1/ERK2 Pathway

The ERK1/ERK2 pathway, also known as the RAS-Raf-MEK-ERK signaling pathway, involves two ERK isoforms: ERK1 (p42) and ERK2 (p44) (Zhang & Dong, 2007). These kinases are crucial for transducing signals related to cell proliferation, differentiation, and survival. RTKs, GPCRs, integrins, and small GTPases at the cell surface regulate ERK1/ERK2 pathway activation. In the canonical ERK1/2 signaling pathway, members of the Raf family, Mos, and Tpl2 constitute the

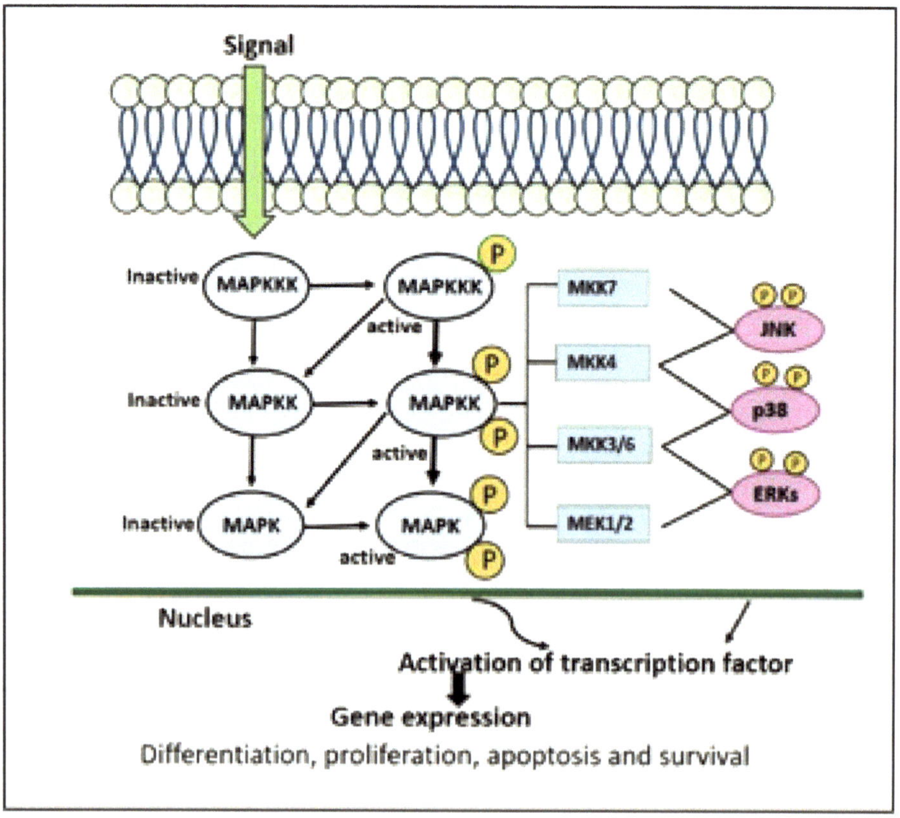

FIGURE 4.11 Various external stimuli can trigger the stimulation of the MAP kinase family after receptor-ligand binding. Members of this family can activate one another through the phosphorylation of serine/threonine amino acids.

MAPKKKs, while MEK1 and MEK2 act as the MAPKKs (Saha & Raghava, 2006). When a ligand, such as growth factor receptor-bound protein 2 (Grb2), binds to RTKS, Grb2 recruits another protein called son of sevenless (SOS), which activates the small G-protein Ras. Ras recruits MAPKKK-Raf, which then activates the MAPKK, MEK. Subsequently, MEK phosphorylates ERK1/2 (the MAPK) on the tyrosine and threonine residues within the TEY motif (Knight & Irving, 2014). Activated ERK1/2 migrates to the nucleus, where it phosphorylates specific transcription factors, affecting the expression of certain genes (Figure 4.12). Dual-specificity MAPK phosphatases (MKPs) can inhibit the pathway. For example, MKP2/4 can dephosphorylate ERK1/2 and other MAPKs, while MKP3 and MKP-X are more ERK-specific (Owens & Keyse, 2007). Additionally, the tyrosine phosphatase SHP2 is a key component of the pathway since it helps modulate the activation of Ras. This intricate network of kinases and phosphatases ensures precise control over cell signaling processes.

4.4.5.3 P38/SAPKs Signaling Pathway

P38/SPAKs are another crucial component of the MAPK signaling cascade. Like the JNK pathway, stress and cytokines trigger the activation of various isoforms of p38 MAPK. These activated isoforms are critical for stimulating inflammatory responses within cells. Each of these isoforms is encoded by a distinct gene and exhibits unique patterns of expression in different tissues (Cuadrado & Nebreda, 2010). Like ERK isoforms, p38 MAPKs undergo sequential activation. Typically, stress or cytokine signals activate a MAPKKK (e.g., MEKK1, ASK1, or TAK1). Proteins like TRAF 2/3/6 or Rho are involved in the activation of this cascade. The activated MAPKKK phosphorylates a MAPKK (such as MKK3 or MKK6), which then activates the Thr-Gly-Tyr (TGY) motif of

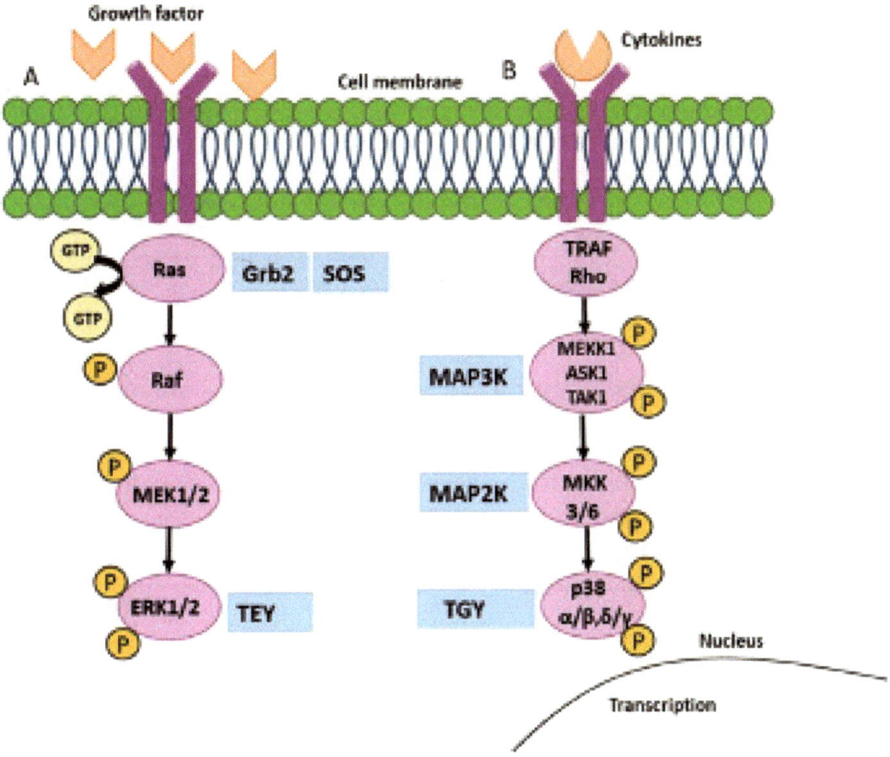

FIGURE 4.12 (A) Extracellular signals activate the ERK1/2 pathway. Once activated, ERK1 translocates to the nucleus, activates transcription of target genes, and regulates various cellular processes. (B) p38 pathway: Several environmental factors like TRAF or Rho activate MAP3K and MAP2K (MKK3/6), followed by p38 activation.

p38. Various dual-specific protein phosphatases are essential for dephosphorylating p38 MAPK isoforms. MPK1 shows higher selectivity in dephosphorylating p38 MAPK, while MPK2/4 dephosphorylate both p38 MAPK isoforms and ERK. Additionally, MPK5/7 may dephosphorylate JNK and p38 MAPKs (Soares-Silva et al., 2016) (Figure 4.12)

4.4.5.4 JNK Signal Transduction Pathway

The JNK pathway is a key signaling cascade within the MAPK signaling pathway, regulating cellular processes such as embryonic development, proliferation, and apoptosis. JNK protein kinases, also known as stress-activated protein kinases (SPAKs), are encoded by three genes: JNK-1, JNK-2, and JNK-3. Alternative splicing generates different isoforms from these genes (Davis, 2000). Various forms of cellular stress activate this pathway, including inflammatory cytokines, heat shock, UV radiation, osmotic stress, and other environmental stressors (Katz et al., 2007). These signals activate RAC1, a small GTPase involved in signal transduction and cell movement. RAC1 serves as an upstream regulator of the JNK pathway, activating an MAPKKK (MEKK or MUK). These kinases then phosphorylate and activate downstream kinases. The activated MAPKKK subsequently phosphorylates MAPKK (either MKK4 or MKK7), activating it. This MAPKK then phosphorylates JNK on specific threonine and tyrosine residues in a TPY motif. This dual phosphorylation fully activates JNK, which then moves to the nucleus, where it phosphorylates transcription factors, such as c-Jun, a component of the activator protein-1 (AP-1) complex. This leads to changes in gene expression associated with stress responses, including apoptosis, differentiation, and inflammation. In essence, the JNK pathway comprises a series of phosphorylation events that transduce stress signals from the cell surface membrane to the nucleus (Figure 4.13).

4.4.5.5 JAK/STAT Pathway

The JAK/STAT pathway plays a pivotal role in RTK signaling transduction. This pathway may be initiated through various interferons, cytokines, and other stimulation factors. More than 50 different cytokines and growth factors regulate the JAK/STAT signaling system. This pathway stimulates different processes, including immunological response, inflammation, proliferation, hematopoiesis, tissue regeneration, differentiation, and apoptosis. There are four members of the JAK family: JAK1, JAK2, JAK3, and TYK2. The STAT family consists of seven members: STAT1, STAT2, STAT3, STAT4, STAT5a, STAT5b, and STAT6 (O'Shea et al., 2015). JAKs are tyrosine kinases typically associated with the cytoplasmic region of type I and type II cytokine receptors (Darnell, 1997). They are activated upon binding of signaling molecules to their corresponding cytokine receptors. These receptors exist as either homodimers (GRH, EPO) or heteromultimers (IFN, IL). Ligand binding induces structural changes in the receptor, causing dimerization of receptor units and triggering JAK activation. Activated JAKs initiate transphosphorylation on tyrosine residues, creating binding sites for STAT recruitment (Levy & Darnell, 2002). This phosphorylation leads to STAT dimerization with other STATs. Activated STAT dimers then translocate to the nucleus, where they attach to specific DNA sequences called STAT response elements (SREs) and gamma-activated sequence (GAS) elements. These dimers then act as transcription factors, regulating gene expression by either activating or suppressing it (Figure 4.14).

4.4.5.6 Phosphoinositide 3-Kinase (PI3K)/Akt Pathway

PI3K is a family of lipid kinases found in the plasma membrane. It consists of three subunits: a p110 catalytic subunit, a p85 regulatory subunit, and a p55 regulatory subunit. These kinases are classified into three types based on their structures and substrates: Class I PI3K (IA and IB), Class II PI3K, and Class III PI3K. Under normal conditions, PI3K is activated by extracellular signals such as growth factors, hormones, and cytokines. Activation catalyzes the phosphorylation of PIP2 to PIP3. PIP3 serves as a second messenger within the cell, recruiting lipid-binding domains, such as pleckstrin-homology (PH), FYVE, Phox (PX), C1, and C2 domains. Several signaling proteins, including PDK1 and Akt, can bind to lipids generated by PI3K, translocate to the cell membrane,

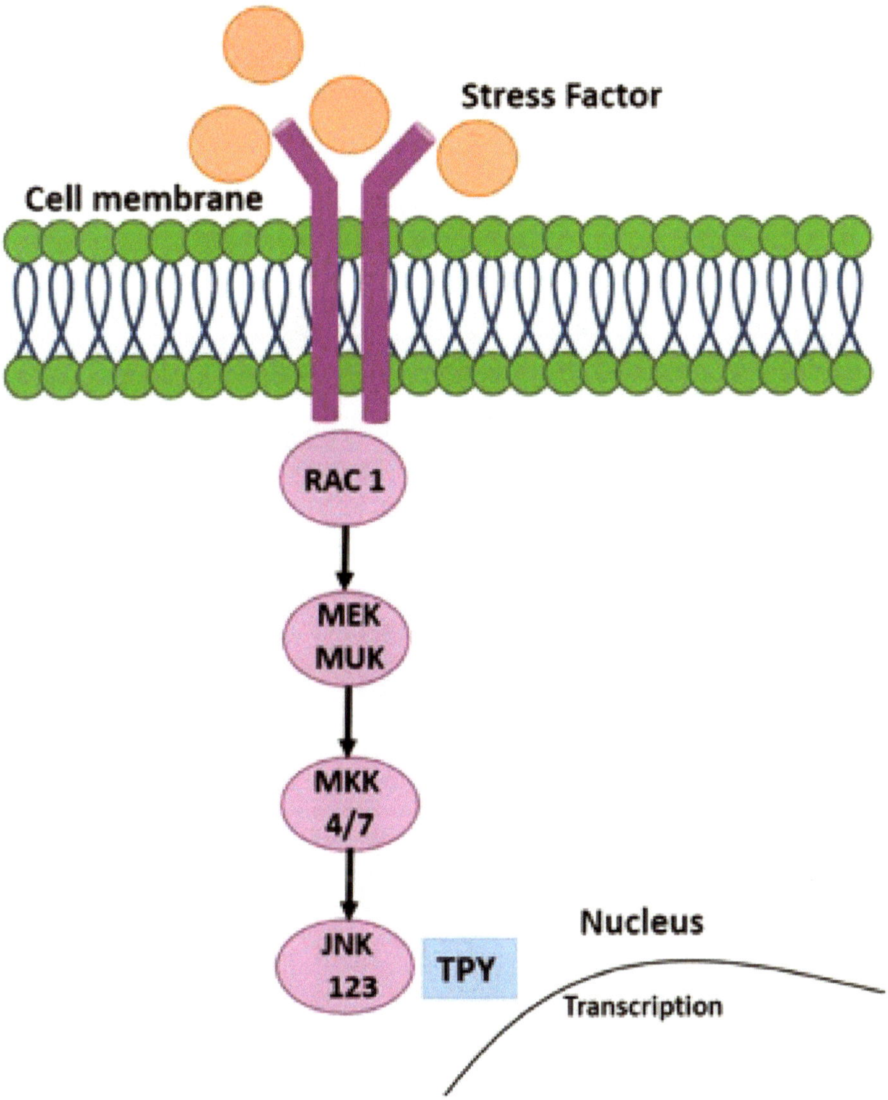

FIGURE 4.13 JNK pathway: Different stress factors activate RAC1, followed by the activation of MAPKKK and MAPKK. Activated MAPKK activates JNK, which moves to the nucleus and phosphorylates various transcription factors.

and activate pathways that promote cell growth and survival. PDK1 and mTORc2 activate Akt, resulting in the phosphorylation of various downstream targets, which control cellular processes. PTEN is involved in regulating this pathway by dephosphorylating PIP3 to PIP2, thereby inhibiting the subsequent signaling processes (Yang et al., 2019) (Figure 4.15).

4.5 MODERN TECHNIQUES AND METHODS FOR ANALYZING AND IDENTIFYING INTEGRAL PROTEIN PROPERTIES

The immense biological diversity of integral proteins, coupled with their heterogeneous nature, requires high-resolution techniques to unravel and understand their association with different

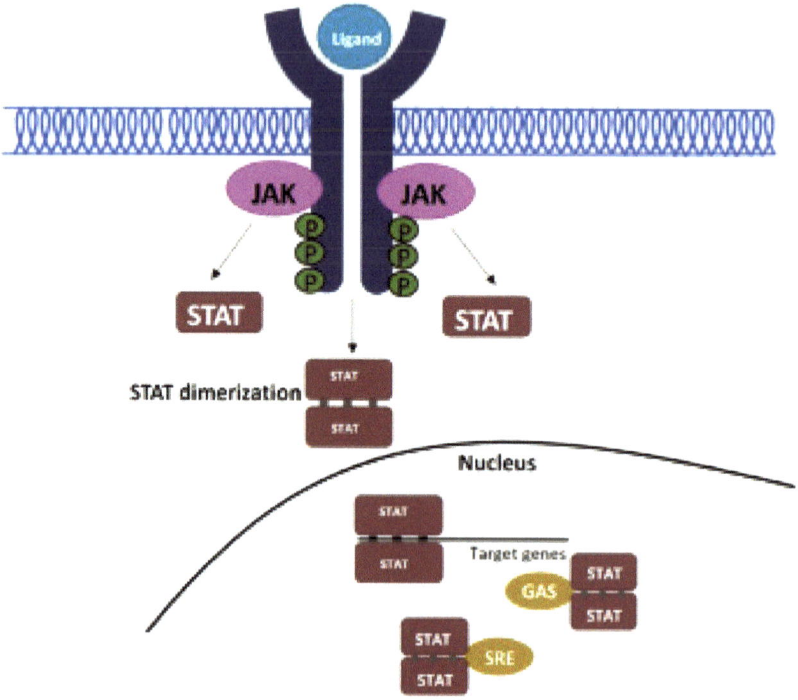

FIGURE 4.14 Schematic representation of JAK/STAT signaling pathway. Cytokines initiate the pathway by binding to receptors, facilitating JAK transactivation. Activated JAKs phosphorylate tyrosine residues on the receptor's intracellular domains, attracting STAT factors and forming STAT dimers. STAT dimers move into the nucleus and bind to SRE and GAS elements to activate gene expression.

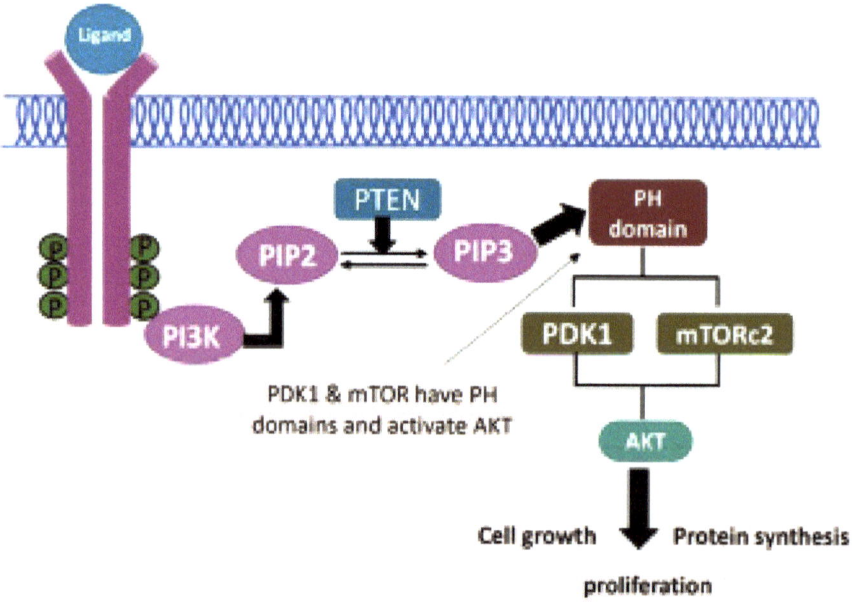

FIGURE 4.15 Schematic illustration of PI3K pathway: RTK recruits PI3K, which phosphorylates PIP2 to PIP3, followed by the activation of Akt with the help of PDK1 and mTORc2.

ligands and environmental stimuli. Researchers employ a wide variety of tools are to understand the complex interactions of these receptors. Understanding their physical dimensions, stability, affinity, and nature helps create a complete picture of these proteins and serves as an invaluable tool for researchers aiming to design therapeutic targets against these receptors.

Techniques for analyzing integral proteins can be classified into four broad categories based on the receptor property they aim to uncover:

1. Structure-based assays: These techniques aim to reveal the receptor's structure in high resolution, so that their stereochemical properties can be properly estimated for better drug design and control.
2. Affinity-based assays: These assess the interaction between receptor and ligand, aiming to estimate the affinity of the two interacting molecular species.
3. Stability-based techniques: These techniques evaluate the stability of the receptors under various conditions, such as different pH levels and temperatures.
4. Signaling-based assays: These determine receptor activation or inactivation by analyzing downstream secondary messengers, helping understand receptor activation under various conditions and in the presence of different molecules (Yen et al., 2023).

The following sections provide a brief overview of these categories and their associated experimental techniques.

4.5.1 Structural Assay

GPCRs don't exist in just two discrete states (on or off). Instead, their output is analogous, with varying levels of activation based on the type of agonists that bind to them. Some agonists might induce maximal efficiency, others partial signaling, and some even cause a decrease in signaling. GPCRs can also have a baseline signal, exhibiting activity even without agonists. The structural conformation of the GPCR is often crucial in determining the magnitude of their signal output, necessitating structural studies to better understand their binding properties (Hilger, 2021). Similar structural insights are needed for ion channels and enzyme-linked receptors.

Cryogenic electron microscopy (Cryo-EM) and X-ray crystallography have been foundational in structural chemistry, particularly for membrane-bound structures. X-ray crystallography, the oldest technique, has undergone numerous developments over time, helping scientists unravel the intricacies of structural biology. Its basic principle involves crystallizing the subject and bombarding it with X-rays; the resulting diffraction pattern allows computer reconstruction of a 3D image. However, not all molecules can be crystallized, leading to the development of alternative techniques like nuclear magnetic resonance (NMR).

In NMR (Asadollahi et al., 2023; Jin et al., 2023), the protein of interest is placed in a solution under strong magnetic fields. Radiofrequency waves bombard the sample, causing atoms to change their energy state. These changes, affected by the molecule's environment, allow derivation of a 3D structure. NMR has limitations, including the need for high sample volumes, which is problematic for proteins that are challenging to purify (Nogales, 2016).

Cryo-EM addresses these issues, requiring neither crystallization nor large sample volumes. This technique involves cry-fixating integral proteins and examining them with a high-resolution electron microscope. Cryofixation rapidly freezes the sample with water at very low temperatures, preserving the protein structure (Gusach et al., 2023). These techniques have proven vital for understanding and uncovering the structure of integral proteins, helping pave the way for the development of advanced therapeutics (Egelman, 2016).

Hydrogen-Deuterium Exchange Mass Spectrometry (HDX-MS) has recently gained popularity as a complementary structural analysis technique. It involves placing the protein in a solution of deuterium oxide, where hydrogen atoms are gradually replaced by deuterium atoms. Mass spectrometry analysis of the changing mass reveals protein structural dynamics under specific conditions (Martens & Politis, 2020; Ozohanics & Ambrus, 2020). Metanalyses show that X-ray crystallography remains the most widely used technique for structural protein analysis. However, Cryo-EM has gained significant traction in recent years and is now a leading technique for studying these proteins (Choy et al., 2021).

4.5.2 Signaling-Based Assays

Another method for determining the properties of integral membrane proteins is to study their output, specifically the secondary signaling molecules they produce. Various techniques focus on assessing the quantitative expression of downstream molecules, helping determine receptor output in the presence of particular agonists and uncovering the relationship between the ligand and the receptor.

For some integral proteins like GPCRs, cyclic AMP is produced as a secondary product, and is quantified to understand GPCR activation (Guo et al., 2022). Similarly, IP3/IP1 and Ca^{2+} are used to estimate GPCR activation involving phospholipase C (Garbison et al., 2012). Another method uses a luciferase reporter gene fused with the downstream signaling product. When expressed, luciferin is produced, emitting light that is used to quantify the receptor signal strength in response to certain ligands.

For ion channels, electrophysiological techniques are primarily used, employing three main methods to extract quantitative data: voltage clamp methods, membrane potential estimations, and absorption spectroscopy (McManus et al., 2012). Voltage clamp methods detect ion channel fluxes but are labor-intensive, although automated methods have been developed. Membrane potential estimation, another electrophysiological technique, can also assess signal strength. Lastly, absorption spectroscopy can be used to determine changing ionic concentrations.

4.5.3 Stability-Based Assays

Understanding the adaptability of integral proteins to various environmental conditions is crucial for researchers. They deploy a wide variety of tools and techniques to gauge different environmental factors' effects on protein stability, which is critical for drug design and therapeutic development.

4.5.3.1 Dynamic Light Scattering (DLS)

DLS assesses the structural integrity of integral proteins based on the Brownian motion of particles in solution, where small particles move faster than larger ones. When agitated, water molecules travel at different speeds and cover different distances based on their size (Lorber et al., 2012). Analyzing the scattering of monochromatic light reveals the particle size distribution. Researchers can alter solution conditions to determine integral proteins' physical characteristics and the effects of various environmental conditions based on size and aggregation property changes. This information is essential for designing therapies that induce or are sensitive to environmental changes (Reis & Moraes, 2020). DLS can also estimate the homogeneity of integral membrane proteins over time.

4.5.3.2 Differential Scanning Fluorimetry (DSF)

In DSF, a fluorescent dye such as SYPRO Orange is paired with the protein of interest, and a temperature gradient is applied. Temperature changes cause conformational changes in the protein of interest, altering the fluorophore's luminescent properties. This can be used to estimate protein stability under various thermal conditions and determine melting temperature, a crucial parameter for protein purification and expression (Boland et al., 2018; Kwan et al., 2022).

Assessing physical characteristics and environmental effects on these proteins can be challenging. Therefore, researchers often combine techniques for better results and understanding. For example, nuclear magnetic resonance spectroscopy can be paired with techniques studying environmental impacts on protein structure, providing a more comprehensive visualization of proteins under various conditions.

4.5.4 Affinity-Based Assays

Affinity-based assays study the binding relationship between a receptor and its ligand, providing crucial information for designing drugs that target integral proteins such as tyrosine kinases, ligand-gated ion channels, and GPCRs. Mass spectrometry, with its diverse applications and high throughput, dominates this field (Meissner et al., 2022).

4.5.4.1 Microbead Assays

In this approach, target integral proteins are attached to the surface of micro-sized beads loaded with fluorescent dyes or fluorophores. These beads provide a solid foundation for attaching proteins and investigating their affinity for different ligands (Zhao et al., 2021). If the ligand attaches to the receptor under study, the beads show a change in fluorescence pattern, which is then detected and analyzed.

4.5.4.2 Membrane Mass Spectrometry

This technique places integral proteins in a simulated or mimetic lipid environment and binds them to their respective ligands. This mimetic approach helps preserve the proteins' natural state and condition, simulating the natural lipid membrane structure. It allows researchers to understand protein structural dynamics and membrane stability (Bolla et al., 2021). A mass spectrometer then analyzes the sample, ionizing constituents and separating objects by mass.

4.5.4.3 Native Mass Spectrometry

Unlike traditional mass spectrometry, which breaks down the sample into smaller fragments, native MS preserves molecular structure (Sahin et al., 2020). It uses gentle ionization technology to maintain larger fragments and a soft desolvation method to transfer components to the gas phase while maintaining their structural integrity (Pan & Vachet, 2022). This can reveal information about ligand-receptor binding affinity and properties.

4.5.4.4 Surface Plasmon Resonance

In surface plasmon resonance, the integral protein under study is immobilized on a metal surface, usually gold. Light directed at the surface causes free electrons to undergo resonant oscillation, due to surface plasmon resonance, inducing a change in the refractive index. This change also depends on the immobilized structure. When a ligand attaches to the receptor, it causes a shift in surface plasmon resonance, which is used to gather data on the binding affinity of the receptor and ligand (Capelli et al., 2020).

4.5.4.5 Microscale Thermophoresis (MST)

In MST, a temperature gradient is applied to a solution containing the integral proteins. The proteins move through the gradient, with their movement affected by their binding state. A receptor with a bound ligand will move differently from an unbound receptor, which can be tracked through fluorescence (Asmari et al., 2023; Nighat et al., 2024). This movement can then be used to estimate the binding kinetics of the two interacting molecules.

4.5.4.6 Fluorescence Resonance Energy Transfer (FRET)

FRET uses an acceptor and a donor molecule. When these molecules are close enough, energy can be transferred from one to another. To study the ligand-receptor interaction, fluorophores are attached to each molecule, which are brought together by their affinity. Light of a specific wavelength corresponding to the donor is directed at the particles. Once excited, the donor transfers the energy to the acceptor, which emits light in a different wavelength. This emission is then recorded (Fang et al., 2023; Jiang et al., 2019).

4.6 CONCLUSION

Modern techniques and methods for analyzing and identifying integral protein properties are essential tools for unraveling the intricacies of these crucial biological molecules. Continued advancement and integration of these techniques promise to deepen our understanding and open new avenues for therapeutic interventions targeting integral proteins.

REFERENCES

Absalom, N. L., Liao, V. W., & Chebib, M. (2020). Ligand-gated ion channels in genetic disorders and the question of efficacy. *The International Journal of Biochemistry & Cell Biology*, *126*, 105806. https://doi.org/10.1016/J.BIOCEL.2020.105806

Alberts, B., Johnson, A., Lewis, J., Raff, M., Roberts, K., & Walter, P. (2002). Membrane proteins. https://www.ncbi.nlm.nih.gov/books/NBK26878/

Alexander, S., Mathie, A., & Peters, J. (2011). Ligand-gated ion channels. *British Journal of Pharmacology*, *164*(Suppl 1), S115. https://doi.org/10.1111/J.1476-5381.2011.01649_4.X

Allen, K. N., Entova, S., Ray, L. C., & Imperiali, B. (2019). Monotopic membrane proteins join the fold. *Trends in Biochemical Sciences*, *44*(1), 7. https://doi.org/10.1016/J.TIBS.2018.09.013

Anderson, K. J., Cormier, R. T., & Scott, P. M. (2019). Role of ion channels in gastrointestinal cancer. *World Journal of Gastroenterology*, *25*(38), 5732–5772. https://doi.org/10.3748/WJG.V25.I38.5732

Angus, M., Channels, P. R.-, & 2019, undefined. (2019). Voltage gated sodium channels in cancer and their potential mechanisms of action. *Taylor & FrancisM Angus, P RubenChannels, 2019•Taylor & Francis*, *13*(1), 400–409. https://doi.org/10.1080/19336950.2019.1666455

Asadollahi, K., Scott, D. J., & Gooley, P. R. (2023). NMR applications to GPCR recognition by peptide ligands. *Current Opinion in Pharmacology*, *70*, 102366. https://doi.org/10.1016/J.COPH.2023.102366

Asmari, M., Michalcová, L., Ibrahim, A. E., Glatz, Z., Wätzig, H., & El Deeb, S. (2023). Studying molecular interactions via capillary electrophoresis and microscale thermophoresis: A review. *Electrophoresis*, *44*(13–14), 1114–1142.

Bachmanov, A. A., & Beauchamp, G. K. (2007). Taste receptor genes. *Annual Review of Nutrition*, *27*, 389. https://doi.org/10.1146/ANNUREV.NUTR.26.061505.111329

Bezanilla, F. (2005). Voltage-gated ion channels. *IEEE Transactions on Nanobioscience*, *4*(1), 34–48. https://doi.org/10.1109/TNB.2004.842463

Blandin, C., Gravez, B., Hatem, S., Cells, E. B., & 2021, undefined. (2021). Remodeling of ion channel trafficking and cardiac arrhythmias. *Mdpi.ComCE Blandin, BJ Gravez, SN Hatem, E BalseCells, 2021•mdpi.Com*. https://doi.org/10.3390/cells10092417

Boland, C., Olatunji, S., Bailey, J., Howe, N., Weichert, D., Fetics, S. K., Yu, X., Merino-Gracia, J., Delsaut, C., & Caffrey, M. (2018). Membrane (and soluble) protein stability and binding measurements in the lipid cubic phase using label-free differential scanning fluorimetry. *Analytical Chemistry*, *90*(20), 12152–12160. https://doi.org/10.1021/ACS.ANALCHEM.8B03176/ASSET/IMAGES/LARGE/AC-2018-03176T_0004.JPEG

Bolla, J. R., Fiorentino, F., & Robinson, C. V. (2021). Mass spectrometry informs the structure and dynamics of membrane proteins involved in lipid and drug transport. *Current Opinion in Structural Biology*, *70*, 53–60.

Bootman, M., Harbor, G. B.-C. S., & 2020, undefined. (2020). Fundamentals of cellular calcium signaling: a primer. *Cshperspectives.Cshlp.OrgMD Bootman, G BultynckCold Spring Harbor Perspectives in Biology, 2020•cshperspectives.Cshlp.Org*. https://doi.org/10.1101/cshperspect.a038802

Bos, J. L. (2006). Epac proteins: multi-purpose cAMP targets. In *Trends in Biochemical Sciences* (Vol. 31, Issue 12, pp. 680–686). Elsevier Ltd. https://doi.org/10.1016/j.tibs.2006.10.002

Breitwieser, G. E. (2004). G protein-coupled receptor oligomerization: Implications for G protein activation and cell signaling. In *Circulation Research* (Vol. 94, Issue 1, pp. 17–27). https://doi.org/10.1161/01.RES.0000110420.68526.19

Bulk, E., Todesca, L. M., & Schwab, A. (2021). Ion channels in lung cancer. *Reviews of Physiology, Biochemistry and Pharmacology*, *181*, 57–80. https://doi.org/10.1007/112_2020_29
Caliz, A. D., Vertii, A., Fisch, V., Yoon, S., Yoo, H. J., Keaney, J. F., & Kant, S. (2022). Mitogen-activated protein kinase kinase 7 in inflammatory, cancer, and neurological diseases. *Frontiers in Cell and Developmental Biology*, *10*, 979673. https://doi.org/10.3389/FCELL.2022.979673/BIBTEX
Capelli, D., Parravicini, C., Pochetti, G., Montanari, R., Temporini, C., Rabuffetti, M., Trincavelli, M. L., Daniele, S., Fumagalli, M., & Saporiti, S. (2020). Surface plasmon resonance as a tool for ligand binding investigation of engineered GPR17 receptor, a G protein coupled receptor involved in myelination. *Frontiers in Chemistry*, *7*, 910.
Catacuzzeno, L., Sforna, L., & Franciolini, F. (2020). Voltage-dependent gating in K channels: Experimental results and quantitative models. *Pflügers Archiv-European Journal of Physiology*, *472*, 27–47.
Catterall, W. A. (2010). Ion channel voltage sensors: Structure, function, and pathophysiology. *Neuron*, *67*(6), 915–928.
Catterall, W. A., Wisedchaisri, G., & Zheng, N. (2017). The chemical basis for electrical signaling. *Nature Chemical Biology*, *13*(5), 455. https://doi.org/10.1038/NCHEMBIO.2353
Cell Membrane (Plasma Membrane). (n.d.). Retrieved February 5, 2024, from https://www.genome.gov/genetics-glossary/Cell-Membrane
Choy, B. C., Cater, R. J., Mancia, F., & Pryor, E. E. (2021). A 10-year meta-analysis of membrane protein structural biology: Detergents, membrane mimetics, and structure determination techniques. *Biochimica et Biophysica Acta (BBA) - Biomembranes*, *1863*(3), 183533. https://doi.org/10.1016/J.BBAMEM.2020.183533
Collingridge, G. L., Olsen, R. W., Peters, J., & Spedding, M. (2009). A nomenclature for ligand-gated ion channels. *Neuropharmacology*, *56*(1), 2–5. https://doi.org/10.1016/J.NEUROPHARM.2008.06.063
Cooper, G. M. (2000a). Structure of the plasma membrane. https://www.ncbi.nlm.nih.gov/books/NBK9898/
Cooper, G. M. (2000b). Structure of the plasma membrane. https://www.ncbi.nlm.nih.gov/books/NBK9898/
Cuadrado, A., & Nebreda, A. R. (2010). Mechanisms and functions of p38 MAPK signalling. *The Biochemical Journal*, *429*(3), 403–417. https://doi.org/10.1042/BJ20100323
Cuevas, B., Abell, A., & Johnson, G. (2007). Role of mitogen-activated protein kinase kinase kinases in signal integration. *Oncogene*, *26*, 3159–3171.
Darnell, J. E. R. (1997). STATs and gene regulation. *Science*, *277*, 1630–1635.
Davis, R. J. (2000). Signal transduction by the JNK group of MAP kinases. *Cell*, *103*, 239–252.
De Lera Ruiz, M., & Kraus, R. L. (2015). Voltage-gated sodium channels: Structure, function, pharmacology, and clinical indications. *Journal of Medicinal Chemistry*, *58*(18), 7093–7118. https://doi.org/10.1021/JM501981G/ASSET/IMAGES/LARGE/JM-2014-01981G_0013.JPEG
De Rubio, R. G., Ransom, R. F., Malik, S., Yule, D. I., Anantharam, A., & Smrcka, A. V. (2018). Phosphatidylinositol 4-phosphate is a major source of GPCR-stimulated phosphoinositide production. *Science Signaling*, *11*(547). https://doi.org/10.1126/scisignal.aan1210
Del-Río-ibisate, N., Granda-Díaz, R., Rodrigo, J. P., Menéndez, S. T., & García-Pedrero, J. M. (2021). Ion channel dysregulation in head and neck cancers: Perspectives for clinical application. *Reviews of Physiology, Biochemistry and Pharmacology*, *181*, 375–428. https://doi.org/10.1007/112_2020_38
Dolphin, A. C., & Lee, A. (2020). Presynaptic calcium channels: Specialized control of synaptic neurotransmitter release. *Nature Reviews Neuroscience 2020 21:4*, *21*(4), 213–229. https://doi.org/10.1038/s41583-020-0278-2
Dowd, F. J., & Abel, P. W. (2017). Pharmacodynamics: Mechanisms of drug action. *Pharmacology and Therapeutics for Dentistry: Seventh Edition*, 2–14. https://doi.org/10.1016/B978-0-323-39307-2.00001-1
Egelman, E. H. (2016). The current revolution in cryo-EM. *Biophysical Journal*, *110*(5), 1008–1012.
Fang, C., Huang, Y., & Zhao, Y. (2023). Review of FRET biosensing and its application in biomolecular detection. *American Journal of Translational Research*, *15*(2), 694. /pmc/articles/PMC10006758/
Gadsby, D. C. (2009). Ion channels versus ion pumps: The principal difference, in principle. *Nature Reviews Molecular Cell Biology*, *10*(5), 344–352. https://doi.org/10.1038/nrm2668
Gamal El-Din, T. M., Lantin, T., Tschumi, C. W., Juarez, B., Quinlan, M., Hayano, J. H., Li, J., Zweifel, L. S., & Catterall, W. A. (2021). Autism-associated mutations in KV7 channels induce gating pore current. *Proceedings of the National Academy of Sciences of the United States of America*, *118*(45), e2112666118. https://doi.org/10.1073/PNAS.2112666118/SUPPL_FILE/PNAS.2112666118.SAPP.PDF
Garbison, K. E., Heinz, B. A., & Lajiness, M. E. (2012). *IP-3/IP-1 Assays*. Eli Lilly & Company and the National Center for Advancing Translational Sciences
Gilman, A. (1987). G proteins: transducers of receptor-generated signals? *Annual Review of Biochemistry?*, *56*, 615.

Guo, S., Zhao, T., Yun, Y., & Xie, X. (2022). Recent progress in assays for GPCR drug discovery. *American Journal of Physiology-Cell Physiology*, *323*(2), C583–C594.

Gusach, A., García-Nafría, J., & Tate, C. G. (2023). New insights into GPCR coupling and dimerisation from cryo-EM structures. *Current Opinion in Structural Biology*, *80*, 102574. https://doi.org/10.1016/J.SBI.2023.102574

Hauser, A. S., Attwood, M. M., Rask-Andersen, M., Schiöth, H. B., & Gloriam, D. E. (2017). Trends in GPCR drug discovery: New agents, targets and indications. *Nature reviews Drug discovery*, *16*(12), 829–842.

Heng, B. C., Aubel, D., & Fussenegger, M. (2013). An overview of the diverse roles of G-protein coupled receptors (GPCRs) in the pathophysiology of various human diseases. *Biotechnology Advances*, *31*(8), 1676–1694. https://doi.org/10.1016/J.BIOTECHADV.2013.08.017

Hilger, D. (2021). The role of structural dynamics in GPCR-mediated signaling. *The FEBS Journal*, *288*(8), 2461–2489. https://doi.org/https://doi.org/10.1111/febs.15841

Hilger, D., Masureel, M., & Kobilka, B. K. (2018). Structure and dynamics of GPCR signaling complexes. *Nature Structural and Molecular Biology*, *25*(1), 4–12. https://doi.org/10.1038/s41594-017-0011-7

Hofmann, L., & Palczewski, K. (2015). The G protein-coupled receptor rhodopsin: A historical perspective. *Methods in Molecular Biology (Clifton, N.J.)*, *1271*, 3. https://doi.org/10.1007/978-1-4939-2330-4_1

Hubbard, S. R. (1999). Structural analysis of receptor tyrosine kinases. *Progress in Biophysics and Molecular Biology*, *71*, 343–358.

Hubbard, S. R., & Till, J. H. (2000). Protein tyrosine kinase structure and function. *Annual Review of Biochemistry*, *69*, 373–398.

Iorio, J., Duranti, C., Lottini, T., Lastraioli, E., Bagni, G., Becchetti, A., & Arcangeli, A. (2020). KV11.1 potassium channel and the Na+/H+ Antiporter NHE1 modulate adhesion-dependent intracellular pH in colorectal cancer cells. *Frontiers in Pharmacology*, *11*. https://doi.org/10.3389/FPHAR.2020.00848/FULL

Jiang, L., Xiong, Z., Song, Y., Lu, Y., Chen, Y., Schultz, J. S., Li, J., & Liao, J. (2019). Protein–protein affinity determination by quantitative FRET quenching. *Scientific Reports*, *9*(1), 1–8. https://doi.org/10.1038/s41598-018-35535-9

Jin, B., Thakur, N., Wijesekara, A. V., & Eddy, M. T. (2023). Illuminating GPCR signaling mechanisms by NMR spectroscopy with stable-isotope labeled receptors. *Current Opinion in Pharmacology*, *72*, 102364. https://doi.org/10.1016/J.COPH.2023.102364

Kamato, D., Thach, L., Bernard, R., Chan, V., Zheng, W., Kaur, H., Brimble, M., Osman, N., & Little, P. J. (2015). Structure, function, pharmacology, and therapeutic potential of the G protein, Gα/q,11. In *Frontiers in Cardiovascular Medicine* (Vol. 2). Frontiers Media S.A. https://doi.org/10.3389/fcvm.2015.00014

Karageorgos, V., Venihaki, M., Sakellaris, S., Pardalos, M., Kontakis, G., Matsoukas, M. T., Gravanis, A., Margioris, A., & Liapakis, G. (2018). Current understanding of the structure and function of family B GPCRs to design novel drugs. *Hormones*, *17*(1), 45–59. https://doi.org/10.1007/S42000-018-0009-5

Kariev, A. M., & Green, M. E. (2012). Voltage gated ion channel function: Gating, conduction, and the role of water and protons. *International Journal of Molecular Sciences*, *13*(2), 1680–1709. https://doi.org/10.3390/IJMS13021680

Katz, M., Amit, I., & Yarden, Y. (2007). Regulation of MAPKs by growth factors and receptor tyrosine kinases. *Biochimica et Biophysica Acta (BBA)-Molecular Cell Research*, *1773*, 1161–1176.

Kefauver, J., Ward, A., Nature, A. P.-, & 2020, undefined. (n.d.). Discoveries in structure and physiology of mechanically activated ion channels. *Nature.ComJM Kefauver, AB Ward, A PatapoutianNature, 2020•nature.Com*. Retrieved January 28, 2024, from https://www.nature.com/articles/s41586-020-2933-1

Khan, S. M., Sleno, R., Gora, S., Zylbergold, P., Laverdure, J. P., Labbé, J. C., Miller, G. J., & Hébert, T. E. (2013). The expanding roles of Gβ γ subunits in G protein-coupled receptor signaling and drug action. *Pharmacological Reviews*, *65*(2), 545–577. https://doi.org/10.1124/pr.111.005603

Kimple, A. J., Bosch, D. E., Giguère, P. M., & Siderovski, D. P. (2011). Regulators of G-protein signaling and their Gα substrates: Promises and challenges in their use as drug discovery targets. *Pharmacological Reviews*, *63*(3), 728–749. https://doi.org/10.1124/pr.110.003038

Knight, T., & Irving, J. A. E. (2014). Ras/Raf/MEK/ERK pathway activation in childhood acute lymphoblastic leukemia and its therapeutic targeting. *Frontiers in Oncology*, *4*, 160.

Köhn, M. (2020). Turn and face the strange: A new view on phosphatases. *ACS Central Science*, *6*(4), 467. https://doi.org/10.1021/ACSCENTSCI.9B00909

Kolch, W. (2005). Coordinating ERK/MAPK signalling through scaffolds and inhibitors. *Nature Reviews Molecular Cell Biology*, *6*, 827–837.

Komolov, K. E., & Benovic, J. L. (2018). G protein-coupled receptor kinases: Past, present and future. In *Cellular Signalling* (Vol. 41, pp. 17–24). Elsevier Inc. https://doi.org/10.1016/j.cellsig.2017.07.004

Kondratskyi, A., Kondratska, K., Skryma, R., Klionsky, D. J., & Prevarskaya, N. (2018). Ion channels in the regulation of autophagy. *Autophagy*, *14*(1), 3. https://doi.org/10.1080/15548627.2017.1384887

Kuhn, M. (2016). Molecular physiology of membrane guanylyl cyclase receptors. *Physiological Reviews*, *96*(2), 751–804. https://doi.org/10.1152/PHYSREV.00022.2015/ASSET/IMAGES/LARGE/Z9J0021627600013.JPEG

Kwan, T. O. C., Kolek, S. A., Danson, A. E., Reis, R. I., Camacho, I. S., Shaw Stewart, P. D., & Moraes, I. (2022). Measuring protein aggregation and stability using high-throughput biophysical approaches. *Frontiers in Molecular Biosciences*, *9*, 890862.

Lemmon, M. A., & Schlessinger, J. (2010). Cell signaling by receptor tyrosine kinases. *Cell*, *141*, 1117–1134.

Levy, D. E., & J., D. J. R. (2002). Stats: Transcriptional control and biological impact. *Nature Reviews Molecular Cell Biology*, *3*, 651–662.

Li, E., & Hristova, K. (2006). Role of receptor tyrosine kinase transmembrane domains in cell signaling and human pathologies. *Biochemistry*, *45*, 6241–6251.

Li, E., & Hristova, K. (2010). Receptor tyrosine kinase transmembrane domains: Function, dimer structure and dimerization energetics. *Cell Adhesion & Migration*, *4*, 249–254.

Li, L., Yao, Y., Wang, Y., Cao, J., Jiang, Z., Yang, Y., Wang, H., & Ma, H. (2024). G protein-coupled estrogen receptor 1 ameliorates nonalcoholic steatohepatitis through targeting AMPK-dependent signaling. *The Journal of Biological Chemistry*, 105661. https://doi.org/10.1016/j.jbc.2024.105661

Lombard, J. (2014). Once upon a time the cell membranes: 175 years of cell boundary research. *Biology Direct*, *9*(1), 1–35. https://doi.org/10.1186/S13062-014-0032-7/FIGURES/11

Lorber, B., Fischer, F., Bailly, M., Roy, H., & Kern, D. (2012). Protein analysis by dynamic light scattering: Methods and techniques for students. *Biochemistry and Molecular Biology Education*, *40*(6), 372–382. https://doi.org/10.1002/BMB.20644

Martens, C., & Politis, A. (2020). A glimpse into the molecular mechanism of integral membrane proteins through hydrogen–deuterium exchange mass spectrometry. *Protein Science*, *29*(6), 1285–1301.

McManus, O. B., Garcia, M. L., Weaver, D., Bryant, M., Titus, S., & Herrington, J. B. (2012). Ion channel screening. *Assay Guidance Manual*, 1–25. https://www.ncbi.nlm.nih.gov/books/NBK100915/

Meissner, F., Geddes-McAlister, J., Mann, M., & Bantscheff, M. (2022). The emerging role of mass spectrometry-based proteomics in drug discovery. *Nature Reviews Drug Discovery*, *21*(9), 637–654.

Miller, E. J., & Lappin, S. L. (2022). Physiology, cellular receptor. *StatPearls*. https://www.ncbi.nlm.nih.gov/books/NBK554403/

Morrison, D. K. (2012). MAP kinase pathways. *Cold Spring Harbor Perspectives in Biology*, *4*, 18.

Moustakas, A., Souchelnytskyi, S., & Heldin, C. (2005). Receptor serine/threonine kinases. *Encyclopedic Reference of Genomics and Proteomics in Molecular Medicine*, 1603–1608. https://doi.org/10.1007/3-540-29623-9_3130

Nighat, N., Roshan, A., Muhammad, A., Kamran, T. M., Zahid, M. M., & Patching, S. G. (2024). *Microscale Thermophoresis Analysis of Membrane Proteins*. HAL open science.

Niswender, C. M., & Conn, P. J. (2010). Metabotropic glutamate receptors: Physiology, pharmacology, and disease. *Annual Review of Pharmacology and Toxicology*, *50*, 295. https://doi.org/10.1146/ANNUREV.PHARMTOX.011008.145533

Nogales, E. (2016). The development of cryo-EM into a mainstream structural biology technique. *Nature Methods*, *13*(1), 24. https://doi.org/10.1038/NMETH.3694

Oelstrom, K., Goldschen-Ohm, M. P., Holmgren, M., & Chanda, B. (2014). Evolutionarily conserved intracellular gate of voltage-dependent sodium channels. *Nature Communications*, *5*. https://doi.org/10.1038/NCOMMS4420

O'shea, J. J., Schwartz, D. M., Villarino, A. V, Gadina, M., Mcinnes, I. B., & Laurence, A. (2015). The JAK-STAT pathway: Impact on human disease and therapeutic intervention. *Annual Review of Medicine*, *66*, 311–328.

Owens, D., & Keyse, S. (2007). Differential regulation of MAP kinase signalling by dual-specificity protein phosphatases. *Oncogene*, *26*, 3203–3213.

Ozohanics, O., & Ambrus, A. (2020). Hydrogen-deuterium exchange mass spectrometry: A novel structural biology approach to structure, dynamics and interactions of proteins and their complexes. *Life*, *10*(11), 286.

Pan, X., & Vachet, R. W. (2022). Membrane protein structures and interactions from covalent labeling coupled with mass spectrometry. *Mass Spectrometry Reviews*, *41*(1), 51–69.

Pavone, P., Pappalardo, X. G., Parano, E., Falsaperla, R., Marino, S. D., Fink, J. K., & Ruggieri, M. (2022). Fever-associated seizures or epilepsy: An overview of old and recent literature acquisitions. *Frontiers in Pediatrics, 10*. https://doi.org/10.3389/FPED.2022.858945/FULL

Pawson, T., Gish, G. D., & Nash, P. (2001). SH2 domains, interaction modules and cellular wiring. *Trends in Cell Biology, 11*, 504–511.

Pitt, G. S. (2016). *Ion channels in health and disease.* Academic Press.

Purves, D., Augustine, G. J., Fitzpatrick, D., Katz, L. C., LaMantia, A.-S., McNamara, J. O., & Williams, S. M. (2001). Receptor types. https://www.ncbi.nlm.nih.gov/books/NBK10989/

Reis, R. I., & Moraes, I. (2020). Probing membrane protein assembly into nanodiscs by in situ dynamic light scattering: A2A receptor as a case study. *Biology, 9*(11), 400.

Rosenbaum, D. M., Rasmussen, S. G. F., & Kobilka, B. K. (2009). The structure and function of G-protein-coupled receptors. In *Nature* (Vol. 459, Issue 7245, pp. 356–363). https://doi.org/10.1038/nature08144

Roux, B. (2017). Ion channels and ion selectivity essays in biochemistry. *Essays in Biochemistry, 61*(2), 201. https://doi.org/10.1042/EBC20160074

Saha, S., & Raghava, G. P. S. (2006). Prediction of continuous B-cell epitopes in an antigen using recurrent neural network. *Proteins: Structure, Function, and Bioinformatics, 65*, 40–48.

Sahin, C., Reid, D. J., Marty, M. T., & Landreh, M. (2020). Scratching the surface: Native mass spectrometry of peripheral membrane protein complexes. *Biochemical Society Transactions, 48*(2), 547–558.

Sassone-Corsi, P. (2012). The cyclic AMP pathway. *Cold Spring Harbor Perspectives in Biology, 4*(12). https://doi.org/10.1101/cshperspect.a011148

Schlessinger, J. (2000). Cell signaling by receptor tyrosine kinases. *Cell, 103*(2), 211-225.

Singer, S. J., & Nicolson, G. L. (1972). The fluid mosaic model of the structure of cell membranes. *Science (New York, N.Y.), 175*(4023), 720–731. https://doi.org/10.1126/SCIENCE.175.4023.720

Soares-Silva, M., Diniz, F. F., Gomes, G. N., & Bahia, D. (2016). The mitogen-activated protein kinase (MAPK) pathway: Role in immune evasion by trypanosomatids. *Frontiers in Microbiology, 7*(February), 179112. https://doi.org/10.3389/FMICB.2016.00183/BIBTEX

Südhof, T. C. (2012). Calcium control of neurotransmitter release. *Cold Spring Harbor Perspectives in Biology, 4*(1). https://doi.org/10.1101/CSHPERSPECT.A011353

Sudhesh Dev, S., Zainal Abidin, S. A., Farghadani, R., Othman, I., & Naidu, R. (2021). Receptor tyrosine kinases and their signaling pathways as therapeutic targets of curcumin in cancer. *Frontiers in Pharmacology, 12*, 772510.

Tóth, K. (2021). Diversity of ion channels. *The Journal of Physiology, 599*(10), 2603–2604. https://doi.org/10.1113/JP281513

Uray, I., sciences, K. U.-I. journal of molecular, & 2021, undefined. (n.d.). Mechanotransduction at the plasma membrane-cytoskeleton interface. *Mdpi.ComIP Uray, K UrayInternational Journal of Molecular Sciences, 2021•mdpi.Com*. Retrieved January 28, 2024, from https://www.mdpi.com/1422-0067/22/21/11566

Van Der Heyden, M. A. G., Delisle, B. P., & Abriel, H. (2018). Editorial: Ion channel trafficking and cardiac arrhythmias. *Frontiers in Physiology, 9*(September), 417846. https://doi.org/10.3389/FPHYS.2018.01254/BIBTEX

Vit, O., & Petrak, J. (2017). Integral membrane proteins in proteomics. How to break open the black box? *Journal of Proteomics, 153*, 8–20. https://doi.org/10.1016/J.JPROT.2016.08.006

Vizurraga, A., Adhikari, R., Yeung, J., Yu, M., & Tall, G. G. (2020). Mechanisms of adhesion G protein–coupled receptor activation. *The Journal of Biological Chemistry, 295*(41), 14065. https://doi.org/10.1074/JBC.REV120.007423

Wang, J., & Xia, Y. (2012). Assessing developmental roles of MKK4 and MKK7 in vitro. *Communicative & Integrative Biology, 5*(4), 319–324. https://doi.org/10.4161/CIB.20216

Wehrman, T., He, X., Raab, B., Dukipatti, A., Blau, H., & Garcia, K. C. (2007). Structural and mechanistic insights into nerve growth factor interactions with the TrkA and p75 receptors. *Neuron, 53*(1), 25–38.

Weis, W. I., & Kobilka, B. K. (2018). The molecular basis of G protein–coupled receptor activation. *Annual Review of Biochemistry, 87*, 897. https://doi.org/10.1146/ANNUREV-BIOCHEM-060614-033910

Wensel, T. (1999). Heterotrimeric G-proteins: Structure, regulation, and signaling mechanisms. *Introduction to Cellular Signal Transduction*, 29–46. https://doi.org/10.1007/978-1-4612-1990-3_3

Wheeler, D. L., & Yarden, Y. (2015). *Receptor Tyrosine Kinases: Family and Subfamilies.* Springer.

Whited, A. M., & Johs, A. (2015). The interactions of peripheral membrane proteins with biological membranes. *Chemistry and Physics of Lipids, 192*, 51–59. https://doi.org/10.1016/J.CHEMPHYSLIP.2015.07.015

Whitelegge, J. P. (2013). Integral membrane proteins and bilayer proteomics. *Analytical Chemistry*, *85*(5), 2558. https://doi.org/10.1021/AC303064A

Xu, Y., & Fisher, G. J. (2012). Receptor type protein tyrosine phosphatases (RPTPs) – roles in signal transduction and human disease. *Journal of Cell Communication and Signaling*, *6*(3), 125. https://doi.org/10.1007/S12079-012-0171-5

Yang, D., Zhou, Q., Labroska, V., Qin, S., Darbalaei, S., Wu, Y., Yuliantie, E., Xie, L., Tao, H., Cheng, J., Liu, Q., Zhao, S., Shui, W., Jiang, Y., & Wang, M. W. (2021). G protein-coupled receptors: Structure- and function-based drug discovery. In *Signal Transduction and Targeted Therapy* (Vol. 6, Issue 1). Springer Nature. https://doi.org/10.1038/s41392-020-00435-w

Yang, J., Nie, J., Ma, X., Wei, Y., Peng, Y., & Wei, X. (2019). Targeting PI3K in cancer: Mechanisms and advances in clinical trials. *Molecular Cancer*, *18*, 1–28.

Yen, H.-Y., Jazayeri, A., & Robinson, C. V. (2023). G protein-coupled receptor pharmacology—insights from mass spectrometry. *Pharmacological Reviews*, *75*(3), 397. https://doi.org/10.1124/pharmrev.120.000237

Zalewska, M., Siara, M., & Sajewicz, W. (2014). G protein-coupled receptors: abnormalities in signal transmission, disease states and pharmacotherapy. *Acta Poloniae Pharmaceutica*, *71*(2), 229–243.

Zhang, L., Malik, S., Pang, J., Wang, H., Park, K. M., Yule, D. I., Blaxall, B. C., & Smrcka, A. V. (2013). Phospholipase Cε hydrolyzes perinuclear phosphatidylinositol 4-phosphate to regulate cardiac hypertrophy. *Cell*, *153*(1), 216–227. https://doi.org/10.1016/j.cell.2013.02.047

Zhang, W., & Liu, H. T. (2002). MAPK signal pathways in the regulation of cell proliferation in mammalian cells. *Cell Research*, *12*, 9–18.

Zhang, Y., & Dong, C. (2007). Regulatory mechanisms of mitogen-activated kinase signaling. *Cellular and Molecular Life Sciences*, *64*, 2771–2789.

Zhao, Q., Shen, Y., Li, X., Tian, F., Yu, X., Yobas, L., Park, H., Duan, Y., & Huang, P. (2021). Analyzing protein–protein interactions in rare cells using microbead-based single-molecule pulldown assay. *Lab on a Chip*, *21*(16), 3137–3149.

Zheng, J., & Trudeau, M. (2023). Textbook of ion channels: Three volume set. https://books.google.com/books?hl=en&lr=&id=qVPKEAAAQBAJ&oi=fnd&pg=PP1&dq=textbook+of+ion+channels+volume+1jie+zheng&ots=EypPNP4WSE&sig=Jt91GUFzkbrX0HPES3b-rJNcJ9k

Zhou, X. E., Melcher, K., & Xu, H. E. (2012). Structure and activation of rhodopsin. *Acta Pharmacologica Sinica*, *33*(3), 291–299. https://doi.org/10.1038/aps.2011.171

5 Antigen-Antibody Interaction in Cell Signaling

Saman Noor, Anam Fatima, Iman Fatima, and Aneela Javed
Department of Biomedicine, Atta ur rahman School of Applied Biosciences (ASAB), National University of Sciences and technology (NUST), H12 Islamabad, Pakistan

5.1 OVERVIEW OF THE IMMUNE SYSTEM

Cell signaling involves communication between cells throughout the body and with the external environment. Antigen-antibody interactions are a crucial component of the immune system in living organisms. Immunity enables the body to identify and eliminate external substances without damaging its own tissues, thereby preventing subsequent infections caused by the same substance.

The immune system plays a crucial role in protecting the skin, respiratory tract, intestinal tract, and other body parts from external antigens such as viruses, bacteria, fungi, parasites, cancer cells, and toxic substances. It comprises two lines of defense mechanisms: innate and adaptive immunity.

Innate immunity is the first line of defense against infections. It includes anatomic barriers such as skin and mucous membrane, physiological barriers such as pH and temperature, as well as chemical mediators and inflammatory barriers. Within hours of encountering an antigen, the host mounts an antigen-independent defense (Turvey & Broide, 2010).

Innate immunity quickly attracts immune cells to infection and inflammation sites using cytokines and chemokines. TNF, IL-1, and IL-6 are the primary inflammatory cytokines generated early in bacterial infection. The complement system identifies and opsonizes bacteria and pathogens. It destroys some infections and infected cells directly and makes others susceptible to phagocytosis. The phagocytic role of the innate immune system eliminates foreign substances, dead cells, and antibody complexes from organs, blood, tissues, and lymph. It also mobilizes and activates antigen-presenting cells (APCs) to activate adaptive immunity. Innate immune cells include phagocytes (macrophages and neutrophils), mast cells, dendritic cells, basophils, natural killer (NK) cells, eosinophils, and innate lymphoid cells (Stone et al., 2010).

Phagocytes are classified as neutrophils or macrophages. Both cells phagocytose and destroy bacteria via several bactericidal mechanisms. Dendritic cells phagocytose and trigger the acquired immune response as APCs, linking the adaptive and innate immune responses. Basophils and mast cells trigger acute inflammatory reactions such as asthma and allergies. Eosinophils, which are phagocyte-like cells, kill parasites that are too large to phagocytose. These cells, along with mast cells and basophils, control allergy and asthma pathways. NK cells are responsible for tumor rejection and virus-infected cell death. They release granzymes and perforins from their granules, which trigger apoptosis in infected cells (Kennedy & DeLeo, 2009). NK cells also play a crucial role in the production of interferon-gamma (IFN-γ), which mobilizes APCs and enhances antiviral immunity (Jonsson & Yokoyama, 2009).

The innate immune system lacks immunological memory and cannot recognize the same pathogen upon future exposure. However, the adaptive immune response depends on the recognition of past encounters with specific antigens, allowing the host to generate a faster and more

 DOI: 10.1201/9781003594901-5

effective immune response. Innate and adaptive immune responses are not mutually exclusive processes of host defense; rather, they are complementary. Any impairments in either innate or adaptive immune systems can result in host vulnerability or improper responses (Bonilla & Oettgen, 2010).

5.2 OBJECTIVES OF THE CHAPTER

This chapter aims to explore the complexities of antigen-antibody interactions in cell signaling, revealing the molecular interactions that occur when the immune system encounters foreign invaders. The primary goals include:

- Investigating the structural characteristics of antigens and antibodies
- Gaining insight into the processes of antigen presentation
- Understanding the signaling cascades triggered by the interaction of antigens with antibodies
- Analyzing the biological reactions elicited by these interactions
- Exploring approaches to control and modulate these interactions for therapeutic applications

5.3 SCOPE AND SIGNIFICANCE

This chapter provides an extensive overview of antigen-antibody interactions, including their fundamental structures and their significant role in cell signaling. By understanding these mechanisms at a molecular and cellular level, we gain knowledge about the coordination of immune responses and their broader implications. This understanding is crucial for developing therapeutic interventions, including targeted treatments and vaccinations that utilize antigen-antibody interactions to provide protection.

As we progress through this chapter, we embark on an exploration of the intricate domain of immune system regulation, where the precise balance of antigen-antibody interactions is crucial for our body's capacity to protect itself and ensure optimal health.

5.4 SIGNIFICANCE OF ANTIGEN-ANTIBODY INTERACTIONS

Antigen-antibody interactions promote the immunological response. The immune system recognizes pathogens by their surface antigens, typically proteins or glycoproteins. Antibodies, produced by specialized white blood cells known as B cells, serve as the immune system's highly specific targeted weapons. Recognition and binding between antigens and antibodies initiate the body's defenses.

This interaction extends beyond pathogen detection. It plays a crucial role in preserving the immune system's balance, preventing autoimmune diseases, and improving immunological memory—the immune system's ability to recall past encounters and generate a faster and more efficient response upon repeated exposure. To ensure sufficient diversity for recognizing various foreign substances, numerous advanced genetic mechanisms have evolved to maximize the variety of immunoglobulin receptors (Geysen, 1985; *An Introduction to Antibodies*, n.d.-b).

5.5 ANTIGENS

Antigens are macromolecules typically composed of proteins present on the outer surfaces of pathogens, including viruses, bacteria, fungi, and various other external substances. When these potentially harmful pathogens enter the body, they initiate an immune response that promotes antibody production.

Antigens are detected by antibodies and T-cell receptors. The immune system produces different antigen receptors to recognize various specific antigens. When an antigen is encountered, only those lymphocytes that can identify that specific antigen are stimulated and undergo expansion, a process referred to as clonal selection (Abbas, 1989).

Host cells can recognize and discriminate antigens with precision due to antigenic specificity. The specificity of antigens is primarily determined by their side-chain conformations.

5.1 Structure of Antigens

Epitopes, also known as antigenic determinants, are the basic components of an antigen. Each antigen possesses several epitopes. An antibody has a minimum of two binding sites capable of attaching to specific epitopes on antigens (Figure 5.1).

Lymphocytes are activated and initiate immunological responses against antigens by binding their receptors to the antigens' surface molecules. This may involve producing antibodies or activating cytotoxic cells. Antibody production following stimulation is highly influenced by the concentration and type of antigen, the specific features of the host, and the route of antigen entry into the body.

5.5.2 Types of Antigens

Antigens can be categorized based on their origin as follows.

5.5.2.1 Exogenous Antigens

These are the antigens that enter from outside through ingestion, inhalation, or injection. Exogenous antigens are endocytosed or phagocytosed by APCs and are fragmented. APCs present these fragments to CD4+ T-helper cells through class II histocompatibility molecules on their surfaces.

5.5.2.2 Endogenous Antigens

These are produced by normal metabolic processes in cells or as a result of infection caused by either bacteria or viruses. After interacting with class I major histocompatibility complex (MHC) molecules, the fragments are displayed on the cell surface. Toxins released by T cells destroy this complex upon recognition by activated T-cytotoxic (TC) cells. These antigens include autologous, xenogenic, idiotypic, and allogenic antigens. Autoimmune diseases can involve host antigens.

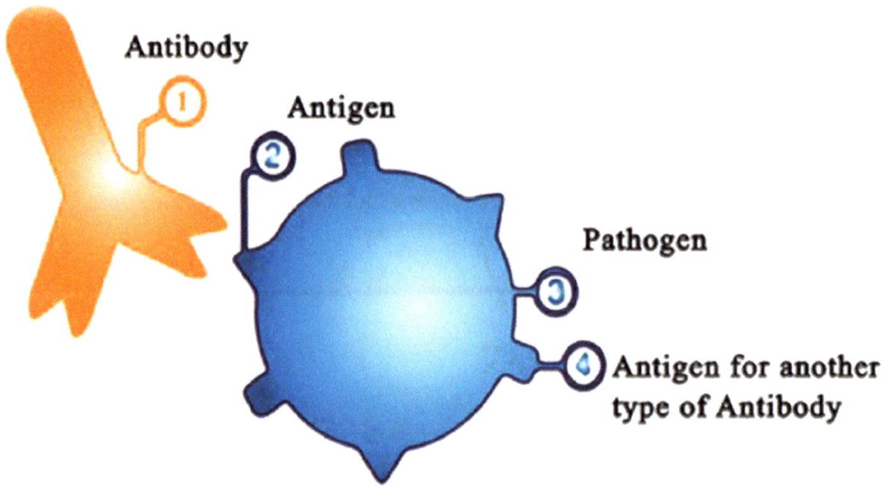

FIGURE 5.1 Antibody-antigen interaction.

5.5.2.3 Autoantigens

These are the antigens that are recognized by the immune system in cases of autoimmune diseases. These are usually protein complexes, self-proteins, RNA, and DNA.

5.5.2.4 Neoantigens

These antigens are absent from normal human DNA. Neoproteins are important for tumor management because central T-cell tolerance does not eliminate the pool of T cells for such antigens. Recently developed technology can measure T-cell reactivity to neoantigens. Neoantigens are quantifiable. Virus-associated malignancies such as cervical and head and neck cancers produce neoantigens from viral open reading frames (Schumacher & Schreiber, 2015).

5.5.2.5 Tumor Antigens

Tumor cells express MHC I or II molecules that present tumor antigens. Tumor-specific antigens (TSAs) are detected only in tumor cells and usually result from a tumor mutation. Cytotoxic T lymphocytes (CTLs) that detect these antigens may kill tumor cells. B cells can identify mutated receptors on tumor surfaces as tumor antigens. DNA mutations create unique peptides (neo-epitopes) in human cancers without a viral cause (Schumacher & Schreiber, 2015).

5.5.2.6 Native Antigens

Antigens that have not been processed by an APC are called native. T cells require APCs for binding with native antigens, but B cells can be stimulated by them directly.

5.5.2.7 Antibodies

An antibody (Ab) is alternatively referred to as an immunoglobulin (Ig). These are large, Y-shaped protein molecules in the blood generated by plasma cells. They attach to foreign substances and invade them (Mian et al., 1991).

5.5.3 Structure of Antibodies

An antibody possesses a Y-shaped configuration, comprising four polypeptide subunits. Each subunit consists of a pair of identical heavy chains and light chains. Antibodies possess three constant regions and a variable region within their heavy chains. The antigen-binding domain is created by binding the N-terminus of each heavy chain with a light chain. The "Y" shape consists of two antigen-binding domains constituting the arms known as "fragment antigen-binding" (Fab) domains.

The Fc domain, located at the C-terminus of the heavy chains, facilitates interactions with effector cells by a process called "fragment crystallization." Disulfide and noncovalent linkages maintain the connection of all four polypeptide subunits (Schroeder & Cavacini, 2010).

Antibodies' variable domains (VH and VL) form Fv (small, oval) with three segments joining β-strands, varying in their sequence and length. V domain complementarity-determining regions (CDRs) form the combining site owing to their closeness. CDRs specifically bind to Ig domain apexes. Antigen recognition relies on the six CDRs' three-dimensional structure binding to the antigen's surface (Schutkowski & Reineke, 2009).

5.5.4 Types of Antibodies

There are five distinct isotypes of antibodies or immunoglobulins. Their classification is based on the characteristics of their H chains (Table 5.1).

5.5.5 Antibody Production and Mechanism

Macrophages capture and break down foreign antigens that the immune system encounters for the first time and present them to B cells. When these antigens are presented, B cells synthesize a

TABLE 5.1
Major Functions of Human Ig Antibodies

Antibody	Function
IgM	• The first antibody generated by B cells in response to a pathogenic attack (primary response). • The largest antibody, present in a pentameric structure. • Present in both the blood and lymph, comprising about 6% of the total antibodies in the serum. • Plays a role in the processes of agglutination and opsonization.
IgG	• The most prevalent isotype found in plasma. • Constitutes around 80% of the overall antibody load in the serum. • Identifies the antibody-antigen complex and removes toxic substances. • Enhances phagocytosis and provides immunity to the growing fetus.
IgA	• Predominantly present in secretions like breast milk, saliva, serum, and intestinal secretions; commonly referred to as secretory immunoglobulin. • Makes up about 13% of the overall antibody load in the serum. • Present in both monomeric and dimeric forms. • Limits infection and serves as the primary barrier against pathogens. • Enhances complement system activity and plays a role in the immune response.
IgD	• Constitutes less than 1% of the overall antibody concentration in serum. • Appears to improve mucosal homeostasis and immune surveillance
IgE	• Found in minimal quantities, approximately 0.02% of the total antibody concentration in the serum. • Present in its monomeric form; found in the linings of the intestinal and respiratory tracts; reacts to allergic responses.

unique antibody with a discrete paratope, allowing it to specifically attach to a particular epitope on the antigen. Each B-cell-associated lymphocyte generates an antibody that targets a specific epitope. After B cells encode, they release antibodies that bind to invaders and eliminate them.

Antigens function by directly attacking pathogens, usually viruses, or by sticking to the pathogen's surface, such as bacteria, and signaling the immune system to eliminate it. These cells remain in the body forever, ready to respond if the same pathogen returns.

Leukocytes aid the immune system in fighting against pathogenic microorganisms and harmful substances. In addition to leukocytes, lymphoid organs, and tissues, antibodies help the immune system mount an effective response (Geysen, 1985).

5.6 STAGES OF ANTIGEN-ANTIBODY INTERACTION

The interactions between Ag and Ab can be divided into three stages:

1. Formation of the Ag-Ab complex
2. Processes such as agglutination and precipitation
3. Destruction or neutralization of Ag

Antigen-antibody affinity describes their strong interaction at specific antigenic regions. The variable antibody "arm" connects with the antigen at multiple regions within each antigenic site via weak noncovalent forces. The level of interaction corresponds with affinity.

The Ab-Ag complex's stability or strength is better measured by avidity. This process is affected by the strength of the Ag-Ab interaction and the number of binding sites present on both the Ag and Ab. These factors influence whether an antibody will bind to an antigen epitope, thus defining its specificity.

Cross-reactivity is the ability of an antibody or set of antibodies to bind to a specific antigen region. This occurs due to the antibody's poor avidity or specificity, or the presence of multiple antigens with identical or similar epitopes. Cross-reactivity can be beneficial for universal binding to several homologous antigens or cross-species labeling, especially when the epitope sequence of an antigen is not well conserved. However, it might lead to overestimation or underestimation of antigen concentration during measurement, hindering immunoassays. Despite the presence of contaminated particles, immunochemical methods use each immunoglobulin's molecular specificity for its antigen. Since most antigens and antibodies have multiple binding sites, they may precipitate (Frank, 2002).

Antibodies may neutralize, opsonize, or fix complement to antigens:

5.6.1 Neutralization

Antibodies neutralize antigens by preventing their action. They may inhibit viruses from entering cells by blocking their binding to cellular receptors. Antibodies may also prevent bacterial toxins from causing harm.

5.6.2 Opsonization

Antibodies help immune cells phagocytose antigens through opsonization. They assist neutrophils and macrophages in detecting and engulfing antigens.

5.6.3 Complement Fixation

The complement system is a set of proteins essential to the immunological response. Antibodies can activate the complement system by binding with antigens and forming a membrane attack complex that lyses the antigen.

5.7 ANTIBODY-ANTIGEN INTERACTION KINETICS

Specific binding of antigens and antibodies involves electrostatic forces, hydrophobic interactions, hydrogen bonding, and Van der Waals forces. These interactions are weak and noncovalent, although certain antigen-antibody interactions may be strong. Antigens may be multivalent, either by having multiple copies of the same epitope or several different epitopes that can be identified by multiple antibodies, similar to antibodies. Multivalent interactions can generate more stable complexes. However, multivalency can also result in steric challenges, thereby decreasing the chances of binding. Antigen-antibody binding is completely reversible and adheres to the fundamental thermodynamic principles of every reversible reaction between two molecules:

$$K_A = \frac{[Ab-Ag]}{[Ab][Ag]}$$

The affinity constant, KA, represents the binding strength of the antibody and antigen. [Ab-Ag] represents the molar concentration of the Ag-Ab complex, whereas [Ag] refers to the molar concentration of free binding sites on the antigen and [Ab] represents the molar concentrations of available binding sites on the antibody.

Optimal equilibrium time depends on the diffusion rate and antibody-antigen affinity and may vary significantly. The KA for Ag-Ab binding might vary dramatically between 105/mol and 1012/

mol. KA can be influenced by factors such as temperature, solvent, and pH. Affinity constants of monoclonal antibodies can be calculated, but those of polyclonal antibodies cannot be determined due to the occurrence of numerous bond forms between polyclonal Abs and their antigens. Equilibrium dialysis quantifies antibody affinity for antigens. Scatchard plots are created by performing many equilibrium dialyses by keeping antibody concentration constant and varying the ligand concentrations. These graphs provide valuable insights into potential cross-reactivity and affinity valence.

Understanding the differences between monoclonal and polyclonal antibodies is vital for creating effective experimental methods, since they determine both the advantages and disadvantages of their application. (Reverberi & Reverberi, 2007; *An Introduction to Antibodies*, n.d.-a)

5.8 NATURE OF ANTIGEN-ANTIBODY BONDS

The F(ab) segment of an antibody contains the hypervariable heavy and light chains' epitope-binding region. This site and the antigen interact through these features and mechanisms:

- Antigens bind with antibody-combining sites via noncovalent interactions, making them easily breakable and repairable. Bonds may include electrostatic, hydrogen, or Van der Waals forces.
- Antibody-antigen interactions are robust and secure due to multiple bond types. The cell's antigenic determinant (epitope) and the antibody's antigen-combining site (paratope) interact selectively and only with short molecular segments, often consisting of a few amino acids.
- Antigen-antibody interactions depend on specific binding between the epitope and paratope, which are initially drawn together through electrostatic and hydrophobic interactions. These forces overcome hydration energy, allowing water molecules to be removed as the two molecules approach. Van der Waals forces bring the epitope and paratope closer, increasing the attraction (Janeway et al., 2001c).

5.9 FACTORS AFFECTING ANTIGEN-ANTIBODY REACTIONS

Various factors may influence antigen-antibody reactions. Several prevalent factors include:

5.9.1 Temperature

The ideal temperature for the antigens-antibodies reactions is determined by the specific bonds present between them and by the chemical nature of the paratope and epitope. Usually, hydrogen bond formation is an exothermic reaction that results in the release of heat energy. These bonds have greater stability at lower temperatures and play a major role in the case of carbohydrate antigens.

5.9.2 pH

The equilibrium constant of the Ag-Ab complex is affected by the pH values ranging from 6.5 to 8.4. Antigen-antibody interaction is significantly suppressed when the pH is below 6.5 or above 8.4. Antibodies can experience structural changes under highly acidic or alkaline conditions, which can disrupt their ability to bind to the antigen.

5.9.3 Ionic Strength

The effect of ionic strength on the antigen-antibody interaction is especially significant in the field of blood group serology. The presence of sodium and chloride ions has a significant effect on the process. In normal saline solution, the Na+ and Cl– ions gather around the complex and partially counteract the charges, which could potentially disrupt the binding of antibodies to antigens. This may pose a challenge when low-affinity antibodies are utilized. Under conditions of extremely low ionic strength, γ-globulins tend to aggregate and form reversible complexes with lipoproteins present in red blood cells, resulting in their sedimentation.

5.10 APPLICATIONS OF ANTIGEN-ANTIBODY INTERACTIONS

Laboratory serological tests for blood compatibility and pathogenic disorders involve antigen-antibody interaction. The simplest test is the determination of the ABO blood group for blood transfusion. More advanced techniques include ELISA, immunofluorescence, immunoelectrophoresis, and enzyme-linked immunospot (Elispot) (Taylor et al., 1983; *Immunochemical Methods | SpringerLink*, n.d.; *Immunofluorescence Techniques - PubMed*, n.d.).

5.10.1 Precipitation Reaction

In an electrolyte at appropriate temperature and pH, soluble antibodies and antigens create an insoluble visible complex, known as a precipitation reaction. This reaction measures antigens and antibodies qualitatively and quantitatively. The interaction of soluble antigens and antibodies forms a lattice. This process has two phases. Initially, the antigens and antibodies establish Ag-Ab complexes within seconds, followed by a slower process in which the Ab-Ag complexes form lattices that precipitate from the solution. A unique ring test may detect anthrax and food adulteration (*An Immuno-Diffusion Assay to Assess the Protective Antigen Content of Anthrax Vaccine - PubMed*, n.d.).

5.10.2 The Agglutination Reaction

The agglutination reaction causes particle clumping by cross-linking particulate antigens with antibodies. Agglutination may be active or passive (Dennison et al., 2012).

5.10.3 Complement Fixation

Complement fixation is a standard method for detecting antibodies in patient serum. This process requires nonspecific, unstable fresh serum complements for red blood cell (RBC) and microbe lysis.

The complement system consists of 11 proteins present in everyone. These proteins bind to the Fc subunit of the Ag-Ab complex. In complement fixation assays, the Ag-Ab complex fixes complement.

The process involves mixing pre-heated (56°C) antiserum and test antigen with a known concentration of complement. This mixture is then incubated at 4°C for 18 hours. Serum containing Ag-specific Ab will form an antigen-antibody complex and eventually result in complement fixation.

5.10.4 Immunofluorescence

Fluorescence is the ability to emit and absorb distinct wavelengths of light. UV-exposed fluorescent dyes produce bright, visible light. In 1942, experiments demonstrated how to link labeled dyes to antibodies to detect antigens. Fluorescein is an organic dye that absorbs blue light with a wavelength of 490 nm and generates intense yellow-green fluorescence with a wavelength of 517 nm. It is the most common immunofluorescence marker. When labeling immunofluorescence, phycoerythrin is employed because it emits red fluorescence and absorbs light (30 times better than fluorescein).

5.10.5 The ELISA Test

Antigens and antibodies labeled with enzymes were first developed in 1971 for antibody and antigen testing.

A ligand that detects Ab is covalently bound to an enzyme like peroxidase, beta-galactosidase, alkaline phosphatase, etc.

ELISA may be classified into three distinct types:

5.10.5.1 Indirect ELISA

The indirect ELISA approach can detect human immunodeficiency virus (HIV). Microtiter plates are covered with envelope proteins produced using recombinant technology. Unbound proteins are removed by the process of washing when test serum is introduced.

5.10.5.2 Sandwich ELISA

The sandwich ELISA technique detects antigens in a sample. After coating the well with Ag-specific antibodies, the test serum is introduced and allowed to undergo a reaction. The unbound antibodies are then washed out using a wash buffer.

A labeled antibody that targets a distinct antigenic epitope is then introduced. After washing out unbound antibodies, a colored substrate is added, and color development occurs. The color intensity closely correlates with the concentration of Ag in the serum.

5.10.5.3 Competitive ELISA

This method determines antigen levels by incubating the antibody with an antigen-containing test sample in a solution.

The antigen-antibody combination is then introduced into a microtiter well that has been coated with antigens. The amount of antigen in the sample inversely affects the quantity of free antibodies available to bind to the antigen-coated well.

In indirect ELISA, a secondary antibody (Ab2) coated with an enzyme and specific to the primary antibody can quantify the amount of bound primary antibody.

5.10.6 Surface Plasmon Resonance (SPR)

SPR detects binding interactions between immobilized biomolecules and free-flowing analytes. Various surface chemistries, such as dextran, carboxylation, and streptavidin coating, can immobilize antigens or antibodies. These methods help prevent binding artifacts and reduce nonspecific binding. SPR offers significant advantages compared to traditional immunofluorescence or ELISA techniques for investigating the interactions between a solution phase analyte and immobilized biomolecules ("Probing Antibody-Antigen Interactions," *PubMed*, n.d.; Krüger et al., 1998).

5.10.7 Fluorescence-Activated Cell Sorting (FACS)

FACS uses flow cytometry to identify antigens on cells or other particles using fluorescently labeled reagents. This method efficiently establishes antigen-antibody interactions without altering or fixing the antigen. Fluorophore-conjugated antibodies mark cells directly or indirectly in FACS.

The tagged cell-antibody combination flows through in a fluid stream, allowing for the individual sorting and analysis of the cells. Laser light focused on the cells examines their scattering properties, revealing granularity, size, and surface antigen presence. Multiple antibodies can simultaneously identify various target antigens on cell membranes. Cells with desirable traits can gathered for further cultivation and characterization (*Flow Cytometry in Biotechnology - PubMed*, n.d.).

5.10.8 Atomic Force Microscopy (AFM)

While the previously described methods scrutinize collective antigen-antibody interactions, investigating individual molecular interactions can be advantageous. The binding interactions between antibodies and antigens are intricate and constantly changing processes. Proteins might have multiple conformations, making large-scale data collection difficult or impossible. AFM effectively quantifies individual antibody-antigen interactions at the molecular level (*Probing Antibody-Antigen Interactions - PubMed*, n.d.).

5.11 T LYMPHOCYTES AND ANTIGEN-PRESENTING CELLS

T cells migrate from bone marrow hematopoietic stem cells to the thymus, where they mature. The T-cell receptor (TCR) binds to several antigens. Each T cell has one TCR form and can rapidly proliferate and differentiate when activated. Cells recognize antigens via APCs such as dendritic cells, macrophages, B cells, fibroblasts, and epithelial cells. APCs have MHC proteins on their surfaces.

When an immature T cell encounters an antigen bound to an MHC molecule, it proliferates and transforms into various unique effector and memory T cells (Janeway et al., 2001a).

5.11.1 Antigen-Presenting Cells

Cytokines released by the T-helper (TH) cells play a crucial role in activating cell-mediated and antibody-mediated humoral responses.

TH cell activation must be carefully controlled to prevent potentially fatal autoimmune effects. To ensure regulated activation, TH cells only detect antigens presented in combination with class II MHC molecules on APC surfaces. These APCs include B lymphocytes, macrophages, and dendritic cells (Figure 5.2).

Class I MHC molecules are present on almost all nucleated cells, whereas class II MHCs are exclusive to APCs. TH cells specifically detect antigenic peptides presented by class II MHC molecules, while TC cells only detect the MHC class I-bound antigenic peptides.

5.11.2 MHC Molecules

MHC molecules are categorized into two classes:

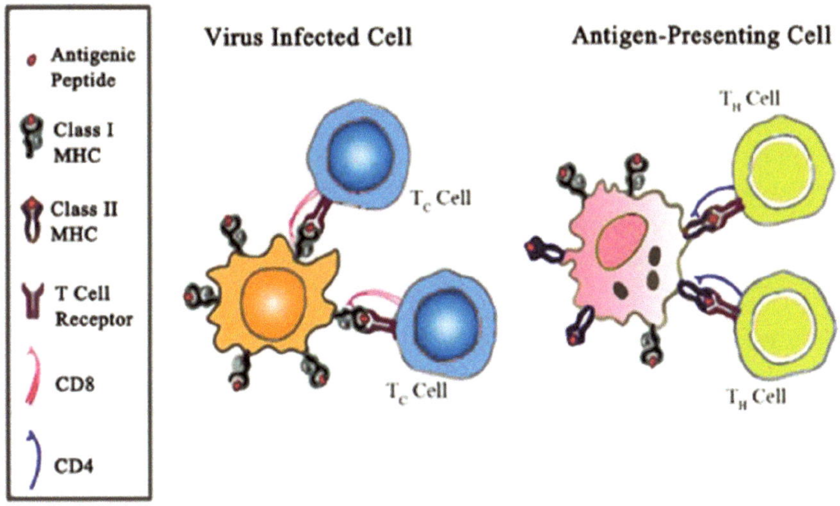

FIGURE 5.2 The role of MHC molecules in the recognition of antigens by T cells.

5.11.2.1 MHC Class I

MHC class I, also known as human leukocyte antigen (HLA) A, B, and C, is expressed on the surface of all nucleated cells. It consists of a heavy chain linked to a small, stable protein called β2-microglobulin.

MHC Class I molecules display endogenous peptides, which originate within the cell.

These include viral proteins produced within infected cells and distinct proteins produced by cancer cells. Endogenous antigens are degraded into peptide segments that attach to MHC class I molecules in the endoplasmic reticulum.

The antigenic peptide-MHC class I complex is then transported to the cell membrane. Given that all MHC Class I molecules are present on all nucleated cells, any cells that generate endogenous antigens utilize this pathway for antigen processing. T lymphocytes expressing CD8 receptors specifically detect antigens bound to MHC Class I molecules, indicating that they are restricted to MHC class I molecules. Cytotoxic T cells selectively target and eliminate cells displaying these complexes (Cruz-Tapias et al., 2013) (Figure 5.3).

Endogenous antigens, generated intracellularly due to viral infection, are degraded in the cytoplasm into peptides. These peptides then migrate to the ER, bind with MHC class I molecules, and transit through the Golgi complex to reach the cell surface.

5.11.2.2 MHC Class II

MHC class II molecules, often referred to as HLA DP, DQ, and DR, are only expressed by APCs such as dendritic cells, macrophages, and B cells. They are composed of one alpha and one beta glycoprotein chain.

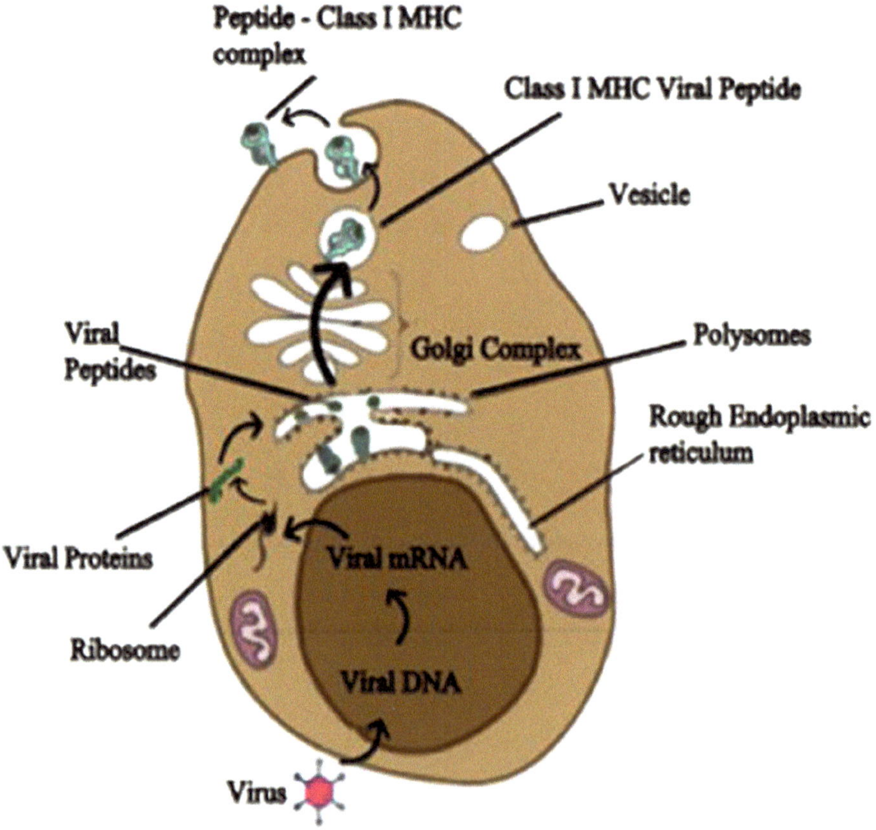

FIGURE 5.3 Processing and display of self-antigens.

MHC class II molecules found on APCs exhibit exogenous peptides, which come from outside the cell. These peptides are derived from antigens in cases of intracellular pathogenic infection or from foreign proteins or organisms engulfed by the cell (Janeway et al., 2001b).

Exogenous antigens enter host cells via endocytosis or phagocytosis. Macrophages, dendritic cells, and B cells break down these antigens into smaller peptide fragments through the endocytic processing pathway. MHC class II molecules in this pathway bind to the antigenic fragments.

Subsequently, the peptide-carrying MHC molecules are transported to the cell surface. Only APCs can present exogenous peptide-class II MHC complexes. CD4+ TH lymphocytes specifically identify antigens bound to class II MHC molecules, making them class II MHC restricted (Sadegh-Nasseri & Kim, 2015; Cruz-Tapias et al., 2013) (Figure 5.4).

Foreign antigens taken up by phagocytosis or endocytosis are degraded into small peptides under acidic conditions. These peptides are then presented on APC membranes with class II MHC molecules.

T cells are activated when they encounter an APC presenting the appropriate antigenic peptide along with its MHC molecule.

The lymphatic system and bloodstream help T cells circulate throughout the body, increasing the likelihood of effective interactions between specific T cells and APCs with the suitable peptide-MHC complex. T cells and APCs in lymph nodes maximize this potential. The MHC-antigen complex activates the TCR, triggering cytokine release and immune response. Antigen presentation differentiates T cells into CD4+ TH or CD8+ TC cells.

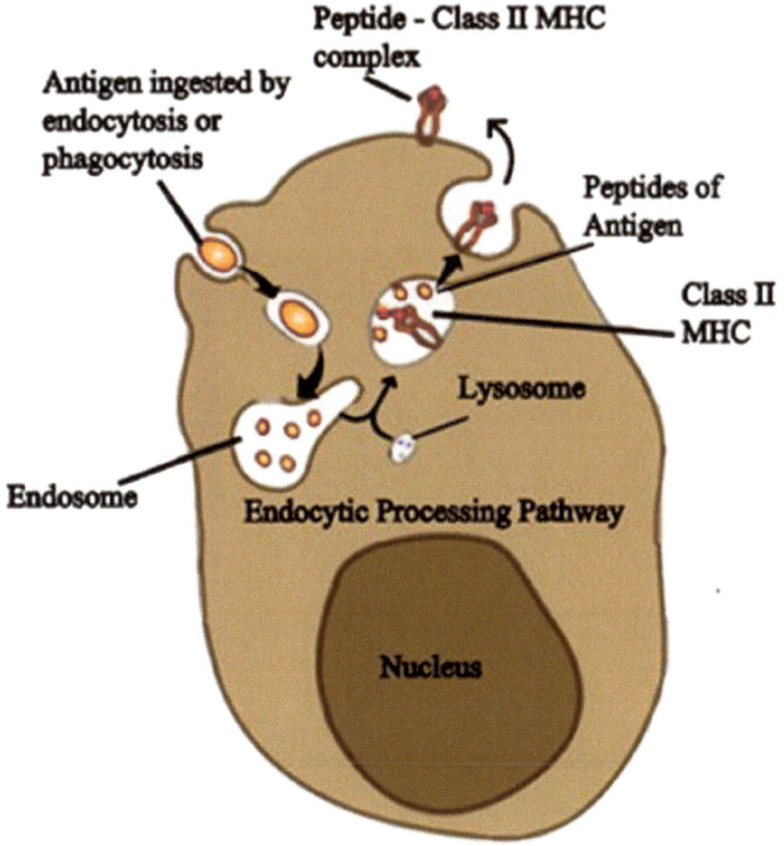

FIGURE 5.4 Processing and display of foreign antigens.

5.11.3 TH and TC Cells

TH and TC cells can be distinguished by the presence of either CD4+ (TH) or CD8+ (TC) membrane glycoproteins on their surface.

5.11.3.1 CD8+ Cytotoxic T Cells

CD8+ cytotoxic T cells play a significant role in destroying cells that have been infected by external agents like viruses, and in destroying tumor cells presenting the appropriate antigens. Upon interaction with an antigenic peptide-class II MHC molecular complex, a TH cell activates, converting into an effector cell that releases cytokines. These cytokines stimulate TC cells, B cells, macrophages, and other immune cells. Variations in cytokine patterns lead to diverse immunological responses. Clonal proliferation of TC cells generates effector cells that secrete chemicals capable of inducing apoptosis in target cells. After infection is resolved, most effector cells undergo apoptosis and are eliminated by phagocytes, but some persist as memory cells, capable of swiftly converting into effector cells following subsequent encounters with the same antigen (Joshi & Kaech, 2008).

5.11.3.2 CD4+ T Cells

When exposed to TH-derived cytokines, a TC cell that detects the antigenic peptide-class I MHC complex undergoes cell division and transforms into an effector T cell known as a cytotoxic T lymphocyte (CTL). These cells play a crucial role in generating and enhancing the immune response. Although CD4+ T cells lack direct cytotoxic or phagocytic functions and cannot directly destroy infected cells or eliminate infections, they act as mediators of the immune response. They direct other cells to perform these functions and regulate the specific type of immunological response that occurs. Upon activation, T cells release cytokines that regulate the activities of other cell types, such as the APCs that caused their activation (König et al., 1996).

5.12 CELL SIGNALING PATHWAYS

Cells communicate through chemical impulses. A transmitting cell produces chemical messengers, such as peptides or other compounds, which it releases into the extracellular space. These signals can then freely move to nearby cells, similar to information enclosed in a container (Bournazos et al., 2017). The significance of the cell signaling process can be understood by considering several key factors:

5.12.1 Homeostasis Maintenance

Cellular communication is crucial in sustaining a living organism's inner equilibrium. Cells communicate to control and synchronize processes, such as regulating humidity, pH, and nutrient levels, to provide optimal circumstances for cellular activity.

5.12.2 Progress and Expansion

Cell signaling plays a crucial role in directing biological events such as cell division, tissue creation, and organ growth. Signaling pathways regulate proliferation and development, guaranteeing the proper development of an organism from a single fertilized egg into a complex, multilayered structure (Nair et al., 2019).

5.12.3 Adaptation to Environmental Changes

Cells must react to alterations in their surroundings, including signals from neighboring cells or stimuli such as light, temperature, or drugs. Cell signaling enables cells to adapt and react appropriately to these environmental cues.

5.12.4 Immune Reaction

Cell signaling is very important for proper immune system function. It enables cells to communicate, identify pathogens, and respond to abnormal cells and diseases. Signaling pathways trigger immune cell activation, enhancing the body's defensive capabilities.

5.12.5 Antibodies as Signaling Molecules

Antibodies not only protect the body from infections but also serve as signaling molecules in certain bodily functions (Yang et al., 2017).

5.12.6 Cell-Surface Receptors

Antibodies can act as cell-surface receptors in the immune system. Key proteins on immune cells, including NK cells, neutrophils, and macrophages, communicate with the Fc (fragment, crystallizable) region of antibodies, leading to immune cell activation and phagocytosis.

5.12.7 FC Receptor Signaling

Immune cell Fc receptors attach to antibody Fc regions, initiating intracellular signaling pathways. This mechanism is essential for immune system functions, including ADCC and ADCP.

5.12.8 Immunoreceptor Tyrosine-Based Activation Motif (ITAM) Signaling

Antibodies trigger signaling pathways through the ITAM. When antibodies bind to Fc receptors, ITAM in the cytoplasmic tails is phosphorylated, activating downstream signaling processes (Bezbradica et al., 2014).

5.12.9 B-cell Receptor Signaling

Antibodies are expressed as B-cell receptors by immune cells. B-cell receptors (BCRs) tell B cells to activate, multiply, and differentiate into antibody-releasing plasma cells upon encountering specific antigens.

5.12.10 Antibodies Signal Intracellularly

Intrinsic antibodies communicate with physiological receptors. This relationship affects cell cycle control and death by modulating signaling pathways.

5.12.11 Signaling Molecule Neutralization

Antibodies can neutralize signaling molecules by attaching to them or blocking receptors. Monoclonal antibodies, which mimic the immune system's capacity for targeting particles, are key therapy tools for treating infectious diseases, cancer, and autoimmune disorders.

5.13. ITAM SIGNALING

The ITAM plays a crucial role in cell regulation, particularly in immune system actions. It is a pattern of peptides found in the cytoplasmic tails of some cell-surface receptors, typically on lymphocytes, B cells, and mast cells. The ITAM functions as a communication module that undergoes phosphorylation upon receptor engagement, facilitating subsequent signaling networks (Abram & Lowell, 2007).

Kinases, like the Src family kinases, and adaptor proteins, like Syk (spleen tyrosine kinase), work with ITAM to transmit signals. These proteins trigger physiological reactions, such as gene stimulation, gene expression alterations, and adjustments in immune cell activities. The ITAM signaling pathway is crucial in synchronizing immune reactions, assisting in the control of both adaptive and innate immunity. Understanding the organization and operation of the ITAM provides insight into the intricate processes that drive immune cell stimulation and its importance in coordinating effective immune responses.

5.14 CELLULAR RESPONSES TO ANTIBODY-ANTIGEN INTERACTION

Antibody-antigen interactions trigger numerous cellular reactions that help the immune system in identifying and eradicating infections. Gene transcription initiation is a vital step in these reactions. When antibodies bind to antigens, a series of internal signaling events activates transcription factors, which then regulate various immune genes (Huebner, 2004).

Beyond antibody production, this activation is important for the overall immune response. Antibody-antigen interactions cause adaptations in cellular behavior and functionality, stimulating the growth or maturation of immune cells like B cells, macrophages, and dendritic cells, increasing resistance to infections. Antibodies also influence processes such as cytokine secretion and phagocytosis, modulating cell behavior. These modifications collectively increase the ability to recognize and eradicate infectious agents, highlighting the diverse and complex nature of the immune system's cellular responses to antibody-antigen interactions.

5.15 CROSSTALK WITH KINASE PATHWAYS AND INTERACTIONS WITH G-PROTEIN-COUPLED RECEPTORS

There is a powerful connection among cell signaling pathways that helps them work together across numerous cellular processes. Interaction with kinase pathways is particularly common. Kinases attach phosphate groups to proteins, playing a role in numerous transmission activities. Antibody-antigen interactions can control kinase pathways by triggering or inhibiting specific kinases, thus regulating downstream signaling cascades. Information is integrated from multiple routes to carry out synchronized reactions within cells.

The antibody-antigen connection not only affects kinase pathways but also G-protein-coupled receptors (GPCRs). These GPCRs are present in cells and perform multiple biological functions. GPCRs spread information by triggering internal G proteins, after which some effectors, including ion channels, are regulated. GPCR signaling can be affected by either antibodies' attachment to GPCRs or by indirectly affecting signaling mechanisms. The reciprocation between various signaling pathways adds a further level of complexity to cellular reactions. Consequently, GPCRs carry out numerous activities, including immunological responses and neurotransmission. Understanding these interactions provides insight into cellular signaling networks and their effect on cell performance (Di Liberto et al., 2019).

5.16 B-CELL ACTIVATION

B cells play an important role in the immune system's adaptive capability. They are known for their ability to recognize and react to foreign substances by carrying out a process known as B-cell activation. An ample amount of BCRs are present on the outer layer of B cells. These BCRs function to bind to specific antigens on the outer layer of cells. B cells perform an important role in the immune system, initiating certain defenses against a comprehensive range of pathogens. B-cell activation holds significance because of its important function in immune system adaptation. The rapid and general reactions of the innate immune system have less specificity than adaptive immunity.

B cell activation is an important part of the adaptive immune response. This induction aids the body's immune system in recognizing and maintaining data regarding specific infections. This process is mandatory for setting up immune memory. Thus, the body generates the capacity to initiate a stronger resistance in response to future exposure to similar infections in the years to come. When B cells contact antigens, they go through a complex sequence of actions (Harwood & Batista, 2010).

Activating B cells involves many complicated chemical reactions, signaling cascades, and physiological processes that lead to the creation of plasma cells. These cells make antibodies and memory B cells. Understanding the intricacies of B cell activation is essential for unraveling the fundamental mechanisms of adaptive immunity and developing efficient strategies to harness the immune system to fight infections and diseases (Shlomchik, 2008).

5.16.1 BCR Engagement

The BCR-target antigen contact initiates B-cell stimulation. BCR chemical composition is key to this binding activity. The Y-shaped BCR is made of membrane-bound immunoglobulin (Ig) molecules with two similar heavy chains and two similar light chains. These chains' polymorphic sections contain antigen-binding sites, giving the BCR specificity. Recombination of genes during B cell development allows for the production of a wide range of BCRs that can recognize different antigens.

The complementary relationship between the antigen and polymorphic BCR sections mediates BCR antibody detection and attachment, which are highly specialized mechanisms. Pathogen-derived antigens have distinctive epitopes that fit perfectly into the BCR's antigen-binding sites. B cells use this lock-and-key strategy to distinguish antigens and launch a pathogen-specific defense (Young et al., 2015).

When a BCR encounters its corresponding antigen, it changes shape, prompting intracellular signaling processes that activate B cells. This interaction also clusters BCRs on the cell's surface, improving antigen responsiveness. The molecular details of BCR interactions show the precise and specific nature of the adaptive immune response, allowing B cells to recognize and respond to a wide range of infections in the immune system's complex terrain.

5.16.2 Clonal Selection and Expansion

The clonal selection hypothesis, an essential concept in immunology, explains the processes that govern the stimulation of B cells and the ensuing immune reaction. According to this theory, the immune system maintains a wide range of B cells, each having a unique BCR that is sensitive to different antigens. When a B cell encounters its matching antigen, it is selected for stimulation based on the uniqueness of its BCR. The selected B cell undergoes clonal expansion, a process that involves rapid multiplication to create a variety of related or clonal daughter cells.

The daughter cells, referred to as effector B cells, maintain the same antigen specificity as the parent B cell. Clonal selection and expansion are crucial processes that ensure the immune system mounts a precise and intensified attack against a specific infection. This creates a group of effector B cells that can differentiate into plasma cells, which make antibodies, and memory B cells, which help the immune system remember a pathogen upon future encounters. The adaptive immune response is based on the coordinated relationship between clonal selection and proliferation, allowing the immune system to adapt to the diverse antigens that an organism faces over its lifetime (Figure 5.5).

5.16.3 Signal Transduction Pathways

The BCR plays a vital role in translating external signals into internal cellular responses by activating signaling pathways. Upon binding to a specific antigen, the BCR undergoes structural changes

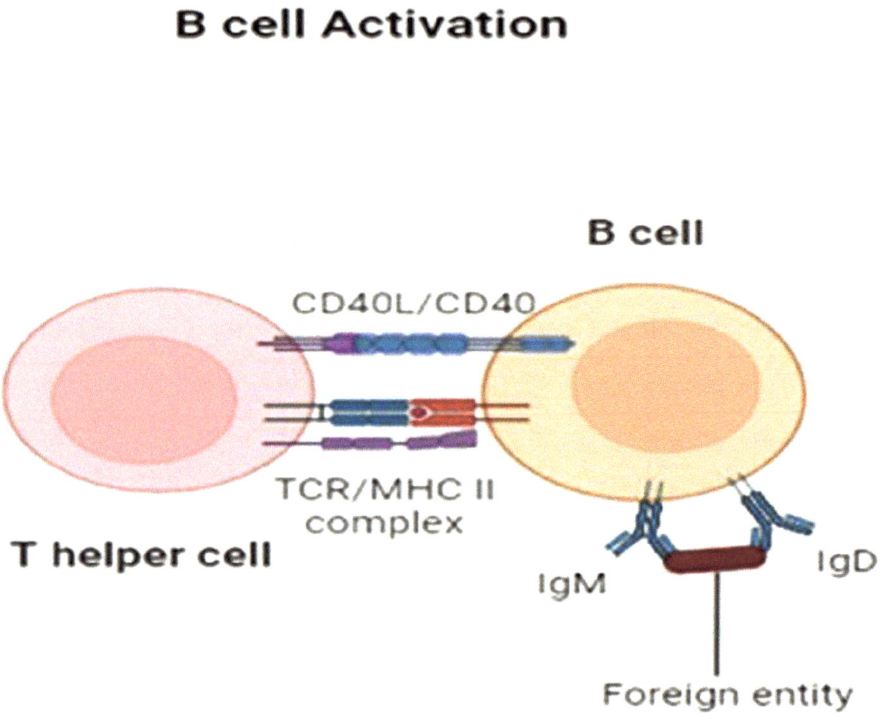

FIGURE 5.5 CD4 T cell and B cell interaction. B cells internalize and process antigens via the MHC-II pathway, breaking down antigens and presenting fragments on their surface with MHC-II. CD4 T cells recognize this complex and produce signals with other surface molecules to activate B cells.

that stimulate internal signaling pathways. This process involves the recruitment and phosphorylation of ITAMs found in the intracellular areas of BCR-associated proteins. This initiates the recruitment and stimulation of tyrosine kinases, especially members of the Src family, such as Lyn. These enzymes phosphorylate signaling molecules further down the pathway, including Syk kinase.

Syk activation triggers a series of intracellular processes, leading to the release of calcium ions, the activation of phospholipase C (PLC), and the production of inositol trisphosphate (IP3) and diacylglycerol (DAG). IP3 stimulates the release of calcium ions from internal storage compartments, further promoting protein kinase C (PKC). DAG activates the Ras-mitogen-activated protein kinase (MAPK) pathway. Additionally, the BCR signaling cascade stimulates the activation of the phosphatidylinositol 3-kinase (PI3K)-Akt pathway. These intricate signaling cascades regulate numerous activities, including B-cell activation, cytoskeleton reorganization, specialization in the adaptive immune response, and the initiation of several cellular processes. Understanding the intricacies of these signaling pathways is important for explaining the molecular processes that form the foundation of B-cell reactions upon their interaction with antigens (Ruland & Mak, 2003).

5.16.4 Role of Cytokines in B-Cell Activation

Cytokines are crucial for the coordination of B-cell initiation and development, influencing immunological responses. These signaling molecules affect B cells' differentiation into memory B cells or into plasma cells. Cytokines exert their effect by binding to specific receptors on the B-cell surface. For example, interleukin-4 (IL-4) promotes the differentiation of B cells into immunoglobulin E (IgE)-producing cells, while interleukin-21 (IL-21) enhances the production of plasma cells and antibody secretion. Cytokines can also guide B cells to produce specific types of antibodies and influence class switching (Vazquez et al., 2015).

Cytokine receptors initiate signaling cascades by activating Janus kinases (JAKs) and signal transducer and activator of transcription (STAT) proteins. These networks regulate gene expression, thereby determining the functional outcome of B cells. The intricate relationship between cytokines and their target cells is important for generating a personalized adaptive immune response to various pathogens. Understanding the mechanisms of cytokine-mediated signaling pathways provides insight into the regulation of B-cell activation and antibody responses.

5.16.5 Class Switching and Antibody Production

Class switch recombination (CSR) is a crucial process in B-cell development, leading to the generation of multiple antibody isotypes while preserving the specificity gained through antigen identification. This transformation occurs at the immunoglobulin heavy chain gene locus and involves the deliberate restructuring of genes in specific regions. CSR allows B cells to switch from producing immunoglobulin M (IgM) and immunoglobulin D (IgD) to making other antibody isotypes such as immunoglobulin G (IgG), immunoglobulin A (IgA), and immunoglobulin E (IgE). Activation-induced cytidine deaminase (AID) initiates this process by creating double-strand DNA breaks at specific sites, which are then rejoined.

This recombination generates antibodies with adapted effector activities, leading to personalized immune responses. The specific isotype produced is determined by the cytokine environment and the microenvironment. Interleukins like IL-4 play a pivotal role in this process. Class witching and subsequent antibody production enhance the immune system's ability to effectively combat a wide range of illnesses by increasing the diversity of available antibodies. Understanding the mechanisms involved in class switching provides insight into the immune system's ability to produce antibodies with diverse functions, demonstrating its versatility and durability (T. Zhang et al., 2010).

5.16.6 Formation of Memory B Cells

Memory B cells play a significant role in the adaptive immune system, providing long-lasting protection against previously encountered infections. Their production is an important result of B-cell activation, known for the continued existence and maintenance of a specific group of B cells with an enhanced lifespan. Some B cells differentiate into memory B cells upon their first encounter with an antigen rather than fully developing into antibody-producing plasma cells.

Memory B cells retain the capacity to recognize the specific antigen that initiated their activation. They demonstrate an accelerated and robust immune response upon subsequent encounters with the same antigen. As a consequence, antibody production is accelerated, leading to quicker microbe eradication. The production and maintenance of memory B cells are integral components of immunological memory, an attribute of the adaptive immune system that ensures a more robust and timely response to previously encountered pathogens.

5.16.7 B-Cell Activation Regulation

A complex array of mechanisms regulates B-cell activation to prevent aberrant immune responses and maintain immune system equilibrium. These mechanisms control B-cell proliferation and prevent the development of autoimmunity. Regulatory T cells (Tregs) inhibit B-cell responses, while inhibitory receptors (CD22 and FcγRIIB) act as barriers to impede communication. Furthermore, the balance between co-stimulatory and co-inhibitory signals is crucial in shaping B-cell fate. This meticulous regulation ensures appropriate responses to pathogens while preventing reactions to self-antigens. A comprehensive understanding of these regulations is critical for unraveling the complexities of immunity and developing treatments for diseases associated with aberrant B-cell activation.

5.17 T-CELL ACTIVATION

T cells play a significant role in coordinating defenses against infection and faulty cells. Their activation is crucial for initiating and regulating immune responses. The two main T-cell subsets are CD4+ and CD8+. CD4+ cells, also called helper T cells, supervise the regulation of immunological reactions by communicating with other immune cells and releasing cytokines (Smith-Garvin et al., 2009).

CD8+ T cells, also known as cytotoxic T cells, possess the ability to rapidly eliminate infected or abnormal cells. T-cell activation is triggered by the binding of TCRs to particular antigens presented by antigen-presenting cells. This interaction, along with co-stimulatory signals, sets off intracellular signaling pathways, leading to T-cell proliferation and differentiation into effector cells. Understanding the intricacies of T-cell activation is crucial for unraveling the complexities of adaptive immunity and developing accurate treatment strategies to fight illnesses, including infections, cancer, and immune-related disorders.

5.17.1 TCR Engagement

T-cell receptors stimulate T-cell proliferation upon antigen binding. T cells display two chains, alpha and beta, which combine to form a heterodimeric complex. This complex consists of variable and constant regions, with the variable region participating in antigen binding. For activation, the peptide fragment must directly contact the TCR's antigen-binding site.

MHC molecules on the surface of cells present these peptides. The binding mechanism is quite specific and crucial for T-cell activation. When TCR molecules interact with MHC molecules, co-receptors like CD4 or CD8 facilitate TCR binding to their respective parts. The specific molecular interactions between the TCR, antigenic peptides, and MHC molecules are essential for T-cell antigen recognition. These interactions trigger signaling mechanisms that ultimately lead to T-cell proliferation and the production of antigen-specific responses.

5.17.2 Co-Stimulation and Co-Inhibition

Co-stimulation and co-inhibition are two major regulatory processes crucial for T-cell activation and immune response regulation, preventing autoimmunity. CD28, a co-stimulatory compound, plays an important role in amplifying T-cell activation signals.

CD28 is found on T cells and connects to ligands on APCs, like B7 molecules, to provide additional signals that T cells need to function optimally. This interaction of co-stimulatory molecules enhances TCR signaling and promotes T-cell proliferation and function. Co-inhibitory signals, such as CTLA-4, play a significant part in regulating T-cell reactions (Chen & Flies, 2013).

CTLA-4 and CD28 compete for attachment to B7. When activated, CTLA-4 sends inhibitory signals that reduce T-cell activation. The balance between co-stimulation and co-inhibition keeps T-cell reactivity under control, preventing immune responses and maintaining immune system homeostasis. Understanding these regulatory mechanisms is crucial for developing therapies that can effectively modulate T-cell responses in autoimmune diseases, organ transplantation, and tumor immunotherapy (Figure 5.6).

5.17.3 Signal Transduction Pathways in T Cells

Intricate signal transduction pathways are activated when TCRs bind to antigen-MHC complexes, resulting in T-cell activation. The TCR complex, comprising CD3 and ζ-chains, contains ITAMs. Key events include the activation of the kinase ZAP-70 and the phosphorylation of ITAMs by the Src family kinase Lck. ZAP-70 then catalyzes the phosphorylation of multiple adapter proteins, resulting in the activation of phospholipase C-γ1 (PLC-γ1). PLC-γ1 hydrolyzes phosphatidylinositol 4,5-bisphosphate (PIP2) into inositol trisphosphate (IP3) and diacylglycerol (DAG). IP3 triggers

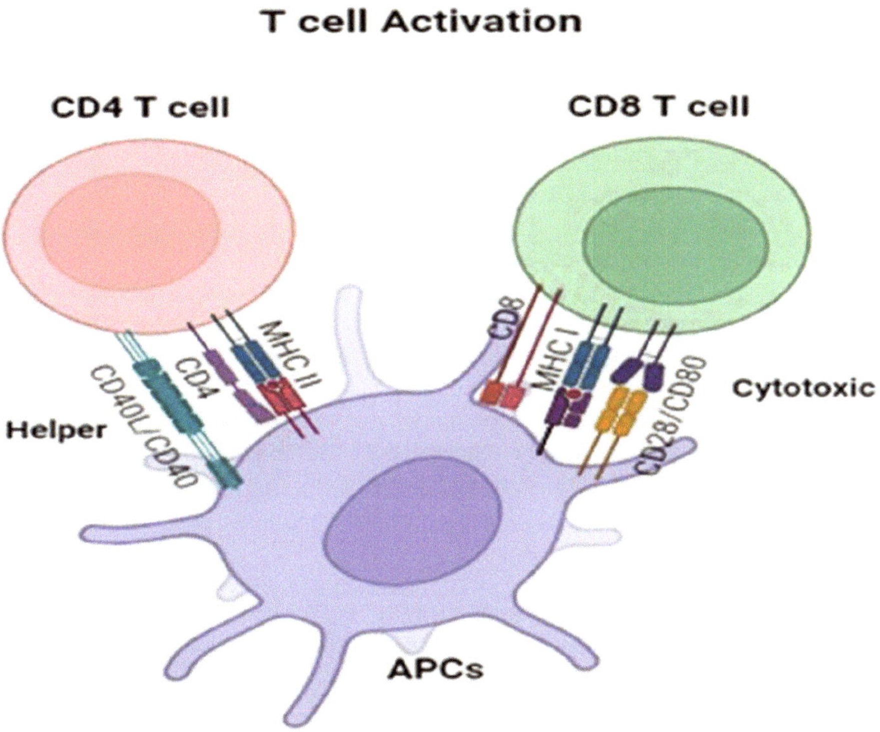

FIGURE 5.6 T-cell activation. A T cell binds to a dendritic cell (DC) presenting its corresponding peptide in an MHC molecule via CD3 and CD4 or CD8. Co-stimulation occurs via DC-bound CD86, CD80, OX40L, and 4-1BBL, resulting in T cell effector function induction and full activation.

calcium release from internal stores, activating calcineurin and NFAT, while DAG activates protein kinase C (PKC) (Arasanz et al., 2017).

The increase in intracellular calcium levels leads to NFAT and PKC activation, resulting in cytokine production and gene transcription. Moreover, the MAPK pathway is stimulated, enhancing T-cell development and activation. This intricate set of intracellular events following TCR stimulation ensures a precise and coordinated T-cell response to antigens. Understanding these signaling pathways is critical for comprehending the complexities of T-cell activation and their role in immune regulation and therapy.

5.17.4 Cytokine Signaling in T-Cell Activation

Cytokines significantly impact T-cell activation, coordinating their development and proliferation upon their exposure to numerous immunological stressors. They also influence the fate of T cells. IL-2, a key cytokine, is pivotal in enhancing the development of T cells. It also helps in the generation of both effector and memory T cells. Specific cytokines direct the differentiation of T cells into distinct subsets. For example, IL-4 directs T cells towards a Th2 phenotype, promoting an antibody response. Meanwhile, interferon-gamma (IFN-γ) promotes a Th1 phenotype, increasing cellular immunity.

Cytokine receptors on the T-cell surface stimulate intracellular signaling cascades, often involving JAKs and STAT proteins. The pathways play a crucial role in shaping gene expression, ultimately determining T-cell fate. Precise regulation of cytokine signaling during T-cell activation is essential to ensure appropriate immune responses to specific pathogens, highlighting the importance of cytokines in immune system modulation and regulation.

5.17.5 T-Cell Differentiation and Effector Functions

T cells experience transformation after activation, leading to the formation of numerous effector subsets. Each subset is designed to function in specific situations. CD4+ T cells can differentiate into Th1 cells, which mediate cell-mediated immunity. Th1 cells secrete interferon-gamma (IFN-γ) to enhance cytotoxicity and phagocytosis. Th2 cells produce interleukins, including IL-4 and IL-5, which modulate humoral immunity, trigger the production of antibodies, and aid in defense against extracellular parasites. IL-17 produced by Th17 cells plays a role in inflammatory reactions and is crucial for immune protection on mucosal surfaces.

Regulatory T cells play a pivotal role in maintaining immune system balance, restricting uncontrolled immune responses, and preventing autoimmune diseases. CD8+ T cells, or cytotoxic T cells, differentiate into effector cells capable of rapidly killing infected or abnormal cells by emitting harmful substances. The differentiation of T cells into specific subsets and their activation to perform distinct functions demonstrates the remarkable adaptability of the immune system. This adaptability enables the immune system to tailor its defenses to various challenges, resulting in an effective and targeted immunological response.

5.17.6 Memory T-Cell Formation

Memory T cells, crucial for immunological memory, are formed following the conclusion of an initial immune response and persist for a prolonged duration. After the eradication of an illness or exposure to an antigen, a subset of activated T cells differentiates into memory T cells. These memory T cells have heightened responsiveness and an enhanced ability to rapidly activate upon re-exposure to the same antigen (Ando et al., 2020).

Memory T cells can persist for an extended period, providing the immune system with a form of immunological "memory." This persistence ensures a faster and more robust reaction when encountering familiar pathogens, thus improving the effectiveness of the adaptive immune system. The generation and maintenance of memory T cells exemplify the dynamic nature of immunological memory, playing a vital role in the body's ability to provide effective and timely immune responses to previously encountered threats.

5.17.7 Regulation of T-Cell Activation

Effective regulation of T-cell activation is necessary to preserve immunological tolerance and prevent the development of autoimmunity. Tregs prevent inappropriate immunological responses. Tregs are characterized by the presence of the transcription factor Foxp3, and can suppress the activation and functions of effector T cells. T-cell functions are further regulated by negative regulators and checkpoints. CTLA-4, present on T cells, interacts with ligands on antigen-presenting cells to deliver an inhibitory signal that regulates the T-cell activation mechanism.

Programmed cell death protein 1 (PD-1) and its ligands also play an essential role in immune regulation by restricting T-cell activation. The interaction between Tregs and antagonistic regulators ensures a harmonized immune response, enabling efficient protection against pathogens while avoiding detrimental reactions towards self-antigens. Understanding these regulatory processes is essential for maintaining immunological balance. This has implications for treating disorders involving abnormal T cells, such as autoimmune diseases, and in cancer immunotherapy.

5.17.8 Integration of T- and B-Cell Responses

The coordination of T- and B-cell responses is a complex and well-organized mechanism in adaptive immunity. T and B cells cooperate through complex pathways to provide an efficient immune defense against infections. Dendritic cells (antigen-presenting cells) deliver antigens to both T and

B cells. Activated helper T cells (CD4+) secrete cytokines that impact the activation and differentiation of B cells. Meanwhile, B cells present antigens to T cells, enhancing T-cell responses.

This collaboration among cell types is crucial for producing high-affinity antibodies, enabling class switching, and forming immunological memory. A coordinated adaptive immune response requires the interplay of signaling molecules, co-stimulatory signals, and feedback mechanisms to ensure synchronized and efficient protection against various pathogens. This cooperation is essential for establishing a strong and rapid response upon re-encountering the same antigens, as well as for clearing up previous infections. This demonstrates the specificity and remarkable flexibility of the immune system (Rawlings et al., 2012).

5.18 SIGNALING MOLECULES

Signaling molecules are essential in antigen-antibody interactions, coordinating and regulating immune responses. The encounter between antigens, which are foreign substances, and antibodies, proteins generated by the immune system, initiates a series of interconnected reactions that ultimately result in the eradication of the invading pathogen (Janeway et al., 2001b). The following are essential signaling molecules involved in antigen-antibody interactions:

5.18.1 Cytokines

Cytokines are a heterogeneous group of compact proteins that have a pivotal function in cellular signaling and intercellular communication within the immune system (Cui et al., 2024). They are synthesized by diverse cell types, including immune cells, and function as signaling molecules to modulate the immune response, inflammation, and other physiological processes (Figure 5.7).

Cytokines can exert autocrine effects by acting on the same cell that secreted them, paracrine effects by acting on neighboring cells, or endocrine effects by acting on distant cells via circulation (J.-M. Zhang & An, 2007). These proteins are initially inactive while circulating, but they can quickly become active to enhance phagocytosis, trigger inflammation, or promote the destruction of

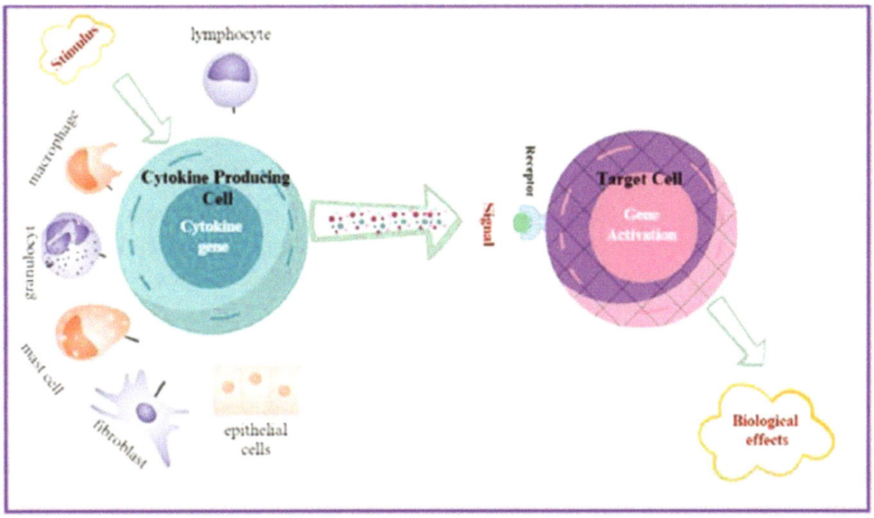

FIGURE 5.7 The main types of cytokine-producing cells. Upon immune stimulation, these cells produce cytokines that bind to specific membrane receptors on target cells. Target cells respond by altering membrane protein expression, proliferation, and effector molecule secretion.

microbial pathogens (Cui et al., 2024). The binding of antigens to antibodies triggers the activation of diverse immune cells, and cytokines play a crucial role in facilitating the communication and control of these immunological reactions (Chaplin, 2010).

Upon antigen recognition, B cells undergo activation. Cytokines, including IL-4, facilitate the differentiation of B cells into plasma cells, which are responsible for antibody production. IL-4, in conjunction with other cytokines, influences the specific class of antibodies produced (e.g., IgM, IgG, IgA, etc.) (Fillatreau, 2018). Lymphocytes rely on cytokines for their activation and differentiation (Cano & Lopera, 2013). Helper T cells, namely those expressing the CD4+ marker, secrete cytokines such as IL-2 and IL-4 (Alberts et al., 2002). These cytokines promote the growth and specialization of B cells into plasma cells, which produce and release antibodies (Janeway et al., 2001a).

Interactions between antigens and antibodies can initiate inflammatory reactions. Cytokines such as TNF and IL-1 are generated during inflammation, facilitating the attraction of immune cells to the site of infection or injury (J.-M. Zhang & An, 2007).

Cytokines also play a role in regulating immune responses. IL-10 and TGF-beta have immunosuppressive activity, which aids in controlling excessive inflammation and preventing damage to host tissues (Taylor et al., 2006). Viruses trigger the activation of many genes responsible for producing cytokine proteins, which are normally inactive. Additionally, viruses can also promote the expression or replication of similar cytokines (Dembic, 2015). Certain cytokines, particularly IL-12, in conjunction with interferon γ, reduce the virulence of viruses. Conversely, cytokines can stimulate or enhance viral pathogenicity in many diseases, such as in *Leishmania major* infection, where IL-4 otherwise has a detrimental impact (Justiz Vaillant & Qurie, 2023).

An appropriate balance between pro-inflammatory and anti-inflammatory cytokines is imperative for establishing a well-regulated effector response by the immune system (C. Liu et al., 2021). Excessive synthesis of pro-inflammatory cytokines and/or insufficient levels of anti-inflammatory cytokines can result in immunological disorders (Kahan, 2008).

The production of cytokines by T-helper (Th) cells plays a crucial role in determining the fate of numerous viral infections. The production of the appropriate set of cytokines in response to an infectious agent can determine survival or mortality (Bretscher, 2022). The Th1/Th2 dichotomy, while an oversimplification, has demonstrated its utility in the examination of immune responses to infections. In certain infectious disorders, such as those caused by *Leishmania* parasites or gastrointestinal helminths, the presence of a specific type of T-helper cell is crucial for eliminating the infection, while the presence of the opposite type of T-helper cell is harmful. In some cases, both Th1 and Th2 responses are necessary at different time intervals to eliminate an infectious pathogen (Tymoshok et al., 2014).

5.18.2 Chemokines

Chemokines, also known as chemotactic cytokines, are a diverse group of small proteins that transmit signals through cell-surface G proteins coupled with hepta-helical chemokine receptors. They play an important role in responses stimulated by antibody-antigen interactions, not by directly forming antigen-antibody complexes but by acting as chemoattractants. They recruit immune cells such as phagocytic cells (neutrophils and macrophages) to sites of infection or inflammation where these complexes form (Hughes & Nibbs, 2018). They also attract T cells, particularly helper T cells, to coordinate immune responses (Alcami & Lira, 2010).

This recruitment occurs due to the expression of specific receptors on immune cell surfaces. Neutrophils express receptors such as CXCR1 and CXCR2, while T cells express receptors like CCR5 and CXCR4 (Reyes-Robles et al., 2013). Chemokines binding to their receptors guide immune cells to the site of antigen-antibody complexes (Figure 5.8).

Chemokines play a central role in shaping and maintaining the immune system, influencing all immune and inflammatory responses, whether protective or destructive (Harvanová et al., 2023). Traditionally recognized for guiding specific chemotactic migration, chemokines can also elicit

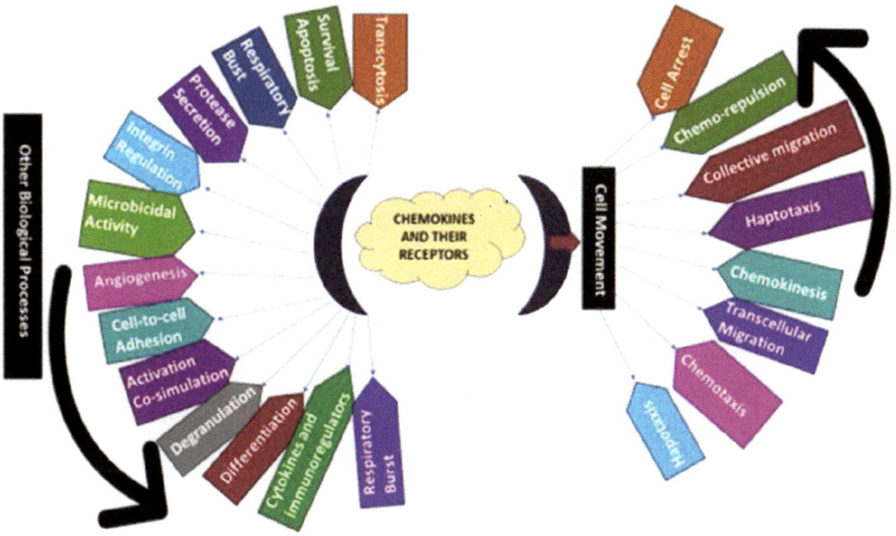

FIGURE 5.8 Biological processes governed by chemokines and their receptors. Processes are displayed in boxes grouped around the central "Chemokines & Receptors" box. The primary activity, "Cell Movement," encompasses various chemokine-mediated migratory behaviors of immune cells, shown separately on the right.

various other forms of directed and undirected migratory responses, including chemokinesis, haptotaxis, and haptokinesis (Hughes & Nibbs, 2018). Additionally, chemokine receptors on leukocytes can impact the biology of numerous non-leukocytic cell types beyond directing migration. These molecules regulate their biology (Laufer & Legler, 2018).

Post-translational modifications greatly affect chemokines by interacting with the extracellular matrix and regulating their localization and quantity through binding with hepta-helical receptors. They also regulate cell adhesion, tissue migration, and immune cell contacts, all crucial for a well-coordinated immune response.

Some chemokines play distinct roles in immune responses. For example, interleukin-8 (CXCL8) is a potent chemokine involved in recruiting neutrophils. Monocyte chemoattractant protein-1 (CCL2) attracts monocytes and T cells (Deshmane et al., 2009). During adaptive immune response, chemokines also contribute to organizing immune cells in lymphoid tissues, facilitating interactions between B and T cells (Cano & Lopera, 2013).

5.18.3 Transforming Growth Factor (TGF)

TGF-beta is a multifunctional pleiotropic cytokine with both anti-inflammatory and immunosuppressive properties involved in regulating immune responses, apoptosis, angiogenesis, cellular proliferation, and tissue regeneration (Xue et al., 2020). There are three isoforms of TGF: TGF-β1, TGF-β2, and TGF-β3. The body's immune cells generate the TGF-β1 isoform, which possesses potent anti-inflammatory properties and acts as a key mediator of the immune response involved in tissue repair and fibrosis (Prud'homme, 2007). TGF-β2 and TGF-β3 play pivotal roles in wound healing, tissue morphogenesis, and embryonic development, and influence cell fate decisions and tissue patterning.

TGF transmits signals via a receptor complex, including TGF-β receptors type I and type II. Downstream Smad proteins (Smad2 and Smad3) are activated and transfer signals to the nucleus to regulate gene expression (Hata & Chen, 2016). Dysregulation of TGF signaling is associated with a range of illnesses, such as fibrotic disorders, autoimmune diseases, and cancer.

TABLE 5.2
Important Interleukins and Their Roles

Function	Interleukin Involved	Role	Reference
B-cell activation and antibody production	IL-2	Upon antigen recognition, B cells undergo activation, and IL-2 supports its proliferation and differentiation into plasma cells responsible for antibody production.	Charles A Janeway et al. (2001a)
Class switching	IL-4	IL-4 and TGF-β stimulate B cells expressing surface IgM (sIgM+) to transform into other forms of antibodies, such as IgE and IgA.	Fujihashi et al. (2008)
Helper T-cell differentiation	IL-12 and IL-4	Involved in differentiating helper T cells (CD4+ T cells) into specific subsets: Th1 cells (stimulated by IL-12) enhance cellular immunity; Th2 cells (stimulated by IL-4) facilitate antibody-mediated immune responses.	Alberts et al. (2002)
Differentiation and maintenance of T follicular helper cells	IL-21	Essential for facilitating B-cell activation and producing high-affinity antibodies.	Crotty (2014)
Inflammation	IL-1 and IL-6	Pro-inflammatory cytokines that contribute to initiating and regulating inflammation by recruiting immune cells to infection or injury sites.	J.-M. Zhang & An (2007)
Immune suppression	IL-10	Suppresses excessive inflammation, preventing damage to host tissues.	Dembic (2015)
Antiviral responses	IL-15	Contributes to NK cell activation and proliferation, playing a role in early defense against viral infections.	Garlanda et al. (2013)

5.18.4 Interleukins

Interleukins are essential for activating and differentiating immune cells in response to any foreign activity, as well as for proliferation, maturation, migration, and adhesion at target sites (Garlanda et al., 2013). Additionally, they possess both pro-inflammatory and anti-inflammatory characteristics (Justiz Vaillant & Qurie, 2023). Interleukins primarily serve to regulate activation, growth, differentiation, and activation in inflammatory and immunological responses. They induce various responses at cellular and tissue levels by binding to high-affinity receptors on cell surfaces (Dembic, 2015). Some important interleukins and their role in different stages of antigen-antibody interactions are discussed in Table 5.2.

5.18.5 The Role of Antigen-Presenting Cells

APCs have a vital function in the immune system as they capture, process, and present antigens to other immune cells, thus initiating and regulating immune responses (Wynne et al., 2016). APCs constantly monitor the body for foreign intruders, especially dendritic cells, which are carefully located in tissues and organs to capture and process antigens, providing constant surveillance (K. Liu, 2016). APCs also contribute to the formation of memory T cells, which are long-lived cells that mount more robust immune responses upon re-exposure to the same antigen (Ratajczak et al., 2018). Moreover, APCs also help maintain immune tolerance by facilitating the distinction between self and non-self, preventing inappropriate immune reactions against the body's own tissues (Nicholson, 2016) (Table 5.3).

TABLE 5.3
Localization and Role of APCs

Antigen-Presenting Cell	Location	Role	Reference
Dendritic cells	Epithelia, central lymphoid organs, bone marrow, thymus.	Antigen capturing and processing, migration, T-cell activation, bridging innate and adaptive immunity.	K. Liu (2016)
Macrophages	Almost all organs: liver, brain, lungs, etc.	Phagocytosis, antigen presentation, cytokine production, tissue repair, immune surveillance.	Hirayama et al. (2017)
B cells	Lymphoid follicles of the spleen and lymph nodes (if not in the gut-associated lymphoid tissue).	Antigen recognition and presentation, T-cell activation, endocytosis, antibody production.	Charles A Janeway et al..(2001a), Nicholson (2016)
Langerhans cells	Skin and mucous membranes	Antigen capture, migration, and T-cell activation; specialized dendritic cells found in skin.	Jaitley & Saraswathi (2012)
Follicular dendritic cells	Lymphoid follicles (especially lymph nodes).	Provide antigens to B cells during germinal center immune responses; B-cell interaction and affinity maturation (not conventional APCs).	Heesters et al. (2014)

5.19 EFFECTOR MECHANISMS

The immune system incorporates many effector mechanisms to protect the body from infections, germs, and other foreign substances (Nicholson, 2016). These systems entail the synchronized activity of several immune cells, chemicals, and processes to eradicate or counteract pathogens. (Dembic, 2015). A concise overview of basic effector mechanisms within the immune system is given in Table 5.4 (Figure 5.9).

5.20 DISABLED INTERACTIONS LEADS TO IMMUNE DISORDERS

The immune system plays a crucial role in safeguarding the body against many diseases and infections. If specific interactions within the immune system become ineffective or compromised, it can lead to various health problems, including chronic diseases.

5.20.1 Immunodeficiency Disorders

Immunodeficiency disorders arise when the immune system is compromised or impaired, resulting in heightened vulnerability to infections. These illnesses can either be present at birth (congenital) or develop later in life (acquired). Immunodeficiency illnesses can impact several components of the immune system, such as B cells, T cells, phagocytes, complement proteins, and other components (Justiz Vaillant & Qurie, 2024).

Two common types of immunodeficiency disorders are as follows:

5.20.1.1 Primary Immunodeficiency Disorder (PID)

PIDs encompass a diverse range of more than 130 disorders that arise from abnormalities in the development and/or functioning of the immune system. They are categorized into two main groups:

TABLE 5.4
Overview of the Effector Mechanisms of Action Against the Immune System

Effector Mechanism	Description	Role	Reference
Antibody-mediated immunity/humoral immunity	Antibodies detect and bind to foreign antigens and neutralize them.	Neutralization of pathogens; opsonization and complement system activation.	Neurath (2008)
Cell-mediated immunity	T lymphocytes recognize and eliminate infected cells and foreign antigens.	Destruction of intracellular pathogen-harboring/ cancerous cells.	Kaufmann (1999)
Phagocytosis	Phagocytic cells engulf and digest pathogens/cellular debris.	Robust and nonspecific elimination of pathogens/ cellular debris.	Rosales & Uribe-Querol (2017)
Complement system	Protein cascade pathways leading to lysis/ opsonization of foreign antigens.	Facilitates phagocytosis; promotes inflammation and cell/pathogen lysis.	Merle et al. (2015)
Cytokines	Signaling proteins regulating immune responses.	Mediate immune cell activation, growth, and proliferation.	Cui et al. (2024)
Inflammation	Localized immune response against infection/tissue damage.	Recruitment of immune cells; enhancement of phagocytosis and prevention of pathogen survival.	J.-M. Zhang & An (2007)
Interferons	Antiviral proteins.	Activate immune cells to disturb viral replication.	Banerjee et al. (2019)
Immunological memory	Remembering earlier encounters with foreign antigens.	Enhancing immune response efficiency upon subsequent exposure.	Ratajczak et al. (2018)

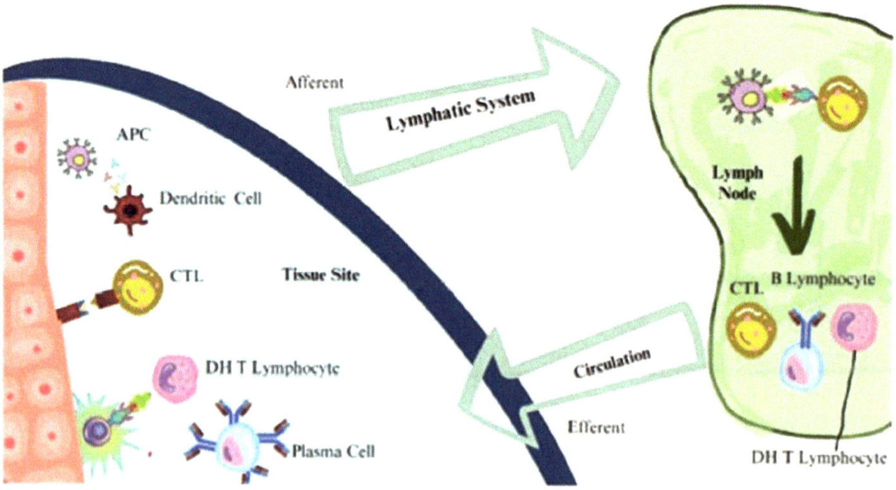

FIGURE 5.9 Effector response pathways in adaptive immunity. The immune response extends beyond the affected tissue, concluding when immune effectors encounter antigens within tissues after release from lymph nodes and circulation. The three primary effector mechanisms include CTLs, delayed hypersensitivity (DH) T lymphocytes (CD4-Th lymphocytes), and antibody-producing plasma cells derived from B lymphocytes.

those affecting adaptive immunity (T-cell, B-cell, or mixed immunodeficiencies) and those affecting innate immunity (including phagocyte and complement disorders) (McCusker & Warrington, 2011). While the clinical presentations of PIDs can vary greatly, most are characterized by an elevated vulnerability to infections. Timely diagnosis and treatment are crucial to prevent serious health complications. Examples include severe combined immunodeficiency (SCID), X-linked agammaglobulinemia (XLA), common variable immunodeficiency (CVID), chronic granulomatous disease (CGD), and DiGeorge syndrome (McCusker et al., 2018).

5.20.1.2 Secondary Immunodeficiency Diseases (SID)

Secondary immunodeficiency diseases are acquired in adulthood and can be caused by various factors, including infections, medications, malnutrition, and certain medical interventions (Chinen & Shearer, 2010). For example, HIV invades and destroys CD4+ T cells, compromising the immune system and making individuals vulnerable to opportunistic infections and malignancies (Okoye & Picker, 2013). Chemotherapy and radiation therapy can suppress the immune system's function, rendering patients susceptible to infections (Li et al., 2023). Malnutrition can weaken the immune system, increasing vulnerability to infections. Chronic diseases such as diabetes, chronic renal disease, and certain autoimmune disorders can gradually impair the immune system (H. Wang et al., 2017).

5.20.2 Allergic Reactions

Allergic reactions occur when the immune system overreacts to harmless substances, known as allergens, such as pollen, dust mites, or specific foods. This overreaction triggers the production of histamines and other substances, causing symptoms such as itching, swelling, hives, and respiratory issues. Allergies may result from a breakdown of immunological tolerance mechanisms.

Initial exposure to an allergen leads to sensitization, increasing the immune system's responsiveness. During this phase, the immune system recognizes the allergen as a foreign agent and generates specialized antibodies known as immunoglobulin E (IgE) (Abbas et al., 2024). Sensitization may be asymptomatic. Upon subsequent exposure, the allergen attaches to IgE antibodies on mast cells and basophils, activating them and leading to the secretion of inflammatory substances like histamine, leukotrienes, and cytokines (Poulsen & Hummelshoj, 2007). This release results in various allergic inflammatory manifestations, such as pruritus, sneezing, lacrimation, nasal discharge, dermatological eruption, or urticaria (Wang et al., 2023).

5.20.3 Autoimmune Diseases

The immune system comprises small molecules, soluble proteins (antibodies, cytokines, chemokines, and complement proteins), cell-surface proteins (receptors, adhesion molecules), and immune cells. This complex system eliminates invading microorganisms, wounded cells, and debris, and monitors cancerous cells. It consists of innate and adaptive immune responses.

Individuals' innate immune responses are similar and target conserved pathogen compounds. The adaptive immune response is altered by foreign chemicals and varies between individuals over time. Both arms of the immune system activate to discriminate invaders from host cells, but immune responses can cause localized damage, potentially leading to inflammation and tissue damage in chronic infections (Anaya et al., 2013).

In autoimmune disorders, the adaptive immune system misidentifies host epitopes as alien, potentially causing permanent damage to target organs. Chronic inflammation can also result from uncontrolled innate immune system stimulation, leading to autoinflammatory disorders. These can be caused by mutations in genes encoding proteins that activate innate immune pathways. Mutations in the adaptive system can also cause autoimmune disorders (Pisetsky, 2023). Over 100

TABLE 5.5
Autoimmune Diseases and Genetic Causes

Pathway/Process	Autoimmune Diseases	Mutated Genes or Signaling Molecules	Reference
Innate immunity and bacterial recognition	Crohn's disease, rheumatoid arthritis, primary biliary cirrhosis, systemic lupus erythematosus	ATG16L1, NOD2, IRF5, ATG5, IFIH1, PRDM1	Anaya et al. (2013), Tobón et al., (2010), Carey et al. (2015), Lahita et al., (2010)
Transcription factors	Multiple sclerosis, systemic sclerosis, rheumatoid arthritis, primary biliary cirrhosis, Crohn's disease	STAT4, STAT3, c-Rel	Anaya et al. (2013), Hilliard et al. (2002), Isomäki et al. (2015), McFarland & Martin (2007), Willson et al. (2012)
Cytokine receptors and cytokines	Primary biliary cirrhosis, Crohn's disease, psoriasis, ankylosing spondylitis, ulcerative colitis	IL23R, IL2RA, IL7R, IL12B, IL10	Braat et al. (2006), Webb et al. (2015), Armstrong et al. (1983), Targan & Karp (2005)
Lymphocyte activation and intracellular signaling	Majority of the autoimmune diseases, including systemic lupus erythematosus and type-I diabetes mellitus	CD40, HLA, CTLA4, TNFAIP3, TAGAP, PTPN2, TYK2, TNIP1, PTPN22, PRKCQ	Taher et al. (2017), Márquez & Martín (2022), Richard-Miceli & Criswell (2012), Paschou et al. (2018)

autoimmune disorders exist. Some autoimmune diseases, along with the affected genes and signaling molecules, are listed in Table 5.5.

REFERENCES

Abbas, A. K. (1989). Antigen presentation by B lymphocytes: Mechanisms and functional significance. *Seminars in Immunology*, *1*(1), 5–12.

Abbas, M., Moussa, M., & Akel, H. (2024). Type I hypersensitivity reaction. In *StatPearls*. StatPearls Publishing. http://www.ncbi.nlm.nih.gov/books/NBK560561/

Abram, C. L., & Lowell, C. A. (2007). The expanding role for ITAM-based signaling pathways in immune cells. *Science's STKE*, *2007*(377), re2. https://doi.org/10.1126/stke.3772007re2

Alberts, B., Johnson, A., Lewis, J., Raff, M., Roberts, K., & Walter, P. (2002). Helper T cells and lymphocyte activation. In *Molecular Biology of the Cell* (4th ed.). Garland Science. https://www.ncbi.nlm.nih.gov/books/NBK26827/

Alcami, A., & Lira, S. A. (2010). Modulation of chemokine activity by viruses. *Current Opinion in Immunology*, *22*(4), 482–487. https://doi.org/10.1016/j.coi.2010.06.004

An immuno-diffusion assay to assess the protective antigen content of anthrax vaccine—PubMed. (n.d.). Retrieved January 25, 2024, from https://pubmed.ncbi.nlm.nih.gov/15908061/

An Introduction to Antibodies: Antibody-Antigen Interaction. (n.d.-b). Retrieved January 23, 2024, from https://www.sigmaaldrich.com/PK/en/technical-documents/technical-article/protein-biology/elisa/antibody-antigen-interaction

Anaya, J.-M., Castiblanco, J., & Lessard, C. J. (2013). Non-HLA genes and autoimmune diseases. In *Autoimmunity: From Bench to Bedside [Internet]*. El Rosario University Press. https://www.ncbi.nlm.nih.gov/books/NBK459446/

Ando, M., Ito, M., Srirat, T., Kondo, T., & Yoshimura, A. (2020). Memory T cell, exhaustion, and tumor immunity. *Immunological Medicine*, *43*(1), 1–9. https://doi.org/10.1080/25785826.2019.1698261

Arasanz, H., Gato-Cañas, M., Zuazo, M., Ibañez-Vea, M., Breckpot, K., Kochan, G., & Escors, D. (2017). PD1 signal transduction pathways in T cells. *Oncotarget*, *8*(31), 51936–51945. https://doi.org/10.18632/oncotarget.17232

Armstrong, R. D., Panayi, G. S., & Welsh, K. I. (1983). Histocompatibility antigens in psoriasis, psoriatic arthropathy, and ankylosing spondylitis. *Annals of the Rheumatic Diseases*, *42*(2), 142–146.

Banerjee, A., Falzarano, D., Rapin, N., Lew, J., & Misra, V. (2019). Interferon regulatory factor 3-mediated signaling limits middle-east respiratory syndrome (MERS) coronavirus propagation in cells from an insectivorous bat. *Viruses*, *11*(2), E152. https://doi.org/10.3390/v11020152

Bezbradica, J. S., Rosenstein, R. K., DeMarco, R. A., Brodsky, I., & Medzhitov, R. (2014). A role for the ITAM signaling module in specifying cytokine-receptor functions. *Nature Immunology*, *15*(4), Article 4. https://doi.org/10.1038/ni.2845

Bonilla, F. A., & Oettgen, H. C. (2010). Adaptive immunity. *The Journal of Allergy and Clinical Immunology*, *125*(2 Suppl 2), S33–S40. https://doi.org/10.1016/j.jaci.2009.09.017

Bournazos, S., Wang, T. T., Dahan, R., Maamary, J., & Ravetch, J. V. (2017). Signaling by Antibodies: Recent Progress. *Annual Review of Immunology*, *35*(1), 285–311. https://doi.org/10.1146/annurev-immunol-051116-052433

Braat, H., Peppelenbosch, M. P., & Hommes, D. W. (2006). Immunology of Crohn's disease. *Annals of the New York Academy of Sciences*, *1072*(1), 135–154. https://doi.org/10.1196/annals.1326.039

Bretscher, P. A. (2022). The role of cytokines in determining the Th1/Th2 phenotype of an immune response: Coherence of the T cell response and the Cytokine implementation hypothesis. *Scandinavian Journal of Immunology*, *95*(1), e13110. https://doi.org/10.1111/sji.13110

Cano, R. L. E., & Lopera, H. D. E. (2013). Introduction to T and B lymphocytes. In *Autoimmunity: From Bench to Bedside [Internet]*. El Rosario University Press. https://www.ncbi.nlm.nih.gov/books/NBK459471/

Carey, E. J., Ali, A. H., & Lindor, K. D. (2015). Primary biliary cirrhosis. *The Lancet*, *386*(10003), 1565–1575. https://doi.org/10.1016/S0140-6736(15)00154-3

Chaplin, D. D. (2010). Overview of the immune response. *The Journal of Allergy and Clinical Immunology*, *125*(2 Suppl 2), S3–23. https://doi.org/10.1016/j.jaci.2009.12.980

Charles A Janeway, J., Travers, P., Walport, M., & Shlomchik, M. J. (2001a). T cell-mediated immunity. In *Immunobiology: The Immune System in Health and Disease* (5th ed.). Garland Science. https://www.ncbi.nlm.nih.gov/books/NBK10762/

Charles A Janeway, J., Travers, P., Walport, M., & Shlomchik, M. J. (2001b). The major histocompatibility complex and its functions. In *Immunobiology: The Immune System in Health and Disease* (5th ed.). Garland Science. https://www.ncbi.nlm.nih.gov/books/NBK27156/

Charles A Janeway, J., Travers, P., Walport, M., & Shlomchik, M. J. (2001c). The structure of a typical antibody molecule. In *Immunobiology: The Immune System in Health and Disease* (5th ed.). Garland Science. https://www.ncbi.nlm.nih.gov/books/NBK27144/

Chen, L., Deng, H., Cui, H., Fang, J., Zuo, Z., Deng, J., Li, Y., Wang, X., & Zhao, L. (2017). Inflammatory responses and inflammation-associated diseases in organs. *Oncotarget*, *9*(6), 7204–7218. https://doi.org/10.18632/oncotarget.23208

Chen, L., & Flies, D. B. (2013). Molecular mechanisms of T cell co-stimulation and co-inhibition. *Nature Reviews Immunology*, *13*(4), Article 4. https://doi.org/10.1038/nri3405

Chinen, J., & Shearer, W. T. (2010). Secondary immunodeficiencies, including HIV infection. *The Journal of Allergy and Clinical Immunology*, *125*(2 Suppl 2), S195–S203. https://doi.org/10.1016/j.jaci.2009.08.040

Crotty, S. (2014). T follicular helper cell differentiation, function, and roles in disease. *Immunity*, *41*(4), 529–542. https://doi.org/10.1016/j.immuni.2014.10.004

Cruz-Tapias, P., Castiblanco, J., & Anaya, J.-M. (2013). Major histocompatibility complex: Antigen processing and presentation. In *Autoimmunity: From Bench to Bedside [Internet]*. El Rosario University Press. https://www.ncbi.nlm.nih.gov/books/NBK459467/

Cui, A., Huang, T., Li, S., Ma, A., Pérez, J. L., Sander, C., Keskin, D. B., Wu, C. J., Fraenkel, E., & Hacohen, N. (2024). Dictionary of immune responses to cytokines at single-cell resolution. *Nature*, *625*(7994), Article 7994. https://doi.org/10.1038/s41586-023-06816-9

Dembic, Z. (2015). Chapter 6 - Cytokines of the immune system: Interleukins. In Z. Dembic (Ed.), *The Cytokines of the Immune System* (pp. 143–239). Academic Press. https://doi.org/10.1016/B978-0-12-419998-9.00006-7

Dennison, S., Moore, M. J., Fahlman, A., Moore, K., Sharp, S., Harry, C. T., Hoppe, J., Niemeyer, M., Lentell, B., & Wells, R. S. (2012). Bubbles in live-stranded dolphins. *Proceedings of the Royal Society B: Biological Sciences*, *279*(1732), 1396–1404. https://doi.org/10.1098/rspb.2011.1754

Deshmane, S. L., Kremlev, S., Amini, S., & Sawaya, B. E. (2009). Monocyte chemoattractant protein-1 (MCP-1): An overview. *Journal of Interferon & Cytokine Research*, *29*(6), 313–326. https://doi.org/10.1089/jir.2008.0027

Di Liberto, V., Mudò, G., & Belluardo, N. (2019). Crosstalk between receptor tyrosine kinases (RTKs) and G protein-coupled receptors (GPCR) in the brain: Focus on heteroreceptor complexes and related functional neurotrophic effects. *Neuropharmacology*, *152*, 67–77. https://doi.org/10.1016/j.neuropharm.2018.11.018

Fillatreau, S. (2018). B cells and their cytokine activities implications in human diseases. *Clinical Immunology*, *186*, 26–31. https://doi.org/10.1016/j.clim.2017.07.020

Flow cytometry in biotechnology—PubMed. (n.d.). Retrieved January 25, 2024, from https://pubmed.ncbi.nlm.nih.gov/11549001/

Frank, S. A. (2002). Specificity and cross-reactivity. In *Immunology and Evolution of Infectious Disease*. Princeton University Press. https://www.ncbi.nlm.nih.gov/books/NBK2396/

Fujihashi, K., Boyaka, P. N., & McGhee, J. R. (2008). 19—Host defenses at mucosal surfaces. In R. R. Rich, T. A. Fleisher, W. T. Shearer, H. W. Schroeder, A. J. Frew, & C. M. Weyand (Eds.), *Clinical Immunology* (3rd ed., pp. 287–303). Mosby. https://doi.org/10.1016/B978-0-323-04404-2.10019-3

Garlanda, C., Dinarello, C. A., & Mantovani, A. (2013). The Interleukin-1 family: Back to the future. *Immunity*, *39*(6), 1003–1018. https://doi.org/10.1016/j.immuni.2013.11.010

Harvanová, G., Duranková, S., & Bernasovská, J. (2023). The role of cytokines and chemokines in the inflammatory response. *Alergologia Polska - Polish Journal of Allergology*, *10*(3), 210–219. https://doi.org/10.5114/pja.2023.131708

Harwood, N. E., & Batista, F. D. (2010). Early events in B cell activation. *Annual Review of Immunology*, *28*(1), 185–210. https://doi.org/10.1146/annurev-immunol-030409-101216

Hata, A., & Chen, Y.-G. (2016). TGF-β signaling from receptors to smads. *Cold Spring Harbor Perspectives in Biology*, *8*(9), a022061. https://doi.org/10.1101/cshperspect.a022061

Heesters, B. A., Myers, R. C., & Carroll, M. C. (2014). Follicular dendritic cells: Dynamic antigen libraries. *Nature Reviews Immunology*, *14*(7), Article 7. https://doi.org/10.1038/nri3689

Hilliard, B. A., Mason, N., Xu, L., Sun, J., Lamhamedi-Cherradi, S.-E., Liou, H.-C., Hunter, C., & Chen, Y. H. (2002). Critical roles of c-Rel in autoimmune inflammation and helper T cell differentiation. *The Journal of Clinical Investigation*, *110*(6), 843–850. https://doi.org/10.1172/JCI15254

Hirayama, D., Iida, T., & Nakase, H. (2017). The phagocytic function of macrophage-enforcing innate immunity and tissue homeostasis. *International Journal of Molecular Sciences*, *19*(1), 92. https://doi.org/10.3390/ijms19010092

Huebner, J. (2004). Antibody-antigen interactions and measurements of immunologic reactions. In *Immunology, Infection, and Immunity* (pp. 207–232). John Wiley & Sons, Ltd. https://doi.org/10.1128/9781555816148.ch9

Hughes, C. E., & Nibbs, R. J. B. (2018). A guide to chemokines and their receptors. *The Febs Journal*, *285*(16), 2944–2971. https://doi.org/10.1111/febs.14466

Immunochemical methods | SpringerLink. (n.d.). Retrieved January 25, 2024, from https://link.springer.com/chapter/10.1007/978-94-011-1552-0_12

Immunofluorescence techniques—PubMed. (n.d.). Retrieved January 25, 2024, from https://pubmed.ncbi.nlm.nih.gov/23299451/

Isomäki, P., Junttila, I., Vidqvist, K.-L., Korpela, M., & Silvennoinen, O. (2015). The activity of JAK-STAT pathways in rheumatoid arthritis: Constitutive activation of STAT3 correlates with interleukin 6 levels. *Rheumatology*, *54*(6), 1103–1113. https://doi.org/10.1093/rheumatology/keu430

Jaitley, S., & Saraswathi, T. (2012). Pathophysiology of Langerhans cells. *Journal of Oral and Maxillofacial Pathology: JOMFP*, *16*(2), 239–244. https://doi.org/10.4103/0973-029X.99077

Jonsson, A. H., & Yokoyama, W. M. (2009). Natural killer cell tolerance licensing and other mechanisms. *Advances in Immunology*, *101*, 27–79. https://doi.org/10.1016/S0065-2776(08)01002-X

Joshi, N. S., & Kaech, S. M. (2008). Effector CD8 T cell development: A balancing act between memory cell potential and terminal differentiation. *Journal of Immunology (Baltimore, Md.: 1950)*, *180*(3), 1309–1315. https://doi.org/10.4049/jimmunol.180.3.1309

Justiz Vaillant, A. A., & Qurie, A. (2023). Interleukin. In *StatPearls*. StatPearls Publishing. http://www.ncbi.nlm.nih.gov/books/NBK499840/

Justiz Vaillant, A. A., & Qurie, A. (2024). Immunodeficiency. In *StatPearls*. StatPearls Publishing. http://www.ncbi.nlm.nih.gov/books/NBK500027/

Kahan, B. D. (2008). 80 - Concepts and challenges in transplantation: Rejection, immunosuppression and tolerance. In R. R. Rich, T. A. Fleisher, W. T. Shearer, H. W. Schroeder, A. J. Frew, & C. M. Weyand (Eds.), *Clinical Immunology* (3rd ed., pp. 1199–1214). Mosby. https://doi.org/10.1016/B978-0-323-04404-2.10080-6

Kaufmann, S. H. E. (1999). Cell-mediated immunity: Dealing a direct blow to pathogens. *Current Biology*, *9*(3), R97–R99. https://doi.org/10.1016/S0960-9822(99)80059-1

Kennedy, A. D., & DeLeo, F. R. (2009). Neutrophil apoptosis and the resolution of infection. *Immunologic Research*, *43*(1–3), 25–61. https://doi.org/10.1007/s12026-008-8049-6

König, R., Fleury, S., & Germain, R. N. (1996). The structural basis of CD4-MHC class II interactions: Coreceptor contributions to T cell receptor antigen recognition and oligomerization-dependent signal transduction. *Current Topics in Microbiology and Immunology, 205*, 19–46. https://doi.org/10.1007/978-3-642-79798-9_2

Krüger, R., Kuhn, W., Müller, T., Woitalla, D., Graeber, M., Kösel, S., Przuntek, H., Epplen, J. T., Schöls, L., & Riess, O. (1998). Ala30Pro mutation in the gene encoding alpha-synuclein in Parkinson's disease. *Nature Genetics, 18*(2), 106–108. https://doi.org/10.1038/ng0298-106

Lahita, R. G., Tsokos, G., Buyon, J. P., & Koike, T. (2010). *Systemic Lupus Erythematosus*. Academic Press.

Laufer, J. M., & Legler, D. F. (2018). Beyond migration—Chemokines in lymphocyte priming, differentiation, and modulating effector functions. *Journal of Leukocyte Biology, 104*(2), 301–312. https://doi.org/10.1002/JLB.2MR1217-494R

Li, Q., Lei, X., Zhu, J., Zhong, Y., Yang, J., Wang, J., & Tan, H. (2023). Radiotherapy/chemotherapy-immunotherapy for cancer management: From mechanisms to clinical implications. *Oxidative Medicine and Cellular Longevity, 2023*, 7530794. https://doi.org/10.1155/2023/7530794

Liu, C., Chu, D., Kalantar-Zadeh, K., George, J., Young, H. A., & Liu, G. (2021). Cytokines: From clinical significance to quantification. *Advanced Science, 8*(15), 2004433. https://doi.org/10.1002/advs.202004433

Liu, K. (2016). Dendritic cells. *Encyclopedia of Cell Biology*, 741–749. https://doi.org/10.1016/B978-0-12-394447-4.30111-0

Mario Geysen, H. (1985). Antigen-antibody interactions at the molecular level: Adventures in peptide synthesis. *Immunology Today, 6*(12), 364–369. https://doi.org/10.1016/0167-5699(85)90096-9

Márquez, A., & Martín, J. (2022). Genetic overlap between type 1 diabetes and other autoimmune diseases. *Seminars in Immunopathology, 44*(1), 81–97. https://doi.org/10.1007/s00281-021-00885-6

McCusker, C., Upton, J., & Warrington, R. (2018). Primary immunodeficiency. *Allergy, Asthma, and Clinical Immunology : Official Journal of the Canadian Society of Allergy and Clinical Immunology, 14*(Suppl 2), 61. https://doi.org/10.1186/s13223-018-0290-5

McCusker, C., & Warrington, R. (2011). Primary immunodeficiency. *Allergy, Asthma, and Clinical Immunology : Official Journal of the Canadian Society of Allergy and Clinical Immunology, 7*(Suppl 1), S11. https://doi.org/10.1186/1710-1492-7-S1-S11

McFarland, H. F., & Martin, R. (2007). Multiple sclerosis: A complicated picture of autoimmunity. *Nature Immunology, 8*(9), Article 9. https://doi.org/10.1038/ni1507

Merle, N. S., Church, S. E., Fremeaux-Bacchi, V., & Roumenina, L. T. (2015). Complement system part I – molecular mechanisms of activation and regulation. *Frontiers in Immunology, 6*. https://www.frontiersin.org/articles/10.3389/fimmu.2015.00262

Mian, I. S., Bradwell, A. R., & Olson, A. J. (1991). Structure, function and properties of antibody binding sites. *Journal of Molecular Biology, 217*(1), 133–151. https://doi.org/10.1016/0022-2836(91)90617-f

Nair, A., Chauhan, P., Saha, B., & Kubatzky, K. F. (2019). Conceptual evolution of cell signaling. *International Journal of Molecular Sciences, 20*(13), Article 13. https://doi.org/10.3390/ijms20133292

Neurath, A. R. (2008). Immune response to viruses: Antibody-mediated immunity. *Encyclopedia of Virology*, 56–70. https://doi.org/10.1016/B978-012374410-4.00591-4

Nicholson, L. B. (2016). The immune system. *Essays in Biochemistry, 60*(3), 275–301. https://doi.org/10.1042/EBC20160017

Okoye, A. A., & Picker, L. J. (2013). CD4+ T cell depletion in HIV infection: Mechanisms of immunological failure. *Immunological Reviews, 254*(1), 54–64. https://doi.org/10.1111/imr.12066

Paschou, S. A., Papadopoulou-Marketou, N., Chrousos, G. P., & Kanaka-Gantenbein, C. (2018). On type 1 diabetes mellitus pathogenesis. *Endocrine Connections, 7*(1), R38–R46. https://doi.org/10.1530/EC-17-0347

Pisetsky, D. S. (2023). Pathogenesis of autoimmune disease. *Nature Reviews Nephrology, 19*(8), Article 8. https://doi.org/10.1038/s41581-023-00720-1

Poulsen, L. K., & Hummelshoj, L. (2007). Triggers of IgE class switching and allergy development. *Annals of Medicine, 39*(6), 440–456. https://doi.org/10.1080/07853890701449354

Probing Antibody-Antigen Interactions—PubMed. (n.d.). Retrieved January 25, 2024, from https://pubmed.ncbi.nlm.nih.gov/26082105/

Prud'homme, G. J. (2007). Pathobiology of transforming growth factor β in cancer, fibrosis and immunologic disease, and therapeutic considerations. *Laboratory Investigation, 87*(11), Article 11. https://doi.org/10.1038/labinvest.3700669

Ratajczak, W., Niedźwiedzka-Rystwej, P., Tokarz-Deptuła, B., & Deptuła, W. (2018). Immunological memory cells. *Central-European Journal of Immunology, 43*(2), 194–203. https://doi.org/10.5114/ceji.2018.77390

Rawlings, D. J., Schwartz, M. A., Jackson, S. W., & Meyer-Bahlburg, A. (2012). Integration of B cell responses through Toll-like receptors and antigen receptors. *Nature Reviews Immunology*, *12*(4), Article 4. https://doi.org/10.1038/nri3190

Reverberi, R., & Reverberi, L. (2007). Factors affecting the antigen-antibody reaction. *Blood Transfusion*, *5*(4), 227–240. https://doi.org/10.2450/2007.0047-07

Reyes-Robles, T., Alonzo, F., Kozhaya, L., Lacy, D. B., Unutmaz, D., & Torres, V. J. (2013). Staphylococcus aureus leukotoxin ED targets the chemokine receptors CXCR1 and CXCR2 to kill leukocytes and promote infection. *Cell Host & Microbe*, *14*(4), 453–459. https://doi.org/10.1016/j.chom.2013.09.005

Richard-Miceli, C., & Criswell, L. A. (2012). Emerging patterns of genetic overlap across autoimmune disorders. *Genome Medicine*, *4*(1), 6. https://doi.org/10.1186/gm305

Rosales, C., & Uribe-Querol, E. (2017). Phagocytosis: A fundamental process in immunity. *BioMed Research International*, *2017*. https://doi.org/10.1155/2017/9042851

Ruland, J., & Mak, T. W. (2003). From antigen to activation: Specific signal transduction pathways linking antigen receptors to NF-κB. *Seminars in Immunology*, *15*(3), 177–183. https://doi.org/10.1016/S1044-5323(03)00034-4

Sadegh-Nasseri, S., & Kim, A. (2015). Exogenous antigens bind MHC class II first, and are processed by Cathepsins Later. *Molecular Immunology*, *68*(2), 81–84. https://doi.org/10.1016/j.molimm.2015.07.018

Schroeder, H. W., & Cavacini, L. (2010). Structure and function of immunoglobulins. *The Journal of Allergy and Clinical Immunology*, *125*(2 Suppl 2), S41–S52. https://doi.org/10.1016/j.jaci.2009.09.046

Schumacher, T. N., & Schreiber, R. D. (2015). Neoantigens in cancer immunotherapy. *Science (New York, N.Y.)*, *348*(6230), 69–74. https://doi.org/10.1126/science.aaa4971

Schutkowski, M., & Reineke, U. (Eds.). (2009). *Epitope mapping protocols* (Vol. 524). Humana Press. https://doi.org/10.1007/978-1-59745-450-6

Shlomchik, M. J. (2008). Sites and stages of autoreactive B cell activation and regulation. *Immunity*, *28*(1), 18–28. https://doi.org/10.1016/j.immuni.2007.12.004

Smith-Garvin, J. E., Koretzky, G. A., & Jordan, M. S. (2009). T cell activation. *Annual Review of Immunology*, *27*(1), 591–619. https://doi.org/10.1146/annurev.immunol.021908.132706

Stone, K. D., Prussin, C., & Metcalfe, D. D. (2010). IgE, mast cells, basophils, and eosinophils. *The Journal of Allergy and Clinical Immunology*, *125*(2 Suppl 2), S73–S80. https://doi.org/10.1016/j.jaci.2009.11.017

Taher, T. E., Bystrom, J., Ong, V. H., Isenberg, D. A., Renaudineau, Y., Abraham, D. J., & Mageed, R. A. (2017). Intracellular B lymphocyte signalling and the regulation of humoral immunity and autoimmunity. *Clinical Reviews in Allergy & Immunology*, *53*(2), 237–264. https://doi.org/10.1007/s12016-017-8609-4

Targan, S. R., & Karp, L. C. (2005). Defects in mucosal immunity leading to ulcerative colitis. *Immunological Reviews*, *206*(1), 296–305. https://doi.org/10.1111/j.0105-2896.2005.00286.x

Taylor, A., Verhagen, J., Blaser, K., Akdis, M., & Akdis, C. A. (2006). Mechanisms of immune suppression by interleukin-10 and transforming growth factor-β: The role of T regulatory cells. *Immunology*, *117*(4), 433–442. https://doi.org/10.1111/j.1365-2567.2006.02321.x

Taylor, C. W., Chakrabarty, S., Schauder, K. S., & Yeoman, L. C. (1983). Identification of cytosolic antigens from GW-39 adenocarcinoma cells by crossed immunoelectrophoresis and immunofluorescence. *Immunological Communications*, *12*(3), 315–329. https://doi.org/10.3109/08820138309050753

Tobón, G. J., Youinou, P., & Saraux, A. (2010). The environment, geo-epidemiology, and autoimmune disease: Rheumatoid arthritis. *Autoimmunity Reviews*, *9*(5), A288–A292. https://doi.org/10.1016/j.autrev.2009.11.019

Turvey, S. E., & Broide, D. H. (2010). Chapter 2: Innate immunity. *The Journal of Allergy and Clinical Immunology*, *125*(2 Suppl 2), S24–S32. https://doi.org/10.1016/j.jaci.2009.07.016

Tymoshok, N. O., Lazarenko, L. M., Bubnov, R. V., Shynkarenko, L. N., Babenko, L. P., Mokrozub, V. V., Melnichenko, Y. A., & Spivak, M. Y. (2014). New aspects the regulation of immune response through balance Th1/Th2 cytokines. *EPMA Journal*, *5*(1), A134. https://doi.org/10.1186/1878-5085-5-S1-A134

Vazquez, M. I., Catalan-Dibene, J., & Zlotnik, A. (2015). B cells responses and cytokine production are regulated by their immune microenvironment. *Cytokine*, *74*(2), 318–326. https://doi.org/10.1016/j.cyto.2015.02.007

Wang, H., Chen, W., Li, D., Yin, X., Zhang, X., Olsen, N., & Zheng, S. G. (2017). Vitamin D and chronic diseases. *Aging and Disease*, *8*(3), 346–353. https://doi.org/10.14336/AD.2016.1021

Wang, J., Zhou, Y., Zhang, H., Hu, L., Liu, J., Wang, L., Wang, T., Zhang, H., Cong, L., & Wang, Q. (2023). Pathogenesis of allergic diseases and implications for therapeutic interventions. *Signal Transduction and Targeted Therapy*, *8*(1), Article 1. https://doi.org/10.1038/s41392-023-01344-4

Webb, G. J., Siminovitch, K. A., & Hirschfield, G. M. (2015). The immunogenetics of primary biliary cirrhosis: A comprehensive review. *Journal of Autoimmunity*, *64*, 42–52. https://doi.org/10.1016/j.jaut.2015.07.004

Willson, T. A., Kuhn, B. R., Jurickova, I., Gerad, S., Moon, D., Bonkowski, E., Carey, R., Collins, M., Xu, H., Jegga, A. G., Guthery, S. L., & Denson, L. A. (2012). STAT3 genotypic variation and cellular STAT3 activation and colon leukocyte recruitment in pediatric crohn disease. *Journal of Pediatric Gastroenterology and Nutrition*, *55*(1), 32–43. https://doi.org/10.1097/MPG.0b013e318246be78

Wynne, J. W., Woon, A. P., Dudek, N. L., Croft, N. P., Ng, J. H. J., Baker, M. L., Wang, L.-F., & Purcell, A. W. (2016). Characterization of the antigen processing machinery and endogenous peptide presentation of a bat MHC class I molecule. *Journal of Immunology (Baltimore, Md.: 1950)*, *196*(11), 4468–4476. https://doi.org/10.4049/jimmunol.1502062

Xue, V. W., Chung, J. Y.-F., Córdoba, C. A. G., Cheung, A. H.-K., Kang, W., Lam, E. W.-F., Leung, K.-T., To, K.-F., Lan, H.-Y., & Tang, P. M.-K. (2020). Transforming growth factor-β: A multifunctional regulator of cancer immunity. *Cancers*, *12*(11), 3099. https://doi.org/10.3390/cancers12113099

Yang, F., Wen, W., & Qin, W. (2017). Bispecific antibodies as a development platform for new concepts and treatment strategies. *International Journal of Molecular Sciences*, *18*(1), Article 1. https://doi.org/10.3390/ijms18010048

Young, R. M., Wu, T., Schmitz, R., Dawood, M., Xiao, W., Phelan, J. D., Xu, W., Menard, L., Meffre, E., Chan, W.-C. C., Jaffe, E. S., Gascoyne, R. D., Campo, E., Rosenwald, A., Ott, G., Delabie, J., Rimsza, L. M., & Staudt, L. M. (2015). Survival of human lymphoma cells requires B-cell receptor engagement by self-antigens. *Proceedings of the National Academy of Sciences*, *112*(44), 13447–13454. https://doi.org/10.1073/pnas.1514944112

Zhang, J.-M., & An, J. (2007). Cytokines, inflammation and pain. *International Anesthesiology Clinics*, *45*(2), 27–37. https://doi.org/10.1097/AIA.0b013e318034194e

Zhang, T., Franklin, A., Boboila, C., McQuay, A., Gallagher, M. P., Manis, J. P., Khamlichi, A. A., & Alt, F. W. (2010). Downstream class switching leads to IgE antibody production by B lymphocytes lacking IgM switch regions. *Proceedings of the National Academy of Sciences*, *107*(7), 3040–3045. https://doi.org/10.1073/pnas.0915072107

6 Immunological and Neuron Synapses

Muhammad Salman Khalid[1], *Somia Mazhar*[1], *Ebtesam A Al-Suhaimi*[3,4], *and Adeeb Shehzad*[2]
[1]Department of Biomedical Engineering and Sciences, School of Mechanical and Manufacturing Engineering, National University of Sciences and Technology, Islamabad
[2]Biodiversity Unit, Research Center, Dhofar University, Oman
[3]Vice presidency for Scientific Research and Innovation, Imam Abdulrahman bin Faisal University Saudi Arabia
[4]King Abdulaziz and his Companions Foundation for Giftedness and Creativity "Mawhiba" Riyad, Saudi Arabia

6.1 INTRODUCTION

A synapse is a stable adhesive connection that allows information transfer between two cells via controlled secretion. These unique cell-surface interactions are used by the immune and nervous systems to directly transmit and transduce highly regulated secretory signals among their component cell populations. Each synapse is based on a microdomain structure consisting of adhesion domains surrounding the primary active zones of endocytosis and exocytosis. The functional status of the synapse is modulated by surface molecules that may be integrated into and surround the active zones [1]. While synapses have been known as the central point of communication between neurons for well over a century, their immunological equivalent has only recently been discovered. Within the immune system, synapses serve to coordinate cell migration and antigen detection during immune response induction, as well as provide specificity to otherwise nonspecific soluble substances by confining them to the synaptic cleft [2, 3].

The term "synapse" refers to the close interaction between individual cells in both immunological and neuronal systems. In the nervous system, chemical synapses are sites where two cells can attach without fusing together. Moreover, they serve as sites for vectorial communication, polarity, and stability. The term "immune synapse," originally used to characterize interactions between T and B cells, components of the adaptive immune system, has now been expanded to include innate immunity cells such as natural killer (NK) cells and, more recently, phagocytes [4]. Specialized signaling areas known as immunological synapses (ISs) are characterized by complex molecular segregation and clustering at the interface of immune system cells. They interact with T cells and form different immunological synapses based on their state of activation and the antigen-presenting cells. These include the T-cell receptor (TCR), adhesion molecules, co-activation molecules, an associated intracellular complex that facilitates signal transmission, and several cytoskeletal components [5].

Synapses are the sites on neurons where connections are made and messages are exchanged. A single neuron can form hundreds of thousands or even millions of synaptic connections with neighboring neurons or other neurons in the same region of the brain. At one end, an electrical signal is converted into a chemical stimulus (neurotransmitter), while the other end has specialized receptors. The neurotransmitter diffuses across the synaptic cleft and binds to the receptor. The type of chemical released and the specific receptor binds to are critical in determining the quality and intensity of information transferred between neurons [6].

 DOI: 10.1201/9781003594901-6

This chapter focuses on immunological and neuron synapses, their structure and composition, and the pathways involved. The main topics discussed are the junctions where immune cells meet with the neurons via soluble mediators and modulators, termed neurotransmitters, cytokines, and chemokines. The chapter also explores how signaling is carried out at these junctions and how aberrant signaling can lead to various diseases related to the central nervous system (CNS).

6.2 IMMUNOLOGICAL SYNAPSES

The immunological synapse (IS) is an interface between T cells and specialized antigen-presenting cells. This structure facilitates the presentation of both self-derived and foreign molecular fragments, known as antigens, via certain surface molecules called major histocompatibility complexes (MHCs). Although T-cell-reliant immune responses help prevent cancer in the host, they can also play a role in harmful autoimmunity. T-cell polarization, TCR engagement, and the development of a specialized junction result from T-cell locomotion suppression. This combination of a specialized junction, cell polarization, and positional stability strongly resembles the conventional nervous system synapse. During migration, T cells can also integrate signals through the antigen receptor; this mobile antigen recognition structure is known as immunological synapse [7, 8].

T cells and antigen-presenting cells (APCs) form a junction characterized by closely spaced membrane appositions separated by "synaptic" regions. When TCRs bind peptide/MHC complexes, signaling occurs via a complex of proteins on T cells, including the transmembrane proteins CD3 and CD4. Additionally, other receptor-ligand pairs, such as CD28-B7 and CD40L-CD40, also concentrate at this interface. Synapses are also formed between NK cells and dendritic cells, as well as between T cells and other T cells. Successive synapses between immune system components likely contribute to the specificity of the immune response, enabling individual cells to eventually produce a "collective" response [8, 9].

6.2.1 Architecture of the Immunological Synapse—First Half of Synapse

The IS framework comprises several elements, including TCR, various cytoskeletal components, an associated intracellular complex, and adhesion molecules. These molecules cluster at the IS through a combination of large-scale molecular segregation at the micrometer scale and smaller-scale submicronic segregation. Many synapses display typical configurations with the TCR and protein kinase C (PKC) grouped in a core cluster surrounded by a ring of adhesion molecules. This arrangement may occasionally be preceded by a synapse with the TCR at the perimeter and integrins grouped in the core. The immunological synapse integrates several major types of receptors, including adhesion receptors, costimulatory receptors, checkpoint inhibitor receptors, and TCR microclusters. The description of these receptors and their associated signal transduction is provided below [10, 11].

6.2.1.1 T Cell Antigen Receptors

Understanding the interaction between TCRs and MHC-peptide complexes is crucial for determining the specificity of the immune response and the origin of antigens in both cellular and humoral responses. TCR genes undergo a rearrangement process that is comparable to that of antibody genes. Naïve T cells, with subunits encoded by the α and β genes, form a diverse repertoire capable of combating pathogens and tumors. T cells with subunits encoded by the γ and δ genes are specialized cells that are not MHC-restricted. On the cell surface, the TCR consists of two distinct antigen recognition subunits, α β or γ δ, which are specific to each T cell. These subunits form a non-covalent complex with CD3ε–CD3δ, CD3ε–CD3γ, and CD3ζ–CD3ζ dimers that are not polymorphic [12]. Understanding the entire complex is crucial for surface expression and signal transduction, as it lacks inherent catalytic activity. The TCR complex contains ten immunotyrosine activation motifs (ITAMs) with paired tyrosine residues. Once phosphorylated, these residues can recruit the non-receptor tyrosine kinase ζ-associated protein of 70 kDa Zeta-chain-associated protein kinase (ZAP-70) [13].

There are two main categories of MHC that regulate antigen recognition. Class I MHC is responsible for presenting peptides produced within the cell by the proteasome to cytotoxic T cells. Class II MHC is responsible for presenting antigens brought into the cell through endocytosis to helper T cells. Cross-presentation, also known as cross-dressing, enables specialized APCs to present external antigens via the class I route. This mechanism is crucial for the development of cellular immunity against cancer [14].

Other MHC-like molecules present non-peptide antigenic structures. These include CDld presenting glycolipids to invariant natural killer T cells, and MR1 presenting bacterial metabolites to mucosa-associated invariant T (MAIT) cells. These naturally occurring T cells may be useful in the immune response against malignancies.

Conventional α β T cells have distinct CD8 and CD4 coreceptors. CD8 interacts with MHC class I conserved areas, whereas CD4 interacts with class II conserved regions. During T-cell development, thymocytes rearrange TCR genes and test their receptors' ability to interact with MHC-presented self-peptides. Strong interactions may induce apoptosis or produce thymic regulatory T (Treg) cells, which are essential in tumor immunology. Treg cells decrease anticancer responses by disrupting T-effector immunological synapses [15].

Conventional α β T cells and Tregs have distinct signaling pathways that may promote regulatory responses in autoimmunity and improve cancer therapy. Thymocyte survival requires a less stringent identification method, which favors the TCR repertoire for CD4 or CD8 MHC molecules. Weak self-recognition prepares T cells to detect foreign peptides by weakly recognizing peptide flanking regions and polymorphic α-helices of MHC proteins. Mature T cells receive survival signals. It is still debated whether germ-line-encoded TCRs have an evolutionary predisposition for identifying MHC molecules in various orientations or whether they learn to recognize them in the thymus. Antigen recognition in tumors may either facilitate or impede triggered responses [4].

Tumor-infiltrating lymphocytes (TILs) are CD8 cytotoxic T cells that can directly destroy tumor cells. The presence of CD8 T lymphocytes infiltrating tumors has been associated with improved prognosis in certain cancer types. Tumors may contain antigen-presenting cells, such as CD11c+ cells, which display characteristics similar to those of dendritic cells. Cancer vaccines aim to promote the development and specialization of T lymphocytes, enabling them to selectively target and eliminate tumors in clinical settings. In adoptive immunotherapy, T-cell specificity can be modified by introducing chimeric antigen receptors into patients' cells [16, 17].

6.2.1.2 Adhesion Receptors

Adhesion molecules play a crucial role in the precise process of antigen recognition necessary for tumor rejection. Several important discoveries were made using immunization and subsequent selection for monoclonal antibodies that inhibit cytotoxic T-cell killing. These include the identification of lymphocyte function-associated antigen-1 (now known as CD11a/CD18, commonly referred to as LFA-1), LFA-2 (now called CD2), and LFA-3 (now known as CD58) [16]. CD2 and CD58 form a heterophilic adhesion receptor pair. These immunoglobulin (Ig) superfamily members are similar in size to the TCR and MHC-peptide complex, suggesting CD2 and the TCR work together. Early T cell-B cell interactions are important for germinal center formation and the establishment of specific T-cell lineages [18].

The LFA-1 molecule has a unique α and β subunit shared by three additional cell-surface heterodimers, each containing a distinct α subunit with individual expression patterns. The first LFA-1 ligand, intercellular adhesion molecule-1 (ICAM-1 or CD54), was discovered in B cells lacking LFA-1. LFA-1 is an integrin, whereas ICAM-1 is an Ig superfamily member. LFA-1 and ICAM-1 form longer intercellular connections than the TCR [19].

Inflammatory cytokines including TNF, interleukin-1β, and IFN-γ can activate integrin ligands in stromal cells, which are found on all nucleated cells. Anti-LFA-1 mAbs are challenging to use in therapy because LFA-1 interacts with endothelial cells, which are required for cell localization and immunological synapses. Anti-LFA-1 mAbs approved for psoriasis have a narrow therapeutic window between disease reduction and infection susceptibility. Unlike antigen and costimulatory receptors, integrins and the signaling lymphocyte activation molecule (SLAM) family use the Src family kinase Fyn rather than Lck (62). LFA-1 and CD2 also interact with immunological synapse ligands at the outer edges. Interestingly, CD2-CD58 interactions are strong enough to be detected in the core region of the TCR [19, 20].

6.2.1.3 Costimulatory Receptors

Costimulatory receptors have minimal signaling or adhesive activity, but they can enhance local adhesion and signaling when combined with other stimuli, such as the TCR. Pathogens or tumors require boost stimulation to elicit a T-cell response. Tumor vaccines employ costimulatory ligands applied directly to tumor cells. When mature T cells are activated without costimulation, clonal deletion protects the host from immune responses to harmless environmental antigens. Antigen recognition without costimulation may result in a state of non-responsiveness, known as anergy [21].

CD28 is the primary costimulatory receptor, an Ig superfamily member with a homodimeric structure. It lacks enzymatic activity in its cytoplasmic domain but contains motifs that recruit and activate Lck. It also indirectly activates PKC-θ, a crucial PKC isoform in T cells. This activation contributes to the activation of nuclear factor kappa-light-chain-enhancer of activated B cells (NF-κB belongs to a family of transcription factors that are primarily involved in cell proliferation, immunity and apoptosis) transcription factors and promotes IL-2 production [22, 23]. CD28 function relies on the increased expression of B7-1 (CD80) and B7-2 (CD86) on APCs and their interaction in the immunological synapse. Stimulation via toll-like receptors (TLRs) and inflammatory cytokines upregulates both CD80 and CD86 on dendritic cells connecting innate and adaptive immunity. Some cancers can develop without activating the production of costimulatory ligands, suppressing tumor-specific T lymphocyte activity [24].

Inducible T-cell costimulator (ICOS) is indeed a secondary costimulatory receptor belonging to the immunoglobulin superfamily. It plays a crucial role in T cell activation and differentiation, particularly in T helper cell responses. Unlike CD28, naïve T cells do not express it, but it increases after T-cell activation and binds to the ICOS ligand (LICOS, CD275). LICOS is highly expressed on dendritic cells and B cells, and it can be induced on stromal cells by inflammatory cytokines. ICOS plays a crucial role in germinal center processes, particularly through its ability to activate phosphatidylinositol-3 kinase. In cancer immunotherapy, ICOS costimulation promotes Th17 differentiation in CD4 T cells. Other significant costimulatory receptors belong to the TNF receptor (TNFR) superfamily, including CD27, GITR (CD357), 4-1BB (CD137), and OX40 (CD134) [11, 25]. These receptors use TNFR-associated factors (TRAFs) and K63 ubiquitination to activate NF-κB2 transcription factors. Except for CD27, which is constitutively expressed, all costimulatory receptors are induced on conventional T cells hours to days after initial activation, participating in T cell effector and memory functions beyond the initial growth phase. Tregs constitutively express these receptors, providing a potential means to regulate negative control in the tumor microenvironment (TME) [26].

6.2.1.4 Checkpoint Inhibitor Receptors

Co-inhibitory or checkpoint receptors complement costimulatory receptors by using diverse negative signaling pathways. These pathways often recruit tyrosine phosphatases, such as SHP1 and SHP2, to mitigate tyrosine kinase cascades or other signaling pathways. Cytotoxic T-lymphocyte antigen-4 (CTLA-4, CD152) is a well-defined checkpoint inhibitor receptor that is quickly activated after T-cell activation. Its cytoplasmic domain contains a constitutively active internalization motif, preventing its accumulation on the cell surface unless it is ligated in the synapse or undergoes phosphorylation of the Y-based internalization motif. CTLA-4 has a higher binding affinity for CD80 than CD86. CTLA-4 is constitutively expressed in Tregs, and its interaction with the endocytic pathway results in the elimination of CD80 and CD86 from APCs, hampering CD28-dependent immune responses in other T cells. CTLA-4 binds to PKCη in Tregs. Anti-CTLA-4 monoclonal antibodies may facilitate melanoma therapy [27].

Programmed cell death protein-1 (PD-1, CD279) is another checkpoint receptor. PD-1 binding to PD-1 ligand 1 and 2 (PD-L1, CD274, and PD-L2, CD273) leads to its recruitment to the immunological synapse, influenced by the strength and quantity of the MHC peptide. Upon ligand interaction, PD-1 recruits SHP-2 tyrosine phosphatase, reducing early TCR signaling. PD-1's effect on immunological synapse stability varies: it destabilizes in autoimmune and delayed-type hypersensitivity responses, but promotes stability in in antiviral reactions. Synapse stabilization leads to immune paralysis, preventing T-cell responses or interactions with other targets. PD-1 is associated with T-cell exhaustion in chronic viral infection, potentially protecting the host from immunopathology caused by chronic immune response toxicity. However, bacteria and tumors may exploit this mechanism to evade detection. PD-1 is also present on Tregs and inhibits their function. Inhibiting PD-1 may enhance Treg activity, supporting the combination of anti-PD-1 and anti-CTLA-4 therapies, as the latter inhibits Treg function. Vista and signal regulatory protein (SIRP is a family of cell surface receptors that are involved in the regulation of immune responses. They initiate phenomenon's like cell adhesion, signal transduction, and immune regulation) are other checkpoint inhibitor receptors [28].

6.2.1.5 TCR Microclusters

The first visible structures at the interface are sub-micron TCR clusters that form in an F-actin-dependent manner when TCRs are ligated by MHC-peptide complexes on cells, anti-CD3 Abs on solid substrates, or MHC-peptide complexes and ICAM-1 on supported planar bilayers. While some argue that TCRs are pre-clustered on T cells, ligand-induced clusters exhibit many unique properties.

Ligand-induced microclusters recruit the Src family kinase Lck to the TCR, phosphorylating ITAMs and recruiting and activating ZAP-70. ZAP-70 phosphorylates the membrane-anchored linker of activated T cells (LAT), recruiting phospholipase C-γ (PLC-γ) and enhancing its activation via inducible T-cell kinase (ITK). PLC-γ produces diacylglycerol and inositol-1,3,5-trisphosphate, activating Ras, PKC, and calcium ion effector pathways, which stimulate T-cell proliferation and regulatory cytokines transcription, like IL-2. The TCR can also use a module that incorporates the guanine nucleotide exchange factor Vav, activating Notch, which directly regulates Myc transcription and T-cell proliferation without relying on IL-2 [29].

Intramolecular interactions regulate Src family kinases Fyn and Lck (the dominant kinase downstream of TCR). All Src family kinases have a C-terminal tyrosine (Y505 in Lck) acting as an intramolecular ligand for its Src homology 2 (SH2) domains. When the SH2 domain binds to this tyrosine and the SH3 domain interacts with the polyproline motif between the kinase and SH2 domains, the kinase is inhibited. C-terminal Src kinase (Csk) phosphorylates the tyrosine when recruited to membrane-bound Src substrates near the active Src kinase. The transmembrane tyrosine phosphatase CD45 counteracts this negative feedback. Csk inhibition leads to ligand-independent TCR signaling that works with CD28 signals. The absence of CD45 suppresses TCR signaling, though CD45 can also counteract Lck- and ZAP-70-mediated activating phosphorylation. Notably, CD45 is absent from TCR microclusters, and this reduction in CD45 density, likely due to the substantial size of its extracellular domain, seems crucial for optimal local phosphatase activity in TCR activation. Any adhesion system of modest size will produce interactions that result in CD45 exclusion. Therefore, it is likely that CD45 exclusion alone is insufficient for complete TCR signaling. However, some degree of exclusion appears necessary [29, 30].

6.2.2 Mechanism Underlying Antigenic Regulation

The microtubule organizing center (MTOC) is a cellular structure that organizes microtubules. In T cells, the MTOC reorients towards the immunological synapse upon antigen recognition, facilitating the polarization of the cell and the efficient delivery of secretory granules to the synapse. Its migration towards the contact site is believed to be crucial for the targeted release of cytotoxic granules by killer cells. After stable contact is established, the uropod disassembles, relocating the MTOC between the nucleus and the contact surface. This reorients the polarity axis towards the surface of the antigen-presenting cell, setting up the motility apparatus for extended contact periods (minutes to hours). While the precise mechanism behind MTOC reorientation remains unclear, recent findings indicate that diacylglycerol (DAG) may play a significant role. Diacylglycerol (DAG) is formed by PIP2 hydrolysis via the activation of phospholipase C. Its localized generation at the IS indicates activated receptor locations at the contact membrane, since DAG is restricted to the plasma membrane and has a very brief half-life. PLC-γ localization to the IS enables local DAG synthesis and potential recruitment of proteins with DAG-binding C1 domains. MTOC reorientation may be initiated only by sudden DAG formation at a specific T cell pole [31].

Microtubule motors are often necessary for microtubule movement. Microtubules grow rapidly at the "plus" end and more slowly at the "minus" end. The minus end is linked to the MTOC, whereas the plus end points towards the plasma membrane. Two groups have demonstrated dynein (a minus-end motor protein) recruitment to the IS and its role in MTOC repolarization. Dynein may be recruited to the IS via binding to the scaffold protein adaptor protein (ADAP is a crucial signaling molecule involved in T cell receptor (TCR) signaling), potentially linking connection dynein recruitment to integrin activation. Dynein induces MTOC reorientation by attracting microtubules and the MTOC towards the IS. The MTOC's proximity to the contact surface plays a crucial role in cytolytic granule release in cytotoxic T lymphocytes (CTLs are a type of T cell that plays a crucial role in the immune response by killing infected cells and cancer cells by releasing cytotoxic granules, which contain perforin and granzymes that targets cell membrane). This indicates that direct contact between the MTOC and the immunological synapse is essential for this process, since it seems to rely solely on end-directed motors [32].

6.2.3 Deciphering Antigen Presentation-Second Half of Synapse

The availability of MHC complexes on APCs is crucial for immunological synapse formation. APCs actively participate in the formation and translocation of complexes, but it remains unclear whether they remain active after MHC complexes bind to cell surfaces. Both T cells and B cells contribute to synapse formation. Each type of APC has unique features and characteristics that influence IS regulation. For example, dendritic cells activate and regulate naïve T lymphocytes, while other APCs are associated with the activation of memory or effector T cells. The uniqueness of each IS corresponds to changes in antigen processing, membrane organization, molecular expression, and active participation of these cells at distinct phases. Antigen presentation occurs via two pathways: MHC class I presents antigens to CD8+ cells, while MHC class II presents to CD4+ cells [33].

6.2.3.1 Role of Dendritic Cells

Dendritic cells (DCs) are antigen-presenting cells (APCs) that play a crucial role in shaping immune responses. DCs encounter antigens in peripheral tissues and migrate to lymph nodes, where they present antigen-derived peptides on major histocompatibility complex (MHC) molecules to T cells, thereby activating the adaptive immune response. Immature DCs retain intact MHC pools but express only a few until encountering a maturation signal. Antigens are invaginated by these DCs through micropinocytosis and via specific Fc and C-type lectin receptors. DCs mature upon sensing bacterial endotoxins (IL-1 and TNF-α), presenting antigens via mature MHC complexes, which move to the cell surface via contact-induced intracellular vesicles (CIIVs). Latrunculin A inhibits this movement, suggesting that CIIVs are constantly associated with the action cytoskeleton, allowing active transport to the surface. Upon maturation, DCs become dynamic, constantly changing their cytoskeleton to enhance motility. Although the exact mechanism of CIIV binding remains debated, T cells are believed to be indirectly involved [33, 34].

6.2.3.2 Role of MHC Complexes

MHC interaction with the cytoskeleton could contribute to motility and proper organization of the immunological synapse. Sub-micronic superstructures of these molecules facilitate efficient antigen presentation to T lymphocytes. Co-clustering of both MHC classes is associated with sessile arrangements of B and T cells in lymphoid organs and blood plasma, primarily contributing to efficient signal transduction and facilitating cell-cell interaction. The abundance of costimulatory peptides and tetraspanin family members further enhances the potential of these complexes in T-cell regulation and interaction. MHC complex oligomers may initially form in vesicles, but CD86 and MHC class I and II appear as colocalized protein aggregates on the DC surface. Cholesterol's role in the formation of these structures is being studied, highlighted by the presence of glycosylphosphatidylinositol (GPI)-anchored proteins (GPI-anchored proteins are involved in the protection from proteolytic cleavage, immune responses, cell proliferation, and cell adhesion) in the sub-cellular compartments, which are crucial for maintaining APC rafts. The tight packing of MHC complexes regulates signaling and ligand binding in APCs. MHC clustering in rafts enhances T cell activation, and upon binding a compatible receptor, these clusters may transduce signals to the APC cytoskeleton. The ligand-induced MHC interactions can dynamically influence cytoskeleton rearrangement, ultimately affecting the stability and formation of IS [33, 35].

6.3 NEURONAL SYNAPSES

Neuronal synapses are specific junctions between neighboring neurons that enable rapid end-to-end communication, allowing neurons to organize into clusters. Synapses are generally of two types: chemical (communication via chemicals) and electrical (communication via ion diffusion). Synapses not only transmit information but also process it. Due to their versatility (each synapse has different properties), predicting synaptic behavior is complex, necessitating a deep understanding of their components. Neurons fire bursts of action potential as spikes that vary in length and frequency. A postsynaptic neuron fires a spike after neurotransmitters bind to specific receptors. Understanding postsynaptic neuron identity is crucial for comprehending plasticity in short and distinct patterns fired from postsynaptic neurons, indicating their role in maintaining short-term plasticity at the junction. Besides synaptic signaling, neurons also engage in non-synaptic signaling mediated by modulatory neuropeptides and diffusible molecules that generally diffuse outside the synaptic cleft [6].

Synapses possess a wide range of properties based on release probability, neurotransmitter types, absence or presence of specific modulators, and postsynaptic receptor composition. Consequently, the brain contains hundreds of synapse types, creating a diverse "synaptome." Axons form synapses arranged like pearls on a string, allowing them to form many synapses with a few neurons or single synapses with many postsynaptic neurons.

6.3.1 Architecture of Neuronal Synapses—Essential Components

6.3.1.1 Presynaptic Terminals

Presynaptic terminals consist of neurons that transmit signals across the synaptic cleft to postsynaptic. They are specialized Ca^{2+}-triggered neurotransmitter release, while postsynaptic terminals contain receptors that bind neurotransmitters and start signaling cascades. Axons primarily form presynaptic terminals, though dendrites can occasionally contribute, as seen in the thalamus and olfactory bulb. Neurotransmitter release is mediated by a canonical machinery comprising SNARE proteins, Munc18, synaptotagmins, and active zone scaffold proteins like Rab3-interaction molecules (RIMs). This machinery facilitates membrane fusion and calcium-dependent release. Even synapses with unusual specializations utilize this canonical machinery [6, 36].

6.3.1.2 Postsynaptic Terminals

Postsynaptic terminals, or "receiving neurons," receive information via neurotransmitters that cross the synaptic cleft and bind to specific receptors. This binding enables postsynaptic neurons to generate their own action potentials. Four main receptor families are present on postsynaptic terminals:

1. Pentameric cys-loop receptors (GABA, nicotine acetylcholine, glycine, and ionotropic receptors)
2. Tetrameric glutamate receptors (NMDARs, AMPARs, and kainate receptors)
3. Metabotropic G protein-coupled receptors
4. Trimeric P2X receptors (ATP receptors)

Synapses primarily utilize pentameric or tetrameric receptors, while ATP receptor is rarely used for neurotransmission [6].

6.3.1.3 Synaptic Cleft

The synaptic cleft is a narrow gap between pre- and postsynaptic membranes. Presynaptic axon terminals contain synaptic vesicles filled with neurotransmitters, which are released into the synaptic cleft. After the release, these neurotransmitters then bind to specific receptors on the postsynaptic membrane. The synaptic cleft maintains separation between neurons while allowing for targeted neurotransmitter release and binding, facilitating signal conduction.

6.3.1.4 Chemical Messengers

Neurotransmitters and neuropeptides are the main chemical messengers in synaptic communication. They act as specific "keys" that unlock signaling cascades after receptor binding. Neurotransmitters are categorized into small molecule neurotransmitters (GABA, glutamate, ATP, adenosine, acetylcholine, and various catecholamines) and neuropeptides (combinations of three or more amino acids, including enkephalins, endorphins, neuropeptide Y, and substance P). These messengers can play excitatory, inhibitory, or modulatory roles depending on the context.

When a neurotransmitter binds to the receiving cell, it causes ion channels to open or close, altering the cell's membrane potential. This can result in either excitation or inhibition. When a cell generates its own action potential, the shift in membrane potential is called an excitatory postsynaptic potential (EPSP). Conversely, when a cell is unable to fire its action potential, it's termed an inhibitory postsynaptic potential (IPSP). Glutamate is the main excitatory neurotransmitter in the CNS, while GABA is associated with synaptic inhibition. EPSPs depolarize the inner membrane by allowing positive ion influx, bringing the membrane potential closer to the firing threshold. IPSPs counteract EPSPs by maintaining the membrane potential below the threshold, canceling the stimulatory effects of EPSPs [37].

6.3.2 Mechanism of Action-Release and Reception

6.3.2.1 Neurotransmitter Release

When an action potential reaches the presynaptic terminal, it opens voltage-gated calcium channels, causing the terminal to fuse with the membrane and releasing neurotransmitters into the synaptic cleft. Prior to release, vesicles are docked and ready to fuse with the membrane, but this process only occurs upon calcium influx. SNARE proteins mediate membrane fusion, regulated by complexin (synapsin) and synaptotagmins. Synaptotagmins are calcium-binding transmembrane proteins that regulate SNARE-mediated fusion, while complexin determines whether fusion proceeds. Calcium binding to synaptotagmins signals complexin to activate vesicle fusion with the presynaptic membrane, releasing neurotransmitters into the synaptic cleft [38].

6.3.2.2 Neurotransmitter Reception

After release into the synaptic cleft, neurotransmitters bind to receptors on the receiving end. This binding induces conformational changes in the receptors, activating signaling cascades that can either inhibit or excite the neuron. There are two main categories of receptors:

1. Ligand-activated ion channels (ionotropic receptors): Directly influenced by neurotransmitter binding (e.g., N-methyl-D-aspartate (NMDA) glutamate receptors)
2. Metabotropic receptors: Indirectly influenced by neurotransmitters, often G-protein coupled receptors (GPCRs) that activate second messenger systems like cAMP and IP3

Ionotropic receptors mediate fast synaptic transmission, while metabotropic receptors typically modulate slower, longer-lasting responses.

Ionotropic receptors are large protein complexes with binding sites for neurotransmitters and segments that form voltage-gated ion channels. They provide rapid physiological responses and ensure the breakdown and removal of neurotransmitters. Metabotropic receptors are only affected indirectly by neurotransmitters, as they are mostly influenced by the composition and activation of molecules, often linked with second messenger pathways. Their response time is slower as multiple signaling pathways are involved.

A classic example of synaptic transmission is the release and uptake of acetylcholine at the neuromuscular junction. This neurotransmitter is released from the presynaptic membrane after the fusion of vesicles to the membrane, and then into the synaptic cleft, where it attaches to specific receptors (either nicotinic or muscarinic) on the postsynaptic terminal of the receiving neuron. This attachment activates signaling cascades involved in muscle contraction through nerve impulses. Neurons also form junctions with astrocytes, glial cells, and macrophages, initiating inflammation and other defense mechanisms.

6.3.3 The Changing Brain-Synaptic Plasticity and Memory Formation

Synaptic plasticity refers to the experience-dependent changes in nerve impulse transmission that strengthen or weaken synapses over time. It is the cellular basis of memory and learning (cognitive functions). Impairment in synaptic plasticity is linked to many psychiatric and neurodegenerative anomalies, including Parkinson's disease, Alzheimer's disease, epilepsy, and autism spectrum disorder (ASD). Synaptic plasticity formation and maintenance primarily depend on the proper functioning of tropomyosin receptor kinase B (TrkB) in neurons, along with other receptors such as NMDA, α-amino-3-hydroxy-5-methyl-4-isoxazolepropionic acid (AMPA), neurotrophin-3 (NT-3), serotonin (5-HT), and brain-derived neurotrophic factor (BDNF). Excitatory synapses are located on the dendritic spines that cover excitatory neuron dendrites. These spines are extremely dynamic, constantly modulated by environmental stimuli, enabling brain adaptation [39].

Cdk5 and TrkB interact during memory formation. TrkB binding to BDNF causes Cdk5-mediated phosphorylation at the junction, activating Rho-GTPase Rac1 via recruitment of exchange factor Tiam1. This process, triggered by the release of glutamate, stimulates actin reorganization. Cdk5 activation also activates S6, a downstream effector of p70S6K, stimulating the formation of new proteins. Additionally, protein kinase A (PKA) plays a critical role in the development of synaptic plasticity. The melanocortin-4 receptor (MC4R), a cognate neuropeptide receptor, maintains hippocampal plasticity through PKA. MC4R activation enhances the regulation of AMPA-subtype glutamate receptors via PKA. Two key concepts in neural plasticity are long-term potentiation (LTP) and long-term depression (LTD) [40].

6.3.3.1 Long-Term Potentiation (LTP)

LTP is a process in which synaptic transmission is strengthened by repeated nerve impulse stimulation. This process is associated with memory formation and learning improvement. LTP lasts longer than LTD and is input-specific, involving a coincidence-detection mechanism. This mechanism relies on NMDA receptors, which regulate calcium ion movement across an established voltage gradient. NMDAR activation occurs only when two conditions are fulfilled: glutamate release from the presynaptic terminal and magnesium ion ejection from the depolarized postsynaptic terminal, which initially blocks the receptor's ionic pores. LTP occurs in the hippocampus (C1 area), activating and releasing enzymes like calcium/calmodulin-dependent kinase-2 (CaMK2) and cyclic AMP (cAMP). These kinases modulate the phosphorylation of proteins based on calcium ion levels, notably activating AMPAR receptors at the postsynaptic neuronal membrane. LTP maintenance involves the formation and modification of new proteins, promoting the expression of key transcription factors [41].

6.3.3.2 Long-Term Depression (LTD)

LTD is the opposite of long-term potentiation, weakening signal transmission across synapses. It is associated with synaptic pruning and memory weakening. LTD involves the activation of mGluRs receptors on the postsynaptic membrane, triggering calcium ion influx and calcineurin phosphorylation. Protein phosphatase activation leads to AMPA receptor internalization, weakening the synapse. The exclusion of AMPA receptors is also observed during LTD, related to the deregulation of key pathways [41].

6.4 INTERSECTION-NEUROIMMUNE INTERFACE

6.4.1 Components of Neuroimmune Synapse

The term "synapse" emphasizes close cell-cell interactions, involving direct communication, soluble mediators, and extracellular vesicles between various cellular systems. In the nervous system, chemical synapses are characterized by their role in maintaining polarity, stability, and vectorial communication, allowing two cells to adhere without fusion. Initially focused on the adaptative immune system, the concept of immune synapses now encompasses T and B cells, as well as NK cells and phagocytes [42]. While these cells form specific synapses at different levels, phagocytic synapses may be the oldest, serving as a template and often used as a general term in sense-based studies (junctions derived from phagocytic receptors).

In addition to immunological components, several neuronal components contribute to the development and stability of neuroimmune synapses, including neurotransmitters (which activate immune cells), cytokines, adhesion molecules, chemokines, and neuropeptides (which enhance immune cell differentiation). The main neurotransmitters involved are glutamate, γ-aminobutyric acid (GABA), and acetylcholine (ACh), while neurokinins and vasoactive intestinal peptides (VIP) are the main neuropeptides [4].

In the nervous system, even the smallest and simplest activities require the synergistic functioning of many parallel synapses in a well-coordinated fashion. Similarly, the immune response against pathogens involves a relay between adaptative and innate immunity. Most immune cells use similar mechanisms to target and kill invading pathogens, employing phosphatase exclusion. While classical neuronal synapses are well-established and well-defined, neuroimmune synapses are more complex due to their robust and heterogeneous nature. A key characteristic of these synapses is that they can form at various locations, notably in the meninges (protective layers on the brain), vascular spaces, blood-brain barrier (BBB), and choroid plexus [4, 43].

6.4.2 Insights into Neuroimmune Interactions

It is now understood that the immune system is not self-propelled but closely linked to the nervous system, with various hormones, neuropeptides, cytokines, and chemokines impacting immune cells. Locally, the nervous system utilizes circulating immune cells for protection against pathogens attempting to enter the brain or to cross the blood-brain barrier, while also benefiting from neuroendocrine mediators. Interactions between nerve endings and immune cells allow the immune system to attract neuronal elements for proper regulation of responses against circulating pathogens. Hormones and other chemical compounds actively participate in neuroimmune interactions, some of which are described below:

6.4.2.1 Role of Lymphoid Organs

Lymphoid organs facilitate direct cell-cell interactions between neurons and immune cells. The autonomic nervous system, particularly the sympathetic nervous system, is involved in immune regulation, forming "neuroeffectors junctions" near T cells. Nerve fibers in close contact with lymphoid organs contain localized neuropeptides like neuropeptide Y, Met-enkephalin, substance P, and neurotensin. In addition to neuronal acetylcholine expression, non-neuronal ACh expression has been reported in lymphocytes, which possess the necessary components for independent, non-neuronal ACh activation and muscarinic receptors. The expression of various components during helper T cell activation suggests that T cells can act in a paracrine or autocrine fashion [44].

The main lymphoid organs involved in neuroimmune response activation are the spleen, thymus, and lymph nodes. They play key roles in immune cell activation and communication, antigen presentation through APCs (modulated by hormone and neuropeptide release), and modulation of local inflammatory responses through cytokine production [45].

6.4.2.2 Role of Catecholamines

The adaptive immune system produces β2-adrenergic receptors. Their activation triggers a signaling cascade involving cAMP formation, which activates guanine exchange proteins or protein kinase A (PKA). PKA signals downstream targets via cytoplasmic kinases, notably p38 mitogen-activated protein kinase and cAMP response element-binding protein (CREB). These receptors can also activate β-arrestins, which connect major signaling pathways during inflammation (NF-κB and mitogen-activated protein kinases (MAPK is a family of kinases that are activated by external factors such as cytokines and stress signals and are involved in cell proliferation and differentiation) regulation). β2 activation is linked to increased retention time of immune cells, mediated by chemokine receptors like CXC and CC for T cells and B cells, respectively.

Chemokines and catecholamines acting on internal cell linings (adhesion molecules) modulate both types of immune responses. High adrenergic receptor tone activated by these chemicals helps prepare the innate immune response for potential infections [46]. Catecholamines, including dopamine, epinephrine, and norepinephrine, serve as immunomodulators. When produced, these neurotransmitters act in a paracrine or autocrine manner via adrenergic receptors. Like T cells, phagocytes can also produce such neurotransmitters, which positively modulate local inflammation and activate NF-κB signaling cascades through the overproduction of these mediators [45].

6.4.2.3 Role of Glucocorticoids

Glucocorticoids can have both anti- and proinflammatory effects in neuroimmune responses. They are typically found in the cytosol near heat-shock proteins, and their action on cells depends on binding to glucocorticoid receptors (GRs). Upon binding, the receptors translocate to the nucleus and form dimers, which mediate a local anti-inflammatory response by binding to response elements such as activator protein 1 (AP-1) and the interferon (IFN) pathway. This downregulates genes involved in regulating adhesion molecules, mediators, and key enzymes. In certain circumstances, glucocorticoids can also have a pro-inflammatory role that is secondary to the selection of corticoid-resistant cells such as Th17 cells, which overexpress multidrug-resistant proteins and IL-23 receptors. This proinflammatory action is also associated with an excessive release of anti-inflammatory factors by the immune system modulators. Thymic-derived glucocorticoids are linked to thymocyte development and the rate of new cell formation. Additionally, intestinal epithelial cells are also involved in extra-adrenal glucocorticoid synthesis, which plays both inhibitory and costimulatory roles in intestinal T-cell activation [45, 47].

6.4.2.4 Other Hormones and Neuropeptides

While IFN-α, ACTH, and endorphins play active roles, other molecules are also involved in neuroimmune crosstalk. Notably, leukocyte-derived adrenocorticotropic hormone (ACTH is a hormone that stimulates the production of cortisol by the adrenal cortex and it regulates blood pressure, immune functions, and metabolism) is produced by leukocyte proopiomelanocortin in the anterior pituitary gland. The signaling pathways of ACTH and opioid receptors are still under investigation, and opioid binding sites on leukocytes have recently been reported. Other precursors and their receptors include enkephalin (activates Met-enkephalin), prodynorphin (source of dynorphin), and preproenkephalin (responsible for Leu-enkephalin). Enkephalins positively regulate immunocytes at baseline and suppress activated immune cells, acting as the immune system's "tuners." During inflammation, opioid-containing immunocytes migrate to the inflamed area, producing β-endorphin, which attenuates pain by activating nerve receptors. Inflammatory mediators like IL-1β and norepinephrine enhance the release of these peptides. Growth hormone and prolactin also have immunomodulatory activities, as demonstrated in pre-clinical trials using rat models. Both B and T lymphocytes have specific binding sites for prolactin and growth hormone [45].

6.4.3 Organizing Information at Neuroimmune Synapses

Both immune and nervous systems utilize receptors in synapses in various ways. In the immune system, two important types of clusters are T-B synapses supported by the LFA-1 ring and integrin bounded by the TCR cluster (Bullseye pattern). Parallel studies of MHC-1 and ICAM-1 (LFA-1 ligand) have demonstrated a bilayer supported by CD2, which generates this pattern through active processes in T cells. LFA-1 forms the peripheral supramolecular structure, while TCR forms the central supramolecular activation cluster. F-actin mediates initial rapid contact through WAVE2, a Rac effector that activates formins and Arp2/3 complexes. Initial TCR and LFA-1 complexes are crucial for assembling supramolecular activation clusters (SMACs). TCR microclusters associate the CD2–CD58 adhesion system and CD28–CD80 costimulatory pathway [42].

Negative regulators, such as CTLA-4 and PD-1, are also part of these clusters in a ligand-dependent manner. Although spatially segregated, LFA-1 interaction improves TCR sensitivity to ligands 100-fold and increases calcium signaling duration. In the immunological synapse, LFA-1 combines with talin (an adaptor protein) in the ring, while TCR microclusters associate with spaces toward the synapse center. This process is mediated by the TSG101, essential for endosomal transport. TSG101 recognizes receptors with a single attached ubiquitin group. After recognition, TCR complexes are degraded by Cbl-b and Cbl-c ubiquitin ligases, activated by agonistic MHC-I class polypeptides. TSG101-supported TCR signaling maintains endosomal complexes essential for transporting the substance across gradients [42, 43].

6.4.4 Neuroimmune Crosstalk and Inflammation

The cholinergic anti-inflammatory reflex activates systemic and local immune responses. This neural reflex was first identified in endotoxemia. Macrophages sense lipopolysaccharides (LPS) secreted by bacteria and, upon activation, secrete a variety of proinflammatory cytokines, such as IL-1β, IL-18, TNF-α, and IL-6. When produced in a large amount, these cytokines can induce a potentially life-threatening cytokine storm. At this point, the nervous system intervenes as a brake. The vagus nerve's sensory afferent portion detects cytokines, and its efferent portion sends the signals back to the brain via a typical brain circuit to halt cytokine production (mainly TNF-α and IL-1β) [48]. Additionally, another complex mechanism activates as ACh attaches to its post-ganglionic neuronal receptor, triggering the release of norepinephrine. In this context, both splenic and vagus nerve stimulation inhibit LPS-induced inflammation. Beyond inflammation, neuroimmune interactions significantly impact intestinal homeostasis, tissue repair, adipose tissue homeostasis, and host defense against pathogens [49].

Vagus nerve activity inhibits several cytokines produced by splenic macrophages, crucial for maintaining immune system homeostasis and bypassing infections. The vagus nerve inhibits TNF-α, which initiates inflammation, via acetylcholine receptors on TNF-α-producing cells in the spleen. Studies have demonstrated that the spleen-vagus nerve connection is regulated by adrenergic neurons produced by the ganglion [46]. Choline-acetyltransferase forms synapses with these adrenergic neurons, and T-cell stimulation through this fusion initiates acetylcholine release, acting on TNF-α-producing cells. These immune synapses have been well documented using electron microscopy, with the synaptic cleft positioned close enough to arrest the freely moving T cells while excluding CD45. Additionally, synapses containing mast cells in combination with N-cadherin are important in allergy studies [42, 49].

Figure 6.1 illustrates the complex bidirectional crosstalk at the neuroimmune synapse. Brain stimulation of the vagus nerve is unidirectional (involving celiac plexus activation), while catecholaminergic and neuropeptide activation of cytokines, chemokines, and immune cells is bidirectional, following the downregulation of Treg cells by leptin from adipocytes.

6.5 CELL SIGNALING AT NEUROIMMUNE SYNAPSE

Neuroimmune synapses are specific areas of bidirectional communication between neurons and immune cells, contributing to homeostasis maintenance and neuronal function modulation. This

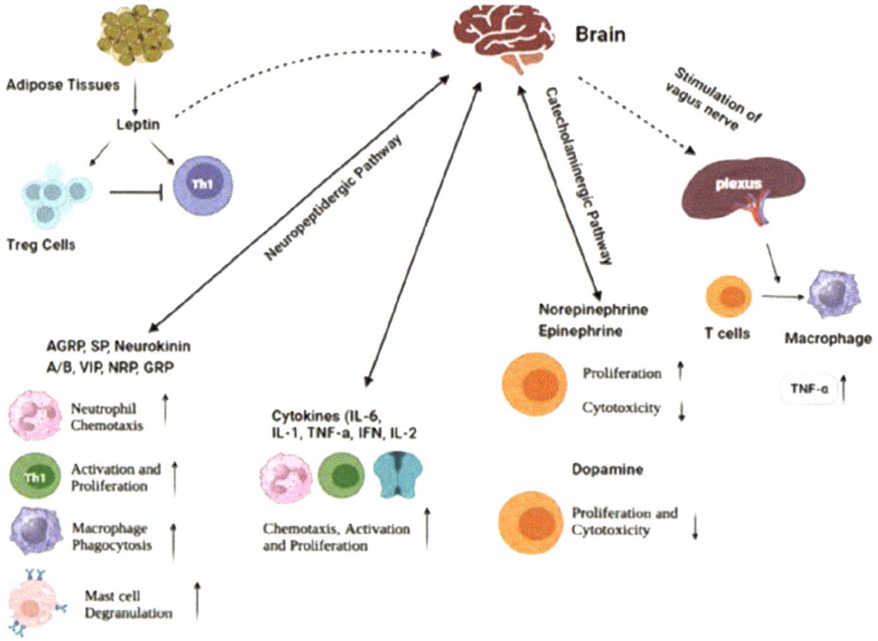

FIGURE 6.1 Crosstalk between neuronal and immune systems. Communication between neuroimmune synapses is bidirectional. The CNS conveys messages to the immune system through vagus nerve stimulation and sympathetic pathway activation (components of the autonomic nervous system), neuropeptides, catecholaminergic pathway, and hormone release. Leptin decreases Treg cell activity and hyperactivates T cells. Catecholamines play both inhibitory and stimulatory roles, while neurotransmitters and neuropeptides positively regulate the immune cells to maintain proper synapses.

communication involves various signaling mechanisms modulated by several chemical moieties, described below.

6.5.1 Direct Cell-Cell Signaling

This signaling mechanism is mediated by various molecules, such as cadherins, integrins, immunoglobulin superfamily members, and tight junction proteins (primarily occludin). A classic example is the interaction between glial cells and astrocytes of the nervous system with dendritic and T cells of the immune system at synapses via various receptors and membrane-bound ligands. These molecules form tight associations that help in information exchange.

Initially, signalosomes connect to the IS via TCR, CXCR4, and LFA-1. TCR activation by MHC complexes and CD4 attachment to Lck initiates phosphorylation of ZAP-70 and CD3 subunit tails. These complexes, initially on the plasma membrane surface, interact with TCR. CXCR4 also activates PLCγl by binding ZAP-70 through G proteins, independent of LAT association, favoring inositol-3-phosphate (IP3) production [50]. IP3 activation then stimulates its specific receptor on the endoplasmic reticulum (ER) surface, allowing calcium flow by increasing intracellular channels. Integrin interaction with Lck and ZAP-70 also activates Kidlin-3 and talin, enhancing Erk1/2 signaling activated by CXRC4 at the neuroimmune synapse. N-cadherins and E-cadherins facilitate cell attachment (responsible for EMT-MET transitions), enabling cells to actively travel and attach themselves to neighbors. This enables proper communication between cells mediated by immunoglobulin, integrin, and cadherin family proteins [51].

Various membrane glycoproteins, including CD47, CD22, and neural cell adhesion molecule (NCAM), also play dominant roles in neuronal-immune cell communication at synapses. CD200, expressed on various cell types, interacts with CD200R, a receptor expressed on immune cells. Blocking this interaction can lead to autoimmune encephalitis and enhanced immune responses. Blocking this receptor causes autoimmune encephalitis and enhanced IL-6 production mediated by IFN-γ activation. CD22, generally expressed in cortical neurons, initiates microglial attachment to neurons via CD45, preventing TNF production. Neurons also inhibit TNF and NO induced by glial cells through NCAM (an IgSF-associated adhesion molecule) [52].

6.5.2 Signaling through Soluble Mediators

Cytokines, neurotransmitters, and chemokines are the dominant soluble mediators actively participating in signal transduction at the neuroimmune junction. These modulators facilitate effective communication between the two systems at various junctions. They can be involved in both cell-cell communication and independent signaling.

6.5.2.1 Neurotransmitter-Mediated Signaling

Dopamine (DA) acts outside the CNS, primarily affecting the immune system by cytokine production and activating receptors D1–5. D2 activation directly affects IL-10 production, D3 activation triggers IFN and TNF-α production, and D5 induces IL-10 secretion in T cells. D4 plays a key role in promoting helper T cell differentiation through enhanced production of GATA3 and STAT5 transcription factors [53]. Macrophagic expression of D1 and D2 receptors induces proinflammatory cytokines production, while D1 activation alone on dendritic cells activate c-Jun N-terminal kinase (JNK is a signaling molecule that is involved in cell proliferation and apoptosis), Extracellular signal-Regulated Kinase (ERK is a key signaling molecule involved in cell proliferation, differentiation, and survival) and NF-κB signaling pathways. Dopamine can be both immunostimulatory and immunoinhibitory. Its immunostimulatory effects include cytokine production and activation of inflammation-related signaling cascades, while its inhibitory effects involve IL-10 and IFN suppression via Krüppel-like factor (KLF-2) hyperactivation and downregulation of ERK1/2, Fyn, Lck, and protein kinases involved in TCR signal transduction [54].

Serotonin also facilitates information exchange in immune networks. Immune cells express almost 14 different 5-HT receptor subtypes, nine of which are GPCRs. 5-HT1a activation positively regulates cell migration, while 5-HT1B, 5-HT1D, and 5-HT2A are inhibitory. 5-HT is primarily released from platelets and acts on lymphoid and myeloid cells during inflammation [55]. 5-HT7 expression increases with T-cell differentiation, leading to increased calcium levels. 5-HT7 activation enhances T-cell proliferation through NF-κB and ERK pathway phosphorylation. 5-HT1B and 5-HT2A cause a massive shift in IFN-γ and IL-2 production and enhance CTL proliferation. 5-HT2A and 5-HT2B transactivation increases IL-17 production in CD4+ cells. 5-HT1A activation is associated with increased CCL2, reduced CCR5, and increased MIP-1α levels, along with enhanced macrophagic phagocytosis. It also triggers cytokine production and release from mature dendritic cells. Serotonin has both immunoinhibitory and immunostimulatory roles [56].

L-glutamate, an abundant neurotransmitter accounting for nearly 90% of synaptic connections, maintains synaptic plasticity after its release from glutamine in the CNS via the glutaminase enzyme. Glu has two receptor families: metabotropic glutamate receptors (eight GPCRs acting through second messenger systems) and ionotropic glutamate receptors (three families allowing ions passage upon activation: α-amino-3-hydroxy-5-methyl-4-isoxazolepropionic acid receptors [AMPARs], kainate receptors [KARs], and NMDARs). The eight metabotropic receptors are further subdivided into three groups: group-1 receptors activate phospholipase C, increasing Ip3 levels, while group-2 and group-3 receptors negatively regulate cAMP levels by downregulating AC [57, 58].

Glutamate signaling is complex, involving GluA3 and GluR5 expressed on naïve T cells. GluA3 activation increases calcium influx, upregulates the IL-2 receptor, and enhances its secretion. This interleukin increase is countered by the mGluR5 and increased cAMP levels, which also downregulate RAS/ERK, NF-κB, and Janus kinase pathways. This inhibition activates C-terminal Src kinase (CSK) [59]. Enhanced GluN1/2B regulation also decreases GluA3 expression after cleavage by granzyme B, while mGluR1/5 activation prevents cell death by downregulating Fas ligand (FasL) expression. Increased mGluR1 activity upregulates P13k and ERK1/2 signaling pathways in mature activated T cells and enhances TNF-α, IFN-γ, and IL-2 expression. DCs expressing mGluR1 effectively release CXCL-1 and IL-8, while mGluR1 and GluN1 activation is linked to CXRC4-mediated migration and integrin-meditated cell-cell adhesion. Overexpression of these Glu receptors is associated with tumor metastasis and progression of prostate cancer, triple-negative breast cancer, and melanoma [56].

6.5.2.2 Cytokine-Mediated Signaling

Immunophysiology posits that immune signals can control the immune system via hormones and peptides, ensuring the downregulation of the neuroendocrine loop involving cytokines. The immune system is regulated by an external loop involving the CNS and its communication with the whole body (apart from the self-regulatory process intrinsic to the immune system). Cytokines play a pivotal role due to their pyrogenic interactions. Pathogen-associated molecular patterns (PAMPs) are recognized by toll-like receptors on monocytes/macrophages [60]. Receptor activation initiates a signaling cascade that produces proinflammatory cytokines. Bacterial LPS binds to LBP receptors with the help of cluster of differentiation 14 (CD14) and myeloid differentiation factor 2 (MD-2). TLR4 activation occurs via two pathways, one dependent on MyD88 and the other independent, resulting in the recruitment of IL-1 receptor-associated kinases (IRAKs) and the main adaptor molecule TNF receptor-associated factor 6 (TRAF6), which activates both Janus kinases (JNK) and p38 [61].

Apart from these, NF-κB and CREB also play active roles in the production of several proinflammatory cytokines. Due to their large size and hydrophilic nature, these cytokines cannot cross the bold-brain barrier and therefore work in areas of the brain lacking these barriers (circumventricular organs). After macrophages produce COX-1 and COX-2 in these areas, astrocytes produce cytokines (such as IL-6) [62]. Similarly, cytokines in the brain produce and release prostaglandin-E2 (PGE2) after microsomal PGE synthase-1 and COX-2 produce the synthesizing enzyme. PGE2 then activates receptors, causing changes in CNS homeostasis and modifying synaptic plasticity and neuronal excitability. This demonstrates the bidirectional role of cytokines in the regulation and maintenance of neurons and immune cells at the neuroimmune interface [45].

6.5.2.3 Chemokine-Mediated Signaling

Chemokines recruit immune cells to sites of inflammation during injury. Neuronal cells secrete these molecules in damaged areas, where they bind to specific GPCR receptors on targeted immune cells. Immune cells are attracted to their respective chemokines via chemotaxis. Chemokine binding activates G proteins, modulating the GPCR and activating various signals simultaneously. This triggers a series of cascades where Rac-1 and Ras homolog gene family (RhoA belongs to the Rho family of GTPases and is primarily involved in cytoskeletal organization and cell adhesion), member A-like GTPases activate tyrosine kinases, which phosphorylate downstream effector proteins. GPCRs and calcium ions activate many presynaptic pathways [63]. GPCR activation also causes G_α and $G_{\beta\gamma}$ dissociation, initiating many other downstream pathways. This calcium ion and receptor stimulation causes intense changes to several potential protein kinases and secondary messengers. Virtually all GPCRs are attached to $G_{i/0}$, G_s, or G_q. G_q acts on synapses by activating PKC and DAG, while G_s activation is linked to PKA. Signaling modulated by $G_{i/0}$ is complex, suppressing PKA while stimulating PKC via its α and β subunits, respectively [62].

These subunits also facilitate crosstalk that activates AC, IP3, and DAG, even in the presence of a single subunit. This signaling mechanism results in the activation of the RAS-RAF pathway, targeting specific proteins, primarily synapsin (MARCKS). Besides protein activation, DAG pathways also activate modulators like Munc13 and Munc18, which strongly affect protein secretion. Direct DAG binding initiates Munc13, while Munc18 activation is mediated by PKC (DAG phosphorylates PKC). A third effector in this pathway, SNAP-25, increases Readily Releasable Pool of Synaptic Vesicles (RRPs are synaptic vesicles that are docked at active zone and are released upon the arrival of action potential) upon phosphorylation [64].

6.6 DISEASES RELATED TO NEUROIMMUNE DYSFUNCTION

The above-mentioned signaling pathways enable effective communication at neuroimmunology synapses. Any dysregulation can result in aberrant signaling, potentially causing various diseases or, in rare cases, multiple diseases simultaneously. Table 6.1 summarizes various diseases that arise from synaptic dysfunction in neuro and immune synapses, along with imaging techniques and optimal treatment options.

6.7 CONCLUSIONS AND FUTURE PROSPECTS

Advancements in imaging and serological titration have enabled researchers to gain a more advanced view of neuroimmune junction interactions, which was previously difficult to interpret. Recent findings reveal that various sub-microscopic structures actively participate in signal transduction at these sites. Scaffold proteins and receptor tyrosine kinases modulate and regulate these microclusters (sub-microscopic structures) via mechanisms such as myosin 2-dependent processes. Although the neuronal and immune systems have varying characteristics and limited component similarities, they function in parallel to maintain homeostasis, protect cells from pathogens, and provoke or inhibit inflammation—a key factor in regulating neuronal immunity. This involves integrating various components from both systems, ultimately activating proper signaling cascades that drive regulation. Synapses are notably dynamic and ever-changing, contributing to learning through a trial-and-error method (reinforced learning) that helps in memory formation and improved immunity at infection sites. A comprehensive understanding of these synapses can help scientists effectively utilize resources to design optimal therapies for diseases caused by aberrant signaling mechanisms.

TABLE 6.1
Anomalies Related to Dysregulation of Neuroimmune Synapses

Disease	Location	Immunophysiology	Diagnosis	Treatment	Ref
Neuromyelitis optica spectrum disorder (NMSOSD)	CNS	Antibody-mediated disease causing complement system activation against neuronal surface molecules, resulting in neuronal loss and demyelination. Antibodies against aquaporin 4 present, B lymphocytes involved	MRI including LETM, OCT, neurological assessment, serological testing	Immunosuppressors, intravenous immunoglobulin (IVIG) therapy, monoclonal antibodies (mAbs)	[65, 66]
Multiple sclerosis (MS)	CNS	Chronic disease causing axonal damage, synapse reduction, demyelination, and decreased resistance to CNS antigens. Involves activation of innate and adaptive immune responses and all major immune cells	Motor function assessment, MRI, cerebrospinal fluid (CSF is a diagnostic tool that used to detect various neurological disorders, such as meningitis, encephalitis, and multiple sclerosis) analysis (high levels of IgG), evoked potentials	Disease modifying therapies (DMTs), rehabilitation, corticosteroids, BTK inhibitors	[66, 67]
MOG-antibody associated disease (MOGAD)	CNS	Autoimmune-mediated inflammation causing damage to oligodendrocytes and myelin sheaths, plus reactive gliosis in gray and white matter. Auto-antibodies target myelin oligodendrocyte glycoprotein (MOG), involving B lymphocytes	Neuroimaging (MRI), CSF analysis, clinical assessment, antibody detection, and OCT	Plasmapheresis, IVIG therapy, corticosteroids, immunosuppressors, PLEX	[65, 68]
Guillain-Barre syndrome (GBS)	PNS	Peripheral neuropathic disease mediated by mimicry of molecule and antibody involvement. Antibodies produced against gangliosides in response to LPS, involving T lymphocytes	CSF analysis, MRI, electromyography (EMG), and nerve conduction tests	Plasma exchange, corticosteroids, IVIG treatment	[65, 69]
Chronic inflammatory demyelinating polyneuropathy (CIDP)	PNS	Polyneuropathy affecting nerve roots and peripheral nerves, stimulated and maintained by humoral and cellular mechanisms. T lymphocytes and macrophages predominate, destroying nerves	EMG and nerve conduction tests, neurological assessments, nerve biopsy	IVIG, plasma exchange, corticosteroid therapy, targeted therapy	[65]

Myasthenia gravis	Neuromuscular junction	Antibodies produced against postsynaptic membrane, impairing junction function. B lymphocytes predominate, producing antibodies against Muscular specific kinase (MuSK is a receptor tyrosine kinase that is involved in the formation and maintenance of neuromuscular junction) or LRP4	EMG, neurological examination, edrophonium test	Acetylcholinesterase inhibitors, immunosuppressors, thymectomy, CAR-T cell therapy	[69]
Lambert-Eaton syndrome	Neuromuscular junction	Auto-antibodies attack presynaptic voltage-gated calcium channels. Involves VGCCs, aberrant ACh release, and glial cell interference	EMG, nerve conduction studies, serological testing, and functional studies	IVIG, immuno-suppressors, DAP treatment	[70]
Alzheimer's disease (AD)	CNS	Formation of amyloid-beta peptides, hyperphosphorylation of tau proteins, and neuroinflammation. Involves microglia, complement, astrocytes, and cytokines	MRI, PET scan, CSF analysis	Cholinesterase inhibitors and memantine	[65, 66]
Autoimmune encephalitis	CNS	Auto-antibodies produced targeting neurons at synapses, affecting ion channels, adhesion molecules, and cell-surface receptors	MRI, EEG, serological testing for specific antibodies	IVIG, corticosteroids, plasmapheresis	[66]
Leucodystrophies (X-linked)	CNS	Impairs motor functions and cognitive progression, arising from accumulation of very long chain fatty acids. Involves large lymphocytic response and monocyte impairment	Measurement of very long-chain fatty acid (VLCFA elevated levels of VLCFAs in the blood can indicate certain genetic disorders, such as peroxisomal disorders, that affect the body's ability to break down these fatty acids) levels, parental testing, and genetic testing	Hormone replacement therapy and hematopoietic stem cell transplantation	[71]
Parkinson's disease (PD)	CNS	Loss of dopaminergic neurons and formation of Lewy bodies. Involves microglia and astrocytes, plus IL-1β gene association	Physical examination, MRI, DAT scans	Dopamine agonists, levodopa treatment, DBS stimulation, rehabilitation	[72]

REFERENCES

1. Dustin, M.L. and D.R.J.S. Colman, *Neural and immunological synaptic relations*. 2002. **298**(5594): p. 785–789.
2. Meier, J., et al., *Fast and reversible trapping of surface glycine receptors by gephyrin*. 2001. **4**(3): p. 253–260.
3. Chockley, P.J., et al., *Synapse-tuned CARs enhance immune cell anti-tumor activity*. 2023. **41**(10): p. 1434–1445.
4. Grakoui, A., et al., *The immunological synapse: a molecular machine controlling T cell activation*. 1999. **285**(5425): p. 221–227.
5. Leithner, A., et al., *Dendritic cell actin dynamics control contact duration and priming efficiency at the immunological synapse*. 2021. **220**(4): p. e202006081.
6. Südhof, T.C.J.J.O.C.B., *The cell biology of synapse formation*. 2021. **220**(7): p. e202103052.
7. Liu, D., et al., *The role of immunological synapse in predicting the efficacy of chimeric antigen receptor (CAR) immunotherapy*. 2020. **18**: p. 1–20.
8. Capitani, N. and C.T.J.F.I.I. Baldari, *The immunological synapse: an emerging target for immune evasion by bacterial pathogens*. 2022. **13**: p. 943344.
9. Davis, D.M. and M.L.J.T.I.I. Dustin, *What is the importance of the immunological synapse?* 2004. **25**(6): p. 323–327.
10. Trautmann, A. and S.J.C.O.I.I. Valitutti, *The diversity of immunological synapses*. 2003. **15**(3): p. 249–254.
11. Thauland, T.J. and D.C.J.I. Parker, *Diversity in immunological synapse structure*. 2010. **131**(4): p. 466–472.
12. Davis, M.M.J.A.R.O.I., *Molecular genetics of the T cell-receptor beta chain*. 1985. **3**(1): p. 537–560.
13. Dustin, M.L., C.T.J.T.I.S.M. Baldari, and Protocols, The immune synapse: past, present, and future. *Methods Mol Biol*. 2017: p. 1–5. doi: 10.1007/978-1-4939-6881-7_1.
14. Wakim, L.M. and M.J.J.N. Bevan, *Cross-dressed dendritic cells drive memory CD8+ T-cell activation after viral infection*. 2011. **471**(7340): p. 629–632.
15. Pilones, K.A., J. Aryankalayil, and S.J.J.O.I.R. Demaria, Invariant NKT cells as novel targets for immunotherapy in solid tumors. *Clin Dev Immunol*. 2012. **2012: p.** 720803. doi: 10.1155/2012/720803. Epub 2012 Oct 17
16. Engelhardt, J.J., et al., *Marginating dendritic cells of the tumor microenvironment cross-present tumor antigens and stably engage tumor-specific T cells*. 2012. **21**(3): p. 402–417.
17. Van Laethem, F., et al., *Lck availability during thymic selection determines the recognition specificity of the T cell repertoire*. 2013. **154**(6): p. 1326–1341.
18. Pasquier, B., et al., *Defective NKT cell development in mice and humans lacking the adapter SAP, the X-linked lymphoproliferative syndrome gene product*. 2005. **201**(5): p. 695–701.
19. Eitler, J., et al., *CAR-mediated targeting of NK cells overcomes tumor immune escape caused by ICAM-1 downregulation*. 2024. **12**(2).
20. Saint-Martin, M. and Y.J.T.F.J. Goda, *Astrocyte–synapse interactions and cell adhesion molecules*. 2023. **290**(14): p. 3512–3526.
21. Maya, J.J.I.J.O.M.S., *Surveying the metabolic and dysfunctional profiles of T cells and NK cells in myalgic encephalomyelitis/chronic fatigue syndrome*. 2023. **24**(15): p. 11937.
22. Cappell, K.M. and J.N.J.N.R.C.O. Kochenderfer, *A comparison of chimeric antigen receptors containing CD28 versus 4-1BB costimulatory domains*. 2021. **18**(11): p. 715–727.
23. González-Fernández, C., et al., *Molecular characterization of the T cell costimulatory receptors CD28 and CTLA4 in the European sea bass*. 2021. **109**: p. 106–115.
24. Wahyuningtyas, R., et al., *Toll-like receptor-mediated immunomodulation of Th1-type response stimulated by recombinant antigen of type 2 porcine reproductive and respiratory syndrome virus (PRRSV-2)*. 2023. **15**(3): p. 775.
25. Park, S., et al., *Force-dependent trans-endocytosis by breast cancer cells depletes costimulatory receptor CD80 and attenuates T cell activation*. 2020. **165**: p. 112389.
26. Lin, M., et al., *The roles of TRAF3 in immune responses*. 2023. **2023**.
27. Yu, J.-L., S.R.-J. Jang, and K.-Y.J.C. Liu, *Exploring the interactions of oncolytic viral therapy and immunotherapy of anti-CTLA-4 for malignant melanoma mice model*. 2023. **12**(3): p. 507.
28. Singh, A., et al., *Role of lymphocytes, macrophages and immune receptors in suppression of tumor immunity*. 2023. **194**: p. 269–310.
29. Jung, Y.J.F.I.I., *Pre-organized landscape of T cell surface*. 2023. **14**: p. 1264721.

30. Lippert, A.H., et al., *Antibody agonists trigger immune receptor signaling through local exclusion of receptor-type protein tyrosine phosphatases.* 2024. **57**(2): p. 256–270. e10.
31. Rodriguez-Rodriguez, C., et al., *Partial loss of sorting nexin 27 resembles age-and down syndrome-associated T cell dysfunctions.* 2024. **21**(1): p. 2.
32. Sebastian, J. and K.J.C.D.T. Rathinasamy, *Microtubules and cell division: potential pharmacological targets in cancer therapy.* 2023. **24**(11): p. 889–918.
33. Bromley, S.K., et al., *The immunological synapse.* 2001. **19**(1): p. 375–396.
34. Alcaraz-Serna, A., et al., *Immune synapse instructs epigenomic and transcriptomic functional reprogramming in dendritic cells.* 2021. **7**(6): p. eabb9965.
35. Vogt, A.B., S. Spindeldreher, and H.J.I.R. Kropshofer, *Clustering of MHC–peptide complexes prior to their engagement in the immunological synapse: lipid raft and tetraspan microdomains.* 2002. **189**(1): p. 136–151.
36. Ma, J.J.J.O.Z.U.-S.A., *Biophysical neurons, energy, and synapse controllability: a review.* 2023. **24**(2): p. 109–129.
37. Teleanu, R.I., et al., *Neurotransmitters—key factors in neurological and neurodegenerative disorders of the central nervous system.* 2022. **23**(11): p. 5954.
38. Carr, C.M. and M.J.E.R. Munson, *Tag team action at the synapse.* 2007. **8**(9): p. 834–838.
39. Goto, A.J.N.r., *Synaptic plasticity during systems memory consolidation.* 2022. **183**: p. 1–6.
40. Richter, J.D., E.J.G. Klann, and development, *Making synaptic plasticity and memory last: mechanisms of translational regulation.* 2009. **23**(1): p. 1–11.
41. Bliss, T.V. and S.F.J.C. Cooke, *Long-term potentiation and long-term depression: a clinical perspective.* 2011. **66**: p. 3–17.
42. Dustin, M.L.J.T.J.O.C.I., *Signaling at neuro/immune synapses.* 2012. **122**(4): p. 1149–1155.
43. Tournier, J.-N. and A.Q.J.T.I.I. Hellmann, *Neuro-immune connections: evidence for a neuro-immunological synapse.* 2003. **24**(3): p. 114–115.
44. Benarroch, E.E.J.N., *Autonomic nervous system and neuroimmune interactions: new insights and clinical implications.* 2019. **92**(8): p. 377–385.
45. Dantzer, R.J.P.R., *Neuroimmune interactions: from the brain to the immune system and vice versa.* 2018. **98**(1): p. 477–504.
46. Batista, F.D., D. Iber, and M.S.J.N. Neuberger, *B cells acquire antigen from target cells after synapse formation.* 2001. **411**(6836): p. 489–494.
47. Walker, D.J., K.A.J.G. Spencer, and c. endocrinology, *Glucocorticoid programming of neuroimmune function.* 2018. **256**: p. 80–88.
48. Rao, J.S., et al., *Neuroinflammation and synaptic loss.* 2012. **37**: p. 903–910.
49. Chu, C., D. Artis, and I.M.J.I. Chiu, *Neuro-immune interactions in the tissues.* 2020. **52**(3): p. 464–474.
50. Suzuki, A., et al., *N-cadherin plays a role in the synapse-like structures between mast cells and neurites.* 2004. **27**(12): p. 1891–1894.
51. Martín-Cófreces, N.B., et al., *Adhesive interactions delineate the topography of the immune synapse.* 2018. **6**: p. 149.
52. Tian, L., et al., *Neuroimmune crosstalk in the central nervous system and its significance for neurological diseases.* 2012. **9**: p. 1–10.
53. Cerovic, M., et al., *Molecular and cellular mechanisms of dopamine-mediated behavioral plasticity in the striatum.* 2013. **105**: p. 63–80.
54. Klein, M.O., et al., *Dopamine: functions, signaling, and association with neurological diseases.* 2019. **39**(1): p. 31–59.
55. Wirth, A., K. Holst, and E.J.P.I.N. Ponimaskin, *How serotonin receptors regulate morphogenic signalling in neurons.* 2017. **151**: p. 35–56.
56. Hodo, T.W., et al., *Critical neurotransmitters in the neuroimmune network.* 2020. **11**: p. 523925.
57. Groc, L. and D.J.S. Choquet, *Linking glutamate receptor movements and synapse function.* 2020. **368**(6496): p. eaay4631.
58. Choudhury, P.R., et al., *Glutamate mediated signaling in the pathophysiology of autism spectrum disorders.* 2012. **100**(4): p. 841–849.
59. Huang, W. and A.J.J.O.L.B. August, *The signaling symphony: T cell receptor tunes cytokine-mediated T cell differentiation.* 2015. **97**(3): p. 477–485.
60. Song, X. and Y.J.C.S. Qian, *IL-17 family cytokines mediated signaling in the pathogenesis of inflammatory diseases.* 2013. **25**(12): p. 2335–2347.
61. Kennedy, R.H. and R. Silver, *Neuroimmune signaling: cytokines and the central nervous system, in neuroscience in the 21st century: from basic to clinical.* 2022, Springer. p. 883–922.

62. Zeytun, A., et al., *Induction of cytokines and chemokines by toll-like receptor signaling: strategies for control of infammation*. 2010. **30**(1).
63. Tripathi, D.K. and K.M.J.F.I.B.-L. Poluri, *Molecular insights into kinase mediated signaling pathways of chemokines and their cognate G protein coupled receptors*. 2020. **25**(7): p. 1361–1385.
64. de Jong, A.P. and M.J.C.O.I.N. Verhage, *Presynaptic signal transduction pathways that modulate synaptic transmission*. 2009. **19**(3): p. 245–253.
65. Winkelmann, A., et al., *Vaccination and immunotherapies in neuroimmunological diseases*. 2022. **18**(5): p. 289–306.
66. Dos Passos, G.R., et al., *Reshaping neuroimmunology: diagnosis and treatment in the era of precision medicine*. 2024. **81**: p. 1125–1133.
67. Boulanger, L.M. and C.J.J.N.R.N. Shatz, *Immune signalling in neural development, synaptic plasticity and disease*. 2004. **5**(7): p. 521–531.
68. Sechi, E., et al., *Myelin oligodendrocyte glycoprotein antibody-associated disease (MOGAD): a review of clinical and MRI features, diagnosis, and management*. 2022. **13**: p. 885218.
69. Nutma, E., et al., *Neuroimmunology–the past, present and future*. 2019. **197**(3): p. 278–293.
70. Titulaer, M.J., B. Lang, and J.J.J.T.L.N. Verschuuren, *Lambert–Eaton myasthenic syndrome: from clinical characteristics to therapeutic strategies*. 2011. **10**(12): p. 1098–1107.
71. Costello, D.J., A.F. Eichler, and F.S.J.T.N. Eichler, *Leukodystrophies: classification, diagnosis, and treatment*. 2009. **15**(6): p. 319–328.
72. Armstrong, M.J. and M.S.J.J. Okun, *Diagnosis and treatment of Parkinson disease: a review*. 2020. **323**(6): p. 548–560.

7 Rhinoviruses Receptors and Their Interactions

Ariba Qaiser, Areej A. Sattar*, and Sobia Manzoor*
Molecular Virology and Viral Oncology Lab, National University of Science & Technology (NUST), Atta-Ur-Rahman School of Applied Biosciences (ASAB), Islamabad, Pakistan
*Equally contributed to this work and are recognized as co-first authors.

7.1 INTRODUCTION

Rhinoviruses (*rhino* means "nose"), the notorious culprit behind the common cold, stand as ubiquitous and highly contagious etiological agents of the upper respiratory tract in human beings (Ong et al., 2023). They exert a profound impact on public health by virtue of their exceptional prevalence and capacity to precipitate a spectrum of illnesses ranging from the common cold to severe respiratory conditions. They cause many of the illnesses in children and neonates (Aizawa et al., 2023). Human rhinovirus (RV) belongs to the *Enterovirus* genus within the *Picornaviridae* family. After the discovery of RVs in 1956, they were genetically subdivided into three species, namely, rhinovirus A (RV-A), rhinovirus B (RV-B), and rhinovirus C (RV-C), consisting of more than 160 types (Falsey & Branche, 2023; Ljubin-Sternak & Meštrović, 2023).

Rhinoviruses are single-stranded, positive-sense RNA viruses of small genome size (approximately 7200 bases) and no envelope. They have a single open reading frame (ORF) that is joined to a short viral priming protein (VPg) and a 50 untranslated region (Ljubin-Sternak & Meštrović, 2023). The ORF acts as a messenger RNA (mRNA) and encodes 11 proteins. Of these proteins, VP1, VP2, and VP3 are responsible for the capsid formation, whereas VP4 helps in the anchorage of ssRNA with the virus capsid (Figure 7.1). The VP1 protein, present at the canyon structures of capsids, facilitates the receptor binding site to the cell surface receptor (Bizot et al., 2021). RVs orchestrate their entry into the host cells through a finely tuned interaction with specific cellular receptors, primarily intercellular adhesion molecule-1 (ICAM-1), cadherin-related family member 3 (CDHR3), and members of the low-density lipoprotein receptor (LDLR) family (Basnet et al., 2019). Elucidating these intricate molecular mechanisms not only enhances the understanding of viral pathogenesis but also provides potential targets for antiviral interventions aimed at disrupting the critical steps in rhinovirus entry and replication within the host respiratory epithelium.

7.2 RHINOVIRUS INTERACTIONS WITH CELLULAR RECEPTORS

Rhinoviruses are classified into subgroups based on their interaction with specific plasma membrane receptors present on the cell surface. Interestingly, RVs have shown different interactions with human respiratory epithelial cells, but some have also shown interaction with other cells, for example, immune cells (Nikonova et al., 2020). The major subgroup of RV comprises RV-A and RV-B that bind with ICAM-1; the minor group includes the subset of RV-A that binds with the LDLR (Shukla et al., 2022). RV-C was recently discovered to show interaction with CDHR3 receptors

DOI: 10.1201/9781003594901-7

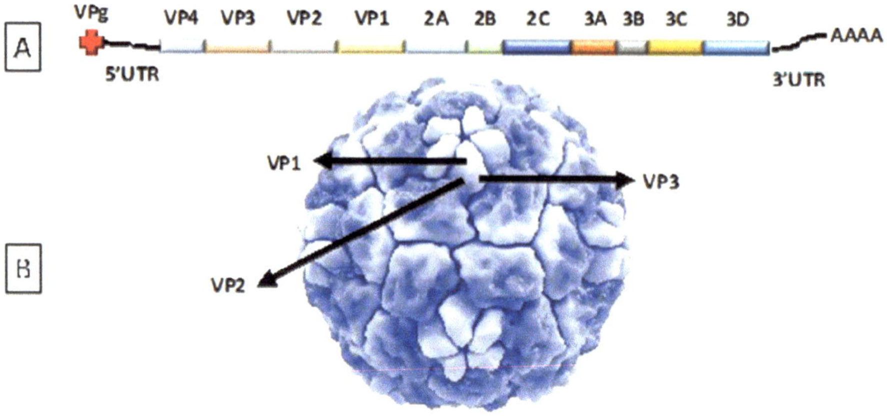

FIGURE 7.1 HRV (human rhinovirus) structure. (A) HRV RNA containing 5′ CAP, 3′ polyadenylated tail, 5′ and 3′ UTR, flanking structural (VP1, VP2, VP3, and VP4) proteins, and nonstructural proteins (2A, 2B, 2C, 3A, 3B, 3C, and 3D). (B) HRV's icosahedron consists of pentameric distribution with VP1, VP2, VP3, and VP4 (not shown) beneath. The capsid encapsulates viral nucleoprotein.

(Bochkov et al., 2015; Watters & Palmenberg, 2018). Some major RVs, especially RV-A and some RV-C, can also use heparan sulfate proteoglycan as a substitute receptor when they are cell culture adapted (Bochkov et al., 2016; Vlasak et al., 2005).

7.2.1 Intercellular Adhesion Molecule 1

Intercellular adhesion molecule 1 (ICAM-1) is a member of the immunoglobulin (Ig) superfamily that was first discovered as a major cellular receptor of RVs. It is a transmembrane glycoprotein that is expressed on the surface of multiple cells, such as leukocytes, endothelial cells, and epithelial cells. These cells express ICAM-1 specifically under stress or as a response to inflammatory stimuli. Enhanced expression of ICAM-1 also appears to be associated with increased susceptibility to infections, especially viruses. However, other bacteria and pathogens also use ICAM-1 as a docking site and invade the target cell.

The structural analysis of ICAM-1 reveals that it has five Ig-like domains, a carboxyl-proximal transmembrane domain, and a short C-terminal cytoplasmic tail (Figure 7.2) (Hubbard & Rothlein, 2000). During the interaction with rhinoviruses, the N-terminal Ig domain is sequestered within the binding pocket formed by the VP1 and VP2 capsid proteins located deep in the canyon (Figure 7.2) (Esneau et al., 2019). As an inherent aspect of its usual functioning, ICAM-1 plays a pivotal role in cellular activation and cell-to-cell interactions, and facilitates the transmigration of leukocytes to the inflammatory site in damaged tissue (Bui et al., 2020).

In the context of prominent RV classification, except RV-B, the hydrophobic pocket within the VP1 is a key player for the docking of VP1 with ICAM-1. It's a structural feature that can assume two different conformations, either filled or empty, which is harbored with a "pocket factor" (a fatty acid-like molecule), which assists in viral entry and uncoating (Ledford et al., 2004). The binding of ICAM-1, for all the major groups of RVs, occurs only when the binding pocket is empty. This happens because the VP1 and VP2 docking residues are present in the vicinity of the pocket pore and at the base of the canyon. The location other than this doesn't facilitate the formation of binding conformation. Once the RV hinges with the ICAM-1, the viral genome is engulfed by the cell, and endocytosis occurs. This results in the lowering of the pH in the endosome, leading toward conformational changes and hence results in the viral RNA translocation in the cytoplasm (Daussy & Wodrich, 2020; Nurani et al., 2003).

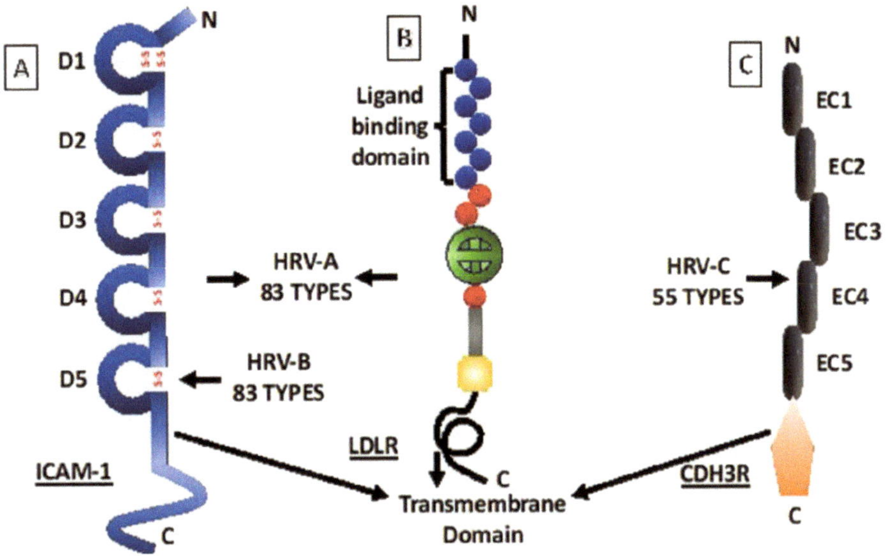

FIGURE 7.2 HRV receptors. (A) ICAM-1 (intracellular adhesion molecule-1) binds to 83 serotypes of both HRV-A and HRV-B. It consists of five domains. The N-terminal gives rise to D1, contains two disulfide bonds, and interacts with VP1 and VP2 of HRV. Below are the other four domains with single disulfide bonds in each, ending on the transmembrane domain and the C terminal. (B) The LDLR (low-density lipoprotein receptor) binds to 83 serotypes of HRV-B. Below the N-terminal are the LDLR type A repeats (in blue), which interact with the ligand (HRV-B); the EFG (epidermal growth factor)-like domain (in red), responsible for acidification in the endosome for the viral genome release; β propeller (in green); O-linked sugar domain (in gray); transmembrane anchor (in yellow); and cytoplasmic domain ending (black) on the C-terminal. (C) CDH3R (cadherin 3) contains five extracellular domains (in gray), out of which EC1 binds to 55 serotypes of HRV-C, then a transmembrane domain, and a cytoplasmic domain.

7.2.2 Low-Density Lipoprotein Receptor Family

The low-density lipoprotein receptor family is another ligand-binding family that was discovered as a basic receptor for the minor group of RVs. This family includes at least three types of receptors that interact with RVs: LDLRs, LDLR-related proteins, and very low-density lipoprotein receptors (VLDLR) (Bochkov & Gern, 2016). Unlike ICAM-1, the minor group of RVs binds to the LDLR present on the star-shaped dome structure at the vertex, outside the canyon region.

The LDLR is responsible for the uptake of cholesterol motifs present in the lipoprotein particles of cells. It is a cell surface receptor that is unique in structure, carrying multiple domains, seven ligand-binding domains (LBDs) (also known as cysteine-rich domains), three epidermal growth factor precursor homology domains (EGFP), followed by a β-propeller domain and O-linked sugar residue. The β-propeller domain contains Tyr-Trp-Thr-Asp residues that are implicated in the low pH-mediated release of ligands in endosomes. They also have a transmembrane domain that is linked to the cytoplasmic tail. Approximately 40 amino acid residues constitute a single LBD, which comprises six cysteine residues involved in the formation of three disulfide bridges. These residues are responsible for the octahedral calcium-binding motif and are involved in the attachment with viruses (Figure 7.2) (Rudenko et al., 2002).

The LDLR naturally uptakes its natural ligands by the process of clathrin-mediated endocytosis. Once the "cargo" is shipped inside the target cell, the pH of the endosome lowers and ruptures to release the molecules in the cytosol. The receptor then recycles back toward the cell membrane (Romagnuolo et al., 2017, p. 9). Members of the LDLR family have a cytoplasmic tail that is utilized in signal transduction pathways to connect with various cytoplasmic adaptors and scaffold proteins

(Bochkov & Gern, 2016). X-ray analysis has revealed that the LBD2 and LBD3 dock with the VP1 capsid protein of RV. When the viral capsid protein hinges with the receptor, no large conformational changes are observed; only local rearrangements are observed at Lys-1228, Asn-1088, and Ala-1084. These residues are directly involved in the attachment with the receptor (Verdaguer et al., 2004). These structural rearrangements lead to the engulfment of the viral particle and ultimately the release of its RNA into the cytoplasm.

7.2.3 Cadherin-Related Family Member 3

Cadherin-related family member 3 (CDHR3) is a transmembrane glycoprotein that is expressed on the surface of respiratory epithelial cells (ECs). They mostly function in calcium-dependent cell-to-cell adhesion, epithelial cell polarity, signal transduction, tissue structure, and organization. While the native function of CDHR3 at the surface is still elusive, these receptors are highly expressed in fallopian tubes, the brain, and respiratory epithelia. These are recognized as the sole receptors of RV-C, a subtype of RV.

The defining characteristic of all cadherin is their extracellular domain (ED), which is arranged in a unique linear tandem repeat, forming a rod-like structure. These EC units of about 110 amino acids long have distinctive sequences, but they frequently take on similar seven-stranded, antiparallel "Greek key" patterns. These repeat units, usually five in number, precede a signal sequence and are followed by a brief transmembrane segment and a cytoplasmic tail (Figure 7.2). Collectively or independently, they mediate cis and trans interactions that are essential for adhesion specificity (Shapiro & Weis, 2009). Two or three calcium ions chelated by multiple acidic residues are incorporated into each EC junction, which is essential in determining the overall rigidity of the stretched, slightly curved rod-like confirmations as well as relative domain orientation. In both cis (parallel orientation, same cell) and trans (antiparallel orientation, opposite cell) configurations, cadherins interact in both homotypic and heterotypic ways. Their adhesion functions are mediated by reciprocal ionic contacts across their different EC surfaces (Perez & Nelson, 2004).

The 885 amino acid residues of the CDHR3 sequence code for six EC repeats with well-defined carboxyl and amino acid extensions. The structurally anticipated contact between EC5 and EC6 is where the white 592/C529 dichotomy is present, and it may have an impact on the calcium stability of that junction. A protein docking model suggests that CDHR3 may only interact with RV-C15A through contact in the first two domains, which is consistent with the most recent structure resolution of this virus (Watters & Palmenberg, 2018). The N-terminal of EC1 repeat binds in a glycosylation-independent and calcium-dependent manner with the HRV-C (Lee et al., 2021).

7.2.4 Uncoating and Endocytosis

Human RVs are internalized in the cells by mostly clathrin-mediated endocytosis. Clathrin is present in the plasma membrane, and it forms a coated pit around the endosome as a result of receptor–ligand interaction. This process involves the assembly of multiple molecules, including AP2 (adaptor complex protein 2), epsin (linker proteins), CALM/AM180, and PIP2 (lipid phosphatidylinositol-4, 5-P2) (Le Roy & Wrana, 2005). Then endophilin, amphiphysin, and GTPase dynamin help in the constriction and fission of coated pits during endocytosis (Vlasak et al., 2005).

Major group viruses exhibit less of the pH-dependent kinetics of RV uncoating than minor group viruses after cellular uptake. Fivefold axis conformational changes bring about a 4% expansion of the capsid proteins and release the N-terminal components of the VP4 and VP1 capsid proteins (Figure 7.3) (Hewat et al., 2002). Sedimentation coefficient evaluations are used to quantify these structural changes in the virus particles. An intermediate subviral particle A (135S) is created when VP4 separates from the entire particle (150S). Subsequently, a subviral particle B (80S) appears, which is distinguished by the lack of RNA (Weiss et al., 2012).

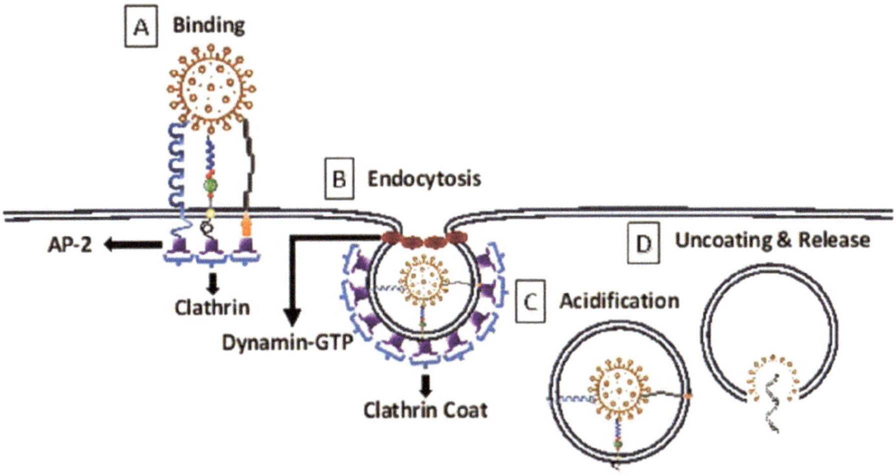

FIGURE 7.3 HRV interaction with the receptors ICAM-1, LDLR, and CDH3R on the plasma membrane. (A) Binding of capsid proteins (VP1 and VP2) with their receptors, which give signals to clathrin that is bound to AP2 (adaptor complex protein-2) to form an endosome. (B) Endocytosis begins when the virus is enclosed in an endosome comprised of a clathrin-coated pit. The endosome excises from the membrane with the help of dynamin GTPase. (C) Acidification inside the endosome occurs when pH drops to >5.6. (D) Uncoating and release of virus into the cytoplasm is a result of acidification.

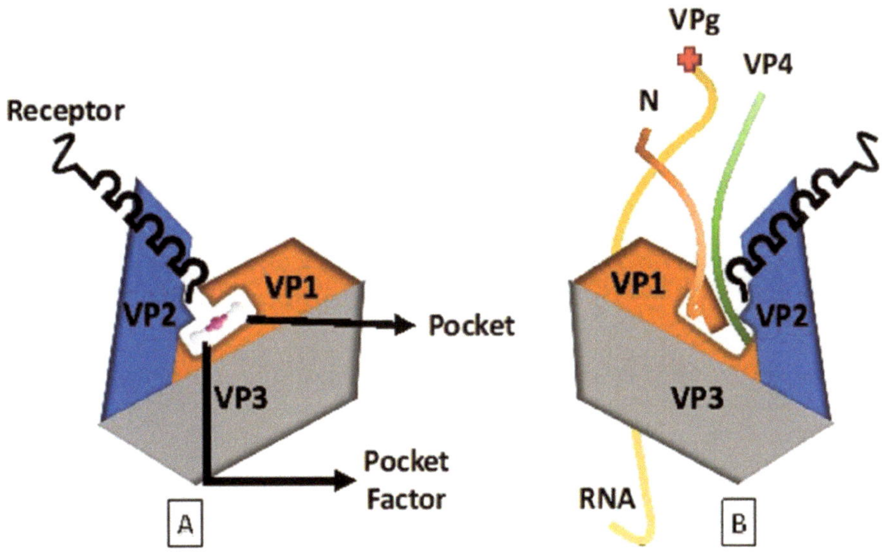

FIGURE 7.4 Release of HRV genome. (A) The two capsid proteins, VP1 and VP2, interact with the receptor above the pocket in VP1. The pocket contains a pocket factor when the genome is still enclosed in the capsid. (B) Receptor–ligand interaction and low pH generate conformational changes in the capsid. The pocket factor is released to empty the pocket allowing the N-terminal of VP1, VP4, and viral RNA to exit the virus.

It is thought that for the minor group of RVs, lowering of pH (>5.6) is the main cause of conformational changes, which in turn causes the viral particle to separate from its LDLR. After that, as soon as ten minutes post-infection, VP4 opens an aqueous pore in the endosomal membrane, allowing the release of viral RNA into the cytoplasm. The capsid proteins are then broken down in the lysosomes (Figure 7.4) (Modrow et al., 2013). The major group of viruses, on the other hand, have a receptor-mediated uncoating mechanism in addition to pH changes. ICAM-1 acts as an uncoating signal and, in certain physiological situations, may also cause endosomal breakdown (Esneau et al., 2019).

7.3 CONCLUSION

RVs are single-stranded, positive-sense RNA viruses that are members of the *Picornaviridae* family. They are highly notorious for causing respiratory illnesses, specifically among children. RVs are further classified into three species—RV-A, RV-B, and RV-C—consisting of more than 160 subtypes on the basis of their attachment with the respective receptors. These viruses have a small genome size (approximately 7200 bp) and are nonenveloped. Viral proteins of RVs interact with specific cellular receptors, primarily ICAM-1, CDHR3, and the LDLR family. The major groups of RVs—RV-A and RV-B—bind with ICAM-1, whereas some minor subgroups of RV-A bind with the LDLR family. However, RV-C shows an interaction with CDHR3. This results in the clathrin-mediated endocytosis of the viral particle that is released in the cytosol upon low pH.

REFERENCES

Aizawa, Y., Ikuse, T., Izumita, R., Habuka, R., Yamanaka, T., & Saitoh, A. (2023). Human rhinovirus as a cause of fever in neonates and young infants during the COVID-19 pandemic, 2020–2022. *The Pediatric Infectious Disease Journal*. https://doi.org/10.1097/INF.0000000000004139

Basnet, S., Bochkov, Y. A., Brockman-Schneider, R. A., Kuipers, I., Aesif, S. W., Jackson, D. J., Lemanske, R. F., Ober, C., Palmenberg, A. C., & Gern, J. E. (2019). CDHR3 asthma-risk genotype affects susceptibility of airway epithelium to rhinovirus C infections. *American Journal of Respiratory Cell and Molecular Biology*, *61*(4), 450–458. https://doi.org/10.1165/rcmb.2018-0220OC

Bizot, E., Bousquet, A., Charpié, M., Coquelin, F., Lefevre, S., Le Lorier, J., Patin, M., Sée, P., Sarfati, E., Walle, S., Visseaux, B., & Basmaci, R. (2021). Rhinovirus: A narrative review on its genetic characteristics, pediatric clinical presentations, and pathogenesis. *Frontiers in Pediatrics*, *9*, 643219. https://doi.org/10.3389/fped.2021.643219

Bochkov, Y. A., & Gern, J. E. (2016). Rhinoviruses and their receptors: Implications for allergic disease. *Current Allergy and Asthma Reports*, *16*(4), 30. https://doi.org/10.1007/s11882-016-0608-7

Bochkov, Y. A., Watters, K., Ashraf, S., Griggs, T. F., Devries, M. K., Jackson, D. J., Palmenberg, A. C., & Gern, J. E. (2015). Cadherin-related family member 3, a childhood asthma susceptibility gene product, mediates rhinovirus C binding and replication. *Proceedings of the National Academy of Sciences of the United States of America*, *112*(17), 5485–5490. https://doi.org/10.1073/pnas.1421178112

Bochkov, Y. A., Watters, K., Basnet, S., Sijapati, S., Hill, M., Palmenberg, A. C., & Gern, J. E. (2016). Mutations in VP1 and 3A proteins improve binding and replication of rhinovirus C15 in HeLa-E8 cells. *Virology*, *499*, 350–360. https://doi.org/10.1016/j.virol.2016.09.025

Bui, T. M., Wiesolek, H. L., & Sumagin, R. (2020). ICAM-1: A master regulator of cellular responses in inflammation, injury resolution, and tumorigenesis. *Journal of Leukocyte Biology*, *108*(3), 787–799. https://doi.org/10.1002/JLB.2MR0220-549R

Daussy, C. F., & Wodrich, H. (2020). "Repair me if you can": Membrane damage, response, and control from the viral perspective. *Cells*, *9*(9), Article 9. https://doi.org/10.3390/cells9092042

Esneau, C., Bartlett, N., & Bochkov, Y. A. (2019). Chapter 1—Rhinovirus structure, replication, and classification. In N. Bartlett, P. Wark, & D. Knight (Eds.), *Rhinovirus Infections* (pp. 1–23). Academic Press. https://doi.org/10.1016/B978-0-12-816417-4.00001-9

Falsey, A. R., & Branche, A. R. (2023). Rhinoviruses. In *Reference Module in Biomedical Sciences*. Elsevier. https://doi.org/10.1016/B978-0-323-99967-0.00053-3

Hewat, E. A., Neumann, E., & Blaas, D. (2002). The concerted conformational changes during human rhinovirus 2 uncoating. *Molecular Cell*, *10*(2), 317–326. https://doi.org/10.1016/S1097-2765(02)00603-2

Hubbard, A. K., & Rothlein, R. (2000). Intercellular adhesion molecule-1 (ICAM-1) expression and cell signaling cascades. *Free Radical Biology and Medicine*, *28*(9), 1379–1386. https://doi.org/10.1016/S0891-5849(00)00223-9

Le Roy, C., & Wrana, J. L. (2005). Clathrin- and non-clathrin-mediated endocytic regulation of cell signalling. *Nature Reviews Molecular Cell Biology*, *6*(2), Article 2. https://doi.org/10.1038/nrm1571

Ledford, R. M., Patel, N. R., Demenczuk, T. M., Watanyar, A., Herbertz, T., Collett, M. S., & Pevear, D. C. (2004). VP1 sequencing of all human rhinovirus serotypes: Insights into genus phylogeny and susceptibility to antiviral capsid-binding compounds. *Journal of Virology*, *78*(7), 3663–3674. https://doi.org/10.1128/JVI.78.7.3663-3674.2004

Lee, W., Frederick, R. O., Tonelli, M., & Palmenberg, A. C. (2021). Solution NMR determination of the CDHR3 rhinovirus-C binding domain, EC1. *Viruses*, *13*(2), Article 2. https://doi.org/10.3390/v13020159

Ljubin-Sternak, S., & Meštrović, T. (2023). Rhinovirus—A true respiratory threat or a common inconvenience of childhood? *Viruses*, *15*(4), 825. https://doi.org/10.3390/v15040825

Modrow, S., Falke, D., Truyen, U., & Schätzl, H. (2013). Viruses with single-stranded, positive-sense RNA genomes. *Molecular Virology*, 185–349. https://doi.org/10.1007/978-3-642-20718-1_14

Nikonova, A., Khaitov, M., Jackson, D. J., Traub, S., Trujillo-Torralbo, M.-B., Kudlay, D. A., Dvornikov, A. S., del-Rosario, A., Valenta, R., Stanciu, L. A., Khaitov, R., & Johnston, S. L. (2020). M1-like macrophages are potent producers of anti-viral interferons and M1-associated marker-positive lung macrophages are decreased during rhinovirus-induced asthma exacerbations. *EBioMedicine*, *54*, 102734. https://doi.org/10.1016/j.ebiom.2020.102734

Nurani, G., Lindqvist, B., & Casasnovas, J. M. (2003). Receptor priming of major group human rhinoviruses for uncoating and entry at mild low-pH environments. *Journal of Virology*, *77*(22), 11985–11991. https://doi.org/10.1128/JVI.77.22.11985-11991.2003

Ong, H. H., Liu, J., Oo, Y., Thong, M., Wang, D. Y., & Chow, V. T. (2023). Prolonged primary rhinovirus infection of human nasal epithelial cells diminishes the viral load of secondary influenza H3N2 infection via the antiviral state mediated by RIG-I and interferon-stimulated genes. *Cells*, *12*(8), Article 8. https://doi.org/10.3390/cells12081152

Perez, T. D., & Nelson, W. J. (2004). Cadherin adhesion: Mechanisms and molecular interactions. *Handbook of Experimental Pharmacology*, *165*, 3–21. https://doi.org/10.1007/978-3-540-68170-0_1

Romagnuolo, R., Scipione, C. A., Marcovina, S. M., Gemin, M., Seidah, N. G., Boffa, M. B., & Koschinsky, M. L. (2017). Roles of the low density lipoprotein receptor and related receptors in inhibition of lipoprotein(a) internalization by proprotein convertase subtilisin/kexin type 9. *PLoS One*, *12*(7), e0180869. https://doi.org/10.1371/journal.pone.0180869

Rudenko, G., Henry, L., Henderson, K., Ichtchenko, K., Brown, M. S., Goldstein, J. L., & Deisenhofer, J. (2002). Structure of the LDL receptor extracellular domain at endosomal pH. *Science (New York, N.Y.)*, *298*(5602), 2353–2358. https://doi.org/10.1126/science.1078124

Shapiro, L., & Weis, W. I. (2009). Structure and biochemistry of cadherins and catenins. *Cold Spring Harbor Perspectives in Biology*, *1*(3), a003053. https://doi.org/10.1101/cshperspect.a003053

Shukla, S. D., Shastri, M. D., Vanka, S. K., Jha, N. K., Dureja, H., Gupta, G., Chellappan, D. K., Oliver, B. G., Dua, K., & Walters, E. H. (2022). Targeting intercellular adhesion molecule-1 (ICAM-1) to reduce rhinovirus-induced acute exacerbations in chronic respiratory diseases. *Inflammopharmacology*, *30*(3), 725–735. https://doi.org/10.1007/s10787-022-00968-2

Verdaguer, N., Fita, I., Reithmayer, M., Moser, R., & Blaas, D. (2004). X-ray structure of a minor group human rhinovirus bound to a fragment of its cellular receptor protein. *Nature Structural & Molecular Biology*, *11*(5), 429–434. https://doi.org/10.1038/nsmb753

Vlasak, M., Goesler, I., & Blaas, D. (2005). Human rhinovirus type 89 variants use heparan sulfate proteoglycan for cell attachment. *Journal of Virology*, *79*(10), 5963–5970. https://doi.org/10.1128/JVI.79.10.5963-5970.2005

Watters, K., & Palmenberg, A. C. (2018). CDHR3 extracellular domains EC1-3 mediate rhinovirus C interaction with cells and as recombinant derivatives, are inhibitory to virus infection. *PLoS Pathogens*, *14*(12), e1007477. https://doi.org/10.1371/journal.ppat.1007477

Weiss, V. U., Subirats, X., Pickl-Herk, A., Bilek, G., Winkler, W., Kumar, M., Allmaier, G., Blaas, D., & Kenndler, E. (2012). Characterization of rhinovirus subviral A particles via capillary electrophoresis, electron microscopy and gas-phase electrophoretic mobility molecular analysis: Part I. *Electrophoresis*, *33*(12), 1833–1841. https://doi.org/10.1002/elps.201100647

8 Role of Ligand- and Voltage-Gated Ion Channels in Cell Signaling

Maham Mehrangaze Khonza[1], *Hira Ali*[1], *Sadaf Mumtaz*[1,2], *Rumeza Hanif*[1], *and Aneela Javed*[1]
[1]Department of Biomedicine, Atta ur Rahman School of Applied Biosciences (ASAB), National University of Sciences and technology (NUST), H12 Islamabad, Pakistan
[2]Department of Physiology, Rawalpindi Medical University, Rawalpindi, Pakistan

8.1 INTRODUCTION

Ion channels are specific protein pores that permit various ions to pass across different membranes, such as organelle inner membranes and the plasma membrane. These transmembrane proteins embedded in the cell membrane are crucial for maintaining cellular balance and are involved in diverse physiological processes such as nerve transmission, muscular contraction, and signal transduction. They selectively allow the flow of ions across cell membranes, governing cell excitability, resting potentials, and responses to stimuli (Hille, 2001). Ion channels can be categorized into several classes based on their ion specificity: potassium, sodium, calcium, and chloride channels. Their regulation is strictly controlled, allowing precise control of ion flow and membrane potential (Alexander et al., 2021) (Figure 8.1).

8.2 TYPES OF ION CHANNELS

Ion channels are classified by how they respond to their environment. For example, voltage-sensitive ion channels open and close in response to changes in voltage across the membrane. Ligand-gated channels form another important class; these ion channels open and close in response to the binding of a ligand molecule, such as a neurotransmitter.

8.2.1 Ligand-Gated Ion Channels

Ionotropic receptors are a ligand-gated ion channel subtype, allowing the binding of an extracellular molecule and transient passage of ions through a transmembrane pore (Rao et al., 2022). Ligand-gated ion channels are cell surface proteins that bind small soluble molecules at their outer surfaces to induce functional changes in their permeability (Schmauder et al., 2023). The receptor channels open, or gate, when a neurotransmitter binds to an orthosteric site, causing a conformational change that leads to the conducting state. Endogenous or exogenous modifiers can modify gating by binding to allosteric sites (S. Alexander et al., 2011). LGICs serve as signal propagating factors, making them an integral part of the rapid intracellular signaling process. These cation channels are crucial in various physiological processes such as neurotransmission and muscle contraction.

 DOI: 10.1201/9781003594901-8

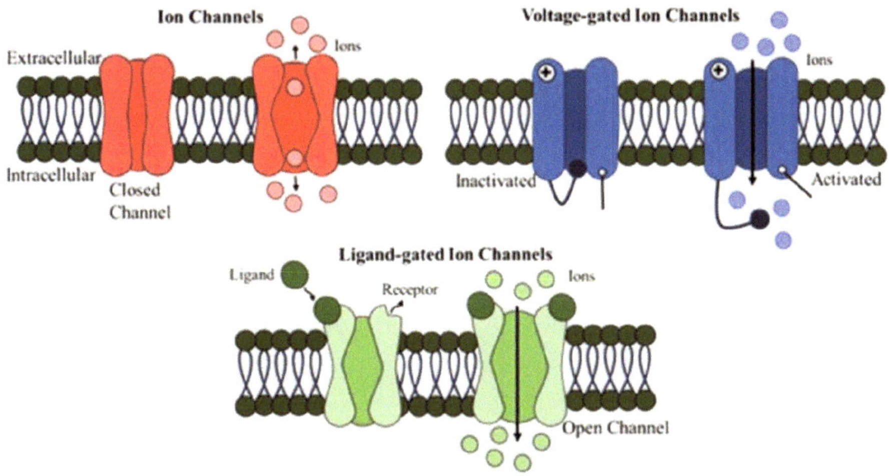

FIGURE 8.1 The two major types of ion channels.

8.2.1.1 Superfamilies of Ligand-Gated Ion Channels

8.2.1.1.1 Cys-Loop Superfamily

The Cys-loop superfamily is one of the largest and most studied LGIC superfamilies. It consists of nAChRs, GABAR-As, glycine receptors, and serotonin type-3 receptors (Changeux & Christopoulos, 2016). These receptors contain a conserved loop between two cysteine residues in the extracellular domain, which plays a key role in ligand binding and channel gating. Recent studies have provided insight into both the structural dynamics and functional aspects of Cys-loop receptors, identifying their role in synaptic transmission (Miller & Aricescu, 2014).

8.2.1.1.1.1 Nicotinic Acetylcholine Receptors (nAChRs)

Nicotinic acetylcholine receptors (nAChR) form a part of the Cys-loop superfamily, along with GABA-A, glycine, and serotonin or 5HT3 ligand-gated ion channels. A multigene family encodes 16 homologous mammalian nAChR subunits. These subunits combine to yield many diverse nAChR subtypes, each with unique patterns of expression, functional properties, and pharmacological features. One of the most studied types of ligand-activated ion channels, nAChRs can be found in both central and peripheral nervous systems (Hogg et al., 2003). The endogenous ligand for nAChRs is acetylcholine, which binds to the receptor and causes structural changes that result in the opening of the ion channel. Pharmacologically, several compounds target nAChRs as agonists and antagonists. For instance, nicotine is an agonist that mimics the actions of Ach, while curare acts as a competitive antagonist, blocking the binding process. (Figure 8.2)

8.2.1.1.1.2 Gamma-Aminobutyric Acid Receptors (GABAARs)

Gamma-aminobutyric acid (GABA) is the primary inhibitory neurotransmitter in the central nervous system (CNS), playing a crucial role in maintaining the balance between neuronal excitation and inhibition (Watanabe et al., 2002). GABA receptors are integral components of the intricate neural signaling network, mediating inhibitory effects through ligand-gated ion channels.

8.2.1.1.1.2.1 GABA Receptors The GABA receptor family includes a major class of ionotropic receptors, namely GABA-A. These receptors are pentameric and have five subunits, which form a chloride ion channel. When GABA binds to the receptor, a series of structural alterations take place in the channel that allows chloride ions to enter neurons. The increased inflow of chloride

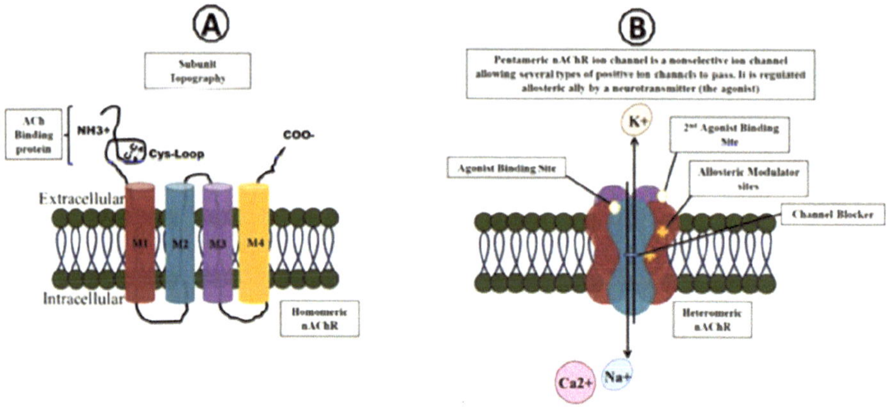

FIGURE 8.2 Structure of the nAChR receptor. (A) Transmembrane view of an nAChR subunit, showing four transmembrane domains (M1–M4), an extracellular amino and carboxy terminus, and an intracellular M3–M4 loop of variable length (B) Schematic diagram of an nAChR showing five subunits forming its ion channel.

ions hyperpolarizes the membrane, decreasing the likelihood of generating an action potential. This hyperpolarization underlies the GABAergic effect on neuronal activity. Various pharmacological substances modify the activity of GABA-A receptors. Benzodiazepines enhance inhibitory effects by increasing chloride channel opening. Barbiturates also act on GABA-A receptors, producing potentiated function and increased inhibitory neurotransmission (Saunders & Ho, 1990) (Figure 8.3).

8.2.1.1.1.2.2 GABA-B Receptors Unlike GABA-A receptors, GABA-B belongs to a family of metabotropic receptors. The activation of GABA-B receptors initiates a complicated intracellular signaling chain through G proteins. The result is the modulation of ion channels, including potassium channels. The opening of potassium channels hyperpolarizes the neuron, causing inhibitory effects. Agonists directly stimulate the receptors, such as baclofen, which is used clinically as an antispastic agent (Pinard et al., 2010). GABA-B receptor antagonists are also being studied for possible therapeutic purposes. Beyond direct agonists and antagonists, modulatory drugs have been developed to fine-tune GABAergic neurotransmission. Benzodiazepines, for example, act as positive allosteric modulators, enhancing the effect of GABA without directly activating the receptor (Bettler et al., 2004) (Figure 8.4).

8.2.1.1.2 Glutamate-Gated Ion Channels Superfamily

Ligand-gated ion channels are membrane proteins that allow ions to pass through the cell membrane in response to the binding of specific ligands. The glutamate-gated ion channels superfamily comprises the major components of excitatory neurotransmission, which are central to many processes in the CNS. The iGluRs belonging to this superfamily are the N-methyl-D-aspartate receptors (NMDAR), alpha-amino-3-hydroxy-5-methyl-4-isoxazole propionic acid receptors, and kainite receptors (Traynelis et al., 2010). Glutamate receptors are involved in synaptic plasticity, learning, and memory mechanisms. Recent studies have aimed at unraveling the structural complexity of these receptors and their long-term subunit composition, as well as providing evidence for their vital role in different neurological disorders (Greger & Mayer, 2019). There are two main types of glutamate receptors that act as ligand-gated ion channels:

8.2.1.1.2.1 Ionotropic Glutamate Receptors (IGluRs)

Ionotropic glutamate receptors (iGluRs) are directly coupled to ion channels. Upon binding of glutamate, these receptors undergo a conformational change that opens the ion channel, allowing

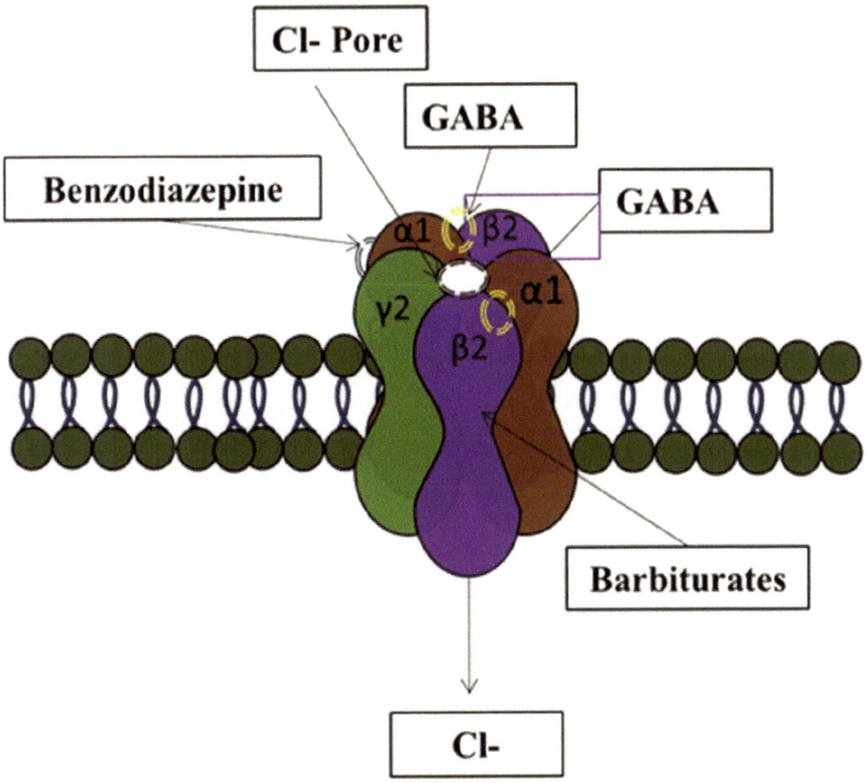

FIGURE 8.3 GABA-A receptor: a heteropentameric GABA-gated chloride ion channel. It is selectively permeable to chloride ions depending on the membrane potential. The receptor is composed of five subunits. Benzodiazepines and barbiturates modulate its pharmacological activity.

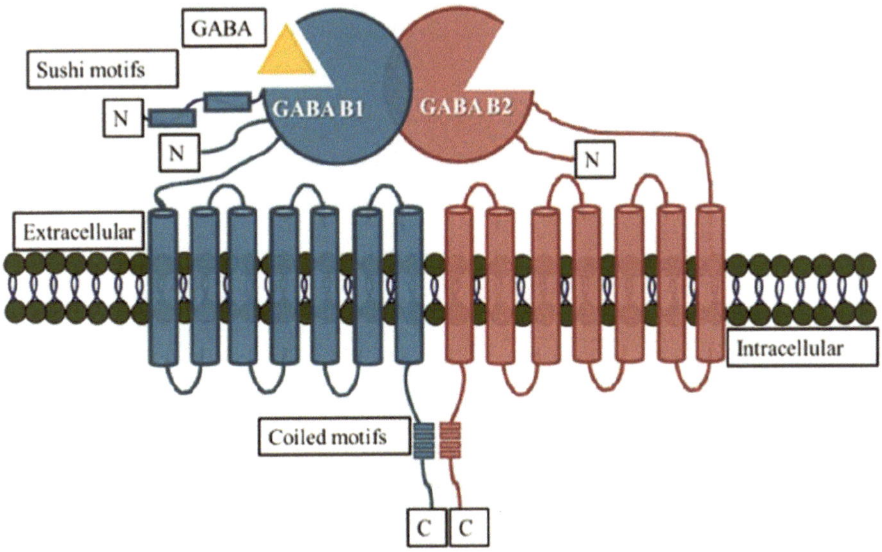

FIGURE 8.4 GABA-B receptor: a heterodimer of GABA-B1 and GABA-B2 subunits. Each has seven transmembrane domains with an extracellular N-terminus and intracellular C-terminus.

ions (Na^+, K^+, Ca^{2+}) to pass through the cell membrane. The three main subtypes of iGluRs are NMDA (N-methyl-D-aspartate) receptors, AMPA (α-amino-3-hydroxy-5-methyl-4-isoxazolepropio nic acid) receptors, and kainite receptors. Activation of these receptors is essential for excitatory neurotransmission in the CNS. They engage in processes such as learning, memory, and synaptic plasticity (Figure 8.5).

8.2.1.1.2.2 Metabotropic Glutamate Receptors (mGluRs)

Metabotropic glutamate receptors (mGluRs) are G protein-coupled receptors that, when activated by glutamate, trigger intracellular signaling cascades through second messengers. Unlike ionotropic receptors, mGluRs do not directly gate ion channels. Instead, they modulate cellular processes through the activation of signaling pathways. There are three groups (Groups I, II, and III) of mGluRs, and they play roles in regulating neurotransmitter release, synaptic transmission, and neuronal excitability (Crupi et al., 2019).

8.2.1.1.3 P2X Receptor Superfamily

The P2X receptors are ion channels activated by extracellular adenosine triphosphate (ATP). They participate in many physiological activities, such as neurotransmission, smooth muscle contraction, and immunity. The P2X receptor superfamily comprises seven subtypes (P2X1–7), and recent studies have shed light on their unique ligand-binding properties and signaling mechanisms. Additionally, P2X receptors have been implicated in several pathological conditions, making them potential therapeutic targets (Burnstock & Kennedy, 2011) (Figure 8.6).

8.2.1.2 Post-Translational Modifications as Crucial Regulators of Ligand-Gated Ion Channels

Ligand-gated ion channels, also known as mechanosensitive gating elements, are instrumental in cellular communication and the transduction of ions by opening channel gates upon binding to specific ligands (Mao et al., 2011).

8.2.1.2.1 Phosphorylation

Phosphorylation, the addition of phosphate groups, is the most prominent post-translational modification method affecting LGIC function. Several kinases (PKA, PKC, and CaMKII) have been associated with LGIC subunit phosphorylation. For instance, the NMDA receptor, which belongs to the ionotropic glutamate receptor class, has multiple phosphorylation sites (Mammen et al., 1999). Phosphorylation mechanisms play a significant role in regulating channel desensitization, receptor trafficking, and conductance rates (Swope et al., 1999). Notably, phosphorylation may either increase or decrease the LGIC function, mirroring changes in cell response pathways.

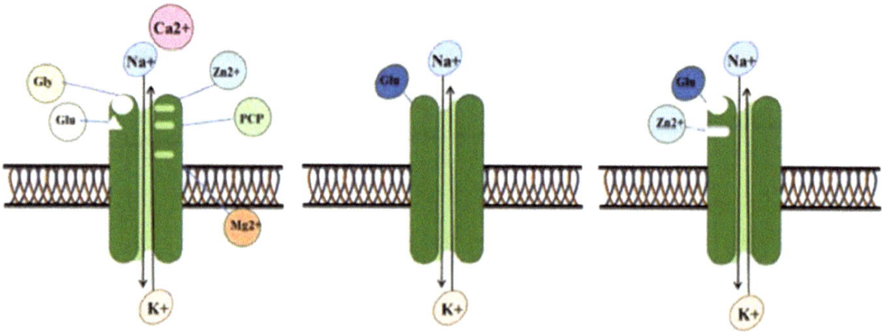

FIGURE 8.5 Glutamate receptors family.

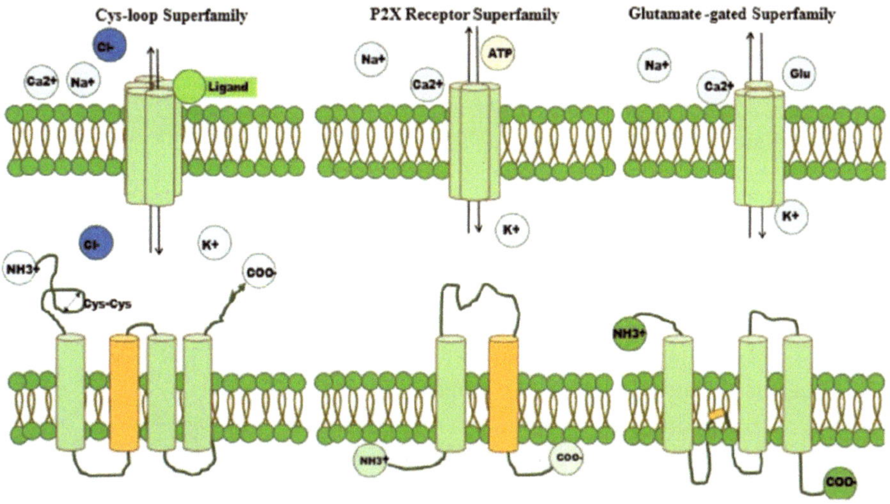

FIGURE 8.6 Ligand-gated ion channel superfamilies. A top view shows ligand preferences and ion selectivity for Cys-loop receptors, P2-type ionotropic purinoreceptors, and glutamate-activated ionotropic receptors. The illustration depicts each of the three receptor superfamilies, with ligand-binding domains highlighted in orange.

8.2.1.2.2 *Ubiquitination*

Ubiquitination, a covalent post-translational modification involving the conjugation of ubiquitin molecules to proteins, has been recognized as an important physiological process regulating LGIC stability and turnover. This system regulates the breakdown of various cellular proteins, including LGICs, via ubiquitin-proteasomes systems (Z. J. Chen & Sun, 2009). Ubiquitination of LGICs facilitates their proteasomal degradation, modulating their presence at the cell membrane surface. This process is vital for managing the LGIC signaling duration and reducing synaptic plasticity. Additionally, accumulating data indicates that ubiquitination may have nondegradative functions in regulating LGIC endocytosis and recycling, further increasing the complexity of this regulation.

8.2.1.2.3 *Glycosylation*

Glycosylation, an important post-translational modification affecting LGIC function, is the enzymatic incorporation of sugar molecules onto proteins. Several LGIC subtypes exhibit N-linked glycosylation on asparagine residues (Arey, 2012). This modification is critical for the folding, trafficking, and surface expression of LGICs. For instance, the glycine receptor, an LGIC subfamily member, requires N-linked glycosylation to achieve proper assembly and functional expression at synapses. Glycosylation also contributes to the stability of LGICs by protecting against early degradation and regulating interaction with other cellular partners.

In summary, post-translational modifications profoundly influence ligand-gated ion channels, altering their functional properties and resulting cellular responses. These regulatory mechanisms include phosphorylation, ubiquitination, and glycosylation, among others. These alterations are vital not only for understanding cell signaling but also for developing therapies. As research in this field advances, deciphering the complex mosaic of post-translational modifications linked to ligand-gated ion channels will undoubtedly expand our understanding of neuronal communication and therapeutic approaches.

8.2.2 Voltage-Gated Ion Channels

Voltage-dependent ion channels are sensitive to changes in the voltage gradient across the cell membrane. Voltage-gated ion channels are specialized transmembrane proteins that act as complex molecular gatekeepers, allowing specific ion permeation through the membrane in response to alterations in membrane potential. They are essential cellular components responsible for controlling various physiological functions. Understanding the molecular characteristics of these channels is important due to their roles in sensing and transmitting electrical impulses in excitable cells, such as neurons and muscles.

The study of voltage-gated ion channels can be traced back to critical research conducted during the mid-twentieth century. Significantly, the work of Hodgkin and Huxley revealed the negative polarity of axonal membrane potential and demonstrated that potentials of more than 100 millivolts were necessary for generating nerve impulses. Their work also helped elucidate the ionic bases of action potentials (Hodgkin & Huxley, 1952). Since this discovery, numerous studies have identified various voltage-gated ion channel types, each with specific traits and functions.

8.2.2.1 Types of Voltage-Gated Ion Channels

Various types of voltage-gated ion channels exist, differentiated by the specific ions they allow passage of. The major types of voltage-gated ion channels are sodium (Na^+), potassium (K^+), calcium (Ca^{2+}), and chloride (Cl^-) channels. Each class has several subtypes that contribute to certain cellular functions. Voltage-gated sodium channels play a vital role in initiating and propagating action potentials in neurons, while voltage-gated potassium channels are responsible for restoring the cell membrane to its resting state and controlling the membrane's resting potential (Figure 8.7).

8.2.2.1.1 Voltage-Gated Potassium Ion Channels

Voltage-gated potassium ion channels, specifically the Kv type, play a critical role in controlling the electrical potential across the cell membrane and maintaining cellular balance. These channels facilitate the specific movement of potassium ions across the cell membrane in response to changes in voltage. Their capacity to switch between open and closed states in response to voltage changes is crucial for the effective function of excitable cells. Recent research has elucidated the complex regulatory mechanisms that control voltage-gated potassium channel activity. Post-translational changes, such as phosphorylation and glycosylation, affect the function and location of these channels. Furthermore, their dynamic regulation is influenced by interactions with auxiliary subunits, intracellular signaling pathways, and environmental variables.

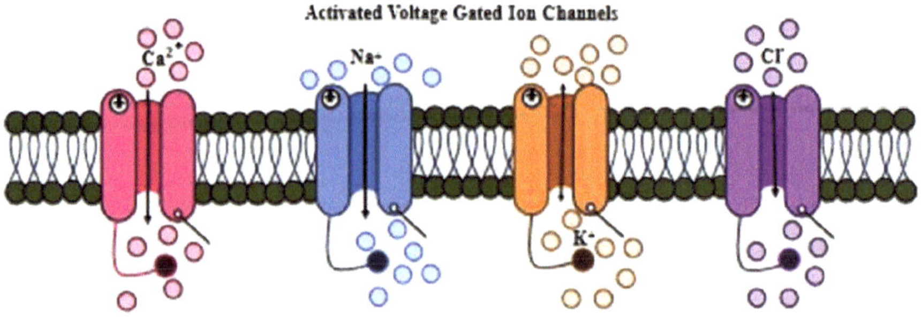

FIGURE 8.7 Types of voltage-gated ion channels.

8.2.2.1.1.1 Structure of Voltage-Gated Potassium Ion Channels

At the molecular level, voltage-gated potassium ion channels consist of multiple conserved domains. These proteins are characterized by their large size and structure, comprising many subunits that span the cell membrane. The main structural elements consist of four alpha subunits, which form the channel through which ions flow, and auxiliary subunits that regulate channel activity. These channels typically have four subunits organized in a tetrameric configuration around a central pore (Catterall, 2010). The voltage-sensory domain consists of four α-helices (S1–S4), with the S4 segment containing positively charged residues that serve as the main voltage sensors. These sensors detect alterations in the membrane potential, initiating conformational changes (Yu & Catterall, 2004). The ion-conducting pathway is formed by the transmembrane domains, consisting of six α-helical segments (S1–S6). The pore domain comprises the S5 and S6 segments. The P-loop region connects the S5 and S6 segments and plays a critical role in determining ion selectivity. Regulatory domains in voltage-gated potassium channels control their activity, often through interactions with cellular signaling molecules or post-translational alterations at the N- and C-terminal domains (Luján, 2010) (Figure 8.8).

8.2.2.1.1.1.1 Functional Diversity Kv channels exhibit remarkable functional diversity, with distinct subtypes expressed in a range of tissues and cell types. Each subtype has unique biophysical characteristics, such as voltage sensitivity, activation and inactivation rates, and sensitivity to pharmacological modulators. The presence of multiple Kv channels enables precise modulation of cellular excitability to meet the distinct requirements of various tissues (Shah & Aizenman, 2014).

Subtype diversity is a key feature of voltage-gated potassium channels. Multiple genes encode different channel isoforms, each with unique biophysical characteristics and expression patterns. For example, the Kv1 subfamily demonstrates rapid activation and inactivation kinetics, whereas the Kv7 subfamily is characterized by sluggish activation and no inactivation. This diversity enables fine-tuning of cellular excitability and contributes to the complexity of physiological processes (Shah & Aizenman, 2014).

The diversity of Kv channels is a result of numerous genes encoding the alpha subunits and distinct splice variants. Different Kv channel subtypes have unique biophysical characteristics, enabling them to specialize in particular cellular tasks. Recent investigations using advanced techniques like cryo-electron microscopy (cryo-EM) and X-ray crystallography have yielded detailed,

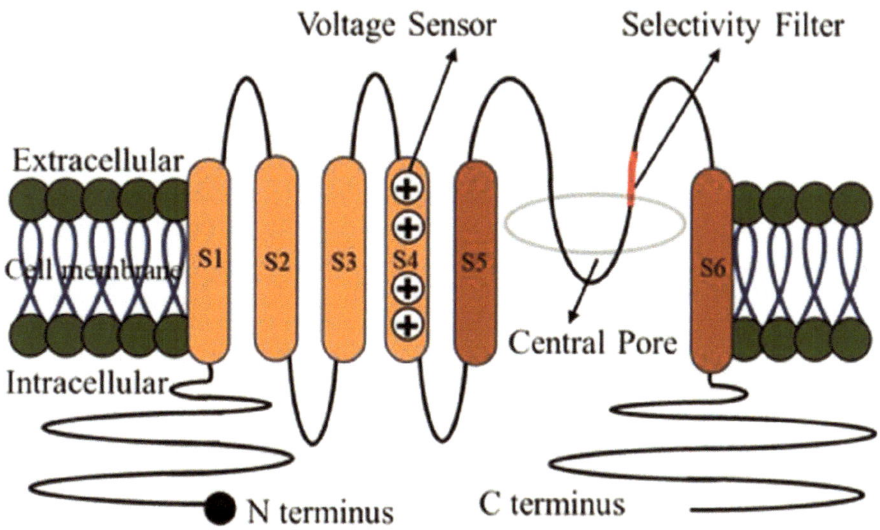

FIGURE 8.8 A voltage-gated potassium channel monomer. It comprises six transmembrane segments (S1–S6). A pore loop (P) between S5 and S6 pore loop contains the selective filter for ions (shown in red). S4 acts as a voltage sensor to detect changes in membrane potential due to stimulus.

high-resolution structural information. Studies of Kv1.2 and Kv1.2/2.1 chimeric channel structures provided a comprehensive understanding of channel design and the complex mechanisms involved in voltage sensing and gating (Lee et al., 2005; Long et al., 2005).

8.2.2.1.1.2 Mechanism of Voltage-Gated Potassium Ion Channel Regulation in Cellular Signaling

8.2.2.1.1.2.1 Voltage-Sensing and Gating Mechanisms Landmark work on potassium channels, such as the Kv1.2 channel, has revealed the details about the ion-channel residue interactions that regulate selectivity (Doyle et al., 1998). Voltage-gated potassium channel function relies on the intricate interaction between voltage sensing and gating. At rest, the channel remains closed, with the voltage sensor in its inactive conformation (Jan & Jan 2012; Jiang et al., 2002). Membrane depolarization causes the voltage sensor to change shape, opening the channel pore. Channel opening, closing, and inactivated states are finely tuned by intracellular and extracellular factors (Jensen et al., 2012). This allows potassium ions to flow out more easily, repolarizing the membrane (Kim & Nimigean, 2016).

Cell depolarization shifts the voltage sensors, initiating structural changes that activate the channel pore. C-type inactivation in specific potassium channels adds complexity to the gating mechanism. Structural analysis of C-type inactivation in Kv1.2 channels (Hoshi & Armstrong, 2013) has significantly advanced our understanding of its molecular basis. Channels can be closed or inactivated through various mechanisms; N-type inactivation, for instance, uses a ball-and-chain mechanism. A period of inactivity called the refractory period follows, during which the channel cannot be reactivated immediately, preventing ongoing depolarization (Payandeh et al., 2011).

To precisely control the duration of the action potential, voltage-gated potassium channels undergo inactivation to prevent prolonged opening. N-type inactivation results from cells obstructing the channel's N terminus, whereas C-type inactivation is the consequence of pore region conformational changes (Hoshi et al., 1990; Zhou et al., 2001).

8.2.2.1.1.3 Physiological Significance of Voltage Gated Potassium Channels

8.2.2.1.1.3.1 Neuronal Excitability Voltage-gated potassium (K^+) channels play a vital role in neuronal excitability. The interaction among various Kv channel subtypes determines the responsiveness of neurons, impacting signal transmission between nerve cells, incoming information integration, and neural network functioning. The channels comprise alpha subunits, which form the opening for potassium ion flow, and auxiliary subunits that regulate channel characteristics. Kv1, Kv2, and Kv3 are among the multiple Kv channel families. Each family has unique characteristics and is found in distinct parts of the nervous system (Hille, 2001).

8.2.2.1.1.3.1.1 Action Potential and Resting Membrane Kv1 channels help control action potential duration and neurotransmitter release. Kv3 channels enable fast-spiking neurons to generate high-frequency firing. Specific Kv7 channel subtypes regulate neuronal excitability and prevent hyperexcitability. Voltage-gated potassium channels primarily serve to uphold the resting membrane potential of neurons, which is determined by the potassium ion concentration gradient across the cell membrane. Potassium channels facilitate potassium ion efflux, which hinders positive charge buildup within the cell. This outflow of K^+ generates a negative electrical charge inside the cell, maintaining a resting membrane potential of –70 mV (Hille, 2001). Disruptions to this delicate balance can affect overall neuronal excitability, potentially causing hyperexcitability or depolarization.

Voltage-gated sodium channels become active during an action potential, allowing sodium ions to enter the cell and depolarize the cell membrane. Voltage-gated potassium channels open during the repolarization phase. As the action potential reaches its maximum, these channels open, enabling potassium ion efflux from the cell. This outflow of potassium ions opposes the depolarization induced by the inflow of sodium ions, restoring the membrane potential to its resting condition. The rapid repolarization by potassium channels enables neurons to generate prolonged action potentials, facilitating efficient signal transmission in the nervous system (Hille, 2001).

The diversity of potassium channel subtypes, each with distinct biophysical characteristics, contributes to variation in action potential shapes across different neuronal populations. Kv1 channels are linked to delayed rectifier currents, which impact the repolarization phase and overall duration of the action potential (Jan & Jan, 2012).

8.2.2.1.1.3.1.2 ***Synaptic Transmission*** Voltage-gated potassium channels affect not only action potentials but also synaptic transmission. These channels, located within the presynaptic terminals, regulate the release of neurotransmitters. After an action potential, the activation of potassium channels causes a hyperpolarization of the presynaptic terminal, restricting the entry of calcium ions and decreasing the likelihood of neurotransmitter release. This feedback mechanism controls the intensity of synaptic transmission, impacting the overall responsiveness of brain circuits.

The malfunctioning of voltage-gated potassium channels has been linked to a range of neurological diseases. Channelopathies can occur when mutations in these channels disrupt their normal functioning, which in turn affects the excitability of neurons. For example, mutations in the KCNQ2 and KCNQ3 genes, which produce Kv7.2 and Kv7.3 potassium channels, are linked to benign familial newborn seizures (Maljevic & Lerche, 2014) (Figure 8.9).

8.2.2.1.1.3.2 Cardiac Function Voltage-gated potassium channels play physiologically significant roles in cardiac function, including regulating contractility, overall electrical stability, and the duration of action potentials. Their importance is highlighted by their role in preserving electrical stability and averting arrhythmias (Sanguinetti & Tristani-Firouzi, 2006). Kv channels are essential for the heart's membrane repolarization after each action potential, helping maintain proper rhythm and synchronization of cardiac contractions (Jan & Jan, 2012). Genetic changes to Kv11.1 (hERG) can cause long QT syndrome, as this gene is crucial for cardiac repolarization. Kv channels also modulate action potential duration and maintain the resting membrane potential in cardiac cells (Sanguinetti & Tristani-Firouzi, 2006).

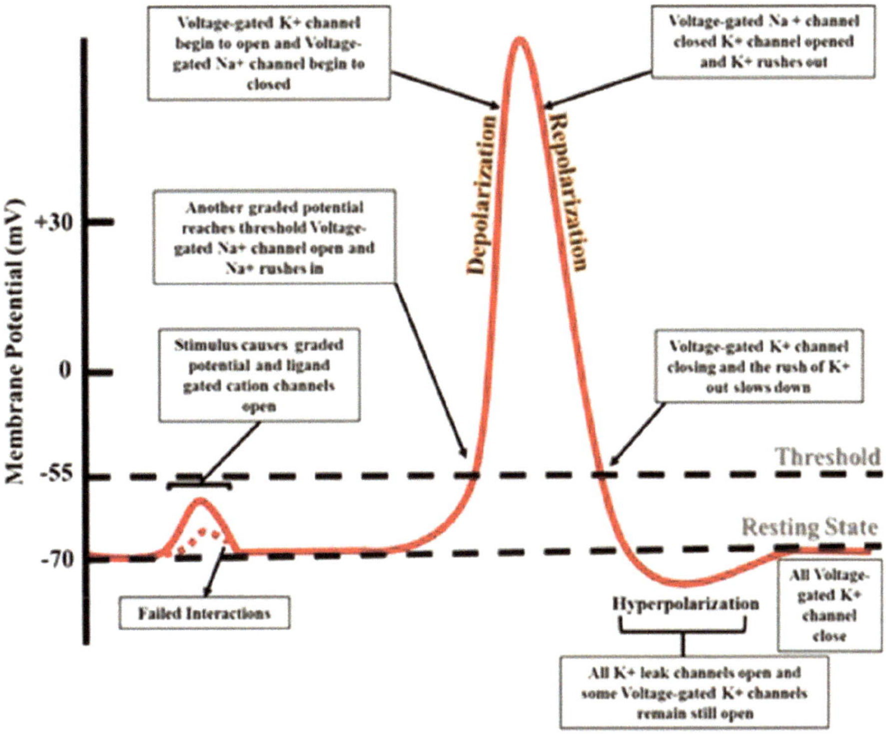

FIGURE 8.9 Changes in axon membrane potential during an action potential.

8.2.2.1.1.3.2.1 Action Potential and Cardiac Repolarization Voltage-gated potassium (K^+) channels are crucial in controlling heart function by modulating action potential repolarization. The heart's action potential is an intricate electrical phenomenon that governs the heart's rhythmic contraction and relaxation. Voltage-gated potassium channels primarily facilitate the restoration of the resting membrane potential during the repolarization phase.

The rapidly activating delayed rectifier potassium current (IKr), mediated by the hERG (human ether-à-go-go-related gene) channel, is crucial for cardiac repolarization. This channel facilitates potassium ion efflux, playing a key role in extending the cardiac action potential repolarization phase. This is essential for maintaining appropriate ventricular contraction and relaxation duration. IKr dysfunction is linked to long QT syndrome, a disorder characterized by a heightened susceptibility to life-threatening arrhythmias (Sanguinetti & Tristani-Firouzi, 2006). Aside from other voltage-gated potassium channels, the slow activating delayed rectifier potassium current (IKs) also plays a significant role in cardiac repolarization. The interaction among these channels ensures precise coordination of ion flows, contributing to heart rhythm stability and consistency (Barhanin et al., 1996).

8.2.2.1.1.3.2.2 Regulation of Excitability and Contractility Voltage-gated potassium channels influence both repolarization and the regulation of excitability and contractility in cardiomyocytes. The transient outward potassium current also contributes to heart excitability, playing a role in initial repolarization and influencing the length of the action potential. Alterations in the expression or function of genes have been linked to several cardiac conditions, such as hypertrophy and heart failure (Nerbonne, 2000). Furthermore, proper Kv channel function is crucial for regulating intracellular calcium levels, a key determinant of cardiac contractility. Inconsistencies in the quantity of potassium currents can result in compromised calcium regulation, diminished contractile performance, and cardiac arrhythmias (Bers, 2002).

8.2.2.1.1.3.3 Cellular Homeostasis Voltage-gated potassium channels regulate cell volume, apoptosis, and cell cycle progression, all crucial for maintaining cellular homeostasis. Cell volume regulation is essential for cellular homeostasis. Cells must balance water absorption and excretion to avoid swelling or shrinking, which can result in cellular malfunction or even cell demise. Osmoregulation, involving ion transport systems, primarily regulates cell volume maintenance. The sodium-potassium pump is a crucial participant in this process, actively transporting sodium ions out of the cell and potassium ions into the cell. This pump, along with various ion channels and transporters, helps regulate the osmotic pressure within the cell, preventing excessive swelling or shrinkage (Alberts et al., 2022).

Apoptosis is a highly controlled process crucial for development, tissue balance maintenance, and removal of damaged or unnecessary cells. Voltage-gated potassium channels contribute to apoptosis by regulating membrane potential and impacting calcium signaling, a crucial factor in controlling apoptotic pathways. These channels help maintain the electrical potential across the mitochondrial membrane and influence the release of chemicals that trigger programmed cell death. Additionally, the outflow of potassium through these channels can initiate cell contraction and the formation of membrane protrusions, which are distinctive characteristics of apoptotic cells (Ghanshani et al., 2000).

Voltage-gated potassium channels also regulate the cell cycle, influencing cell growth and differentiation. Certain potassium channels are crucial during the G1 phase for sustaining a hyperpolarized membrane potential, facilitating the advancement of the cell cycle. Furthermore, alterations in the expression and functioning of potassium channels are detected during the shift from the G1 to the S phase, impacting the process of DNA synthesis and cellular division (Urrego et al., 2014).

8.2.2.1.1.4 Classification of Voltage-Gated Potassium Channel

Voltage-gated potassium channels are divided into three superfamilies, which are further classified into many subtypes, as shown in Figure 8.10 and in Table 8.1.

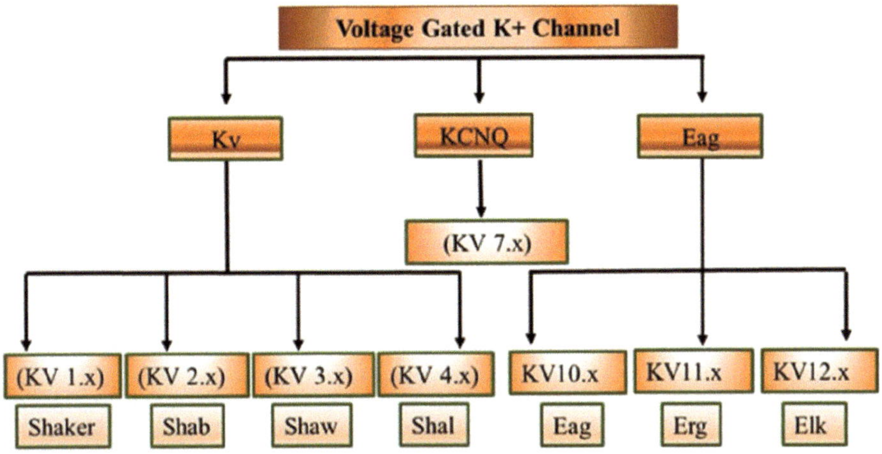

FIGURE 8.10 Classification of voltage-gated potassium channels.

TABLE 8.1
Various subtypes of potassium channels along with their respective genes, localization within the body, physiological functions, and associated channelopathies

Type	Gene	Location	Physiological Function	Channelopathy
Kv1.1	KCNA1	Axon	Helps in repolarization and shaping of action potential	Myokymia & periodic ataxia
Kv1.2	KCNA2	Axon	Regulates pre- and postsynaptic membrane excitability	Pharmacoresponsive epilepsy, episodic ataxia
Kv1.3	KCNA3	Axon	Essential for T-cell proliferation and activation	Episodic ataxia type 1
Kv1.4	KCNA4	Axon	Regulates fast-repolarizing phase of cardiac action potentials, influencing duration	Neuropathic pain, myasthenic syndromes, paroxysmal dyskinesias
Kv1.5	KCNA5	Brain	Restores resting membrane potential of beta cells after depolarization, contributing to insulin secretion regulation	Familial atrial fibrillation type 7
Kv1.6	KCNA6	Axon	Important for ocular neurovascular beds, interacts with other proteins	Infantile epilepsy, neurodevelopment anomalies
Kv1.7	KCNA7	Skeletal muscle, heart, kidney	Increases negative membrane potential, raising threshold for triggering action potential	Inherited cardiac disorders
Kv1.8	KCNA10	Brain	Regulated by cGMP, mediates effects of substances that increase intracellular cGMP	Diabetic neuropathy, post-herpetic neuralgia, other pain syndromes
Kv2.1	KCNB1	Soma, proximal dendrites	Determines intrinsic neuronal excitability	Autism, epilepsy, intellectual disability
Kv2.2	KCNB2	Soma, dendrites	Controls hyperpolarization, after potentials, excitability	Alzheimer's disease, Parkinson's disease, amyotrophic lateral sclerosis

(Continued)

TABLE 8.1 (CONTINUED)

Various subtypes of potassium channels along with their respective genes, localization within the body, physiological functions, and associated channelopathies

Type	Gene	Location	Physiological Function	Channelopathy
Kv3.1	KCNC1	Axon	Involved in rapid firing of neurons	Myoclonic epilepsy in infants, episodic ataxia
Kv3.2	KCNC2	Soma, dendrites	Rapid spiking of interneurons	Various types of epilepsy
Kv3.3	KCNC3	Axon	Important for endocytosis of synaptic vesicles	Spinocerebellar ataxia type 13
Kv3.4	KCNC4	Axon	Generates atypical voltage-dependent transient current important for neuronal excitability	Congenital disorder of glycosylation, autism spectrum disorder, progressive myoclonus epilepsy
Kv4.1	KCND1	Soma, dendrites	Functions in repolarization of membrane action potentials; regulates neurotransmitter release, heart rate, insulin secretion, smooth muscle contraction	Epilepsy and Cardiac Arrhythmias
Kv4.2	KCND2	Soma, dendrites	Mediates a rapidly inactivating, A-type outward potassium current which is not under the control of the N terminus	Early myoclonic encephalopathy
Kv4.3	KCND3	Soma, dendrites	Contributes to cardiac transient outward potassium current and neuronal potassium current	Spino-cerebellar ataxia type 19, Burgada syndrome 9, hereditary ataxia
Kv5.1	KCNF1	Brain	No specific function as voltage-gated ion channel reported yet	Arts syndrome, fragile X syndrome
Kv6.1	KCNG1	Brain	Forms heterotetramer with KV2.1 ion channel	Coumarin resistance, fragile X syndrome
Kv6.2	KCNG2	Brain	Regulates neurotransmitter release, heart rate, neuronal excitability, smooth muscle contraction, cell volume	Sick sinus syndrome, pyromania, opioid abuse
Kv6.3	KCNG3	Brain	No independent function; modulates activation and inactivation of KV2.1 ion channel	Retinal cone dystrophy 3b
Kv6.4	KCNG4	Brain	No specific activity alone; forms with other channels for various functions	Night blindness, congenital stationary type 1d
Kv7.1	KCNQ1	Various tissues	Functions in colon, inner ear, heart; mediates current kinetics	Long QT syndrome (LQT1), short QT syndrome 2 (SQT2), atrial fibrillation
Kv7.2	KCNQ2	Axon	Increases sub-threshold current to delay inactivation process of neuronal excitation	Developmental and epileptic encephalopathy, febrile seizures
Kv7.3	KCNQ3	Axon, soma, dendrites	Determines neuronal responsiveness to synaptic stimuli and sub-threshold for excitability	Benign neonatal seizures

(Continued)

TABLE 8.1 (CONTINUED)

Various subtypes of potassium channels along with their respective genes, localization within the body, physiological functions, and associated channelopathies

Type	Gene	Location	Physiological Function	Channelopathy
Kv7.4	KCNQ4	Axon	Regulates excitability cochlear sensory cells	Rare genetic deafness, autosomal dominant non-syndromic deafness 2a
Kv7.5	KCNQ5	Soma, dendrites	Regulates neuronal excitability	Autosomal intellectual development disorder
Kv8.1	KCNV1	Brain	Modulates Kv 2.1 & 2.2 activity by shifting threshold to more negative values, delaying inactivation	Familial adult myoclonic epilepsy, atrial septal defect 5
Kv9.1	KCNS1	Brain	Modulates activation and inactivation of Kv 2.1 and 2.2 channels	Marsili syndrome, pain disorder, causalgia
Kv9.2	KCNS2	Brain	Modulates Kv2.1 and Kv 2.2 ion channel functions by forming heterotetrameric channels	Hereditary essential tremor 1, 4,5
Kv9.3	KCNS3	Brain	Forms heterotetrameric channel with Kv2.1, increasing action potential amplitude and delaying inactivation	Generalized epilepsy with febrile seizures type 4, developmental and epileptic encephalopathy
Kv10.1	KCNH1	Brain	Mediates potassium current in myoblasts; involved in cellular differentiation and proliferation	Temple-Baraitser syndrome, Zimmermann-Laband syndrome 1
Kv10.2	KCNH5	Brain	Produces non-inactivating outward rectifying current	Early infantile epileptic encephalopathy, developmental and epileptic encephalopathy
Kv11.1	KCNH2	Brain	No specific channel activity alone; modulates other channels by forming heterotetramers	LQT1, SQT1
Kv11.2	KCNH6	Brain	Produces slowly activating rectifying current	LQT2, congenital hypomyelinating, autosomal recessive neuropathy
Kv11.3	KCNH7	Brain	Pore-forming unit of ion channel; no other specific activity alone	Autosomal recessive deafness 1a
Kv12.1	KCNH8	Brain	Elicits slowly activating rectifying current in action potential	Epilepsy, progressive myoclonic 4 with and without renal failure
Kv12.2	KCNH3	Brain	Produces outward current with rapid inactivation	Autosomal dominant deafness 2a, chronic wasting disease
Kv12.3	KCNH4	Brain	Elicits outward current without inactivation	Epilepsy and cardiac arrhythmias

8.2.2.1.2 Voltage-Gated Sodium Ion Channels

Voltage-gated sodium ion channels (VGSCs) facilitate action potential generation and other vital physiological processes by enabling the specific migration of sodium ions across the cell membrane in neurons and muscle cells. These channels consist of a prominent α-subunit that contains the voltage-sensing and pore-forming regions, along with smaller β-subunits that regulate channel activity. Sodium channel activation causes a fast entry of sodium ions, leading to the depolarization of the

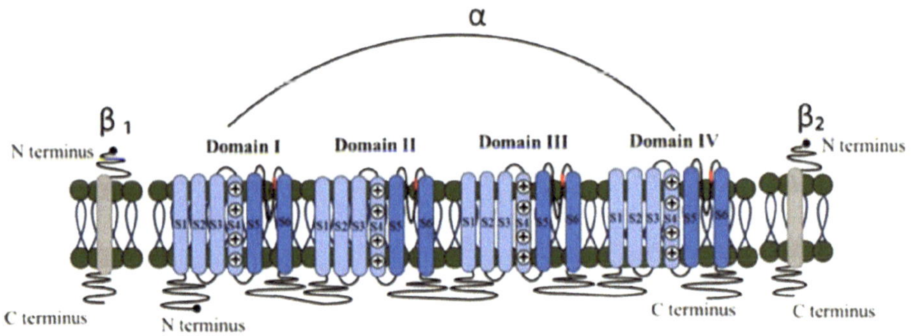

FIGURE 8.11 Schematic representation of a voltage-gated sodium channel. The α subunit consists of four homologous domains (I–IV) with six transmembrane segments (S1–S6) per domain. The pore region is located between segments S5 and S6. Sodium channels in the central nervous system (CNS) associate with two β subunits (β1 and β2). The β1 subunit is noncovalently linked to the α subunit, while β2 is covalently attached by a disulfide bond. All β subunits are structurally similar, including a small cytoplasmic region, a transmembrane carboxy-terminal region, and a large external amino-terminal region containing an immunoglobulin domain. The α subunits control channel opening and closing, while the β subunits modulate channel properties.

cell membrane (Payandeh et al., 2011). Recent structural investigations, particularly those employing cryo-EM (cryo-electron microscopy), have provided detailed and high-resolution insights into the dynamic conformational alterations that occur during channel activation and inactivation (Shen et al., 2017).

8.2.2.1.2.1 Structure of Voltage-Gated Sodium Ion Channels
The alpha subunit, consisting of four homologous domains (I–IV), is the fundamental structural component of sodium channels. Each domain contains six transmembrane segments (S1–S6) that form the voltage-sensing and ion-conducting pore (De Lera Ruiz & Kraus, 2015). The S1–S4 segments create voltage-sensing domains to detect changes in membrane potential. An ion-conducting pore that facilitates ion passage is created by the pore loop that lies between the S5 and S6 segments, influencing ion permeability and selectivity (Rems et al., 2020). In addition to α subunits, sodium channels frequently form associations with auxiliary β subunits, which regulate channel kinetics, expression, and the positioning of the channels on the cell surface. These subunits facilitate the many processes by which sodium channels operate in different organs and tissues (Bouza & Isom, 2017) (Figure 8.11).

8.2.2.1.2.2 Function of Voltage-Gated Sodium Ion Channels
8.2.2.1.2.2.1 Action Potential Function in Excitable Cells Cell communication mostly occurs through action potentials, which are quick and transient changes in the membrane potential of excitable cells. Neurons produce action potentials in response to stimuli, which then propagate along the axon via voltage-gated sodium channels, ensuring fast and effective communication (Drukarch et al., 2018). Sodium and potassium channels help maintain the resting state by selectively permitting or limiting ion movement. At rest, sodium channels remain closed. However, when the cell membrane depolarizes, these channels change shape, allowing sodium ions to enter. Upon sufficient stimulation, voltage-gated sodium channels activate, causing a fast sodium ion influx into the cell (Alberts et al., 2022). This depolarization propagates throughout the cell, enabling swift signal conduction.

The sudden increase in sodium permeability causes depolarization, reducing the membrane potential's negative charge. Sodium influx continues until the membrane potential reaches its optimal level, often around +40 mV. At this point, sodium channels initiate inactivation, halting the sodium influx. Simultaneously, voltage-gated potassium channels open, resulting in the release of potassium ions. After repolarization, the membrane potential may briefly hyperpolarize because of

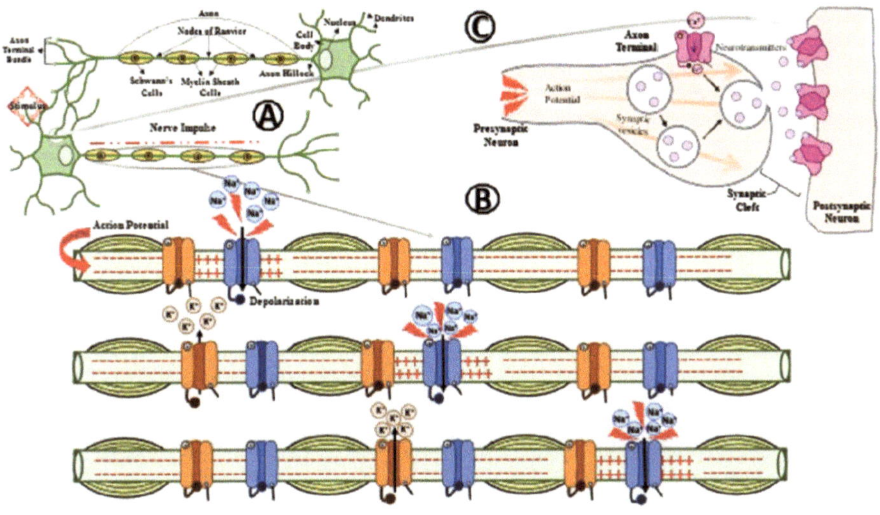

FIGURE 8.12 Action potential propagation and synaptic transmission. (A) A nerve impulse is triggered by a stimulus, causing a change in the electrical charge across the neuron's membrane. (B) An action potential triggers the opening of voltage-gated Na^+ channels along the axon. Myelin insulation increases conduction speed. Nodes of Ranvier, rich in Na^+ and K^+ channels, allow ion flow every 1–2 mm (Poliak & Peles, 2003). (C) At synapses, presynaptic membrane depolarization opens voltage-gated Ca^{2+} channels. Ca^{2+} influx causes synaptic vesicle fusions neurotransmitter release. Neurotransmitters bind to postsynaptic ligand-gated ion channels, causing local depolarization or hyperpolarization.

the ongoing efflux of potassium ions. This hyperpolarization maintains sodium channels in an inactive state during the refractory period, preventing immediate initiation of another action potential (Grider et al., 2024) (Figure 8.12).

Various neurological illnesses exemplify the significance of sodium channels in governing neuronal activity. Nav1.7 and Nav1.2 channels are essential for the generation of the action potentials in the neurons. These channels are associated with disorders such as epilepsy and neuropathic pain when mutated (Wang et al., 2017). Dravet syndrome, a severe form of childhood epilepsy, is associated with genetic abnormalities in the SCN1A gene, which encodes the Nav1.1 channel (Bender et al., 2012).

Action potential propagation is made possible by the successive opening and closing of sodium channels along the axon. Myelination by Schwann cells in the peripheral nervous system and oligodendrocytes in the central nervous system enhances conduction speed via saltatory conduction. At synapses, action potentials trigger neurotransmitter release, facilitating signal transmission between neurons (Seidl, 2014).

Normal neuromuscular system function relies on sodium channels in excitable cells and their involvement in action potentials. Coordinated ion transport through these channels is crucial for cellular signaling, signal transmission, and various physiological processes. In skeletal muscle fibers, sodium channels are critical for excitation-contraction coupling, promoting depolarization and initiating muscle contraction by triggering calcium release from the sarcoplasmic reticulum (Bean, 2007).

At the neuromuscular junction, where motor neurons interact with muscle fibers, voltage-gated sodium channels activate when an action potential arrives. This activation depolarizes the muscle cell membrane, propagating to the sarcoplasmic reticulum and triggering calcium ion release, which initiates muscle fiber contraction. Sodium channels thus play a crucial role in excitation-contraction coupling in muscle cells (Hille, 2001).

In the heart, sodium channels are vital for initiating and transmitting electrical signals that result in synchronized muscle contractions. Nav1.5, the primary cardiac sodium channel, is responsible for the swift increase in the electrical potential during the contraction of heart muscle cells.

Malfunctioning of sodium channels in the heart can result in arrhythmias. For example, mutations in the SCN5A gene, which codes for Nav1.5, have been linked to long QT syndrome and Brugada syndrome (Veerman et al., 2015). Understanding the complex role of sodium channels in cardiac electrophysiology is crucial for comprehending and treating cardiac arrhythmias.

Mutations in sodium channel genes can also cause disorders such as myotonia and periodic paralysis (George, 2005). These conditions highlight the importance of sodium channels in the precise regulation of muscle contraction and relaxation (Catterall, 2012).

8.2.2.1.2.3 Regulation of Voltage-Gated Sodium Ion Channel in Cellular Signaling

Before examining sodium channel regulation, it is important to understand their basic function and structure. Sodium channels are transmembrane proteins with various domains, such as voltage-sensing and pore-forming domains. The channels exhibit distinct conformational states, including closed, open, and inactivated, which are regulated by alterations in membrane potential. Sodium channel regulation is complex, involving voltage-dependent activation and inactivation mechanisms, regulatory proteins, post-translational modifications, and pharmacological modulation (Marban et al., 1998).

8.2.2.1.2.3.1 Voltage-Gating and Inactivation Mechanism The S4 segments of voltage sensors consist of positively charged amino acids that react to alterations in membrane potential. Membrane depolarization initiates structural alterations in the voltage-sensing domains, activating the channel. The velocity and likelihood of activation significantly influence cellular excitability. Sodium channels experience swift inactivation, becoming unresponsive to subsequent activation. This guarantees appropriate action potential duration and prevents prolonged depolarization, maintaining optimal cellular function (Bezanilla, 2000).

Sodium channels demonstrate rapid inactivation, which is essential for preventing continuous sodium influx. The inactivation gate, situated within the intracellular loop connecting domains III and IV, effectively blocks the ion-conducting channel. Inactivation occurs via two primary mechanisms: fast and slow inactivation. Fast inactivation occurs when the channel pore is quickly obstructed by the inactivation gate. Slow inactivation refers to gradual conformational changes affecting channel availability over a longer duration (Motoike et al., 2004).

8.2.2.1.2.3.2 Regulatory Proteins Several auxiliary and regulatory proteins regulate sodium channel function. Beta subunits, a well-known group of proteins, affect channel-gating properties and membrane expression. They also contribute to channel trafficking and positioning within the cell membrane. Other proteins, such as fibroblast growth factors and neuronal cell adhesion molecules, can also influence sodium channels (Arikkath & Campbell, 2003; Isom, 1994).

8.2.2.1.2.3.3 Post-Translational Modifications Sodium channels experience diverse post-translational modifications that fine-tune their function. Phosphorylation, an enzymatic process facilitated by protein kinases, is an essential controller of sodium channels. Protein kinase A, protein kinase C, and other kinases phosphorylate residues on the channels, modifying their activity and responsiveness to voltage fluctuations. Protein phosphatases, such as protein phosphatase 1 and protein phosphatase 2, can remove phosphate groups from sodium channels, serving as a dynamic regulatory mechanism.

8.2.2.1.2.3.4 Pharmacological Modulation Various pharmacological agents target sodium channels for therapeutic purposes. In clinical settings, lidocaine and bupivacaine are widely used local anesthetics to block sodium channels for pain management. Antiarrhythmic drugs, such as flecainide and mexiletine, modulate sodium channel activity to treat cardiac arrhythmias (Sheets et al., 2010). Developing selective sodium channel modulators remains an active area of research for various neurological and cardiovascular disorders.

8.2.2.1.2.4 Voltage-Gated Sodium Ion Channel Subtypes

Nine subtypes of the α subunit (Nav1.1–Nav1.9) have different expression patterns and physiological functions. For example, Nav1.1 is highly expressed in the central nervous system and is associated with disorders such as schizophrenia (Mantegazza & Catterall, 2012). Nav1.5 is highly expressed in the heart and is implicated in cardiovascular disease (Abriel, 2010). Nav1.7 is highly expressed in tissues enriched with sensory neurons and plays an important role in pain perception (Cox et al., 2006).

Recent studies have demonstrated the importance of VGSC subtypes in health and disease. For example, Nav1.8 is associated with inflammatory and neuropathic pain (Nassar et al., 2004), while Nav1.9 is associated with pain perception and intestinal motility (Leipold et al., 2013). Understanding the role of different VGSC subtypes is crucial for developing targeted therapies against diseases related to sodium channel activity (Table 8.2).

8.2.2.1.3 Voltage-Gated Calcium Ion Channels

Voltage-gated calcium ion channels (VGCCs) have pivotal functions in various physiological processes, including neurotransmitter release, muscle contraction, and gene expression. Accurate

TABLE 8.2

Roles of different sodium channel subtypes in physiological processes and their implications in various channelopathies

Type	Gene	Location	Physiological Function	Channelopathy
Nav1.1	SCN1A	Axon initial segment (AIS)	Crucial role in generation and propagation of action potentials, enabling communication within the nervous system	Seizures, epilepsy
Nav1.2	SCN2A	Axons of cortex, thalamus, hippocampus	Permits sodium influx from extracellular space into cytosol after nerve membrane depolarization	Autism, episodic ataxia, epilepsy
Nav1.3	SCN3A	Dorsal root ganglion (DRG) sensory neurons, fetal brain	Generation and conduction of action potential in excitable cells	Hemiplegic migraine, epilepsy
Nav1.4	SCN4A	Central nervous system, skeletal muscles	Generation and propagation of action potentials in neurons and muscles	Myotonia and periodic paralysis disorders
Nav1.5	SCN5A	Intercalated disc and lateral membrane of cardiac cells	Crucial for cardiac electrophysiology; major role in cardiac sodium current	Long QT syndrome, Brugada syndrome, atrial fibrillation, dilated cardiac myopathy
Nav1.6	SCN8A	Nodes of Renvier, axon initial segment	Regulation of neuronal excitability	Intellectual disability, movement disorders, epileptic encephalopathy
Nav1.7	SCN9A	Peripheral somatic, visceral sensory neurons	Pain signaling; initiation and maintenance of action potentials in nociceptive (pain-sensing) neurons	Pain syndrome
Nav1.8	SCN10A	Trigeminal ganglia, dorsal root ganglia, C- Fibers	Detection of painful stimuli (nociception); detection of extreme cold, heat, chemical, and mechanical signals	Human lingual neuromas, small fiber neuropathy (SFN), inborn errors of metabolism (IEM)
Nav1.9	SCN11A	Myenteric neurons, small diameter DRG neurons, TG neurons	Pain perception	Severe paroxysmal pain, congenital inability to sense pain

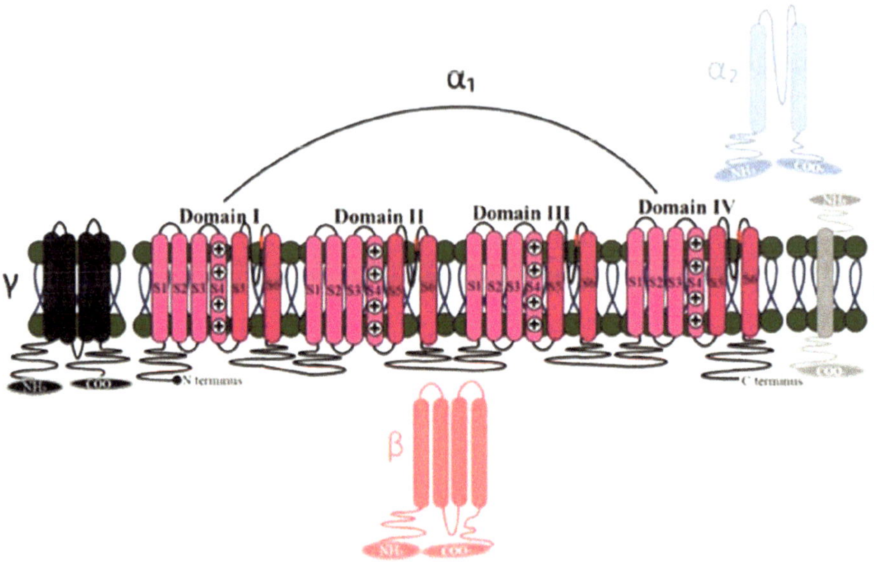

FIGURE 8.13 Voltage-gated calcium channel structure. The α1 subunit resembles the voltage-gated sodium channel α subunit, with four homologous domains (I–IV), each containing six transmembrane segments (S1–S6). The pore region (P) in the α1 subunit triggers the gating mechanism. Most calcium channels contain an intracellular β-subunit (pink), an extracellular α2-subunit (blue) linked by a disulfide bond to a transmembrane δ-subunit (gray), and a transmembrane or associated γ-subunit (black). These accessory subunits modulate channel properties.

targeting of medicinal compounds to VGCCs depends on understanding their biological structure. This study provides an in-depth examination of the intricate molecular components and structural features of VGCCs (Catterall, 2011).

8.2.2.1.3.1 Structure of Voltage-Gated Calcium Ion Channels

VGCCs comprise several subunits, including the pore-forming α1 subunit, the auxiliary α2δ subunit, the intracellular β subunit, and the transmembrane γ subunit (Dolphin, 2016). (Figure 8.13)

8.2.2.1.3.1.1 Structural Features of the α1 Subunit The α1 subunit is the primary and indispensable component of VGCCs, responsible for their structure, function, and pore formation. It controls the channel's biophysical functioning. This subunit consists of four homologous domains (I–IV), each containing six transmembrane segments (S1–S6). The S4 segments in each domain function as voltage sensors, detecting changes in membrane potential and facilitating channel activation. During membrane depolarization, these channels open, allowing Ca^{2+} ions to enter the cell. This calcium influx initiates cellular processes and activates subsequent signaling cascades. The S5 and S6 segments, along with the S5-S6 loop, form the pore-forming region that governs ion selectively and permeability properties. Different α1 subunit isoforms produce distinct VGCCs, including L-type, N-type, P/Q-type, R-type, and T-type channels (Catterall, 2000). Research has shown that different versions of the α1 subunit are present in all tissues, contributing to VGCC diversity. For example, the Cav1 subfamily (Cav1.1, Cav1.2, and Cav1.3) is primarily found in excitable cells like neurons and muscle cells.

8.2.2.1.3.1.2 Structural Features of the Auxiliary (α2δ) Subunit VGCCs are extraordinarily complex proteins composed of several subunits, each of which plays a role in the channel's structure and function. The S1–S6 subunits are involved in channel gating and interactions with auxiliary subunits. The α2δ subunit, composed of α2 and δ proteins, increases channel surface expression and affects the timing of channel activation and inactivation. It is crucial for normal trafficking, membrane localization, and regulation of the α1 subunit. This glycoprotein binds to the α1 subunit during channel complex formation. Recent research suggests that α2δ subunits may have additional functions in synaptogenesis and neuronal cell development beyond their role in VGCC assembly (Campiglio & Flucher, 2015; Dolphin, 2016).

8.2.2.1.3.1.3 Structural Features of the β Subunit The intracellular β subunit regulates VGCCs, altering the behavior of the channels by modulating their kinetics and voltage dependence of activation and inactivation. It engages with the α1 subunit's cytoplasmic region, influencing channel gating properties. Additionally, the β subunit plays a role in signal transduction pathways, increasing the complexity of cellular responses mediated by VGCCs (Herlitze et al., 1996). Multiple β subunit isoforms exist, with tissue-specific expression patterns. This diversity contributes to the functional variability of VGCCs in various cell types (Dolphin, 2012). The primary function of β subunits is to adjust the voltage dependency, kinetics, and pharmacological characteristics of VGCCs. Their integration with the α-subunit affects channel trafficking, membrane expression, and sensitivity to different ligands (Dolphin, 2016).

8.2.2.1.3.1.4 Structural Features of the γ Subunit The transmembrane γ subunit is a distinctive characteristic of VGCCs, although it has been less extensively studied. It has a role in regulating channel function and can potentially influence the speed of the calcium current. Although the γ submit is less specialized compared to other subunits, recent research suggests it has a significant impact on channel assembly and stabilization. Initially discovered in skeletal muscle, ongoing investigations aim to elucidate its function in various tissues and identify clinical disorders caused by aberrant calcium mobilization (Davies et al., 2007).

8.2.2.1.3.2 Regulation Mechanisms of Voltage-Gated Calcium Ion Channels in Cellular Signaling VGCCs play a critical role in cellular physiology by regulating calcium ion entry, which leads to changes in the membrane potential. These channels are crucial for neurotransmitter release, muscle contraction, and gene expression. Precise Ca^{2+} influx into cells is regulated by coordinated voltage sensing and gating mechanisms. This analysis examines the intricate mechanisms that govern VGCC voltage sensing and gating, focusing on their structural basis and various forms of gating.

8.2.2.1.3.2.1 Neurotransmission VGCCs play a significant role in neurotransmission. Their depolarization-induced activation increases intracellular calcium levels, triggering neurotransmitter release from synaptic vesicles in neurons. This mechanism is fundamental for intracellular communication in the nervous system. Recent research has identified multiple VGCC subtypes, including N-type, P/Q-type, and L-type, which play varying roles in neurotransmission (Catterall, 2011). VGCCs also contribute to long-term potentiation (LTP) and long-term depression (LTD), key processes linked to learning and memory. They work in conjunction with NMDA receptors to facilitate the entry of calcium ions into postsynaptic neurons (Liu et al., 2007).

8.2.2.1.3.2.2 Muscle Contraction The process of excitation-contraction coupling in muscle cells relies heavily on VGCCs. During depolarization, sarcolemmal VGCCs allow Ca^{2+} entry into the cytoplasm, triggering the release of stored calcium from the sarcoplasmic reticulum. The resulting increase in cytoplasmic calcium ions stimulates the proteins responsible for muscle contraction (Cain & Snutch, 2011). VGCC dysfunction is associated with various muscular diseases, underscoring their importance in muscle physiology (Figure 8.14).

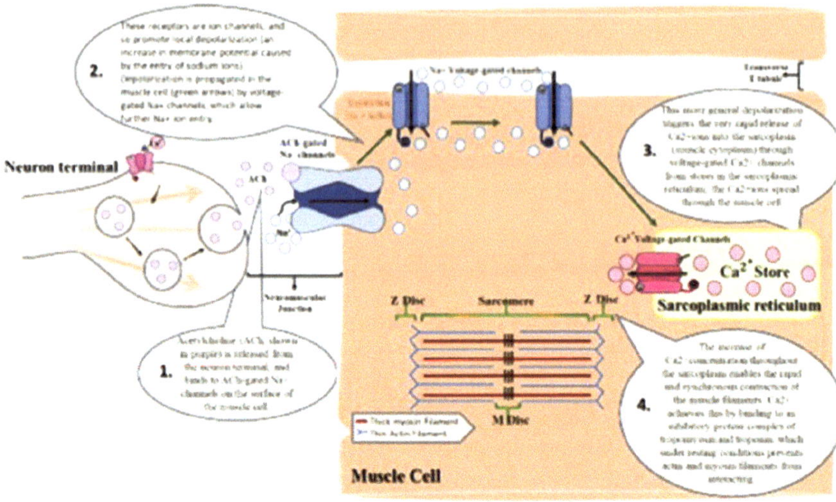

FIGURE 8.14 Calcium ions in skeletal muscle contraction synchronization.

8.2.2.1.3.2.3 Cellular Proliferation and Apoptosis VGCCs also play a role in controlling cellular growth and apoptosis (programmed cell death). VGCC activation leads to the Ca^{2+} influx, initiating signaling pathways that regulate the cell cycle. VGCCs are crucial in regulating the G1-S phase transition of the cell cycle, affecting cell proliferation (Monteith et al., 2007). Conversely, sustained increases in intracellular Ca^{2+} levels due to VGCC activity can trigger apoptosis by activating various apoptotic pathways, including the mitochondrial pathway and endoplasmic reticulum stress (Berridge et al., 2000). VGCC-mediated calcium signaling carefully balances cell survival and apoptosis.

8.2.2.1.3.2.4 Gene Expression Calcium regulates many aspects of cellular life. Intracellular Ca^{2+} levels increase through activation of intracellular Ca^{2+} or release from the endoplasmic reticulum. However, Ca^{2+} signaling must be tightly regulated to accomplish its task. Calmodulin (CaM) is an important Ca^{2+}-binding protein in the central nervous system. It contains four domains, each with a Ca^{2+} binding site. When Ca^{2+} binds to CaM, it changes conformation, allowing the CaM-Ca^{2+} complex to interact with and modify target proteins. The main CaM targets are Ca^{2+}/CaM-dependent kinases (CaMKs), which are involved in many neuronal regulatory processes, from synaptic plasticity to gene transcription (Soderling, 1999). The Ca^{2+}/calmodulin-dependent protein kinase (CaMK) pathway, which is linked to VGCC-mediated calcium entry, plays a role in the activation of transcription factors such as CREB (cAMP response element-binding protein) (Catterall, 2011). (Figure 8.15)

8.2.2.1.3.2.5 Voltage Sensing The intricate molecular interplay of VGCCs, including the coordination of charged residues in specific positions and the lipid environment, is crucial for voltage-dependent activation. Changes in membrane potential cause movement of these charged residues, resulting in structural changes that facilitate or hinder calcium ion entry. Positively charged residues such as arginine and lysine play a key role in this process, which can activate or deactivate the ion channel (Bezanilla, 2008). A study conducted by Catterall (2011) identified the crystal structure of a voltage-gated sodium channel as a key factor in voltage sensing, demonstrating the conservation of this voltage-sensing mechanism across the ion channel family.

8.2.2.1.3.2.6 Gating Mechanisms: Activation and Inactivation VGCC activation is a multi-stage process. The voltage sensor domains in the $\alpha 1$ subunit undergo a conformational change, influenced by interactions between the S4-S5 linker and the intracellular loop. This regulates the transition of

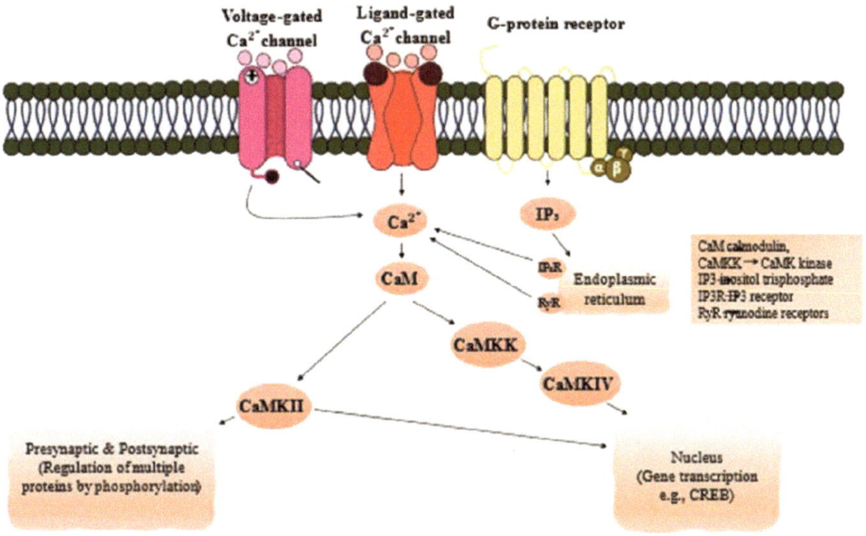

FIGURE 8.15 Calcium signaling and CaMK pathway activation. Intracellular Ca^{2+} increases via voltage-gated channels, ligand-gated ionotropic channels, G-protein receptors, or the release of IP3R or RyR from the endoplasmic reticulum. Ca^{2+} /calmodulin activates CaMKII directly or CaMKIV via CaMKK phosphorylation.

the channel pore from the inactivated to the activated state, leading to calcium ion influx and subsequent physiological reactions. Inactivation of VGCCs occurs to regulate excessive calcium influx. There are two primary methods of inactivation: fast and slow. The inactivation gate, a specific segment in the intracellular loop connecting domains III and IV, facilitates rapid inactivation. Cryo-electron microscopy experiments by Catterall et al. (2020) have identified the factors that contribute to the structural basis of fast inactivation. Slow inactivation involves gradual VGCC restructuring. The kinetics of VGCC activation and inactivation are crucial for cellular physiology. For example, L-type channels in excitable cells, including neurons and muscle cells, exhibit rapid activation and inactivation, while T-type channels in pacemaker cells display delayed activation and lack deactivation. Understanding these functional distinctions is important due to the diverse roles of VGCCs in different cell types (Striessnig et al., 2014).

8.2.2.1.3.2.7 Modulation of VGCCs VGCC modulation is complex and diverse, involving many cellular mechanisms and external factors. Auxiliary subunits, post-translational changes, and interactions with other cellular components all contribute to VGCC regulation. Additionally, neurotransmitters, hormones, and second messengers can modulate VGCC activity, allowing cells to adjust calcium transmission based on certain biological signals. Pharmacological regulation of VGCCs has therapeutic potential, with channel blockers targeting VGCCs to treat conditions such as hypertension, angina, and cardiac arrhythmias (Dolphin, 2016).

8.2.2.1.3.2.7.1 Post-Translational Modifications The $\alpha 1$ subunit of VGCCs is a key target for regulation. Post-translational modifications, including phosphorylation, glycosylation, and palmitoylation, influence channel function. Phosphorylation has been extensively studied as a means of regulating VGCCs. Intracellular signaling pathways, modal modulation, and auxiliary subunits all contribute to VGCC regulation. Phosphorylation can alter channel activity, influencing its voltage sensitivity and response to other modulators. For example, phosphorylation of certain amino acid residues by kinases like PKA and PKC can modify channel activity channel and voltage responsiveness. The $\alpha 2\delta$ subunit interacts with the $\alpha 1$ subunit to influence VGCC expression and pharmacological characteristics, expanding the functional repertoire of these channels (Zamponi et al.,

2015). Various kinases, including protein kinase A (PKA), protein kinase C (PKC), and calcium/calmodulin-dependent protein kinase II (CaMKII), have been linked to the phosphorylation of VGCC. PKA phosphorylation of the α1 can increase channel activity by elevating open probability (Catterall, 2011).

8.2.2.1.3.2.7.2 G Protein Modulation of VGCCs G protein-coupled receptors (GPCRs) can modulate VGCC activity through intracellular signaling cascades. Interactions between G proteins and VGCCs are well-established mechanisms for regulating calcium channel activity in response to various monoamine neurotransmitters and hormones (Zamponi & Currie, 2013). Activation of Gαq/11-coupled receptors, such as muscarinic acetylcholine receptors, activates phospholipase C (PLC). PLC generates IP3 and DAG. IP3 triggers intracellular calcium release, while DAG activates PKC, leading to a subsequent modification of VGCC activity (Zamponi & Snutch, 1998).

8.2.2.1.3.2.7.3 Pharmacological Modulation of VGCCs Various medicines and substances have been developed to manipulate VGCCs for medicinal purposes. These agents selectively target subtypes of VGCCs and can either increase or decrease channel activity. Calcium channel blockers (CCBs) are commonly used to treat cardiovascular problems. These medications inhibit VGCCs, reducing calcium ion influx. This leads to smooth muscle relaxation, and blood vessel dilation (Chakraborty & Hamilton, 2024).

8.2.2.1.3.3 Types of Voltage-Gated Calcium Ion Channels

Voltage-gated calcium ion channels are transmembrane proteins that facilitate calcium ion entry into cells in response to alterations in membrane potential. These channels are categorized into various subtypes, including L-type, N-type, P/Q-types, R-type, and T-type. Each type possesses unique functional characteristics and distinct tissue locations (Table 8.3).

8.2.2.1.4 Voltage-Gated Chloride Ion Channels

Voltage-gated chloride ion channels are membrane proteins that regulate the movement of chloride ions across cell membranes in response to variations in membrane voltage. They are present in all

TABLE 8.3

Different subtypes of calcium channels, their genetic basis, localization, physiological significance, and associated disorders

Type	Channel	Gene	Location	Physiological Function
L-Type	Cav1.1, Cav1.2, Cav1.3, Cav1.4	CACNA1, CACNA1F	Invaginations of cellular membranes (T-tubules)	Provide Ca^{2+} entry pathway in cardiac, skeletal, and smooth muscles
N-Type	Cav2.1	CACNA1B	Synaptic nerve terminals in dorsal horn	Facilitate binding of key synaptic transmission proteins
P/Q-Type	Cav2.2	CACNA1A	Somatodendritic compartments of central and peripheral neurons, axon terminals	Contribute to synaptic vesicle release
R-Type	Cav2.2	CACNA1E	Neurons	Provide dendritic Ca^{2+} source, regulate neuronal functions
T-Type	Cav3.1, Cav3.2, Cav3.2	CACNA1, CACNA1, CACNA1I	Cardiac myocytes, pacemaker cells, fibroblasts, neurons	Generate rhythmic bursts in heart, control SA node activity

tissues and primarily contribute to the maintenance of cellular ionic integrity and osmoregulation (Jentsch, 2008). Extensive research has revealed the broad implications of the CLC family, which includes both voltage-dependent and non-voltage-gated chloride channels (Koch et al., 1992).

8.2.2.1.4.1 Structure of Voltage-Gated Chloride Ion Channels

Voltage-gated Cl^- channels belong to the CLC family, which includes CLC-K and CLC-1. The CLC homodimer consists of two monomers, each containing two transmembrane domains and a large cytoplasmic domain, following a typical CLC structure. Each subunit contains 18 transmembrane helices. The pore-forming region determines which ions can pass through and when the channel opens and closes, while the transmembrane helices help maintain channel integrity.

8.2.2.1.4.1.1 Transmembrane Domains Voltage-gated chloride ion channels consist of several transmembrane domains, which form the channel's structural foundation. These domains contain α-helical segments that contribute to pore formation and overall structure. The central structure comprises helices S1–S6, with helix S4 serving as the voltage sensor (Jentsch, 2008).

8.2.2.1.4.1.2 Pore Region The pore region plays a critical role in determining ion selectivity and permeability. Specific amino acid residues lining the pore helices determine chloride ion preference over other anions (Dutzler et al., 2003). The voltage-sensing domains, particularly the S4 segment, are important in channel activation. During depolarization, the S4 helical region undergoes structural changes, shifting to open the channel pore.

8.2.2.1.4.1.3 Intracellular Loops Intracellular loops serve to connect transmembrane domains and frequently contain regulatory sites. Phosphorylation of certain residues within these loops can alter channel function, providing a mechanism for cellular regulation of chloride ion transit.

8.2.2.1.4.1.4 Extracellular Domains Extracellular domains of voltage-gated chloride ion channels can interact with ligands and other proteins, modulating channel performance. The diversity of extracellular regulatory components reflects the complexity of channel regulation (Figure 8.16).

8.2.2.1.4.2 Physiological Significance of Voltage-Gated Chloride Ion Channel Regulation in Cellular Signaling

Voltage-gated chloride ion channels perform diverse functions across various tissues and cell types. They play roles in various physiological processes, such as regulating cell volume, controlling neuronal excitability, and maintaining acid-base balance (Jentsch, 2008). CLC-2 regulates the electrical charge across the membrane and influences neuronal activation. Epithelial chloride channels, such as CLC-K, are essential for preserving electrolyte equilibrium in the kidneys (Devuyst et al., 2005).

8.2.2.1.4.2.1 Gating Mechanisms The conformational changes that occur during voltage-dependent gating in CLC channels are highly intricate. Similar to voltage sensors and other ion channels, the S4 helix experiences movements triggered by alterations in membrane potentials (Dutzler et al., 2002).

8.2.2.1.4.2.2 Regulation by Voltage Sensing Voltage-gated chloride ion channels uniquely respond to changes in membrane potential. Charged residues in the transmembrane domains detect voltage changes by sensing the electric field across the membrane (Jentsch, 2008).

8.2.2.1.4.2.3 Cell Volume Regulation Voltage-activated chloride ion channels primarily regulate cell volume. Ion movement across the cell membrane maintains proper osmotic balance, preventing excessive cell swelling or shrinkage. These channels allow chloride ions to move in or out of the cell, affecting intracellular osmotic pressure. Chloride channels play crucial roles in volume control

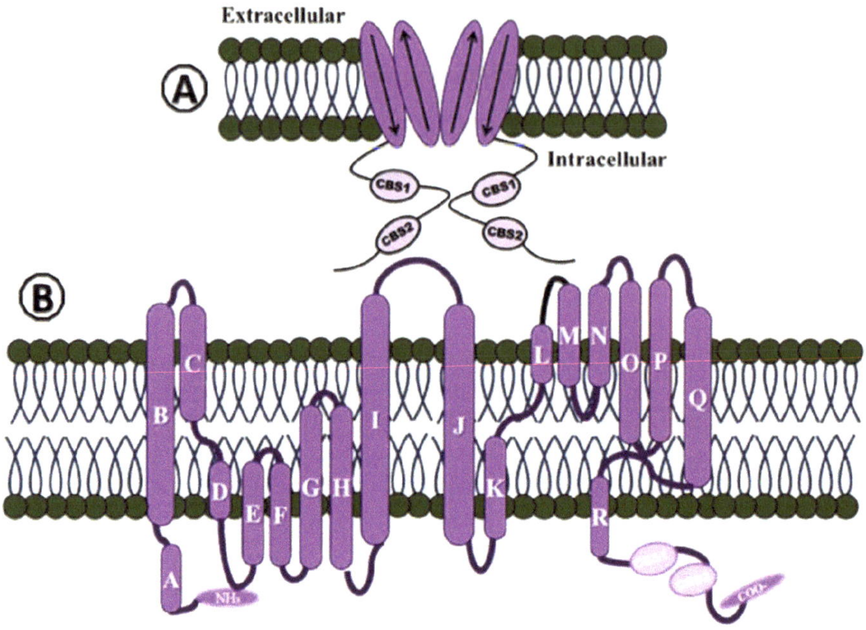

FIGURE 8.16 (A) Basic structure of voltage-gated chloride channel. (B) One subunit of a CLC chloride channel. It comprises 18 α-helical segments (A–R). The amino-terminal and carboxy-terminal halves are homologous but oppositely oriented in the membrane. They surround a central selective filter. The functional channel is a dimer with two independent pores.

for several cell types, including neurons, epithelial cells, and red blood cells (Lambert et al., 2008). For example, in neurons, alterations in the concentration of solutes outside the cell trigger chloride ion channel opening, preventing cell swelling or dehydration. Additionally, voltage-gated chloride channels influence other ion channels, such as potassium and sodium channels, impacting overall ion equilibrium. This contributes to optimized cell volume regulation during physiological changes (Poroca et al., 2017).

8.2.2.1.4.2.4 Neuronal Excitability

8.2.2.1.4.2.4.1 Action Potential Regulation Chloride influx in excitatory cells can lead to membrane depolarization, increasing action potential likelihood. Conversely, chloride ion efflux in inhibitory neurons leads to membrane hyperpolarization, decreasing action potential probability.

8.2.2.1.4.2.4.2 Synaptic Inhibition Voltage-gated chloride channels also contribute to synaptic inhibition. GABA-A and glycinergic synapses use chloride channels to maintain inhibitory postsynaptic potentials. This inhibitory mechanism is crucial for regulating physiological neuronal excitation (Farrant & Kaila, 2007).

8.2.2.1.4.2.5 Acid-Base Balance

8.2.2.1.4.2.5.1 pH Regulation Voltage-gated chloride channels help regulate intracellular pH, affecting the balance of acidity and alkalinity (Weinreich & Jentsch, 2001). Chloride ion movement drives proton redistribution across the cell membrane due to electrochemical gradients. In renal tubular cells, chloride channels participate in uric acid accumulation and proton transport, which is crucial for maintaining a stable acid-base balance in the body.

TABLE 8.4

Various subtypes of chloride channels along with their respective genes, localization within the body, physiological functions, and associated channelopathies

Subtype	Gene	Localization	Physiological Function	Channelopathy
CLC-1	CLCN1	Skeletal muscle membranes	Responsible for 80% of resting membrane conductance in inactive muscle fibers	Myotonia congenita
CLC-2	CLCN2	Apical membrane of lung and intestinal epithelia	Improves bowel movement by inducing passive movement of sodium and water into intestinal lumen	Leukoencephalopathy, cystic fibrosis, retinal degeneration, azoospermia
CLC-3	CLCN3	Synaptic vesicles and postsynaptic plasma membranes	Cell volume regulation and proliferation	Intellectual disability, epilepsy, global development delay, psychiatric conditions
CLC-4	CLCN4	Brain, kidney, liver	Regulares transepithelial transport, electrical excitability, and ion homeostasis	Dent's disease
CLC-5	CLCN5	Kidneys, particularly proximal tubules	Albumin endocytosis	Dent's disease
CLC-6	CLCN6	Golgi bodies of vascular smooth muscle cells	Endosomal acidification	Lysosomal storage disease
CLC-7	CLCN7	Osteoclast resorption lacuna, lysosomes	Lysosomal enzyme activation, protein degradation, receptor trafficking	Autosomal dominant osteopetrosis, visual loss, bone marrow failure
CLC-Ka	CLCNKA	Kidney epithelial cell basolateral membrane, inner ear	Renal chloride re-absorption, collecting duct potential generation, blood pressure regulation	Bartter syndrome 4
CLC-Kb	CLCNKB	Thin and thick ascending limbs of Henle's loop, marginal cells of stria vascularis of inner ear	Corticomedullary osmotic gradient, potassium handling, acid-base balance maintenance	Bartter Syndrome 3

8.2.2.1.4.2.5.2 Cellular pH Sensing Chloride channels also function as cellular pH sensors. Alterations in intracellular pH can impact the activity of these channels, regulating the feedback mechanisms that stabilize pH fluctuations.

8.2.2.1.4.3 Classification of Voltage-Gated Chloride Ion Channels

The CLC channel family is well-studied and known for its structural diversity and role in fundamental cellular functions. These channels are widely expressed in various tissues, and their dysregulation has been linked to a range of human diseases (Table 8.4).

8.3 PATHOPHYSIOLOGY OF ION CHANNELS

Dysregulation of ion channels is implicated in a myriad of pathological conditions. Mutations affecting ion channel properties can cause channelopathies, a group of diseases characterized by ion channel dysfunction (George, 2005).

8.3.1 Neurological Disorders

In neurological disorders, dysfunctional ion channels contribute to conditions such as epilepsy and neurodegenerative diseases, where aberrant ion channel activity contributes to pathogenesis. Voltage-gated sodium channels play a significant role in the development and propagation of neuronal activity, and mutations or genetic changes in these channels are associated with various types of epilepsy. This highlights the importance of these channels in the regulation of neuronal excitability. Additionally, altered calcium channel function has been implicated in the pathogenesis of neurodegenerative diseases like Alzheimer's and Parkinson's (Surmeier et al., 2017). Recent studies have focused on developing selective sodium channel modulators as potential antiepileptic drugs. Calcium channels, particularly L-type calcium channels, are also implicated in neurodegenerative diseases, making them attractive targets for drug intervention (Catterall, 2012; Zamponi et al., 2015).

Alzheimer's disease (AD) is a progressive neurodegenerative disorder characterized by the deposition of beta-amyloid plaques and intracellular fibrillary tangles in neurons. Recent findings indicate that abnormalities in ligand-gated ion channels, particularly N-methyl-D-aspartate (NMDA) receptors, contribute to the pathophysiology of AD. Hyperactivation of these receptors results in excitotoxicity, leading to neuronal cell loss and the memory and cognition decline observed in patients with AD (Palop & Mucke, 2010).

Epilepsy is a chronic neurological disease that causes recurring seizures. Ligand-gated ion channels like GABA type A (GABA-A) receptors and nicotinic acetylcholine receptors specifically regulate neuronal excitability. In certain forms of epilepsy, mutations have been identified in genes encoding these receptors, suggesting a role for ligand-gated ion channels in disease pathogenesis (Macdonald et al., 2010). GABA, the main inhibitory neurotransmitter in the brain, modulates neuronal excitability through GABAergic inhibitory signaling. Dysfunctional GABAergic inhibition, whether from receptor alterations or decreased GABA synthesis, can result in increased excitability and seizures. Glutamate, a major excitatory neurotransmitter in the brain, can provoke hyperexcitability and seizures when over-released or improperly cleared. Alterations in glutamate receptors, such as NMDA and α-amino-3-hydroxy-5-methyl-4-isoxazolepropionic acid (AMPAR) receptors, are implicated in epilepsy (T.-S. Chen et al., 2023).

Glutamate receptor dysfunction has been associated with various neurological disorders. Consequently, pharmacological modulation of these receptors offers a promising target for drug development. Research is focused on developing drugs targeting glutamate receptors to regulate synaptic transmission (Johnson et al., 2009). For instance, NMDA receptor antagonists may be used in neuroprotective approaches, and mGluR subtype agonists are being studied as potential treatments for psychiatric disorders (Hovelso et al., 2012). Glutamate receptors play major roles in synaptic transmission and neuronal communication. Pharmacological agents that specifically modulate these receptors enable researchers to study their functions in health and disease (Traynelis et al., 2010). Overall, glutamate receptor pharmacology in the context of ligand-gated ion channels is pivotal to understanding neurotransmission and presents several therapeutic implications for many neurological and psychiatric disorders.

8.3.2 Neuromuscular Disorders

Myasthenia gravis (MG) is an autoimmune disease causing muscle weakness and fatigue. In MG, autoantibodies target and destroy acetylcholine receptors (AChRs) at the neuromuscular junction, blocking signal transmission. This dysfunction in ligand-gated ion channels, especially nicotinic AChRs, results in muscle weakness and fatigue due to neuromuscular transmission in myasthenia gravis patients (Dresser et al., 2021). Congenital myasthenic syndromes (CMS) are a group of genetic disorders characterized by impaired neuromuscular transmission. Mutations have been identified in genes encoding ligand-gated ion channels, including AChRs and endplate acetylcholinesterase, in CMS patients. These mutations alter the stability and function of ion channels, causing muscular weakness and fatigue (Engel et al., 2015).

8.3.3 Sensory Disorders

Retinitis Pigmentosa (RP) is a group of inherited retinal disorders leading to progressive vision loss. Neurotransmission in the retina relies on ligand-gated ion channels, particularly ionotropic glutamate receptors. Dysregulation of these receptors has been linked to retinal cell degeneration in RP. Targeting ligand-gated ion channels may offer potential therapeutic approaches for preserving eyesight in people with RP (Parmeggiani et al., 2011).

Ligand-gated ion channels play a vital role in auditory signal transduction in the inner ear. Various mutations in genes encoding these channels, including those for glutamate receptors, have been linked to sensorineural hearing loss (Roux et al., 2006). Understanding the molecular mechanisms involved in these disorders is crucial as it could provide a basis for developing hearing function restoration therapies.

8.3.4 Cardiovascular Diseases

Abnormalities in potassium channels, for example, have been linked to arrhythmias. Antiarrhythmic drugs often target these channels to restore normal cardiac rhythm. Moreover, voltage-gated calcium channels are implicated in regulating cardiac contractility, and drugs targeting these channels are used to manage conditions such as hypertension and angina (Zamponi et al., 2015; De Lera Ruiz & Kraus, 2015).

8.3.5 Cancer

The role of ion channels in cancer progression is a growing area of research. Potassium channels have been found to influence cell proliferation, migration, and apoptosis in various cancer types. Targeting these channels has shown promise in inhibiting cancer cell growth. Calcium channels also play a role in cancer, with some studies suggesting that their modulation could be a potential strategy for cancer therapy (Arcangeli et al., 2009; Prevarskaya et al., 2018).

8.3.6 Pain Management

Chronic pain conditions, such as neuropathic pain, are often associated with dysregulated ion channel activity. Voltage-gated sodium channels have been identified as key players in pain signal transmission. Sodium channel blockers are being explored for their potential in pain management. Additionally, ligand-gated ion channels, such as TRP channels, have emerged as potential targets for analgesic drug development (Cortright et al., 2007; Dib-Hajj et al., 2010).

8.3.7 Inflammatory Disorders

Ion channels are involved in controlling immune responses and inflammatory processes. For example, chloride channels, such as CFTR, are involved in the regulation of inflammatory mediators. Modulating these channels has been explored as a potential therapeutic approach in inflammatory diseases like cystic fibrosis. Similarly, potassium channels are also involved in the function of immune system cells, and their regulation in conditions such as arthritis is being studied (Canning BJ, 2014; Moran, 2018).

REFERENCES

Abriel, H. (2010). Cardiac sodium channel Nav1.5 and interacting proteins: Physiology and pathophysiology. *Journal of Molecular and Cellular Cardiology*, *48*(1), 2–11. https://doi.org/10.1016/j.yjmcc.2009.08.025

Alberts, B., Heald, R., Johnson, A., Morgan, D., Raff, M., Roberts, K., & Walter. Peter. (2022). *Molecular Biology of the Cell / Bruce Alberts, Alexander Johnson, Julian Lewis, David Morgan, Martin Raff, Keith Roberts, Peter Walter; with Problems by John Wilson, Tim Hunt* (7th Edition). W. W. Norton & Company.

Alexander, S. P. H., Mathie, A., & Peters, J. A. (2011). Ligand-gated ion channels. *British Journal of Pharmacology*, *164*(s1). https://doi.org/10.1111/j.1476-5381.2011.01649_4.x

Alexander, S. P. H., Mathie, A., Peters, J. A., Veale, E. L., Striessnig, J., Kelly, E., Armstrong, J. F., Faccenda, E., Harding, S. D., Pawson, A. J., Southan, C., Davies, J. A., Aldrich, R. W., Attali, B., Baggetta, A. M., Becirovic, E., Biel, M., Bill, R. M., Catterall, W. A., … Zhu, M. (2021). The concise guide to pharmacology 2021/22: Ion channels. *British Journal of Pharmacology*, *178*(S1). https://doi.org/10.1111/bph.15539

Arcangeli, A., Crociani, O., Lastraioli, E., Masi, A., Pillozzi, S., & Becchetti, A. (2009). Targeting ion channels in cancer: A novel frontier in antineoplastic therapy. *Current Medicinal Chemistry*, *16*(1), 66–93. https://doi.org/10.2174/092986709787002835

Arey, B. J. (2012). The role of glycosylation in receptor signaling. In *Glycosylation*. InTech. https://doi.org/10.5772/50262

Arikkath, J., & Campbell, K. P. (2003). Auxiliary subunits: Essential components of the voltage-gated calcium channel complex. *Current Opinion in Neurobiology*, *13*(3), 298–307. https://doi.org/10.1016/S0959-4388(03)00066-7

Barhanin, J., Lesage, F., Guillemare, E., Fink, M., Lazdunski, M., & Romey, G. (1996). KvLQT1 and IsK (minK) proteins associate to form the IKS cardiac potassium current. *Nature*, *384*(6604), 78–80. https://doi.org/10.1038/384078a0

Bean, B. P. (2007). The action potential in mammalian central neurons. *Nature Reviews Neuroscience*, *8*(6), 451–465. https://doi.org/10.1038/nrn2148

Bender, A. C., Morse, R. P., Scott, R. C., Holmes, G. L., & Lenck-Santini, P.-P. (2012). SCN1A mutations in Dravet syndrome: Impact of interneuron dysfunction on neural networks and cognitive outcome. *Epilepsy & Behavior*, *23*(3), 177–186. https://doi.org/10.1016/j.yebeh.2011.11.022

Berridge, M. J., Lipp, P., & Bootman, M. D. (2000). The versatility and universality of calcium signalling. *Nature Reviews Molecular Cell Biology*, *1*(1), 11–21. https://doi.org/10.1038/35036035

Bers, D. M. (2002). Cardiac excitation–contraction coupling. *Nature*, *415*(6868), 198–205. https://doi.org/10.1038/415198a

Bettler, B., Kaupmann, K., Mosbacher, J., & Gassmann, M. (2004). Molecular structure and physiological functions of $GABA_B$ receptors. *Physiological Reviews*, *84*(3), 835–867. https://doi.org/10.1152/physrev.00036.2003

Bezanilla, F. (2000). The voltage sensor in voltage-dependent ion channels. *Physiological Reviews*, *80*(2), 555–592. https://doi.org/10.1152/physrev.2000.80.2.555

Bezanilla, F. (2008). How membrane proteins sense voltage. *Nature Reviews Molecular Cell Biology*, *9*(4), 323–332. https://doi.org/10.1038/nrm2376

Bouza, A. A., & Isom, L. L. (2017). *Voltage-Gated Sodium Channel β Subunits and Their Related Diseases* (pp. 423–450). https://doi.org/10.1007/164_2017_48

Burnstock, G., & Kennedy, C. (2011). *P2X Receptors in Health and Disease* (pp. 333–372). https://doi.org/10.1016/B978-0-12-385526-8.00011-4

Cain, S. M., & Snutch, T. P. (2011). Voltage-gated calcium channels and disease. *BioFactors*, *37*(3), 197–205. https://doi.org/10.1002/biof.158

Campiglio, M., & Flucher, B. E. (2015). The role of auxiliary subunits for the functional diversity of voltage-gated calcium channels. *Journal of Cellular Physiology*, *230*(9), 2019–2031. https://doi.org/10.1002/jcp.24998

Canning BJ, C. A. B. D. S. J. M. S. M. L. C. E. C. Panel. (2014). Anatomy and neurophysiology of cough. *Chest*, *146*(4), 1S. https://doi.org/10.1378/chest.1464s1

Catterall, W. A. (2000). Structure and regulation of voltage-gated Ca^{2+} channels. *Annual Review of Cell and Developmental Biology*, *16*(1), 521–555. https://doi.org/10.1146/annurev.cellbio.16.1.521

Catterall, W. A. (2010). Ion channel voltage sensors: Structure, function, and pathophysiology. *Neuron*, *67*(6), 915–928. https://doi.org/10.1016/j.neuron.2010.08.021

Catterall, W. A. (2011). Voltage-gated calcium channels. *Cold Spring Harbor Perspectives in Biology*, *3*(8), 1–23. https://doi.org/10.1101/cshperspect.a003947

Catterall, W. A. (2012). Voltage-gated sodium channels at 60: Structure, function and pathophysiology. *Journal of Physiology*, *590*(11), 2577–2589. https://doi.org/10.1113/jphysiol.2011.224204

Catterall, W. A., Lenaeus, M. J., & Gamal El-Din, T. M. (2020). Structure and pharmacology of voltage-gated sodium and calcium channels. *Annual Review of Pharmacology and Toxicology*, *60*(1), 133–154. https://doi.org/10.1146/annurev-pharmtox-010818-021757

Chakraborty, R. K., & Hamilton, R. J. (2024). *Calcium Channel Blocker Toxicity*. StatPearls Publishing.

Changeux, J.-P., & Christopoulos, A. (2016). Allosteric modulation as a unifying mechanism for receptor function and regulation. *Cell*, *166*(5), 1084–1102. https://doi.org/10.1016/j.cell.2016.08.015

Chen, T.-S., Huang, T.-H., Lai, M.-C., & Huang, C.-W. (2023). The role of glutamate receptors in epilepsy. *Biomedicines*, *11*(3), 783. https://doi.org/10.3390/biomedicines11030783

Chen, Z. J., & Sun, L. J. (2009). Nonproteolytic functions of ubiquitin in cell signaling. *Molecular Cell*, *33*(3), 275–286. https://doi.org/10.1016/j.molcel.2009.01.014

Cortright, D. N., Krause, J. E., & Broom, D. C. (2007). TRP channels and pain. In *Biochimica et Biophysica Acta - Molecular Basis of Disease* (Vol. 1772, Issue 8, pp. 978–988). https://doi.org/10.1016/j.bbadis.2007.03.003

Cox, J. J., Reimann, F., Nicholas, A. K., Thornton, G., Roberts, E., Springell, K., Karbani, G., Jafri, H., Mannan, J., Raashid, Y., Al-Gazali, L., Hamamy, H., Valente, E. M., Gorman, S., Williams, R., McHale, D. P., Wood, J. N., Gribble, F. M., & Woods, C. G. (2006). An SCN9A channelopathy causes congenital inability to experience pain. *Nature*, *444*(7121), 894–898. https://doi.org/10.1038/nature05413

Crupi, R., Impellizzeri, D., & Cuzzocrea, S. (2019). Role of metabotropic glutamate receptors in neurological disorders. *Frontiers in Molecular Neuroscience*, *12*. https://doi.org/10.3389/fnmol.2019.00020

Davies, A., Hendrich, J., Van Minh, A. T., Wratten, J., Douglas, L., & Dolphin, A. C. (2007). Functional biology of the α2δ subunits of voltage-gated calcium channels. *Trends in Pharmacological Sciences*, *28*(5), 220–228. https://doi.org/10.1016/j.tips.2007.03.005

De Lera Ruiz, M., & Kraus, R. L. (2015). Voltage-gated sodium channels: Structure, function, pharmacology, and clinical indications. *Journal of Medicinal Chemistry*, *58*(18), 7093–7118. https://doi.org/10.1021/jm501981g

Devuyst, O., Jouret, F., Auzanneau, C., & Courtoy, P. J. (2005). Chloride channels and endocytosis: New Insights from dent's disease and ClC-5 knockout mice. *Nephron Physiology*, *99*(3), p69–p73. https://doi.org/10.1159/000083210

Dib-Hajj, S. D., Cummins, T. R., Black, J. A., & Waxman, S. G. (2010). Sodium channels in normal and pathological pain. *Annual Review of Neuroscience*, *33*, 325–347. https://doi.org/10.1146/annurev-neuro-060909-153234

Dolphin, A. C. (2012). Calcium channel auxiliary α2δ and β subunits: Trafficking and one step beyond. *Nature Reviews Neuroscience*, *13*(8), 542–555. https://doi.org/10.1038/nrn3311

Dolphin, A. C. (2016). Voltage-gated calcium channels and their auxiliary subunits: Physiology and pathophysiology and pharmacology. *The Journal of Physiology*, *594*(19), 5369–5390. https://doi.org/10.1113/JP272262

Doyle, D. A., Cabral, J. M., Pfuetzner, R. A., Kuo, A., Gulbis, J. M., Cohen, S. L., Chait, B. T., & MacKinnon, R. (1998). the structure of the potassium channel: Molecular basis of K^+ conduction and selectivity. *Science*, *280*(5360), 69–77. https://doi.org/10.1126/science.280.5360.69

Dresser, L., Wlodarski, R., Rezania, K., & Soliven, B. (2021). Myasthenia gravis: Epidemiology, pathophysiology and clinical manifestations. *Journal of Clinical Medicine*, *10*(11), 2235. https://doi.org/10.3390/jcm10112235

Drukarch, B., Holland, H. A., Velichkov, M., Geurts, J. J. G., Voorn, P., Glas, G., & de Regt, H. W. (2018). Thinking about the nerve impulse: A critical analysis of the electricity-centered conception of nerve excitability. *Progress in Neurobiology*, *169*, 172–185. https://doi.org/10.1016/j.pneurobio.2018.06.009

Dutzler, R., Campbell, E. B., & MacKinnon, R. (2003). Gating the selectivity filter in ClC chloride channels. *Science*, *300*(5616), 108–112. https://doi.org/10.1126/science.1082708

Dutzler, R., Campbell, E. B., Cadene, M., Chait, B. T., & MacKinnon, R. (2002). X-ray structure of a ClC chloride channel at 3.0 Å reveals the molecular basis of anion selectivity. Nature, 415(6869), 287-294.

Engel, A. G., Shen, X.-M., Selcen, D., & Sine, S. M. (2015). Congenital myasthenic syndromes: Pathogenesis, diagnosis, and treatment. *The Lancet Neurology*, *14*(4), 420–434. https://doi.org/10.1016/S1474-4422(14)70201-7

Farrant, M., & Kaila, K. (2007). *The Cellular, Molecular and Ionic Basis of GABAA Receptor Signalling* (pp. 59–87). https://doi.org/10.1016/S0079-6123(06)60005-8

George, A. L. (2005). Inherited disorders of voltage-gated sodium channels. *Journal of Clinical Investigation*, *115*(8), 1990–1999. https://doi.org/10.1172/JCI25505

Ghanshani, S., Wulff, H., Miller, M. J., Rohm, H., Neben, A., Gutman, G. A., Cahalan, M. D., & Chandy, K. G. (2000). Up-regulation of the IKCa1 potassium channel during T-cell activation. *Journal of Biological Chemistry*, *275*(47), 37137–37149. https://doi.org/10.1074/jbc.M003941200

Greger, I. H., & Mayer, M. L. (2019). Structural biology of glutamate receptor ion channels: Towards an understanding of mechanism. *Current Opinion in Structural Biology*, *57*, 185–195. https://doi.org/10.1016/j.sbi.2019.05.004

Grider, M. H., Jessu, R., & Kabir, R. (2024). *Physiology, Action Potential*. StatPearls Publishing LLC.

Herlitze, S., Garcia, D. E., Mackie, K., Hille, B., Scheuer, T., & Catterall, W. A. (1996). Modulation of Ca2+ channels β γ G-protein py subunits. *Nature*, *380*(6571), 258–262. https://doi.org/10.1038/380258a0

Hille, B. (2001). *Ion Channels of Excitable Membranes*. Sinauer Associates.

Hodgkin, A. L., & Huxley, A. F. (1952). A quantitative description of membrane current and its application to conduction and excitation in nerve. *The Journal of Physiology*, *117*(4), 500–544. https://doi.org/10.1113/jphysiol.1952.sp004764

Hogg, R. C., Raggenbass, M., & Bertrand, D. (2003). Nicotinic acetylcholine receptors: From structure to brain function. In *Reviews of physiology, biochemistry and pharmacology* (pp. 1–46). Springer Berlin Heidelberg. https://doi.org/10.1007/s10254-003-0005-1

Hoshi, T., & Armstrong, C. M. (2013). C-type inactivation of voltage-gated K+ channels: Pore constriction or dilation? *Journal of General Physiology*, *141*(2), 151–160. https://doi.org/10.1085/jgp.201210888

Hoshi, T., Zagotta, W. N., & Aldrich, R. W. (1990). Biophysical and molecular mechanisms of *Shaker* potassium channel inactivation. *Science*, *250*(4980), 533–538. https://doi.org/10.1126/science.2122519

Hovelso, N., Sotty, F., P. Montezinho, L., S. Pinheiro, P., F. Herrik, K., & Mork, A. (2012). Therapeutic potential of metabotropic glutamate receptor modulators. *Current Neuropharmacology*, *10*(1), 12–48. https://doi.org/10.2174/157015912799362805

Isom, L. (1994). Auxiliary subunits of voltage-gated ion channels. *Neuron*, *12*(6), 1183–1194. https://doi.org/10.1016/0896-6273(94)90436-7

Jan, L. Y., & Jan, Y. N. (2012). Voltage-gated potassium channels and the diversity of electrical signalling. *The Journal of Physiology*, *590*(11), 2591–2599. https://doi.org/10.1113/jphysiol.2011.224212

Jensen, M. Ø., Jogini, V., Borhani, D. W., Leffler, A. E., Dror, R. O., & Shaw, D. E. (2012). Mechanism of voltage gating in potassium channels. *Science*, *336*(6078), 229–233. https://doi.org/10.1126/science.1216533

Jentsch, T. J. (2008). CLC chloride channels and transporters: From genes to protein structure, pathology and physiology. *Critical Reviews in Biochemistry and Molecular Biology*, *43*(1), 3–36. https://doi.org/10.1080/10409230701829110

Jiang, Y., Lee, A., Chen, J., Cadene, M., Chait, B. T., & MacKinnon, R. (2002). The open pore conformation of potassium channels. *Nature*, *417*(6888), 523–526. https://doi.org/10.1038/417523a

Johnson, K., Conn, P., & Niswender, C. (2009). Glutamate receptors as therapeutic targets for parkinsons disease. *CNS & Neurological Disorders - Drug Targets*, *8*(6), 475–491. https://doi.org/10.2174/187152709789824606

Kim, D. M., & Nimigean, C. M. (2016). Voltage-gated potassium channels: A structural examination of selectivity and gating. *Cold Spring Harbor Perspectives in Biology*, *8*(5), a029231. https://doi.org/10.1101/cshperspect.a029231

Koch, M. C., Steinmeyer, K., Lorenz, C., Ricker, K., Wolf, F., Otto, M., Zoll, B., Lehmann-Horn, F., Grzeschik, K.-H., & Jentsch, T. J. (1992). The skeletal muscle chloride channel in dominant and recessive human myotonia. *Science*, *257*(5071), 797–800. https://doi.org/10.1126/science.1379744

Lambert, I. H., Hoffmann, E. K., & Pedersen, S. F. (2008). Cell volume regulation: Physiology and pathophysiology. *Acta Physiologica*, *194*(4), 255–282. https://doi.org/10.1111/j.1748-1716.2008.01910.x

Lee, S.-Y., Lee, A., Chen, J., & MacKinnon, R. (2005). Structure of the KvAP voltage-dependent K + channel and its dependence on the lipid membrane. *Proceedings of the National Academy of Sciences*, *102*(43), 15441–15446. https://doi.org/10.1073/pnas.0507651102

Leipold, E., Liebmann, L., Korenke, G. C., Heinrich, T., Gießelmann, S., Baets, J., Ebbinghaus, M., Goral, R. O., Stödberg, T., Hennings, J. C., Bergmann, M., Altmüller, J., Thiele, H., Wetzel, A., Nürnberg, P., Timmerman, V., De Jonghe, P., Blum, R., Schaible, H.-G., … Kurth, I. (2013). A de novo gain-of-function mutation in SCN11A causes loss of pain perception. *Nature Genetics*, *45*(11), 1399–1404. https://doi.org/10.1038/ng.2767

Liu, Y., Wong, T. P., Aarts, M., Rooyakkers, A., Liu, L., Lai, T. W., Wu, D. C., Lu, J., Tymianski, M., Craig, A. M., & Wang, Y. T. (2007). NMDA receptor subunits have differential roles in mediating excitotoxic neuronal death both *In Vitro* and *in Vivo*. *The Journal of Neuroscience*, *27*(11), 2846–2857. https://doi.org/10.1523/JNEUROSCI.0116-07.2007

Long, S. B., Campbell, E. B., & MacKinnon, R. (2005). Crystal structure of a mammalian voltage-dependent *Shaker* family K + channel. *Science*, *309*(5736), 897–903. https://doi.org/10.1126/science.1116269

Luján, R. (2010). Organisation of potassium channels on the neuronal surface. *Journal of Chemical Neuroanatomy*, *40*(1), 1–20. https://doi.org/10.1016/j.jchemneu.2010.03.003

Macdonald, R. L., Kang, J.-Q., & Gallagher, M. J. (2010). Mutations in GABA $_A$ receptor subunits associated with genetic epilepsies. *The Journal of Physiology*, *588*(11), 1861–1869. https://doi.org/10.1113/jphysiol.2010.186999

Maljevic, S., & Lerche, H. (2014). *Potassium Channel Genes and Benign Familial Neonatal Epilepsy* (pp. 17–53). https://doi.org/10.1016/B978-0-444-63326-2.00002-8

Mammen, A. L., Kamboj, S., & Huganir, R. L. (1999). *[19] Protein Phosphorylation of Ligand-Gated Ion Channels* (pp. 353–370). https://doi.org/10.1016/S0076-6879(99)94022-3

Mantegazza, M., & Catterall, W. A. (2012). *Voltage-Gated Na+ Channels: Structure, Function, and Pathophysiology*. Oxford University Press.

Mao, L.-M., Guo, M.-L., Jin, D.-Z., Fibuch, E. E., Choe, E. S., & Wang, J. Q. (2011). Post-translational modification biology of glutamate receptors and drug addiction. *Frontiers in Neuroanatomy*, *5*. https://doi.org/10.3389/fnana.2011.00019

Marban, E., Yamagishi, T., & Tomaselli, G. F. (1998). Structure and function of voltage-gated sodium channels. *The Journal of Physiology*, *508*(3), 647–657. https://doi.org/10.1111/j.1469-7793.1998.647bp.x

Miller, P. S., & Aricescu, A. R. (2014). Crystal structure of a human GABAA receptor. *Nature*, *512*(7514), 270–275. https://doi.org/10.1038/nature13293

Monteith, G. R., McAndrew, D., Faddy, H. M., & Roberts-Thomson, S. J. (2007). Calcium and cancer: Targeting Ca2+ transport. *Nature Reviews Cancer*, *7*(7), 519–530. https://doi.org/10.1038/nrc2171

Moran, M. M. (2018). 58:309-30 First published as a review in advance on. *Annual Review of Pharmacology and Toxicology*. https://doi.org/10.1146/annurev-pharmtox

Motoike, H. K., Liu, H., Glaaser, I. W., Yang, A.-S., Tateyama, M., & Kass, R. S. (2004). The Na+ channel inactivation gate is a molecular complex. *The Journal of General Physiology*, *123*(2), 155–165. https://doi.org/10.1085/jgp.200308929

Nassar, M. A., Stirling, L. C., Forlani, G., Baker, M. D., Matthews, E. A., Dickenson, A. H., & Wood, J. N. (2004). Nociceptor-specific gene deletion reveals a major role for Na $_v$ 1.7 (PN1) in acute and inflammatory pain. *Proceedings of the National Academy of Sciences*, *101*(34), 12706–12711. https://doi.org/10.1073/pnas.0404915101

Nerbonne, J. M. (2000). Molecular basis of functional voltage-gated K $^+$ channel diversity in the mammalian myocardium. *The Journal of Physiology*, *525*(2), 285–298. https://doi.org/10.1111/j.1469-7793.2000.t01-1-00285.x

Palop, J. J., & Mucke, L. (2010). Amyloid-β–induced neuronal dysfunction in Alzheimer's disease: From synapses toward neural networks. *Nature Neuroscience*, *13*(7), 812–818. https://doi.org/10.1038/nn.2583

Parmeggiani, F., S. Sorrentino, F., Ponzin, D., Barbaro, V., Ferrari, S., & Di Iorio, E. (2011). Retinitis pigmentosa: Genes and disease mechanisms. *Current Genomics*, *12*(4), 238–249. https://doi.org/10.2174/138920211795860107

Payandeh, J., Scheuer, T., Zheng, N., & Catterall, W. A. (2011). The crystal structure of a voltage-gated sodium channel. *Nature*, *475*(7356), 353–359. https://doi.org/10.1038/nature10238

Pinard, A., Seddik, R., & Bettler, B. (2010). *GABAB Receptors: Physiological Functions and Mechanisms of Diversity* (pp. 231–255). https://doi.org/10.1016/S1054-3589(10)58010-4

Poliak, S., & Peles, E. (2003). The local differentiation of myelinated axons at nodes of Ranvier. *Nature Reviews Neuroscience*, *4*(12), 968–980. https://doi.org/10.1038/nrn1253

Poroca, D. R., Pelis, R. M., & Chappe, V. M. (2017). ClC channels and transporters: Structure, physiological functions, and implications in human chloride channelopathies. *Frontiers in Pharmacology*, *8*. https://doi.org/10.3389/fphar.2017.00151

Prevarskaya, N., Skryma, R., & Shuba, Y. (2018). Ion channels in cancer: Are cancer hallmarks oncochannelopathies? *Physiological Reviews*, *98*, 559–621. https://doi.org/10.1152/physrev.00044.2016.-Genomic

Rao, R., Shah, S., Bhattacharya, D., Toukam, D. K., Cáceres, R., Pomeranz Krummel, D. A., & Sengupta, S. (2022). Ligand-gated ion channels as targets for treatment and management of cancers. *Frontiers in Physiology*, *13*. https://doi.org/10.3389/fphys.2022.839437

Rems, L., Kasimova, M. A., Testa, I., & Delemotte, L. (2020). Pulsed electric fields can create pores in the voltage sensors of voltage-gated ion channels. *Biophysical Journal*, *119*(1), 190–205. https://doi.org/10.1016/j.bpj.2020.05.030

Roux, I., Safieddine, S., Nouvian, R., Grati, M., Simmler, M.-C., Bahloul, A., Perfettini, I., Le Gall, M., Rostaing, P., Hamard, G., Triller, A., Avan, P., Moser, T., & Petit, C. (2006). Otoferlin, defective in a human deafness form, is essential for exocytosis at the auditory ribbon synapse. *Cell*, *127*(2), 277–289. https://doi.org/10.1016/j.cell.2006.08.040

Sanguinetti, M. C., & Tristani-Firouzi, M. (2006). hERG potassium channels and cardiac arrhythmia. *Nature*, *440*(7083), 463–469. https://doi.org/10.1038/nature04710

Saunders, P. A., & Ho, I. K. (1990). Barbiturates and the GABAA receptor complex. In *Progress in Drug Research / Fortschritte der Arzneimittelforschung / Progrès des recherches pharmaceutiques* (pp. 261–286). Birkhäuser Basel. https://doi.org/10.1007/978-3-0348-7128-0_7

Schmauder, R., Eick, T., Schulz, E., Sammler, G., Voigt, E., Mayer, G., Ginter, H., Ditze, G., & Benndorf, K. (2023). Fast functional mapping of ligand-gated ion channels. *Communications Biology*, *6*(1), 1003. https://doi.org/10.1038/s42003-023-05340-w

Seidl, A. H. (2014). Regulation of conduction time along axons. *Neuroscience*, *276*, 126–134. https://doi.org/10.1016/j.neuroscience.2013.06.047

Shah, N. H., & Aizenman, E. (2014). Voltage-gated potassium channels at the crossroads of neuronal function, ischemic tolerance, and neurodegeneration. *Translational Stroke Research*, *5*(1), 38–58. https://doi.org/10.1007/s12975-013-0297-7

Sheets, M. F., Fozzard, H. A., Lipkind, G. M., & Hanck, D. A. (2010). Sodium channel molecular conformations and antiarrhythmic drug affinity. *Trends in Cardiovascular Medicine*, *20*(1), 16–21. https://doi.org/10.1016/j.tcm.2010.03.002

Shen, H., Zhou, Q., Pan, X., Li, Z., Wu, J., & Yan, N. (2017). Structure of a eukaryotic voltage-gated sodium channel at near-atomic resolution. *Science*, *355*(6328). https://doi.org/10.1126/science.aal4326

Soderling, T. R. (1999). The Ca2+–calmodulin-dependent protein kinase cascade. *Trends in Biochemical Sciences*, *24*(6), 232–236. https://doi.org/10.1016/S0968-0004(99)01383-3

Striessnig, J., Pinggera, A., Kaur, G., Bock, G., & Tuluc, P. (2014). L-type Ca $^{2+}$ channels in heart and brain. *Wiley Interdisciplinary Reviews: Membrane Transport and Signaling*, *3*(2), 15–38. https://doi.org/10.1002/wmts.102

Surmeier, D. J., Halliday, G. M., & Simuni, T. (2017). Calcium, mitochondrial dysfunction and slowing the progression of Parkinson's disease. In *Experimental Neurology* (Vol. 298, pp. 202–209). Academic Press Inc. https://doi.org/10.1016/j.expneurol.2017.08.001

Swope, S. L., Moss, S. J., Raymond, L. A., & Huganir, R. L. (1999). *3 Regulation of Ligand-Gated Ion Channels by Protein Phosphorylation* (pp. 49–78). https://doi.org/10.1016/S1040-7952(99)80005-6

Traynelis, S. F., Wollmuth, L. P., McBain, C. J., Menniti, F. S., Vance, K. M., Ogden, K. K., Hansen, K. B., Yuan, H., Myers, S. J., & Dingledine, R. (2010). Glutamate receptor ion channels: Structure, regulation, and function. *Pharmacological Reviews*, *62*(3), 405–496. https://doi.org/10.1124/pr.109.002451

Urrego, D., Tomczak, A. P., Zahed, F., Stühmer, W., & Pardo, L. A. (2014). Potassium channels in cell cycle and cell proliferation. *Philosophical Transactions of the Royal Society B: Biological Sciences*, *369*(1638), 20130094. https://doi.org/10.1098/rstb.2013.0094

Veerman, C. C., Wilde, A. A. M., & Lodder, E. M. (2015). The cardiac sodium channel gene SCN5A and its gene product NaV1.5: Role in physiology and pathophysiology. *Gene*, *573*(2), 177–187. https://doi.org/10.1016/j.gene.2015.08.062

Wang, J., Ou, S.-W., & Wang, Y.-J. (2017). Distribution and function of voltage-gated sodium channels in the nervous system. *Channels*, *11*(6), 534–554. https://doi.org/10.1080/19336950.2017.1380758

Watanabe, M., Maemura, K., Kanbara, K., Tamayama, T., & Hayasaki, H. (2002). *GABA and GABA Receptors in the Central Nervous System and Other Organs* (pp. 1–47). https://doi.org/10.1016/S0074-7696(02)13011-7

Weinreich, F., & Jentsch, T. J. (2001). Pores formed by single subunits in mixed dimers of different CLC chloride channels. *Journal of Biological Chemistry*, *276*(4), 2347–2353. https://doi.org/10.1074/jbc.M005733200

Yu, F. H., & Catterall, W. A. (2004). The VGL-Chanome: A protein superfamily specialized for electrical signaling and ionic homeostasis. *Science's STKE*, *2004*(253). https://doi.org/10.1126/stke.2532004re15

Zamponi, G. W., & Currie, K. P. M. (2013). Regulation of CaV2 calcium channels by G protein coupled receptors. *Biochimica et Biophysica Acta (BBA) - Biomembranes*, *1828*(7), 1629–1643. https://doi.org/10.1016/j.bbamem.2012.10.004

Zamponi, G. W., & Snutch, T. P. (1998). Modulation of voltage-dependent calcium channels by G proteins. *Current Opinion in Neurobiology*, *8*(3), 351–356. https://doi.org/10.1016/S0959-4388(98)80060-3

Zamponi, G. W., Striessnig, J., Koschak, A., & Dolphin, A. C. (2015). The physiology, pathology, and pharmacology of voltage-gated calcium channels and their future therapeutic potential. *Pharmacological Reviews*, *67*(4), 821–870. https://doi.org/10.1124/pr.114.009654

Zhou, Y., Morais-Cabral, J. H., Kaufman, A., & MacKinnon, R. (2001). Chemistry of ion coordination and hydration revealed by a K+ channel–Fab complex at 2.0 Å resolution. *Nature*, *414*(6859), 43–48. https://doi.org/10.1038/35102009

9 Hormone and Receptor Interactions in Cell Signaling

Tehmina Bibi[1,2], *Shahar Bano*[1,2],
Eun Kyoung Seo[3], *and Salman Khan*[1,2]
[1]Pharmacological Sciences Research Lab, Department of Pharmacy, Faculty of Biological Sciences, Quaid-i-Azam University, Islamabad 45320, Pakistan
[2]Department of Pharmacy, Faculty of Biological Sciences, Quaid-i-Azam University, Islamabad, 45320, Pakistan
[3]College of Pharmacy, Ehwa Woman University, Seoul, Korea

9.1 INTRODUCTION

Many molecules are involved in triggering the process of cell signaling and signal transduction. The process is mandatory for cell–cell communication through signaling molecule secretion. Each molecule is released in response to changes in the body and transmits particular information between the cells. Moreover, every molecule in the process acts as a ligand. In the cell–cell communication, endocrine signaling is one of the four types of signaling forms in multicellular organisms. Endocrine signaling involves signal reception from distant cells. It shows delayed response; however, the effect produced as a result of endocrine signaling lasts longer (Cooper and Hausman, 2000). Moreover, molecules for signaling are formed at different sites but show effects on distant parts of the body. In the endocrine system, these cell signaling molecules are termed hormones. Hormones are formed far away from their site of action in the specialized cells of endocrine glands and secreted in the blood circulation. Molecules have to go with the stream and cover long distances between cells from the site of action and their formation site. As a result, hormones undergo the process of dilution and become low in concentration when they reach their destination.

9.2 HORMONE RECEPTORS

Hormones bind to specific receptors present on cells at the target site. Every cell expresses specific kinds of receptors on its surface. These receptors are protein in nature. Hormones move through the whole body via blood circulation, and they affect specific cells. Due to the specificity of receptors, hormones bind to their respective receptor and show their specific response. In addition, various cells may contain receptors for the particular hormone. For instance, thyroid hormone receptors are found in different tissues of the body. Thyroid hormones bind to these receptors and regulate various developmental processes and metabolic activities in the body (Anyetei-Anum et al., 2018). On the other hand, the receptor may be limited to a small specialized population of cells. Moreover, variation in the functions and structure is due to various types of signaling molecules involved in signal transmission. Single cells may contain receptors for more than one hormone. Therefore, a cell may receive various signals and then integrate the information. Resultantly, the response may not be restricted to the receiving cell; however, it may send signals to other cells.

The amplitude of cellular response toward a certain signaling molecule depends on the number of receptors for a hormone present in the cell. The cellular response toward any hormone may vary with time. This change is based on the number of responding receptors out of the total number

DOI: 10.1201/9781003594901-9

of receptors on a cell. It is a vital mechanism in the assessment of the dose–response relationship of hormones (Andersen and Barton, 1999). The mechanism behind the change in response is the upregulation and downregulation of hormone receptors. Upregulation of receptors involves the enhancement of receptor numbers that resultantly enhances the cellular response toward hormones. Contrary to this, a persistent increase in hormonal levels leads to the downregulation of receptors, i.e., a decrease in responding receptors. This reduces the cell response toward hormonal signals. Mostly, receptors of hormones are regulated by their ligands. For instance, the estrogen hormone induces upregulation of estrogen receptors in the case of immature uteri. However, when there is a need to reduce the response in mature uteri, estrogen initiates the downregulation of estrogen receptors (Kiang et al., 1989).

9.3 TYPES AND LOCATIONS OF HORMONE RECEPTORS

Hormone receptors have different locations over the cell and within the cell. This is due to the variation, such as binding hormones to the different receptors. Two major categories of hormones are lipophilic and hydrophilic hormones. Lipophilic hormones are lipid soluble; therefore, these hormones can easily pass through the cell membrane. Examples of lipophilic hormones are steroids and thyroid hormones. These hormones cross the cell membrane due to their lipophilicity and bind to their intracellular receptors. Upon binding to their specific binding site, signal transmission initiates by conformational alterations, and the response will appear after the transcription of genes (Reichel and Jacob, 1993). In contrast, hydrophilic hormones cannot cross the lipid bilayer of the cell membrane. Therefore, hydrophilic hormones bind to receptors that are present on the cell membrane outside the cell. Peptide or protein hormones and insulin are examples of hydrophilic hormones that bind to cell membrane-bound receptors. Upon binding with the receptor, hormones initiate signal transduction, transcription, and translation of proteins (Seifter et al., 2021). Based on the nature of hormones, receptors may be categorized as cell membrane-bound receptors, cytoplasmic receptors, and nuclear receptors.

9.3.1 Membrane-Bound Receptors

Membrane-bound receptors are present on the cell membrane and possess the extracellular and intracellular domains of a receptor. These receptors are glycoproteins with high molecular weight. Hormones, having a hydrophilic nature, circulate with the blood and come in contact with their receptors. The outer extracellular domain of the hormone receptor possesses a binding site for the particular hormone. A hormone with a specific chemical structure binds to the receptor binding site. The binding of hormones with their receptors is mostly reversible but very structure-specific. As a result, a hormone–receptor complex is formed. These complexes initiate conformational changes in the internal domain of the receptor. This leads to signal transduction within the cell via intracellular regulator proteins (Kahn, 1976). Signal transduction downflow starts with the effector's interactions with the activated hormone–receptor complex. The interaction between the hormone–receptor complex and effectors is based on the relative mobility of the cell membrane due to fluidity. The theory was first coined by Singer and Nicolson (1972), and further studies authenticated the concept. Therefore, the receptors are further categorized based on effector interactions and initiation of cell signaling. Two main kinds are G protein-coupled receptors and enzyme-linked receptors.

9.3.1.1 G Protein-Coupled Receptors

G protein-coupled receptors are responsible for molecule binding and signal transduction of peptide hormones, calcitonin, glucagon, and parathyroid hormones. These receptors have two main domains that are involved in signal transmission within the cell (Figure 9.1). The ligand binding domain is referred to as the extracellular domain (ECD), while the part of the receptor that interacts with the effectors is known as the transmembrane domain (TMD). The outer ECD has a specific binding site for hormones and shows affinity toward ligands. Therefore, the ECD is a fundamental

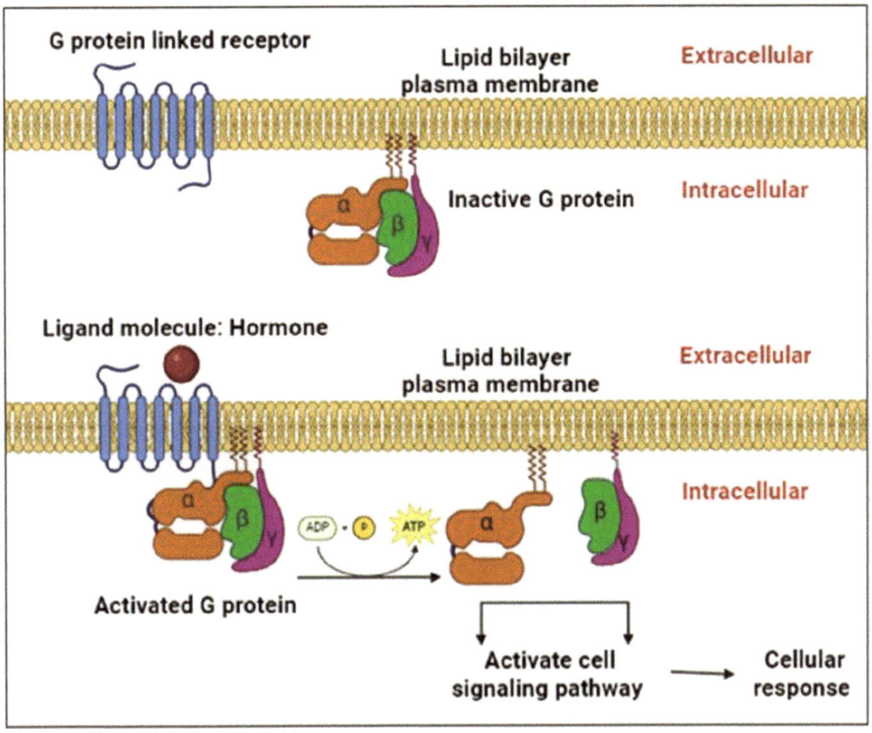

FIGURE 9.1 The signaling molecule, i.e., hormone, binds to the G protein-coupled receptors embedded in the lipid bilayer plasma membrane. G protein gets activated by the exchange of GDP by GTP, and the activated intracellular α, β, and γ subunits further transduce signals within the cell that ultimately produce the response.

part of recognizing ligands. Whenever a hormone binds to the ECD, it activates the TMD, which is composed of seven helices. The resultant complex initiates the coupling of the cytosolic domain of the receptor with the inner trimeric G protein that further transmits signals downstream toward effector and regulator proteins (Pal et al., 2012; Alberts et al., 1994).

9.3.1.2 Enzyme-Linked Receptors

Enzyme-linked receptors are important receptors for ligand recognition and cell signaling of growth hormones (Figure 9.2). The response through enzyme-linked receptors is mostly slow, as there is the involvement of multiple cell signaling proteins until the signal reaches the gene expression level. Enzyme-linked receptors also have an extracellular domain for binding and recognition of hormonal signals. Once the hormone binds to the binding site, the cytosolic transmembrane domain shows enzymatic activity, or it links directly with the enzyme (Alberts et al., 1994).

9.3.2 Cytoplasmic Receptors

Cytoplasmic receptors are present within the cytoplasm of the cell (Figure 9.3A). Lipophilic hormones bind to the cytoplasmic receptors, for instance, steroid hormones. Lipophilic hormones are released from endocrine glands in the bloodstream. The carrier proteins carry the lipophilic hormones from the site of production and along the bloodstream to the site of action. These carrier proteins are mandatory to make lipophilic hormones soluble in blood circulation. The hormones are detached from the carrier proteins at the site of action. Due to lipophilicity, steroid hormones cross the lipid bilayer of the cell membrane through the process of diffusion. In the cytoplasm of

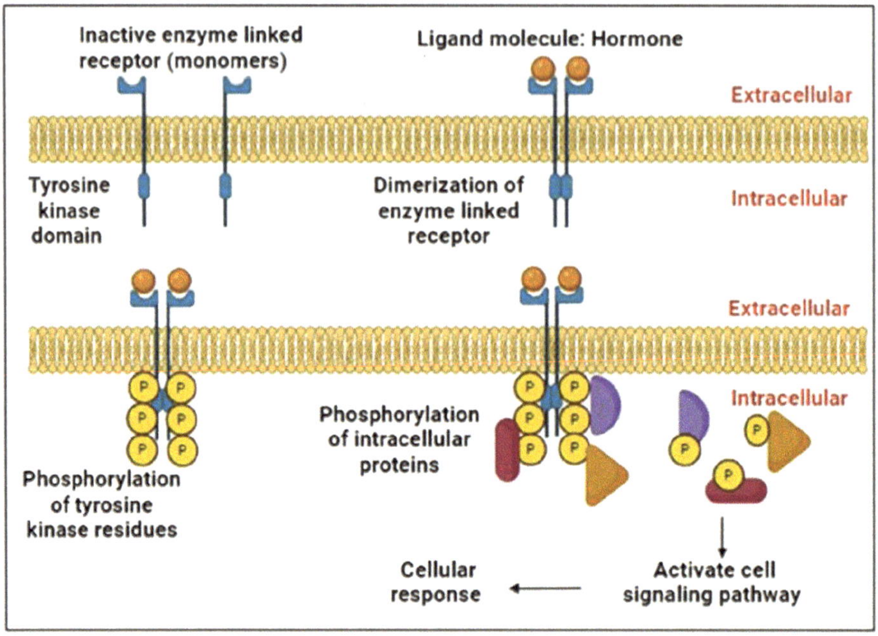

FIGURE 9.2 The signaling molecule, i.e., hormone, binds to the enzyme-linked receptors embedded in the lipid bilayer plasma membrane that causes dimerization of receptors. Dimerization leads to the phosphorylation of tyrosine kinase residue on the intracellular domain of receptors. The regulator proteins get phosphorylated and activated through the intracellular tyrosine domain and cause signal transduction within the cell that ultimately leads to cellular response.

the cell, receptors and hormones bind to form a receptor–hormone complex. The receptor–hormone complex then initiates the process of cell signaling in response to signals produced by the hormone. The complex regulates gene transcription in response to signal recognition and transmission. The formation of mRNA of a particular gene is increased or decreased accordingly. This alteration in gene expression corresponds to the synthesis of protein in response to hormonal signals received by the cell (Karlson, 1979; Molnar and Gair, 2015).

9.3.3 Nuclear Receptors

Nuclear receptors are proteins in nature. Hormones, such as thyroid hormones, transmit signals by binding to the nuclear receptors (Sever and Glass, 2013). These receptors are monomeric proteins that are a superfamily of transcription factors (Figure 9.3B). The ligands of nuclear receptors are mostly low-affinity molecules. Some of the nuclear receptors are referred to as orphan receptors, as these receptors do not have known ligands. Nuclear receptors have two domains: an N-terminal domain and a core domain. Transcription factors bind to the activation function 1 (AF1) of the N-terminal domain and alter the activity of receptors. However, the hormone response element (HRE), a portion of a gene, binds to the core domain that is further involved in DNA recognition (Aranda and Pascual, 2001).

9.4 REGULATION OF HORMONE PRODUCTION

The essential feature of stopping the stimulating components from acting unchecked must be a part of all regulatory systems. Otherwise, metabolic processes might go out of balance, which could

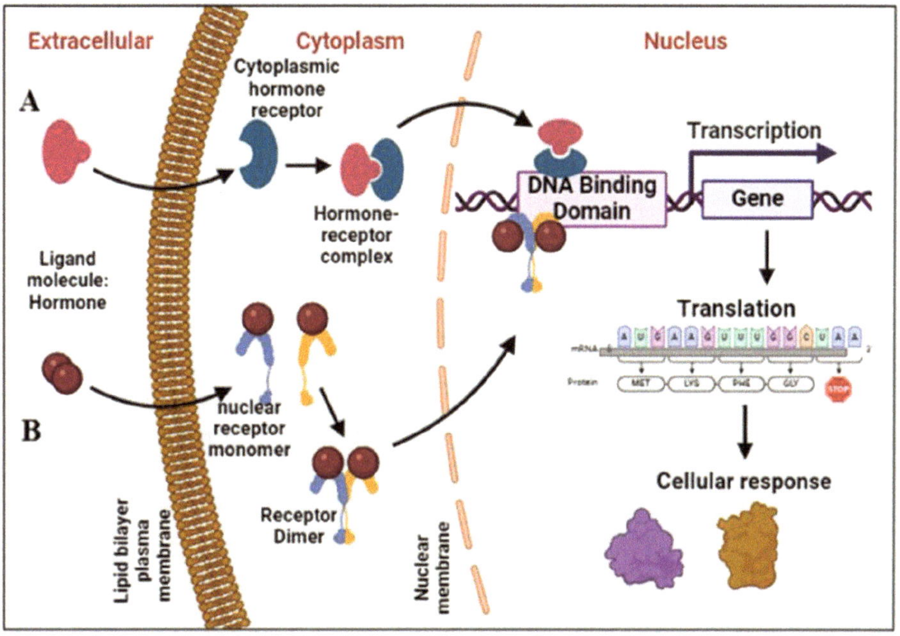

FIGURE 9.3 Signaling molecules, i.e., hormones, diffuse through the lipid bilayer plasma membrane. (A) The hormone binds to the cytoplasmic receptor that forms the hormone–receptor complex. The complex translocates in the nucleus and binds to the DNA binding site that causes specific gene transcription and translation. This results in the formation of a protein that is a cellular response against particular signals. (B) Hormone molecules bind to the monomers of nuclear receptors. The receptors undergo dimerization and the receptor dimer moves into the nuclear membrane. Within the nucleus, dimer binds to the DNA binding site and initiates gene transcription and translation for the manifestation of cellular response.

lead to problems in the case of the hormones, which are known to have stimulatory qualities. To avoid metabolic dysfunction brought on by excessive hormone activity, the hormone is either rapidly broken down to limit its duration of effect or its level in the bloodstream is managed. For instance, because it can be fatal in high amounts due to its capacity to elevate blood pressure and stimulate the heart, epinephrine is rapidly metabolized. Other hormones, such as adrenocortical hormones, thyroid hormones, and sex hormones, suppress the production of their own hormones. These hormones govern the endocrine glands. The most well-known organ where this type of regulation occurs is the human pituitary gland, which is located behind the nasal root near the base of the brain. This gland could be thought of as the central nervous system for controlling the body's hormones (Schauer, 1972).

9.5 PEPTIDE HORMONE INSULIN

The endocrine pancreas secretes the powerful peptide hormone insulin. It is essential to the control of glucose homeostasis. Moreover, it is vital to controlling the metabolisms of proteins, lipids, and carbohydrates. Insulin is supplied to the body continuously to eliminate extra glucose from the blood, except for metabolic diseases like metabolic syndrome and diabetes mellitus. Insulin also has anabolic effects on the body. Insulin contributes to the metabolism of carbohydrates by raising the rate at which glucose is burned in tissues and by facilitating the entry of glucose into cells. Additionally, it promotes liver and muscle production of glycogen from lactate and carbohydrates (DeFronzo and Tripathy, 2009; Iliya et al., 2016).

9.6 CELLULAR PATHWAY OF INSULIN SIGNALING

The target tissues of insulin are the liver, muscles, and adipose tissues. Insulin facilitates the passage of glucose, amino acids, and electrolytes across cell membranes. The binding of the hormone to certain receptors known as insulin receptors (INSR) on the surface of the cell membrane of all insulin target tissues is the first step in the cellular processes by which insulin starts its stimulating effects on glucose metabolism (Kumar and O'Rahilly, 2005; Moore et al., 2012). The typical INSR is a heterotetrameric of glycoproteins comprising two α and two β subunits connected by disulfide links to form a heterotetrameric complex (Figure 9.4).

Both of the α subunits are fully extracellular and have the domain responsible for binding insulin. The β subunits, conversely, possess an internal domain (tyrosine kinase C-terminal domain) that produces insulin-stimulated kinase activity directed toward its tyrosine residues, as well as an extracellular domain and a transmembrane domain (Whitehead et al., 2000).

Following its binding to the α subunit, insulin transduces signals across the plasma membrane, unsuppressing the activity of the β subunit. This promotes autophosphorylation of the β subunit, which in turn phosphorylates the tyrosine residues, leading to a spike in tyrosine kinase function (Iliya et al., 2016). After activation, a sequence of intramolecular transphosphorylation processes is triggered by the INSR tyrosine kinase, wherein a β subunit phosphorylates its neighboring partner on particular tyrosine residues (Hirabara et al., 2012), which afterward undergoes tyrosine phosphorylation in regions containing particular amino acid sequence motifs, which when phosphorylated, act as recognition sites for proteins containing the Src homology 2 (SH2) domains. These intracellular proteins are known as docking proteins (Draznin, 2006). These tyrosine residues can be autophosphorylated, which increases the receptor tyrosine kinase's catalytic activity and attracts docking proteins (Hirabara et al., 2012). Two proteins with Src homology (SH2 domains), Shc and Grb2, and the insulin receptor substrate (IRS) proteins have been conventionally identified among

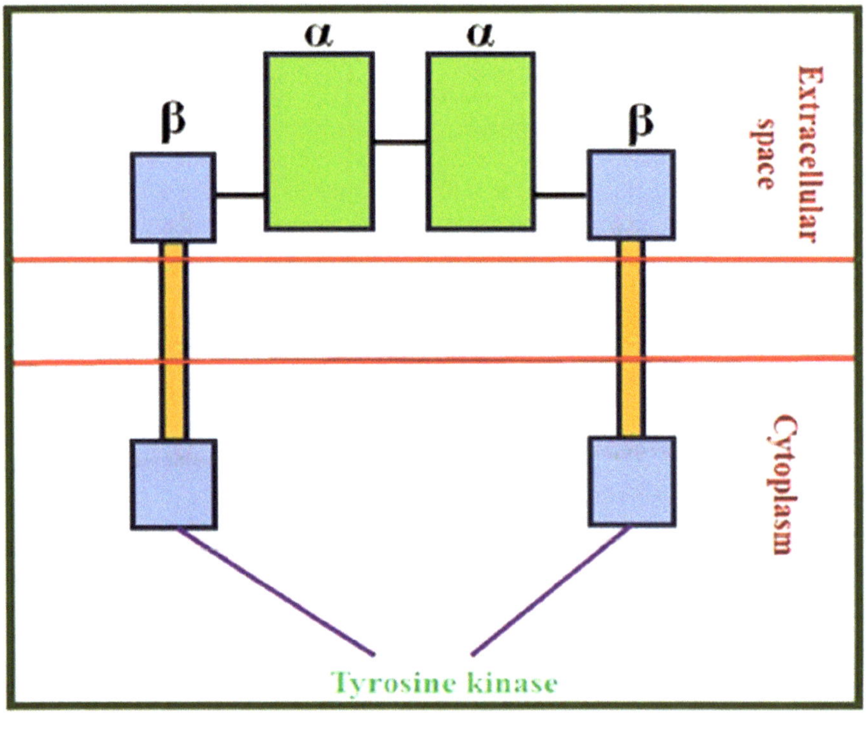

FIGURE. 9.4 Basic structure of insulin receptor.

these intracellular docking proteins (Hirabara et al., 2012; Draznin, 2006). The primary docking protein in muscle tissue that interacts with the IR tyrosine kinase domain is IRS-1 (Kovacs et al., 2003). However, the primary docking protein in the liver is IRS-2. By inducing the intracellular enzyme phosphatidylinositol-3-kinase (Pi3K), both of these docking proteins undergo tyrosine phosphorylation and modulate the effect of insulin on hepatic glucose synthesis, gluconeogenesis, and glycogen formation (Hirabara et al., 2012; Draznin, 2006). The interaction between the phosphorylated tyrosine residues of the IRS proteins and the regulatory subunit of Pi3K (85 kDa) results in the activation of the enzyme (DeFronzo and Tripathy, 2009; Iliya et al., 2016; Bradshaw and Dennis, 2009).

The Pi3K enzyme comprises a 110 kDa catalytic unit and an 85 kDa regulatory unit. Phosphatidylinositol-4,5-biphosphate and phosphatidylinositol-3,4,5-triphosphate (PIP3) are formed when certain phosphoinositide in the plasma membrane undergoes 3′ phosphorylation, which is catalyzed by the catalytic unit (DeFronzo and Tripathy, 2009).

9.6.1 Phosphatidylinositol 3-Kinase Pathway

The IRS proteins' tyrosine residues produce PIP3 and activate Pi3K, which in turn induces the phosphoinositide to activate phosphoinositide-dependent kinase-1 (PDK-1), which is a serine/threonine kinase. Once activated, PDK-1 phosphorylates another serine/threonine kinase, which is also known as protein kinase B (PKB or AKT). A PH domain found in PKB interacts directly with PIP3 (Hill et al., 1999). Activated PKB then proceeds to connect the insulin route to the glucose transporter (GLUT2 or GLUT4). When the signaling pathway is connected to these insulin-dependent GLUT proteins, the transporter is activated, moving to the cell membrane and bringing glucose into the intracellular space of the cell (Iliya et al., 2016; Hill et al., 1999). After entering the cell, glucose is processed by several different enzymes, three of which have been shown to use glucose intracellularly in a classical manner. Glucokinase, a kind of hexokinase present in the liver, is one such enzyme. It regulates how glucose is phosphorylated within the cell to glucose-6-phosphate (G6P). Next, phosphofructokinase, a crucial regulatory enzyme, phosphorylates G6P to fructose-1,6-bisphosphate, controlling the rate-limiting step into the glycolytic process. Glycogen synthetase handles excess glucose by converting it to glycogen, which is the body's stored form of glucose (Martin et al., 2000).

9.6.2 Cbl/Cbl-Associated Protein Pathway

Cbl-associated protein pathway can cause GLUT2 or GLUT4 translocation to the cell membrane since it is a compartmentalized system that functions in connection to the Pi3K/PKB pathway (Kimura et al., 2002; Smart et al., 1999). According to some research, the specific portions of the plasma membrane that are rich in cholesterol, sphingolipids, glycosylphosphatidylinositol-anchored proteins, glycolipids, and lipid-modified signaling proteins are where insulin signaling via this pathway starts. Lipid rafts or caveolin are the names given to these enhanced, specialized areas (Smart et al., 1999). Other investigations have provided evidence that at least a portion of the INSR resides in these lipid rafts (Ferre and Foufelle, 2010; Inokuchi, 2010). The tyrosine phosphorylation of the proto-oncogenes Cbl-c and Cbl-b is stimulated upon activation of the INSR in these subdomains of the plasma membrane. Cbl must be drawn to the adapter protein APS, which has PH and SH2 domains, in order for this phosphorylation step to occur (Kumar and O'Rahilly, 2005; Kimura et al., 2001; Liu et al., 2002). Along with Cbl, the Cbl-associated protein (CAP) is drawn to the INSR/APS complex (Liu et al., 2002). CAP is a bifunctional adaptor protein that possesses three SH3 domains in its COOH-terminus and an NH2-terminal region known as the sorbin homology (SoHo) domain that bears resemblance to the gut peptide sorbin (Ferre and Foufelle, 2010).

The level of insulin sensitivity and CAP expression are highly correlated. Protein production increases following activation of nuclear receptor peroxisome proliferator-activated receptor γ, which is the receptor for the thiazolidinedione class of insulin-sensitizing medicines. The protein is primarily present in insulin-sensitive tissues (Yu et al., 2002). These proteins are constitutively linked because the PXXP motif in the Cbl and the COOH terminal SH3 domain of CAP connect. CAP interacts with the lipid raft domain protein flotillin through its SoHo domain after being recruited to the INSR (Liu et al., 2005). Tyrosine phosphorylation causes Cbl to connect with Crk-II, an adaptor protein that contains SH2/SH3 (Akagi et al., 2000). Through its SH2 domain, Crk-II interacts with certain phosphorylation sites on Cbl (Watanabe et al., 2006) and, through its SH3 domain, is constitutively linked to the nucleotide exchange factor C3G or GRF2 (Knudsen et al., 1994). The small G protein TC10 is then activated by the C3G/Crk-II complex (Chiang et al., 2002). When TC10 is activated, it interacts with several molecules that could be effectors. A splice variant of the adaptor protein CIP4 and Exo70, a part of the exocyst complex, are two examples of these. The tethering or docking of secretory vesicles has been linked to the exocyst complex (Ewart et al., 2005). After becoming active, the GLUT will translocate to the plasma membrane to import glucose since these two proteins are involved in membrane fusion and vesicular trafficking (Akagi et al., 2000).

9.6.3 Mitogen-Activated Protein Kinase Pathway

Some other signal transduction proteins engage with IRS, such as Grb2, an adaptor protein with SH2 domains that binds to the nucleotide exchange factor Son of Sevenless. This interaction triggers the activation of the mitogen-activated protein kinase (MAPK) cascade, which in turn triggers mitogenic reactions (Whitehead et al., 2000; Asnaghi et al., 2004). Another insulin substrate is shc. Shc phosphorylates and binds to Grb2, which enables it to trigger the MAPK pathway without the help of IRS. The production of transcription factors and the stimulation of cell division, growth, and proliferation are both facilitated by the MAPK pathway (Figure 9.5). Therefore, it has been

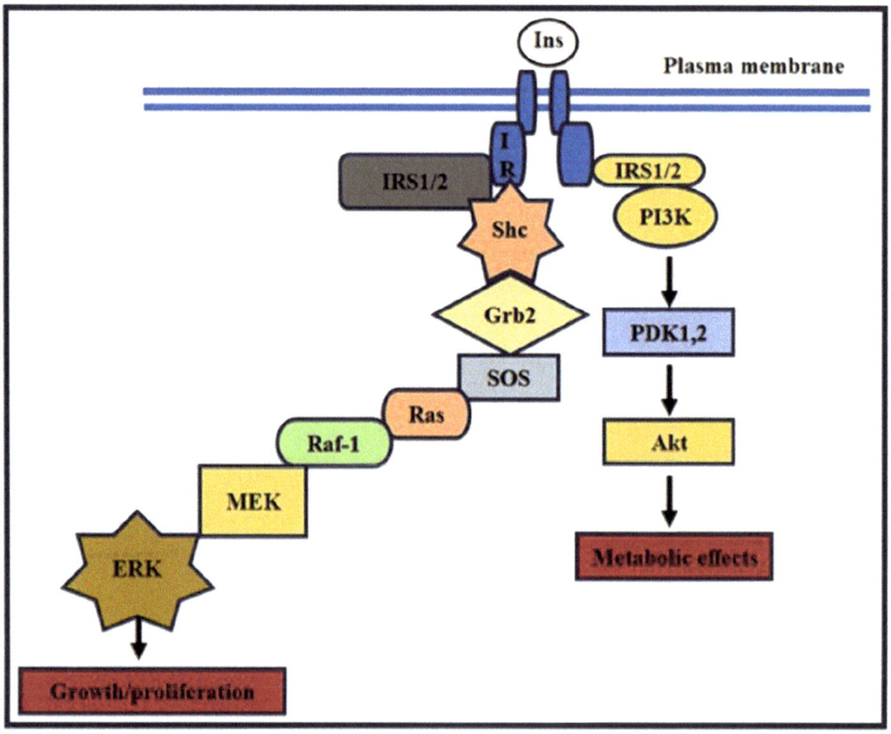

FIGURE 9.5 Insulin-associated MEK/ERK and PI3K/Akt signaling pathways.

demonstrated that blocking this route stops insulin from stimulating cell growth but has no effect on the metabolic effects of insulin (Whitehead et al., 2000; Asnaghi et al., 2004; Iliya et al., 2016).

9.7 GROWTH HORMONE RECEPTOR

As a member of the class I cytokine receptor family, the growth hormone receptor (GHR) is related to more than 30 other receptors, including the prolactin receptor (PRLR), erythropoietin receptor (EPOR), thrombopoietin receptor (TPOR), granulocyte-macrophage colony-stimulating factor receptor, and interleukin-3, interleukin-6, and interleukin-7 receptors (Waters and Brooks, 2015; Dehkhoda et al., 2018). Since GHR was the first cytokine receptor to be cloned and have its ECD crystal structure determined, it has been regarded as the prototypical class I cytokine receptor (De Vos et al., 1992).

9.7.1 Growth Hormone Receptor Structure and Mechanism of Action

One GH molecule binds two GHR molecules, according to the crystal structure of the GH bound to the GHR ECD (De Vos et al., 1992). According to biophysical research, GH first attaches to a single receptor via its "site 1" motif before using its "site 2" interactions to connect to a second receptor (Cunningham et al., 1991; Wells, 1996). Tryptophan 104 and tryptophan 169 substantially contributed to site 1 interactions in the receptor hydrophobic patch, which was shown to include the primary energy-contributing residues of the hormone–receptor complex. Moreover, tryptophan 104 is essential for the weaker site 2 interaction (Clackson et al., 1998; Walsh et al., 2003). In a study employing a hybrid receptor of the GHR ECD fused to the base of the CRH domain of granulocyte colony-stimulating factor receptor (G-CSFR), the remaining fibronectin type III homology domains (FNIII), transmembrane, and ICD supported this dimerization-induced activation model (Figure 9.6). Monoclonal antibodies to the GHR ECD activated the hybrid receptor, while monovalent

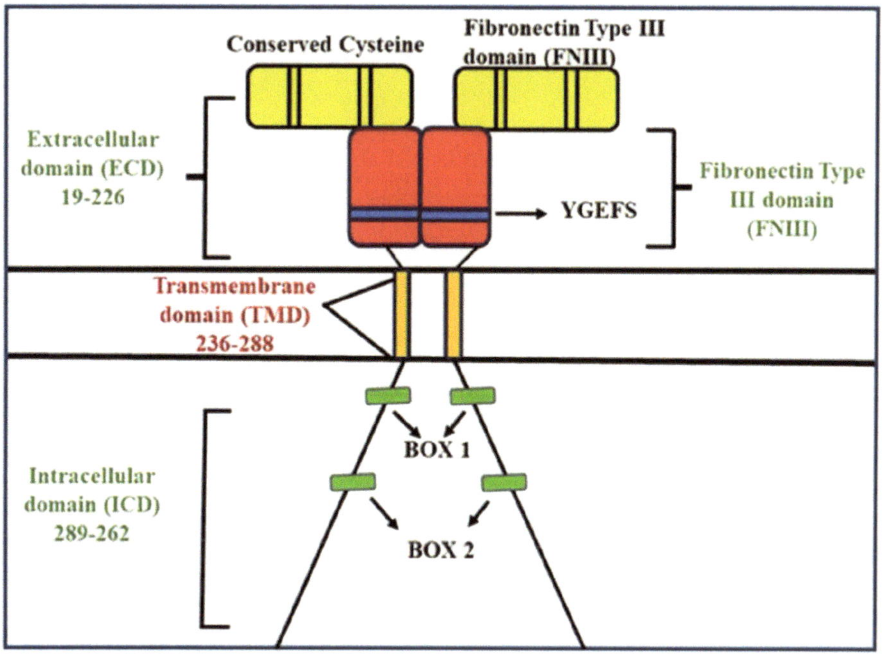

FIGURE 9.6 Growth hormone receptor domain organization.

fragments did not; however, these data may have been influenced by the additional FNIII domains (Fuh et al., 1992).

9.7.2 Mechanism of JAK2 Activation by the GHR

It was discovered that the Box2 motifs of the intracellular receptor domains spread apart when the receptor was activated by GH binding. Following this discovery, a new model for JAK2 activation by cytokine binding was needed, according to which a JAK2 PKD in trans inhibits JAK2 KD in the cytokine-free receptor. The PKD of a separate JAK2 molecule attached to the opposing receptor interacts with the KD of one JAK2 molecule bound to a Box1–Box2 motif from one receptor in this model to inhibit the KD within the receptor homodimer. The receptor transmembrane domain interaction shifts from a parallel form to a left-handed crossover interaction through a structural alteration in the ECD caused by GH binding. This structural change separates the JAK2 kinase-pseudo kinase trans connection, pulls the KDs closer together, enabling transphosphorylation and activation, and separates the ICD, at least to the Box1–Box2 motif (Waters and Brooks, 2015; Brooks et al., 2014).

9.7.3 Signal Transducers and Activators of Transcription (STATs)

When cells were treated with interferon, STATs were first discovered as cytokine-induced transcription factors (Schindler et al., 1992; Darnell Jr et al., 1994; Darnell Jr, 1997). Seven distinct STAT family members have been found, including STAT1–4, 5a, 5b, and 6. Except for sections at the N- and C-termini, two distinct genes with substantial sequence homology encode STAT5a and 5b (Leonard, 2001). Proteolytic cleavage and posttranslational modification have also been demonstrated to generate different versions of STAT1 and STAT3 (Schaefer et al., 1995). While all members of the STAT family are activated by different cytokines, IFN-α and IL-4 primarily activate STAT2 and STAT6, respectively (Hou et al., 1994).

Six domains make up STAT proteins: the transactivation domain, linker domain, DNA-binding domain, coiled-coil domain, N-terminal domain, and SH2 domain (Darnell Jr, 1997). By phosphorylating a conserved tyrosine, a JAK can activate each STAT member. Once in the nucleus, activated STATs take on the role of transcriptional regulators. The primary transcription factor that causes GH-induced cell proliferation and activities is STAT5, even though JAK2 activates STAT members 1, 3, and 5 after GH induction.

Homodimers and STAT1–STAT3 heterodimers are formed by active STAT1, STAT3, and activated STAT5 (Brooks and Waters, 2010; Gouilleux et al., 1995; Han et al., 1996).

9.7.4 GHR-Mediated Src Family Kinase (SFK) Signaling

The nine members of the Src family of tyrosine kinases are non-receptor kinases. They are Src, Lck, Hck, Fyn, Blk, Lyn, Fgr, Yes, and Yrk. The members of the family are distinguished from one another by a "unique" region near the N-terminus and a shared multidomain topology that consists of consecutive SH3, SH2, and SH1 KDs. Through processes including phosphorylation, dephosphorylation, and intricate intramolecular interactions that result in the displacement of the SH3 and SH2 domains, the regulation of SFKs is achieved (Boggon and Eck, 2004; Engen et al., 2008). The control of immune system activity and cellular homeostasis depends on SFK members (Latour and Veillette, 2001). Unlike other members of this family, Lyn serves as both a signal amplifier and a signal inhibitor, making it stand apart from the others in the family (Hibbs and Harder, 2006).

Lyn is expressed in every blood cell except T lymphocytes, and its primary roles have been linked to blood cell formation and hematopoietic growth factor receptor signaling (Lowell, 2004). Research has demonstrated that GH can activate SFKs without any JAK2 activation (Rowlinson et al., 2008). When employing JAK2-deficient γ2A cells, it was still demonstrated that GH activates the SFK member Lyn, which is linked to the membrane-proximal 150 residues of the cytoplasmic

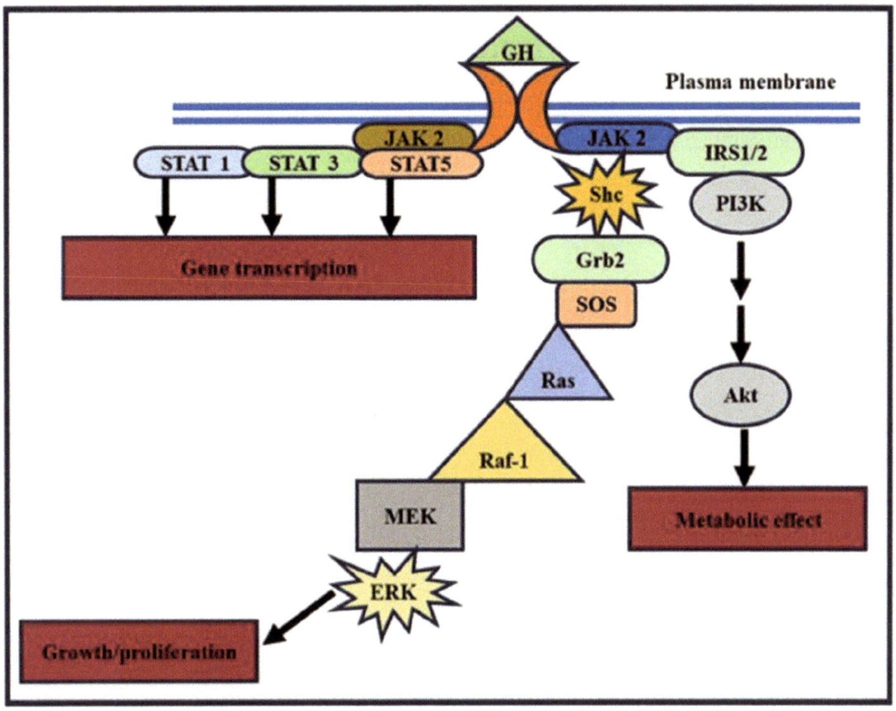

FIGURE 9.7 The principal intracellular signaling pathways that GH activates. The activated GHR/JAK2 complex primarily activates the MEK/ERK, PI3K/Akt, and STAT signaling pathways.

domain of the human GHR. Additionally, a GHR Box 1 mutant that lost GHR-mediated JAK2-STAT signaling was found to still activate SFKs and ERK1/2 in response to GH stimulation (Rowlinson et al., 2008; Barclay et al., 2010). From these investigations, it was suggested that this new SFK-dependent pathway starts with PLCγ activation, which then releases diacylglycerol and inositol 1,4,5-triphosphate (IP3) (Figure 9.7). This increases the amount of cytoplasmic calcium ions, which activates the Ras guanine nucleotide exchange factor RasGRP1. The ERK1/2 signaling pathway is started and RAS is activated as a result of this activity (Rowlinson et al., 2008). Using growth hormone as a ligand, related investigations verified that this JAK2-independent route can activate ERK1/2 via activating Rapl and Rap2, as well as Ras-like small GTPases RalA and RalB in NIH-3T3 cells (Zhu et al., 2002; Pessentheiner et al., 2013). Impaired ERK1/2 activation was the outcome of mutations made in a loop of the GHR ECD that modifies GH-induced conformational changes but not JAK2 or STAT5 activation. It was demonstrated that GHR communicates with Lyn (an SFK) and that the decreased ERK1/2 activation was caused by a deficit in Lyn's activation (Rowlinson et al., 2008). These investigations demonstrated that conformational changes in the ECD regulate the amount of JAK2 or SFK signaling. It's interesting to note that exon 3 removed GHR, which causes the ECD to lose 22 amino acids, and shows normal STAT5 and AKT signaling but changed ERK signaling. This work demonstrated the physiological significance of modifying these pathways' signaling intensity because exon 3-deleted GHR subjects had longer lifespans and were taller (Ben-Avraham et al., 2017).

9.8 DEFECTS IN TRANSDUCTION WITHIN THE GH SIGNAL

Cellular GH signaling can be altered or improved by variations in post-GHR signaling molecules and GHR availability or abundance. $GHR^{-/-}$ mice are viable and nearly normal in size when they are born, but they acquire only about half of the average adult body weight because they mature later than their normal siblings (Zhou et al., 1997). GHR gene variations, including nonsense, splicing, frameshift, and nonsense mutations, have been identified to date; GH resistance has been associated with a variety of GHR mutations. These mutations may impact GH binding, GH signaling, or the expression of the GHR gene. GH variation in a short-statured infant revealed reduced ERK1/2 activation but normal STAT5 activation (Lewis et al., 2004). A distinct mutant GHR found in a family with notably shorter stature showed higher GH activation of STAT5 and only mild activation of the ERK1/2 pathway when compared to wild-type GHR (Metherell et al., 2001).

9.9 INSULIN-MEDIATED GHR LEVEL MAINTENANCE

It appears that proper hepatic growth hormone response depends on physiological insulin levels, which rise temporarily after meals. Reduced levels of GH binding protein (GHBP), which is released into the bloodstream after being cleaved from cell surface GHR, have been associated with insulin deficit in type 1 diabetics. Streptozotocin decreases plasma IGF-1, GH binding, and liver GHR levels. It is the same chemical that causes type 1 diabetes in animals. IGF-1 and GH binding are increased by insulin therapy (Xu and Messina, 2009). Insulin is known to stimulate the growth of the fetus and the newborn (Messina, 2010).

9.10 CONCLUSION

Cell signaling is a process of transmitting information from one part of the body to another. To convey information across the body, cell–cell communication is a fundamental process. In this regard, the endocrine system is one of the main systems that is involved in cell–cell communication via distant signal transmission. Hormones, the signaling molecules, are released from endocrine glands and flow with the bloodstream to reach the site of action. Wherever the action is desired, hormones bind to their specific receptors and cause activation of signaling molecules downstream in the cell. The regulator proteins get activated and initiate gene transcription and translation. Resultantly, the specific response appears in the form of protein formation. The hormonal response is not limited to the cells on which they bind. They regulate their release, and the responding cell may further affect other cells.

REFERENCES

Akagi, T., Shishido, T., Murata, K. & Hanafusa, H. 2000. v-Crk activates the phosphoinositide 3-kinase/AKT pathway in transformation. *Proceedings of the National Academy of Sciences*, 97, 7290–7295.

Alberts, B., Bray, D., Lewis, J., Raff, M., Roberts, K. & Watson, J. D. 1994. *Molecular biology of the cell.* Garland New York.

Andersen, M. E. & Barton, H. A. 1999. Biological regulation of receptor-hormone complex concentrations in relation to dose-response assessments for endocrine-active compounds. *Toxicological Sciences*, 48, 38–50.

Anyetei-Anum, C. S., Roggero, V. R. & Allison, L. A. 2018. Thyroid hormone receptor localization in target tissues. *Journal of Endocrinology*, 237, R19–R34.

Aranda, A. & Pascual, A. 2001. Nuclear hormone receptors and gene expression. *Physiological Reviews*, 81, 1269–304.

Asnaghi, L., Bruno, P., Priulla, M. & Nicolin, A. 2004. mTOR: A protein kinase switching between life and death. *Pharmacological Research*, 50, 545–549.

Barclay, J. L., Kerr, L. M., Arthur, L., Rowland, J. E., Nelson, C. N., Ishikawa, M., D'aniello, E. M., White, M., Noakes, P. G. & Waters, M. J. 2010. In vivo targeting of the growth hormone receptor (GHR) Box1 sequence demonstrates that the GHR does not signal exclusively through JAK2. *Molecular Endocrinology*, 24, 204–217.

Ben-Avraham, D., Govindaraju, D. R., Budagov, T., Fradin, D., Durda, P., Liu, B., Ott, S., Gutman, D., Sharvit, L. & Kaplan, R. 2017. The GH receptor exon 3 deletion is a marker of male-specific exceptional longevity associated with increased GH sensitivity and taller stature. *Science Advances*, 3, e1602025.

Boggon, T. J. & Eck, M. J. 2004. Structure and regulation of Src family kinases. *Oncogene*, 23, 7918–7927.
Bradshaw, R. A. & Dennis, E. A. 2009. *Handbook of cell signaling*. Academic Press.
Brooks, A. J., Dai, W., O'mara, M. L., Abankwa, D., Chhabra, Y., Pelekanos, R. A., Gardon, O., Tunny, K. A., Blucher, K. M. & Morton, C. J. 2014. Mechanism of activation of protein kinase JAK2 by the growth hormone receptor. *Science*, 344, 1249783.
Brooks, A. J. & Waters, M. J. 2010. The growth hormone receptor: Mechanism of activation and clinical implications. *Nature Reviews Endocrinology*, 6, 515–525.
Chiang, S.-H., Hou, J. C., Hwang, J., Pessin, J. E. & Saltiel, A. R. 2002. Cloning and functional characterization of related TC10 isoforms, a subfamily of rho proteins involved in insulin-stimulated glucose transport. *Journal of Biological Chemistry*, 277, 13067–13073.
Clackson, T., Ultsch, M. H., Wells, J. A. & De Vos, A. M. 1998. Structural and functional analysis of the 1: 1 growth hormone: Receptor complex reveals the molecular basis for receptor affinity. *Journal of Molecular Biology*, 277, 1111–1128.
Cooper, G. M. & Hausman, R. J. T. C. N. E. S. 2000. *A Molecular Approach*. Sinauer Associates.
Cunningham, B. C., Ultsch, M., De Vos, A. M., Mulkerrin, M. G., Clauser, K. R. & Wells, J. A. 1991. Dimerization of the extracellular domain of the human growth hormone receptor by a single hormone molecule. *Science*, 254, 821–825.
Darnell JR, J. E. 1997. Stats and gene regulation. *Science*, 277, 1630–1635.
Darnell JR, J. E., Kerr, L. M. & Stark, G. R. 1994. Jak-Stat pathways and transcriptional activation in response to IFNs and other extracellular signaling proteins. *Science*, 264, 1415–1421.
De Vos, A. M., Ultsch, M. & Kossiakoff, A. A. 1992. Human growth hormone and extracellular domain of its receptor: Crystal structure of the complex. *Science*, 255, 306–312.
Defronzo, R. A. & Tripathy, D. 2009. Skeletal muscle insulin resistance is the primary defect in type 2 diabetes. *Diabetes Care*, 32, S157.
Dehkhoda, F., Lee, C. M., Medina, J. & Brooks, A. J. 2018. The growth hormone receptor: mechanism of receptor activation, cell signaling, and physiological aspects. *Frontiers in Endocrinology*, 9, 297357.
Draznin, B. 2006. Molecular mechanisms of insulin resistance: Serine phosphorylation of insulin receptor substrate-1 and increased expression of p85α: The two sides of a coin. *Diabetes*, 55, 2392–2397.
Engen, J. R., Wales, T. E., Hochrein, J. M., Meyn, M. R., Banu Ozkan, S., Bahar, I. & Smithgall, T. E. 2008. Structure and dynamic regulation of Src-family kinases. *Cellular and Molecular Life Sciences*, 65, 3058–3073.
Ewart, M.-A., Clarke, M., Kane, S., Chamberlain, L. H. & Gould, G. W. 2005. Evidence for a role of the exocyst in insulin-stimulated Glut4 trafficking in 3T3-L1 adipocytes. *Journal of Biological Chemistry*, 280, 3812–3816.
Ferre, P. & Foufelle, F. 2010. Hepatic steatosis: A role for de novo lipogenesis and the transcription factor Srebp-1c. *Diabetes, Obesity and Metabolism*, 12, 83–92.
Fuh, G., Cunningham, B. C., Fukunaga, R., Nagata, S., Goeddel, D. V. & Wells, J. A. 1992. Rational design of potent antagonists to the human growth hormone receptor. *Science*, 256, 1677–1680.
Gouilleux, F., Pallard, C., Dusanter-Fourt, I., Wakao, H., Haldosen, L., Norstedt, G., Levy, D. & Groner, B. 1995. Prolactin, growth hormone, erythropoietin and granulocyte-macrophage colony stimulating factor induce MGF-Stat5 DNA binding activity. *The EMBO Journal*, 14, 2005–2013.
Han, Y., Leaman, D. W., Watling, D., Rogers, N. C., Groner, B., Kerr, I. M., Wood, W. I. & Stark, G. R. 1996. Participation of JAK and stat proteins in growth hormone-induced signaling (*). *Journal of Biological Chemistry*, 271, 5947–5952.
Hibbs, M. L. & Harder, K. W. 2006. The duplicitous nature of the Lyn tyrosine kinase in growth factor signaling. *Growth Factors*, 24, 137–149.
Hill, M. M., Clark, S. F., Tucker, D. F., Birnbaum, M. J., James, D. E. & Macaulay, S. L. 1999. A role for protein kinase Bβ/Akt2 in insulin-stimulated Glut4 translocation in adipocytes. *Molecular and Cellular Biology*, 19, 3208–3219.
Hirabara, S. M., Gorjao, R., Vinolo, M. A., Rodrigues, A. C., Nachbar, R. T. & Curi, R. 2012. Molecular targets related to inflammation and insulin resistance and potential interventions. *BioMed Research International*, 2012, 1–13.
Hou, J., Schindler, U., Henzel, W. J., Ho, T. C., Brasseur, M. & Mcknight, S. L. 1994. An interleukin-4-induced transcription factor: IL-4 Stat. *Science*, 265, 1701–1706.
Iliya, I., Mohammed, B., Akuyam, S., Nok, A., Bauchi, Z., Tanko, M., Timbuak, J. & Yusuf, B. 2016. Molecular mechanisms of insulin signaling. *Sub-Saharan African Journal of Medicine*, 3, 3–7.
Inokuchi, J.-I. 2010. Membrane microdomains and insulin resistance. *FEBS Letters*, 584, 1864–1871.

Kahn, C. R. 1976. Membrane receptors for hormones and neurotransmitters. *Journal of Cell Biology*, 70, 261–86.
Karlson, P. 1979. [Hormone receptors and hormone action (author's transl)]. *Klinische Wochenschrift*, 57, 607–12.
Kiang, D. T., Kollander, R. E., Thomas, T. & Kennedy, B. J. C. R. 1989. Up-regulation of estrogen receptors by nonsteroidal antiestrogens in human breast cancer. *Cancer Research*, 49, 5312–5316.
Kimura, A., Baumann, C. A., Chiang, S.-H. & Saltiel, A. R. 2001. The sorbin homology domain: a motif for the targeting of proteins to lipid rafts. *Proceedings of the National Academy of Sciences*, 98, 9098–9103.
Kimura, A., Mora, S., Shigematsu, S., Pessin, J. E. & Saltiel, A. R. 2002. The insulin receptor catalyzes the tyrosine phosphorylation of caveolin-1. *Journal of Biological Chemistry*, 277, 30153–30158.
Knudsen, B. S., Feller, S. M. & Hanafusa, H. 1994. Four proline-rich sequences of the guanine-nucleotide exchange factor C3G bind with unique specificity to the first Src homology 3 domain of Crk. *Journal of Biological Chemistry*, 269, 32781–32787.
Kovacs, P., Hanson, R. L., Lee, Y.-H., Yang, X., Kobes, S., Permana, P. A., Bogardus, C. & Baier, L. J. 2003. The role of insulin receptor substrate-1 gene (IRS1) in type 2 diabetes in Pima Indians. *Diabetes*, 52, 3005–3009.
Kumar, S. & O'rahilly, S. 2005. *Insulin Resistance: Insulin Action and Its Disturbances in Disease*. John Wiley & Sons.
Latour, S. & Veillette, A. 2001. Proximal protein tyrosine kinases in immunoreceptor signaling. *Current Opinion in Immunology*, 13, 299–306.
Leonard, W. J. 2001. Role of Jak kinases and STATs in cytokine signal transduction. *International Journal of Hematology*, 73, 271–277.
Lewis, M. D., Horan, M., Millar, D. S., Newsway, V., Easter, T. E., Fryklund, L., Gregory, J. W., Norin, M., Del Valle, C. B.-J. & López-Siguero, J. P. 2004. A novel dysfunctional growth hormone variant (Ile179Met) exhibits a decreased ability to activate the extracellular signal-regulated kinase pathway. *The Journal of Clinical Endocrinology & Metabolism*, 89, 1068–1075.
Liu, J., Deyoung, S. M., Zhang, M., Dold, L. H. & Saltiel, A. R. 2005. The stomatin/prohibitin/flotillin/HflK/C domain of flotillin-1 contains distinct sequences that direct plasma membrane localization and protein interactions in 3T3-L1 adipocytes. *Journal of Biological Chemistry*, 280, 16125–16134.
Liu, J., Kimura, A., Baumann, C. A. & Saltiel, A. R. 2002. APS facilitates c-Cbl tyrosine phosphorylation and GLUT4 translocation in response to insulin in 3T3-L1 adipocytes. *Molecular and Cellular Biology*, 22, 3599–3609.
Lowell, C. A. 2004. Src-family kinases: rheostats of immune cell signaling. *Molecular Immunology*, 41, 631–643.
Martin, S., Millar, C. A., Lyttle, C. T., Meerloo, T., Marsh, B. J., Gould, G. W. & James, D. E. 2000. Effects of insulin on intracellular GLUT4 vesicles in adipocytes: Evidence for a secretory mode of regulation. *Journal of Cell Science*, 113, 3427–3438.
Messina, J. L. 2010. Insulin as a growth-promoting hormone. *Comprehensive Physiology*, 783–811.
Metherell, L. A., Akker, S. A., Munroe, P. B., Rose, S. J., Caulfield, M., Savage, M. O., Chew, S. L. & Clark, A. J. 2001. Pseudoexon activation as a novel mechanism for disease resulting in atypical growth-hormone insensitivity. *The American Journal of Human Genetics*, 69, 641–646.
Molnar, C. & Gair, J. 2015. *Concepts of Biology*, BCcampus.
Moore, M. C., Coate, K. C., Winnick, J. J., An, Z. & Cherrington, A. D. 2012. Regulation of hepatic glucose uptake and storage in vivo. *Advances in Nutrition*, 3, 286–294.
Pal, K., Melcher, K. & Xu, H. E. 2012. Structure and mechanism for recognition of peptide hormones by Class B G-protein-coupled receptors. *Acta Pharmacologica Sinica*, 33, 300–11.
Pessentheiner, A. R., Pelzmann, H. J., Walenta, E., Schweiger, M., Groschner, L. N., Graier, W. F., Kolb, D., Uno, K., Miyazaki, T. & Nitta, A. 2013. NAT8L (N-acetyltransferase 8-like) accelerates lipid turnover and increases energy expenditure in brown adipocytes. *Journal of Biological Chemistry*, 288, 36040–36051.
Reichel, R. R. & Jacob, S. T. 1993. Control of gene expression by lipophilic hormones. *FASEB Journal*, 7, 427–36.
Rowlinson, S. W., Yoshizato, H., Barclay, J. L., Brooks, A. J., Behncken, S. N., Kerr, L. M., Millard, K., Palethorpe, K., Nielsen, K. & Clyde-Smith, J. 2008. An agonist-induced conformational change in the growth hormone receptor determines the choice of signaling pathway. *Nature Cell Biology*, 10, 740–747.
Schaefer, T. S., Sanders, L. K. & Nathans, D. 1995. Cooperative transcriptional activity of Jun and Stat3 beta, a short form of Stat3. *Proceedings of the National Academy of Sciences*, 92, 9097–9101.

Schauer, R. 1972. The mode of action of hormones. *Angewandte Chemie International Edition in English*, 11, 7–16.

Schindler, C., Shuai, K., Prezioso, V. R. & Darnell JR, J. E. 1992. Interferon-dependent tyrosine phosphorylation of a latent cytoplasmic transcription factor. *Science*, 257, 809–813.

Seifter, J. L., Walsh, E. & Sloane, D. E. 2021. *Integrated Physiology and Pathophysiology: Integrated Physiology and Pathophysiology E-Book*. Elsevier Health Sciences.

Sever, R. & Glass, C. K. 2013. Signaling by nuclear receptors. *Cold Spring Harbor Perspectives in Biology*, 5, a016709.

Singer, S. J. & Nicolson, G. L. 1972. The fluid mosaic model of the structure of cell membranes. *Science*, 175, 720–731.

Smart, E. J., Graf, G. A., Mcniven, M. A., Sessa, W. C., Engelman, J. A., Scherer, P. E., Okamoto, T. & Lisanti, M. P. 1999. Caveolins, liquid-ordered domains, and signal transduction. *Molecular and Cellular Biology*, 19, 7289–7304.

Walsh, S. T., Jevitts, L. M., Sylvester, J. E. & Kossiakoff, A. A. 2003. Site2 binding energetics of the regulatory step of growth hormone–induced receptor homodimerization. *Protein Science*, 12, 1960–1970.

Watanabe, T., Tsuda, M., Makino, Y., Ichihara, S., Sawa, H., Minami, A., Mochizuki, N., Nagashima, K. & Tanaka, S. 2006. Adaptor molecule Crk is required for sustained phosphorylation of Grb2-associated binder 1 and hepatocyte growth factor–induced cell motility of human synovial sarcoma cell lines. *Molecular Cancer Research*, 4, 499–510.

Waters, M. J. & Brooks, A. J. 2015. JAK2 activation by growth hormone and other cytokines. *Biochemical Journal*, 466, 1–11.

Wells, J. A. 1996. Binding in the growth hormone receptor complex. *Proceedings of the National Academy of Sciences*, 93, 1–6.

Whitehead, J. P., Clark, S. F., Ursø, B. & James, D. E. 2000. Signalling through the insulin receptor. *Current Opinion in Cell Biology*, 12, 222–228.

Xu, J. & Messina, J. L. 2009. Crosstalk between growth hormone and insulin signaling. *Vitamins & Hormones*, 80, 125–153.

Yu, C., Chen, Y., Cline, G. W., Zhang, D., Zong, H., Wang, Y., Bergeron, R., Kim, J. K., Cushman, S. W. & Cooney, G. J. 2002. Mechanism by which fatty acids inhibit insulin activation of insulin receptor substrate-1 (IRS-1)-associated phosphatidylinositol 3-kinase activity in muscle. *Journal of Biological Chemistry*, 277, 50230–50236.

Zhou, Y., Xu, B. C., Maheshwari, H. G., He, L., Reed, M., Lozykowski, M., Okada, S., Cataldo, L., Coschigamo, K. & Wagner, T. E. 1997. A mammalian model for Laron syndrome produced by targeted disruption of the mouse growth hormone receptor/binding protein gene (the Laron mouse). *Proceedings of the National Academy of Sciences*, 94, 13215–13220.

Zhu, T., Ling, L. & Lobie, P. E. 2002. Identification of a JAK2-independent pathway regulating growth hormone (GH)-stimulated p44/42 mitogen-activated protein kinase activity: GH activation of ral and phospholipase D is Src-dependent. *Journal of Biological Chemistry*, 277, 45592–45603.

10 Signaling Networks in Development and Differentiation

Aroosa Younis Nadeem[1], *Areej Fatima*[1], *Maleeha Azhar*[1], *and Adeeb Shehzad*[1,2]
[1]Department of Biomedical Sciences, School of Mechanical and Manufacturing Engineering, National University of Sciences & Technology, H-12 Islamabad, Pakistan
[2]Biodiversity Unit, Research Center, Dhofar University, Salalah 211, Oman
*Department of Biomedical Engineering & Sciences, School of Mechanical and Manufacturing Engineering, National University of Sciences and Technology, Islamabad, Pakistan

10.1 INTRODUCTION

Developing embryonic cells are perpetually in contact with their neighboring cells, and the molecules they employ to transmit and acquire signals are vital to embryonic development. Researchers have discovered a multitude of intracellular signaling pathways, of which a subset is stimulated by secreted growth regulators. Morphogens are chemicals that establish a concentration gradient and determine cell fate in response to growth factor levels. Fibroblast growth factor (FGF), retinoic acid (RA), sonic hedgehog (SHH), and wingless (WNT) are examples of such pathways. These pathways regulate proliferation and its sustenance, in addition to controlling the final state of cells. In contrast, cells utilize planar cell polarity (PCP) pathways to regulate cell morphology and polarity and to interpret their orientation within the tissue plane. Adjacent cells utilize the Notch signaling pathway to relay information and regulate binary cell fate determinations, as well as the development of complex tissue boundaries and patterns. This chapter provides a detailed description of chosen pathways, illustrating with specific examples how a single signaling pathway can be utilized iteratively for multiple purposes throughout embryogenesis, how cells perceive and integrate subtle variations in signal strength, how signaling pathways are regulated and interconnected, and how morphogenesis and cell fate are regulated.

10.2 BRIEF REVIEW OF THE CELLULAR DEVELOPMENTAL SIGNAL TRANSDUCTION CASCADE

10.2.1 Decoding the Insights and Implications of Receptor Tyrosine Kinases

Receptor tyrosine kinases (RTKs) phosphorylate tyrosine residues. Researchers have identified a total of 58 RTKs, categorized into 20 classes. In numerous circumstances, distinct secretory ligands may stimulate specific RTKs [1]. As a developmentally relevant signaling module, RTK signaling serves a wide variety of purposes. Consequently, mutations in RTK signaling pathways are linked to a wide range of developmental disorders and diseases [2]. Developmental biologists are especially interested in a subset of RTKs because their absence causes noticeable developmental abnormalities. Examples of RTKs that can be activated by various growth factors include epidermal

 DOI: 10.1201/9781003594901-10

growth factor (EGF), vascular endothelial growth factor (VEGF), platelet-derived growth factor (PDGF), and glial cell line-derived neurotrophic factor (GDNF), which are all produced from glial cell lines. Although it would be impractical to include a comprehensive list and explanation of all RTK ligands, Table 10.1 provides examples of ligands that can stimulate RTKs with implications for embryogenesis.

10.2.1.1 RTK-Mediated Intracellular Signal Transduction Mechanisms

A typical RTK has both an intracellular tyrosine kinase domain and an external ligand-binding domain. Receptor activation occurs when they bind to their surface-bound cognate ligands. This results in the stabilization of a receptor dimer or oligomer with active tyrosine kinase activity, regardless of the chemical mechanisms by which various ligands activate specific receptors. Complementary molecules control the expression of certain receptors. For example, heparan sulfate proteoglycans (HSPGs) latch to and interact with both the receptor and the ligand, thereby stabilizing FGF:FGFR complexes [23]. The catalytic activity of activated receptor complexes is enhanced through transphosphorylation, which involves phosphorylating one or more tyrosine residues [24]. Phosphotyrosine residues serve as binding sites for downstream signaling molecules through their Src homology-2 (SH2) or phosphotyrosine binding (PTB) domains. Docking proteins, such as FRS2, IRS1, or Gab1 (Grb2-associated binder), are among the first to be recruited. Proteins with SH2 and SH3 domains have many functional modules that work together for interactions. Multimolecular signaling complexes form when these molecules interact with one another after recruitment (Figure 10.1).

Figure 10.1 shows that several downstream signaling pathways are activated by multimolecular complexes. These pathways include the Ras/MAPK, PLCγ, and PI3K pathways. The PI3K pathway has long been linked to cell survival, the Ras/MAPK pathway to increased cell proliferation, and Rho/Rac activity to alterations in cell shape and movement. It remains unclear how distinct growth factors selectively activate downstream cascades or to what degree this specificity is achieved. Xenopus has a mechanism that uses negative feedback to target specific pathways. For example, one study reported that Sprouty proteins block PLCγ-dependent pathways and morphogenesis, while Spred family feedback antagonists selectively decrease MAPK signaling and mesoderm induction [25]. Another factor that could determine the downstream pathway and cellular response is the location of the initial signaling complex within the cell, whether on the cytoplasmic membrane or an intracellular membrane like an endosome [26]. However, it is now known that several RTK signaling pathways engage in substantial crosstalk, resulting in complex signaling networks. Figure 10.1 shows how extracellular signal-regulated kinase (ERK) signaling can be enhanced when active PKC activates Raf1 downstream of the PLCγ pathway [27].

In vitro experimental investigations have provided a significant portion of our understanding of the molecular mechanisms behind RTK signal transduction. Many of these in vivo discoveries have been confirmed by gene knockout experiments in mice. For instance, due to defects in FGF-dependent processes like gastrulation, embryos that are FRS2α-deficient do not develop past E7.5 [28]. In vivo investigations have revealed considerable functional differences between the mammalian RTK signaling components. For example, while $Erk2^{-/-}$ mice are viable with only mild anomalies, $Erk1^{-/-}$ embryos die soon after implantation due to deficiencies in the production of extraembryonic tissues [29].

10.2.1.2 Morphogenetic Roles and Receptor Mechanisms of WNT Signaling Cascade

Numerous studies have reported that WNT ligands are secreted morphogens with lipid modifications that play important roles in developmental phases [30].

The signaling process takes place via Frizzled (Fz) receptors, which are seven-pass transmembrane proteins [31]. There are 10 Fz genes and 19 WNT genes in the human and mouse genomes. In Xenopus embryos, canonical WNT ligands like WNT1, WNT3a, and WNT8 can induce a secondary axis, while noncanonical WNTs like WNT4, WNT5a, and WNT11 induce convergent extension,

TABLE 10.1
Ligands That Can Stimulate RTKs

Receptor Tyrosine Kinase	Ligands	Cofactors	Pathways	Functions	References
Fibroblast growth factor receptor (FGFR1–4)	FGF1–10, FGF15–20, FGF22 (canonical FGFs)	Heparan sulfate proteoglycan	MAPK/ERK pathway PI3K/AKT pathway PLCγ pathway	Stimulates cell proliferation, differentiation, and migration	[3]
	FGF15/19	b-Klotho	MAPK/ERK pathway PI3K/AKT pathway	Stimulates glucose uptake in adipose tissue and skeletal muscle	[4]
	FGF23	a-Klotho		Regulates phosphate homeostasis	[5]
	FGF21 (hormone-like/endocrine FGFs)	Unknown		Regulates glucose and lipid metabolism	[6]
Rearranged during transfection (RET)	GDNF	GFRa1	PI3K/AKT and MAPK/ERK pathways	Promote cell survival, proliferation, and differentiation of enteric nervous system (ENS)	[7]
	Neurturin	GFRa2		Development and maintenance of neurons in the peripheral nervous system (PNS)	[8]
	Persephin	GFRa3		Development and maintenance of neurons in the enteric nervous system (ENS) and peripheral nervous system (PNS)	[9]
	Artemin	GFRa4		Development and maintenance of sensory and sympathetic neurons	[10]
EGFR (ErbB1)	EGF, TGFa, HB-EGF, amphiregulin	EGF receptor substrate 15 (EPS15)	MAPK/ERK pathway PI3K/AKT pathway	Induces cell proliferation, migration, and survival	[11]
	Betacellulin	Shc adaptor protein		Involved in tissue development and regeneration.	[12]
	Epiregulin	Grb2		Contributes to tissue development and wound healing	[13]
	Epigen			Implicated in cancer progression and metastasis	[14]
VEGFR1-3	VEGF-A (isoforms VEGF121, 165, 189, and 206)	NRP1 NRP2 HSPGs	MAPK/ERK pathway PI3K/AKT pathway PLCγ	Induces angiogenesis, enhances vascular permeability	[15]
	VEGF-B	NRP1 NRP2	MAPK/ERK pathway PI3K/AKT	Promotes endothelial cell survival and migration	[16]
	VEGF-C	NRP2		Induces lymph angiogenesis	[17]
	VEGF-D	NRP2		Promotes lymphatic metastasis in cancer	[18]
	Placental growth factor (PGF)	NRP1 NRP2		Enhances arteriogenesis	[19]
Erythroblastic leukemia viral oncogene homologue ErbB2 ErbB3	Heregulin		MAPK/ERK pathway PI3K/AKT pathway PLCγ	Plays crucial roles in tissue development; implicated in neuronal and development, synaptic plasticity	[20]
ErbB4	Neuregulin			Contributes to cardiomyocyte proliferation, myocardial growth, and maintenance of cardiac contractility	[21]
PDGFR	PDGF-A, B, C, D	(HSPGs)		Plays essential roles in tissue development, particularly in embryogenesis, angiogenesis, and wound healing	[22]

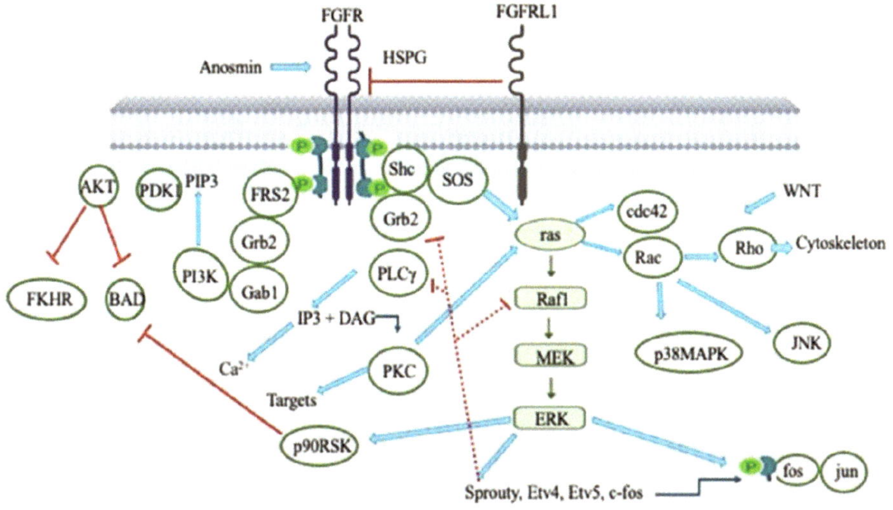

FIGURE 10.1 RTK signaling is depicted by FGF signaling cascades. The transphosphorylation of FGFRs is facilitated by the interaction of heparan sulfate proteoglycan (HSPG) and FGF ligand in the form of multimeric complexes [TK] motifs. Concurrence between TK-deficient FGFRL1 molecules may impede stimulation of receptors. Many cofactors, including anosmin, have been identified as potential enhancers of FGFR activation. Phosphorylated FGFRs facilitate the recruitment of adapter molecules, including Shc and FRS2, which in turn recruit Grb2 and SOS to initiate the Ras /MAPK pathway, and Gabl to initiate the PI3K cascade. Ras is activated by the guanine exchange factor SOS, facilitating the transmission of the signal to Raf1, MEK, and ERK. From there, the signal translocates to the nucleus, where it modulates gene expression in conjunction with transcriptional cofactors. Feedback inhibitors of the pathway are encoded by Sprouty genes (broken lines). C-fos, an essential indicator of signal potency, is transcriptionally and post-translationally regulated by ERK. Recruitment and activation of AKT are correlated with PI3K pathway activation via PIP3 production and PDK1 activity. AKT obstructs cellular death pathways that are controlled by BAD and forkhead transcription factors (FKHR). Additionally, ERK can impede cell death via the mechanism mediated by p90RSK on BAD. Recruitment to activated RTKs induces the activation of PLCγ, which subsequently releases the second messengers IP3 and DAG. IP3 stimulates the release of intracellular Ca2+ and DAG-activated-PKC, which activates Raf1, alters gene expression downstream, and contributes to the Ras/MAPK pathway. Upon activation, Ras can stimulate the activity of additional small GTPases, including cdc42, Rho, Rac, and p38MAPK or JNK, which modify the cytoskeleton's stability. In addition to Rac and rho activation, noncanonical WNT signaling contributes to this pathway by influencing the cytoskeleton and activating JNK.

depending on the circumstances [32]. Researchers have identified three main WNT signaling pathways: the canonical WNT/β-catenin, noncanonical WNT/Ca^{2+}, and WNT/PCP pathways.

10.2.1.3 Canonical WNT/β-Catenin Signaling Pathway

The stabilization of the downstream effector molecule, β-catenin, is the central event in canonical signaling; thus, it is also called the WNT/β-catenin pathway. When WNT signaling is absent, β-catenin is broken down by a complex system, including Axin, adenomatous polyposis coli (APC), and glycogen synthase kinase 3 (GSK-3). New recognition sites for the F-box/WD repeat protein β-TrCP arise when GSK-3 phosphorylates β-catenin at multiple Ser/Thr domains. According to Shimizu et al. (2017), β-TrCP is a component of an E3 ubiquitin ligase complex that specifically targets β-catenin for annihilation [33].

Members of the low-density lipoprotein receptor-related protein (LRP) family, specifically LRP5 and LRP6, are crucial for canonical WNT signaling [34]. Researchers found that Axin is recruited away from the disintegration complex when phosphorylated LRP5/6 generates a novel binding site for Axin. Furthermore, studies have shown that Disheveled (Dsh) can prevent the destruction complex from functioning properly by phosphorylating and inactivating GSK3β through Frat/GBP [35]. Interestingly, animals lacking Frat show no abnormalities, indicating that Frat is not necessary for normal WNT signaling on its own. When the degradation complex is inhibited, robust, nonphosphorylated β-catenin can accumulate. This β-catenin then moves to the nucleus and interacts with T-cell factor/lymphoid enhancer factor (TCF/LEF) complexes to alter gene regulation and expression [36].

10.2.1.4 Noncanonical Routes of WNT Signal Transduction

Across lower vertebrates and *Drosophila*, researchers have identified a noncanonical Fz/Ca2+ pathway. Ca2+ is released from intracellular stores, and NFAT (nuclear factor associated with T cells) is activated when Fz-mediated heterotrimeric G-proteins are activated [37]. Recent research has supported the importance of this pathway during organogenesis in mice.

A third system, the Fz/PCP pathway in *Drosophila*, regulates PCP and involves the activation of Fz receptors, which do not require WNT ligands to operate. The function of the Fz/PCP pathway in mammals is unclear, in contrast to its well-established role in *Drosophila*. Some vertebrates have been found to induce PCP signals upon Fz ligation using ligands such as WNT5A, WNT5B, and WNT11 [38]. Cell polarity is affected by the stimulation of components of this pathway, such as Van Gogh-like (Vangl), Prickle, and Celsr (Flamingo). This activation is Dsh-dependent and is mediated by Rho A and JNK, while Nemo-like kinase is Dsh-independent. The asymmetric localization of Vangl/Pk complexes in the cell may result from Dsh's association with the ubiquitin ligase Smurf, which is necessary for mammalian PCP and targets Prickle for degradation. Tang et al. (2018) reported that PCP signaling is disrupted in Smurf1 and Smurf2 knockout mice embryos [39].

The atypical RTKs Ryk and Ror2 can bind to and be activated by WNT ligands. These receptors can activate other downstream signaling molecules, such as SRC kinase (Ryk) and JNK (Ror2), or they can function as Fz coreceptors. A recent study demonstrated that Ror2 senses the concentration of WNT ligand and controls PCP in developing limb chondrocytes by forming a complex with Vangl2 in a WNT-dependent manner [40].

In summary, the WNT/β-catenin, WNT/PCP, WNT/Ca2+, WNT/Ryk/Src, and WNT/Ror2/JNK pathways are interconnected. Many of these pathways likely interact with one another and are regulated by each other.

10.2.1.5 Molecular Mechanisms and Cellular Effects of the SHH Signal Transduction Pathway

In humans and mouse models, three overlapping hedgehog ligands have been identified: desert hedgehog (DHH), Indian hedgehog (IHH), and sonic hedgehog (SHH). SHH is crucial for the proper formation of numerous organs and structures during embryonic development. Embryos lacking SHH exhibit distinct abnormalities, including holoprosencephaly and the absence of distal limb structures and ventral cell types in the neural tube. Throughout development, SHH serves as a classical morphogen, and its effect depends on both its concentration and duration of signaling. The SHH pathway involves crucial stages, including ligand transformation, release, and receptor binding [41]. The seven-pass transmembrane molecule Smoothened is relieved of repression by Patched on the cell surface. Smoothened translocation into the primary cilium results in the accumulation of full-length Gli activator proteins within the nucleus. Three distinct Gli proteins operate in the downstream pathway of SHH signaling during mammalian development. The precise functions and interrelationships of these factors continue to be debated. In the neural tube, the most thoroughly studied system, high levels of Gli1 gene expression occur in regions where SHH signaling is active. During development, Gli1 and Gli2 predominantly operate as transcriptional activators. Posttranslational regulation governs Gli3 activity: regions with low SHH signaling contain a higher proportion of the cleaved, repressor form of Gli3, whereas SHH signals inhibit the processing of the full-length Gli3 protein, which typically functions as a transcriptional activator, into the cleaved repressor form [42]. Future research must investigate how alterations in cell fate correlate with distinct levels of Gli activity.

10.2.1.6 Insights into the Complexity of the BMP Signal Transduction Pathway

The 20 or so bone morphogenetic proteins (BMPs) identified to date are members of the transforming growth factor-β (TGF-β) family of signaling molecules. BMPs are categorized into the BMP-2/4, BMP-5/6/7/8, growth and differentiation factor (GDF)-5/6/7, and BMP-9/10 subfamilies, respectively. Disruption of BMP signaling in mice results in numerous defects. BMP is a crucial modulator of cellular differentiation and proliferation in these contexts. BMP ligands also serve as morphogens during mammalian development [43].

TGF-β ligands activate two subtypes of Ser kinase receptors: type I and type II (Figure 10.4). Both receptor types are required for in vivo BMP activity. Seven type I receptors (Activin-like kinase [ALK]1–7) have been identified. ALK1, ALK2, ALK3 (BMPR-IA), and ALK6 (BMPR-IB) mediate the activation of signaling downstream from BMPs, while the remaining receptors (also known as TβRI receptors) function downstream from alternative TGF-β ligands. Mammalian type II receptors that serve as BMP ligands include BMPR-II, ActR-II, and ActR-IIB, of which ActR-IIB is specific to BMPs [44]. GPI-anchored proteins of the repulsive guidance molecule (RGM), DRAGON, SARA, and CD44 are among the coreceptors and modulators that have been identified. Ligand binding of homodimers or heterodimers stabilizes and enhances oligomer assembly, allowing the presence of type I and II receptors on cell surfaces. Heterodimerization triggers type I receptor transphosphorylation, a conformational change, and activation, as the intracellular serotonin kinase domain of the type II receptor is constitutively active. These receptors phosphorylate and activate receptor-regulated SMADs (R-SMADs). SMAD1, 5, and 8 operate as R-SMADS for BMP signaling. Activated R-SMADs form a complex with SMAD4, which then translocates to the nucleus and modifies gene expression. Despite the apparent pivotal significance of SMAD4 in the transduction of BMP signals, its absence does not significantly impede the progression of certain developmental processes in mice, as demonstrated by the deletion of SMAD4. SMADs, frequently in conjunction with other transcriptional regulators like Runx and MyoD, regulate gene transcription by binding directly to GC-rich regulatory elements [45] (Figure 10.2).

10.2.1.7 Regulatory Mechanisms Governing the Notch Signaling Cascade

The activation of Notch signaling occurs when transmembrane receptors corresponding to the Delta/Serrate and Jagged families adhere to the transmembrane Notch receptor located on neighboring cells or, in certain instances, the same cell. Notch intracellular domain (NICD) is released upon proteolytic cleavage in response to Notch activation; the NICD then translocates to the nucleus, where it modifies gene expression. Notch signaling is dependent on the enzymes that facilitate Notch processing; mutations in these enzymes result in mutants that lack the functionality of Notch. In conjunction with transcriptional coactivators, NICD regulates gene expression, most prominently RBP-Jκ, which is the mammalian homolog of Su(H). The Notch signaling pathway primarily targets the Hairy/enhancer of split (Hes) genes. Hes family transcriptional regulators inhibit the expression of Notch ligands [46]. Therefore, when Notch signaling is activated within a cell, ligand expression is downregulated; conversely, lack of Notch signaling leads to ligand upregulation. The potential of Notch signaling to amplify gene expression disparities between adjacent cells through an approach known as lateral inhibition is supported by this feed-forward mechanism. In the mammalian cochlea, binary cell fate decisions, such as becoming a supporting versus sensory hair cell, require Notch-mediated lateral inhibition. It has been observed that Notch signaling plays a critical role in determining the fate of stem cells by inhibiting the expression of neurogenic bHLH factors such as Hes1. An essential mechanism by which asymmetric cell divisions are regulated during stem cell self-renewal is the asymmetric segregation of proteins such as Numb, which functions as a Notch antagonist to demonstrate that the Notch signaling pathway is activated as a downstream

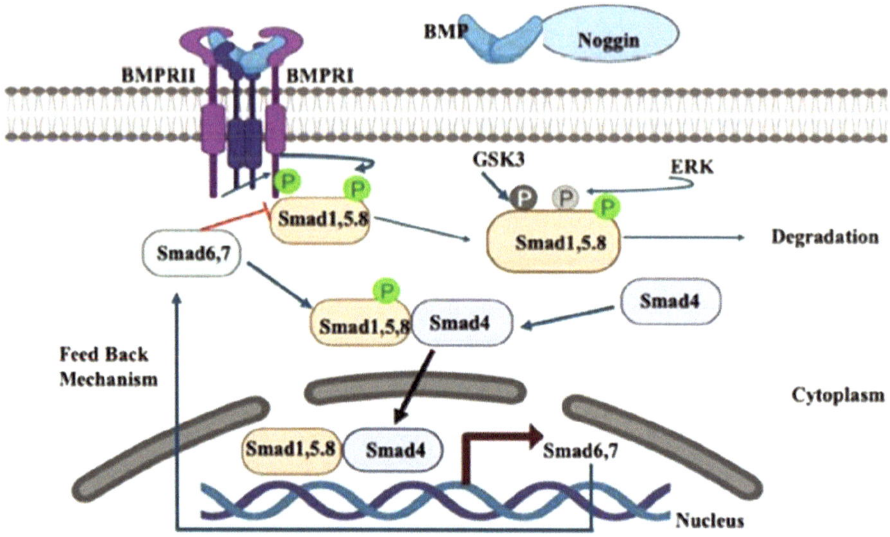

FIGURE 10.2 Modulation of SMAD activation and the BMP signaling cascade. The dimerization of BMPRI and phosphorylation of receptor (R-)Smads (Smad1, 5, or 8) in close proximity to the carboxyl terminus occur upon the binding of BMP ligand to BMPRII. The stimulated R-Smads form an intricate complex with a Co-Smad (Smad4), and the Smad complex relocates to the nucleus, where it regulates gene transcription through interactions with other transcriptional regulators. One instance of negative feedback mechanism involves the activation of SMAD6, 7, which engages in competition with SMAD1, 5, 8 for BMPRI binding. It has been demonstrated that the WNT and RTK signaling pathways inhibit BMP signaling via Smad1. Phosphorylation of the linker region centered adjacent to the amino and carboxy-terminal domains of Smad by ERK prepares Smad for GSK3 to phosphorylate it further. Smad that has been dually phosphorylated is set up for degradation. Noggin and other secreted BMP antagonists bind to and encapsulate BMP ligands in the extracellular environment.

effector by the cell polarity pathways regulated by Par proteins and the cellular machinery that controls the orientation of the cell spindle in order to maintain a balance between differentiation and proliferation. The Notch pathway, apart from its function in lateral inhibition, is also necessary for the acquisition of developmental competency. This function seems to be accomplished through a lateral induction process, wherein Notch signaling stimulates the production of the Jag1 ligand [47].

10.3 CELLULAR DIFFERENTIATION REGULATORY CASCADE

Several crucial signaling pathways are utilized repeatedly throughout development to control a multitude of cellular processes that contribute to the formation of the developing organism. Determining how fundamental cellular mechanisms, including migration, adhesion, proliferation, fate determination, differentiation, apoptosis, and morphological changes, are orchestrated during morphogenesis is a formidable obstacle.

10.3.1 Molecular Control of Cell Proliferation by FGF Signaling Pathway

By governing the expression of cell cycle regulators, the Ras/MAPK pathway is widely recognized for its capacity to stimulate cell proliferation in embryonic cell types. Stimulation of tissue progenitor expansion is thus a significant function of FGF signaling. This function is highly evident during the process of cardiac development. Various progenitor populations of embryos provide the cells that regulate the development of the cardiac muscle. The fundamental heart field supplies cells to the left ventricle, while the anterior or secondary heart field, SHF, specifies the progenitors of the outflow tract and right ventricle. The expansion of the SHF population in the early embryo is regulated autonomously by FGF8 cells; FGF8 hypomorphic fetuses exhibit shortened right ventricles and outflow tracts. DiGeorge and Alagille syndromes are distinguished by these cardiovascular abnormalities, and the fact that FGF8 expression is downregulated in mouse models of these syndromes demonstrates the criticality of preserving adequate levels of FGF signaling in order to avert developmental disorder [3].

Alisotropic growth and the development of organs and tissues with well-defined three-dimensional structures may be facilitated by intricate regulation of the orientations of cell divisions in a developing tissue, as opposed to simply propelling the size of a particular population of primitive cells or an organ. In fact, research has demonstrated that oriented cell divisions take place during processes such as appendage outgrowth and epithelial tube elongation. The orientation of cell divisions in developing tissues has been demonstrated to be altered by mutations that impact PCP pathways. Recent research indicates that the disruption of oriented cell division also occurs when the FGF pathway is overactivated in the embryonic lung epithelium through the removal of Sprouty antagonists [48]. The precise mechanisms by which dysregulated Ras/MAPK signaling regulates or modifies the orientation of the mitotic spindle remain unknown.

10.3.2 Cell Fate Determination Orchestrated by FGF Signaling Pathway

The regulation of distinctive cellular identities is ultimately determined by epigenetic mechanisms, which involve the consistent activation of predetermined transcriptional networks. An essential overlooked issue is how cell fates are initially determined during embryogenesis by establishing distinct transcription factor patterns in distinct groups of progenitors. The specification of distinct neuronal identities in the neural tube in relation to the intensity and duration of the SHH signal is arguably the most thoroughly investigated system in this regard.

It has been demonstrated that RTK signaling determines cell fate in multiple contexts. A widely recognized role of FGF signaling in the early embryo is to regulate gastrulation-related mesoderm specification. FGF8 signaling via FGFR1 regulates mesoderm formation by controlling the expression of T (Brachyury) and Tbx6 genes, in addition to rendering cell migration away from the primitive streak [49].

An additional extensively researched instance of FGF signals influencing cell fate determination is the fidelity of cranial ectoderm cells to the otic fate. Multiple FGF ligands repeatedly execute this mechanism in the mouse. Cross-regulation between FGF and WNT signals during this process limits otic fate induction to a specific embryonic location, which is an essential aspect to consider.

In a recent study, it was demonstrated that the outcome of neuroepithelial cells in the embryonic cortex is influenced by their proximity to the commissural plate, a critical signaling center and the source of FGF8. Along the anterior–posterior axis of the telencephalon, FGF8 can indeed function as a secreted morphogen to determine cell fate, as demonstrated by this study. These results have significant ramifications for neurodevelopmental disorders that are linked to altered signaling levels [50].

10.3.3 FGF Signaling Pathway in Modulation of Apoptosis in Developmental Processes

A significant proportion, if not the majority, of the cells in the developing embryo are actively kept alive through the transmission of particular signals. Thus, the function of a specific signaling pathway in cellular survival can only be discerned through an increase in apoptosis in response to its inhibition or loss. In recent years, numerous contexts have been identified in which FGF signals prevent apoptosis in cells [51]. Conditional deletion of FGF8 in the IsO induces extensive cell death shortly after the FGF8 signal is withdrawn. When FGF signaling is not entirely inhibited but rather significantly diminished, the cells farthest from the IsO, the origin of the survival signal, are the first to exhibit signs of apoptosis. According to these findings, cell mortality is exclusively induced when FGF signaling diminishes to a specific level. An analogous finding was documented in the context of limb development, wherein a decline in FGF secretion from the distal tip led to cellular mortality in the proximal limb bud [52]. Recent intricate experiments conducted in the Hébert laboratory demonstrate that FGF signals generated by a signaling center in the forebrain also prevent apoptosis in telencephalon cells. The regulation of FGF expression in this particular instance is influenced by the WNT/β-catenin signaling pathway. The antiapoptotic effects of FGF signaling are achieved by inhibiting proapoptotic signals that are downstream of BMP/SMAD4 signaling.

By regulating the withdrawal of a survival factor, developing structures can be transformed. An exemplary instance pertains to interdigital cell death that occurs during limb development; this process is indispensable for the final physical detachment of the developing digits. High FGF8 signaling in the distal limb bud prevents apoptosis in cells, despite the RA-induced expression of cell death molecules such as Bax2, according to a recent study. Nevertheless, with the emergence of digits, there is a progressive downregulation of FGF8 expression in the interdigital regions; this cessation of FGF signaling facilitates the advancement of apoptosis [53]. Interdigital webbing was observed in response to an increase in FGF4 and FGF8 expression following the deletion of BMPR1a from the apical ectodermal ridge (AER), which is consistent with these results.

The PI3K pathway, which has been linked to cell survival, particularly in the prospect of cancer, is activated by FGF signaling. Epiblast cell survival in embryoid bodies and the blastocyst is contingent upon the activation of PI3K signaling. Conversely, investigations conducted in vitro indicate that cellular survival is regulated in a context-dependent fashion by both the Ras/MAPK and PI3K pathways. It is unknown whether PI3K is required for the transduction of survival signals downstream of FGF and to what degree dysregulated PI3K signaling in the embryo directly contributes to apoptosis upon FGF withdrawal [54].

10.3.4 Integrated Signaling Networks Orchestrating Cell Adhesion and Migration Processes

FGF8 signaling is indispensable not only for determining the fate of mesodermal cells during gastrulation but also for facilitating the migration and epithelial-to-mesenchymal transition of mesodermal cells. The aforementioned transition is correlated with the cessation of E-cadherin expression and cellular adhesion [55].

10.3.5 Cell Differentiation Regulatory Signaling Cascade

The concluding phases of cellular differentiation are frequently accompanied by discernible changes in the morphology of cells and are inextricably linked with cell cycle termination. The regulation of lens fibers' notable elongation during their terminal differentiation in the eye is accomplished through the interaction between FGF signaling and other pathways, including WNT and TGF-β. Recent studies demonstrated that distinct FGF ligands can control the formation of various forms of synapses during neuronal differentiation. The precise mechanisms by which FGF signaling facilitates cell differentiation, despite its critical and subtle nature, remain inadequately understood [56]

10.4 REGULATION AND INTERPRETATION OF SIGNAL STRENGTH AND DURATION IN CELLULAR SIGNALING

The magnitude and duration of the signal, as well as the cellular context, influence the cellular response to a given stimulus. There may be a subtle distinction between RTK signals that promote cell proliferation and those that facilitate differentiation. However, as signals progress through the signaling cascade, minute variations in the initial signal may become magnified. A fundamental component in comprehending the mechanisms underlying cell fate determination during development is the ability to decipher how cells respond to signals of a qualitatively distinct nature, such as lineage selection or fate, despite the signals being quantitatively distinct in intensity or duration. An intricate mechanism that governs RTK signaling and allows cells to differentiate between transient and persistent signals entails the control of the transcription factor c-fos. The fundamental nature of this pathway is c-fos multilevel regulation by ERK signaling. From the outset, it has been established that ERK signaling targets the c-fos gene immediately and early via transcription [57]. Nevertheless, c-fos protein degradation occurs at a rapid rate and is relatively unstable. c-fos is stabilized and primed for further ERK phosphorylation, thereby improving its transcriptional activity, through phosphorylation of serine residues on c-fos mediated by ERK. Therefore, a transient RTK signal will only induce transient c-fos activity; it will not induce downstream gene expression in a significant way. On the contrary, sustained activation of ERK leads to the buildup of active c-fos. Therefore, during ERK signaling in cells, c-fos serves as a crucial sensor [58].

10.4.1 Regulatory Feedback Loops in RTK Signaling Cascade

A substantial percentage of the signaling molecules that are secreted during embryonic development serve as morphogens; the extent to which these factors influence cell fate varies with their concentration. The presence of both positive and negative feedback mechanisms is critical for the regulation of signaling levels. Furthermore, the embryo utilizes these to discern minute variations in signal intensity or duration and convert them into pivotal developmental switches.

10.4.1.1 Role of Positive Feedback in Signal Transduction Pathways

A positive feedback mechanism frequently noticed in carcinogenesis is an upregulation of ligand or receptor expression, which can augment signaling via autocrine means. There is a profusion of paracrine positive feedback mechanisms in embryonic development. For instance, reciprocal signaling between epithelial and mesenchymal cells is essential for organogenesis and the continual proliferation of stem/progenitor cells; it is frequently mediated by combinations of FGF ligands and receptors expressed by epithelial and mesenchymal cells. These mechanisms tend to establish positive feedback loops in which an upregulation of FGF ligands in response to elevated FGF signaling in one tissue induces a corresponding increase in intensity of signaling in the other tissue. In order to disrupt these feed-forward mechanisms, negative feedback mechanisms are frequently implemented; removing negative regulators can have severe repercussions on development [59].

The distinct impacts of NGF or EGF on PC12 cells can be rationalized through the plausible mechanism of positive feedback. This mechanism has been utilized as an in vitro model to demonstrate how distinct cell fates transpire from varying signal intensities. Stimulation with NGF, but not EGF, activates a positive feedback loop from MAPK to Raf. Positive feedback has been shown to significantly alter the kinetics of RTK signaling in Xenopus oocytes; specifically, activation of ERK triggers a robust positive feedback loop that is facilitated by Mos [60]. Consequently, the cells exhibit bimodal ERK activation and subsequent effects, which can be described as either ON or OFF $Mos^{-/-}$ mouse oocytes, which are consistent with the results obtained in Xenopus, do not induce MAPK activation during meiosis and are linked to metaphase II arrest [61]. Therefore, cells utilize positive feedback mechanisms to differentiate between various growth factors and transform variations in signal intensity or duration into unique developmental switches, thereby ensuring that responses to secreted morphogens throughout embryonic development are robust.

10.4.1.2 Role of Negative Feedback in Signal Transduction Pathways

Numerous negative feedback mechanisms have been identified to restrict the intensity, time frame, and/or scope of growth factor signaling throughout the developmental process. A mechanism by which signaling molecules such as PKC can directly alter EGF receptors to prevent them from binding ligands with high affinity may permit transmodulation between receptors. Receptor ubiquitination, internalization, and subsequent degradation are facilitated by the recruitment and stimulation of molecules with ubiquitin ligase activity, such as Cbl. This leads to the expeditious elimination of activated RTKs from the exterior of cells. Slightly delayed mechanisms entail the activation of additional feedback regulators, which could attach to and retain the ligand or impede a ligand–receptor coupling that would otherwise be productive. Additional intracellular molecules that have the potential to impede intracellular signal transduction consist of proteins belonging to the MAPK phosphatases; transcriptional repressors; and Sef, Spred, and Sprouty families [62].

Subsequent research conducted in Xenopus has demonstrated that in particular circumstances, the cellular response can be altered and fine-tuned through the superimposition of feedback mechanisms that selectively and efficiently target downstream signaling cascades. These negative feedback mechanisms limit signaling to small, distinct regions in developing organs in mice, enabling exquisite control over cellular-level cell fate decisions.

Collectively, minute variations in signaling properties translate into conspicuous distinctions in cellular responses via a variety of mechanisms involving the stabilization of transcription factors, activation of positive and negative feedback loops, and modification of the dynamics and stability of signaling complexes [63].

10.4.2 Coordination of Signaling Pathways: Positive and Negative Regulators in WNT, SHH, and BMP Signaling Networks

All the pathways that have been examined thus far have positive and negative regulators identified. Secreted Fz-related proteins (sFRPs) bind directly to WNT ligands during WNT signaling in order to either enhance the WNT signaling range or compete with Fz for binding. Several WNT coreceptors and inhibitors, including SOST, Dickkopf (Dkk), R-spondin, and Wise, have been documented to engage in direct interactions with LRP5/6, which is consistent with the pivotal role of LRP5/6 in WNT/β-catenin signaling [64].

Several SHH pathway modulators, including Cdo, Boc, LRP2 (Megalin), and Gas1, have been identified. Particularly at relatively small doses of SHH, these molecules appear to bind to SHH and augment SHH signaling. The interpretation of the SHH gradient requires knowledge of feedback regulation of SHH signaling, particularly via upregulation of Ptch1 [65].

The regulation of BMP signaling is carried out by extracellular agonists, such as Chordin, Noggin, Gremlin, and Follistatin, which bind to BMPs and impede their receptor activation. Negative feedback mechanisms, such as upregulation of the inhibitory (I-SMAD) proteins SMAD6 and SMAD7, which hinder the activity of stimulated type I receptors by binding to them, are also triggered by BMP activation. Moreover, the activity of R-SMADS can be impeded by I-SMADs through the formation of complexes. In addition to interacting with Smurfs, which are Smad ubiquitin regulatory factors, I-SMADS recruits ubiquitin ligases to the receptors, resulting in the ubiquitination and degradation of the receptors. SMAD1/5 degradation can also be modulated directly by Smurfs in response to BMPs [66].

10.5 COORDINATED SIGNALING NETWORKS IN CELLULAR DEVELOPMENTAL PROCESSES

The functions of individual signaling pathways in regulating cellular behavior throughout embryonic development have been the focus of the discussions. Evidently, this reductionist strategy—concentrating on the role of a singular pathway, process, or gene in the development of a particular structure—is indispensable for elucidating fundamental molecular mechanisms. All biologists must acknowledge, however, that the reality is considerably more intricate. Multiple signaling pathways are known to be encountered and responded to concurrently or sequentially by cells, which assimilate this information in some pattern. Comprehension of the mechanisms by which cells integrate information during development and the way the morphogenetic processes that culminate in the formation of a fully functional organism are controlled by the combined action of multiple signaling factors on a tissue constitute a formidable obstacle.

A limited selection of the numerous instances of cross-regulation between distinct pathways have been observed. Studies suggested that the activation of AKT after RTK signaling could amplify WNT/β-catenin signaling via the inhibition of GSK3β, thereby leading to improved β-catenin persistence and functionality. WNT ligands in the roof plate of the embryonic neural tube regulate Gli3 expression, which in turn inhibits the expression of the SHH target gene in the dorsal neural tube. Crosstalk between the Notch and RTK signaling pathways has been observed in *Drosophila* on multiple occasions. Moreover, crucial downstream targets of the Notch pathway, including genes associated with Hes, are additionally regulated by alternative pathways, including WNT [67].

The burgeoning discipline of "systems biology" seeks to provide an explanation for the interconnections and mutual influences among various molecules and pathways within a given system. Nevertheless, the significance of particular interactions might differ depending on the developmental procedure or period in question. Moreover, numerous crucial interactions in the field of developmental biology occur through intercellular communication rather than being cell autonomous. Consequently, it will be necessary to validate each of these interactions and evaluate their significance or influence on a specific cellular process. A comprehensive signaling network may ultimately require intracellular and intercellular signals from each cell population at each developmental time point in order to comprehend the integration of signaling information throughout development.

10.5.1 Integrative Signaling Pathway-Mediated Intracellular Signal Transduction Mechanisms

Certain molecules have been identified as "nodes" in networks of protein interactions. It is anticipated that these molecules will play a crucial role in integrating numerous signaling pathways. Gaining insight into the molecular mechanisms that govern the activity of these critical molecules is thus an essential prerequisite for comprehending molecular signal integration.

One of the most extensively researched instances of signal integration at the molecular level is SMAD1, which serves as a mediator of BMP signals throughout the process of development. The phosphorylation of carboxy-terminal residues of SMAD1 is facilitated by BMP signaling, thereby enhancing the nuclear translocation of SMAD1. The "linker" region of SMAD1 is situated between the carboxy- and amino-terminal regions of the protein. Several residues in this linker region are susceptible to reversible modification by additional signaling cascades. The proteasomal degradation of SMAD1 by SMURF1 is enhanced through phosphorylation by ERK and GSK3. Therefore, it can be observed that the FGF/Ras/MAP kinase pathway inhibits BMP signals, while canonical WNT signaling promotes BMP signaling through its inhibition of SMAD1 phosphorylation mediated by GSK3. More recently, it was demonstrated that RA can disrupt BMP signaling via SMAD1 by increasing MAPK activation. Thus, an understanding unfolds in which SMAD1 serves as an essential locus upon which numerous signaling cascades converge, in addition to its function of transducing BMP signals to the nucleus [68].

10.5.2 Regulatory Mechanism of Interconnected Signaling Networks in Behavior of Cell Populations

The emergence of a complex morphogenetic process can also be attributed to the independent effects of distinct signaling networks on a population of cells, in addition to the direct intersection of two pathways within the same cell. An exemplary depiction of this concept can be observed in developing limb outgrowth.

The initial morphogenetic event that occurs during limb formation is the limb bud protruding along its proximal-distal axis. This process has been linked to both FGF and WNT signals, and mutations that impair the expression of FGF8 and WNT5a have been linked to malformed, emaciated extremities. The precise mechanisms by which these factors regulate limb outgrowth and the interrelationships between their respective functions remain unknown. Although cell proliferation is an unquestionably necessary condition for growth, research indicates that variations in proliferation levels throughout the bud do not produce anisotropic growth. Conversely, polarization of mesenchymal cells and nonrandom migration and division of cells occur within the developing limb bud. By integrating time-lapse imaging with a potent genetic technique for arbitrarily labeling a limited number of cells with green fluorescent protein (GFP), it was discovered that cells situated in the dorsal and ventral regions of limb buds lacking Wnt5a ceased to migrate in a directed manner toward the ectoderm. When WNT5a was reintroduced into mutant buds, their coordinated movement was restored. Oriented cell divisions necessitated activation of the JNK signaling pathway, as demonstrated by pharmacological inhibitor experiments [69]. Inhibition of the FGF/Ras/MAPK pathway, on the other hand, did not affect oriented cell division but decreased the speed of cell migration. In contrast, cell migration was enhanced when this pathway was stimulated via constitutively active MEK1 expression or FGF8 application. The correlation between ERK activity and cell movement in the normal limb bud, where both are greatest in proximity to the source of FGF signals in the ectoderm, is well reflected by these experimental results. It is noteworthy that modifications in FGF/MAPK signaling did not influence the orientation of cell division, and modulation of WNT/JNK signaling had no effect on the velocity of cell movement. The researchers hypothesized, in light of these results, that the anisotropic growth of the limb bud is facilitated by the combined activity of these two pathways [70].

10.6 CONCLUSION

In conclusion, an all-encompassing comprehension of the way in which cells interact with their surroundings and exchange information while in embryonic development will be necessary. This will necessitate an in-depth understanding of the mechanisms by which these signals govern fundamental biological processes, including oriented cell division, directed cell migration, and fate determination, as well as the regulation and function of intracellular pathways, and secreted morphogens and other signaling molecules. Significant progress has been made in recent years regarding the influence of various signaling pathways and their interlink with morphogenesis and cellular fate determinations, which aids in our investigation into the mechanisms governing embryonic development. While considerable progress has been made in knowledge over the past few decades, these discoveries have also prompted numerous intriguing inquiries. By effectively integrating mechanical biological technologies and emergent systems to manipulate genetic material in mouse embryos, substantial progress can be made toward comprehending the ways in which genetic and epigenetic mechanisms govern morphogenesis and cell fate.

REFERENCES

1. Lam, I., C.M. Pickering, and F. Mac Gabhann, Context-dependent regulation of receptor tyrosine kinases: Insights from systems biology approaches. *Wiley Interdiscip Rev Syst Biol Med*, 2019. **11**(2): p. e1437.
2. McDonell, L.M., et al., Receptor tyrosine kinase mutations in developmental syndromes and cancer: two sides of the same coin. *Hum Mol Genet*, 2015. **24**(R1): p. R60–R66.
3. Xie, Y., et al., FGF/FGFR signaling in health and disease. *Signal Transduct Target Ther*, 2020. **5**(1): p. 181.

4. Jones, S.A., Physiology of FGF15/19. *Adv Exp Med Biol*, 2012. **728**: p. 171–182.
5. Erben, R.G., Physiological actions of fibroblast growth factor-23. *Front Endocrinol*, 2018. **9**.
6. Phan, P., et al., The saga of endocrine FGFs. *Cells*, 2021. **10**(9).
7. Kawai, K. and M. Takahashi, Intracellular RET signaling pathways activated by GDNF. *Cell Tissue Res*, 2020. **382**(1): p. 113–123.
8. Samiotaki, G., et al., Enhanced delivery and bioactivity of the neurturin neurotrophic factor through focused ultrasound—mediated blood—brain barrier opening in vivo. *J Cereb Blood Flow Metab*, 2015. **35**(4): p. 611–622.
9. Sidorova, Y.A. and M. Saarma, Small molecules and peptides targeting glial cell line-derived neurotrophic factor receptors for the treatment of neurodegeneration. *Int J Mol Sci*, 2020. **21**(18): p. 6575.
10. Ilieva, M., et al., Artemin and an artemin-derived peptide, artefin, induce neuronal survival, and differentiation through ret and NCAM. *Front Mol Neurosci*, 2019. **12**.
11. Singh, B., G. Carpenter, and R.J. Coffey, EGF receptor ligands: Recent advances. *F1000Res*, 2016. **5**.
12. Chava, S., et al., Betacellulin promotes tumor development and EGFR mutant lung cancer growth by stimulating the EGFR pathway and suppressing apoptosis. *iScience*, 2022. **25**(5): p. 104211.
13. Martin, L.J., et al., Epiregulin and EGFR interactions are involved in pain processing. *J Clin Invest*, 2017. **127**(9): p. 3353–3366.
14. Srinivasan, S., et al., Ligand-induced transmembrane conformational coupling in monomeric EGFR. *Nat Commun*, 2022. **13**(1): p. 3709.
15. Peach, C.J., et al., Molecular pharmacology of VEGF-A isoforms: Binding and signaling at VEGFR2. *Int J Mol Sci*, 2018. **19**(4): p. 1264.
16. Kuonqui, K., et al., Dysregulation of lymphatic endothelial VEGFR3 signaling in disease. *Cells*, 2024. **13**(1): p. 68.
17. Lin, Q.Y., et al., VEGF-C/VEGFR-3 axis protects against pressure-overload induced cardiac dysfunction through regulation of lymphangiogenesis. *Clin Transl Med*, 2021. **11**(3): p. e374.
18. Eroğlu, A., et al., Vascular endothelial growth factor (VEGF)-C, VEGF-D, VEGFR-3 and D2-40 expressions in primary breast cancer: Association with lymph node metastasis. *Adv Clin Exp Med*, 2017. **26**(2): p. 245–249.
19. Chau, K., A. Hennessy, and A. Makris, Placental growth factor and pre-eclampsia. *J Hum Hypertens*, 2017. **31**(12): p. 782–786.
20. Momeny, M., et al., Heregulin-HER3-HER2 signaling promotes matrix metalloproteinase-dependent blood-brain-barrier transendothelial migration of human breast cancer cell lines. *Oncotarget*, 2015. **6**(6): p. 3932–3946.
21. Wang, J., et al., Neuregulin 1/ErbB4 signaling contributes to the anti-epileptic effects of the ketogenic diet. *Cell & Biosci*, 2021. **11**(1): p. 29.
22. Basciani, S., et al., Expression of platelet-derived growth factor-A (PDGF-A), PDGF-B, and PDGF receptor-alpha and -beta during human testicular development and disease. *J Clin Endocrinol Metab*, 2002. **87**: p. 2310–2319.
23. Maruyama, I.N., Mechanisms of activation of receptor tyrosine kinases: Monomers or dimers. *Cells*, 2014. **3**(2): p. 304–330.
24. Seif, F., et al., The role of JAK-STAT signaling pathway and its regulators in the fate of T helper cells. *Cell Commun Signal*, 2017. **15**(1): p. 23.
25. Basson, M.A., Signaling in cell differentiation and morphogenesis. *Cold Spring Harb Perspect Biol*, 2012. **4**(6).
26. Nair, A., et al., Conceptual evolution of cell signaling. *Int J Mol Sci*, 2019. **20**(13).
27. Dillon, M., et al., Progress on Ras/MAPK signaling research and targeting in blood and solid cancers. *Cancers (Basel)*, 2021. **13**(20).
28. Neben, C., et al., Feedback regulation of RTK signaling in development. *Dev Biol*, 2017. **447**.
29. Satoh, Y., et al., Deletion of ERK1 and ERK2 in the CNS causes cortical abnormalities and neonatal lethality: Erk1 deficiency enhances the impairment of neurogenesis in Erk2-deficient mice. *J Neurosci*, 2011. **31**(3): p. 1149–1155.
30. Parchure, A., N. Vyas, and S. Mayor, Wnt and hedgehog: Secretion of lipid-modified morphogens. *Trends Cell Biol*, 2018. **28**(2): p. 157–170.
31. Milhem, R.M. and B.R. Ali, Disorders of FZ-CRD; Insights towards FZ-CRD folding and therapeutic landscape. *Mol Med*, 2019. **26**(1): p. 4.
32. Qin, K., et al., Canonical and noncanonical Wnt signaling: Multilayered mediators, signaling mechanisms and major signaling crosstalk. *Genes Dis*, 2024. **11**(1): p. 103–134.

33. Shimizu, K., et al., The SCFβ-TRCP E3 ubiquitin ligase complex targets Lipin1 for ubiquitination and degradation to promote hepatic lipogenesis. *Sci Signal*, 2017. **10**(460).
34. Ren, Q., J. Chen, and Y. Liu, LRP5 and LRP6 in Wnt signaling: Similarity and divergence. *Front Cell Dev Biol*, 2021. **9**.
35. Sharma, M., et al., Dishevelled: A masterful conductor of complex Wnt signals. *Cell Signal*, 2018. **47**: p. 52–64.
36. Liu, J., et al., Wnt/β-catenin signalin: Function, biological mechanisms, and therapeutic opportunities. *Signal Transduct Target Ther*, 2022. **7**(1): p. 3.
37. Pan, H.Y., et al., Role of nuclear factor of activated T cells (NFAT) pathway in regulating autophagy and inflammation in retinal pigment epithelial cells. *Int J Mol Sci*, 2021. **22**(16).
38. Yang, Y. and M. Mlodzik, Wnt-Frizzled/planar cell polarity signaling: Cellular orientation by facing the wind (Wnt). *Annu Rev Cell Dev Biol*, 2015. **31**: p. 623–646.
39. Tang, Y., et al., Generation of Smurf2 conditional knockout mice. *Int J Biol Sci*, 2018. **14**(5): p. 542–548.
40. Green, J., R. Nusse, and R. Amerongen, The role of ryk and ror receptor tyrosine kinases in wnt signal transduction. *Cold Spring Harb Perspect Biol*, 2013. **6**.
41. Sasai, N., M. Toriyama, and T. Kondo, Hedgehog signal and genetic disorders. *Front Genet*, 2019. **10**: p. 1103.
42. Sabol, M., et al., Role of GLI transcription factors in pathogenesis and their potential as new therapeutic targets. *Int J Mol Sci*, 2018. **19**(9).
43. Miyazono, K., Y. Kamiya, and M. Morikawa, Bone morphogenetic protein receptors and signal transduction. *J Biochem*, 2009. **147**: p. 35–51.
44. Heldin, C.H. and A. Moustakas, Signaling receptors for TGF-β family members. *Cold Spring Harb Perspect Biol*, 2016. **8**(8).
45. Dituri, F., et al., The interactivity between TGF-β and BMP signaling in organogenesis, fibrosis, and cancer. *Cells*, 2019. **8**(10).
46. Zhou, B., et al., Notch signaling pathway: Architecture, disease, and therapeutics. *Signal Transduct Target Ther*, 2022. **7**(1): p. 95.
47. Haan, S.D., et al., Jag1 represses Notch activation in lateral supporting cells and inhibits an outer hair cell fate in the medial compartment of the developing cochlea. *bioRxiv*, 2024: p. 2024.02.02.577075.
48. Chaffee, B., et al., FGFR and PTEN signaling interact during lens development to regulate cell survival. *Dev Biol*, 2016. **410**.
49. Mathieu, J., et al., Nodal and FGF pathways interact through a positive regulatory loop and synergize to maintain mesodermal cell populations. *Development*, 2004. **131**(3): p. 629–641.
50. Harish, R.K., et al., Real-time monitoring of an endogenous Fgf8a gradient attests to its role as a morphogen during zebrafish gastrulation. *Development*, 2023. **150**(19).
51. Bock, F.J., et al., Apoptotic stress-induced FGF signaling promotes non-cell autonomous resistance to cell death. *Nat Commun*, 2021. **12**(1): p. 6572.
52. Sountoulidis, A., et al., A topographic atlas defines developmental origins of cell heterogeneity in the human embryonic lung. *Nature Cell Biology*, 2023. **25**(2): p. 351–365.
53. Farooq, M., et al., The role of fibroblast growth factor (FGF) signaling in tissue repair and regeneration. *Cells*, 2021. **10**(11).
54. Rodgers, S.J., et al., Regulation of PI3K effector signaling in cancer by the phosphoinositide phosphatases. *Biosci Rep*, 2017. **37**(1).
55. Hao, Y., et al., FGF8 induces epithelial-mesenchymal transition and promotes metastasis in oral squamous cell carcinoma. *Int J Oral Sci*, 2021. **13**(1): p. 6.
56. Kumar, V., et al., Functional roles of FGF signaling in early development of vertebrate embryos. *Cells*, 2021. **10**(8).
57. Yokoyama, K., et al., C-Fos regulation by the MAPK and PKC pathways in intervertebral disc cells. *PLoS One*, 2013. **8**(9): p. e73210.
58. Lara Aparicio, S.Y., et al., Current opinion on the use of c-Fos in neuroscience. *NeuroSci*, 2022. **3**(4): p. 687–702.
59. Gattiglio, M., M. Protzek, and C. Schröter, Population-level antagonism between FGF and BMP signaling steers mesoderm differentiation in embryonic stem cells. *Biology Open*, 2023. **12**(8).
60. Hennig, A., et al., Feedback activation of neurofibromin terminates growth factor-induced Ras activation. *Cell Commun Signal*, 2016. **14**(1): p. 5.
61. Gelaude, A., et al., Hydrogen Sulfide Impairs meiosis resumption in xenopus laevis oocytes. *Cells*, 2020. **9**(1): p. 237.

62. Lake, D., S.A.L. Corrêa, and J. Müller, Negative feedback regulation of the ERK1/2 MAPK pathway. *Cell Mol Life Sci*, 2016. **73**(23): p. 4397–4413.
63. Sonnen, K.F. and C.Y. Janda, Signaling dynamics in embryonic development. *Biochem J*, 2021. **478**(23): p. 4045–4070.
64. Jeong, W. and E.-H. Jho, Regulation of the low-density lipoprotein receptor-related protein LRP6 and its association with disease: Wnt/β-Catenin signaling and beyond. *Front Cell Dev Biol*, 2021. **9**.
65. Christ, A., K. Herzog, and T.E. Willnow, LRP2, an auxiliary receptor that controls sonic hedgehog signaling in development and disease. *Dev Dyn*, 2016. **245**(5): p. 569–579.
66. Miyazawa, K. and K. Miyazono, Regulation of TGF-β family signaling by inhibitory smads. *Cold Spring Harb Perspect Biol*, 2017. **9**(3).
67. LaFoya, B., et al., Src kinase phosphorylates Notch1 to inhibit MAML binding. *Sci Rep*, 2018. **8**(1): p. 15515.
68. Luo, K., Signaling cross talk between TGF-β/Smad and other signaling pathways. *Cold Spring Harb Perspect Biol*, 2017. **9**(1).
69. Gros, J., et al., WNT5A/JNK and FGF/MAPK pathways regulate the cellular events shaping the vertebrate limb bud. *Current Biology : CB*, 2010. **20**: p. 1993–2002.
70. Jin, L., et al., Fibroblast growth factor 10 and vertebrate limb development. *Front Genet*, 2019. **9**.

11 Cell Signaling and Metabolism

Saadia Zahid[1], Sanila Amber[1,2],*
Laila tul Qadar[1], and Akasha Khurshid[1]
[1]Neurobiology Research Laboratory, Department of Healthcare Biotechnology, Atta-ur-Rahman School of Applied Biosciences, National University of Sciences and Technology, Islamabad, Pakistan
[2]Department of Biotechnology, University of Mianwali, Mianwali-Pakistan
*Department of Healthcare Biotechnology, Atta-ur-Rahman School of Applied Biosciences, National University of Sciences and Technology, Islamabad, Pakistan

11.1 INTRODUCTION

In order to communicate with each other and with the outer environment, cells use various signaling pathways. The signals that a cell receives and those it generates are essential for the primary mechanism of cellular life, fate, and behavior. Receptor molecules are the transmembrane proteins that accept the signaling molecules and start a signaling cascade inside the cell by activating multiple proteins like cyclic adenosine monophosphate (cAMP) and calcium ions that act as second messengers. Multiple signaling pathways are identified and discussed to be involved in basic cellular processes, like protein and lipid synthesis, energy generation and storage, cell survival, inflammation, cell division, proliferation, differentiation, and apoptosis (Valls and Esposito 2022). Important signaling pathways include PI3K/Akt signaling, which is known as the pro-survival pathway. mTOR signaling also works along with it and is involved in proliferation, growth, and survival. AMP-activated protein kinase (AMPK) signaling is known for energy generation. p53 is activated in response to errors in replication and mutations, and may lead to senescence, apoptosis, or cell cycle arrest. They are also called tumor suppressor genes and have a pivotal role in cancer. NF-κB signaling is involved in innate and adaptive immune responses and is important in autoimmunity. Toll-like receptor (TLR) signaling is activated in response to the attack of pathogenic viruses and bacteria. Notch signaling is important in deciding the fate of cells at embryonic stages, while Wnt signaling is involved in adult as well as embryogenic differentiation of cells. Wnt and Notch signaling pathways also have an important role in metastatic cancers. Vascular endothelial growth factor (VEGF) signaling is responsible for angiogenesis, particularly in embryos but to some extent in adults during an injury. It has a well-studied role in cancer as well. Moreover, insulin signaling maintains glucose homeostasis and impacts multiple pathways in different metabolic disorders (Wu et al. 2021).

11.2 CELL METABOLISM

Cell metabolism is crucial for generating energy and sustaining cellular functions. Metabolism consists of two major processes: anabolism and catabolism. Anabolism is an energy-storing procedure, whereas catabolism is the process of obtaining energy or energy expenditure. Mitochondria are the hub of cellular metabolism and are considered the powerhouse of the cell where oxidative phosphorylation takes place, which generates ATP. Multiple nutrient molecules like glucose, lipids, and proteins undergo metabolic processing and yield energy-rich compounds like $FADH_2$ and NADPH that enter the electron transport chain to yield ATPs (Xiao et al. 2018).

DOI: 10.1201/9781003594901-11

11.3 SIGNAL TRANSDUCTION AND REGULATION OF CELL METABOLISM

Energy storage and its expenditure is a vital mechanism for life. Mammalian cells need external growth factors to activate the receptor-mediated signal transduction for regulation of the cell cycle. Multiple signal transduction pathways regulate the continuous energy and nutrient requirement by growing and proliferating cells and for sustainable production of new biomolecules. These highly intricate pathways are integrated for generation, storage, and expenditure of energy in expendable form, i.e., ATP through carbohydrates, fats, and protein metabolism, to maintain a systematic energy balance within the cell (Ward and Thompson 2012).

11.3.1 P13K/Akt Signaling

The phosphoinositide 3-kinase (PI3K)/Akt pathway is a highly conserved signal transduction pathway initiated by extracellular growth factors and has a diversified role in glucose metabolism (Figure 11.1). P13K is activated following the stimulation of G-protein-coupled receptors (GPCRs),

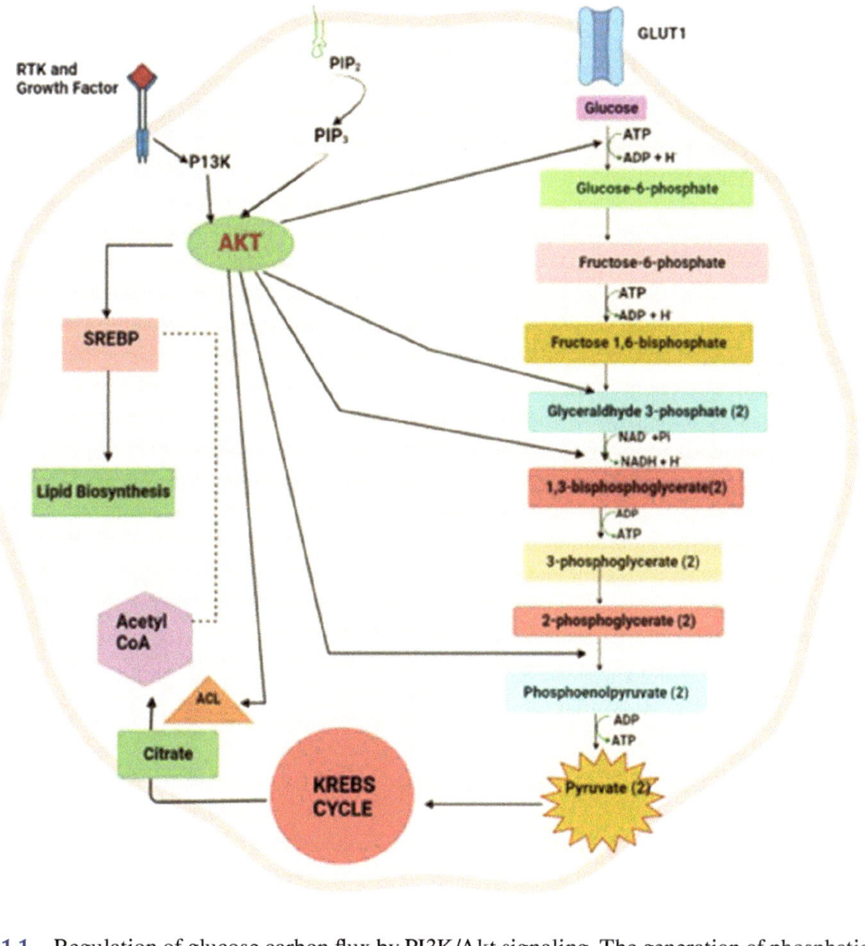

FIGURE 11.1 Regulation of glucose carbon flux by PI3K/Akt signaling. The generation of phosphatidylinositol 3,4,5-trisphosphate (PIP3) from phosphatidylinositol 4,5-bisphosphate (PIP2) and activation of Akt by PIP_3 influence a variety of proteins, thus increasing the rate of glycolysis. Akt also promotes lipogenesis from glucose carbon and can indirectly increase the proteolytic release of sterol regulatory element-binding protein transcription factors (SREBPs), which subsequently mediate the expression of genes implicated in the fatty acids, cholesterol, and isoprenoid synthesis pathways.

cell membrane receptor tyrosine kinases (RTKs), and cytokine receptors by extracellular growth factors. The membrane phosphatidylinositol lipids are phosphorylated by the activated P13k that sequentially causes the recruitment and stimulation of additional kinases, specifically Akt. Various downstream targets involved in essential biological functions for cell survival can subsequently be activated or deactivated by Akt (Figure 11.1). A notable impact of Akt on glucose metabolism is the enhanced glucose uptake through upregulating the glucose transporter 1 (GLUT1) expression on the cell surface. This phenomenon not only regulates the translocation of GLUT1 but also promotes the translation of GLUT1 mRNA through the activation of mammalian target of rapamycin complex 1 (mTORC1) and 4E binding protein (4EBP), which typically sequester eIF4E and obstruct cap-dependent mRNA translation. mTOR is an evolutionary conserved protein and belongs to the PI3K-related kinase family. At least two different multiprotein complexes, mTORC1 and mTORC2, are nucleated by mTOR (Chandel 2021).

The activity of multiple glycolytic enzymes is also influenced by Akt signaling. The association of hexokinase with mitochondria for an increased rate of glycolysis is facilitated by the activation of Akt. It indirectly activates phosphofructokinase-1 (PFK1), a glycolysis rate-regulating enzyme, by directly phosphorylating phosphofructokinase-2 (PFK2) (Simons et al. 2012; Gottlob et al. 2001). In mammalian insulin-dependent tissues such as fat and muscles, the PI3K/Akt pathway is responsible for the absorption of glucose, representing a downstream effect on insulin signaling. The movement of the glucose transporter 4 (GLUT4) to the cell surface is also promoted by the PI3/Akt pathway (Kohn et al. 1996; Hemmings and Restuccia 2012).

11.3.2 HIF1 Signaling

Hypoxia-inducible factor 1 (HIF1) signaling pathways play a significant part in regulating glucose metabolism within cells (Figure 11.2). HIF1 is a transcription factor that is essential to the adaptive response of cells under low oxygen levels. HIF1-mediated activation of glycolytic enzymes balances the energy production during oxidative phosphorylation when there is a depletion of oxygen in mitochondria. This intricate metabolic adjustment lowers the mitochondrial activity to control the supply and demand chain of oxygen in the tissue. HIF1 acts as a rate-limiting factor for pyruvate entry into the mitochondria by upregulating the expression of pyruvate dehydrogenase kinase-1 (PDK1) and inhibiting the action of mitochondrial pyruvate dehydrogenase complex (PDH) that is responsible for pyruvate conversion to acetyl CoA, and consequently, a greater amount of lactate is produced from pyruvate. The newly converted lactate is then transported to the extracellular spaces, and NAD is regenerated for continued glycolysis (Papandreou et al. 2006). However, increased HIF1 signaling has been observed in proliferating cells even at normal oxygen levels. Numerous pathways, such as Akt activation through mTORC1, enhanced cellular food absorption, and elevated reactive oxygen species (ROS) levels, can increase the translation of HIF1α mRNA. Subsequently, the elevated ROS increases the fumarate and succinate levels in addition to being strong inhibitors of prolyl hydroxylase domain enzymes (PHDs) that target HIF1. The inhibitory effects of ROS on PHDs initiate a feedback mechanism that reduces the amount of glucose carbon flow into the mitochondria, whereas succinate and/or fumarate counteracts ROS formation through mitochondrial metabolism, prevents further accumulation of ROS inside the cell, and aids in the regulation of HIF1α breakdown (Selak et al. 2005; Laplante and Sabatini 2012). Recently, it has been shown that during prenatal hypoxic conditions, the activity of HIF1α increased in the hippocampus resulting in an imbalance between the pentose phosphate pathway and glycolysis followed by oxidative stress (Vetrovoy et al. 2023).

11.3.3 Myc Signaling

The absorption and metabolism of amino acids through growth factor signaling-mediated mechanisms liberate nitrogen that is in turn utilized by cells for protein and nucleotide biosynthesis. Glutamine metabolism is regulated by the growth factor Myc downstream signaling. In the proliferating cells, Myc proto-oncogene is under stringent regulation downstream of several receptor

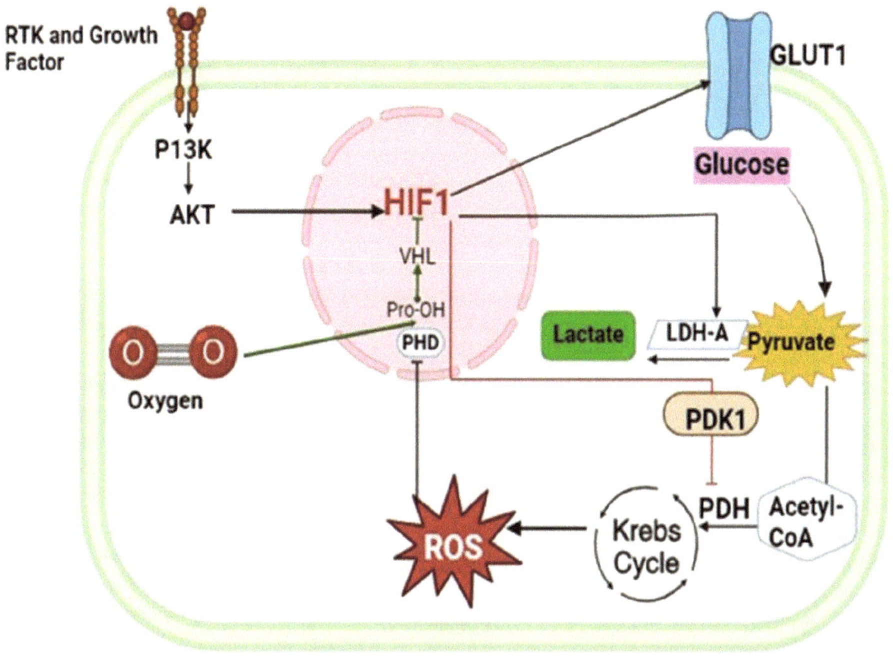

FIGURE 11.2 HIF1-mediated regulation of glucose metabolism. The regulation of cellular stability is mediated by HIF1 signaling via hydroxylation of its proline residues on α-subunits. Prolyl hydroxylase domain enzymes (PHDs) target HIF1α for recognition by the VHL ubiquitin E3 ligase and degradation by the proteasome. HIF1 is positively regulated by growth factor signaling and promotes the expression of genes encoding glucose transporters and glycolytic enzymes. The inhibition of glucose carbon flow in the mitochondria by HIF1 mediates the expression of pyruvate dehydrogenase kinase (PDK).

signaling pathways, including WNT, Hedgehog, Notch, TGFβ, and tyrosine kinases (Kumar et al. 2021; Chachoua et al. 2022). The stimulation of glutamine uptake by Myc, either by increased expression of glutaminase or glutamine transporter proteins such as SLC1A5, SLC7A11, and GLS1, results in elevated glutamine metabolism (Li et al. 2024). Myc facilitates the SLC7A5/SLC43A1-mediated essential amino acid uptake, while SLC7A5/SLC43A1 increases its mRNA translation through the elevation of essential amino acid. This regulatory mechanism stops whenever there is a depletion of SLC7A5 and SLC43A1, thus an overall inhibition of glucose and glutamine metabolism (Yue et al. 2017).

Similarly, anaplerosis, or the replenishment of the supply of intermediates in the Krebs cycle, depends on the carbon backbone of glutamine, which is achieved through the conversion of glutamine to glutamate, and that is later converted to α-ketoglutarate, a potent intermediate of the Krebs cycle. This Myc-mediated glutamine breakdown yields amide and amino groups required for non-essential amino acid synthesis. Interestingly, glutamine itself acts as a relevant regulator of c-Myc protein expression as observed in various studies where glutamine replenishment restored Myc expression (Dejure et al. 2017). In addition to generating precursors for nucleotide and nicotinamide biosynthesis, Myc also upregulates the expression of numerous key enzymes responsible for nucleotide biosynthesis (Tong et al. 2009; Lane and Fan 2015).

11.3.4 mTOR Signaling

mTOR is a major PI3K-related kinase involved in cellular proliferation and growth. It plays a pivotal role in maintaining the energy demand and supply for various cellular processes like protein and

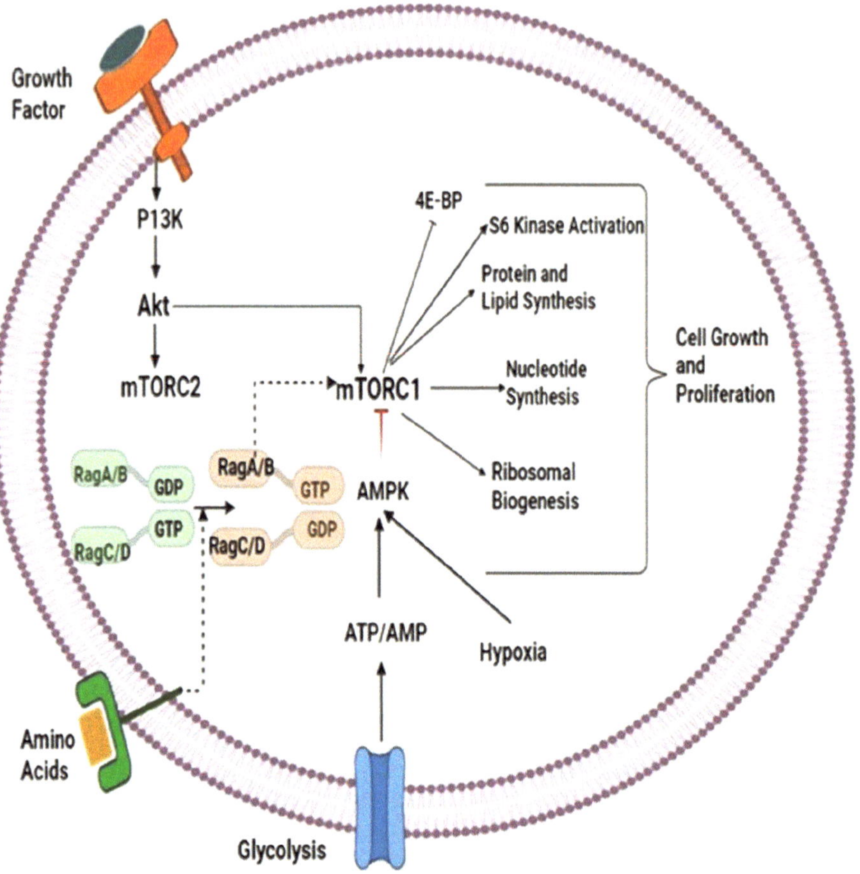

FIGURE 11.3 mTOR-mediated pyruvate kinase and amino acid uptake and metabolism. The mechanism is regulated by mTOR and Myc. Pyruvate kinase catalyzes the conversion of phosphoenolpyruvate (PEP) to pyruvate and generates ATP. Fructose 1,6-bisphosphate activates the M2 isoform Myc pyruvate kinase, while tyrosine kinase signaling downstream to the growth factor receptors releases F1,6P2 that diminishes the enzymatic activity of pyruvate kinase. This inhibition promotes the redistribution of glycolytic intermediates into anabolic pathways like nucleotide biosynthesis. Myc also regulates PKM2 expression via the upregulation of RNA splicing factors and glutamine uptake and metabolism. mTORC1 activates S6 kinase that inhibits the eukaryotic initiation factor 4EBP and promotes ribosomal protein expression.

lipid biosynthesis and regulating insulin sensitivity (Hu et al. 2016). It is also crucial for pyruvate kinase and amino acid uptake and metabolism (Figure 11.3). Silva et al. (2023) demonstrated that the fatty acids like oleic acid present in the maternal circulation stimulate the uptake of amino acid in primary human trophoblast cells, which is substantially facilitated by phosphatidic acid and mTOR signaling. Likewise, it was observed that the treatment of human cytotrophoblast cells with vasoactive intestinal peptide (VIP) increased mTOR expression and mTOR/S6 phosphorylation. In the case of VIP silencing, there was a marked reduction in mTOR mRNA and protein expression. The study further emphasized that VIP is substantially involved in glucose and neutral amino acids transport in cytotrophoblast cells through mTOR-regulated pathways, which further highlights the crucial role of mTOR signaling in cell metabolism (Merech et al. 2019).

The mTOR signaling cascade is composed of two distinct multisubunit complexes known as mTOR complex 1/2 (mTORC1/2). Both complexes establish a crucial negative feedback mechanism where the activation of mTORC1 inactivates mTORC2 and vice versa. This negative feedback mechanism ensures the maintenance of proper signaling within the system (Pena-Leon et al. 2020). mTORC1 promotes protein synthesis through phosphorylation and suppresses the activity of 4EBP. Similarly, it also regulates the translation of 5′ TOP mRNAs and ribosome synthesis by transcribing rRNA via activated polymerase I (Ward and Thompson 2012). Recent data showed that during the pharmacological inhibition of mTOR, 4EBP1/2 dominantly regulates the repression of translation of both 5′ TOP and canonical mRNAs (Hochstoeger et al. 2024).

Interestingly, AMPK can inhibit mTORC1 through various pathways, while leucine and other amino acids can activate it by stimulating the Rag GTPase complex on lysosomal membranes (Sancak et al. 2008). For ideal mTOR activation in proliferating HepG2 cells, RagC, Rheb, Rab1A, Rab5, and Arf1 GTPases are required, whereas in the senescence-like counterparts, the activation is mediated by RagC and Rheb only (Jiang et al. 2022).

Studies suggest that cellular leucine absorption and glutamine efflux are related, linking mTORC1 regulation to Myc-regulated glutamine uptake (Nicklin et al. 2009). Here, SLC1A5, a glutamine transporter, mediates reduced mTORC1 signaling and is targeted by MYC, signifying its regulatory impact on transcriptional feedback loops that modulate upstream signaling pathways like mTOR (Zhao et al. 2019). In contrast, high glutamine-induced mTORC1 hyperactivation resulted in the suppression of mTORC2, which regulates metabolic functions and growth factor signaling mostly through phosphorylation and activation of Akt (Saxton and Sabatini 2017). Conversely, reduced glutamine conditions inhibit mTORC1 hyperactivation that subsequently increases mTORC2 activation and RUNX2 stabilization (Gayatri et al. 2022).

11.3.5 Insulin Signaling

Insulin signaling is one of the crucial pathways responsible for energy production and expenditure (Figure 11.4). The uptake of glucose by GLUT receptors of the beta cells results in energy

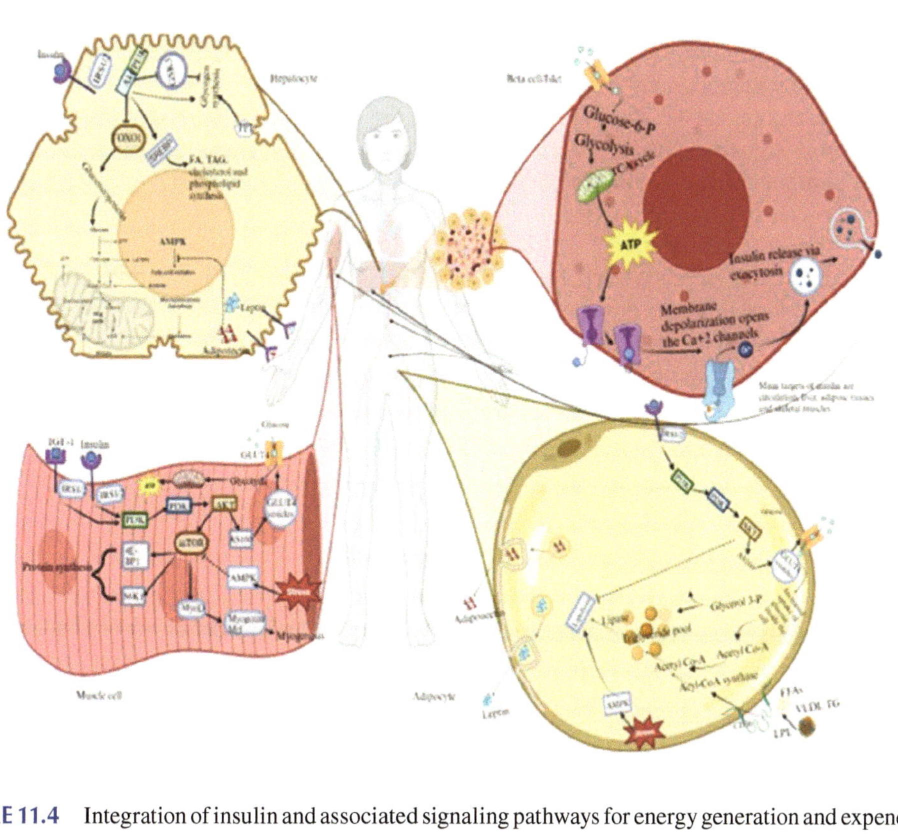

FIGURE 11.4 Integration of insulin and associated signaling pathways for energy generation and expenditure.

production, causing the leaky K^+ channels to close tightly, which mediates partial membrane depolarization. As a consequence, the calcium channels remain open and allow the entry of calcium that releases insulin from insulin-containing vesicles. When insulin enters the bloodstream, it acts on multiple cells and organs to perform various metabolic functions like glycolysis, glycogenesis, lipogenesis, and protein synthesis (Röder et al. 2016).

The binding of insulin to the insulin receptor in hepatocytes initiates the PI3K/Akt signaling cascade. Akt significantly suppresses gluconeogenesis via inhibiting the forkhead box subfamily O-1 (FOXO1) responsible for direct regulation of the de novo synthesis of glucose. The inhibition of GSK3, an important inhibitor of glycogen synthesis by Akt, asserts a promoting effect on glycogen synthesis in the liver (Molinaro et al. 2019).

Insulin, when acting on skeletal muscle cells, accelerates glycolysis and glucose metabolism, contributing to the energy production necessary for muscle contractions. While it actively promotes protein synthesis in muscles via the mTOR signaling pathway where Akt promotes mTOR, it also induces the expression of myogenin/mrf responsible for myogenesis (Nakashima et al. 2020; Sylow et al. 2021).

Additionally, the regulation of insulin by adiponectin and leptin in adipocytes maintains the sensitivity of these cells toward insulin. The action of insulin on white adipocytes initiates lipogenesis and facilitates the normal uptake of glucose by adipose tissues (Stern et al. 2016).

11.3.6 AMPK Pathway and Stress Signaling

Under stress, the body releases specific substances involved in energy expenditure like catecholamines (epinephrine and norepinephrine) and cortisol. Catecholamines are involved in glycolysis, gluconeogenesis, glycogenolysis, and proteolysis, and serve as neuromodulators in the central nervous system (Kruk et al. 2020). AMPK senses the lower cellular energy levels and inhibits the energy-consuming pathways, simultaneously promoting the energy-generating mechanisms. Also, an inhibition of AMPK activation minimizes glucose metabolism and cell survival. As observed in *Echinococcus granulosus* sensu stricto, when the EgAMPK/EgGLUT1 pathway is suppressed it results in limited glucose uptake and metabolism (Yan et al. 2024).

During experimental studies of animal models for depression and anxiety, it was found that AMPK/SIRT1 signaling, which is crucial for maintaining mitochondrial balance, is impaired. Reduced phosphorylated AMPK levels during anxiety may inhibit AMPK activation, impeding downstream factors like SIRT1. This inhibition may reduce mitochondrial biosynthesis, impair mitochondrial structure repair, and ultimately disrupt mitochondrial balance (Zhao et al. 2023).

Stress affects the HPA axis and leads to an increase in stress hormones like epinephrine and cortisol. The interaction of cortisol with endoplasmic reticulum (ER) stress and hypoxia is evident through the regulation of mRNA expression of ER stress-dependent genes (XBP1, DNAJB9) as well as the insulin receptor gene (Minchenko et al. 2024). Moreover, in stress-induced colitis models treated with vitamin D and lactoferrin, a significant upregulation of AMPK expression was observed along with inhibition of mTOR-STAT3 signaling, suggesting augmented colonic autophagy and indicating the imperative role of the AMPK pathway during stress (Abdelmalak et al. 2023).

11.3.7 Triacylglycerol and Cholesterol Metabolism

Triacylglycerol (TAG) metabolism is a dynamic process that enables lipid constituents to perform a variety of interconnected functions inside cells (Figure 11.5). The cellular TAG accumulated in lipid droplets yield fatty acids (FAs) that function as signals to affect transcriptional regulation of gene expression and are specifically routed to β-oxidation. The FA storage within TAG is imperative to preventing the membrane complexes from FA detergent-like properties. These stored FAs also sequester essential fatty acids and phospholipid precursors like diacylglycerol (DAG). Lipid intermediates such as lysophosphatidic acid (LPA), phosphatidic acid (PA), and DAG are produced

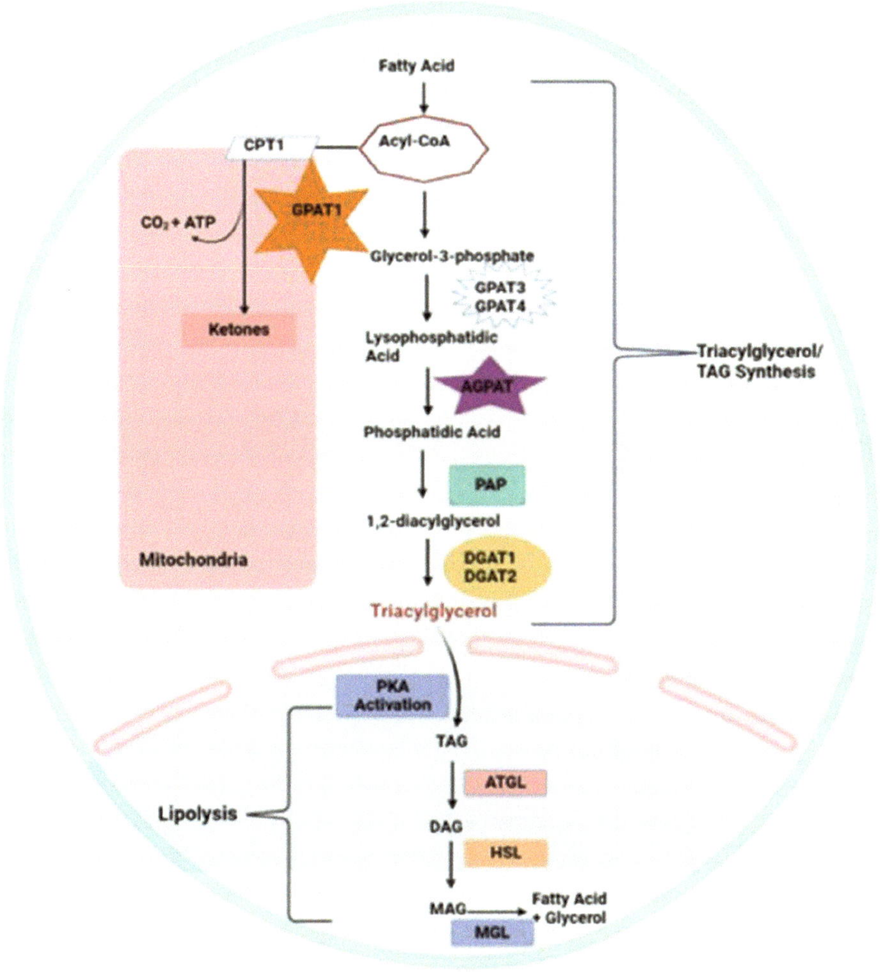

FIGURE 11.5 TAG synthesis and lipolysis oxidation of acyl-CoA. Following their entry in the mitochondrial matrix as acyl-carnitines by CPT1, acyl-CoAs are esterified to glycerol-3-phosphate by a GPAT isoform that produces LPA. The isoform of the enzyme AGPAT utilizes another acyl-CoA for the esterification of LPA to synthesize PA, which is dephosphorylated by PAP to yield DAG. DGAT isozymes utilize the last acyl-CoA to synthesize TAG from DAG. PLIN1 is phosphorylated in response to lipolytic stimulation and PKA activation. After a cascade of intermediatory events involving multiple enzymes, as a final step, MGL hydrolyzes MAG, which yields glycerol and fatty acids.

during the synthesis and breakdown of TAG that may in turn activate or inhibit signaling pathways governed by PPARγ, mTOR, or protein kinase C isoforms (Coleman and Mashek 2011). In white adipocytes, lipolysis directly results in allosteric activation of AMPK through the generation of LC-acyl-CoAs (Rahman et al. 2024). Promotion of lipolysis by leucine is also dependent on the mTOR-SIRT1 signaling pathway along with acetylated S6K1 as observed in finishing pigs, suggesting the imperative involvement of signaling pathways in lipid breakdown and energy production (Yin et al. 2023). Table 11.1 provides a brief summary of the enzymes involved in TAG metabolism and their possible roles in signal transduction.

Cholesterol, a dominant sterol, is a major constituent of mammalian cell membranes. There are two proposed overlapping pathways for cholesterol production from lanosterol (Figure 11.6). In the Bloch pathway, cholesta-5,24-dienol (desmosterol) is the route through which cholesterol synthesis

TABLE 11.1

Major Enzymes of TAG Metabolism Involved in Signal Transduction Pathways

	Enzymes Involved in Triglyceride Metabolism	Expression Sites	Functions	References
1	ATGL (adipose triglyceride lipase)	Brown and white adipose tissues Minimal expression in skeletal and cardiac muscles	Oxidative gene expression Control of fatty acid catabolism	Ahmadian et al. 2009; Zimmermann et al. 2004
2	PNPLA3/adiponutrin	Mainly adipose tissues	TAG hydrolase, transacylase, and phospholipase A2 activities	Jenkins et al. 2004; Lake et al. 2005
3	HSL (hormone-sensitive lipase)	Adipose tissues Adrenal tissues Gonads Placenta Muscles	Lipolysis regulation Control of fatty acid catabolism Responsive to epinephrine and norepinephrine signaling	Coleman and Mashek 2011
4	MGL	Testis Kidneys Adipose tissues	Endocannabinoid signaling	Schlosburg et al. 2010
5	TGH (triacylglycerol hydrolase)	Endoplasmic reticulum of hepatocytes	Lipolysis	Schlosburg et al. 2010; Wang et al., 2007

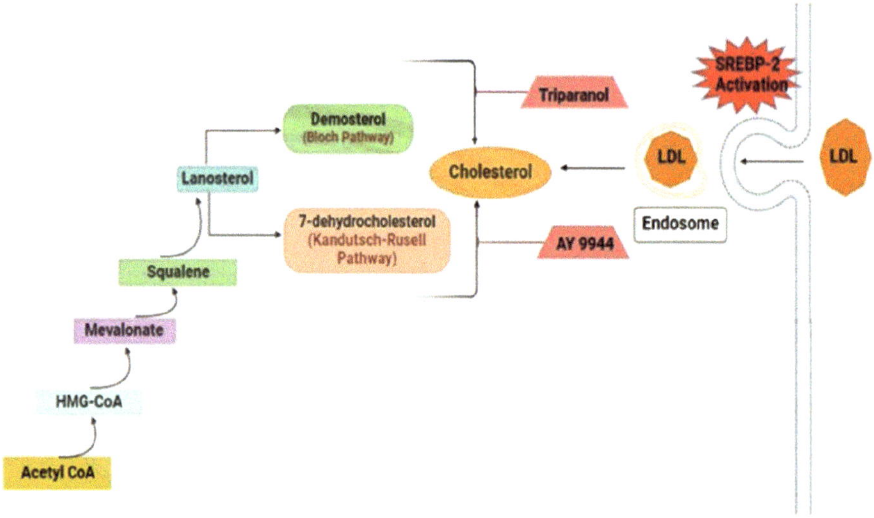

FIGURE 11.6 Mechanistic overview of cholesterol biosynthesis depicting Kandutsch–Russell and Bloch pathways. Both pathways are interconnected by the 24-DHCR enzyme and use acetate as a common precursor but later split post lanosterol synthesis. The inclusion of a double bond at position 24 in the alkyl side chain of the sterol intermediates of the Bloch pathway is the only distinction between the corresponding sterol intermediates of the two pathways. Inhibitors block cholesterol biosynthesis at various stages. Triparanol is a potent inhibitor of the 24-DHCR enzyme and demisterol, while AY 9944 inhibits 7-dehydrocholesterol.

continues if the C24 double bond reduction is maintained until the final reaction. Conversely, in the Kandutsch–Russell pathway, cholesta-5,7-dienol (7-dehydrocholesterol) and cholesterol are produced via early Δ24-reduction involving lanosterol. The C24–C25-terminal intermediates facilitate the synthesis process in the skin and intestines and have greater sterol C24-reductase activity than in the liver or brain (Shrivastava et al. 2023; Nes 2011).

Cholesterol trafficking within the cell organelles is also regulated by vesicular or non-vesicular transport mechanisms. Cellular cholesterol can be synthesized by de novo cholesterol biosynthesis or can be imported into the cells through LDL cholesterol depending on the cellular requirements, controlled by SREBP2 transcription factor. Following the synthesis, the first step of cholesterol metabolism is its oxidation to oxysterol as it can efficiently pass through the membranes than cholesterol. A significant portion of cholesterol is also converted to bile acids predominantly in the hepatic cells. Both oxysterol and bile acids act as principle signaling molecules (van den Boomen et al. 2020; Griffiths and Wang 2021). A recent study highlights the ameliorative effect of polymethoxyflavones on hypercholesterolemia by facilitating the entry of cholesterol into the liver via the reverse cholesterol transport (RCT) pathway followed by excretion via metabolism pathways of cholesterol and bile acid. The study provides further clues on the mechanistic importance of various pathways involved in cholesterol metabolism (Du et al. 2024).

11.4 CONCLUSION

Cell signaling plays a pivotal role in maintaining cellular responses and sensitivity toward altered levels of intracellular metabolites. Aberrations in key metabolic pathways, due to several metabolite-derived modifications distract cellular processing and hinder the cell's feedback control on signal transduction mechanisms. Dysregulation of these mechanisms interrupt cellular growth and division mostly through deprivation of energy reserves. Signaling molecules and hormones that regulate signaling pathways may serve as therapeutic targets for metabolic disorders and associated consequences.

REFERENCES

Abdelmalak, M. F. L., Abdelrahim, D. S., George Michael, T. M. A., Abdel-Maksoud, O. M., & Labib, J. M. W. (2023). Vitamin D and lactoferrin attenuate stress-induced colitis in Wistar rats via enhancing AMPK expression with inhibiting mTOR-STAT3 signaling and modulating autophagy. *Cell Biochemistry and Function*, *41*(2), 211–222. https://doi.org/10.1002/cbf.3774

Ahmadian, M., Duncan, R. E., Varady, K. A., Frasson, D., Hellerstein, M. K., Birkenfeld, A. L., Samuel, V. T., Shulman, G. I., Wang, Y., Kang, C., & Sul, H. S. (2009). Adipose overexpression of desnutrin promotes fatty acid use and attenuates diet-induced obesity. *Diabetes*, *58*(4), 855–866. https://doi.org/10.2337/db08-1644

Chachoua, I., Tzelepis, I., Dai, H., Lim, J. P., Lewandowska-Ronnegren, A., Casagrande, F. B., ... & Göndör, A. (2022). Canonical WNT signaling-dependent gating of MYC requires a noncanonical CTCF function at a distal binding site. *Nature communications*, *13*(1), 204.

Chandel, N. S. (2021). Signaling and metabolism. *Cold Spring Harbor Perspectives in Biology*, *13*(2), a040600.

Coleman, R. A., & Mashek, D. G. (2011). Mammalian triacylglycerol metabolism: synthesis, lipolysis, and signaling. *Chemical Reviews*, *111*(10), 6359–6386. https://doi.org/10.1021/cr100404w

Dang, C. V. (2012). MYC on the path to cancer. *Cell*, *149*(1), 22–35.

Dejure, F. R., Royla, N., Herold, S., Kalb, J., Walz, S., Ade, C. P., Mastrobuoni, G., Vanselow, J. T., Schlosser, A., Wolf, E., et al. (2017). The MYC mRNA 3'-UTR couples RNA polymerase II function to glutamine and ribonucleotide levels. *EMBO Journal*, *36*, 1854–1868. https://doi.org/10.15252/embj.201796662

Du, Y., Su, J., Yan, M., Wang, Q., Wang, T., Gao, S., Tian, Y., Wang, Y., Chen, S., Lv, G., & Yu, J. (2024). Polymethoxyflavones in citrus extract has a beneficial effect on hypercholesterolemia rats by promoting liver cholesterol metabolism. *Journal of Ethnopharmacology*, *322*, 117644. https://doi.org/10.1016/j.jep.2023.117644

Gayatri, M. B., Gajula, N. N., Chava, S., & Reddy, A. B. M. (2022). High glutamine suppresses osteogenesis through mTORC1-mediated inhibition of the mTORC2/AKT-473/RUNX2 axis. *Cell Death Discovery*, *8*(1), 277. https://doi.org/10.1038/s41420-022-01077-3

Gottlob, K., Majewski, N., Kennedy, S., Kandel, E., Robey, R. B., & Hay, N. (2001). Inhibition of early apoptotic events by Akt/PKB is dependent on the first committed step of glycolysis and mitochondrial hexokinase. *Genes & Development*, *15*(11), 1406–1418.

Griffiths, W. J., & Wang, Y. (2021). Sterols, oxysterols, and accessible cholesterol: Signalling for homeostasis, in immunity and during development. *Frontiers in Physiology*, 12, 723224. https://doi.org/10.3389/fphys.2021.723224

Hemmings, B. A., & Restuccia, D. F. (2012). Pi3k-Pkb/Akt pathway. *Cold Spring Harbor Perspectives in Biology*, *4*(9), a011189.

Hochstoeger, T., Papasaikas, P., Piskadlo, E., & Chao, J. A. (2024). Distinct roles of LARP1 and 4EBP1/2 in regulating translation and stability of 5'TOP mRNAs. *Science Advances*, *10*(7), eadi7830. https://doi.org/10.1126/sciadv.adi7830

Hu, F., Xu, Y., & Liu, F. (2016). Hypothalamic roles of mTOR complex I: Integration of nutrient and hormone signals to regulate energy homeostasis. *American Journal of Physiology-Endocrinology and Metabolism*, *310*(11). https://doi.org/10.1152/ajpendo.00121.2016

Jenkins, C. M., Mancuso, D. J., Yan, W., Sims, H. F., Gibson, B., & Gross, R. W. (2004). Identification, cloning, expression, and purification of three novel human calcium-independent phospholipase A2 family members possessing triacylglycerol lipase and acylglycerol transacylase activities. *The Journal of Biological Chemistry*, *279*(47), 48968–48975. https://doi.org/10.1074/jbc.M407841200

Jiang, W., Ou, Z., Zhu, Q., & Zai, H. (2022). RagC GTPase regulates mTOR to promote chemoresistance in senescence-like HepG2 cells. *Frontiers in Physiology*, *13*, 949737. https://doi.org/10.3389/fphys.2022.949737

Kohn, A. D, Summers, S. A., Birnbaum, M. J., & Roth, R. A. (1996). Expression of a constitutively active Akt Ser/Thr Kinase in 3T3-L1 adipocytes stimulates glucose uptake and glucose transporter 4 translocation. *Journal of Biological Chemistry*, *271*(49), 31372–31378.

Kruk, J., Kotarska, K., & Aboul-Enein, B. H. (2020). Physical exercise and catecholamines response: benefits and health risk: possible mechanisms. *Free Radical Research*, *54*(2–3), 105–125. https://doi.org/10.1080/10715762.2020.1726343

Kumar, V., Vashishta, M., Kong, L., Wu, X., Lu, J. J., Guha, C., & Dwarakanath, B. S. (2021). The role of notch, hedgehog, and wnt signaling pathways in the resistance of tumors to anticancer therapies. *Frontiers in Cell and Developmental Biology*, *9*, 650772. https://doi.org/10.3389/fcell.2021.650772

Lake, A. C., Sun, Y., Li, J. L., Kim, J. E., Johnson, J. W., Li, D., Revett, T., Shih, H. H., Liu, W., Paulsen, J. E., & Gimeno, R. E. (2005). Expression, regulation, and triglyceride hydrolase activity of Adiponutrin family members. *Journal of Lipid Research*, *46*(11), 2477–2487. https://doi.org/10.1194/jlr.M500290-JLR200

Lane, A. N., & Fan, T. W. (2015). Regulation of mammalian nucleotide metabolism and biosynthesis. *Nucleic acids Research*, *43*(4), 2466–2485.

Laplante, M., & Sabatini, D. M. (2012). MTOR signaling. *Cold Spring Harbor Perspectives in Biology*, *4*(2), a011593.

Li, X., Peng, X., Li, Y., Wei, S., He, G., Liu, J., Li, X., Yang, S., Li, D., Lin, W., Fang, J., Yang, L., & Li, H. (2024). Glutamine addiction in tumor cell: Oncogene regulation and clinical treatment. *Cell Communication and Signaling : CCS*, *22*(1), 12. https://doi.org/10.1186/s12964-023-01449-x

Merech, F., Soczewski, E., Hauk, V., Paparini, D., Ramhorst, R., Vota, D., & Pérez Leirós, C. (2019). Vasoactive Intestinal Peptide induces glucose and neutral amino acid uptake through mTOR signalling in human cytotrophoblast cells. *Scientific Reports*, *9*(1), 17152. https://doi.org/10.1038/s41598-019-53676-3

Minchenko, D. O., Khita, O. O., Viletska, Y. M., Sliusar, M. Y., Rudnytska, O. V., Kozynkevych, H. E., Bezrodnyi, B. H., Khikhlo, Y. P., & Minchenko, O. H. (2024). Cortisol controls endoplasmic reticulum stress and hypoxia dependent regulation of insulin receptor and related genes expression in HEK293 cells. *Endocrine Regulations*, *58*(1), 1–10. https://doi.org/10.2478/enr-2024-0001

Molinaro, A., Becattini, B., Mazzoli, A., Bleve, A., Radici, L., Maxvall, I., Rotter Sopasakis, V., Molinaro, A., Bäckhed, F., & Solinas, G. (2019). Insulin-driven PI3K-Akt signaling in the hepatocyte is mediated by redundant PI3KA and PI3Kβ activities and is promoted by Ras. *Cell Metabolism*, *29*(6). https://doi.org/10.1016/j.cmet.2019.03.010

Nakashima, K., Ishida, A., Shimamoto, S., Ijiri, D., & Ohtsuka, A. (2020). Insulin stimulation of protein synthesis and mTOR signaling in chick myotube cultures. *The Journal of Poultry Science*, *57*(3), 205–209. https://doi.org/10.2141/jpsa.0190082

Nes W. D. (2011). Biosynthesis of cholesterol and other sterols. *Chemical Reviews*, *111*(10), 6423–6451. https://doi.org/10.1021/cr200021m

Nicklin, P., Bergman, P., Zhang, B., Triantafellow, E., Wang, H., Nyfeler, B., Yang, H., Hild, M., Kung, C., & Wilson, C. (2009). Bidirectional transport of amino acids regulates MTOR and autophagy. *Cell*, *136*(3), 521–534.

Papandreou, I., Cairns, R. A., Fontana, L., Lim, A. L., & Denko, N. C. (2006). HIF-1 mediates adaptation to hypoxia by actively downregulating mitochondrial oxygen consumption. *Cell Metabolism*, *3*(3), 187–197.

Pena-Leon, V., Perez-Lois, R., & Maria Seoane, L. (2020). MTOR pathway is involved in energy homeostasis regulation as a part of the gut–Brain axis. *International Journal of Molecular Sciences*, *21*(16), 5715. https://doi.org/10.3390/ijms21165715

Rahman, A. A., Butcko, A. J., Songyekutu, E., Granneman, J. G., & Mottillo, E. P. (2024). Direct effects of adipocyte lipolysis on AMPK through intracellular long-chain acyl-CoA signaling. *Scientific Reports*, *14*(1), 19. https://doi.org/10.1038/s41598-023-50903-w

Röder, P. V., Wu, B., Liu, Y., & Han, W. (2016). Pancreatic regulation of glucose homeostasis. *Experimental & Molecular Medicine*, *48*(3), e219. https://doi.org/10.1038/emm.2016.6

Sancak, Y., Peterson, T. R., Shaul, Y. D., Lindquist, R. A., Thoreen, C. C., Bar-Peled, L., & Sabatini, D. M. (2008). The Rag GTPases bind raptor and mediate amino acid signaling to mTORC1. *Science*, *320*(5882), 1496-1501.

Saxton, R. A., & Sabatini, D. M. (2017). MTOR signaling in growth, metabolism, and disease. *Cell*, *168*(6), 960–976.

Schlosburg, J. E., Blankman, J. L., Long, J. Z., Nomura, D. K., Pan, B., Kinsey, S. G., Nguyen, P. T., Ramesh, D., Booker, L., Burston, J. J., Thomas, E. A., Selley, D. E., Sim-Selley, L. J., Liu, Q. S., Lichtman, A. H., & Cravatt, B. F. (2010). Chronic monoacylglycerol lipase blockade causes functional antagonism of the endocannabinoid system. *Nature Neuroscience*, *13*(9), 1113–1119. https://doi.org/10.1038/nn.2616

Selak, M. A., Armour, S. M., MacKenzie, E. D., Boulahbel, H., Watson, D. G., Mansfield, K. D., Pan, Y., Simon, M. C., Thompson, C. B., & Gottlieb, E. (2005). Succinate links TCA cycle dysfunction to oncogenesis by inhibiting HIF-α Prolyl hydroxylase. *Cancer Cell*, *7*(1), 77–85.

Shrivastava, S., Paila, Y. D., & Chattopadhyay, A. (2023). Role of cholesterol and its biosynthetic precursors on membrane organization and dynamics: A fluorescence approach. *The Journal of Membrane Biology*, *256*(2), 189–197. https://doi.org/10.1007/s00232-023-00278-w

Silva, E., Ferchaud-Roucher, V., Kramer, A., Madi, L., Pantham, P., Chassen, S., Jansson, T., & Powell, T. L. (2023). Oleic acid stimulation of amino acid uptake in primary human trophoblast cells is mediated by phosphatidic acid and mTOR signaling. *FASEB bioAdvances*, *6*(1), 1–11. https://doi.org/10.1096/fba.2023-00113

Simons, A. L., Orcutt, K. P., Madsen, J. M., Scarbrough, P. M., & Spitz, D. R. (2012). The role of Akt pathway signaling in glucose metabolism and metabolic oxidative stress. *Oxidative Stress in Cancer Biology and Therapy*, 21–46.

Stern, J. H., Rutkowski, J. M., & Scherer, P. E. (2016). Adiponectin, leptin, and fatty acids in the maintenance of metabolic homeostasis through adipose tissue crosstalk. *Cell Metabolism*, *23*(5), 770–784. https://doi.org/10.1016/j.cmet.2016.04.011

Sylow, L., Tokarz, V. L., Richter, E. A., & Klip, A. (2021). The many actions of insulin in skeletal muscle, the paramount tissue determining glycemia. *Cell Metabolism*, *33*(4), 758–780. https://doi.org/10.1016/j.cmet.2021.03.020

Tong, X., Zhao, F., & Thompson, C. B. (2009). The molecular determinants of de novo nucleotide biosynthesis in cancer cells. *Current Opinion in Genetics & Development*, *19*(1), 32–37.

Valls, P. O., & Esposito, A. (2022). Signalling dynamics, cell decisions, and homeostatic control in health and disease. Current Opinion in Cell Biology, 75, 102066.

van den Boomen, D. J. H., Volkmar, N., & Lehner, P. J. (2020). Ubiquitin-mediated regulation of sterol homeostasis. *Current Opinion in Cell Biology*, *65*, 103–111. https://doi.org/10.1016/j.ceb.2020.04.010

Vetrovoy, O., Stratilov, V., Potapova, S., & Tyulkova, E. (2023). Oxidative stress accompanies HIF1-dependent impairment of glucose metabolism in the hippocampus of adult Rats survived prenatal severe hypoxia. *Developmental Neuroscience*. https://doi.org/10.1159/000535326

Wang, H., Gilham, D., & Lehner, R. (2007). Proteomic and lipid characterization of apolipoprotein B-free luminal lipid droplets from mouse liver microsomes: Implications for very low density lipoprotein assembly. *The Journal of Biological Chemistry*, *282*(45), 33218–33226. https://doi.org/10.1074/jbc.M706841200

Ward, P. S., & Thompson, C. B. (2012). Signaling in control of cell growth and metabolism. *Cold Spring Harbor Perspectives in Biology*, *4*(7), a006783.

Wu, F., Yang, J., Liu, J., Wang, Y., Mu, J., Zeng, Q., Deng, S., & Zhou, H. (2021). Signaling pathways in cancer-associated fibroblasts and targeted therapy for cancer. *Signal Transduction and Targeted Therapy*, *6*(1), 1–35. https://doi.org/10.1038/s41392-021-00641-0

Xiao, W., Wang, R. S., Handy, D. E., & Loscalzo, J. (2018). NAD (H) and NADP (H) redox couples and cellular energy metabolism. *Antioxidants & Redox Signaling*, *28*(3), 251–272.

Yan, M., Liu, H., Su, Y., Bi, X., Yang, N., Lin, R., & Lü, G. (2024). Inhibition of AMPK activation in Echinococcus granulosus sensu stricto limits the parasite's glucose metabolism and survival. *Antimicrobial Agents and Chemotherapy*, e0120223. https://doi.org/10.1128/aac.01202-23

Yin, Y., Gong, S., Han, M., Wang, J., Shi, H., Jiang, X., Guo, L., Duan, Y., Guo, Q., Chen, Q., & Li, F. (2023). Leucine regulates lipid metabolism in adipose tissue through adipokine-mTOR-SIRT1 signaling pathway and bile acid-microbe axis in a finishing pig model. *Animal Nutrition (Zhongguo xu mu shou yi xue hui)*, *16*, 158–173. https://doi.org/10.1016/j.aninu.2023.10.005

Yue, M., Jiang, J., Gao, P., Liu, H., & Qing, G. (2017). Oncogenic MYC activates a feedforward regulatory loop promoting essential amino acid metabolism and tumorigenesis. *Cell Reports*, *21*(13), 3819–3832. https://doi.org/10.1016/j.celrep.2017.12.002

Zhao, A. R., Li, J., Wang, S. Q., Bian, L. H., Li, W. J., & Guo, J. Y. (2023). Stress can affect mitochondrial energy metabolism and AMPK/SIRT1 signaling pathway in rats. *Brain Research Bulletin*, *203*, 110770. https://doi.org/10.1016/j.brainresbull.2023.110770

Zhao, X., Petrashen, A. P., Sanders, J. A., Peterson, A. L., & Sedivy, J. M. (2019). SLC1A5 glutamine transporter is a target of MYC and mediates reduced mTORC1 signaling and increased fatty acid oxidation in long-lived Myc hypomorphic mice. *Aging Cell*, *18*(3), e12947. https://doi.org/10.1111/acel.12947

Zimmermann, R., Strauss, J. G., Haemmerle, G., Schoiswohl, G., Birner-Gruenberger, R., Riederer, M., Lass, A., Neuberger, G., Eisenhaber, F., Hermetter, A., & Zechner, R. (2004). Fat mobilization in adipose tissue is promoted by adipose triglyceride lipase. *Science (New York, N.Y.)*, *306*(5700), 1383–1386. https://doi.org/10.1126/science.1100747

12 Inflammatory Signaling and Immune Response

Naseem Zehra[1], Zia Uddin[1],
Muhammad Imran Amir Zada[1], and Muhammad Ikram[1]*
Department of Pharmacy, COMSATS University
Islamabad, Abbottabad Campus, Pakistan

12.1 FUNDAMENTALS OF THE IMMUNE SYSTEM

From birth, our bodies are attacked by pathogens that aim to survive and reproduce in a favorable environment. Some microorganisms are not harmful, like microbiota that have a mutually beneficial relationship with our gut [1]. The immune system possesses complex mechanisms to identify and eliminate or nullify harmful microorganisms that cause disease [2]. However, severe inflammation or tissue damage brought on by an immune reaction that targets the body's own cells, substances, or environmental substances can potentially cause negative effects or organ loss. Numerous protective mechanisms have evolved over time to limit the adverse effects of developing immunological responses, especially to protect vertebrates from reacting to their own cells and to restrict inflammation caused by the immune system.

An essential component in maintaining the proper immune system function is immunological tolerance, achieved through the actions of a certain group of $CD4^+$ T lymphocytes known as regulatory T cells (Tregs). These Tregs possess potent immunosuppressive abilities and consistently produce the transcription factors (FOXP3, IRF4, and SMAD4). They also express high levels of the interleukin-2 (IL-2) receptor α-chain (CD25) [3]. Treg-mediated suppression is a vital mechanism for negatively regulating immune-mediated inflammation and plays a prominent role in autoimmune diseases, allergies, acute and chronic infections, as well as cancer. Additionally, Tregs are important for tissue healing and inflammation resolution.

The recognition of the affiliation between Tregs and various human diseases has prompted extensive investigations into their potential therapeutic manipulation. Emerging approaches for Treg-based therapies are context-specific, aiming to increase Treg numbers and enhance their functional activity in inflammatory, allergic, and autoimmune diseases, while inhibiting or depleting these cells' suppressive function in cancer [3].

The concept of immune senescence refers to age-related changes in immune responses, encompassing the alterations in both cellular and serological mechanisms that impact the generation of specific responses to foreign and self-antigens. As individuals age, their immune system gradually declines, manifesting in increased vulnerability to infectious diseases, reduced vaccine efficacy, and a higher prevalence of cancer, autoimmune disorders, and other chronic illnesses. To combat external threats, various cells throughout the organism engage in intricate interactions, either through direct contact or via a diverse array of soluble mediators. This elaborate network aims to ensure a comprehensive defense mechanism against foreign invasions, all while maintaining the delicate balance of cell proliferation within the body [4].

In ancient times, inflammation was identified through five primary visual manifestations: redness (rubor), swelling (tumor), heat (specifically in the body's extremities), pain, and loss of function. The significant milestone in medicine can be attributed to Celsus, an ancient Roman physician, who, between 30 and 38 BC, named the first four of these indications [5].

DOI: 10.1201/9781003594901-12

The inflammatory response's main objective is to combat infection and tissue damage through complex interactions between various immune cells and molecules. Innate immune cells, including macrophages, fibroblasts, mast cells, and dendritic cells, as well as circulating leukocytes like monocytes and neutrophils, are strategically located in different tissues to detect signs of cell damage or pathogen invasion. These cells possess pattern recognition receptors (PRRs), found either intracellularly or on their surfaces. PRRs can identify molecular patterns associated with damage (DAMPs) or pathogen-associated molecular patterns (PAMPs). PAMPs encompass microbial nucleic acids, lipoproteins, and carbohydrates, while DAMPs refer to molecules released from injured cells.

When PRRs recognize PAMPs or DAMPs, they undergo oligomerization and assembly, forming large multi-subunit complexes. This molecular rearrangement activates signaling cascades that initiate a series of events aimed at resolving infection or tissue damage. One key outcome is the release of factors promoting leukocyte recruitment to the affected area, including chemotactic molecules like chemokines. Additionally, cytokines and growth factors are produced, amplifying the inflammatory response and facilitating tissue repair. Leukocyte recruitment is crucial for pathogen clearance and inflammation resolution. Neutrophils, often the first responders, play a crucial role in phagocytosing and killing invading microbes. Monocytes differentiate into macrophages, which are highly efficient phagocytes and contribute to tissue repair. The coordinated efforts of these immune cells, along with additional immune system elements like complementary proteins, ensure an effective and controlled inflammatory response [6].

In the late 19th century, microbiologists initially believed that leukemic cells facilitated the spread of infectious agents throughout the host organism. However, in 1882, Eli Metchnikoff discerned that white blood cells were actually engaged in eradicating engulfed pathogens. To describe this cellular phenomenon, Metchnikoff coined the term "phagocyte," derived from the Greek words "phagein" (to consume) and "cyte" (cell). This groundbreaking discovery marked the identification of what would eventually be recognized as innate immunity.

Innate immunity, the more ancient arm of the immune system, consists of various components such as physical barriers (e.g., the skin), small molecules like complement, and specialized cells, including macrophages and dendritic cells. The term "innate" reflects the fact that this defense system provides protection against pathogens without requiring any preconditioning from the environment. In other words, when confronted with a pathogen, the innate defense mechanism promptly initiates a response aimed at either eliminating or expelling it from the host organism [7].

12.1.1 Organs of the Immune System

While immune cells are found throughout the body, the immune system has specialized immune organs. This intricate organization facilitates a regulated immune response that efficiently generates a substantial number of cells capable of halting the progression of a spreading infection. These immune reservoirs can release immune cells and molecules that infiltrate virtually any tissue in the body. Although all blood cells originate in the bone marrow, their maturation and residence sites vary significantly [8]. Many innate immune cells originate in the bone marrow and establish themselves in the bloodstream and various tissues. For acquired immune cells, T and B cells undergo recombination, modifying their immune receptors in the thymus and bone marrow, respectively. These specialized organs, referred to as primary immune organs, are crucial for the development and maturation of T and B cells. Secondary immune organs are the sites where lymphocytes reside after maturation. Examples of secondary immune organs include the appendix, tonsils, adenoids, spleen, lymph nodes, and other mucus-associated lymphoid tissues [7].

12.1.2 Primary Immune Organs

12.1.2.1 Thymus

The thymus, an essential lymphoid organ, plays a crucial role in T-cell development and maturation. T cells serve as the initial defense against invading pathogens, immune disorders, and cancerous cells. However, with advancing age, the thymus undergoes involution or shrinkage (Li et al., 2003), characterized by decreased size and cellularity. This involution reduces T-cell production and the capacity for naïve T cells to migrate to peripheral tissues. Thymic involution also significantly alters the composition and functionality of the circulating T-cell pool. Specifically, there is an increase in memory T cells, which have previously encountered specific antigens, and changes in the T-cell receptor (TCR) repertoire [9]. These alterations have important implications for overall immune function and responsiveness.

The impact of thymic involution goes beyond individual immune health. The decline in T-cell production and the subsequent alterations in the T-cell pool increase susceptibility to infectious diseases, such as COVID-19. Individuals experiencing thymic involution are at a higher risk of developing autoimmune disorders and cancer due to decreased immune surveillance. Thymic involution negatively impacts T cell growth, selection, and tolerance maintenance. Processes affecting the thymus's structural components (cortex and medulla), gradual genetic mutations, and gene expression changes may culminate in immune senescence. Understanding the molecular mechanisms of thymic involution is crucial in identifying diagnostic biomarkers and therapeutic targets to alleviate associated complications. This analysis examines the consequences of thymic involution in infections, immune disorders, and diseases, highlighting potential obstacles and strategies to support or restore thymus function, specifically in elderly and immunocompromised populations [10] (Figure 12.1).

12.1.2.2 Bone Marrow

While bone marrow is primarily known as a major organ of hematopoiesis, research shows it is also the site of active immune cell function and trafficking. These immune cells include mesenchymal stem cells, B cells, regulatory T cells, dendritic cells, natural killer T cells, neutrophils,

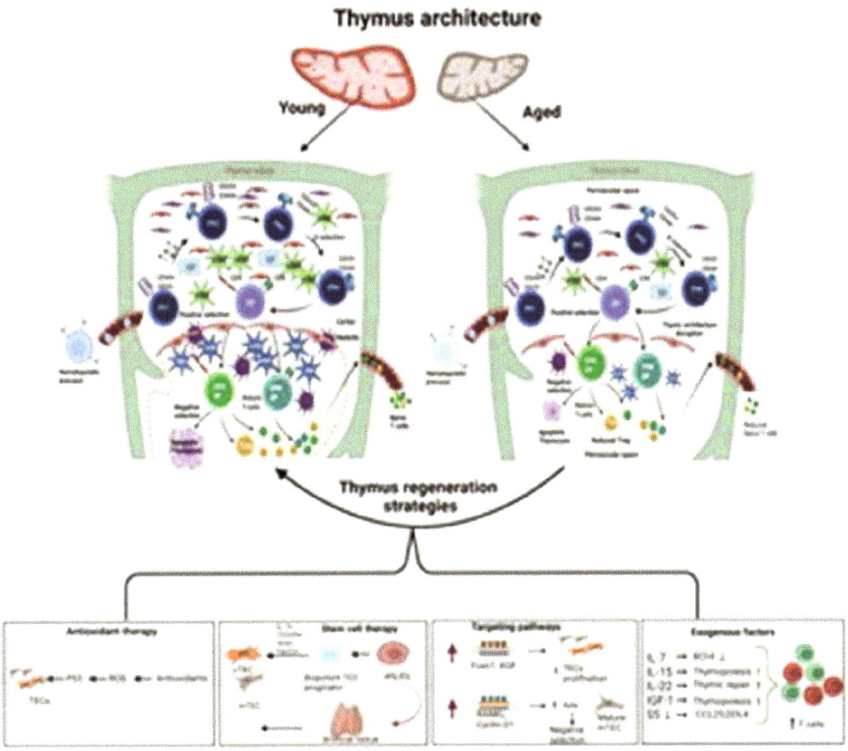

FIGURE 12.1 *Thymus structure. This image is reproduced from [10] with permission from Taylor & Francis (license number 5723031184526).*

and myeloid-derived suppressor cells [11]. Moreover, bone marrow is a preferred site of metastatic spread for several human tumors. It functions as an immune regulatory organ that may finely regulate immunity and could be a therapeutic target for immune vaccination and immunotherapy [12].

12.1.3 Secondary Immune Organs

12.1.3.1 The Spleen

The spleen, the body's largest secondary lymphoid organ, functions as both a component of the immune system and a blood filter. It consists of three primary histological components: the marginal zone, red pulp, and white pulp. The spleen's main activities are mostly restricted to specific anatomical regions. The splenic red pulp filters blood by removing unnecessary erythrocytes and platelets [13]. Red pulp macrophages also fight blood-borne infections. The marginal zone and white pulp are the primary sites of innate and adaptive immune responses. The marginal zone, located at the interface of red and white pulp, is dominated by B cells, dendritic cells, and macrophages. These cells are crucial for both innate immunity and antigen presentation and absorption, which triggers the adaptive immune response [14].

12.1.3.2 Lymph Nodes

Lymph nodes (LNs) are key sites where the body's defense against dangerous antigens begins. The lymphoid system comprises three main kinds of lymphoid tissues: primary, secondary, and tertiary. The primary lymphoid organs are the thymus and bone marrow. Secondary lymphoid organs include the LNs, spleen, and Peyer's patches (PPs). Tertiary lymphoid tissues are diverse structures that develop during inflammation [15].

LNs are distributed throughout mammalian bodies. They share a common design and population of cell types. Their role is to filter the lymph that exits the draining region and detect antimicrobials in the lymph. This process either leads to tolerance development for benign antigens or triggers an immune response to pathogenic antigens. Briefly put, antigen-loaded dendritic cells migrate from the draining area via afferent lymphatics and present antigens to T lymphocytes in T-cell region or paracortex. Activated T cells proliferate and develop T-cell receptor specificity for the presented antigens. Some activated T lymphocytes, called T helper cells, move into the cortex or B-cell region to support B cells. These antigen-specific B cells differentiate into plasma cells that efficiently produce antibodies. All activated effector cells, including plasma cells and $CD4^+$ or $CD8^+$ T cells, migrate to the medulla and exit the LN via efferent lymphatics or the bloodstream to reach the inflamed or threatened region of their specific draining area [16].

12.1.4 Components of the Immune System

The primary host defense mechanism in all multicellular organisms is the immune response, which has evolved to shield the host against environmental toxins and infection agents such as bacteria, fungi, viruses, and parasites. The immune system is broadly categorized into innate immunity and adaptive immunity [17].

12.1.4.1 The Innate Immune Response and Components of Innate Immune System

Compared to the adaptive immune response, the innate immune system is more evolved. The adaptive immune response builds immune memory. Components of the innate defense system include anatomical and physical barriers, effector cells, soluble mediators, cell receptors, and antimicrobial peptides. The mucosa and skin act as effective immunological barriers, with the skin providing both physical and chemical protection [18].

Keratinocytes, the main component of the outermost epidermal layer, are encased in extracellular matrix proteins and tightly connected by desmosomes. Beyond their physical barrier function, keratinocytes possess PRRs and can generate antimicrobial peptides and cytokines, which cause microbial death and trigger inflammatory cascades, respectively [19]. Furthermore, sebaceous glands attached to hair follicles secrete fatty acids, making the surrounding environment acidic and unfriendly to microorganisms. The continuous epithelium of the respiratory, digestive, and genitourinary systems prevents microbes from entering the host. These epithelial cells also produce defensins and other antimicrobial peptides. Inflammatory cytokines like interleukin-1 (IL-1) and tumor necrosis factor-alpha (TNF-α), secreted by immune cells such as macrophages in response to invasive infections, further stimulate defensin synthesis.

Granulocytes, fast-responding innate immune cells, serve as the first line of defense against infectious pathogens. The innate immune system also initiates adaptive immunity, which provides long-lasting immunological memory, to mount an efficient defense against pathogenic agents [20].

12.1.4.2 Effector Cells of Innate Immunity

Cellular components include phagocytic cells, endothelial and epithelial cells, platelets, innate lymphoid cells, and NK cells. Phagocytic cells comprise granulocytes, monocytes, macrophages, eosinophils, mast cells, and basophils. These cells not only perform phagocytosis but also participate in inflammation. Most cell components have microbicidal capabilities because they can express PRRs and produce cytokines. These cells with innate immunity effector pathways impact both innate and adaptive immune systems [21].

12.1.4.3 Granulocytes

Granulocyte effector cells dominate the early or acute phase of the innate immune response. Their primary job is to recognize, absorb, and eliminate microbial pathogens using oxidative processes, enzymes (e.g., lysozyme, collagenase, and elastase), and receptors. This cell type includes mast cells, basophils, eosinophils, and neutrophils [22].

12.1.4.4 Neutrophils

Neutrophils are prevalent and active during phagocytosis and inflammation. Polymorphonuclear neutrophils (PMNs) are the first cell line recruited to inflammation sites after chemotactic stimulation [23]. These stimuli include leukotrienes (e.g., L-B4), which have autocrine and paracrine effects on other neutrophils, chemokines like IL-8, and complement factors like C5a. These compounds are recognized by certain receptors or PRRs, facilitating migration to damage sites.

Neutrophils possess Fc or complement receptors (FcRs or CRs) that recognize complement factors or immunoglobulin Fc fractions, respectively. This enables phagocytosis of bacteria opsonized (tagged) by complement (mostly C3b or iC3b) or antibodies (primarily Immunoglobulin G (IgG is a type of antibody that constitutes approximately 75% of the total antibodies in human serum. It plays a crucial role in the immune response by binding to pathogens such as viruses and bacteria, facilitating their neutralization and elimination)). Neutrophil granules contain various enzymes that activate microbicidal systems and damage microbial pathogens through oxygen-dependent and -independent pathways.

Granulocyte colony-stimulating factor (G-CSF) is a pro-inflammatory cytokine crucial for PMN recruitment and activation. In mice, three distinct PMN subsets have been identified based on surface antigen expression and cytokine production, each associated with macrophage activation [24].

12.1.4.5 Eosinophils

Eosinophils, less common than neutrophils, are found in the respiratory, urinary, and gastrointestinal systems [25]. Their effector action is facilitated by the degranulation and release of histamine, cationic proteins, major basic proteins, sulfatases, and chemotactic agents, including prostaglandins and leukotrienes. Immunoglobulin E (IgE is a class of antibody that plays a critical role in the immune response, particularly in mediating allergic reactions and defending against parasitic infections) or other chemotactic proteins, such as IL-5, promote degranulation. Eosinophils primarily eliminate microbiological infections, especially parasites, and work closely with mast cells to trigger allergic reactions [25].

12.1.4.6 Mast Cell and Basophils

While not naturally phagocytotic, these cells carry several receptors, including the IgE receptor. Basophils are present in lower percentages in blood compared to other granulocytes. Mast cell granules contain histamine, serotonin, and heparin, and are primarily found in tissue mucosa [26]. These cells can also produce a range of cytokines that intensify the inflammatory process, particularly in the early stages. They play a role in allergic reactions and viral infections. Most mast cells are found in connective tissue. Their cell membranes express TLR-1, 2, 4, and 6, CRs, and mannose receptors. They also generate numerous inflammatory mediators, including platelet activator factors, proteases, IL-8, TNF-α, and antimicrobial peptides [27].

12.1.4.7 Monocytes/Macrophages

Macrophages and monocytes are important components of both innate and adaptive immunity. Monocytes in peripheral blood are mobile and can travel to inflammation sites, where they differentiate into tissue macrophages [28]. Once within the tissue, macrophages perform several functions: (i) phagocytic microbicidal mechanisms that are both oxygen-dependent and -independent; (ii) antigen presentation and lymphocytes activation; (iv) immune response regulation; (v) synthesis of matrix metalloproteinases, collagen, and other extracellular matrix proteins, facilitating tissue reorganization; and (vi) production of cytotoxic factor involved in immunity against tumors [29].

12.1.4.8 Dendritic Cells (DCs)

Dendritic cells are professional antigen-presenting cells (APCs) and innate components of the immune system [30]. They activate T-cell responses to provide protection against microbial infection through various mechanisms. They patrol mucosal and skin surfaces. Antigen uptake involves numerous processes, such as phagocytosis, endocytosis, pinocytosis, and micropinocytosis. Dendritic cells can transport antigens from the periphery to the major lymph nodes, the primary site of antigen presentation. These APCs bridge the innate and adaptive immune responses by processing and presenting antigens via major histocompatibility complex (MHC) class II molecules. Dendritic cells also regulate T-cell immune responses, help establish peripheral immunological tolerance, and act as effector cells in innate defense against a range of microbial diseases. These diverse roles depend on the variety of dendritic cell subsets, which include progenitors (pre-dendritic cells) and immature dendritic cells. There are two types of precursor dendritic cells [31].

12.1.4.9 Natural Killer Cells

Natural killer (NK) cells play an immunomodulatory role in cell-mediated immune responses due to their cytotoxic action [32]. They also contribute to immunological surveillance and antimicrobial defenses by regulating tumor development and maintaining immunological homeostasis. These cells use a "negative recognition" mechanism. While T or B cells recognize antigens via MHC, NK cell activation occurs when an antigen is not recognized in this manner. NK cell receptors recognize MHC antigens and inhibit the lytic activity of these cells, functioning as inhibitory receptors. These cells identify malignant or infected (primarily virus-infected) whose expression of MHC molecules has diminished, changed, or stopped.

NK cells distinguish normal host cells through CD94-NKG2A inhibitory receptors and killer cell immunoglobulin-like receptors (KIRs), which recognize MHC class I molecules on normal cell surfaces. Binding of these receptors prevents NK cells from lysing and secreting cytokines. Furthermore, NK cells express PRRs such as TLR-2, 3, 4, 5, 7, and 8 and possess granules containing granzymes and perforins that induce target cell lysis or death. Activated NK cells release chemokines, interferons (IFN-γ is a group of signaling proteins produced by host cells in response to pathogens, particularly viruses. They play a crucial role in modulating the immune response, enhancing the ability of cells to resist infections and activating immune cells to combat viral and bacterial threats), TNF-α growth factors, and IL-5, 10, and 13 [33].

12.2 PHAGOCYTOSIS AND COMPLEMENT SYSTEM

Macrophages, as professional phagocytes with strong phagocytosis ability, pro-inflammatory cytokine generation, and antigen presentation capabilities, occupy a unique position between innate and adaptive immunity components. Phagocytosis, facilitated by host-generated opsonin like complement and antibodies, is an essential mechanism for eliminating bacteria, dying cells, and other foreign particles [34]. Early phagocytosis studies established that complement receptor-mediated phagocytosis was less pro-inflammatory than FcγR-mediated uptake in macrophages [35].

12.2.1 Complement System Receptor

Our immune system and physiology are intricately linked to the complement system. Complement activation produces several chemicals that converge on target opsonization for phagocytosis and immune system activation through soluble anaphylatoxins [36]. This reaction is crucial not only for controlling bacteria and eliminating dead cells but also for promoting tissue repair and adaptive defense. CRs, found on leukocytes and parenchymal cells, are essential for many of these actions because they recognize complement-derived molecules produced to facilitate leukocyte recruitment, pathogen phagocytosis, and immune complex removal [34].

12.2.1.1 The Complement Cascade Pathway

The NF-κB pathway requires the degradation of IκBs, particularly IκBα, for the translocation of NF-κB complexes to the nucleus. IκB kinase (IKK), a trimeric complex composed of IKKα, IKKβ, and IKKγ, mediates this degradation. There are other ways to activate specific NF-κB members. The non-canonical NF-κB pathway processes p100 to activate the RelB/p52 NF-κB complex, rather than degrading IκBα. This pathway is linked to lymphoid cancers and controls crucial biological processes. Understanding the regulation of non-canonical NF-κB activation has therapeutic value. The non-canonical NF-κB pathway differs significantly from the canonical pathway in its signaling mechanisms [37].

Figure 12.2 illustrates the three primary pathways of the human complement system: the lectin, alternative, and classical pathways [38]. Scissors are markers of proteolytic cleavage. Abbreviations include mannose-binding lectin (MBL), factor B (FB), factor D (FD), membrane attack complex (MAC), and MBL-associated serum protease (MASP) [34].

12.2.2 Complement Receptors' Function in Leukocyte Attraction

The host's acute inflammation response is triggered by microbial infection or tissue injury [39]. The inflammatory response aims to neutralize harmful substances and restore equilibrium, often proving advantageous in treating such pathological conditions. When cells that reside in the tissue near the inflammation site identify DAMPs or PAMPs, they produce and release inflammatory chemical mediators. These mediators, including prostaglandins, cytokines, chemokines, histamine, and leukotrienes, not only cause typical inflammatory symptoms (erythema, heat, pain, and edema) but also cause leukocyte extravasation and chemotaxis within the nearby inflammatory tissue. Upon reaching the inflammatory site, leukocytes neutralize the trigger through mechanisms such as phagocytosis, reactive oxygen species (ROS) production, and degranulation [40].

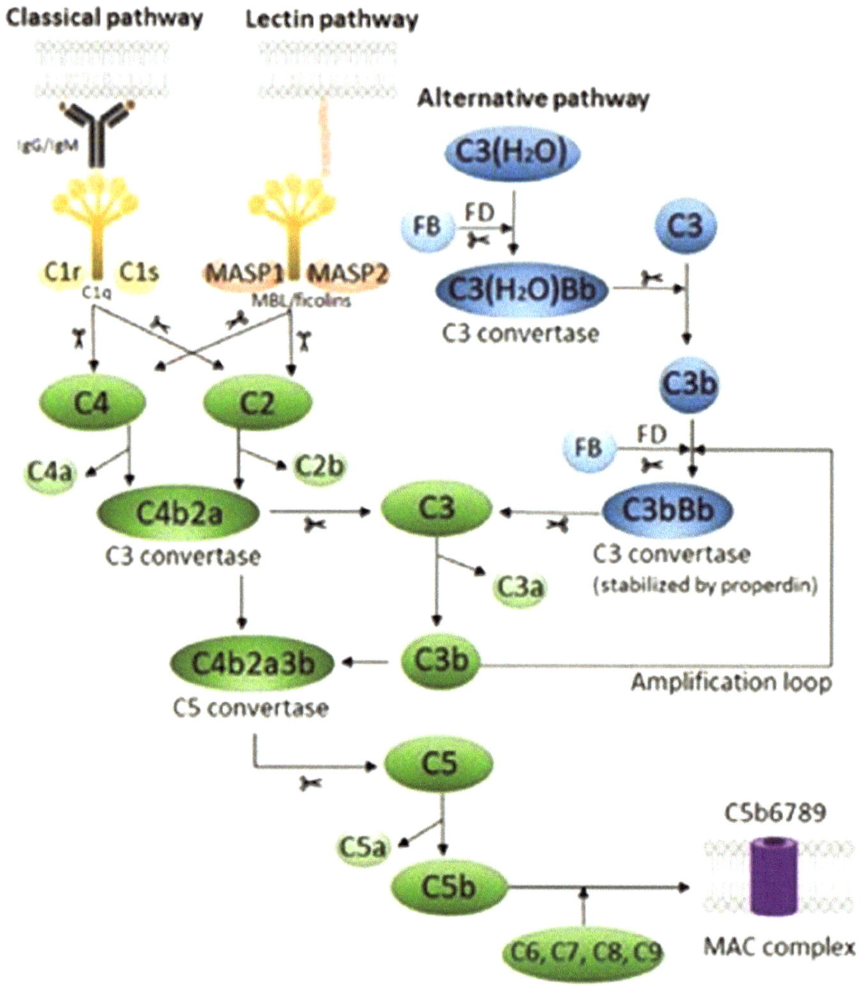

FIGURE 12.2 Complement cascade pathways. This image is reproduced from [34] with an open access mandate, and published under the Creative Commons Attribution license (CC-BY).

12.2.3 Leukocyte Adhesion During Resolving Inflammation

Tissue trauma or infection triggers the quick start of an acute and healing inflammation response in vertebrates. The composition of leukocytes in the tissue infiltrate changes over time, matching the specialized functions of various leukocyte subsets with the needs for tissue repair and adaptive immune response development. Molecular mechanisms facilitate leukocyte recruitment in post-capillary venules, the primary site of leukocyte recruitment during acute inflammation. When endothelial cells (ECs are specialized cells that line blood vessels and play a crucial role in vascular biology, including the regulation of blood flow, barrier function, and immune responses. They are involved in various physiological processes such as inflammation, angiogenesis, and the maintenance of homeostasis within the vascular system) are activated by local inflammatory mediators like histamine or cytokines (e.g., TNF-α), they express specific adhesion that forms temporary interactions with counter-ligands on the surface of leukocytes. E- and P-selectin and vascular adhesion molecule-1 (VCAM-1), members of the immunoglobulin superfamily (IgSF), are the main

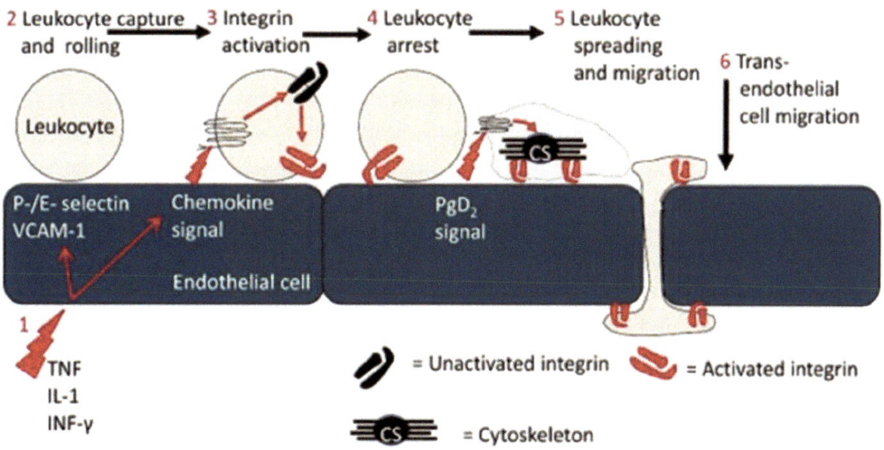

FIGURE 12.3 Leukocyte adhesion during resolving inflammation. This image is distributed under the terms of the Creative Commons Attribution license and reproduced with permission from Taylor & Francis.

receptors that capture leukocytes from rapidly flowing blood. These bonds form and dissolve successively, supporting rolling, a distinct and dynamic type of adhesion that doesn't require leukocyte activation. However, leukocyte migration through the vessel wall and into the inflammatory tissue requires an activating stimulus. Rolling leukocytes are primarily activated by chemokine family peptide agonists, which serve to stabilize adhesion [41] (Figure 12.3).

(1) Inflammatory cytokines activate endothelial cells at inflammation sites, resulting in the expression of adhesion molecules and chemokines that recruit leukocytes. (2) Selectin receptors and VCAM-1 recruit leukocytes from blood, supporting rolling adhesion. (3) Chemokine signals activate integrins on rolling leukocytes. (4) Adhesion is stabilized, and leukocytes firmly attach to endothelial cells. (5) Additional signals cause leukocyte cytoskeleton remodeling, leading to shape change and migration. (6) Leukocytes migrate through endothelial cells and into tissue [41].

12.3 ADAPTIVE IMMUNE RESPONSE

The adaptive immune response is facilitated by the production of pathogen-specific (antigen-specific) B and T lymphocytes [42]. This process involves gene rearrangement. Over time, researchers have identified and characterized various classes of PRRs. These PRRs, which include DNA receptors (cytosolic sensors for DNA), NOD-like receptors (NLRs), RIG-I-like receptors (RLRs), and toll-like receptors (TLRs), are crucial for detecting both extracellular and intracellular pathogens. They can detect a broad range of substances, including nucleic acid proteins, lipids, and carbohydrates [43].

Dendritic cells undergo rapid differentiation upon encountering various stimuli, which encompass microbes, endogenous "alarmins," and other lymphocytes. These stimuli trigger an innate response characterized by the release of numerous cytokines and chemokines. The mechanisms underlying dendritic cells' robust reactivity and the conversion of their prompt innate responses into more sustained adaptive responses, including memory establishment, remain enigmatic. Future could focus on elucidating the events following the interaction between dendritic cells and T cells, particularly when dendritic cells begin expressing cytokines and employing molecules like CD40 and CD70 to foster potent adaptive resistance. Dendritic cells comprise distinct subsets, each with unique receptors for antigen uptake and signaling, different antigen processing pathways, and diverse functional outcomes. Research in this area may uncover whether dendritic cell subsets have evolved to effectively combat specific categories of infections and other antigens [44].

T cells play crucial roles in immune system protection against pathogenic infections and tumor growth [45]. To produce the proper immune response, T cells must be activated by the T-cell receptor, which recognizes specific antigens presented in the context of the MHC. This differentiation also enables T-cell responses to distinguish between "self" and "non-self." Throughout a T cell's lifespan, TCR interactions control the cell's ability to survive, proliferate, and/or differentiate. As such, TCR signaling has important implications for both the appropriate course and effectiveness of the immune system.

Various ligands, such as angiotensin II, TGF-β, insulin, and Epidermal Growth Factor (EGF is a 53-amino acid polypeptide that stimulates cell growth, proliferation, and differentiation by binding to its receptor, EGFR. It plays a crucial role in various biological processes, including tissue repair and development, and is involved in signaling pathways that regulate cellular functions), can stimulate cognate receptors and stimulate the intracellular production of ROS, which are vital secondary messengers required for ligand-mediated regulation of gene expression, protein kinase activation, and proliferative responses. Growing evidence also indicates that ROS generation stimulates cell growth and transformation. Recent research on the redox regulation of T-cell signaling and responses has expanded our understanding that T cells also generate ROS in response to TCR stimulation. Furthermore, these reactive species are essential for controlling T-cell activity, gene expression, and signal transduction [46].

T lymphocytes can be divided into two categories based on their cell surface receptors: $CD4^+$ and $CD8^+$ [47]. The TCR and MHC molecules on lymphocytes and target cells, respectively, form an immunological synapse that depends on the $CD4^+$ and $CD8^+$ receptors. A common term for $CD8^+$ T cells is "cytotoxic T cell." Upon antigen-induced activation, helper T cells differentiate into subgroups, including type 1 helper T (Th1) cells and type 2 helper T (Th2) cells. These subgroups produce different cytokines, providing additional control over $CD4^+$ T cells. Th1 cells secrete interferon-gamma, which stimulates cellular immune responses and inhibits Th2 cells. On the other hand, Th2 cells secrete interleukin-4 (involved in the activation of B cells), transforming growth factor beta (TGF-β), and interleukin-10 (which inhibits Th1 cells). Another subgroup, regulatory T cells, known as type 1 regulatory T (Tr1) cells or type 3 regulatory T (Tr3) cells, secrete immunosuppressive cytokines such as interleukin-10 and TGF-β.

The development of Th1, Th2, and other cytokine-secreting T cells following antigen activation plays a significant role in regulating T-cell function. Collectively, these cytokine-secreting T cells form a subset of typical T cells that control immunological reactions to both foreign and self-antigens. They are antigen-specific but mediate nonspecific suppression through the action of their cytokines. By controlling the magnitude and class of immune responses, these cells contribute to the nonspecific suppression of pathogenic autoimmunity [48].

12.3.1 TCR-Mediated Signaling: Antigen Receptor and Tyrosine Kinases

TCR signaling is initiated through the recruitment and activation of protein tyrosine kinases (PTKs) belonging to the Src, SYK, and thymic epithelial cells (TEC) families [49]. This facilitates the connection of the SYK family member ZAP70 to the immunoreceptor tyrosine-based activation motifs (ITAM) via its tandem Src-homology-2 (SH2) domains. Src PTKs (likely Lck in T cells) phosphorylate the ITAMs of signal transduction antigen receptor subunits. After binding to the ITAM, ZAP70 undergoes tyrosine phosphorylation and activation, subsequently phosphorylating ZAP70 substrates like the adapters SLP76 and LAT (transmembrane protein that plays a crucial role in T-cell receptor (TCR) signaling by serving as a docking site for various signaling molecules, facilitating the propagation of signals essential for T-cell activation and function) [50].

It is crucial to understand that ITAM phosphorylation is a continuous event that can be inhibited by tyrosine phosphatases such as CD45, rather than a binary process. TCR triggering is thought to alter this equilibrium, excluding phosphatase molecules from the TCR-ligand complex. The TCR-MHC interaction is believed to span 150 nm, which determines the distance between T cells and APCs around the TCR-MHC complex. This spatial arrangement excludes large extracellular domain membrane molecules from the TCR-MHC interaction, while allowing smaller costimulatory receptors, like CD28, to participate.

In addition to maintaining baseline levels of active Lck, CD45 is a strong negative regulator of TCR signaling. Due to its large extracellular domain, CD45 is excluded from the proximity of a TCR-MHC complex based on size constraints. T-cell activation can be viewed as a process that disrupts the existing balance between phosphorylation and dephosphorylation through cell-to-cell interaction. This disruption ultimately leads to the activation of cytosolic tyrosine kinases [51].

12.3.2 Signaling After Tyrosine Kinases in T Cells

The primary substrates for antigen receptor-regulated tyrosine kinases are adaptors like SLP-76 (a 76,000 MW protein associated with the SH-2 domain on lymphocytes) and LAT (the linker of activated T cells) [50, 52]. These adaptors create scaffolds that all signal transduction molecules to perform their effector functions, either directly or through the allosteric control of co-assembled regulatory proteins. Tyrosine phosphorylation of adaptors during T-cell activation links antigen receptors to a series of signaling pathways. The most important of these are the GTPase signaling networks of the Ras and Rho families. Additionally, antigen receptor-regulated tyrosine kinases control inositol phospholipid metabolism, which governs intracellular calcium and the activity of several serine/threonine kinases, such as phosphatidyl inositide-3 kinase (PI-3K) and members of the PKC family [53].

12.3.3 The TCR Signaling Network

TCR signaling is essential for the adaptive immune response. Due to its complex connectivity, the TCR signaling network can serve as a model system to study a variety of physiological responses, including cytoskeletal remodeling and gene regulation. Recent years have seen significant advances in understanding TCR signaling, particularly in elucidating the networks connecting early signaling events to subsequent transcriptional responses. Additionally, our knowledge of how other cell-surface receptors influence TCR signaling and T-cell effector activities has greatly improved.

Research on the cell biology context of TCR signaling has become increasingly interesting, providing an overview of the essential components of TCR signaling networks. Early TCR signaling influences the development of the immunological synapse, a stable and specialized junction between T cells and APCs. This process also involves costimulatory receptors of the CD28 family. Furthermore, early TCR signaling is linked to gene expression.

TCR signaling causes the microtubule-organizing center (MTOC) to polarize toward the immunological synapse. This MTOC reorientation appears to require the negatively directed microtubule motor dynein. Microtubules radiate from the MTOC with their positive ends facing inward and negative ends facing outward. As a result, dynein at the immunological synapse can attach to the negative ends of microtubules and "reel" the MTOC toward itself. However, the exact mechanism by which dynein is recruited to the immunological synapse remains unclear. A recent study suggests that dynein may bind directly to ADAP, although this finding seems to be specific to human T cells [54]. Other actin cytoskeleton regulators, such as IQ-motif-containing GTPase-activating protein homolog (IQGAP) and formin diaphanous (Dia), have also been linked to MTOC reorientation [55].

12.3.4 TCR Signal Transduction

The four main modules of TCR-mediated signaling are as follows:

(i) Initial signaling characterized by the recruitment and phosphorylation of lymphocyte-specific PTKs

(ii) The formation of signalosome complexes that include CD6 (T-cell differentiation antigen), LAT (linker for activation of T cells), or both
(iii) The activation of crucial mediators such as protein kinase C θ (PKCθ)
(iv) Downstream signaling via inhibitor of NF-κB kinases (IKK) kinases, mitogen-activated protein kinases (MAPKs), and Ca2+-mediated signaling activates inducible transcription factors, which then promote changes in gene expression.

NFAT (nuclear factor of activated T cells), NF-κB, AP-1 (activator protein 1), and CREB (cyclic adenosine 3′,5′-monophosphate response element-binding protein) are the main transcription factors that enhance TCR-inducible gene expression. Robust. T-cell activation and differentiation result from an appropriate balance in the activation of these transcription factors. Conversely, abnormalities could lead to a compromised immune system, resulting in fatigue or cell death.

For example, NF-κB and NFAT are transcription factors of the Rel family and may compete for binding motifs on DNA targets. The interaction between NFAT and NF-κB affects the cell phenotype. When NF-κB activity is higher than that of NFAT, the cell exhibits a pro-inflammatory response, whereas an anti-inflammatory response is elicited when NF-κB activity is absent. β-catenin and serum response factor (SRF) are two additional transcription factors that respond to TCR activation [56].

12.3.5 Components and Structure of TCR Complex

The core TCR complex comprises six Cluster of Differentiation 3 (CD3) chains and two TCR chains [57]. Additional components include kinases, ligands, and coreceptors. Four TCR genes—TCRα, TCRβ, TCRγ, and TCRδ—are expressed in the human genome and combine to form two different heterodimers: TCRα/TCRβ or TCRγ/TCRδ. A small percentage (0.5–5%) of T lymphocytes (γ δ T cells) express TCRγ and TCRδ isoforms. The bulk of mature T cells, known as α β T cells, express TCRα and TCRβ isoforms.

TCR chains consist of a shortened cytoplasmic tail, an extracellular portion, and a transmembrane region. The extracellular region contains a constant immunoglobulin-like (C) domain, a variable immunoglobulin-like (V) domain, and a connecting peptide. The V domain, which functions as the antigen recognition site, is assembled by RAG1 and RAG2 recombinases. The C domain mediates interactions with CD3 chains. The C domain and connecting peptide differ significantly between α β and γ δ chains, reflected in surface shape, charge distribution, and TCR complex formation.

Both α β and γ δ complexes contain three dimers of the CD3 proteins: a ζ ζ homodimer, a γ ε heterodimer, and a δ ε heterodimer. These CD3 proteins are necessary for the full localization of TCR on the cell surface and interact with TCR through non-covalent hydrophobic contacts [58].

12.4 B Cells

B cells circulate through the bloodstream and lymph, monitoring the body for indications of illness. Their antigen recognition receptor, the B-cell receptor (BCR), is essentially an antibody that is attached to a cell. When a B cell is activated by its targeted protein, it differentiates into a plasma cell and starts to secrete large quantities of antibodies that can bind to and neutralize the target protein. B cells are further categorized into the B1 and B2 subpopulations.

B cells are generally considered positive regulators of immune responses [59]. They act as essential cellular adjuvants promoting optimum $CD4^+$ T-cell activation, function as effector cells giving rise to antibody-producing plasma cells, and mediate various other immune functions. However, B cells also produce autoreactive antibodies, which can lead to harmful outcomes like systemic lupus erythematosus. Additionally, B cells contribute to immunoregulation by producing cytokines such as IL-4, IL-6, IL-10, interferon-γ (IFN-γ), and TGF-β [60].

The thymus is the primary immune organ responsible for producing self-restricted, self-tolerant T lymphocytes. It consists of developing T cells (thymocytes) supported by a complex network of resident cells, including mesenchymal, vascular, dendritic, and thymic epithelial cells. These cell types form several functional microenvironments that guide and support thymocyte development from immature progenitors into mature cells, restricting immigrating lymphoid progenitor cells (LPCs) to the T-cell fate. This process shapes the emerging repertoire, making it both self-restricted (recognizes peptide antigens in the context of "self" MHC) and self-tolerant (does not attack the body's own cells) [61].

Thymic involution compromises immunological function, resulting in the collapse of the peripheral TCR repertoire and decreased production of new naïve T cells. The age-related decline in T-cell production caused by thymic involution is a significant risk factor for numerous infectious diseases and cancers in humans. Furthermore, thymic involution is associated with heightened vulnerability to several diseases, such as cancer, infections, and autoimmune disorders [62].

12.4.1 Receptor Signaling in B Cells

Over the past three decades, extensive research has been conducted on receptor signaling in B cells, yet a comprehensive understanding of the intricate mechanisms underlying this process remains incomplete. Numerous researchers have conducted comprehensive analyses and evaluations of this phenomenon [63].

The BCR activation process begins with antigen binding. This triggers the initial kinase in the BCR signaling cascade, LYN (LYN is a member of the Src family of protein tyrosine kinases primarily expressed in hematopoietic cells, where it plays a crucial role in regulating cell activation and signaling pathways. It is involved in both promoting and inhibiting immune responses, depending on the context of activation, and is essential for processes such as B cell receptor signaling and myeloid cell function), a member of the Src-kinase family, to phosphorylate the ITAM. Subsequently, SYK is recruited to the phosphorylated Igα–Igβ heterodimer via its SH2 domain. The aggregation of ligand-coupled BCR molecules can enhance their engagement with Src-family PTKs, potentially impacting downstream signaling [64]. Following SYK activation, a group of proteins associated with the adaptor protein B-cell linker (BLNK, SLP-65) relay the BCR signal. The phosphorylated form of BLNK serves as a scaffold for the integration of other components, including phospholipase C-gamma 2 (PLCγ2), VAV 1, and Bruton's tyrosine kinase (BTK). The BCR signal is indirectly controlled by at least two molecules not connected with the receptor: C-terminal Src tyrosine kinase (CSK) and B220 [65].

Furthermore, after BCR ligation, tyrosines in the cytoplasmic tail of CD19 are phosphorylated by LYN, creating binding sites for the SH2 domains of the p85 adaptor subunit of PI-3K, as well as other effectors that contain SH2 domains. Activated PLCγ2 cleaves the phosphoinositide PI(4,5)P2 into two second messengers: I(1,4,5)P3 and diacylglycerol (DAG is a glycerolipid consisting of two fatty acid chains esterified to a glycerol backbone, serving as an important second messenger in cellular signaling pathways. It plays a crucial role in activating protein kinase C (PKC) and is involved in various metabolic processes and signal transduction mechanisms). I(1,4,5)P3 production induces the release of calcium ions from intracellular and extracellular compartments. Calcium ion signaling is necessary for protein kinase C (PKC)-facilitated activation of transcription factors, including NF-κB and NFAT [66].

DAG functions as a model activator of PKC isotypes that control the functions of the extracellular signal-regulated kinase (ERK), c-Jun NH2-terminal kinase (JNK/SAPK), and p38 MAPK family members. These complex processes collectively result in B cell activation, antigen presentation, cytokine synthesis, and cell proliferation and differentiation. BCR signaling has significant repercussions throughout the course of B cell maturation and following antigen encounter [67].

12.5 CYTOKINES AS HORMONES OF THE IMMUNE SYSTEM

As soluble factors generated by one cell and acting on another to alter the target cell's function, cytokines can be considered the "hormones" of the inflammatory and immune systems [68]. However, some cytokine characteristics do not fit this description. For instance, in lymphoid organs, cytokines primarily have a local role in paracrine and juxtracrine activities within the humoral and cell-mediated immune response to antigens. Additionally, while cytokines are produced by most cells, hormones are the primary product of specific tissues or cells.

Nevertheless, cytokines exhibit endocrine-like effects during the systemic inflammatory response, activating various processes such as the synthesis of hepatic acute-phase proteins and the mobilization of bone marrow-derived leukocytes, both occurring far from infection or inflammation sites. Cytokines are significantly more potent than hormones on a molar basis. For instance, the cytokine IL-1 can promote cyclooxygenase-2 (COX-2) gene expression and synthesis at concentrations of 10 pM, whereas cytokine IL-12 can produce IFN-γ at 20 pM [69].

12.5.1 CYTOKINES PRODUCED BY THl OR TH2 LYMPHOCYTES

T lymphocytes are essential for regulating immune responses. They identify antigens complexed with MHC molecules or APCs through their TCRs, which comprise two distinct (heterodimer) glycoprotein chains. The TCR-associated CD3 complex, along with other molecules such as CD2, $CD4^+$, and $CD8^+$, physically connect to the TCR and transduce the intracellular signaling pathway that activates resting T cells. TCR binding triggers the activation of associated molecules, leading to the production of many genes, including growth factor receptors (e.g., the IL-2 receptor).

Several cytokines mediate the activation and expansion of T cells. These can be separated into two groups: cytokines that directly stimulate T-cell growth and proliferation (like IL-2, IL-4, and IL-7) and those necessary for optimum stimulation but not directly causing T-cell proliferation and expansion (like IL-la/fl, IL-6, TNFα, TNFβ, and GM-CSF). TH2 cells secrete IL-4, IL-5, IL-6, and IL-10, while TH1 lymphocytes produce IFNγ and IL-2. While examples of both TH1 and TH2 human T-cell subtypes exist, most human T-cell clones do not fall into these categories.

The exact factors causing TH precursors to differentiate into TH1 or TH2 cells are unknown, but they may be related to the type of antigenic trigger. Allergens promote TH2 differentiation, whereas tuberculin favors a TH1 response. TH1 promotes the activation of certain cytotoxic $CD8^+$ lymphocytes and NK cells by producing IL-2 and IFNγ, resulting in delayed hypersensitivity reactions. TH1-released TNFβ and IFNγ enhance macrophages' capacity to kill bacteria and tumors. TH2 cells produce IL-4, 5, and 10, which strongly stimulate B lymphocyte differentiation and proliferation. TH1-derived IFNs inhibit IgE synthesis, whereas IL-4 promotes it. IL-2 and IFNγ can also stimulate B lymphocytes cell proliferation and differentiation, while encouraging IgG2a production.

TH1 cells and TH2 lymphocytes mutually regulate each other. TH2 cells prevent TH1 and NK cell cytokine production by releasing IL-4 and IL-10. Conversely, TH1 cells inhibit eosinophil and mastocyte proliferation and differentiation through IFNγ production, countering the effects of TH2-produced IL-3, 4, 5, and 10 [70].

12.5.2 CYTOKINE TYPES

IL-1, originally known as lymphocyte activation factor, is a macrophage product that stimulates T and B cells. IL-2, previously called T-cell growth factor, enhances T-cell proliferation. The term "interleukin" was first used in 1979 to describe two leukocyte products with molecular weights between 15,000 Daltons and 17,000 Daltons [71]. This term was chosen because these chemicals were both produced by and affected leukocytes.

IL-2 has a specific biological function, directly stimulating T cells, while IL-1 is more generic, mediating several aspects of the systemic acute-phase response. IL-1 affects numerous non-leukocytic targets, including the liver, pancreas, bone, cartilage, muscle, synovial fibroblasts, and brain, making it more than just an interleukin. IL-2 is a single molecule and a true growth factor with the apparent exclusive function of increasing T-cell populations, while IL-1 is part of a polypeptide family [69].

12.5.2.1 Interleukin-l

Microbial invasion, injury, immunological reactions, neoplastic changes, and inflammatory processes constantly threaten host survival. The host's response to these stimuli is commonly known as the acute-phase response. These reactions are distinguished by alterations in metabolic, endocrinologic, neurological, and immune systems. The acute-phase spectrum of changes includes increased hepatic acute-phase protein production, leukocytosis (characterized by elevated circulating immature neutrophils), reduced plasma iron and zinc levels, negative nitrogen balance, and various endocrinologic abnormalities. Neurologic abnormalities include fever, increased lethargy or drowsiness, and decreased appetite.

Aspects of dysregulated immunity vary; polyclonal B-cell activation is the most common, but there may also be an increase in suppressor cells and anorexia. Erythrocyte sedimentation rates are enhanced due to hypergammaglobulinemia and high hepatic proteins. Patients with concealed infections or chronic conditions such as cancer, inflammatory bowel disease, or rheumatoid arthritis also present with acute-phase alterations. The production of IL-1, a family of polypeptides crucial in generating elements of the acute-phase response either directly or indirectly, is undoubtedly the foremost event in initiating the reaction. Phagocytic cells are the primary source of IL-1, which is produced, transported, and affects distant organ systems. As a result, IL-1 serves as a hormone mediating the host's reactions to infection and inflammation.

Blood monocytes and phagocytic lining cells in the liver, spleen, and other organs are the main producers of IL-1. Furthermore, specialized cells, including brain astrocytes, brain microglia, keratinocytes, B cells, large granular lymphocytes, gingival and corneal epithelial cells, renal mesangial cells, and serosal lining cells, produce IL-1. The IL-1 produced by these specialized cells likely has its primary effects in those tissues, contributing to the pathogenic mechanisms of many disease processes [72].

12.6 MOLECULAR MECHANISMS OF INFLAMMATORY SIGNALING

12.6.1 Intracellular Signaling Pathways

Cells use signaling pathways to sense changes in their surroundings, process internal or external signals, and adjust transcriptional activity, metabolism, or other regulatory processes in response. These pathways are essential not only for differentiation and cell fate determination but also for adaptability and survival in a variety of environments. Signaling pathways are crucial for brain plasticity, oncogenesis, and development in multicellular animals. Common elements of signaling pathways include receptors, ERK or MAPK cascades, G proteins, and small G proteins. The architecture of these components appears to be the same across all kingdoms of life [73] (Figure 12.4).

12.6.1.1 NF-κB Signaling Pathway

NF-κB, a specific transcription factor, oversees the regulation of gene expression for several immune system components [74]. These include adhesion molecules, pro-inflammatory cytokines, chemokines, and inducible enzymes that control the innate immune response, like inducible nitric oxide synthase and cycloxygenase-2. Furthermore, they include costimulatory molecules essential for the induction of specialized immunity, as well as proteins that control specific immune responses, such as MHC [75].

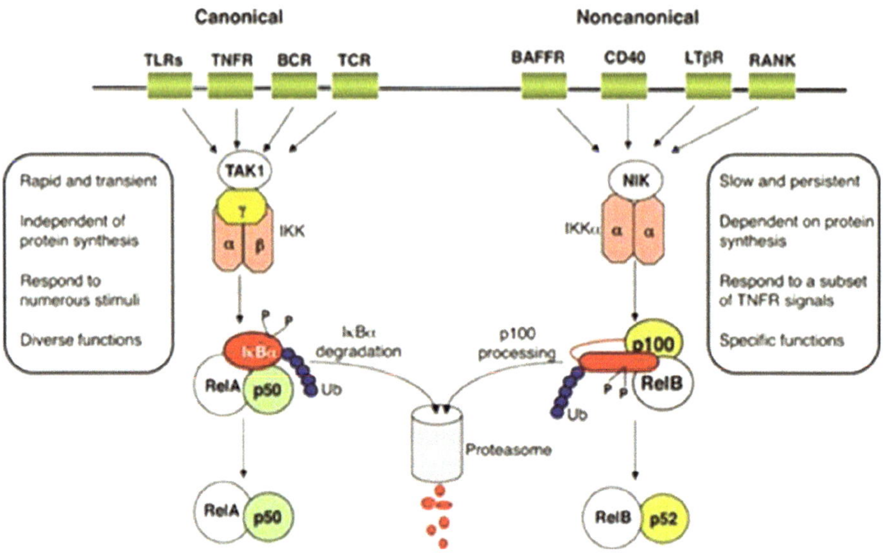

FIGURE 12.4 *Signaling pathway of nuclear factor. This image is reproduced from [37] with permission from Springer Nature (license number 5723091079381).*

Mollusks and Drosophila exemplify organisms that developed the NF-κB signaling pathway early in their evolutionary history. Drosophila activates NF-κB-like transcription factors via the Toll and immune deficiency (Imd) pathways to combat infections. Toll receptors function as cytokine receptors, specifically activated by fungal infections, while the IMd pathway is triggered by Gram-negative bacteria infections. Mammals have evolved to preserve both the elements of the NF-κB immune response function and signaling pathway.

The N-terminal NF-κB/Rel homology domain (RHD) of five distinct NF-κB DNA-binding subunits controls DNA binding, dimerization, nuclear translocation, and interaction with inhibitory IκB proteins. P50 and P5 are the other two family members [37].

The canonical and non-canonical NF-κB signaling pathways are activated by different signals. The canonical pathway involves the activation of the IKK complex and the degradation of IκBα. The non-canonical pathway relies on p100 processing triggered by signaling from TNF receptor (TNFR) family members. This pathway depends on NIK and IKKα and leads to the activation of the RelB/p52 complex.

12.6.1.2 Induced Activation of NF-κB p65:p50

The predominant form of NF-κB activated through the canonical pathway is the heterodimer of p50 and p65. IκB protein phosphorylation releases the NF-κB p65:p50 dimers from their inhibitory complex, and proteasomes subsequently ubiquitinate and destroy them. When IKK (IκB kinase) is activated in a NEMO-dependent manner, nuclear translocation can occur via the arginine- and lysine-rich NLS of the p50 and p65 subunits and association with importin α/β heterodimers. Mechanistically, the α protein of importin typically binds the NLS first, then enlists the help of importin β to move the cargo via nuclear pore complexes (NPCs) into the nucleus. However, importin β directly binds to the p65 NLS, facilitating its nuclear translocation. Within the nucleus, Ran-GTP binds to importin β, releasing it from the complex with p65.

Free p65 then binds to specific nucleotide sequences of target genes, promoting tightly regulated transcriptional programming that shows a high level of tissue specificity. P65's interaction with several components of the basic transcription machinery, including TFIIB (transcription factor IIB), TBP (TATA-binding protein), and coactivators, such as p300 and Tip60, significantly enhances its sequence-specific DNA binding.

Exportin proteins facilitate p65's nuclear export by binding to both Ran-GTP and p65's nuclear export region. After crossing the NPC into the cytoplasm, p65 dissociates due to the Ran-GTPase-activating protein hydrolyzing Ran-GTP to Ran-guanosine diphosphate (GDP is a nucleotide composed of guanine, ribose, and two phosphate groups, serving as an essential energy carrier in cellular processes. It plays a crucial role in signal transduction and is involved in the regulation of G-proteins, which mediate various physiological responses). Structural studies of the NF-κB:IκB complex revealed that the IκB proteins mask the NLS of the p65 subunit but not that of p50. Under physiological conditions, importin α/β dimers can bind the p50 NLS, facilitating the nuclear import of p50:p65 dimers. In the nucleus, IκBα's nuclear export sequence interacts with p65 to create a complex that binds to exportin proteins for nuclear export (Sun, 2011).

NF-κB target genes are implicated in the initiation and progression of inflammation. Upon activation, NF-κB can initiate the transcription of numerous genes, thereby managing the inflammation process. NF-κB also plays a role in regulating cell proliferation, apoptosis, morphogenesis, and differentiation, while specifically addressing inflammation by upregulating the production of inflammatory cytokines, chemokines, and adhesion molecules [76].

12.6.2 MAPK Pathway

ERK1/2 is a protein kinase belonging to the MAPK family [77]. This protein family is essential in signaling cascades, transferring signals from the external environment to intracellular destinations. MAPK cascades play a crucial role in controlling fundamental processes like cell division, proliferation, and stress responses [78]. These cascades transmit signals through sequential activation of three to five layers of protein kinases, known as MAPK kinase kinase kinase (MAP4K), MAPK kinase kinase (MAP3K), MAPK kinase (MAPKK), MAPK, and MAPK-activated protein kinases (MAPKAPK) [79]. The first three central layers are considered the core unit, while the last two layers can vary among cells and stimuli and may be present in certain cascades [80].

12.6.2.1 The MAPK Pathway and its Regulation

The MAPK pathways comprise four distinct signaling families: Big MAP kinase-1 (BMK-1), JNK, p38, and the conventional MAPK/ERK pathway [81]. These families share a fundamental arrangement consisting of a dual-specificity threonine/tyrosine kinase and two serine/threonine kinases [82]. These enzymes are classified according to their proximity to the initial signaling event, with those closest to the nucleus designated as MAP3K, MAPKK, and MAPK [83]. The canonical MAPK/ERK pathway consists of three classes of MAP3K: A-RAF, B-RAF, and C-RAF (or RAF-1) kinases. B-RAF is the most commonly altered gene at this point in human cancer development. One tier below are the MAPKKs, which include MEK1 and MEK2. Further downstream are ERK1 and ERK2, the ultimate effector molecules of the MAPK pathway [84].

ERK activation through phosphorylation leads to the activation of numerous substrates crucial for cell proliferation. The spatial localization of ERK determines its specific substrates and subsequent influence on various cellular processes. In the cytoplasm, ERK phosphorylates cytoskeletal proteins, affecting cell movement, trafficking, metabolism, cell adhesion, and various signaling pathways [85]. Cytoplasmic targets include ribosomal S6 kinases (RSKs), which regulate glycogen synthase kinase 3 (GSK3) involved in metabolism, and L1 adhesion molecule, a neural protein involved in cellular adhesion [86].

Minutes after MAPK/ERK activation, ERK dissociates from its cytoplasmic anchoring proteins and relocates to the nucleus, where it regulates transcription. In the nucleus, active ERK phosphorylates and activates numerous transcription factors, including carbamoyl phosphate synthetase II (CPS-II), involved in DNA synthesis, and p90RSK, which promotes cell cycle progression. These two mechanisms largely drive MEK/ERK-induced cell proliferation. Moreover, activated ERK contributes to the innate immune response at several phases of the inflammatory cascade, boosting the production of inducible nitric oxide synthase (iNOS) and TNF-α [87]. Figure 12.5

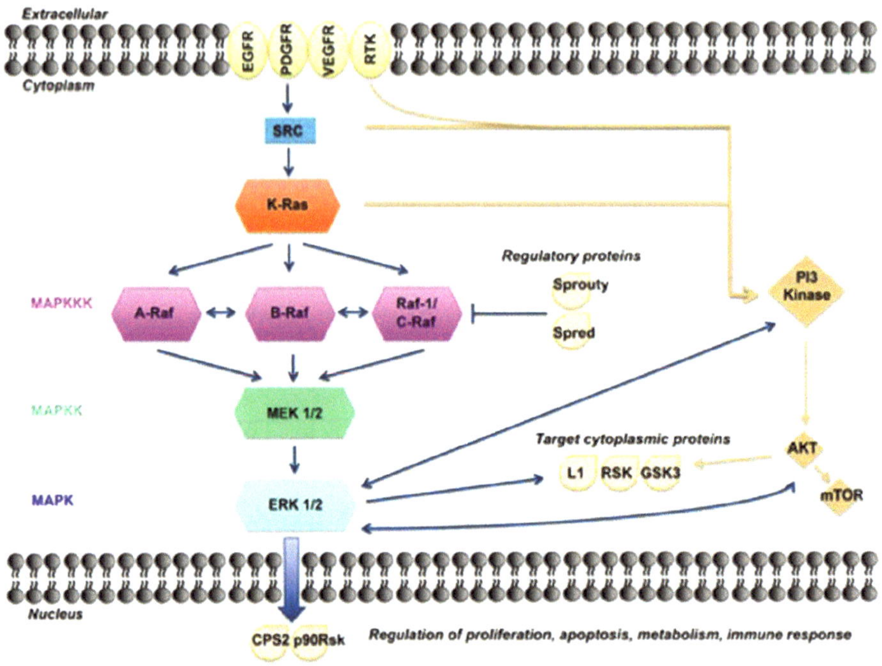

FIGURE 12.5 *MAPK kinases cascade signaling transduction pathways. This image is reproduced from [87]. It is distributed under the Creative Commons Attribution license as an open access image.*

depicts the MAPK/ERK pathway. Adaptor proteins recruit Ras proteins to activate the pathway, culminating in ERK activation. The signal is amplified through successive phosphorylation steps, Raf → MEK → ERK, until ERK activates its targets in the cytoplasm and/or nucleus.

12.6.3 JAK/STAT Pathway

Cytokines regulate numerous significant cellular functions. Upon ligand-receptor binding, the JAK/STAT pathway, an essential signaling cascade downstream of cytokine receptors, is activated [88]. This activation leads to the activation of receptor-associated JAKs. Subsequently, JAK kinases phosphorylate tyrosine kinases residues in STAT proteins to initiate their dimerization and nuclear translocation. These proteins then alter target gene expression in the nucleus [89].

12.6.3.1 JAK/STAT Signaling Pathway

Investigations into IFN (type I interferon) signaling led to the discovery of the JAK/STAT signaling pathway [90]. In 1990, researchers discovered that the IFN-α-responsive transcription factor, interferon-stimulated gene factor 3 (ISGF3), comprises multiple interacting polypeptide chains (48, 84, 91, and 113 kDa) [90]. In 1992, Fu reported that the ISGF3α proteins of ISGF3 (113, 91, and 84 kDa) contain conserved SH2 and SH3 domains, and that a distinct cytoplasmic tyrosine kinase stimulated by IFN-α can phosphorylate and activate ISGF3α. Fu proposed a direct effector model to explain IFN-α-triggered signal transduction, elucidating the mechanism of the JAK/STAT signaling pathway [91].

Later research revealed that ISGF3 consists of IRF9, STAT1, and STAT2. Between 1993 and 1995, STAT3, STAT4, STAT5a, STAT5b, and STAT6 were identified by various laboratories [92, 93]. JAKs were discovered between 1989 and 1994. In 1989, Wilks et al. discovered a tyrosine kinase with a distinct kinase domain and a pseudo kinase domain. They identified two more such tyrosine kinases, JAK1 and JAK2, in 1991 [94, 95]. The remaining two JAKs, tyrosine kinase 2 (TYK2) and JAK3, were identified in 1990 and 1994, respectively [96].

The connection between JAKs and STATs was first identified in 1992 by Velazquez et al., who discovered that TYK2 is crucial for IFN-α/β signaling. Subsequently, Müller et al. demonstrated that JAKs are necessary for STAT phosphorylation in IFN-dependent signaling [97]. Thus, the components and architecture of the JAK/STAT signaling pathway were defined between the late 1980s and early 1990s. Ongoing research into additional proteins and JAK/STAT pathway functions has further expanded our understanding of this signaling landscape [98].

12.6.4 Dysregulation of Inflammatory Signaling

12.6.4.1 Autoimmune Disorders

Macroscopic indicators of inflammation, such as erythema, heat, swelling, and pain, result from dilated capillaries and enhanced permeability. These changes in blood vessels ultimately facilitate the transmission of serum proteins to the site of damage. At the cellular level, this phenomenon is regulated by a complex network of cytokines, chemokines, growth factors, and the nervous system [99]. The term "inflammation" has been used since antiquity and comes from the Latin word "inflammare," meaning to burn or ignite. This analogy is apt, as inflammation can be triggered to defend against various threats and can spontaneously subside once the threat is neutralized. However, inflammation presents three primary challenges. Firstly, not all threats require an inflammatory reaction. Severe ischemia-reperfusion injury, exposure to toxins, trauma, crystal particles, and inflammatory diseases can all contribute to adverse health outcomes. Secondly, inflammation is indiscriminate, damaging both healthy and diseased tissues. Finally, like a smoldering fire, there is always a risk of unchecked and persistent inflammatory spread.

It is crucial to emphasize that this perspective on inflammation is not comprehensive, as detailed information can be found elsewhere. Instead, this overview aims to provide a broad definition and contextualization of inflammation, highlight what inflammation is not, and ultimately improve the management of inflammation [100].

12.6.4.2 Inflammation and Cancer

Cancer is defined as a group of malignant neoplasms characterized by the growth and spread of cancer cells to other parts of the body [101]. It can affect nearly any organ or tissue and is influenced by numerous variables, including environmental stress and genetic instability. Clinical and epidemiological research has established the connection between inflammation and cancer. For example, there is a notable increased risk of colorectal cancer in those with ulcerative colitis, Crohn's disease, and inflammatory bowel disease. Furthermore, studies have shown that taking specific anti-inflammatory drugs reduces the risk of colon cancer in people with colitis.

Evidence suggests that the intensity and duration of inflammatory illnesses in the respiratory system strongly correlate with cancer risk. For example, Epstein-Barr virus is associated with dysplastic progression in nasopharyngeal cancer. Microbial infections and non-infectious physical and chemical irritants can also cause inflammation. Helicobacter pylori infection primarily causes mucosal-associated lymphoid tissue lymphoma and adenocarcinoma in the gastrointestinal tract. Cholangiocarcinoma of the bile tract is linked to persistent inflammatory infiltration brought on by Clonorchis sinensis infection. Chronic hepatitis caused by hepatitis B and C viruses is a significant risk factor for hepatocellular carcinoma, the third most common cause of cancer-related deaths globally [102]. Human papillomavirus infection is the primary cause of penile and anogenital malignancies. Other diseases, like human herpesvirus type 8 and schistosomiasis, can increase the risk of bladder cancer and Kaposi's sarcoma, respectively. Cancer development can also be facilitated by non-infectious chronic inflammation. Pancreatic cancer, esophageal cancer, and gallbladder cancer have all been linked to inflammatory disorders such as esophagitis, Barrett's metaplasia, and chronic pancreatitis.

There are also possible links between Marjolin's ulcer and skin cancer; mesothelioma and asbestos exposure; bronchial cancer and silica exposure; lung, skin, and liver cancer and chronic asthma;

ulcerative lichen planus and verrucous carcinoma; penile cancer and foreskin inflammation/phimosis; and ovarian epithelium inflammation or pelvic inflammatory diseases and chronic prostatitis, whether caused by persistent bacterial infection or non-infective stimuli, have been linked to prostate cancer. Thus, there is increasing evidence supporting the association between chronic inflammation and cancer development [101].

12.6.4.3 Inflammation Contributes to the Complications of Atherosclerosis

Sudden thrombotic episodes, one of the most severe consequences of atherosclerosis, are also influenced by inflammatory and immunological processes. Most myocardial ischemia and certain ischemic strokes are caused by plaque rupture and subsequent thrombus formation, which does not necessitate a constricted artery [103]. The clinical implications of mature atherosclerotic plaque are significantly influenced by its microanatomy. The highly thrombogenic lipid core of the plaque is usually covered by a fibrous cap that shields it from circulating coagulation factors. Plaques with thin fibrous caps are typically associated with rupture and fatal coronary artery thromboses. The fibrous cap of the plaque maintains its biomechanical integrity due to interstitial collagen. This extracellular matrix macromolecule is resistant to most proteinases. Collagen is a tightly wrapped triple helix polymer that can only be initially degraded by a small set of enzymes called matrix metalloproteinases (MMPs), which initiate the catabolic process. Macrophages can accelerate the breakdown of collagen in inflamed atheroma by expressing members of the MMP family, activated by T-cell-originated CD40L or other pro-inflammatory cytokines.

Insufficient collagen synthesis induced by pro-inflammatory cytokines interacting with smooth muscle cells could further undermine the stability of the fibrous cap. To illustrate, IFN released from T cells inhibits the transcription of interstitial collagen mRNA, reducing the amount of collagen synthesized by cultured vascular smooth muscle cells. These T-cell factors regulate certain plaque components associated with its susceptibility to rupture by enhancing collagen breakdown through CD40-mediated MMP production by macrophages and decreasing collagen synthesis due to IFN-mediated smooth muscle activity. Moreover, activated T cells may contribute to arterial thrombosis by releasing CD40L, which induces macrophages to release tissue factor, a protein known for its high prothrombotic activity [104]

12.6.4.4 Inflammation and Arthritis NF-κB

Immune cells infiltrate the synovium of rheumatoid arthritis (RA) patients' joints, causing persistent inflammation, pannus development, and eventual permanent joint and cartilage destruction [105]. The RA synovium primarily consists of B cells, dendritic cells, other immune cells, synovial cells such as macrophage and endothelium (30–40%), T cells (approximately 30%), and synovial fibroblasts [106]. Numerous effector molecules have been detected synovial fluid of RA patients, including MMPs (MMP-1, -3, -9, and -13), pro-inflammatory chemokines (IL-8, IP-10, MCP-1, MIP-1, and RANTES), pro-inflammatory cytokines (IL-1β, IL-6, TNF-α, and IL-18), and metabolic proteins (Cox-1, Cox-2, and iNOS). These may initiate a vicious cycle of pro-inflammatory signals, resulting in chronic and persistent inflammation due to their intricate interactions.

TNF-α is the primary mediator of inflammation and apoptosis inducer. Crucially, recent studies have demonstrated that NF-κB transcription factors regulate TNF-α genes and numerous other factors, suggesting that NF-κB serves as a principal controller of inflammatory cytokine production in RA [107]. Active NF-κB transcription factors have been found in both human arthritic joints and artificially induced RA in the joints of animals, as demonstrated by synovial fibroblast cultures [108].

12.6.4.5 Systemic Lupus Erythematosus and the Immune System

Systemic lupus erythematosus (SLE) affects 20–50 out of every 100,000 patients and is a systemic autoimmune condition of unknown cause. While the early stages of the illness often affect the skin and joints, chronic inflammation-related cardiovascular events and damage to key organs (especially the kidneys, nervous system, and hematological organs), as well as infections caused by immunosuppressive treatments, are typically linked to disease morbidity and mortality.

SLE is characterized by numerous immune system abnormalities, such as the formation of pathogenic autoantibodies, immune complex deposition, immune system infiltration into injured organs, and inflammation. Inflammatory mediators and various immune cells, particularly defective T and B cells, have been proven to be detrimental components of SLE. The breakdown of B- or T-cell tolerance toward self-antigens is associated with hormonal, environmental, and genetic factors, which in turn activate both the innate and adaptive immune systems. The development of SLE is commonly linked to abnormal adaptive immunological responses, particularly B-cell responses, due to the presence of self-reactive IgG antibodies [109].

12.7 RECEPTORS INVOLVED IN INFLAMMATORY SIGNALING

Inflammatory signaling involves numerous receptors in an intricate and tightly controlled process essential for detecting and responding to a wide variety of stimuli. TLRs are particularly important as triggers of the innate immune response. These transmembrane proteins are expressed on the surface of immune cells and can identify specific molecular patterns linked to infections, such as viral nucleic acids or bacterial lipopolysaccharides. TLR activation initiates downstream signaling cascades, resulting in the production of chemokines and pro-inflammatory cytokines. Additionally, intracellular receptors such as NOD-like receptors (NLRs) function within the cytoplasm to recognize endogenous danger signals and microorganisms [110].

12.7.1 Toll-Like Receptors

TLRs belong to the PRR family and play a critical role in innate immunity through their capacity to detect tissue injury through DAMPs and infections via PAMPs. TLRs 1, 2, 4, 5, 6, and 11 are located on the plasma membrane, whereas TLRs 3, 7, 8, and 9 are found in intracellular compartments. Except for TLR3, microbial components trigger TLR signaling through a MyD88 (myeloid differentiation factor)-dependent pathway, which activates the transcription factors NF-κB and AP-1 and causes the release of inflammatory cytokines. TLR interaction also activates MAPKs, such as extracellular signal-regulated kinases (ERK1/2), JNK, and p38. TLR3 and TLR4 activation also triggers an IFN-β response via a mechanism that is dependent on TRIF (TIR domain-containing adaptor-inducing interferon) but independent of MyD88 [111].

TLR activation not only produces inflammatory responses but also leads to the formation of antigen-specific adaptive immunity. The adaptor protein MyD88 mediates a common signaling pathway essential for the TLR-induced inflammatory response. However, evidence suggests other pathways mediate biological responses unique to TLR ligands [112].

12.7.1.1 Structure of TLRs

Two TLR molecules are joined by the ligands, creating a dimer between their extracellular domains. This dimerization brings together the cytoplasmic TIR (toll/IL-1R) domains, according to the crystallographic arrangement of TLR receptors or the complexes associated with their ligands [79]. All TLRs belong to the type I transmembrane glycoprotein receptor family. These receptors have molecular weights ranging from 90 to 115 kilodaltons and contain 16–28 leucine-rich repeats (LRRs) in their extracellular domains [113]. The transmembrane α-helix segment and the intracellular conserved domain exhibit homology with corresponding regions in the human interleukin-18 receptor (IL-18R) and interleukin-1 receptor (IL-1R) [114].

The presence of the LRR domain in both plant and animal proteins indicates a conserved function as a molecular PRR throughout evolution. Notably, TLRs 1–9 are expressed in both humans and mice. TLR10 is expressed in humans but not in mice, while TLRs 11–13 are expressed in mice but absent in humans [115]. TLRs recognize highly conserved patterns in microorganisms such as viruses, bacteria, fungi, and protozoans, known as PAMPs. Certain TLRs (for example, TLR1, TLR2, TLR4, TLR5, and TLR6) are situated on the surface of cells, while others (including TLR3, TLR7, TLR8, and TLR9) are found within intracellular compartments. TLRs are characterized by LRRs that exhibit significant variations in sequence and length compared to the TLR consensus [116]. Each LRR consists of a 24–29 amino acid sequence characterized by the leucine-rich motif XLXXLXLXX and the conserved motif XΦXXΦX4FXXLX. In these motifs, X represents any amino acid, and Φ represents a hydrophobic amino acid. LRRs comprise a β-strand and an α-helix, which are connected by loops [117].

12.7.1.2 Members of TLR Family

Thirteen mouse TLR members and ten human TLR member families have been identified, although TLR10 is non-functional in mice. A mouse strain resistant to lipopolysaccharides (LPS) was found to have a point mutation in the TLR4 gene, validating the role of TLR4, the first toll protein homolog identified in humans. TLR4 has been shown to activate genes linked to inflammatory responses. Functionally, TLRs can be divided into two categories:

1. Cell surface-expressed TLRs: These include heterodimers of TLR2 with TLR1 and TLR6, as well as TLR4, TLR5, and TLR10.
2. Nucleic acid sensors: Also known as intracellular TLRs (TLR3, TLR7, TLR8, and TLR9), these are present in the endoplasmic reticulum (ER), endosomes, and lysosomes [118].

12.7.1.3 Signaling Pathways of TLRs

Ligands binding to TLRs triggers specific intracellular downstream signaling cascades that initiate host defense responses. These PAMP-RRR interactions generate a type 1 interferon and pro-inflammatory cytokines, which direct immune responses against specific microorganisms. TLR signaling is influenced by the type of stimulus, the activated TLR, and the downstream adaptor molecule. Key compounds include Interleukin-1 receptor-associated kinases (IRAKs are a family of serine-threonine kinases that play a crucial role in the signaling pathways of Toll-like receptors (TLRs) and interleukin-1 receptors (IL-1Rs), mediating inflammatory responses. They are essential for the activation of downstream signaling cascades that lead to the expression of inflammatory genes in immune cells), the adaptor molecule MyD88, TAK1-binding protein 1 (TAB1) and 2 (TAB2), TGF-β-activated kinase (TAK1), and tumor necrosis factor receptor-associated factor 6 (TRAF6).

There are two main TLR signaling mechanisms: the MyD88-dependent pathway, which is used by all TLRs except TLR3, produces inflammatory cytokines. When IL-1R or TLR is activated, the adaptor protein MyD88 is drawn to the receptor complex, facilitating its association with IL-1R-associated kinases IRAK4 and IRAK1. IRAK4 activation leads to TRAF6 interaction and IRAK-1 hyperphosphorylation. The IRAK-4·IRAK-1·TRAF6 interaction causes conformational changes, leading to their disengagement from the receptor complex. This activates TAK1 in the cytoplasm, which then triggers IKK. Inactive IKK sequesters NF-κB in the cytoplasm, but when activated, it phosphorylates IκB, leading to its degradation and NF-κB release. TAK1 activation also activates MAPKs and c-Jun NH2-terminal kinases [119] (Figure 12.6).

12.7.1.4 Function of TLRs

The primary function of TLRs is to identify a wide range of pathogens. Each immune system cell expresses a specific subset of TLRs with unique roles in modulating immune responses and identifying PAMPs and DAMPs. TLRs induce antimicrobial actions and produce inflammatory cytokines, which are important aspects of innate immunity. TLR activation by DAMPs results in the recruitment of IRAK 1, 2, 4, and M, TAB2, and TRAF6. This recruitment process ultimately leads to the translocation of pro-inflammatory transcription factors, particularly NF-kB, AP-1, and interferon regulatory factor 3 (IRF3). These transcription factors are responsible for encoding various molecules, such as type 1 interferons (IFN-α and β), chemokines (CXCL8 and CXCL10), pro-inflammatory cytokines (TNF-α, IL-1β, and IL-6), and antimicrobial peptides [120].

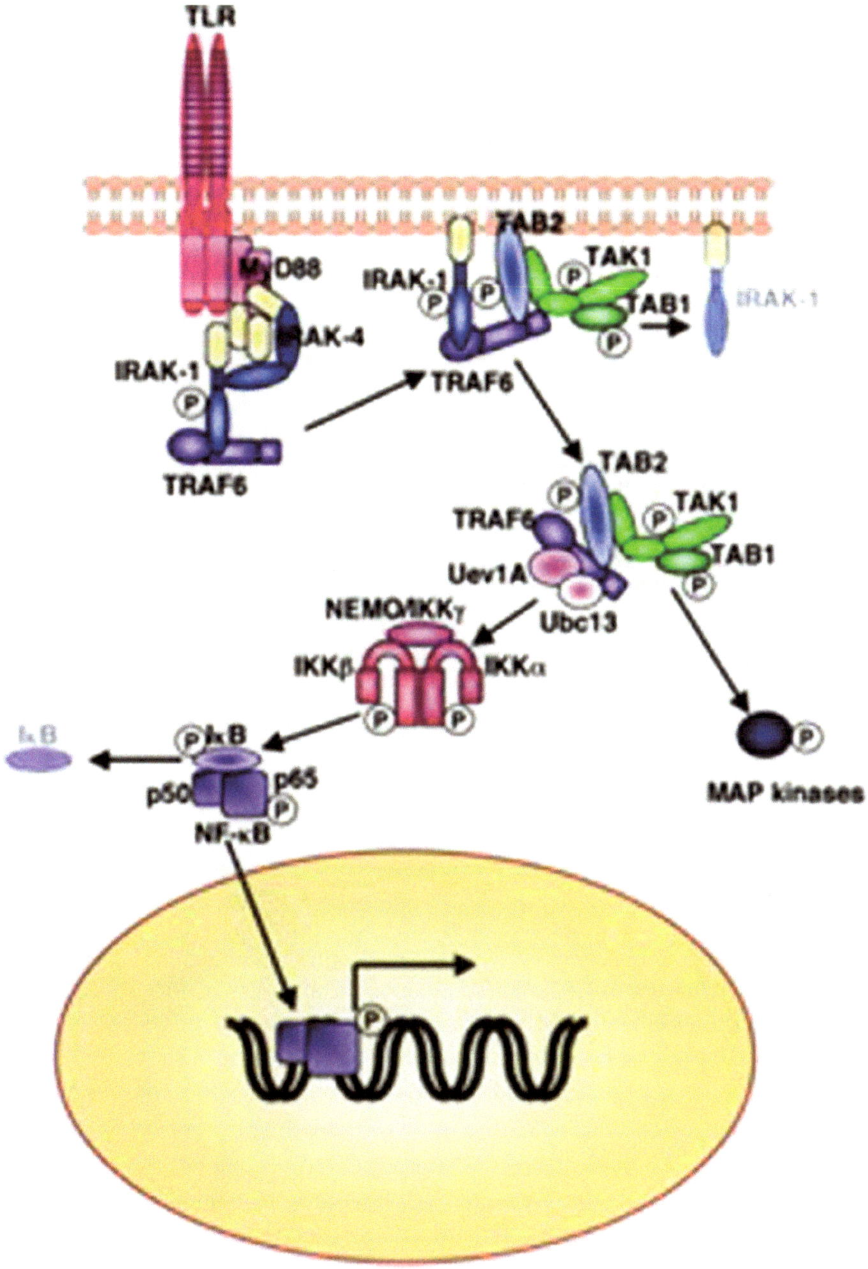

FIGURE 12.6 Toll receptor signaling pathway. This image is reproduced from [119], an open access article under the Creative Commons CC-BY license, which allows unrestricted use, distribution, and reproduction in any medium, provided the original work is properly cited.

12.7.2 Cytokine Receptors

Cytokines are vital immune system communication mediators essential for the host's defense against infections. Cytokines are released by cells and can also be generated or released by a broad range of cells in response to cellular damage or infection, as in the case of interferons. Cytokines regulate many cellular aspects, including survival, growth, and development. Cells in the local tissue environment can be reprogrammed to boost specific immune responses. Cytokines function via a range of conserved signaling routes that regulate cellular metabolism and transcriptional programming.

Germline mutations affecting cytokines, cytokine receptors, or signal transduction mediators are the root cause of many primary immunodeficiency disorders. Excess cytokine production or response can lead to inflammatory and autoimmune disorders [121]. Cytokines can have an autocrine effect on the cells from which they are released, a paracrine effect on nearby cells, or an endocrine effect on distant target cells.

Class I cytokines activate class I cytokine receptors through their four-α-helix bundle protein architectures. Hematopoietic cytokines fold in an "up-up-down-down" pattern with four α-helices, where the two initial helices (A and B) are parallel to each other and antiparallel to the subsequent two (C and D), joined by two long, overhand loops. These helical structures are divided into "short-chain" and "long-chain" groups based on their length. This family includes granulocyte-macrophage colony-stimulating factor (GM-CSF), macrophage colony-stimulating factor (M-CSF), and IL-2, 3, 4, 5, 7, 9, and 13. Long-chain cytokines lack the β-sheets found in short-chain cytokines.

Initially, interferons (IFNs) and IL-10 were the only recognized class II cytokines. However, the identification of other cytokines has led to a reclassification. Class II cytokines, including the IL-10 family of cytokines and type I, II, and III IFNs, are now classified into four distinct categories. Despite their similar structural characteristics, these cytokines elicit different and varied reactions in their target cells [122].

12.7.3 Nucleotide-Binding and Oligomerization Domain-Like Receptors

Autoimmune diseases are chronic immunological disorders typified by immune system dysregulation, leading to a breakdown in the body's ability to tolerate self-antigens. Nucleotide-binding and oligomerization domain (NOD)-like receptors (NLRs) are crucial in managing various autoimmune disorders, including multiple sclerosis (MS), RA, psoriasis, SLE, and inflammatory bowel disease (IBD). NLR proteins comprise a central nucleotide-binding domain, an N-terminal effector domain, and a C-terminal LRR. These proteins form a class of PRRs that mediate immune responses by detecting cellular PAMPs or DAMPs and inducing multiple signaling pathways, such as MAPK, RIP2 kinase, caspase-1, NF-κB, and so forth [123].

12.7.4 NOD-Like Receptors

The NLR family was initially discovered through the nematode homolog CED-4 and its mammalian counterpart, apoptotic protease activating factor 1 (APAF1). These proteins regulate embryonic and p53-dependent apoptosis, also known as programmed cell death. Subsequently, database searches for APAF1/CED-4 homologs led to the identification of two NOD-containing molecules, NOD1 (caspase recruitment domain 4, CARD4) and NOD2 (CARD15). There are currently at least 34 NLR genes in mice and 23 proteins that make up the human NLR family [124].

12.7.4.1 NOD1 and NOD2 Activation and Signaling

NOD1 and NOD2 are the most extensively studied members of the NLR family. NOD1 expression is ubiquitous, while NOD2 expression is only seen in intestinal Paneth cells, monocytes, macrophages, and dendritic cells. Initial biochemical characterization showed that both NOD1 and NOD2 caused TLR-independent NF-κB activation. Further analysis revealed that NOD1 and NOD2 recognize distinct structural core motifs [125].

12.8 INFLAMMATION AND INFLAMMATORY SIGNALING PATHWAYS

Inflammation is a major category of immunological reactions. It is essential for protection against numerous potentially dangerous stimuli, both endogenous and external, and plays a crucial role in both innate and adaptive defense mechanisms [126]. During the inflammatory phase, various immune cells, including leucocytes, and plasma proteins such as cytokines and complement proteins, migrate to the region of inflammation or tissue injury and are activated [127]. These blood-derived immune system components mediate inflammation to drive out invasive pathogens (including fungi, viruses, and bacteria) and promote tissue healing [128].

The immune system has evolved to identify molecular structures of both endogenous and foreign substances, such as LPS and heat shock proteins (HSPs), via receptors expressed by immune cells including macrophages, DCs, endothelial cells, B cells, and T cells [126]. During the inflammatory response, immune cells release various inflammatory mediators, including chemokines and pro-inflammatory cytokines [127]. The most significant pro-inflammatory cytokines in the immune system include IL-1β, IL-6, IL-8, IL-12, IFN-γ, and TNF-α. These cytokines interact with their receptors on immune cell surfaces, triggering positive feedback loops in inflammatory signaling cascades. The three primary signaling systems that mediate the inflammatory response in immune cells are the NF-κB signaling system, the JAK/STAT signaling network, and the MAPK signaling pathway [129].

In immune-mediated inflammatory diseases, these systems, which tightly regulate inflammatory signaling pathways, malfunction in conditions such as MS, RA, SLE, atherosclerosis, and IBD [130, 131]. Consequently, modulating these inflammatory signaling pathways and their signaling molecules may provide an effective strategy for producing anti-inflammatory effects and treating inflammatory diseases [132].

12.8.1 Disruption of NF-κB, MAPK, or JAK/STAT

Inflammatory pathways impact numerous chronic diseases and involve common mediators and regulatory processes. Inflammatory stimuli trigger intracellular signaling pathways, leading to the synthesis of inflammatory mediators. Primary inflammatory stimuli regulate inflammation by interacting with TNFR, IL-1R, IL-6R, and TLRs. Examples of these stimuli include microbiological products and cytokines like TNF-α, IL-1β, and IL-6 [133].

Receptor activation initiates significant intracellular signaling pathways, such as MAPK, JAK/STAT, and NF-κB [134]. Dysregulation of JAK/STAT, MAPK, or NF-κB activity is linked to cancer, autoimmune, and inflammatory illnesses. Transcription factor-mediated signaling results in cytokine secretion [135]. Various transcription factors control many inflammatory genes, including IL-1, TNF-α, IL-6, interferons, TGF, CSF, and chemokines. Disruption of these processes can result in unchecked, persistent chronic inflammation. Chronic inflammation occurs when acute inflammatory systems fail to completely resolve tissue damage [136]. This may result in a variety of illnesses, including cancer, atherosclerosis, type 2 diabetes, RA, and cardiovascular disease [137, 138].

12.8.2 Management Strategies for Conditions Associated with Inflammation

12.8.2.1 Nonsteroidal Anti-Inflammatory Drugs

The view that all nonsteroidal anti-inflammatory drugs (NSAIDs) work solely by preventing prostaglandin synthesis has been challenged. NSAIDs have been found to influence a broad range of cellular processes essential to both their therapeutic effects and adverse consequences. The different actions of both novel and well-established NSAIDs on cyclooxygenase enzymes (COXs), which regulate the synthesis of physiologically significant (COX-1) and inflammatory (COX-2) prostaglandins, have implications for both intrinsic physiological and inflammatory mechanisms.

Some well-known NSAIDs have a minimal risk of gastrointestinal (GI) ulcers and bleeding due to their distinct pharmacokinetic characteristics (e.g., pro-drugs and protein binding). These NSAIDs do not significantly affect GI mucosal prostaglandin synthesis and have unique physicochemical properties that minimize harm to mucosal membranes or non-prostaglandin-related cellular events essential for mucosal defense.

Some novel COX-2 selective medications with methyl or amino-sulfonyl moieties have comparatively high pKa values and share characteristics with well-known NSAIDs that have low GI ulcerogenicity. These physical and chemical characteristics might explain the new COX-2 selective medications' decreased irritancy, even if they don't affect COX-1 in the GI mucosa. Current research is investigating whether COX-2 alternative therapies may have a lower risk of cartilage degradation in osteoarthritis (OA) by circumventing the suppressive impact on cartilage proteoglycan metabolism seen with drugs like indomethacin and salicylates. While meloxicam doesn't seem to inhibit proteoglycan metabolism, it's unclear if this will improve radiographic or magnetic resonance imaging evaluation of OA-related joint abnormalities [139].

12.8.2.2 Corticosteroids

Corticosteroids, also known as glucocorticosteroids, glucocorticoids, or steroids, are widely used medications that demonstrate efficacy in numerous inflammatory and immune disorders. They are primarily used to manage conditions such as asthma, while their effectiveness is relatively limited in other ailments like chronic obstructive pulmonary disease (COPD) [140]. Chronic inflammation is characterized by increased expression of inflammatory genes regulated by transcription factors such as NF-κB and AP-1. These transcription factors attach to coactivator molecules, causing core histone acetylation and initiating gene transcription. Corticosteroids inhibit this process in chronic inflammatory diseases like asthma by reversing histone acetylation of activated inflammatory genes. This suppression is primarily achieved through the recruitment of histone deacetylase-2 (HDAC2) to the active transcription complex and the binding of liganded glucocorticoid receptors (GR) to coactivators. At higher concentrations, corticosteroids cause GR homodimers to engage with DNA recognition sites, activating anti-inflammatory genes and inhibiting genes linked to adverse effects. Significant progress has been made in understanding the molecular mechanisms by which inhaled corticosteroids effectively suppress inflammation in asthma and other inflammatory ailments. These developments are supported by recent advances in our comprehension of the basic principles underlying gene transcription and cell signaling in inflammation [141].

12.9 DISRUPTION OF JAK/STAT SIGNALING IN CANCER

Preclinical research has extensively examined the significance of JAK/STAT equilibrium and the effects of disturbances. Mice lacking JAK1 exhibit aberrant lymphocyte formation, reduced cytokine responses, and prenatal death [142]. It is often suggested that mutations resulting in gain or loss of function (GOF and LOF, respectively) and promoting carcinogenesis are the cause of JAK/STAT dysfunction in several malignancies [143]. Many of these mutations share altered immune function, such as loss of NK cell cytotoxicity, aberrant Th1 differentiation, IFN dysregulation, aberrant myeloproliferation, autoimmunity, and aberrant cytokine signaling and response [144].

Numerous hematological cancers have been linked to somatic mutations in STAT3 [145]. Mutations in the STAT3 gene's SH2 domain have been found in up to 72% of individuals with T-cell large granular lymphocyte leukemia (T-LGL), primarily affecting the Y640F GOF gene. This mutation is linked to cytopenia, aberrant myeloproliferation, and persistent overexpression of Ifngr2, Irf9, Irf7, and Bcl2l1 genes, as well as suppression of the NFκB negative regulator Birc3. Similarly, abnormal $CD4^+$ T-cell growth has been found in up to 55% of patients with STAT5b mutations associated with CD4+ T-LGL oncogenesis [146].

The JAK2 V617F mutation is linked to significantly poorer cumulative survival in myeloproliferative malignancy, with affected individuals having a three-fold higher probability of early death compared to those without the mutation. This mutation has also been linked to non-small cell lung cancer, with GOF mutations connected to changes in tumor cell PD-L1 expression, although JAK/STAT mutations are less common in solid tumors [147].

STAT1 misregulation has been extensively studied in the cancer and immunology context, where greater STAT3 activation is linked to tumor growth. Diminished IFN-γ sensitivity in human melanoma and lung cancer models has been reported due to faulty JAK signal transduction, and altered IFN-γ-induced STAT1 phosphorylation results in decreased MHC inducibility [148]. Loss of STAT1 in mammary epithelium has been linked to spontaneous breast tumor growth in BALB/C mice and neu-driven carcinogenesis, associated with defective T-cell infiltration and killing, as well as loss of epithelial IRF1. The BRCA1 (breast cancer type 1) gene, commonly altered in ovarian and breast cancer, may co-activate type I IFN signaling. It has been suggested that BRCA1 directly regulates GAS binding by complexing with nuclear STAT1 homodimers, an IFN-γ-dependent process that upregulates Immunity Related Guanosine Triphosphatases (IRGs are a family of proteins activated during the early immune response, primarily involved in defending against intracellular pathogens. They function by localizing to and disrupting phagocytic vacuoles, thereby enhancing pathogen clearance) like Irf7. Therefore, BRCA1 loss due to germline mutation in cancerous cells may facilitate immune evasion and subsequent tumor growth [149].

12.10 IMMUNOTHERAPY IMPACT ACROSS VARIOUS CANCERS

Many malignancies are primarily detected in elderly individuals, and chronic inflammation significantly affects the general well-being and immune function of older cancer patients [150]. Inflammatory signaling triggered by cytoplasmic DNA initiates immune cell activation [151]. Therefore, immune system clearance is a crucial hurdle that tumor cells must overcome during the tumorigenic process [152].

Immunotherapy focuses on both innate and adaptive immune responses of the host. Immunotherapies fall into two categories: (i) active immunotherapy, like vaccine treatment, which induces the patient's immune system to produce certain immune effectors, including T cells and antibodies; and (ii) passive (or adoptive) immunotherapy, in which the patient is given ex vivo-generated molecules, like antibodies. Passive immunotherapy does not elicit an immune response.

Antibody-based therapies, such as anti-cytokines, have been successfully used to treat many inflammatory chronic ailments, including RA, IBD, Crohn's disease, and psoriasis [153]. Immunotherapy is a form of cancer therapy that utilizes the body's capacity to recognize and destroy cancer cells. Unlike traditional cancer therapies like chemotherapy and radiation therapy, which attack cancer cells directly, immunotherapy aims to enhance the body's natural defenses against the disease. The immune system, a sophisticated network of cells, tissues, and organs, is the body's defense system against pathogens, viruses, and cancerous cells.

While cancer cells can sometimes elude detection and continue to grow and spread, the immune system can normally recognize and eliminate them as they develop. Immunotherapy functions by improving the immune system's capacity to recognize and fight against malignant cells more effectively. Various forms of immunotherapies include:

1. Checkpoint inhibitors: These prevent certain proteins on cancer cells from evading immune system detection and attack.
2. Adoptive cell transfer: Immune cells are extracted from the patient, modified in a lab to increase their cancer-fighting capacity, then reinfused into the patient.
3. Cancer vaccines: These expose the immune system to cancer cell fragments or related materials, enhancing its ability to recognize and fight cancerous cells.

Immunotherapy has shown promise in treating numerous cancers, including melanoma, lung cancer, bladder cancer, and others. While it can be more effective and have fewer side effects than conventional cancer therapies like radiation and chemotherapy, immunotherapy is not without risks [154].

12.10.1 Types and Features of Vaccines Against Cancer Amino Acid Sequences Derived from Antigens Linked with Tumors (TAAs)

12.10.1.1 Peptide Vaccines

Peptide vaccines are potential cancer immunotherapy strategies that use small amino acid sequences to trigger specific immune responses against cancerous cells [155].

12.10.1.2 DNA/RNA-Based Vaccines

The basic idea behind DNA/RNA-based tumor vaccines is to introduce genetic material encoding tumor antigens into host cells. This strategy triggers the immune system to recognize and target cancer cells, resulting in an adaptive immunological response. These vaccines are delivered using various vectors, including lipid nanoparticles, viral vectors, and bare nucleic acids. They offer several advantages over conventional methods, such as safety, ease of production, strong immune response induction, and adaptability. Additionally, they provide opportunities for personalization to match each patient's specific tumor profile. Despite their potential, challenges remain, including effective DNA/RNA absorption and transfer into cells, potential autoimmune reactions, and financial and time constraints associated with personalization [156].

12.10.1.3 Viral-Vector-Based Vaccines

Viral-vector-based tumor vaccines represent a promising development in cancer immunotherapy. These vaccines exploit viruses' natural ability to enter host cells and effectively deliver tumor antigens, inducing a strong, focused immune response. After infection, this method causes host cells to produce tumor-associated or specific antigens, which T cells then recognize and mount a strong defense against. A key advantage of these vaccines is their ability to elicit robust cellular and humoral immunological responses. They can also be engineered to express a wide variety of tumor antigens, increasing the immune response's effectiveness and breadth. However, challenges remain, including the impact of pre-existing viral immunity and practical difficulties in large-scale production [157].

12.10.1.4 Dendritic-Cell-Based Vaccines

Dendritic cells (DCs) are used extensively in cancer immunotherapy strategies due to their crucial role in antigen presentation and subsequent T-cell activation. DCs can also modulate the immune response by bridging the gap between innate and adaptive immunity. DC-based cancer vaccines use DCs laden with tumor-associated antigens to effectively induce an immune response against cancer cells. These DCs can be loaded with TAAs through various methods, including synthetic peptides, tumor lysates, and mRNA-expressing tumor antigens. Clinical trials have demonstrated the potential of these DC-based vaccines against a range of malignancies. Despite this tremendous promise, there are still obstacles to overcome, such as the immunosuppressive tumor microenvironment, variations in vaccine potency, technical difficulties in DC vaccine production, and a lack of reliable patient biomarkers [158].

12.10.1.5 Whole-Cell-Based Vaccines

Whole-cell-based vaccines offer a broad approach to cancer immunotherapy by combining a wide range of tumor-associated antigens to elicit a strong immune response. These vaccines use irradiated autologous or allogeneic tumor cells to expose the immune system to the tumor's entire antigenic repertoire, triggering targeted and multivalent defenses against a variety of tumor antigens. This method eliminates the need to identify specific antigens for each patient, which can be costly and time-consuming, and addresses the issue of antigen deletion or downregulation, a common tumor escape strategy. However, limitations exist. Producing autologous whole-cell-vaccines can be labor-intensive and complex, requiring isolation and culture of each patient's tumor cells. The immunosuppressive tumor microenvironment may reduce vaccine effectiveness, necessitating the use of adjuvants or immunomodulatory drugs to enhance immunogenicity. Concerns also persist regarding potential autoimmunity triggered by self-antigens in the vaccine formulation. While whole-cell-based vaccines present a promising avenue for cancer immunotherapy, further optimization and refinement are needed to overcome these obstacles and constraints [159].

12.11 COMBINATION THERAPIES

12.11.1 Immune Checkpoint Inhibitors and Cancer Vaccines

To increase the effectiveness of cancer vaccines, they may be used in combination with other treatments. Pembrolizumab (Keytruda), nivolumab (Opdivo), and ipilimumab (Yervoy) are examples of immune checkpoint inhibitors that block immunological checkpoints, unleashing a more powerful assault on malignant cells. This combination aims to enhance immune response (via checkpoint inhibitors) and improve the detection of cancer cells (via vaccination). By stimulating tumor-specific T cells and preventing their depletion, this combination strategy has shown potential in early clinical studies and preclinical models. Cancer vaccines aim to make tumor antigens more visible to T cells; however, the tumor's defense mechanisms can weaken this immune response. Checkpoint inhibitors block T cells' inhibitory checkpoints (PD-1 and CTLA-4), "releasing the brakes" and enhancing their anti-tumor efficacy.

Combining these inhibitors, which ensure T cells are not repressed, with cancer vaccines, which increase the number of T cells recognizing tumors, results in a cooperative enhancement of the anti-tumor immune response. Initial investigations suggest that this combination amplifies the attack on tumors, potentially leading to improved patient outcomes [160].

12.11.1.2 Cancer Vaccines and Chemotherapy

Chemotherapy impairs cancerous cells by inhibiting their proliferation and division. However, it may inadvertently damage rapidly dividing normal cells, such as those in the skin, digestive tract, and bone marrow. The potential synergy between chemotherapy and cancer vaccines stems from certain chemotherapy drugs that cause immunogenic cell death. This process may increase the efficacy of cancer vaccines by making dying cancer cells more visible to the immune system.

Several chemotherapeutic drugs have shown potential to increase the effectiveness of cancer vaccines due to their immunomodulatory actions. For instance, temozolomide and cyclophosphamide can reduce immunosuppressive regulatory T cells (Tregs), making the tumor microenvironment more receptive to vaccination. Docetaxel can improve antigen presentation in cancers such as breast and prostate malignancies, helping the immune system recognize tumor cells more effectively. Additionally, gemcitabine specifically reduces myeloid-derived suppressor cells (MDSCs). When used in conjunction with cancer vaccines, these drugs may alter the tumor environment, reduce immune system inhibition, or boost anti-tumor defenses. However, choosing the right combination depends on various factors, including the patient's condition and cancer type [161].

12.11.1.3 Radiotherapy and Cancer Vaccines

High-energy particles or waves such as protons, electron beams, gamma rays, or X-rays are used in radiotherapy to destroy or damage tumors. This radiation induces DNA damage in cells, preventing their proliferation and division and ultimately resulting in their demise. Radiotherapy may work in concert with cancer vaccines to increase their effectiveness. Research shows that radiation therapy affects the tumor's microenvironment in both cytotoxic and immunomodulatory ways. It not only causes direct damage to tumor cells but also induces immunogenic cell death, releasing DAMPs.

These DAMPs act as "danger signals," stimulating dendritic cell activity and fostering anti-tumor immune responses. At the same time, radiation therapy damages blood vessels within the tumor, increasing their permeability by directly affecting endothelial cells and by increasing VEGF release from radiation-damaged tumor cells. This may lead to greater immune cell infiltration into the tumor and temporary increases in oxygen and nutrient delivery [162].

12.12 ADVANCES IN IMMUNOLOGICAL RESEARCH

12.12.1 Omics Technology and Personalized Medicine in Immunology

The advancement of laboratory technologies and computational biomedicine has facilitated the production and analysis of extensive data. These developments will pave the way for the transformation of medicine, shifting from a uniform approach to a more comprehensive patient classification and tailored treatment strategy [163]. Precision medicine involves customizing medical therapy according to each patient's unique characteristics. This does not mean developing drugs exclusively for individual patients, but rather classifying patients into subpopulations that respond differently to specific treatments. The focus is on recognizing which treatments are effective for specific patients based on their individual genetic and epigenetic attributes. For instance, Herceptin is only effective in breast cancer patients who overexpress HER2. However, the term "precision" can be misleading as it implies a level of accuracy that medicine may lack [164].

The transition to individualized therapy requires a comprehensive, unbiased examination of genomic and molecular characteristics in individuals with specific diseases. This aims to identify dependable patient-specific biomarkers linking genotypes, molecular profiles/endotypes, disease development, and "Omics" data. These biomarkers are then computationally processed to identify personalized markers.

Endotyping (categorizing individuals based on underlying biological mechanisms), genotyping (analyzing individual genetic makeup), and theratyping (tailoring therapeutic interventions based on individual characteristics) have gained substantial attention in precision medicine. Precision medicine aims to revolutionize patient care and develop more effective prevention and therapy plans. Precision health introduces methods for achieving and maintaining good health and preventing allergic diseases and asthma. Allergy plays a prominent role in this area, as allergen-specific immunotherapy is naturally one of the major approaches to personalized/precision medicine, providing the potential for individualized care instead of conventional symptomatic treatments [165].

Big data collection and computational analysis will enable medicine to move from a "one-size-fits-all" approach to precise patient classification and therapy. Notable advancements have been made in personalized oncology using "Omics" data. In cancer, personalized medicine uses proteomic, immunological, and genetic profiling data to guide treatment decisions and offer patient-specific prognostic insights.

The advent of checkpoint inhibitors in immuno-oncology has highlighted that most patients may not respond clinically to these treatments but risk serious immunologic adverse effects. This emphasizes the need for biomarkers to predict patient response and disease progression. Recent success in this area enables the identification of patients likely to benefit from these medications, a first step toward truly personalized immuno-oncology [166].

12.13 CONCLUSION

This chapter emphasizes the significance of the complex mechanisms that control the functions of the immune system. From the building blocks of the immune system to the complex molecular features of inflammatory signaling pathways. Beginning with the basic features of the immune defense system and the major secondary immune organs, we've explored the important functions performed by vital organs like the bone marrow, spleen, thymus, and LNs.

The immune system's components have been examined, explaining the different roles played by innate immune responses and important cellular players, such as NK cells, mast cells, granulocytes, neutrophils, eosinophils, basophils, monocytes/macrophages, and dendritic cells. The thymus and spleen play a crucial role in training immune cells, which are critical for the body's defense mechanism and for fighting against pathogens as well as different types of tumors. Innate and adaptive immunity, key components of the immune system, have been discussed, highlighting how innate immune cells respond rapidly and how adaptive immunity builds immunological memory.

TCR activation is essential for optimal T cell function. T cells are key participants in immune responses and tumor defense. The complex TCR signaling network, which includes tyrosine kinases and adaptive immunological reactions, controls T-cell survival, proliferation, and differentiation. The elements, including $CD4^+$ and $CD8^+$ receptors, as well as the signaling cascade, are essential to comprehending the intricacies of T-cell-mediated immunological reactions.

Inflammatory signaling networks, such as the JAK/STAT, MAPK, and NF-κB pathways, have been explored. The JAK/STAT pathway, frequently triggered by the cytokines produced during inflammation, plays a part in controlling immunological responses, including immune cell activation and the production of inflammatory mediators. The MAPK pathway, a key component of the inflammatory response, controls the expression of genes connected to apoptosis, inflammation, and cell division. Numerous stimuli, including pro-inflammatory cytokines, can activate MAPK.

The dysregulation of inflammatory signaling, an imbalance or breakdown in the complex pathways that control the body's inflammatory response, has been discussed. The molecular pathways, cytokine networks, and cellular interactions have revealed the impressive complexity of the immune system. Inflammatory signaling is a double-edged sword that, when dysregulated, can both contribute to pathological states and promote defensive responses. When these signaling pathways are dysregulated, it might lead to numerous health problems and different diseases like SLE, arthritis, cancer, autoimmune illnesses, and consequences from atherosclerosis.

Finally, the architecture, signaling routes, and activities of receptors implicated in inflammatory signaling, including TLRs, cytokine receptors, and NLRs, have been examined. The field of medical treatments is constantly changing, as seen by the investigation of cancer vaccines and their varieties, combination therapy, and the potential benefits of corticosteroids and NSAIDs.

REFERENCES

1. Gombart, A.F., A. Pierre, and S. Maggini, A review of micronutrients and the immune system–working in harmony to reduce the risk of infection. *Nutrients*, 2020. **12**(1): p. 236.
2. Cruvinel, W.D.M., et al., Immune system: Part I. Fundamentals of innate immunity with emphasis on molecular and cellular mechanisms of inflammatory response. *Revista brasileira de reumatologia*, 2010. **50**: p. 434–447.
3. Plitas, G. and A.Y. Rudensky, Regulatory T cells: differentiation and function. *Cancer Immunology Research*, 2016. **4**(9): p. 721–725.
4. Castelo-Branco, C. and I. Soveral, The immune system and aging: a review. *Gynecological Endocrinology*, 2014. **30**(1): p. 16–22.
5. Punchard, N.A., C.J. Whelan, and I. Adcock, The journal of inflammation. *BioMed Central*. 2004. **1**: p. 1–4.
6. Newton, K. and V.M. Dixit, Signaling in innate immunity and inflammation. *Cold Spring Harbor Perspectives in Biology*, 2012. **4**(3): p. a006049.
7. McComb, S., et al., Introduction to the immune system. *Immunoproteomics: Methods and Protocols*, 2013: p. 1–20.
8. Liang, Y., T. Zhang, and M. Tang, Toxicity of quantum dots on target organs and immune system. *Journal of Applied Toxicology*, 2022. **42**(1): p. 17–40.
9. Li, L., et al., Cellular mechanism of thymic involution. *Scandinavian Journal of Immunology*, 2003. **57**(5): p. 410–422.
10. Gulla, S., et al., Role of thymus in health and disease. *International Reviews of Immunology*, 2023. **42**(5): p. 347–363.

11. Schirrmacher, V., Bone marrow: the central immune system. *Immuno*, 2023. **3**(3): p. 289–329.
12. Zhao, E., et al., Bone marrow and the control of immunity. *Cellular & Molecular Immunology*, 2012. **9**(1): p. 11–19.
13. Cesta, M.F., Normal structure, function, and histology of the spleen. *Toxicologic Pathology*, 2006. **34**(5): p. 455–465.
14. Mebius, R.E. and G. Kraal, Structure and function of the spleen. *Nature Reviews Immunology*, 2005. **5**(8): p. 606–616.
15. Gasteiger, G., M. Ataide, and W. Kastenmüller, Lymph node–an organ for T-cell activation and pathogen defense. *Immunological Reviews*, 2016. **271**(1): p. 200–220.
16. Buettner, M. and U. Bode, Lymph node dissection–understanding the immunological function of lymph nodes. *Clinical & Experimental Immunology*, 2012. **169**(3): p. 205–212.
17. Medzhitov, R. and C.A. Janeway Jr, Innate immunity: impact on the adaptive immune response. *Current Opinion in Immunology*, 1997. **9**(1): p. 4–9.
18. Vasselon, T. and P.A. Detmers, Toll receptors: a central element in innate immune responses. *Infection and Immunity*, 2002. **70**(3): p. 1033–1041.
19. Coates, M., S. Blanchard, and A.S. MacLeod, Innate antimicrobial immunity in the skin: a protective barrier against bacteria, viruses, and fungi. *PLoS Pathogens*, 2018. **14**(12): p. e1007353.
20. Aristizábal, B. and Á. González, Innate immune system. In *Autoimmunity: From Bench to Bedside [Internet]*. El Rosario University Press, 2013.
21. Mai, J., et al., An evolving new paradigm: endothelial cells–conditional innate immune cells. *Journal of Hematology & Oncology*, 2013. **6**(1): p. 1–13.
22. Haslett, C., Granulocyte apoptosis and its role in the resolution and control of lung inflammation. *American Journal of Respiratory and Critical Care Medicine*, 1999. **160**(supplement_1): p. S5–S11.
23. Jaillon, S., et al. Neutrophils in innate and adaptive immunity. In *Seminars in Immunopathology*. Springer, 2013.
24. Segal, A.W., How neutrophils kill microbes. *Annual Review of Immunology*, 2005. **23**: p. 197–223.
25. Hogan, S.P., et al., Eosinophils: biological properties and role in health and disease. *Clinical & Experimental Allergy*, 2008. **38**(5): p. 709–750.
26. Otsuka, A. and K. Kabashima, Mast cells and basophils in cutaneous immune responses. *Allergy*, 2015. **70**(2): p. 131–140.
27. Arinobu, Y., H. Iwasaki, and K. Akashi, Origin of basophils and mast cells. *Allergology International*, 2009. **58**(1): p. 21–28.
28. Germic, N., et al., Regulation of the innate immune system by autophagy: monocytes, macrophages, dendritic cells and antigen presentation. *Cell Death & Differentiation*, 2019. **26**(4): p. 715–727.
29. Prame Kumar, K., A.J. Nicholls, and C.H. Wong, Partners in crime: neutrophils and monocytes/macrophages in inflammation and disease. *Cell and Tissue Research*, 2018. **371**: p. 551–565.
30. Kushwah, R. and J. Hu, Complexity of dendritic cell subsets and their function in the host immune system. *Immunology*, 2011. **133**(4): p. 409–419.
31. Hespel, C. and M. Moser, Role of inflammatory dendritic cells in innate and adaptive immunity. *European Journal of Immunology*, 2012. **42**(10): p. 2535–2543.
32. Freud, A.G. and M.A. Caligiuri, Human natural killer cell development. *Immunological Reviews*, 2006. **214**(1): p. 56–72.
33. Zitti, B. and Y.T. Bryceson, Natural killer cells in inflammation and autoimmunity. *Cytokine & Growth Factor Reviews*, 2018. **42**: p. 37–46.
34. Vandendriessche, S., et al., Complement receptors and their role in leukocyte recruitment and phagocytosis. *Frontiers in Cell and Developmental Biology*, 2021. **9**: p. 624025.
35. Acharya, D., et al., Complement receptor-mediated phagocytosis induces proinflammatory cytokine production in murine macrophages. *Frontiers in Immunology*, 2020. **10**: p. 3049.
36. Rus, H., C. Cudrici, and F. Niculescu, The role of the complement system in innate immunity. *Immunologic Research*, 2005. **33**: p. 103–112.
37. Sun, S.-C., Non-canonical NF-κB signaling pathway. *Cell Research*, 2011. **21**(1): p. 71–85.
38. Kościelska-Kasprzak, K., et al., The complement cascade and renal disease. *Archivum immunologiae et therapiae experimentalis*, 2014. **62**: p. 47–57.
39. Medzhitov, R., Origin and physiological roles of inflammation. *Nature*, 2008. **454**(7203): p. 428–435.
40. Mitroulis, I., et al., Leukocyte integrins: role in leukocyte recruitment and as therapeutic targets in inflammatory disease. *Pharmacology & Therapeutics*, 2015. **147**: p. 123–135.
41. Ed Rainger, G., et al., The role of platelets in the recruitment of leukocytes during vascular disease. *Platelets*, 2015. **26**(6): p. 507–520.

42. Schenten, D. and R. Medzhitov, The control of adaptive immune responses by the innate immune system. *Advances in Immunology*, 2011. **109**: p. 87–124.
43. Kumar, H., T. Kawai, and S. Akira, Pathogen recognition by the innate immune system. *International Reviews of Immunology*, 2011. **30**(1): p. 16–34.
44. Steinman, R.M., Dendritic cells: versatile controllers of the immune system. *Nature Medicine*, 2007. **13**(10): p. 1155–1159.
45. Slaney, C.Y., M.H. Kershaw, and P.K. Darcy, Trafficking of T cells into tumors. *Cancer Research*, 2014. **74**(24): p. 7168–7174.
46. Williams, M.S. and J. Kwon, T cell receptor stimulation, reactive oxygen species, and cell signaling. *Free Radical Biology and Medicine*, 2004. **37**(8): p. 1144–1151.
47. Zuckermann, F.A., Extrathymic CD4/CD8 double positive T cells. *Veterinary Immunology and Immunopathology*, 1999. **72**(1–2): p. 55–66.
48. Jiang, H. and L. Chess, Regulation of immune responses by T cells. *New England Journal of Medicine*, 2006. **354**(11): p. 1166–1176.
49. Acuto, O. and D. Cantrell, T cell activation and the cytoskeleton. *Annual Review of Immunology*, 2000. **18**(1): p. 165–184.
50. Myung, P.S., N.J. Boerthe, and G.A. Koretzky, Adapter proteins in lymphocyte antigen-receptor signaling. *Current Opinion in Immunology*, 2000. **12**(3): p. 256–266.
51. Cantrell, D.A., T-cell antigen receptor signal transduction. *Immunology*, 2002. **105**(4): p. 369.
52. Zhang, W. and L.E. Samelson. The role of membrane-associated adaptors in T cell receptor signalling. In *Seminars in Immunology*. Elsevier, 2000.
53. Mustelin, T., et al., Protein tyrosine phosphorylation in T cell signaling. *Front Biosci*, 2002. **7**: p. d918–d969.
54. Combs, J., Kim, S. J., Tan, S., Ligon, L. A., Holzbaur, E. L., Kuhn, J., and Poenie, M. Recruitment of dynein to the Jurkat immunological synapse. *Proceedings of the National Academy of Sciences*, 2006. **103**(40):p. 14883–14888.
55. Huse, M. The T-cell-receptor signaling network. *Journal of Cell Science*, 2009. **122**(9): p. 1269–1273.
56. Rodríguez-Jorge, O., et al., Cooperation between T cell receptor and Toll-like receptor 5 signaling for CD4+ T cell activation. *Science Signaling*, 2019. **12**(577): p. eaar3641.
57. Guy, C.S. and D.A. Vignali, Organization of proximal signal initiation at the TCR: CD3 complex. *Immunological Reviews*, 2009. **232**(1): p. 7–21.
58. Huse, M., The T-cell-receptor signaling network. *Journal of Cell Science*, 2009. **122**(9): p. 1269–1273.
59. Hoffman, W., F.G. Lakkis, and G. Chalasani, B cells, antibodies, and more. *Clinical Journal of the American Society of Nephrology: CJASN*, 2016. **11**(1): p. 137.
60. Bouaziz, J.D., K. Yanaba, and T.F. Tedder, Regulatory B cells as inhibitors of immune responses and inflammation. *Immunological Reviews*, 2008. **224**(1): p. 201–214.
61. Manley, N.R., et al., Structure and function of the thymic microenvironment. *Front Biosci (Landmark Ed)*, 2011. **16**(7): p. 2461–2477.
62. Liang, Z., et al., Age-related thymic involution: Mechanisms and functional impact. *Aging Cell*, 2022. **21**(8): p. e13671.
63. Dal Porto, J.M., et al., B cell antigen receptor signaling 101. *Molecular Immunology*, 2004. **41**(6–7): p. 599–613.
64. Pierce, S.K. and W. Liu, The tipping points in the initiation of B cell signalling: how small changes make big differences. *Nature Reviews Immunology*, 2010. **10**(11): p. 767–777.
65. Packard, T.A. and J.C. Cambier, B lymphocyte antigen receptor signaling: initiation, amplification, and regulation. *F1000prime Reports*, 2013. **5**.
66. Su, T.T., et al., PKC-β controls IκB kinase lipid raft recruitment and activation in response to BCR signaling. *Nature Immunology*, 2002. **3**(8): p. 780–786.
67. Yam-Puc, J.C., et al., Role of B-cell receptors for B-cell development and antigen-induced differentiation. *F1000Research*, 2018. **7**.
68. Straub, R.H., et al., Cytokines and hormones as possible links between endocrinosenescence and immunosenescence. *Journal of Neuroimmunology*, 2000. **109**(1): p. 10–15.
69. Dinarello, C.A., Historical insights into cytokines. *European Journal of Immunology*, 2007. **37**(S1): p. S34–S45.
70. Nicod, L., Cytokines. 1. Overview. *Thorax*, 1993. **48**(6): p. 660.
71. Garlanda, C., C.A. Dinarello, and A. Mantovani, The interleukin-1 family: back to the future. *Immunity*, 2013. **39**(6): p. 1003–1018.
72. Dinarello, C. and J. Mier, Interleukins. *Annual Review of Medicine*, 1986. **37**(1): p. 173–178.

73. Klipp, E. and W. Liebermeister, Mathematical modeling of intracellular signaling pathways. *BMC Neuroscience*, 2006. **7**: p. 1–16.
74. Li, Q. and I.M. Verma, NF-κB regulation in the immune system. *Nature Reviews Immunology*, 2002. **2**(10): p. 725–734.
75. Moynagh, P.N., The NF-κB pathway. *Journal of Cell Science*, 2005. **118**(20): p. 4589–4592.
76. Liu, T., et al., NF-κB signaling in inflammation. *Signal Transduction and Targeted Therapy*, 2017. **2**(1): p. 1–9.
77. Molina, J.R. and A.A. Adjei, The ras/raf/mapk pathway. *Journal of Thoracic Oncology*, 2006. **1**(1): p. 7–9.
78. Plotnikov, A., et al., The MAPK cascades: signaling components, nuclear roles and mechanisms of nuclear translocation. *Biochimica et Biophysica Acta (BBA)-Molecular Cell Research*, 2011. **1813**(9): p. 1619–1633.
79. Wortzel, I. and R. Seger, The ERK cascade: distinct functions within various subcellular organelles. *Genes & Cancer*, 2011. **2**(3): p. 195–209.
80. Guo, Y.J., et al., ERK/MAPK signalling pathway and tumorigenesis. *Experimental and Therapeutic Medicine*, 2020. **19**(3): p. 1997–2007.
81. Cossa, G., et al., Modulation of sensitivity to antitumor agents by targeting the MAPK survival pathway. *Current Pharmaceutical Design*, 2013. **19**(5): p. 883–894.
82. Dhanasekaran, N. and E.P. Reddy, Signaling by dual specificity kinases. *Oncogene*, 1998. **17**(11): p. 1447–1455.
83. De Luca, A., et al., The RAS/RAF/MEK/ERK and the PI3K/AKT signalling pathways: role in cancer pathogenesis and implications for therapeutic approaches. *Expert Opinion on Therapeutic Targets*, 2012. **16**(2): p. S17–S27.
84. Robinson, M.J. and M.H. Cobb, Mitogen-activated protein kinase pathways. *Current Opinion in Cell Biology*, 1997. **9**(2): p. 180–186.
85. Ma, L., et al., Phosphorylation and functional inactivation of TSC2 by Erk: implications for tuberous sclerosisand cancer pathogenesis. *Cell*, 2005. **121**(2): p. 179–193.
86. Chu, B., et al., Sequential phosphorylation by mitogen-activated protein kinase and glycogen synthase kinase 3 represses transcriptional activation by heat shock factor-1. *Journal of Biological Chemistry*, 1996. **271**(48): p. 30847–30857.
87. Burotto, M., et al., The MAPK pathway across different malignancies: a new perspective. *Cancer*, 2014. **120**(22): p. 3446–3456.
88. Leonard, W., Type 1 cytokines and interferons and their receptors. *Fundamental Immunology*. 2003. **5**: 1–4.
89. Imada, K. and W.J. Leonard, The jak-STAT pathway. *Molecular Immunology*, 2000. **37**(1–2): p. 1–11.
90. Fu, X.-Y., et al., ISGF3, the transcriptional activator induced by interferon alpha, consists of multiple interacting polypeptide chains. *Proceedings of the National Academy of Sciences*, 1990. **87**(21): p. 8555–8559.
91. Fu, X.-Y., A direct signaling pathway through tyrosine kinase activation of SH2 domain-containing transcription factors. *Journal of Leucocyte Biology*, 1995. **57**(4): p. 529–535.
92. Zhong, Z., Z. Wen, and J.E. Darnell Jr, Stat3 and Stat4: members of the family of signal transducers and activators of transcription. *Proceedings of the National Academy of Sciences*, 1994. **91**(11): p. 4806–4810.
93. LIu, X., et al., Cloning and expression of Stat5 and an additional homologue (Stat5b) involved in prolactin signal transduction in mouse mammary tissue. *Proceedings of the National Academy of Sciences*, 1995. **92**(19): p. 8831–8835.
94. Wilks, A.F., Two putative protein-tyrosine kinases identified by application of the polymerase chain reaction. *Proceedings of the National Academy of Sciences*, 1989. **86**(5): p. 1603–1607.
95. Wilks, A.F., et al., Two novel protein-tyrosine kinases, each with a second phosphotransferase-related catalytic domain, define a new class of protein kinase. *Molecular and Cellular Biology*, 1991. **11**(4): p. 2057–2065.
96. Krolewski, J., et al., Identification and chromosomal mapping of new human tyrosine kinase genes. *Oncogene*, 1990. **5**(3): p. 277–282.
97. Müller, M., et al., The protein tyrosine kinase JAK1 complements defects in interferon-α/β and-γ signal transduction. *Nature*, 1993. **366**(6451): p. 129–135.
98. Hu, X., et al., The JAK/STAT signaling pathway: from bench to clinic. *Signal Transduction and Targeted Therapy*, 2021. **6**(1): p. 402.

99. Baum, C.L. and C.J. Arpey, Normal cutaneous wound healing: clinical correlation with cellular and molecular events. *Dermatologic Surgery*, 2005. **31**(6): p. 674–686.
100. Oronsky, B., S. Caroen, and T. Reid, What Exactly Is Inflammation (and What Is It Not?). *International Journal of Molecular Sciences*, 2022. **23**(23): p. 14905.
101. Lu, H., W. Ouyang, and C. Huang, Inflammation, a key event in cancer development. *Molecular Cancer Research*, 2006. **4**(4): p. 221–233.
102. Lu, T., et al., Prevention of hepatocellular carcinoma in chronic viral hepatitis B and C infection. *World Journal of Gastroenterology: WJG*, 2013. **19**(47): p. 8887.
103. Rafieian-Kopaei, M., et al., Atherosclerosis: process, indicators, risk factors and new hopes. *International Journal of Preventive Medicine*, 2014. **5**(8): p. 927.
104. Libby, P., Inflammatory mechanisms: the molecular basis of inflammation and disease. *Nutrition Reviews*, 2007. **65**(suppl_3): p. S140–S146.
105. Müller-Ladner, U., et al., Mechanisms of disease: the molecular and cellular basis of joint destruction in rheumatoid arthritis. *Nature Clinical Practice Rheumatology*, 2005. **1**(2): p. 102–110.
106. Feldmann, M. and R.N. Maini, Anti-TNFα therapy of rheumatoid arthritis: what have we learned? *Annual Review of Immunology*, 2001. **19**(1): p. 163–196.
107. May, M.J. and S. Ghosh, Signal transduction through NF-κB. *Immunology Today*, 1998. **19**(2): p. 80–88.
108. Simmonds, R.E. and B.M. Foxwell, Signalling, inflammation and arthritis: NF-κ B and its relevance to arthritis and inflammation. *Rheumatology*, 2008. **47**(5): p. 584–590.
109. Herrada, A.A., et al., Innate immune cells' contribution to systemic lupus erythematosus. *Frontiers in Immunology*, 2019. **10**: p. 772.
110. Yeung, Y.T., et al., Signaling pathways in inflammation and anti-inflammatory therapies. *Current Pharmaceutical Design*, 2018. **24**(14): p. 1449–1484.
111. Tanti, J.-F., et al., Implication of inflammatory signaling pathways in obesity-induced insulin resistance. *Frontiers in Endocrinology*, 2013. **3**: p. 181.
112. Takeda, K., T. Kaisho, and S. Akira, Toll-like receptors. *Annual Review of Immunology*, 2003. **21**(1): p. 335–376.
113. Matsushima, N., et al., Comparative sequence analysis of leucine-rich repeats (LRRs) within vertebrate toll-like receptors. *BMC Genomics*, 2007. **8**: p. 1–20.
114. Beutler, B., et al., Genetic analysis of host resistance: Toll-like receptor signaling and immunity at large. *Annual Review of Immunology*, 2006. **24**: p. 353–389.
115. Chaturvedi, A. and S.K. Pierce, How location governs Toll-like receptor signaling. *Traffic*, 2009. **10**(6): p. 621–628.
116. Ishii, K.J., S. Uematsu, and S. Akira, 'Toll'gates for future immunotherapy. *Current Pharmaceutical Design*, 2006. **12**(32): p. 4135–4142.
117. Chang, Z., Important aspects of Toll-like receptors, ligands and their signaling pathways. *Inflammation Research*, 2010. **59**: p. 791–808.
118. El-Zayat, S.R., H. Sibaii, and F.A. Mannaa, Toll-like receptors activation, signaling, and targeting: an overview. *Bulletin of the National Research Centre*, 2019. **43**(1): p. 1–12.
119. Akira, S., Toll-like receptor signaling. *Journal of Biological Chemistry*, 2003. **278**(40): p. 38105–38108.
120. Blasius, A.L. and B. Beutler, Intracellular toll-like receptors. *Immunity*, 2010. **32**(3): p. 305–315.
121. O'Shea, J.J., M. Gadina, and R.M. Siegel, Cytokines and cytokine receptors. In Clinical Immunology. Elsevier, 2019. p. 127–155. e1.
122. Brooks, A.J., F. Dehkhoda, and B.B. Kragelund, Cytokine receptors. *Princ Endocrinol Horm Action*. 2016. **4**: 1–20.
123. Chen, L., et al., NOD-like receptors in autoimmune diseases. *Acta Pharmacologica Sinica*, 2021. **42**(11): p. 1742–1756.
124. Irving, A.T., et al., The immune receptor NOD1 and kinase RIP2 interact with bacterial peptidoglycan on early endosomes to promote autophagy and inflammatory signaling. *Cell Host & Microbe*, 2014. **15**(5): p. 623–635.
125. Shaw, M.H., et al., NOD-like receptors (NLRs): bona fide intracellular microbial sensors. *Current Opinion in Immunology*, 2008. **20**(4): p. 377–382.
126. Bianchi, M.E., DAMPs, PAMPs and alarmins: all we need to know about danger. *Journal of Leucocyte Biology*, 2007. **81**(1): p. 1–5.
127. Dinarello, C.A., Proinflammatory cytokines. *Chest*, 2000. **118**(2): p. 503–508.
128. Medzhitov, R., Inflammation 2010: new adventures of an old flame. *Cell*, 2010. **140**(6): p. 771–776.
129. Kyriakis, J.M. and J. Avruch, Sounding the alarm: protein kinase cascades activated by stress and inflammation. *Journal of Biological Chemistry*, 1996. **271**(40): p. 24313–24316.

130. Abou-Raya, A. and S. Abou-Raya, Inflammation: a pivotal link between autoimmune diseases and atherosclerosis. *Autoimmunity Reviews*, 2006. **5**(5): p. 331–337.
131. Brydges, S. and D. Kastner, The systemic autoinflammatory diseases: inborn errors of the innate immune system. In Current Concepts in Autoimmunity and Chronic Inflammation. Springer, 2006. p. 127–160.
132. Kahkhaie, K.R., et al., Curcumin: a modulator of inflammatory signaling pathways in the immune system. *Inflammopharmacology*, 2019. **27**: p. 885–900.
133. Kaminska, B., MAPK signalling pathways as molecular targets for anti-inflammatory therapy—from molecular mechanisms to therapeutic benefits. *Biochimica et Biophysica Acta (BBA)-Proteins and Proteomics*, 2005. **1754**(1–2): p. 253–262.
134. Hendrayani, S.-F., et al., The inflammatory/cancer-related IL-6/STAT3/NF-κB positive feedback loop includes AUF1 and maintains the active state of breast myofibroblasts. *Oncotarget*, 2016. **7**(27): p. 41974.
135. Iwasaki, A. and R. Medzhitov, Toll-like receptor control of the adaptive immune responses. *Nature Immunology*, 2004. **5**(10): p. 987–995.
136. Lintermans, L.L., et al., T cells in vascular inflammatory diseases. *Frontiers in Immunology*, 2014. **5**: p. 504.
137. Sugimoto, M.A., et al., Resolution of inflammation: what controls its onset? *Frontiers in Immunology*, 2016. **7**: p. 160.
138. Chen, L., et al., Inflammatory responses and inflammation-associated diseases in organs. *Oncotarget*, 2018. **9**(6): p. 7204.
139. Rainsford, K., Profile and mechanisms of gastrointestinal and other side effects of nonsteroidal anti-inflammatory drugs (NSAIDs). *The American Journal of Medicine*, 1999. **107**(6): p. 27–35.
140. Van Staa, T., et al., Use of oral corticosteroids and risk of fractures. *Journal of Bone and Mineral Research*, 2000. **15**(6): p. 993–1000.
141. Barnes, P.J., How corticosteroids control inflammation: quintiles prize lecture 2005. *British Journal of Pharmacology*, 2006. **148**(3): p. 245–254.
142. Decker, T. and A. Meinke, Jaks, Stats and the immune system. *Immunobiology*, 1997. **198**(1–3): p. 99–111.
143. Villarino, A.V., Y. Kanno, and J.J. O'Shea, Mechanisms and consequences of Jak–STAT signaling in the immune system. *Nature Immunology*, 2017. **18**(4): p. 374–384.
144. Rodig, S.J., et al., Disruption of the Jak1 gene demonstrates obligatory and nonredundant roles of the Jaks in cytokine-induced biologic responses. *Cell*, 1998. **93**(3): p. 373–383.
145. Shahmarvand, N., et al., Mutations in the signal transducer and activator of transcription family of genes in cancer. *Cancer Science*, 2018. **109**(4): p. 926–933.
146. Andersson, E.I., et al., High incidence of activating STAT5B mutations in CD4-positive T-cell large granular lymphocyte leukemia. *Blood, The Journal of the American Society of Hematology*, 2016. **128**(20): p. 2465–2468.
147. Li, S.D., et al., Cancer gene profiling in non-small cell lung cancers reveals activating mutations in JAK2 and JAK3 with therapeutic implications. *Genome Medicine*, 2017. **9**: p. 1–11.
148. Castro, F., et al., Interferon-gamma at the crossroads of tumor immune surveillance or evasion. *Frontiers in Immunology*, 2018. **9**: p. 847.
149. Owen, K., N. Brockwell, and B. Parker, JAK-STAT signaling: a double-edged sword of immune regulation and cancer progression. *Cancers (Basel)*. 2019.
150. Quinn, K.M., et al., Impact of age-, cancer-, and treatment-driven inflammation on T cell function and immunotherapy. *Journal of Leucocyte Biology*, 2020. **108**(3): p. 953–965.
151. Din, S., et al., Mutational analysis identifies therapeutic biomarkers in inflammatory bowel disease–associated colorectal cancers. *Clinical Cancer Research*, 2018. **24**(20): p. 5133–5142.
152. Chen, M., R. Linstra, and M.A. van Vugt, Genomic instability, inflammatory signaling and response to cancer immunotherapy. *Biochimica et Biophysica Acta (BBA)-Reviews on Cancer*, 2022. **1877**(1): p. 188661.
153. Arora, S., et al., Potential application of immunotherapy for modulation of pulp inflammation: opportunities for vital pulp treatment. *International Endodontic Journal*, 2021. **54**(8): p. 1263–1274.
154. Bashir, D., *Immunotherapy for Cancer: Harnessing the Power of the Immune System*. MDPI, 2023.
155. Kumai, T., et al., Peptide vaccines in cancer—old concept revisited. *Current Opinion in Immunology*, 2017. **45**: p. 1–7.
156. Johnson, P.A., et al., Advances in DNA-and RNA-based oncolytic viral therapeutics and immunotherapies. *Applied Microbiology*, 2022. **2**(2): p. 319–329.

157. Harrop, R., J. John, and M.W. Carroll, Recombinant viral vectors: cancer vaccines. *Advanced Drug Delivery Reviews*, 2006. **58**(8): p. 931–947.
158. Santos, P.M. and L.H. Butterfield, Dendritic cell–based cancer vaccines. *The Journal of Immunology*, 2018. **200**(2): p. 443–449.
159. Liu, D., et al., tumor vaccines: Unleashing the power of the immune system to fight cancer. *Pharmaceuticals*, 2023. **16**(10): p. 1384.
160. Zhao, J., et al., Safety and efficacy of therapeutic cancer vaccines alone or in combination with immune checkpoint inhibitors in cancer treatment. *Frontiers in Pharmacology*, 2019. **10**: p. 1184.
161. Weir, G.M., R.S. Liwski, and M. Mansour, Immune modulation by chemotherapy or immunotherapy to enhance cancer vaccines. *Cancers*, 2011. **3**(3): p. 3114–3142.
162. Hannani, D., et al., Prerequisites for the antitumor vaccine-like effect of chemotherapy and radiotherapy. *The Cancer Journal*, 2011. **17**(5): p. 351–358.
163. Delhalle, S., et al., A roadmap toward personalized immunology. *NPJ Systems Biology and Applications*, 2018. **4**(1): p. 9.
164. Litman, T., Personalized medicine—concepts, technologies, and applications in inflammatory skin diseases. *Apmis*, 2019. **127**(5): p. 386–424.
165. Kucuksezer, U.C., et al., Precision/personalized medicine in allergic diseases and asthma. *Archivum immunologiae et therapiae experimentalis*, 2018. **66**: p. 431–442.
166. Gwin, W.R., M.L. Disis, and E. Ruiz-Garcia, Immuno-oncology in the era of personalized medicine. *Translational Research and Onco-Omics Applications in the Era of Cancer Personal Genomics*, 2019: p. 117–129.

13 Cell Signaling in Cancer

Khoula Sharif Mughal[1], *Nabi Shah*[1], *Muhammad Naeem*[2], *and Muhammad Ikram*[*1]
[1]Department of Pharmacy, COMSATS University Islamabad, Abbottabad Campus, Pakistan
[2]Department of Biological Sciences, National University of Medical Sciences. Rawalpindi, Pakistan

13.1 INTRODUCTION

Cancer is a major cause of death worldwide and a serious public health concern (Peng et al. 2022). According to the American Cancer Society's 2017 report, breast, lung, colon, and pancreatic cancers account for 46% of all cancer deaths (Li et al. 2017). Cancer results from uncontrolled cell growth and division due to genetic and epigenetic modifications that disrupt normal cell function (L. Xia et al. 2018). These disruptions cause malignant cells to proliferate and avoid processes that normally restrict their growth, division, and migration (Vaghari-Tabari et al. 2021). Many of these disruptions correspond to alterations in cell signaling pathways, impacting processes like apoptosis, invasiveness, metastasis, growth, and proliferation (Vaghari-Tabari et al. 2021).

Signaling pathways are interconnected into complex networks in the intricate landscape of cellular communication. Cells receive signals are received from a variety of sources, including cell-matrix interactions, growth factor receptors, and cell-cell connections. They integrate this information to control vital functions such as protein synthesis, cell division, motility, cellular architecture, polarity, and programmed cell death. During tumor development, these signaling pathways are carefully controlled and can produce different results depending on the cell type. Understanding the changes in cancer cells and deciphering the operation of these very complicated signaling networks in vivo present enormous intellectual challenges (Martin 2003).

Cancer cell aggressiveness, which leads to cancer development and metastasis, involves:

1. Detachment from primary tumor,
2. Attachment to and breakdown of the extracellular matrix (ECM)
3. Invasion of the lymphatic or blood systems
4. Tumor angiogenesis
5. Proliferation at secondary sites to form metastases

This complex process characterizes tumor malignancy and poses a challenge to clinical treatment success. A critical therapeutic goal is to target mechanisms that limit or prevent tumor spread. Cancer stem cells (CSCs) play an important role in cancer genesis, recurrence, and metastasis. However, no treatments targeting CSCs are currently in clinical use. Key signaling pathways involved in CSC progression include Hedgehog, Notch, Wnt, Hippo, and autophagy signaling pathways (Yue Yang et al. 2020).

This chapter examines the complex interaction between cell signaling and cancer, focusing on crucial signaling pathways implicated in cancer advancement and the agents of irregular signaling pivotal in various cancer types.

DOI: 10.1201/9781003594901-13

13.2 THE RELATIONSHIP BETWEEN CELL SIGNALING AND CANCER

13.2.1 Cell Signaling Pathway and Cancer

Cell signaling allows cells to coordinate, communicate, and respond appropriately to specific stimuli. It influences cell structure and function (Shahi 2023), enabling cells to adapt to their surroundings. Signaling pathways are essential for communication between the cell surface, nucleus, and extracellular matrix (Lansdowne 2018). These signaling molecules can affect gene expression over time by either directly or indirectly regulating the activity of transcription factors, even if they are not directly involved in transcription [9].

Numerous proteins called kinases are involved in cell signaling pathways. These pathways allow transmembrane proteins, like receptor tyrosine kinases (RTKs), to send important signals to the nucleus for various functions, including metabolism, DNA replication, apoptosis, cell division, proliferation, and repair (Roy and Ghosh 2021). RTKs, characterized by unique structural and biological features, function as specialized receptors on the cell surface. Upon receiving ligands, they initiate specific signaling pathways within tumor cells. RTKs are vital in modulating downstream signaling such as mitogen-activated protein kinase (MAPK), phosphatidylinositol 3-kinase (PI3K)/Akt, and Janus-activated kinase (JAK)/signal transducer and activator of transcription (STAT), which play pivotal roles in regulating cancer stemness, angiogenesis, and metastasis (Butti et al. 2018).

When cells disregard the typical signals that indicate it's time to halt growth or undergo cell death, it leads to abnormalities in cell signaling pathways, which are pivotal in cancer onset and advancement ("Cell Signaling and How It Relates to Cancer," 2024). Cancer cells undergo various alterations during tumor progression, gaining the ability to multiply autonomously, invade neighboring tissues, metastasize to distanced sites, cause angiogenesis, and resist mechanisms that limit cell proliferation and survival. These changes are driven by modifications in cellular signaling pathways that normally govern cell proliferation, motility, and survival in healthy cells. Signaling pathways contain several proteins that have been explored as possible targets for cancer therapy (Martin 2003).

Mutations in signaling proteins enable cells to disregard immune signals, leading to unregulated cell division, genetic instability, hypoxia-induced angiogenesis, immune system evasion, invasion, and migration of neighboring healthy cells. These alterations, along with genetic and/or epigenetic mutations, contribute to cancer development (Roy and Ghosh 2021).

13.2.2 Dysregulated Signaling in Cancer

Disruption of a single signaling pathway can have a large impact on broader signaling networks, promoting cancer growth and metastasis. In cancer, abnormal cell signaling is responsible for a variety of distinctive characteristics in tumor cells, termed the "hallmarks of cancer," which distinguish them from normal cells (Lansdowne 2018).

Individual proteins act as nodes in intracellular signaling networks, responding to diverse inputs and regulating several outputs. Recent significant advancements have improved our understanding of how numerous signaling proteins contain modular regions aiding protein-protein interactions. These modules play crucial roles in directing proteins to substrates or particular locations, responding to post-translational modifications, and assembling multiprotein complexes and pathways (Pawson and Nash 2003). The same modules control intramolecular interactions that affect signaling function. Understanding the architecture of the signaling network demands identifying all these inter- and intramolecular interactions (Martin 2003).

13.3 NORMAL STEM CELLS AND CSCS

Stem cells regulate tissue growth and homeostasis and have the power to self-renew throughout a person's life. Precise control over stem cell functions is critical for appropriate biological functioning.

Cancer begins with cells that create tumors and encourage their growth. These cells also carry oncogenic and tumor suppressor mutations, identifying cancer as a genetic disease (Hanahan and Weinberg 2011).

CSCs are linked to various cancer-related phenomena such as metastasis, recurrence, heterogeneity, and resistance to various drugs and radiation because of their capacity to self-renew and differentiate. Important signaling and developmental pathways that regulate stem cell properties like survival, self-renewal, proliferation, and differentiation include nuclear factor-kappa B (NF-κB), phosphatidylinositol 3-kinase (PI3K)/Akt/mammalian target of rapamycin (mTOR), phosphatase and tensin homolog (PTEN), Hedgehog (Hh), Wnt, JAK/STAT, and Notch (Roy and Ghosh 2021). Dysregulation of these pathways is frequent in cancer. Emerging evidence indicates that dysregulated signaling within and between pathways can sustain the survival of CSCs, resulting in the development of serially transplantable heterogeneous tumors across different cancer types (Matsui 2016a).

13.4 CANCER-RELATED SIGNALING PATHWAYS

The expanding array of impaired signaling pathways in CSCs includes NF-κB, Hh, Wnt, Notch, PI3K/Akt/mTOR/PTEN, JAK/STAT, and others (K. Chen, Huang, and Chen 2013). The exact mechanisms underlying malfunctioning CSC signaling pathways remain unknown. The following section aims to enhance understanding of the molecular intricacies of the signaling pathways linked with cancer, as illustrated in Figure 13.1.

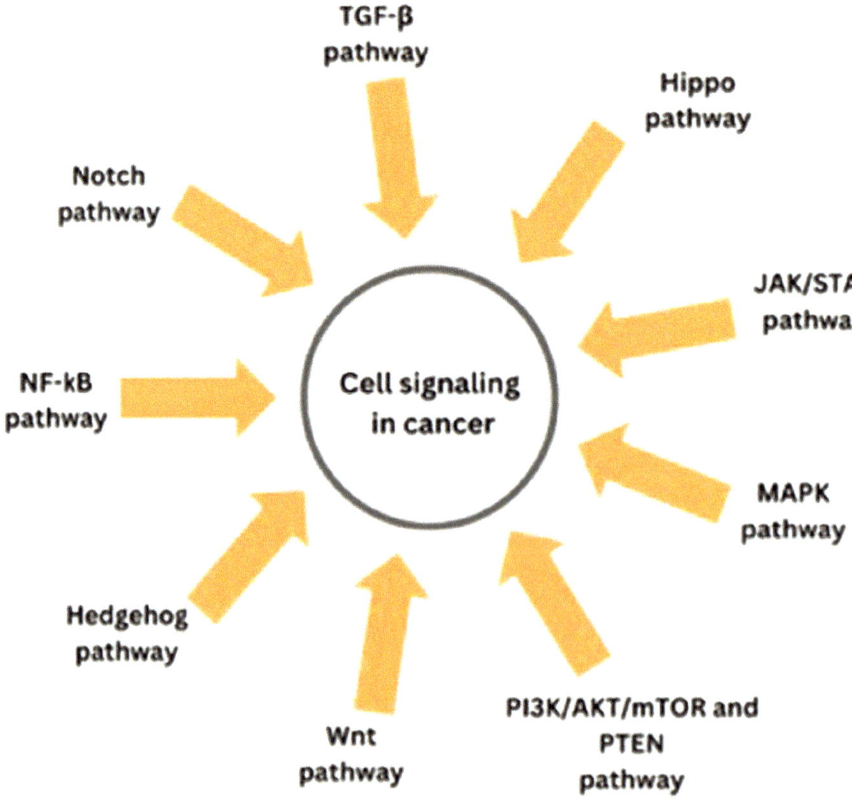

FIGURE 13.1 Signaling pathways related to cancer. Created with canva.com based on information from K. Chen, Huang, and Chen (2013), Roy and Ghosh (2021), and Farooqi, Naureen, and Attar (2022).

13.4.1 JAK/STAT Signaling Pathway

The JAK/STAT pathway is involved in stem cell maintenance, hematopoiesis, and inflammatory responses, among other activities regulated during embryonic development. This system transmits signals from growth factors, interleukins, and cytokines through several families of transmembrane receptors (Thomas et al. 2015). It is particularly integral to immune regulation as a key component of cytokine signaling. Inflammatory mediators serve as important junctions with other signaling pathways due to their activation and role in carcinogenesis. Cancer development is largely influenced by immune system dysregulation and persistent inflammation (Dutta and Li 2013).

13.4.1.1 Regulation of the JAK/STAT Pathway

The human JAK family comprises four members: JAK1, JAK2, JAK3, and TYK2. Each JAK member consists of numerous domains, including the N-terminal FERM domain (protein 4.1, ezrin, radixin, and moesin), which facilitates receptor complex interactions, and the SH2 domain (Src homology 2), which enables phosphotyrosine residue binding. Regulation occurs via a pseudokinase domain and C-terminal protein tyrosine kinase (PTK) domain that phosphorylate specific downstream substrates (Gutiérrez-Hoya and Soto-Cruz 2020; Thomas et al. 2015).

Thomas et al. (2015) identified seven members of the STAT family in humans: STAT1, STAT2, STAT3, STAT4, STAT5A, STAT5B, and STAT6. These proteins regulate numerous biological processes, including apoptosis, inflammation, cell division, and proliferation. When phosphorylated, STAT proteins translocate to the nucleus, form homo/heterodimers, and act as transcription factors. By targeting unique genes, different sets of homo- or heterodimers regulate various biological processes (Gutiérrez-Hoya and Soto-Cruz 2020). In general, STAT3 and STAT5A/B promote oncogenesis, whereas STAT1 activation has been associated with opposite effects (Thomas et al. 2015). Activation involves receptor oligomerization, JAK protein aggregation, subsequent phosphorylation, and stimulation of JAK proteins. Once stimulated, JAK proteins phosphorylate the receptor's cytoplasmic domain, recruiting STAT proteins. Phosphorylated STATs form dimers, translocate to the nucleus, and initiate transcription of target genes (Figure 13.2) (Matsui 2016a). This transcription factor regulates numerous cellular processes, including invasion, angiogenesis, apoptosis, proliferation, and metastasis (Barreto Pires et al. 2016). Negative regulation involves various cytoplasmic proteins that inhibit STAT binding to JAKs, promote STAT dephosphorylation, or sequester STATs from the nucleus (Matsui 2016a).

Under normal physiological conditions, various inhibitory proteins regulate cytokine signaling via the JAK/STAT pathway. Three types of inhibitors are important in this mechanism: inhibitors of stimulated STATs (PIAS), suppressors of cytokine signaling (SOCS), and protein tyrosine phosphatases (SHPs), which include Src homology 2 (SH2). It's interesting to note that SOCS are proteins themselves targets for cytokine signaling; the JAK/STAT pathway increases their levels, which results in a classic negative feedback loop that impedes JAK/STAT signaling (Dutta and Li 2013). STATs' positive regulation of SOCS gene transcription generates a negative feedback loop, which requires precise control throughout the pathway (Thomas et al. 2015).

13.4.1.2 Role of JAK/STAT Signaling in Cancer

Growing evidence indicates that overexpression and heightened activity of JAK/STAT signaling components are linked to the onset of various cancers, including lung, breast, liver, and pancreatic cancers, as well as autoimmune diseases. These associations have garnered increased attention, leading to a heightened focus on both basic and clinical research in these areas (Gutiérrez-Hoya and Soto-Cruz 2020; Hu et al. 2021). The JAK/STAT pathway, a remarkably conserved signaling system, governs various cellular processes linked to the development of diverse diseases (Hu et al. 2021). Fundamental molecular changes, such as gain-of-function mutations affecting different components (JAK, STAT), can lead to abnormal activation of the signaling cascade (Hu et al. 2021). JAK/STAT signaling plays an important role in cancer development by regulating immune

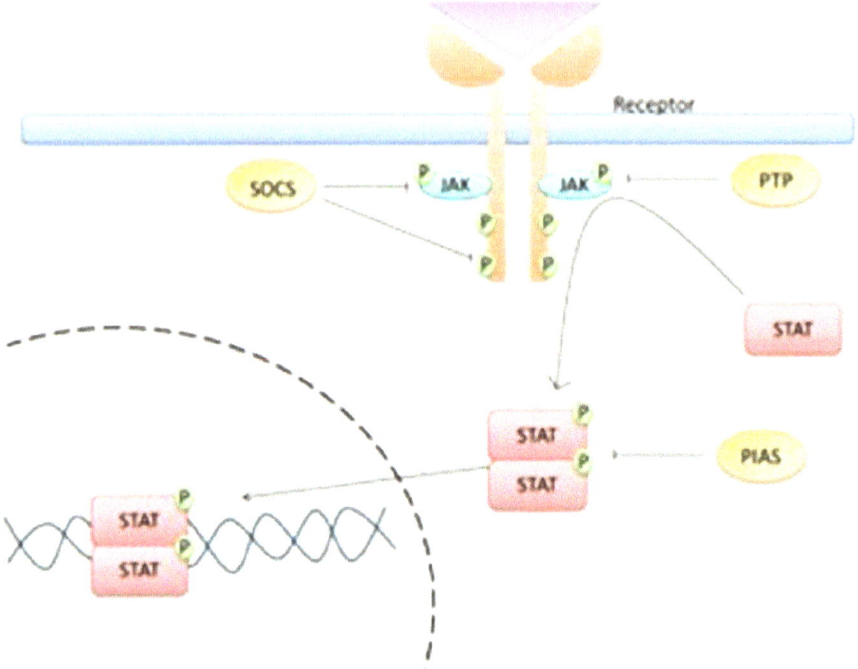

FIGURE 13.2 The JAK/STAT pathway involves membrane cytokine receptors with cytoplasmic tails that associate with inactive JAKs. Cytokine or growth factor binding induces receptor dimerization/oligomerization, leading to JAK autophosphorylation or transphosphorylation. Phosphorylated receptors then recruit phosphorylated cytoplasmic STATs, which are caused by JAKs. Phosphorylated STATs translocate to the nucleus and form dimers, which initiates gene transcription. Negative regulation is mediated by protein tyrosine phosphatases (PTPs), SOCS, and protein inhibitors of activated STAT (PIAS). Reproduced from Gutiérrez-Hoya and Soto-Cruz (2020) (CC BY-NC 4.0).

surveillance and acting as a tumor-specific driver of cancer growth and metastasis (Brooks and Putoczki 2020). Disruption of immune system regulation and the presence of chronic inflammation are critical factors in cancer development. Consequently, research into the JAK/STAT pathway in cancer has become a focal point (Dutta and Li 2013).

Clinical and experimental data indicate that STAT3 plays a role in the development, proliferation, metastasis, and chemoresistance of breast cancer. It is also an early diagnostic marker and a promoter of breast cancer malignancy (Ma, Qin, and Li 2020). Aberrations in JAK/STAT function lead to immune disorders and cancer (Toss et al. 2017). In breast cancer, constitutively activated STAT3 plays an important part in cancer development and spread, as well as breast cancer stem cell (BCSC) development. Its persistent activity influences stem cell-like properties during metastasis, and it is crucial for inflammatory signaling and BCSC maintenance, contributing to a poor prognosis and indicating therapeutic potential (Barreto Pires et al. 2016).

STAT3 activation is triggered by various inflammatory cytokines, carcinogens, growth factors, and oncogenes. STAT3 remains persistently activated and is essential for transformation and prevention of cell death. Overexpression of STAT3 has been observed in numerous tumors, including non-small cell lung cancer (NSCLC). STAT3 activation likely contributes significantly to NSCLC carcinogenesis, with its overexpression correlating with NSCLC cell proliferation, survival, and radioresistance (Alam et al. 2023).

Gain-of-function mutations in JAKs cause the JAK/STAT system to become active in hepatocellular carcinomas (HCCs), promoting cell migration, proliferation, differentiation, and apoptosis.

HCC is often associated with inactive JAK-binding proteins and an overactive JAK/STAT pathway (Dimri and Satyanarayana 2020). STAT3, acting as a transcription factor, enhances HCC angiogenesis, metastasis, proliferation, and survival (Farzaneh et al. 2021). The activation state, rather than increased gene mutations, appears to be crucial for both the Ras and JAK/STAT pathways in HCC. This imbalance underscores the pivotal role of these pathways in HCC (Höpfner, Schuppan, and Scherübl 2008).

13.4.2 MAPK Signaling Pathway

MAPKs govern significant signaling pathways controlling a wide variety of biological processes, including survival, differentiation, apoptosis, and stress responses (Strumberg 2002; Guo et al. 2020). The Ras superfamily of GTPases, comprising over 150 small G-proteins such as HRAS, KRAS, and NRAS, initiates the phosphorylation cascade as the first group of cytosolic intermediates in the MAPK pathway (Braicu et al. 2019). Various elements, including cytokines, viruses, G-protein-coupled receptor ligands, and oncogenes, activate and regulate the MAPK signaling pathway (Guo et al. 2020). Phosphorylation of target proteins like Ras, Raf, MEK, and ERK can either promote or inhibit the signaling pathway. The distinct roles of these target protein locations contribute to the intricate regulation of the Ras/Raf/MEK/ERK signaling pathway, which is crucial for cellular processes such as proliferation, differentiation, and survival (S. Yang and Liu 2017).

13.4.2.1 Regulation of Ras/Raf/MEK/ERK Pathway

MAPK, a serine/threonine protein kinase abundant in eukaryotic cells, consists of four main members: extracellular receptor kinase (ERK), c-Jun N-terminal kinase (JNK), ERK5, and p38 MAPK. These are activated by various stimuli, such as G protein-coupled receptors and RTKs, as illustrated in (Figure 13.3). The ERK cascade regulates processes like differentiation, proliferation, and apoptosis (Iyoda et al. 2003; Huynh et al. 2003; Strumberg 2002). The Raf/MEK/ERK pathway

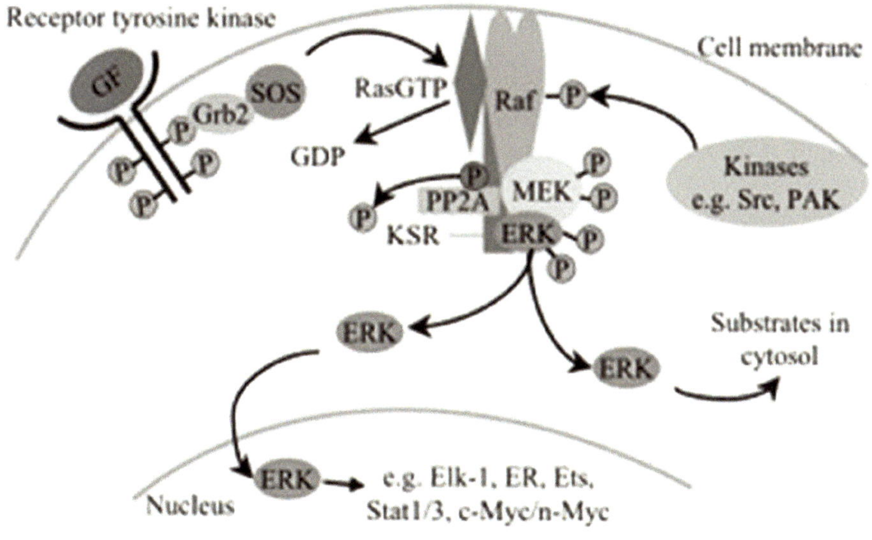

FIGURE 13.3 A MAPK cascade involves three tiers of protein kinases: Extracellular signals (GF) bind to receptor tyrosine kinases, activating Raf via the G-protein Ras. This leads to the activation of MEK, which activates ERK. ERK1/2 phosphorylates proteins involved in cytoskeletal regulation, metabolism, chromatin remodeling, and various transcription factors. The kinase suppressor of Ras (KSR) acts as a scaffold upstream of MEK, crucial for the ERK/MAPK module activation. Reproduced from Strumberg (2002), with permission from Karger Publishers.

is crucial for cell division, apoptosis, and senescence (Choudhary et al. 2023). Mammalian cells employ three distinct MAPK signaling pathways—p38, JNK, and ERK—to activate response genes and transmit extracellular signals to the nucleus. While growth factors primarily trigger the ERK pathway, JNK and p38 signaling respond to diverse stimuli such as cytokines and environmental challenges (Moon and Ro 2021). In summary, Ras recruits and activates protein kinase Raf, which in turn activates dual-specificity protein kinase and ERK1/2. Activated ERK1/2 phosphorylates various substrates, influencing gene expression by controlling distinct transcription factors (Lei Li et al. 2016; Chung and Kondo 2011).

13.4.2.2 Role of MAPK Signaling in Cancer

Most types of cancer are primarily triggered by a malfunction in the Ras-ERK pathway (Guo et al. 2020). Human malignancies frequently involve mutations of Ras proteins, which are proto-oncogenes (Prior, Lewis, and Mattos 2012). Genetic and epigenetic modifications lead to changes in signaling cascades in numerous diseases, including cancer (Braicu et al. 2019). A multitude of events can initiate the Ras-Raf-MEK-ERK signaling cascade. Mutations in Ras and other upstream receptor genes cause aberrant stimulation of the Ras/Raf/MEK/ERK pathway (Roy and Ghosh 2021). While the Ras pathway has a specific function in cancer development, point mutations in each pathway component have diverse effects: MEK and ERK mutations are poor prognostic markers, while Raf and Ras mutations are tumor-causing (Degirmenci, Wang, and Hu 2020). Tumors are caused by abnormal activation of the Ras-Raf-MEK-ERK cascade. Ras mutations are present in about 33% of cancers, while Raf mutations are present in 8% of tumors (Y. Song et al. 2023).

Mutations affecting various components of the MAPK pathway are prevalent in over 50% of human malignancies. Among these, thyroid cancer is particularly associated with mutations that activate the MAPK pathway, constituting approximately 70% of cases. Key mutations driving thyroid carcinoma include B-Raf Proto-Oncogene, Serine/Threonine Kinase (BRAF) and Ras point mutations, alongside Rearranged during Transfection (RET)/PTC rearrangements. Interestingly, the glial-derived neurotrophic factor (GDNF) family activates the transmembrane tyrosine kinase RET receptor, which is essential for the growth, maturation, and maintenance of various tissues, such as the kidney, nervous system, and spermatogonial stem cells. These mutations are exclusive to thyroid tumors, underscoring their reliance on MAPK activation. Although these genetic alterations may activate other pathways like the PI3K-Akt axis, their primary carcinogenic effect on the thyroid predominantly stems from MAPK pathway activation (Zaballos and Santisteban 2017).

During the transformation of colorectal cancer, cells lose normal growth regulation and acquire invasive capabilities. Ras mutations occur in about 36% of cases, with early KRAS proto-oncogene mutations being particularly significant. Heightened Ras activity results in enhanced ERK activity, creating an imbalance between ERK and JNK, a dynamic that holds significant relevance in human colorectal tumorigenesis (Fang and Richardson 2005).

Activation of MAPK signaling has been demonstrated in aggressive breast cancer (BC) and BCSCs. One mechanism initiating MAPK signaling in BCSCs involves the interaction between human epidermal growth factor receptor-2 (HER2) and epidermal growth factor receptor (EGFR). Activated ERK triggers MAPK interacting kinase (MNK) signaling, which activates XIAP, an apoptosis inhibitor protein elevated in aggressive and inflammatory breast cancer. This indicates a link between MAPK and NF-κB signaling (Yousefnia et al. 2020). MAPK activation can lead to uncontrolled cell cycle progression, apoptosis resistance, and therapy insensitivity. RTKs like EGFR, platelet-derived growth factor receptor (PDGFR), and FGFR activate Ras upon interaction with ligands, initiating a cascade where Raf phosphorylates and activates MEK (MEK1 and MEK2), which then phosphorylates ERK or MAPK. This prompts nuclear translocation and release of transcription factors regulating oncogene expression for cell survival and proliferation. Notably, MAP3K1 (8%) and MAP2K4 (4%) are primary targets of major somatic alterations in the MAPK cascade (Toss et al. 2017).

Glioma development involves molecular abnormalities, including mutations in key regulatory networks. MAPKs are crucial in controlling cell proliferation, survival, and differentiation. Alterations in the MAPK pathway, particularly p38 MAPK activation, are common in glial tumors and associated with tumor invasion, metastasis, and higher tumor grade. These changes contribute to brain tumorigenesis and chemotherapy resistance (Grave et al. 2022). Ras overactivation, linked to mutations in the Ras oncogene or its upstream regulators, is notable in gliomas and involves three axes: Raf/MEK/ERK, PI3K/Akt, and Rac1/Rho pathways. High-grade astrocytoma and glioblastoma consistently show Ras overexpression, with approximately 88% of glioblastomas exhibiting this characteristic (Harrison and De Groot 2017).

Research has shown that Raf/MEK/ERK signal transduction actively influences both small cell lung cancer (SCLC) and NSCLC. This pathway plays a significant role in cell development, multiplication, differentiation, inflammation, apoptosis, and senescence. Depending on the type of cell and the stimulus, MAPKs can either stimulate or inhibit apoptosis. While ERKs are crucial for cell survival, JNKs and p38-MAPKs control apoptosis (Abolfathi, Arabi, and Sheikhpour 2023).

The constitutive stimulation of the Raf/MEK/ERK pathway in HCC raises the possibility that this signaling plays a role in cancer development (Whittaker, Marais, and Zhu 2010). Generally, oncogenic mutations in the Ras gene generate constitutive c-Raf (RAF proto-oncogene serine/threonine-protein kinase) activation, and an imbalance in the overexpression of growth factors and their receptors causes constitutive c-Raf activation. These two processes are the main causes of Raf/MEK/ERK signaling in solid tumors (Dimri and Satyanarayana 2020). It is also thought that Raf-1 overexpression is a separate indicator of an early tumor recurrence and a poor diagnosis. According to MEK/ERK expression and phosphorylation, nearly all patients with advanced-stage HCC and around 50% of patients with early-stage HCC have activated MAPK/ERK signaling (Moon and Ro 2021). Likewise, HCC patients' tumors show higher amounts of MAPK, phosphorylated MAPK, and phosphorylated MEK1/2 (Huynh et al. 2003).

13.4.3 PI3K/Akt/mTOR and PTEN Signaling Pathway

All higher eukaryotic cells have a highly conserved main transduction network called PI3K/Akt/mTOR (PAM). This cell signaling system supports survival, growth, proliferation, motility, and immune response modulation (Figure 13.4) (Ortega et al. 2020).

13.4.3.1 Regulation of PI3K/Akt/mTOR Pathway

Growth factors initiate PI3K/Akt/mTOR signal transduction upon binding to cell surface RTKs such as HER2, insulin-like growth factor receptor (IGFR), PDGFR, and vascular endothelial growth factor receptor (VEGFR). The p85 (regulatory) subunit of PI3K is recruited to the plasma membrane by binding to the RTK upon ligand engagement with these receptors. PI3K then undergoes phosphorylation and dimerization. Interaction with the regulatory p85 subunit exposes the catalytic (p110) subunit of PI3K, which then phosphorylates phosphatidylinositol 4,5-bisphosphate (PIP2) to generate phosphatidylinositol 3,4,5-trisphosphate (PIP3) (Choudhary et al. 2023).

PIP3, acting as a secondary messenger, recruits Akt towards the plasma membrane, where it is phosphorylated and stimulated by 3-phosphoinositide-dependent kinase 1 (PDK1) and mTOR complex-2. Stimulated Akt translocates to the cytoplasm, influencing cell proliferation, growth, and survival by phosphorylating downstream effectors (mTOR complex-1, 4EBP1, p70S6K) and regulating prosurvival (XIAP, Bcl-2) and proapoptotic proteins. PTEN typically controls this process, and its dysregulation can contribute to cancer progression and exacerbation (Choudhary et al. 2023).

13.4.3.2 Dysregulation of the PAM Pathway

PAM pathway aberration is recognized as a key promoter of cancer onset and spread. In fact, PAM pathway aberration is the most active route in human cancer and occurs in about 50% of malignancies (Glaviano et al. 2023). Aberrant activation of the PI3K/Akt/mTOR pathway is an important part

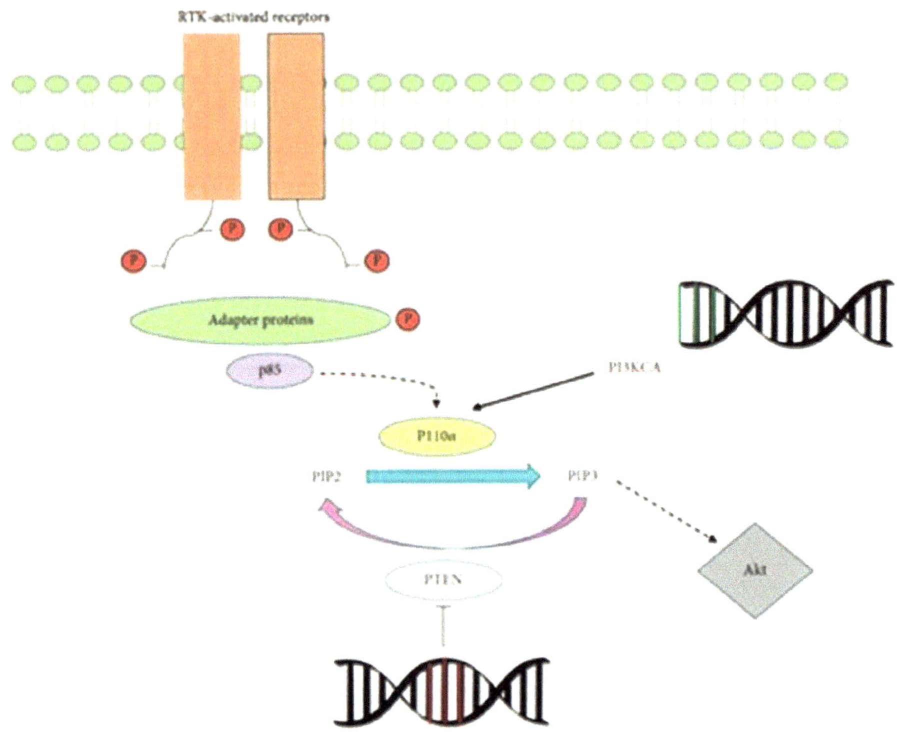

FIGURE 13.4 PI3K and PTEN function via the following mechanism: Upon activation of RTK receptors, self-phosphorylation occurs along with phosphorylation of other adapter proteins, such as IRS-1. The regulatory subunit p85 binds to these phosphorylated sites, releasing the catalytic component p110α. p110α then phosphorylates PIP2 to produce PIP3, which triggers Akt. PTEN dephosphorylates PIP3 to counteract this activation. This pathway can be activated, which leads to the development of cancer, by stimulating mutations in PI3KCA (shown in green) or a loss of function in PTEN (shown in red). Reproduced from Ortega et al. (2020) (CC BY-NC 4.0).

of carcinogenesis (Peng et al. 2022). This pathway has also been linked to a wide range of illnesses and conditions, including vascular disorders, Parkinson's disease, and tuberous sclerosis (Ortega et al. 2020).

Cancer can be initiated and advanced by upregulation of upstream oncogenes, activation of membrane receptors (RTK or GPCR), mutations in PIK3CA, diminished PTEN expression, and gain-of-function Akt mutations. Excessive PAM signaling activity initiates the epithelial-mesenchymal transition (EMT) and metastasis by influencing cell migration. PTEN terminates PI3K/PIP3 signaling by dephosphorylating PIP3 to PIP2, but PTEN loss leads to prolonged pathway activation (Glaviano et al. 2023).

The PAM axis is notable for its role in resistance to anticancer therapy. Critical pathway malfunctions such as Akt gain of function, PTEN loss of function, and increased PI3K activity are well-known catalysts for treatment resistance and cancer progression (Glaviano et al. 2023; Ortega et al. 2020).

Several tumor types exhibit activating mutations in PIK3CA, including breast, uterine corpus endometrial, cervical, colorectal, esophageal, gastric, gallbladder, non-small cell lung, and ovarian carcinomas (Peng et al. 2022). Prostate cancer exhibits synergistic PIK3CA mutation and PTEN loss, accelerating carcinogenesis via PAM pathway hyperactivation. Lung, liver, ovarian, and colorectal cancers frequently feature mutations in PIK3CA, PIK3R1, and PIK3R2. Colorectal cancer often shows PIK3CA mutations linked with KRAS mutations. Renal cancer displays genetic alterations in PIK3CA, PIK3R1, and PIK3R2, with PIK3R1 regulating EMT and stem-like traits. Esophageal squamous-cell carcinoma and poorly differentiated thyroid cancer present genetic changes in PI3K family genes and PTEN. Mutations in the PAM pathway disrupt signaling by affecting RTKs, growth factors, Ras, and PI3K p110 subunits (Glaviano et al. 2023; Ortega et al. 2020).

Akt1, Akt2, and Akt3, members of the AGC kinase family, are downstream effectors of the PI3K signaling pathway and function as serine/threonine protein kinases (Peng et al. 2022). The inner plasma membrane contains phosphorylated phosphatidylinositol lipids that can directly bind intracellular proteins. PIP3 specifically binds Akt and PDK1, which causes them to accumulate close to the membrane. After activation, Akt relocates from the plasma membrane into the nucleus and cytoplasm, where a variety of substrates are found (Glaviano et al. 2023).

Akt2 activation through amplification or overexpression occurs in various malignant tumors, contributing to tumor invasion and metastasis. Akt3 amplification is frequently observed in ovarian epithelial tumors, breast, endometrial, melanoma, cholangiocarcinoma, and NSCLC (Peng et al. 2022). Elevated Akt1 activity is detected in 40% of breast cancer patients, and Akt overactivation is observed in several different cancer types (Glaviano et al. 2023).

The mTOR gene encodes a serine-threonine kinase known as the mammalian target of rapamycin, or mTOR. This member of the serine-threonine kinase family has attracted significant attention due to its crucial role in cancer initiation (Farooqi et al. 2021). Along with regulating cell responses to stressors including growth hormones, starvation, and DNA damage, it also controls tumor development, survival, and metabolic signaling (Peng et al. 2022).

Akt is mainly responsible for controlling cell growth by activating the protein kinase mTORC1. TSC2 within the TSC complex is phosphorylated in an inhibitory manner by nutrients and Akt, which activates mTORC1. mTORC1 acts as both a promoter in the PAM signaling pathway and an inhibitory regulator with potent negative feedback effects on Akt activation by cell surface receptors (RTKs or GPCRs) (Glaviano et al. 2023). Mutations in Akt or upstream PAM pathway genes, particularly PI3K genes and PTWN, or gene amplifications can lead to cancer cells through overactivation of mTORC1.This results in increased cell growth, survival, proliferation, and metabolism. Additionally, altered mTORC1 activation leads to increased signals for tumor cells to spread and infiltration of healthy organs. Mutations in the PAM pathway can cause Rheb to become GTP-loaded when released from the TSC complex, activating mTORC1, which is then attracted to Rag proteins (Glaviano et al. 2023).

PTEN acts as a crucial negative regulator of PI3K/Akt/mTOR signaling, impacting cell survival and proliferation by converting PIP3 to PIP2 as a membrane-bound phosphatase (Peng et al. 2022). Dysregulation of PTEN enhances cell viability, growth, and proliferation. Tumor events contributing to malignancy include mutations, promoter methylation, loss of heterozygosity (LOH), and post-translational PTEN inhibition. PTEN loss may increase genomic instability (Glaviano et al. 2023), leading to the accumulation of harmful mutations. PTEN mutations are common in various cancers, such as prostate adenocarcinoma, glioma, and endometrial carcinoma (Peng et al. 2022).

13.4.3.3 Role of PI3K/Akt/mTOR and PTEN Signaling in Cancer

Research on the PI3K/Akt/mTOR pathway has been integral to cancer studies, with alterations observed in nearly all human tumors, including breast cancer, where approximately 60% of tumors exhibit various hyperactivating variations in this pathway. Growth factors, integrins, cytokines, and other diverse ligands trigger PI3K signaling, an evolutionarily conserved pathway that controls various cellular activities, including cell cycle progression, growth, and survival. This pathway is crucial for self-renewal in mouse embryonic stem cells and contributes to the development and lineage specification of hematopoietic stem cells, and potentially regulates intestinal and neural stem cells (Ortega et al. 2020). In both primary and metastatic breast cancer, PTEN loss frequently results in the hyperactivation of PAM signaling that promotes cell proliferation (Glaviano et al. 2023).

PAM signaling, crucial in central nervous system (CNS) cancers, undergoes frequent genetic and epigenetic alterations. These include oncogenic mutations in PI3K isoforms leading to persistent pathway activation, loss-of-function mutations in the tumor suppressor PTEN, and gain-of-function mutations and amplifications in RTK genes (e.g., EGFR). Aberrant PAM signaling facilitates cell invasion and metastasis in CNS cancers (Crespo, Kind, and Arcaro 2016). The PI3K-Akt pathway and PTEN loss of function are both common and vital in malignant brain tumors (Xu et al. 2008). Genetic changes in the PTEN tumor suppressor gene are associated with dysregulation of the PI3K signaling pathways in at least 60% of glioblastomas. Loss of PTEN function from a mutation or LOH is linked to poor survival in anaplastic astrocytoma and glioblastoma, impacting patient prognosis (Koul 2008). Dysregulation of the PI3K pathway, initiated by cell surface RTKs and oncogenic proteins like Ras, activates PI3K at the cell membrane, triggering phosphorylation and activation of Akt. Akt then activates downstream targets such as CREB, FOXO, and mTOR, regulating cell growth, division, and apoptosis (Harrison and De Groot 2017).

The PI3K/Akt/mTOR pathway plays a critical role in the regulation and progression of lung cancers, particularly in NSCLC. It influences tumorigenesis by regulating cellular processes such as proliferation, survival, angiogenesis, migration, metastasis, genomic integrity, and stem cell self-renewal in lung cancer (Alam et al. 2023). Tumor suppressor protein PTEN often blocks Akt activation by inversely regulating PIP3, which halts the PI3K/Akt/mTOR pathway (Choudhary et al. 2023). Mutations in the PTEN gene accelerate lung cancer progression by activating the PI3K-Akt-NF-κB signaling pathway, promoting cancer cell growth and invasion (Abolfathi, Arabi, and Sheikhpour 2023). PI3K/Akt/mTOR pathway dysregulation in NSCLC is primarily caused by inactivating mutations in PTEN and overexpression of PIK3CA, especially with activating mutations in its helical and kinase domains, leading to overactivity of the p110α catalytic subunit and contributing to lung cancer progression. Numerous microRNAs also regulate PTEN, adding to the complexity of its dysregulation in lung cancer (Choudhary et al. 2023).

Activated PI3K/Akt/mTOR pathway has emerged as a crucial factor in HCC tumorigenesis (Höpfner, Schuppan, and Scherübl 2008). Around 50% of HCC cases exhibit hyperactivation of this signaling pathway, often associated with PTEN gene mutations or decreased expression (Dimri and Satyanarayana 2020). Loss of heterozygosity in the PTEN locus (10q) is commonly observed in HCC (Villanueva et al. 2007). Both mTORC1 and mTORC2 are vital components of this pathway (Sia and Villanueva 2011), with mTORC1 controlling protein synthesis and cell cycle progression via S6 kinase activation. Elevated phospho-mTOR expression is found in 15% of HCCs (Dimri and Satyanarayana 2020). mTOR plays a key role in regulating cell division, growth, and nutritional sensing. Enhanced activation of RTKs, such as IGF receptor and EGFR, leads to overstimulation of PI3K/Akt/mTOR signaling in HCC (Villanueva et al. 2007). Upregulation of both IGF and its receptor is observed in HCC and cirrhotic liver, stimulating this pathway along with the Raf/MEK/ERK and Wnt/β-catenin pathways. The role of Akt in HCC remains uncertain, but phosphorylated Akt in resected HCCs has been associated with a significantly increased risk of early recurrence (Whittaker, Marais, and Zhu 2010).

13.4.4 Wnt Signaling Pathway

The Wnt pathway is a sophisticated and evolutionarily conserved signaling network involving 19 Wnt ligands and over 15 receptors (Roy and Ghosh 2021), important for embryonic development as well as adult tissue homeostasis (Liu et al. 2022). It regulates various cellular processes such as proliferation, differentiation, survival, apoptosis, and cell motility (Tennis, Van Scoyk, and Winn 2007). Wnt signaling begins with communication between secreted Wnt proteins and co-receptors Frizzled (FZD) along with low-density lipoprotein receptor-related protein 5 and 6 (LRP5/6), activating intracellular pathway cascades through canonical (β-catenin dependent) or non-canonical (β-catenin independent) pathways (Ehmsen and Ditzel 2021). The canonical pathway, mediated by the transcriptional regulator β-catenin, controls cell fate determination (Katoh and Katoh 2007), while the non-canonical pathway regulates cell migration and tissue polarity independently of β-catenin (Figure 13.5) (Roy and Ghosh 2021). Canonical Wnt pathways are necessary not only for

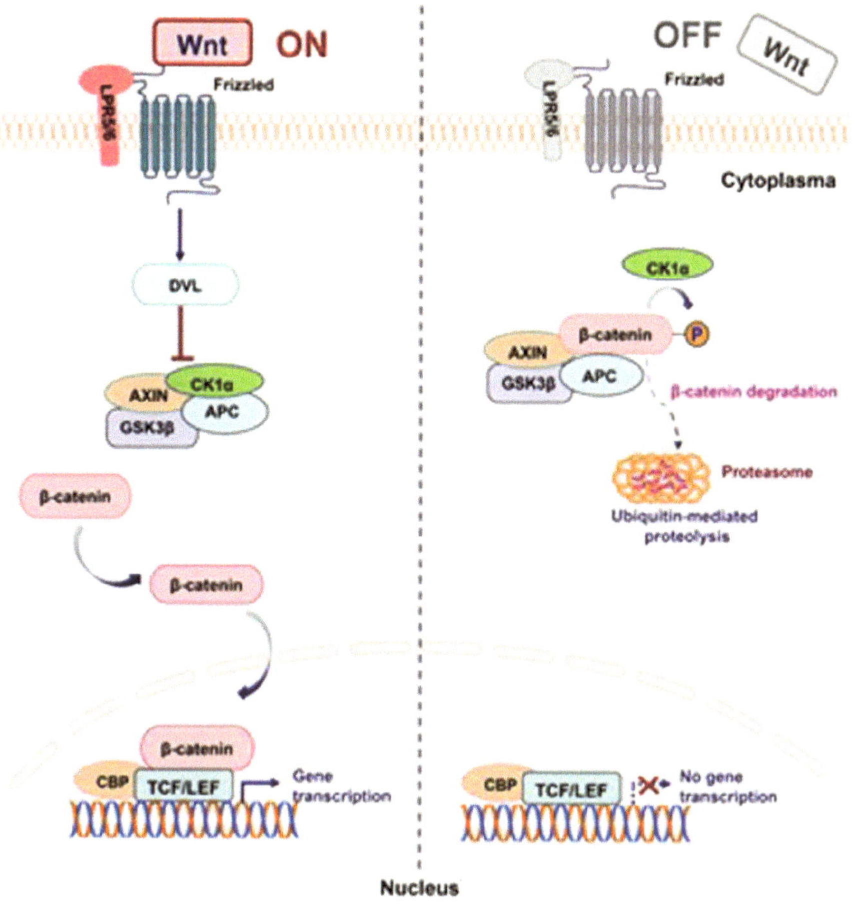

FIGURE 13.5 Illustration of Wnt/β-catenin pathway states: "Wnt ON": Wnt binding to receptors activates DVL, inhibiting GSK3β, stabilizing β-catenin, which translocates into the nucleus, stimulating TCF/LEF transcription factors. "Wnt OFF": When Wnt is absent, the destruction complex (β-catenin, AXIN, CK1α, GSK3β, and APC) phosphorylates β-catenin, leading to its degradation. Reproduced from Y. Zhang and Wang (2020) (CC BY-NC 4.0).

embryonic development but also for tissue homeostasis. Whereas the conventional system governs proliferation, survival, and cell fate, the non-canonical pathway governs asymmetric cell division, polarity, and migration. In postnatal tissues such as the kidney, colon, mammary gland, hair follicle, central nervous system, and epidermis, canonical Wnt signaling interacts with normal stem cells, as evidenced by lineage tracing investigations (Kwon et al. 2011). Dysregulation of the Wnt pathway is associated with various cancer types, including colon, lung, breast, thyroid, and prostate cancers, potentially leading to carcinogenesis (Tennis, Van Scoyk, and Winn 2007). The canonical route is particularly recognized for its significant contribution to cancer development (Roy and Ghosh 2021).

13.4.4.1 Regulation of Wnt/β-Catenin Pathway

The canonical Wnt signaling pathway is activated when Wnt ligands bind to FZD receptors and/or LRP5 and LRP6 co-receptors (Zhan, Rindtorff, and Boutros 2017). There are ten members of the FZD proteins family (Zhao et al. 2022). In the absence of Wnt ligands, a complex including glycogen synthase kinase 3β (GSK3β), casein kinase I (CK I), Axin, and adenomatous polyposis coli (APC) phosphorylates cytoplasmic β-catenin, leading to its degradation (Zhao et al. 2022). Upon

Wnt ligand binding, the Axin/GSK3/APC complex is disrupted, preventing β-catenin degradation. This allows β-catenin to accumulate in the cytoplasm and translocate to the nucleus, where it forms a complex with transcription factor TCF/LEF1, initiating the expression of Wnt target genes (Zhao et al. 2022). R-spondin ligands can further augment the Wnt pathway by interacting with G-protein-coupled receptors containing leucine-rich repeats (Lgrs). The R-spondin receptor Lgr5 serves as a marker for intestinal stem cells and can initiate tumor growth upon APC deletion in these cells (Zhan, Rindtorff, and Boutros 2017).

Non-canonical Wnt signaling operates independently of β-catenin and is initiated by the binding of Wnt ligands to Frizzled receptors and co-receptors like ROR1, ROR2, or RYK. This pathway activates planar cell polarity (PCP), RTK, or Ca++ signaling cascades (Ordaz-Ramos, Tellez-Jimenez, and Vazquez-Santillan 2023) and regulates both transcriptional and non-transcriptional responses in cells. The two primary β-catenin-independent Wnt pathways are PCP and Wnt/Ca2+ (Zhao et al. 2022). The Wnt/β-catenin pathway disrupts other signaling pathways and promotes EMT and fibrosis in breast and intestinal tumors through transcriptional synergy with TGF-β. Additionally, Wnt/β-catenin interacts with Notch during developmental processes and cancer progression (Kwon et al. 2011). In colorectal cancers, the Wnt pathway is frequently activated, controlling intestinal crypt cell division and stem cell niche survival and maintenance by suppressing GSK3β activity, leading to enhanced nuclear β-catenin (Zhao et al. 2022).

13.4.4.2. Dysregulated Wnt/β-Catenin Signaling

Aberrant Wnt/β-catenin signaling influences carcinogenesis by promoting the renewal, proliferation, and differentiation of CSCs (Zhao et al. 2022; Ehmsen and Ditzel 2021). It is involved in colorectal, esophageal, papillary thyroid, invasive ductal breast carcinoma, leukemia, and melanoma cancers (Roy and Ghosh 2021). Mutations in Axin, APC, and β-catenin can cause dysregulation of the signaling system, resulting in hyperactivation. Most colon tumors contain altered APC genes, which eventually lead to cancer. However, colorectal cancer is not exclusively caused by APC deactivation. In fact, 92% of spontaneous colorectal cancers include at least one mutation in a regulator of the Wnt pathway (White, Chien, and Dawson 2012). Targeting the Wnt pathway's components may be a promising therapeutic approach, as it is activated in many breast cancers. Nuclear β-catenin is elevated in several breast cancer subtypes that have overexpressed canonical Wnt receptors and ligands (Roy and Ghosh 2021). In recent years, our understanding of the role of Wnt signaling in hematopoiesis and leukemia has expanded. Normal hematopoietic stem cells (HSC) require a carefully tuned degree of Wnt signaling for long-term survival; however, Wnt activity is significantly elevated in most leukemias (Zhan, Rindtorff, and Boutros 2017). Many malignancies, such as medulloblastoma, lymphoma, leukemia, breast, gastric, and colorectal cancer, have mutations in genes linked to Wnt pathways. Wnt signaling affects stem cell regulation, tumor formation, and metastasis in many tissues, making it a crucial component in the overall development of cancer.

13.4.4.4 Role of Wnt/β-Catenin Signaling in Cancer

Aberrant Wnt signaling significantly contributes to the development, maintenance, and progression of various cancers due to its ability to alter CSC behavior (Duchartre, Kim, and Kahn 2016). Overexpression of the Wnt/β-catenin pathway promotes tumor growth, with evidence suggesting positive regulation in malignant cells across different cancer types, along with Notch signaling overexpression in mammary gland cells (Germán et al. 2019). Misexpression of Wnt ligands induces mammary adenocarcinomas (Kahn and Kim 2014), and Wnt signaling is frequently constitutively active in breast cancer (Barreto Pires et al. 2016), triggering downstream events like β-catenin translocation into the nucleus to regulate oncogenes (Kahn and Kim 2014). Approximately 50% of breast cancer cases display stabilized β-catenin levels, with dysregulation of key regulators (Dvl) and inhibitors (FRP1, DKK1) correlating with poor prognosis. Triple-negative, HER2-driven, and basal-like breast tumors often show increased Wnt/β-catenin pathway activity. Reactivation of silenced Wnt inhibitors has shown efficacy in reducing tumor growth. Furthermore, Wnt pathway activation is implicated in radiation and anticancer drug resistance, potentially through the regulation of stem and progenitor cells **(Figure 13.6).**

Both canonical and non-canonical Wnt signaling affect the "stemness" of CSCs in breast cancer, with CD44 as the target (K. Song and Farzaneh 2021). Mutations in Wnt signaling molecules, predominantly affecting activators or inhibitors, are common (Ordaz-Ramos, Tellez-Jimenez, and Vazquez-Santillan 2023). The Wnt/β-catenin pathway is associated with mammosphere formation,

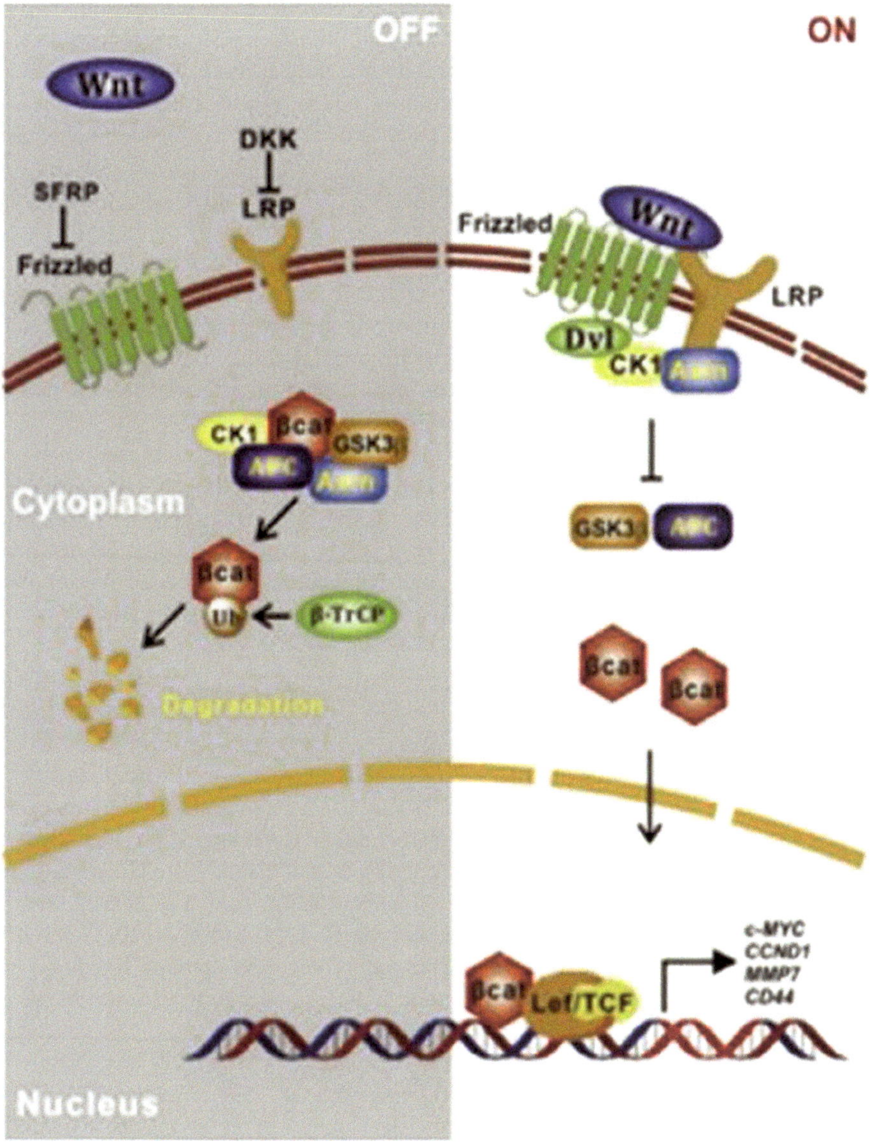

FIGURE 13.6 Breast cancer and mammary gland development are closely associated with the Wnt/β-catenin pathway. Wnt ligands bind to Frizzled and LRPs, preventing β-TrCP from ubiquitinating β-catenin. As β-catenin accumulates, it translocates towards the nucleus and initiates transcription downstream. Wnt signaling inhibitors, including DKKs and SFRPs, act as tumor suppressors by promoting β-catenin degradation and blocking the transcription of oncogenes that target β-catenin. Reproduced from Feng et al. (2018b) (CC BY-NC 4.0).

self-renewal in breast CSCs, cell motility, invasion, and resistance to radiation therapy and apoptosis (Ehmsen and Ditzel 2021). Wnt1 activation enhances stem cell traits and promotes tumorigenesis by targeting the stem/progenitor population. Inhibiting WNT1 alters cell phenotype and reduces tumor formation, suggesting therapeutic potential in modulating Wnt signaling for breast cancer treatment (Feng et al. 2018a).

The Wnt pathway is important in controlling cell fate and differentiation in early development due to the presence of numerous compounds in the developing lung. Components of the Wnt pathway are essential for lung cancer (Kciuk et al. 2023). NSCLC exhibits overexpression of Wnt 1 and Wnt 2, and inhibiting them induces apoptosis. Notably, Wnt 7a re-expression inhibits the development of NSCLC cell lines via activating peroxisome proliferator-activated receptor gamma in an ERK5-dependent manner. This pathway is crucial for maintaining normal epithelial differentiation, representing non-canonical Wnt signaling. Overexpressed disheveled proteins in NSCLC serve as a route for Wnt/FZD activation of Rac and Rho, known to be involved in lung carcinogenesis (Tennis, Van Scoyk, and Winn 2007).

Dysregulated β-catenin, a key component of the Wnt pathway, initiates early events in HCC via the Wnt/β-catenin signaling pathway. High cytoplasmic and nuclear β-catenin levels are observed in 50–70% of HCC tumors, promoting cell division and inhibiting tumor cell differentiation (Dimri and Satyanarayana 2020). Up to 30% of HCC patients exhibit Wnt pathway upregulation (Sia and Villanueva 2011). Wnt/β-catenin signaling additionally regulates angiogenesis in liver cancers. Various triggers, including TGF-β, HGF, and environmental stresses like hypoxia, can stimulate β-catenin. Targeting receptor-ligand interactions or downstream effectors in this pathway could modify its activation in HCC (Farzaneh et al. 2021).

Dysregulated regulation of the transcription factor β-catenin, an essential component of the Wnt signaling pathway, appears to be an important and early phase in the carcinogenesis of HCC. In addition to direct β-catenin mutations, HCC may exhibit other abnormal activations of the Wnt pathway (Whittaker, Marais, and Zhu 2010). Within liver tumors, Wnt ligands are produced by HCC cells and macrophages. Moreover, some environmental risk factors can cause mutations that alter many components of the Wnt pathway, increasing Wnt signaling in HCC (Farzaneh et al. 2021).

13.4.5 Hedgehog Signaling Pathway

The Hedgehog (Hh) pathway, first discovered in Drosophila, is now well recognized for its critical function in human development and pathology (Jiang 2022; Jeng et al. 2019). This evolutionarily conserved pathway transmits signals from the cell membrane to the nucleus. The Hh ligand, Patched (Ptch), and Smoothened (Smo) are the three essential proteins that activate Hh signaling (Skoda et al. 2018). The Hh pathway is essential for several basic functions, such as tissue homeostasis and embryonic development (Pietrobono, Gagliardi, and Stecca 2019; Sari et al. 2018). It is vital for the differentiation of cell and organ formation during embryonic development. In adults, it regulates stem cell maintenance and proliferation in certain organs, contributing to tissue homeostasis and playing a role in carcinogenesis (Di Magliano and Hebrok 2003).

13.4.5.1 Regulation of Hedgehog Signaling Pathway

The Hh signaling pathway, one of the primary developmental pathways, is a complex signaling network comprising both canonical and non-canonical signaling (Gonnissen, Isebaert, and Haustermans 2015). In mammals, there are three distinct types of Hh ligands associated with this pathway: Sonic (SHh), Indian (IHh), and Desert (DHh) (Jeng et al. 2019; Di Magliano and Hebrok 2003). Hh signaling requires two membrane proteins, Smo and Ptch, as well as Gli transcription factors (Gli1, Gli2, Gli3). Additionally, the principal cilia (PC), a tubulin-scaffold protrusion, acts as a unique interface for the transmission of Hh signaling, allowing dynamic coordination of signaling proteins inside this organelle (J. Wang et al. 2023).

Notably, Hh signaling functions via a series of inhibitory mechanisms. Once generated, the three Hh ligands are limited in their dispersion by their interactions with the transmembrane receptors Hipl, Ptch1, and Ptch2 on Hh-responsive cells (Di Magliano and Hebrok 2003). Ptch localizes at the base of the PC and inhibits Smo translocation into the PC, thus decreasing its activity when the Hh ligand is absent. Consequently, GSK3, casein kinase 1 (CK1), and protein kinase A (PKA) phosphorylate Gli repressor (GliR), which is produced when full-length glioma-associated oncogene (GliFL) is proteolytically cleaved. GliR binds to Hh target gene promoters to prevent their activation (Figure 13.7A).

When Hh binds to the Ptch1 protein, the Hh signaling cascade starts. After binding, the Hh-Ptch combination internalizes, and both proteins are then broken down in lysosomes. This connection relieves the inhibition on Smo, allowing the Hh signal to be transmitted downstream via a cytoplasmic protein complex made up of GliFL, Suppressor of fused (Sufu), and kinesin protein (Kif7). Sufu receives a signal from Smo translocating to the PC tip to release the Gli activator (GliA). GliA then migrates to the nucleus and initiates the expressions of target genes (Figure 13.7B). The canonical Hh signaling pathway involves endocrine control of the Gli transcription factor family, while the non-canonical Hh signaling pathway transmits signals via Gli-independent processes (Skoda et al. 2018).

13.4.5.2 Role of the Hh Signaling in Cancer

Hh signaling plays a crucial role in tumor development and/or its progression to more advanced stages (Gonnissen, Isebaert, and Haustermans 2015). Aberrant activation of the Hh pathway is linked to a broad spectrum of tumors, encompassing those affecting the skin, brain, breast, lung, pancreas, gastrointestinal tract, prostate, and ovary. In certain cases, the overproduction of Hh ligands, as observed in prostate cancer, promotes the activation of Hh-Gli signaling or occurs through genetic changes such as Ptch1 loss-of-function or Smo gain-of-function mutations. These mutations are common in diseases such as Shh-subtype medulloblastoma, basal cell carcinoma (BCC), and rhabdomyosarcoma (Pietrobono, Gagliardi, and Stecca 2019). Current estimates suggest that the Hh signaling pathway is involved in the genesis of approximately one-third of all malignant tumors

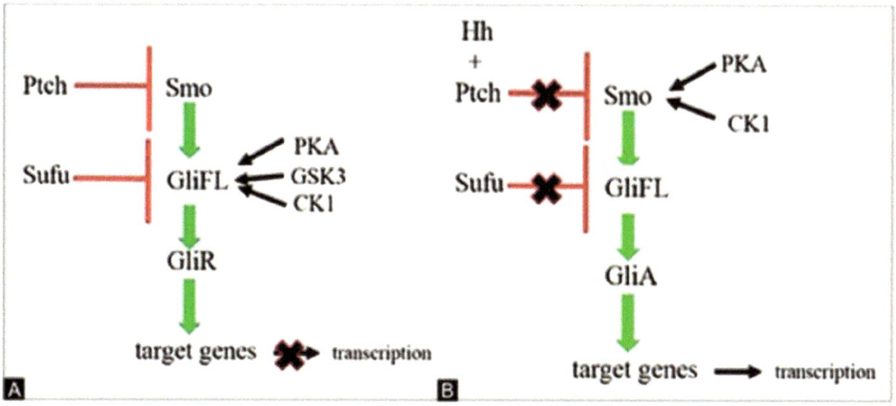

FIGURE 13.7 Hedgehog (Hh) signaling pathway in mammalian cells: (A) Without Hh ligands, Ptch inhibits Smo activity and cilia translocation. Gli2 and Gli3 are phosphorylated, forming repressor forms (Gli2, 3R) that enter the nucleus. (B) Upon Hh ligand binding, Ptch undergoes degradation. Smo accumulates in cilia, causing disassociation of the SUFU/Gli complex. Activator forms of Gli (GliA) then enter the nucleus, initiating target gene transcription. Green arrows indicate activation, whereas red symbols indicate inhibition. Abbreviations: Hh: Hedgehog; Ptch: Patched; Smo: Smoothened; Gli: oncogene associated with gliomas; GSK3: glycogen synthase kinase-3; CK1: casein kinase 1; PKA: protein kinase A; GliFL: full-length Gli; GliA: Gli activator; GliR: Gli repressor. Reproduced from Skoda et al. (2018) (CC BY-NC 4.0).

(Figure 13.8). Three proposed mechanisms outline how the Hh pathway can be aberrantly activated in numerous cancer types (Skoda et al. 2018). These are as follows:

1. **Hh ligand-independent signaling** occurs through mutations activating Smo or deactivating Ptch1 or Sufu, leading to continuous Hh pathway activation even without ligand presence. Somatic mutations in genes related to the Hh pathway are rarely found in certain medulloblastomas (MBs) and basal cell carcinomas. Sporadic BCC, the most prevalent type of cancer in humans, is mostly brought on by mutations that activate the constitutive Hh signaling system (J. Wang et al. 2023).
2. **Hh-dependent oncogenic signaling** operates through autocrine/juxtacrine mechanisms. Following Hh stimulation, this pathway activates, fostering tumor development and progression. Overexpression is noted across diverse tumor types, including stomach, colorectal, esophageal, pancreatic, ovarian, endometrial, breast, prostate, lung, melanomas, gliomas, and other extracutaneous tumors (Skoda et al. 2018; J. Wang et al. 2023).
3. **Hh-dependent paracrine mechanism (ligand-dependent and uses paracrine signaling).** Normal tissue growth and embryonic development depend on paracrine Hh signaling. However, its activation in stromal cells via paracrine processes has been linked to several malignancies, including colon, pancreatic, and prostate cancers. In this scenario, tumor cells release Hh ligands that bind to Ptch1 receptors on stromal cells, activating the Hh pathway (Skoda et al. 2018).

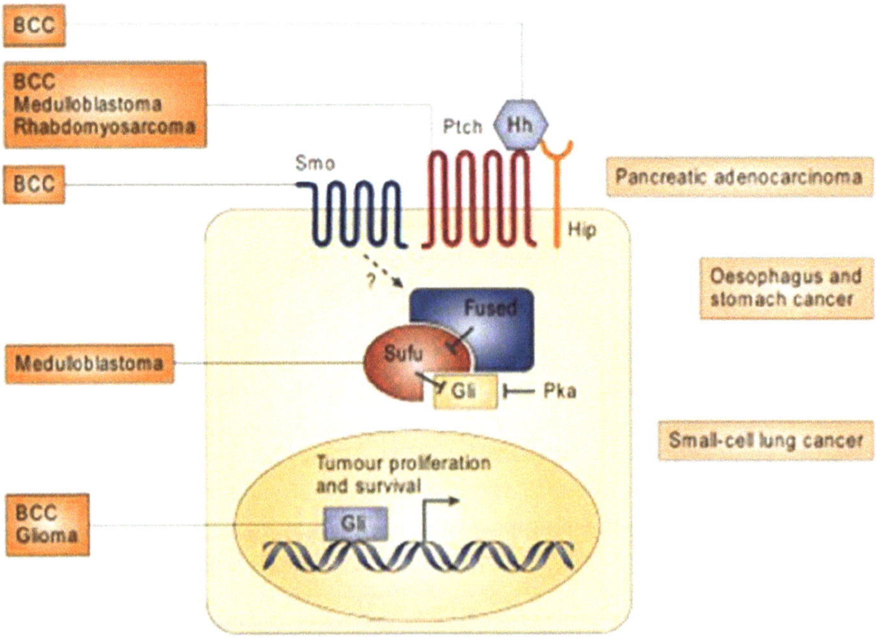

FIGURE 13.8 Dysregulation of Hedgehog (Hh) signaling is implicated in cancer across various tissues. Ptch mutations, often loss-of-function alleles, are linked to rhabdomyosarcoma, medulloblastoma, and BCC, identifying Ptch as a tumor suppressor. Similar loss-of-function mutations in the Suppressor of Fused (Sufu) are seen in medulloblastoma cells. Constitutively active Smo forms, independent of Ptch binding, drive oncogenic BCC. Oncogenic Shh forms are associated with BCC, and ectopic Gli expression induces glioma. Gli inhibition by protein kinase A (Pka) is disrupted in some cases. Reproduced from Di Magliano and Hebrok (2003), with permission from Springer Nature.

Many adult tissues, including the brain and internal organ epithelia, depend on the Hh pathway for the preservation of stem/progenitor cells (Jiang 2022). Hh signaling contributes to tumorigenesis (Sari et al. 2018); consequently, it is anticipated that certain cancers responsive to Hh signaling may exhibit activating mutations in pathway components (Di Magliano and Hebrok 2003). Current estimates suggest that Hh signaling is implicated in the genesis of approximately one-third of all malignant tumors (Skoda et al. 2018).

Several lines of evidence indicate significant involvement of the Hh pathway in breast cancer. Examples include the prolonged stimulation of Smo and increased signaling caused by overexpression of Hh or mutations/polymorphisms in Ptch1. Furthermore, these changes correlate with an aggressive phenotype, particularly in triple-negative breast cancer. The Hh pathway is necessary for preserving the phenotype of BCSCs and controlling their ability to self-renew (Barreto Pires et al. 2016).

Basal cell carcinoma is among the most prevalent types of skin cancer in the Western world. The discovery that germline mutations in Ptch1 cause Gorlin syndrome, also known as nevoid BCC syndrome or Nevoid Basal Cell Carcinoma Syndrome (NBCCS), which is characterized by a propensity for BCCs and other malignancies, first connected aberrant Hh signaling to BCC (Jiang 2022).

Hh signaling is essential for tissue polarity modulation as it influences ductal morphogenesis and proliferation in prostate development. Recent research has highlighted Shh signaling's critical function in endoderm-derived tissue formation, particularly in prostate development, where Shh expression during branching morphogenesis suggests involvement in the onset of androgen-dependent prostate growth (Antón Aparicio et al. 2007).

Hh is extensively studied in CNS development and has a critical function in several important developmental processes, including controlling differentiation, proliferation, and morphogenesis (Clark et al. 2007a). The Sonic Hedgehog (Shh) pathway regulates both developmental processes and cancer (Shahi, Rey, and Castresana 2012). In medulloblastoma, mutations in the Ptch gene disrupt Hh signaling feedback mechanisms (Clark et al. 2007a). Dysregulation of this system leads to severe embryonic defects, potentially induced by teratogens or mutations (Shahi, Rey, and Castresana 2012). Gliomas rely on Shh signaling for CSC self-renewal and tumor growth (Xu et al. 2008). Inhibiting Smo signaling reduces brain tumor stem cell (BTSC) proliferation and tumorigenesis in gliomas (Clark et al. 2007a).

In lung cancer, this pathway promotes cell proliferation, sustains cancer stem cells, and activates cancer-associated fibroblasts (CAFs). Notably, stimulation of Hh signaling is induced by the direct interaction between CAFs and NSCLC cells, which increases the potential for metastatic spread. While SCLC lacks known mutations or amplifications of genes in the Hh pathway, studies using a xenograft model of SCLC cells in nude mice showed that the Hh pathway was particularly activated in SCLC cells that produced Shh. This suggests that SCLC may have an autocrine and/or juxtracrine loop (Abe and Tanaka 2016).

Hh signaling has been demonstrated to be aberrantly active in HCC and is crucial for both development and invasion (Dimri and Satyanarayana 2020). Following injury, hepatocytes and Kupffer cells in the liver can release Shh ligands (Farzaneh et al. 2021). A recent study found that 60% of HCCs had aberrant Shh expression levels (Sia and Villanueva 2011). Moreover, c-myc overexpression, a crucial factor in hepatocarcinogenesis, can be mediated by tumorigenic activation of Smo (Villanueva et al. 2007). Shh can stimulate EMT and metastasis in HCC through binding to the Notch, Wnt, or TGF-β pathways (Farzaneh et al. 2021).

13.4.6 NF-κB Signaling Pathway

The nuclear factor kappa B (NF-κB) family of transcription factors plays a crucial role in the body's response to cellular stresses and limits the correlation between important genes and immunology, inflammation, cell death, and proliferation (Zinatizadeh et al. 2021). The NF-κB family comprises five conserved proteins: NF-κB1 (p105/p50), NF-κB2 (p100/p52), RelA (p65), c-Rel, and RelB (L.

Xia et al. 2018). According to Barreto Pires et al. (2016), the conserved Rel homology domain (RHD) found in all these proteins is responsible for dimerization, DNA binding, and interaction with the repressor protein inhibitor of kappa B (IκB). NF-kB protein is located in the cytoplasm and can be stimulated by numerous cellular stimuli (Zinatizadeh et al. 2021).

13.4.6.1 Proteins and Structure of NF-κB Family

NF-κB proteins contain an N-terminal RHD for dimerization, nuclear localization, and DNA binding. RelB, RelA, and c-Rel contain a C-terminal transactivation domain (TAD) for coactivator binding and gene transcription. Notably, p52 and p50 lack a TAD, inhibiting gene activation in their dimers. Precursors p100 and p105 are cleaved to form active p50 and p52. NF-κB family members form distinct homo- or heterodimers, with p50/RelA and p52/RelB being prominent in canonical and non-canonical pathways. C-Rel/p50 heterodimers are crucial for the innate immune response and pathogen elimination (Ordaz-Ramos, Tellez-Jimenez, and Vazquez-Santillan 2023).

13.4.6.2 Regulation of NF-κB Signaling Pathway

A key hallmark of NF-κB transcription factors is their association with members of the IκB inhibitor family present in the cell cytoplasm. The IκB kinase protein family contains three members (IκBα, IκBβ, and IκBε) with very similar structures (Zinatizadeh et al. 2021). During cell resting states, complexes of NF-κB and IκB dynamically travel between the nucleus and cytoplasm. After stimulation through extracellular signals like lipopolysaccharide, ionizing radiation, TNFα, viral double-stranded RNA, and IL1, NF-κB is activated and translocates inside the nucleus to bind to target genes. The NF-κB-signaling pathways are categorized as canonical or non-canonical upon activation (Figure 13.9) (L. Xia et al. 2018).

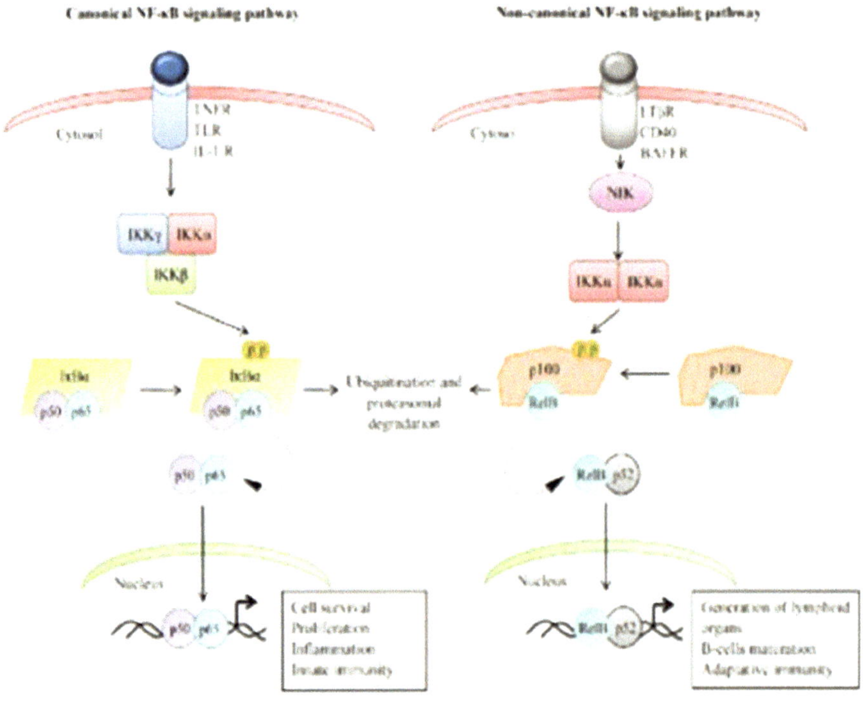

FIGURE 13.9 Initiation of the NF-κB cascade through the canonical and non-canonical signaling pathways. Reproduced from Trocoli and Mergny (2011) (CC BY-NC 4.0).

Canonical signaling is a rapidly induced signaling system that does not depend on protein synthesis; it is linked to functions in inflammation and immunity (Ordaz-Ramos, Tellez-Jimenez, and Vazquez-Santillan 2023). Phosphorylation of IκBα results from the IKK complex (IKKγ, IKKα, and IKKβ) being activated by pro-inflammatory receptors, leading to proteasomal degradation via the canonical NF-κB pathway. This causes dimers (p65/p50) to translocate into the nucleus and bind to target genes' promoter regions.

In the non-canonical NF-κB pathway, stimulation of CD40, LTβR, and BAFF-R stabilizes NIK and activates IKKα homodimers. Active IKKα phosphorylates the NF-κB2/p100 family member, causing proteasomal processing into p52. As a result, the p52/Rel B dimer can translocate to the nucleus and control target gene transcription (Trocoli and Mergny 2011). Activation in these two pathways occurs through the phosphorylation of IκB proteins, leading to the release of inhibition on IκB proteins from NFκB dimers (L. Xia et al. 2018).

13.4.6.3 Role of NF-κB Signaling in Cancer

Many human malignancies exhibit constitutive NF-κB activation caused by an inflammatory microenvironment and carcinogenic changes. This activation induces the epithelial-to-mesenchymal transition, which leads to distant metastasis, in addition to stimulating tumor cell proliferation, suppressing apoptosis, and promoting angiogenesis (Y. Xia, Shen, and Verma 2014). NF-κB, serving as a central coordinator, facilitates reciprocal interactions between inflammation and cancer across various dimensions. Pro-inflammatory cytokines in tumor settings with high NF-κB activity contribute to a tumor-promoting microenvironment. Immune cells infiltrating the gastrointestinal mucosa in inflammatory bowel disease (IBD) emit pro-tumorigenic cytokines such as TNF-α, IL-1, and IL-17, which enhance NF-κB activation and increase the risk of colon cancer (Y. Xia, Shen, and Verma 2014).

Most malignancies activate NF-κB, generally due to increased stimulation of the NF-κB pathway, such as elevated TNF-α and IL-1 in the tumor microenvironment (T. Zhang et al. 2021). Currently, cancer research focuses on the function of the NF-κB pathway in cell biological activities (L. Xia et al. 2018). Abnormal NF-κB pathway activation is related to different malignancies, including blood cancers and solid tumors in the gastrointestinal, genitourinary, gynecological, thoracic, head, neck, and breast tissues (Matsui 2016a). Targeted inhibition of inhibitory kappa B kinase beta (IKKβ) led to sustained activation of NF-κB in intestinal epithelial cells, demonstrating its positive influence on colitis-associated colon cancer (L. Xia et al. 2018).

Persistent activation of NF-κB has been documented across various cancer types, including breast cancer. Continuous stimulation of NF-κB is linked to increased angiogenesis, cellular proliferation, and apoptosis resistance in breast tumors, particularly in triple-negative and HER2/neu subtypes (Barreto Pires et al. 2016). IKKα mutations and gene fusions leading to IKKα activation were found in breast cancer. Persistent IKKα activation in breast cancer contributes to cancer progenitor self-renewal and is implicated in progesterone-induced tumor promotion (L. Xia et al. 2018).

NF-κB serves as a regulatory component in TNF-α signaling and plays a direct role in cancer, contributing to various inflammatory responses and tumor development. Canonical IKKs, such as IKKα/β, are important to NF-κB signaling governing growth, cell survival, and cancer development. IKK and NF-κB are consistently activated in various tumors, promoting drug resistance through antiapoptotic mechanisms. Elevated NF-κB expression is observed in histologically diverse lung cancers, including SCLC and NSCLC. The IKK/NF-κB signaling pathway is linked to early contributions and worse disease-specific endurance in lung tumor patients (Alam et al. 2023).

13.4.7 Notch Signaling Pathway

This pathway's evolutionarily conserved mechanism controls both normal embryonic growth and adult tissue homeostasis (Anusewicz, Orzechowska, and Bednarek 2021a; Bolós, Grego-Bessa, and De La Pompa 2007). It serves as a crucial regulator of differentiation and self-renewal in several tissues and cell types (Lobry et al. 2014). During development, Notch is pivotal for lateral inhibition, a noteworthy mechanism influencing cell fate determination and proliferation (Clark et al. 2007b). Canonical Notch signaling begins when a Notch ligand binds to a Notch transmembrane receptor on a nearby cell's surface. Five ligands comprise the human Notch pathway: three Delta-like proteins (DLL-1, 2, and 4) and two Jagged proteins (Jagged-1 and Jagged-2). Additionally, it has four similar receptors: Notch1–Notch4 (Anusewicz, Orzechowska, and Bednarek 2021a).

13.4.7.1 Regulation of Notch Pathway

Following ligand binding to Notch receptors, a two-step proteolytic cleavage occurs. The initial cleavage, facilitated by A Disintegrin and Metalloproteinase (ADAM) metalloproteases, eliminates the Notch extracellular domain, which is then targeted for lysosomal destruction. After the initial cleavage, the residual segment, Notch extracellular truncated (NEXT), goes through a second cleavage with the assistance of gamma secretase. This complex is made up of the enzymes Presenilin, Nicastrin, PEN2, and APH1. As a result, the intracellular domain (NICD) and transmembrane domain are separated. NICD enters the nucleus and initiates the transcription of Notch target genes by forming an activator complex with the DNA-binding protein CSL (CSL stands for C-promoter binding factor 1 (CBF1) in humans and mice, Suppressor of Hairless (Su(H)) in *Drosophila melanogaster*, and LAG-1 in *Caenorhabditis elegans*) and a transcriptional coactivator from the mastermind-like (MAML) family (Figure 13.10) (J. Li et al. 2022; Anusewicz, Orzechowska, and Bednarek 2021b).

13.4.7.2 Role of Notch Signaling in Cancer

Notch signals have marked non-cell-autonomous and cell-autonomous roles in cancer, both oncogenic and tumor-suppressive (Aster, Pear, and Blacklow 2017). This pathway functions in a wide array of developmental processes (Zhou et al. 2022; Bray 2006), and its dysfunction is linked to several facets of cancer biology, including metastasis, tumor immune evasion, angiogenesis, and CSC phenotype (X. Li et al. 2023). Notch signaling exhibits dual roles in cancer, impacting tumor cells in leukemia, glioblastoma, breast, colon, pancreas, and lung. Depending on the tissue type, it can have a complex and context-dependent role in CSCs as either a promoter or a suppressor. T cell acute lymphoblastic leukemia is mostly caused by Notch1 mutations, while Notch suppression in intestinal crypts encourages differentiation. In pancreatic CSCs, Notch is crucial for stemness, while its inhibition reduces the CSC subpopulation. In esophageal adenocarcinoma, Notch activation prevails, and inhibiting it reduces primary tumor growth and CSC population (Matsui 2016b). Further investigation revealed increased Gli2 and free β-catenin in Notch1-null mice's epidermis, suggesting that the Wnt and Hedgehog pathways are important players in carcinogenesis when Notch signaling is absent (Matsui 2016b).

Notch signaling governs breast epithelial cell differentiation in normal development and significantly contributes to breast cancer progression through dysregulated expression of its upstream and downstream effectors (Yousefi et al. 2022). According to (Bolós, Grego-Bessa, and De La Pompa 2007), the self-renewal of normal mammary stem cells is linked to the Notch signaling system. These single transmembrane proteins—which comprise five ligands: Delta-like (Dll) 1, 3, and 4, and Jagged (JAG) 1 and 2—have several Epidermal Growth Factor (EGF)-like repeats and a DSL domain that facilitates receptor binding. Interestingly, a cysteine-rich domain absent from Delta-like ligands is present in Jagged ligands. These ligands' cytoplasmic regions are poorly understood, although their C-terminal domain has a PDZ-binding motif (Kamdje et al. 2014). Genes under Notch signaling (NS) regulation impact cell development, cell cycle, apoptosis, and metabolism. Approximately 20% of mammary gland tumors stem from Notch4 aberrations, while over 50% result from Notch1 aberrations. NS engages in crosstalk with other signaling transduction pathways. For example, through interactions with the Wnt/β-catenin and Hh pathways, NS activates GSK-3β and increases apoptosis resistance in mammary epithelial cells. NS also involves crosstalk with the Wnt pathway, regulating cell cycle and differentiation events for lineage decisions. Disruption of

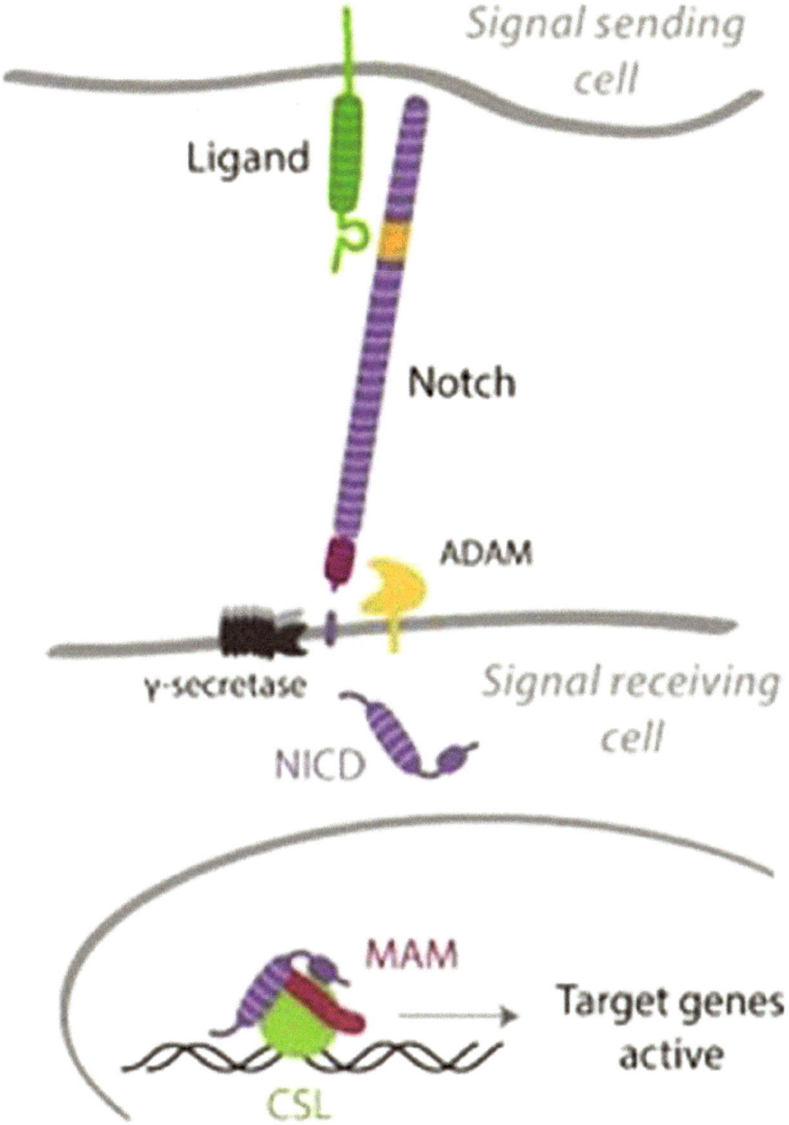

FIGURE 13.10 Summary of the Notch pathway: The Notch receptor undergoes two cleavage events upon contact with its ligand at the cell surface, facilitated by γ-secretase and ADAM metalloprotease. After release from the membrane, the Notch intracellular domain interacts with the coactivator Mastermind (Mam) and DNA-binding CSL (CBF1, Su(H), and LAG-1) transcription factor to activate target gene transcription. Reproduced from Falo-Sanjuan and Bray (2020), with permission from permission from Wiley-VCH.

this communication leads to breast cancer, where a group of cancer cells gradually transforms into cancer progenitor cells (Germán et al. 2019).

Several phases of lung organogenesis depend on Notch signaling (Anusewicz, Orzechowska, and Bednarek 2021b). The Notch signaling pathway's activation status significantly impacts SCLC's complex and diverse biological profile. Notch signaling is believed to affect EMT, neuroendocrine development, cell adhesion, and proliferation, among other aspects of clinical behavior in SCLC. Notably, the Notch3 receptor may play a valuable role in the carcinogenesis of SCLC (Marignol 2017).

Notch signaling, another pathway intricately linked to liver development, governs the differentiation of hepatic progenitor cells and serves as a pivotal regulator in bile duct formation. In HCC, the intracellular domain of Notch3 and one of its potential ligands, JAG1, exhibit upregulation. Experimental models of HCC have demonstrated that inhibiting Notch3 exerts antiproliferative effects (Sia and Villanueva 2011). In HCC, the Notch pathway has disputed implications. This pathway is involved in angiogenesis through the Vascular Endothelial Growth Factor (VEGF) signaling pathway, HCC proliferation through PI3K and mTOR pathways, and CSC maintenance through Wnt and Shh pathways. The Notch pathway is active at different levels in different HCC cell lines, correlating with their invasive characteristics. For example, compared to the HepG2 cell line, the invasive MHCC97 cell line has higher Notch signaling (Farzaneh et al. 2021).

13.4.8 Hippo Signaling Pathway

The Hippo pathway, also known as the Salvador-Warts Hippo pathway, governs cell proliferation and apoptosis and is crucial for the regulation of tissue and organ growth (Teoh and Das 2016). Key elements of the mammalian Hippo pathway include transcriptional enhanced associated domain (TEAD) family, Salvador homolog 1 protein (SAV1), large tumor suppressor kinase 1/2 (LATS1/2), Yes-associated protein 1 (YAP), mammalian STE20-like kinase 1/2 (MST1/2), MOBKL1A/B (MOB1A/B), and WW-domain-containing transcription regulator 1 (TAZ) (Figure 13.11) (Xiao and Dong 2021; Fu et al. 2022), and the transcription factors' downstream interaction with the TEAD family (Masliantsev, Karayan-tapon, and Guichet 2021). YAP and TAZ activation in human cancers promote tumorigenesis by enhancing cancer stem cells, proliferation, metastasis, and chemoresistance in solid tumors (Snigdha et al. 2019).

13.4.8.1 Regulation of Hippo Signaling Pathway

While the Hippo pathway has many known activators, cell contact is the primary trigger. Ras-GTP-binding proteins Ras associated domain family 1/2 (RASSF1/2) and novel Ras/Rap effector 1 (NOREl) undergo autophosphorylation. The Neurofibromin type 2 (NF2) gene produces

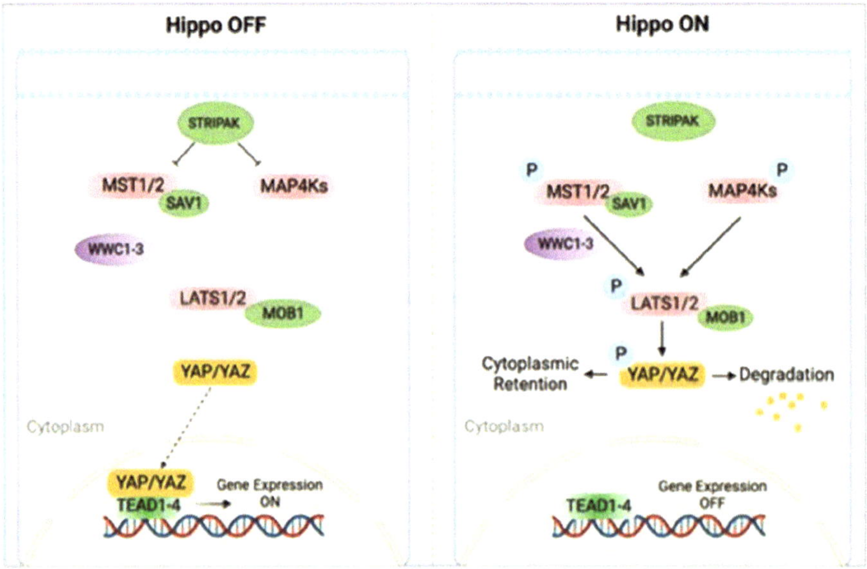

FIGURE 13.11 The core Hippo pathway in mammals (Fu et al. 2022).

the membrane-associated FERM domain protein Merlin (Mer), which is a member of the ezrin-radixin-moesin-related protein family. Kidney and brain (KIBRA) activates Mstl/2 kinases. Mstl/2 phosphorylates RASSF2 to protect Mst2 from degradation and turnover, forming an active complex with Mst2 and RASSF2. Cell contact inhibition may be influenced by Mer's interaction with the cell surface hyaluronan receptor CD44, which activates the Hippo pathway. Independent of Mstl/2, KIBRA was shown to phosphorylate and activate Latsl/2 (Teoh and Das 2016). The pathway, which consists of two antagonistic modules, is essential for both tumor-suppressive and cancer-causing activities, typically regulating a wide variety of genes involved in numerous biological processes (Cunningham and Hansen 2022). Key kinases of the Hippo pathway, including MST1/2 and LATS1/2, function as tumor suppressors in various human malignancies (Ouyang et al. 2020).

13.4.8.2 Role of Hippo Signaling in Cancer

The Hippo pathway contains tumor suppressors and oncogenes. While the core components and upstream regulators mainly function as tumor suppressors, TAZ, YAP, and TEADs are primarily implicated in oncogenic processes (Noorbakhsh et al. 2021; Chan et al. 2011). This pathway has recently been shown to play a role in the development and spread of various malignancies, including ovarian, colorectal, lung, and breast cancers (Harvey, Zhang, and Thomas 2013; Masliantsev, Karayan-tapon, and Guichet 2021). The YAP/TAZ-TEAD transcription factor complex is a typical target of oncogenic transformation (Pan 2010; Calses et al. 2019). YAP/TAZ becomes hyperactive when the Hippo pathway is dysregulated, encouraging the transcription of genes linked to drug resistance, immunosuppression, invasion, migration, cancer survival, proliferation, and other activities (Xiao and Dong 2021). YAP gene locus amplification has been documented at varying frequencies in a range of human and mouse malignancies, including oral squamous-cell carcinomas, medulloblastomas, and lung, pancreatic, esophageal, liver, and mammary gland carcinomas. One study found TAZ overexpression in 21% of primary breast malignancies (Pan 2010).

In HCC, tumor development requires dysregulation of the Hippo pathway (Dimri and Satyanarayana 2020). Several growth factors, including Wnt, Notch, EGF, and Shh, can activate the YAP pathway. Upon activation, YAP translocates into the nucleus and combines with TEAD, PDZ-binding motif (TAZ), and a transcriptional coactivator. In HCC, this interaction inhibits autophagy and apoptosis while promoting proliferation and spread. YAP or TAZ expression is noticeably increased in primary liver tumor samples and HCC cell lines, including hepatic leukemia factor (HLF) and HepG2 (Farzaneh et al. 2021).

The Hippo pathway affects lung malignancies by regulating cell proliferation, organ growth, apoptosis, and carcinogenesis (Teoh and Das 2016). Abundant research supports the importance of dysregulated Hippo signaling in lung cancer (Y. Wang et al. 2015). While many of the Hippo pathway's downstream regulators are known to be tumor suppressors, TAZ and YAP are known oncogenes (Teoh and Das 2016). Most importantly, TAZ and YAP are proven to be oncogenes in lung cancer patients (Y. Wang et al. 2015). TAZ overexpression is reported in more than 60% of NSCLC cases, which is significantly associated with adenocarcinoma (ADC), metastasis, poor differentiation, and poor prognosis. Furthermore, 60–70% of NSCLC patients have overexpressed YAP, strongly correlated with p-TNM stage, lymph node metastases, poor prognosis, and shorter survival time (Yeung, Yu, and Yang 2016). LATS1 is a lung cancer tumor suppressor. MST1 and MST2, previously thought to be proapoptotic cytoplasmic kinases that regulated cell growth, division, and apoptosis, now have a redundant tumor-suppressive function (Y. Wang et al. 2015).

Although the Hippo pathway is well known to function in solid tumors, less is known about its role in gliomas (Masliantsev, Karayan-tapon, and Guichet 2021). An increasing number of studies have shown that Hippo/YAP signaling is crucial for central nervous system development and brain tumors, such as gliomas (Ouyang et al. 2020). Research suggests that patients with glioblastoma (GBM) with higher levels of TAZ overexpression have a worse prognosis. Moreover, abnormal osteoblastic and chondrocyte formation through the mesenchymal transition can be caused by the TEAD2 transcription factor in conjunction with overexpression of TAZ in normal CNS stem cells and proneural glioma stem cells (GSCs) (Masliantsev, Karayan-tapon, and Guichet 2021).

The essential elements of Hippo signaling play an important role in the colonization of metastatic breast tumors both within and beyond breast tissue (Wei, Wang, and Li 2018). In mammary tissue, dysregulation of transcriptional coactivators (TAZ and YAP) and regulatory kinases (MST1, MST2, LATS1, LATS2, SAV1, MOB kinase 1A, MOB kinase 1B) promotes cell proliferation and tissue enlargement. The activated kinase cascade phosphorylates TAZ and YAP, preventing their nuclear entry (Yousefnia et al. 2020). However, their nuclear localization activates target genes linked to invasion, migration, cell proliferation, and chemotherapy resistance in breast cancer cells. TAZ/YAP overexpression in breast cancer stem cells boosts MDR protein activity and interacts with the extracellular matrix (ECM) (Yousefnia et al. 2020). YAP deficiency reduces the incidence of lung metastases in an engineered mouse model of breast cancer (Wei, Wang, and Li 2018).

13.4.9 TGF-B Signaling Pathway

Transforming growth factor (TGF-β) signaling is conserved between flies and humans and regulates diverse functions (Padua and Massagué 2009). It regulates biological processes, including proliferation, differentiation, angiogenesis, inflammation, immunology, autophagy, apoptosis, and EMT (Yun Yang et al. 2021; Elliott and Blobe 2005; Y. Shi and Massagué 2003). TGF-β is a class of structurally similar proteins, part of a superfamily of cytokines that include the TGF-βs (TGF-β1, 2, and 3), activins, anti-mullerian hormone (AMH), inhibins, bone morphogenic protein (BMPs), GDNFs, growth differentiation factors (GDFs), and nodals (Syed 2016; Gu and Feng 2018; Massagué 2008). There are three TGF-β receptors (TGF-βR1, 2, and 3) (Vander Ark, Cao, and Li 2018).

TGF-β is known for its ability to induce EMT in neoplastic epithelial and embryonic cells. While crucial for normal embryonic development, this process is believed to enhance tumor invasion and metastasis during cancer progression (Barcellos-Hoff and Akhurst 2009; Seoane and Gomis 2017a; Colak and ten Dijke 2017). TGF-β has been associated with numerous human diseases, including autoimmune conditions, vascular diseases, and cancer (Syed 2016). TGF-β's complex role in cancer leads to a broad range of gene responses; in early-stage tumors, these include cytostatic and apoptotic tumor-suppressive effects, while in advanced cancer, they include proliferative, invasive, angiogenic, and oncogenic responses (Seoane and Gomis 2017a).

13.4.9.1 Regulation of TGF-β Pathway

TGF-β initiates downstream signaling by binding its ligand to TGF-βR3, which then presents it to TGF-βR2. Subsequently, TGF-βR2 recruits and phosphorylates TGF-βR1, activating its kinase activity. This interaction initiates signaling through the phosphorylation of Smad proteins, leading to the creation of Smad complexes. These complexes are then transported to the nucleus, where they control the transcription of various target genes by binding to specific DNA sequences (Figure 13.12) (Baba et al. 2022).

When any of the three ligands bind to TGF-βR2, TGF-βR1 is recruited and phosphorylated, initiating canonical TGF-β signaling. Subsequently, downstream Smad2 and Smad3 (mothers against decapentaplegic homolog 2 and 3) are phosphorylated by phosphorylated TGF-βR1, each at a serine residue at its carboxy terminus (pSmad2/3C). These phosphorylated Smads then recruit Smad4 and enter the nucleus to control the transcription of TGF-β target genes (Vander Ark, Cao, and Li 2018).

TGF-β can also initiate non-canonical pathways. Smad7, a Smad-specific E3 ligase, may aid in the degradation of activated TGFβRs or pSmad2/3 complexes. Additionally, when TGF-β binds to its receptors, non-canonical signaling can activate a range of kinases, depending on the proteins in the ligand-receptor complex, including Rho-associated protein kinase (ROCK), PI3K/Ack, and MAPKs (ERK, P38, JNK) (Huang and Blobe 2016; P. Y. Chen, Qin, and Simons 2023; Colak and ten Dijke 2017; Seoane and Gomis 2017b; Vander Ark, Cao, and Li 2018).

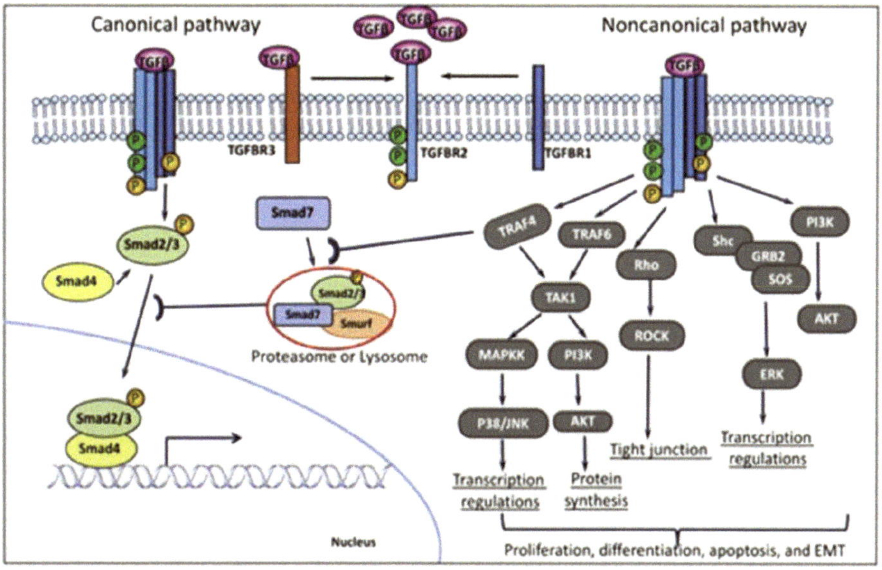

FIGURE 13.12 TGF-β canonical and non-canonical signaling. Reproduced from Vander Ark, Cao, and Li (2018), with permission from Elsevier.

13.4.9.2 TGF-β Dysregulation

Numerous illnesses, including cancer, can result from the activation or inactivation of TGF-β and its downstream mediators (Baba et al. 2022). In early-stage cancer, TGF-β promotes apoptosis and inhibits the progression of the cell cycle, suppressing tumor growth. However, in later stages, TGF-β promotes tumor invasiveness and metastasis (Syed 2016; M. Zhang et al. 2021; Gu and Feng 2018; X. Shi et al. 2022). Antiproliferative responses involve a signaling system with cell surface receptors and intracellular Smad proteins. Tumor cells can lose these responses through molecular mechanisms such as mutations in signaling components and inhibition of the Smad pathway by abnormal regulatory molecule activities (Kretzschmar 2000).

13.4.9.3 Role of TGF-β Signaling in Cancer

TGF-β controls post-lactation involution, ductal and alveolar development, and mammary epithelial proliferation (Elliott and Blobe 2005). Oncogenic mutations, epigenetic modifications, and tumor suppressor gene inactivation redirect the TGF-β signaling pathway from growth inhibition to boosting motility, invasion, and metastasis (Barcellos-Hoff and Akhurst 2009). TGF-β signaling can function as both a tumor suppressor and a carcinogenesis promoter by suppressing proliferation or causing apoptosis in cancer cells, or by promoting angiogenesis, metastasis, and immune evasion to mediate oncogenic effects (Pietrobono, Gagliardi, and Stecca 2019; Yun Yang et al. 2021). TGF-β loss results in growth inhibition, while conversion is linked to elevated TGF-β expression and the advancement of malignant transformation in multiple cancer types, such as gastric, ovarian, cervical, breast, glioma, and melanoma (Barcellos-Hoff and Akhurst 2009). TGF-β plays an important role in carcinogenesis and facilitates tumor growth by inducing EMT, cell division, invasion, metastasis, angiogenesis, and immune surveillance evasion (Baba et al. 2022).

In early breast cancer stages, hyperplastic breast ducts lacking TGF-β receptor type 2 (TβRII) expression have an elevated risk of progressing to invasive breast cancer. Conversely, in advanced stages, TGF-β directly promotes tumorigenesis by stimulating invasion, tumor cell migration, and activating the tumor stroma (de Kruijf et al. 2013). Breast cancer stem cells exhibit EMT characteristics, and as TGF-β affects both EMT and cell stemness, TGF-β signaling likely plays a significant role in the regulation of breast cancer stem cells (Imamura, Hikita, and Inoue 2012). In advanced malignancies, breast tumors, including metastases, often exhibit nuclear phosphorylated Smad2, indicating active TGF-β signaling. TGF-β signaling in breast cancer may also be negatively regulated by estrogens, and evidence points to epigenetic control of pathway elements during critical malignant progression transitions (Barcellos-Hoff and Akhurst 2009). The developmental transcription factor SOX4 can modulate TGF-β activity and facilitate tumor progression, EMT, and metastasis (Yun Yang et al. 2021).

TGF-β is essential for normal lung development and homeostasis. However, an abnormal TGF-β pathway is crucial for the development of lung cancer and pulmonary fibrosis (Saito et al. 2018). TGF-β plays a crucial role in epithelial-mesenchymal interactions during lung development. It is the primary inducer of epithelial-mesenchymal transition in NSCLC and is crucial for shaping the tumor-promoting microenvironment in lung cancer tissue (Saito, Horie, and Nagase 2018). Both SCLC and NSCLC tumors often show upregulation of TGF-β expression, and higher plasma levels of TGF-β are associated with a poorer prognosis for lung cancer patients (Elliott and Blobe 2005).

TGF-β is also highly expressed in HCC, and elevated TGF-β1 levels are linked to disease progression and unfavorable outcomes (Kim et al. 2021). In hepatocytes or stromal cells, CAFs are the main source of TGF-β secretion in hepatocytes or stromal cells. TGF-β signaling promotes HCC growth, induces VEGF production, and attracts endothelial cells to the tumor site while maintaining the CSC population. TGF-β improves mesenchymal characteristics in HCC cell lines in conjunction with the EGF, Wnt, and Shh pathways. TGF-β also transforms tumor-associated macrophages (TAM) into M2-like macrophages, promoting the growth, metastasis, and neo-angiogenesis of HCC (Farzaneh et al. 2021).

TGF-β plays a dual and complex role as both a tumor suppressor and promoter in the initiation, progression, development, and malignancy of brain tumors. The intricate downstream signaling network, involving growth factor pathways, Smad proteins, and other cellular components (Luwor, Kaye, and Zhu 2008), underscores the complexity of tumor progression. Sox9, a transcription factor in the Sox family, is upregulated and negatively correlated with poor prognosis in clinical gliomas. Sox9 functions downstream of the TGF-β cascade, which inhibits its proteasomal degradation in glioma cells. The TGF-β pathway contributes to glioma development by increasing Sox9 expression and accelerating the disease progression (Chao et al. 2021).

Smad mutations are identified in colon cancer cells. In the TGF-β/Smad signaling pathway, Smad4 deletion may indicate poor response to chemotherapy, while Smad7 deletion is a favorable prognostic marker in colorectal cancer patients. Smad4 deletion is known for its regulatory effect on cell growth inhibition and death (Soleimani et al. 2019). Mutations in TGF-βR2, influenced by a deficiency in mismatch repair, lead to colorectal cancers (CRCs) characterized by microsatellite instability. Interestingly, this condition is associated with relatively better survival rates (Itatani, Kawada, and Sakai 2019). Elevated TGF-β expression correlates with a worse prognosis in pancreatic cancer (Kim et al. 2021). A recent exome sequencing study revealed TGF-βR2 as one of the 16 most frequently mutated genes in pancreatic cancer (Javle et al. 2014). While approximately 30% of prostate cancer specimens show reduced or absent TGF-βR2 expression, most prostate tumors gain resistance to TGF-β's antiproliferative actions, but without mutations or deletions in the conventional Smad-dependent signaling pathway (Elliott and Blobe 2005).

REFERENCES

Abe, Yoshinori, and Nobuyuki Tanaka. 2016. "The Hedgehog Signaling Networks in Lung Cancer: The Mechanisms and Roles in Tumor Progression and Implications for Cancer Therapy." *BioMed Research International*. Hindawi Limited. https://doi.org/10.1155/2016/7969286.

Abolfathi, Hanie, Mohadeseh Arabi, and Mojgan Sheikhpour. 2023. "A Literature Review of MicroRNA and Gene Signaling Pathways Involved in the Apoptosis Pathway of Lung Cancer." *Respiratory Research*. BioMed Central Ltd. https://doi.org/10.1186/s12931-023-02366-w.

Alam, Manzar, Gulam Mustafa Hasan, Sayed M. Eldin, Mohd Adnan, Muhammad Bilal Riyaz, Asimul Islam, Ilyas Khan, and Md Imtaiyaz Hassan. 2023. "Investigating Regulated Signaling Pathways in Therapeutic Targeting of Non-Small Cell Lung Carcinoma." *Biomedicine and Pharmacotherapy.* Elsevier Masson s.r.l. https://doi.org/10.1016/j.biopha.2023.114452.

Antón Aparicio, L. M., R. G. García Campelo, J. C. Cassinello Espinosa, M. V. Valladares Ayerbes, M. R. Reboredo López, S. Díaz Prado, and G. A. Aparicio Gallego. 2007. "Prostate Cancer and Hedgehog Signalling Pathway." *Clinical and Translational Oncology* 9 (7): 420–28. https://doi.org/10.1007/s12094-007-0080-x.

Anusewicz, Dorota, Magdalena Orzechowska, and Andrzej K Bednarek. 2021a. "Notch Signaling Pathway in Cancer-Review with Bioinformatic Analysis." https://doi.org/10.3390/cancers.

Anusewicz, D., Orzechowska, M., and Bednarek, A. K. 2021b. "Notch signaling pathway in cancer—review with bioinformatic analysis." *Cancers* 13 (4): 768.

Ark, Alexandra Vander, Jingchen Cao, and Xiaohong Li. 2018. "TGF-β Receptors: In and beyond TGF-β Signaling." *Cellular Signalling.* Elsevier Inc. https://doi.org/10.1016/j.cellsig.2018.09.002.

Aster, Jon C., Warren S. Pear, and Stephen C. Blacklow. 2017. "The Varied Roles of Notch in Cancer." *Annual Review of Pathology: Mechanisms of Disease.* Annual Reviews Inc. https://doi.org/10.1146/annurev-pathol-052016-100127.

Baba, Abdul Basit, Bilal Rah, Gh Rasool Bhat, Ifra Mushtaq, Sabra Parveen, Rukhsana Hassan, Mahrukh Hameed Zargar, and Dil Afroze. 2022. "Transforming Growth Factor-Beta (TGF-β) Signaling in Cancer-A Betrayal Within." *Frontiers in Pharmacology.* Frontiers Media S.A. https://doi.org/10.3389/fphar.2022.791272.

Barcellos-Hoff, Mary H., and Rosemary J. Akhurst. 2009. "Transforming Growth Factor-β in Breast Cancer: Too Much, Too Late." *Breast Cancer Research.* https://doi.org/10.1186/bcr2224.

Barreto Pires, Bruno Ricardo, Ísis Salviano Soares De Amorim, Layane Duarte Souza, Juliana Alves Rodrigues, and Andre Luiz Mencalha. 2016. "Targeting Cellular Signaling Pathways in Breast Cancer Stem Cells and Its Implication for Cancer Treatment." *Anticancer Research.* International Institute of Anticancer Research. https://doi.org/10.21873/anticanres.11151.

Bolós, Victoria, Joaquín Grego-Bessa, and José Luis De La Pompa. 2007. "Notch Signaling in Development and Cancer." *Endocrine Reviews.* https://doi.org/10.1210/er.2006-0046.

Braicu, Cornelia, Mihail Buse, Constantin Busuioc, Rares Drula, Diana Gulei, Lajos Raduly, Alexandru Rusu, et al. 2019. "A Comprehensive Review on MAPK: A Promising Therapeutic Target in Cancer." *Cancers.* MDPI AG. https://doi.org/10.3390/cancers11101618.

Bray, Sarah J. 2006. "Notch Signalling: A Simple Pathway Becomes Complex." *Nature Reviews Molecular Cell Biology.* https://doi.org/10.1038/nrm2009.

Brooks, Andrew J., and Tracy Putoczki. 2020. "Jak-Stat Signalling Pathway in Cancer." *Cancers.* MDPI AG. https://doi.org/10.3390/cancers12071971.

Butti, Ramesh, Sumit Das, Vinoth Prasanna Gunasekaran, Amit Singh Yadav, Dhiraj Kumar, and Gopal C. Kundu. 2018. "Receptor Tyrosine Kinases (RTKs) in Breast Cancer: Signaling, Therapeutic Implications and Challenges." *Molecular Cancer.* BioMed Central Ltd. https://doi.org/10.1186/s12943-018-0797-x.

Calses, Philamer C., James J. Crawford, Jennie R. Lill, and Anwesha Dey. 2019. "Hippo Pathway in Cancer: Aberrant Regulation and Therapeutic Opportunities." *Trends in Cancer.* Cell Press. https://doi.org/10.1016/j.trecan.2019.04.001.

"Cell Signaling and How It Relates to Cancer." 2024. Accessed January 1, 2024. https://pancan.org/view-points-position-papers/.

Chan, Siew Wee, Chun Jye Lim, Liming Chen, Yaan Fun Chong, Caixia Huang, Haiwei Song, and Wanjin Hong. 2011. "The Hippo Pathway in Biological Control and Cancer Development." *Journal of Cellular Physiology.* https://doi.org/10.1002/jcp.22435.

Chao, Min, Nan Liu, Zhichuan Sun, Yongli Jiang, Tongtong Jiang, Meng Xv, Lintao Jia, Yanyang Tu, and Liang Wang. 2021. "TGF-β Signaling Promotes Glioma Progression Through Stabilizing Sox9." *Frontiers in Immunology* 11 (February). https://doi.org/10.3389/fimmu.2020.592080.

Chen, Ke, Ying Hui Huang, and Ji Long Chen. 2013. "Understanding and Targeting Cancer Stem Cells: Therapeutic Implications and Challenges." *Acta Pharmacologica Sinica.* https://doi.org/10.1038/aps.2013.27.

Chen, Pei Yu, Lingfeng Qin, and Michael Simons. 2023. "TGFβ Signaling Pathways in Human Health and Disease." *Frontiers in Molecular Biosciences.* Frontiers Media S.A. https://doi.org/10.3389/fmolb.2023.1113061.

Choudhary, Neeraj, Sweta Bawari, Jack T. Burcher, Dona Sinha, Devesh Tewari, and Anupam Bishayee. 2023. "Targeting Cell Signaling Pathways in Lung Cancer by Bioactive Phytocompounds." *Cancers*. Multidisciplinary Digital Publishing Institute (MDPI). https://doi.org/10.3390/cancers15153980.

Chung, Eva, and Motonari Kondo. 2011. "Role of Ras/Raf/MEK/ERK Signaling in Physiological Hematopoiesis and Leukemia Development." *Immunologic Research*. https://doi.org/10.1007/s12026-010-8187-5.

Clark, Paul A., Daniel M. Treisman, Johnathan Ebben, and John S. Kuo. 2007a. "Developmental Signaling Pathways in Brain Tumor-Derived Stem-like Cells." *Developmental Dynamics*. https://doi.org/10.1002/dvdy.21381.

Colak, Selcuk, and Peter ten Dijke. 2017. "Targeting TGF-β Signaling in Cancer." *Trends in Cancer*. Cell Press. https://doi.org/10.1016/j.trecan.2016.11.008.

Crespo, Silvia, Marcus Kind, and Alexandre Arcaro. 2016. "The Role of the PI3K/AKT/MTOR Pathway in Brain Tumor Metastasis." *Journal of Cancer Metastasis and Treatment* 2 (3): 80. https://doi.org/10.20517/2394-4722.2015.72.

Cunningham, Richard, and Carsten Gram Hansen. 2022. "The Hippo Pathway in Cancer: YAP/TAZ and TEAD as Therapeutic Targets in Cancer." *Clinical Science*. Portland Press Ltd. https://doi.org/10.1042/CS20201474.

Degirmenci, Ufuk, Mei Wang, and Jiancheng Hu. 2020. "Targeting Aberrant RAS/RAF/MEK/ERK Signaling for Cancer Therapy." *Cells*. NLM (Medline). https://doi.org/10.3390/cells9010198.

Dimri, Manali, and Ande Satyanarayana. 2020. "Molecular Signaling Pathways and Therapeutic Targets in Hepatocellular Carcinoma." *Cancers*. MDPI AG. https://doi.org/10.3390/cancers12020491.

Duchartre, Yann, Yong Mi Kim, and Michael Kahn. 2016. "The Wnt Signaling Pathway in Cancer." *Critical Reviews in Oncology/Hematology*. Elsevier Ireland Ltd. https://doi.org/10.1016/j.critrevonc.2015.12.005.

Dutta, Pranabananda, and Willis X Li. 2013. "Role of the JAK-STAT Signalling Pathway in Cancer." In *Encyclopedia of Life Sciences*. Wiley. https://doi.org/10.1002/9780470015902.a0025214.

Ehmsen, Sidse, and Henrik J. Ditzel. 2021. "Signaling Pathways Essential for Triple-Negative Breast Cancer Stem-like Cells." *Stem Cells*. Wiley-Blackwell. https://doi.org/10.1002/stem.3301.

Elliott, Rebecca L., and Gerard C. Blobe. 2005. "Role of Transforming Growth Factor Beta in Human Cancer." *Journal of Clinical Oncology*. https://doi.org/10.1200/JCO.2005.02.047.

Falo-Sanjuan, Julia, and Sarah J. Bray. 2020. "Decoding the Notch Signal." *Development Growth and Differentiation*. Blackwell Publishing. https://doi.org/10.1111/dgd.12644.

Fang, Jing Yuan, and Bruce C. Richardson. "The MAPK Signalling Pathways and Colorectal Cancer." *The lancet oncology* 6, no. 5 (2005): 322-327.

Farooqi, Ammad Ahmad, Humaira Naureen, and Rukset Attar. 2022. "Regulation of Cell Signaling Pathways by Circular RNAs and MicroRNAs in Different Cancers: Spotlight on Wnt/β-Catenin, JAK/STAT, TGF/Smad, SHH/GLI, NOTCH and Hippo Pathways." *Seminars in Cell and Developmental Biology*. Elsevier Ltd. https://doi.org/10.1016/j.semcdb.2021.04.002.

Farooqi, Ammad Ahmad, Humaira Naureen, Rabbia Zahid, Lara Youssef, Rukset Attar, and Baojun Xu. 2021. "Cancer Chemopreventive Role of Fisetin: Regulation of Cell Signaling Pathways in Different Cancers." *Pharmacological Research*. Academic Press. https://doi.org/10.1016/j.phrs.2021.105784.

Farzaneh, Zahra, Massoud Vosough, Tarun Agarwal, and Maryam Farzaneh. 2021. "Critical Signaling Pathways Governing Hepatocellular Carcinoma Behavior; Small Molecule-Based Approaches." *Cancer Cell International*. BioMed Central Ltd. https://doi.org/10.1186/s12935-021-01924-w.

Feng, Yixiao, Mia Spezia, Shifeng Huang, Chengfu Yuan, Zongyue Zeng, Linghuan Zhang, Xiaojuan Ji, et al. 2018a. "Breast Cancer Development and Progression: Risk Factors, Cancer Stem Cells, Signaling Pathways, Genomics, and Molecular Pathogenesis." *Genes and Diseases*. Chongqing University. https://doi.org/10.1016/j.gendis.2018.05.001.

Feng, Y., Liang, J., Li, Z., Yi, P., Xu, L., and Zhang, Z. 2018b. "Transcription of oncogenes that target β-catenin." *Molecular Cancer* 17 (1): 1–14

Fu, Minyang, Yuan Hu, Tianxia Lan, Kun Liang Guan, Ting Luo, and Min Luo. 2022. "The Hippo Signalling Pathway and Its Implications in Human Health and Diseases." *Signal Transduction and Targeted Therapy*. Springer Nature. https://doi.org/10.1038/s41392-022-01191-9.

Germán, Saucedo-Correa, Bravo-Patiño, A., Nuñez-Anita, R. E., Oviedo-Boyso Javier, Valdez-Alarcon Juan José, and Baizabal-Aguirre Víctor Manuel. 2019. "The Notch Signaling Pathway and Breast Cancer: The Importance of Balance and Cellular Self-Control." *Current Signal Transduction Therapy* 16 (1): 38–50. https://doi.org/10.2174/1574362414666190916120659.

Glaviano, Antonino, Aaron S. C. Foo, Hiu Y. Lam, Kenneth C. H. Yap, William Jacot, Robert H. Jones, Huiyan Eng, et al. 2023. "PI3K/AKT/MTOR Signaling Transduction Pathway and Targeted Therapies in Cancer." *Molecular Cancer.* BioMed Central Ltd. https://doi.org/10.1186/s12943-023-01827-6.

Gonnissen, Annelies, Sofie Isebaert, and Karin Haustermans. 2015. "Targeting the Hedgehog Signaling Pathway in Cancer: Beyond Smoothened." *Oncotarget.* Vol. 6. www.impactjournals.com/oncotarget/.

Grave, Nathália, Thamiris Becker Scheffel, Fernanda Fernandes Cruz, Liliana Rockenbach, Márcia Inês Goettert, Stefan Laufer, and Fernanda Bueno Morrone. 2022. "The Functional Role of P38 MAPK Pathway in Malignant Brain Tumors." *Frontiers in Pharmacology.* Frontiers Media S.A. https://doi.org/10.3389/fphar.2022.975197.

Gu, Shuchen, and Xin Hua Feng. 2018. "TGF-β Signaling in Cancer." *Acta Biochimica et Biophysica Sinica.* Oxford University Press. https://doi.org/10.1093/abbs/gmy092.

Guo, Yan-Jun, Wei-Wei Pan, Sheng-Bing Liu, Zhong-Fei Shen, Ying Xu, and Ling-Ling Hu. 2020. "ERK/MAPK Signalling Pathway and Tumorigenesis (Review)." *Experimental and Therapeutic Medicine* (January). https://doi.org/10.3892/etm.2020.8454.

Gutiérrez-Hoya, Adriana, and Isabel Soto-Cruz. 2020. "Role of the JAK/STAT Pathway in Cervical Cancer: Its Relationship with HPV E6/E7 Oncoproteins." *Cells.* NLM (Medline). https://doi.org/10.3390/cells9102297.

Hanahan, Douglas, and Robert A. Weinberg. 2011. "Hallmarks of Cancer: The next Generation." *Cell* 144 (5): 646–74. https://doi.org/10.1016/j.cell.2011.02.013.

Harrison, Rebecca A., and John F. De Groot. 2017. "Cell Signaling Pathways in Brain Tumors." *Topics in Magnetic Resonance Imaging.* Lippincott Williams and Wilkins. https://doi.org/10.1097/RMR.0000000000000112.

Harvey, Kieran F., Xiaomeng Zhang, and David M. Thomas. 2013. "The Hippo Pathway and Human Cancer." *Nature Reviews Cancer.* https://doi.org/10.1038/nrc3458.

Höpfner, Michael, Detlef Schuppan, and Hans Scherübl. 2008. "Growth Factor Receptors and Related Signalling Pathways as Targets for Novel Treatment Strategies of Hepatocellular Cancer." *World Journal of Gastroenterology.* Baishideng Publishing Group Co. https://doi.org/10.3748/wjg.14.1.

Hu, Xiaoyi, Jing li, Maorong Fu, Xia Zhao, and Wei Wang. 2021. "The JAK/STAT Signaling Pathway: From Bench to Clinic." *Signal Transduction and Targeted Therapy.* Springer Nature. https://doi.org/10.1038/s41392-021-00791-1.

Huang, Jennifer J., and Gerard C. Blobe. 2016. "Dichotomous Roles of TGF-β in Human Cancer." *Biochemical Society Transactions* 44 (5): 1441–54. https://doi.org/10.1042/BST20160065.

Huynh, Hung, Thi Thanh, Tuyen Nguyen, Kah-Hoe Pierce Chow, Puay Hoon Tan, Khee Chee Soo, and Evelyne Tran. 2003. "Over-Expression of the Mitogen-Activated Protein Kinase (MAPK) Kinase (MEK)-MAPK in Hepatocellular Carcinoma: Its Role in Tumor Progression and Apoptosis." http://www.biomedcentral.com/1471-230X/3/19.

Imamura, Takeshi, Atsuhiko Hikita, and Yasumichi Inoue. 2012. "The Roles of TGF-β Signaling in Carcinogenesis and Breast Cancer Metastasis." *Breast Cancer* 19 (2): 118–24. https://doi.org/10.1007/s12282-011-0321-2.

Itatani, Yoshiro, Kenji Kawada, and Yoshiharu Sakai. 2019. "Transforming Growth Factor-β Signaling Pathway in Colorectal Cancer and Its Tumor Microenvironment." *International Journal of Molecular Sciences.* MDPI AG. https://doi.org/10.3390/ijms20235822.

Iyoda, Kenya, Yutaka Sasaki, Masayoshi Horimoto, Takashi Toyama, Takayuki Yakushijin, Mitsuru Sakakibara, Tetsuo Takehara, et al. 2003. "Involvement of the P38 Mitogen-Activated Protein Kinase Cascade in Hepatocellular Carcinoma." *Cancer* 97 (12): 3017–26. https://doi.org/10.1002/cncr.11425.

Javle, Milind, Yanan Li, Dongfeng Tan, Xiaoqun Dong, Ping Chang, Siddhartha Kar, and Donghui Li. 2014. "Biomarkers of TGF-β Signaling Pathway and Prognosis of Pancreatic Cancer." *PLoS One* 9 (1). https://doi.org/10.1371/journal.pone.0085942.

Jeng, Kuo Shyang, Chi Juei Jeng, Wen Juei Jeng, I. Shyan Sheen, Shih Yun Li, Chuen Miin Leu, Yeou Guang Tsay, and Chiung Fang Chang. 2019. "Sonic Hedgehog Signaling Pathway as a Potential Target to Inhibit the Progression of Hepatocellular Carcinoma (Review)." *Oncology Letters* 18 (5): 4377–84. https://doi.org/10.3892/ol.2019.10826.

Jiang, Jin. 2022. "Hedgehog Signaling Mechanism and Role in Cancer." *Seminars in Cancer Biology.* Academic Press. https://doi.org/10.1016/j.semcancer.2021.04.003.

Kahn, Michael, and Yong-mi Kim. 2014. "The Role of the Wnt Signaling Pathway in Cancer Stem Cells: Prospects for Drug Development." *Research and Reports in Biochemistry* (July 1). https://doi.org/10.2147/rrbc.s53823.

Katoh, Masuko, and Masaru Katoh. 2007. "Wnt Signaling Pathway and Stem Cell Signaling Network." *Clinical Cancer Research*. https://doi.org/10.1158/1078-0432.CCR-06-2316.

Kciuk, M., Alam, M., Ali, N., Rashid, S., Głowacka, P., Sundaraj, R., ... and Kontek, R. 2023. "Epigallocatechin-3-gallate therapeutic potential in cancer: mechanism of action and clinical implications." *Molecules* 28 (13): 5246.

Kim, Byung Gyu, Ehsan Malek, Sung Hee Choi, James J. Ignatz-Hoover, and James J. Driscoll. 2021. "Novel Therapies Emerging in Oncology to Target the TGF-β Pathway." *Journal of Hematology and Oncology*. BioMed Central Ltd. https://doi.org/10.1186/s13045-021-01053-x.

Koul, Dimpy. 2008. "PTEN Signaling Pathways in Glioblastoma." *Cancer Biology and Therapy*. Landes Bioscience. https://doi.org/10.4161/cbt.7.9.6954.

Kretzschmar, Marcus. 2000. "Transforming Growth Factor-β β and Breast Cancer Transforming Growth Factor-β β/Smad Signaling Defects and Cancer." http://breast-cancer-research.com/content/2/2/107.

Kruijf, E. M. de, T. J. A. Dekker, L. J. A. C. Hawinkels, H. Putter, V. T. H. B. M. Smit, J. R. Kroep, P. J. K. Kuppen, et al. 2013. "The Prognostic Role of TGF-β Signaling Pathway in Breast Cancer Patients." *Annals of Oncology* 24 (2): 384–90. https://doi.org/10.1093/annonc/mds333.

Kwon, Chulan, Paul Cheng, Isabelle N. King, Peter Andersen, Lincoln Shenje, Vishal Nigam, and Deepak Srivastava. 2011. "Notch Post-Translationally Regulates β-Catenin Protein in Stem and Progenitor Cells." *Nature Cell Biology* 13 (10): 1244–51. https://doi.org/10.1038/ncb2313.

Lansdowne, Laura Elizabeth. "Cell Signaling in Cancer" Technology Networks,: December 17, 2018.

Li Jiaxing, Xiao Wang, Jiayu Ding, Yasheng Zhu, Wenjian Min, Wenbing Kuang, Kai Yuan, Chengliang Sun, and Peng Yang. 2022. "Development and Clinical Advancement of Small Molecules for Ex Vivo Expansion of Hematopoietic Stem Cell." *Acta Pharmaceutica Sinica B*. Chinese Academy of Medical Sciences. https://doi.org/10.1016/j.apsb.2021.12.006.

Li, Lei, Guo Dong Zhao, Zhe Shi, Li Li Qi, Li Yuan Zhou, and Ze Xian Fu. 2016. "The Ras/Raf/MEK/ERK Signaling Pathway and Its Role in the Occurrence and Development of HCC (Review)." *Oncology Letters*. Spandidos Publications. https://doi.org/10.3892/ol.2016.5110.

Li, Li, Ping Tang, Shun Li, Xiang Qin, Hong Yang, Chunhui Wu, and Yiyao Liu. 2017. "Notch Signaling Pathway Networks in Cancer Metastasis: A New Target for Cancer Therapy." *Medical Oncology*. Humana Press Inc. https://doi.org/10.1007/s12032-017-1039-6.

Li Xinxin, Xianchun Yan, Yufeng Wang, Balveen Kaur, Hua Han, and Jianhua Yu. 2023. "The Notch Signaling Pathway: A Potential Target for Cancer Immunotherapy." *Journal of Hematology and Oncology*. BioMed Central Ltd. https://doi.org/10.1186/s13045-023-01439-z.

Liu, Jiaqi, Qing Xiao, Jiani Xiao, Chenxi Niu, Yuanyuan Li, Xiaojun Zhang, Zhengwei Zhou, Guang Shu, and Gang Yin. 2022. "Wnt/β-Catenin Signalling: Function, Biological Mechanisms, and Therapeutic Opportunities." *Signal Transduction and Targeted Therapy*. Springer Nature. https://doi.org/10.1038/s41392-021-00762-6.

Lobry, Camille, Philmo Oh, Marc R Mansour, A Thomas Look, and Iannis Aifantis. 2014. "Notch Signaling: Switching an Oncogene to a Tumor Suppressor." https://doi.org/10.1182/blood-2013-08.

Luwor, Rodney B., Andrew H. Kaye, and Hong Jian Zhu. 2008. "Transforming Growth Factor-Beta (TGF-β) and Brain Tumours." *Journal of Clinical Neuroscience*. https://doi.org/10.1016/j.jocn.2008.01.003.

Ma, Jia Hui, Li Qin, and Xia Li. 2020. "Role of STAT3 Signaling Pathway in Breast Cancer." *Cell Communication and Signaling*. BioMed Central Ltd. https://doi.org/10.1186/s12964-020-0527-z.

Magliano, Marina Pasca Di, and Matthias Hebrok. 2003. "Hedgehog Signalling in Cancer Formation and Maintenance." *Nature Reviews Cancer*. European Association for Cardio-Thoracic Surgery. https://doi.org/10.1038/nrc1229.

Marignol, Laure. 2017. "Notch Signalling: The True Driver of Small Cell Lung Cancer?" *Translational Cancer Research*. AME Publishing Company. https://doi.org/10.21037/tcr.2017.09.22.

Martin, G. Steven. 2003. "Cell Signaling and Cancer." *Cancer Cell* 4 (3): 167–74. https://doi.org/10.1016/S1535-6108(03)00216-2.

Masliantsev, Konstantin, Lucie Karayan-tapon, and Pierre Olivier Guichet. 2021. "Hippo Signaling Pathway in Gliomas." *Cells*. MDPI. https://doi.org/10.3390/cells10010184.

Massagué, Joan. 2008. "TGFβ in Cancer." *Cell*. Elsevier B.V. https://doi.org/10.1016/j.cell.2008.07.001.

Matsui, William H. 2016a. "Cancer Stem Cell Signaling Pathways." *Medicine (United States)*. Lippincott Williams and Wilkins. https://doi.org/10.1097/MD.0000000000004765.

Matsui, W. H. 2016b. "Cancer stem cell signaling pathways." *Medicine* 95(1S): S8–S19.

Moon, Hyuk, and Simon Weonsang Ro. 2021. "Mapk/Erk Signaling Pathway in Hepatocellular Carcinoma." *Cancers*. MDPI. https://doi.org/10.3390/cancers13123026.

Noorbakhsh, Negar, Bentolhoda Hayatmoghadam, Marzieh Jamali, Maryam Golmohammadi, and Maria Kavianpour. 2021. "The Hippo Signaling Pathway in Leukemia: Function, Interaction, and Carcinogenesis." *Cancer Cell International*. BioMed Central Ltd. https://doi.org/10.1186/s12935-021-02408-7.

Nwabo Kamdje, Armel Herve, Paul Faustin Seke Etet, Lorella Vecchio, Jean Marc Muller, Mauro Krampera, and Kiven Erique Lukong. 2014. "Signaling Pathways in Breast Cancer: Therapeutic Targeting of the Microenvironment." *Cellular Signalling*. Elsevier Inc. https://doi.org/10.1016/j.cellsig.2014.07.034.

Ordaz-Ramos, Alejandro, Olivia Tellez-Jimenez, and Karla Vazquez-Santillan. 2023. "Signaling Pathways Governing the Maintenance of Breast Cancer Stem Cells and Their Therapeutic Implications." *Frontiers in Cell and Developmental Biology*. Frontiers Media SA. https://doi.org/10.3389/fcell.2023.1221175.

Ortega, Miguel A., Oscar Fraile-Martínez, Ángel Asúnsolo, Julia Buján, Natalio García-Honduvilla, and Santiago Coca. 2020. "Signal Transduction Pathways in Breast Cancer: The Important Role of PI3K/Akt/MTOR." *Journal of Oncology*. https://doi.org/10.1155/2020/9258396.

Ouyang, Taohui, Wei Meng, Meihua Li, Tao Hong, and Na Zhang. 2020. "Recent Advances of the Hippo/YAP Signaling Pathway in Brain Development and Glioma." *Cellular and Molecular Neurobiology*. Springer. https://doi.org/10.1007/s10571-019-00762-9.

Padua, David, and Joan Massagué. 2009. "Roles of TGFβ in Metastasis." *Cell Research*. https://doi.org/10.1038/cr.2008.316.

Pan, Duojia. 2010. "The Hippo Signaling Pathway in Development and Cancer." *Developmental Cell*. https://doi.org/10.1016/j.devcel.2010.09.011.

Pawson, Tony, and Piers Nash. 2003. "Assembly of Cell Regulatory Systems Through Protein Interaction Domains." *Science* 300 (5618): 445–52. https://doi.org/10.1126/SCIENCE.1083653.

Peng, Yan, Yuanyuan Wang, Cheng Zhou, Wuxuan Mei, and Changchun Zeng. 2022. "PI3K/Akt/MTOR Pathway and Its Role in Cancer Therapeutics: Are We Making Headway?" *Frontiers in Oncology*. Frontiers Media S.A. https://doi.org/10.3389/fonc.2022.819128.

Pietrobono, Silvia, Sinforosa Gagliardi, and Barbara Stecca. 2019. "Non-Canonical Hedgehog Signaling Pathway in Cancer: Activation of GLI Transcription Factors beyond Smoothened." *Frontiers in Genetics*. Frontiers Media S.A. https://doi.org/10.3389/fgene.2019.00556.

Prior, Ian A., Paul D. Lewis, and Carla Mattos. 2012. "A Comprehensive Survey of Ras Mutations in Cancer." *Cancer Research*. https://doi.org/10.1158/0008-5472.CAN-11-2612.

Roy, Indrajit, and Goutam Ghosh. 2021. "Journal of Cellular Signaling Review Article Https://Www.Scientificarchives.Com/Journal/Journal-of-Cellular-Signaling Aberrant Signaling Pathways in Cancer Cells: Application of Nanomaterials." *Journal of Cellular Signaling* 2. https://www.scientificarchives.com/journal/journal-of-cellular-signaling.

Saito, Akira, Masafumi Horie, and Takahide Nagase. 2018. "TGF-β Signaling in Lung Health and Disease." *International Journal of Molecular Sciences*. MDPI AG. https://doi.org/10.3390/ijms19082460.

Saito, Akira, Masafumi Horie, Patrick Micke, and Takahide Nagase. 2018. "The Role of TGF-β Signaling in Lung Cancer Associated with Idiopathic Pulmonary Fibrosis." *International Journal of Molecular Sciences*. MDPI AG. https://doi.org/10.3390/ijms19113611.

Sari, Ita Novita, Lan Thi Hanh Phi, Nayoung Jun, Yoseph Toni Wijaya, Sanghyun Lee, and Hyog Young Kwon. 2018. "Hedgehog Signaling in Cancer: A Prospective Therapeutic Target for Eradicating Cancer Stem Cells." *Cells*. MDPI. https://doi.org/10.3390/cells7110208.

Seoane, Joan, and Roger R. Gomis. 2017a. "TGF-b Family Signaling in Tumor Suppression and Cancer Progression." *Cold Spring Harbor Perspectives in Biology*. Cold Spring Harbor Laboratory Press. https://doi.org/10.1101/cshperspect.a022277.

Seoane, J., & Gomis, R. R. 2017b. "TGF-β family signaling in tumor suppression and cancer progression." *Cold Spring Harbor Perspectives in Biology 9* (12): a022277.

Shahi, Mehdi H. 2023. "Introduction to Cell Signaling Pathways and Brain Tumors." In *Role of Signaling Pathways in Brain Tumorigenesis*, 3–12. https://doi.org/10.1007/978-981-15-8473-2_1.

Shahi, Mehdi H., Juan A. Rey, and Javier S. Castresana. 2012. "The Sonic Hedgehog-GLI1 Signaling Pathway in Brain Tumor Development." *Expert Opinion on Therapeutic Targets*. https://doi.org/10.1517/14728222.2012.720975.

Shi, Xueke, Jin Yang, Shuzhi Deng, Hongdan Xu, Deyang Wu, Qingxiang Zeng, Shimeng Wang, Tao Hu, Fanglong Wu, and Hongmei Zhou. 2022. "TGF-β Signaling in the Tumor Metabolic Microenvironment and Targeted Therapies." *Journal of Hematology and Oncology*. BioMed Central Ltd. https://doi.org/10.1186/s13045-022-01349-6.

Shi, Yigong, and Joan Massagué. 2003. "Mechanisms of TGF-Beta Signaling from Cell Membrane to the Nucleus." *Cell* 113: 25.

Sia, Daniela, and Augusto Villanueva. 2011. "Signaling Pathways in Hepatocellular Carcinoma." *Oncology*. https://doi.org/10.1159/000333254.

Skoda, Ana Marija, Dora Simovic, Valentina Karin, Vedran Kardum, Semir Vranic, and Ljiljana Serman. 2018. "The Role of the Hedgehog Signaling Pathway in Cancer: A Comprehensive Review." *Bosnian Journal of Basic Medical Sciences*. Association of Basic Medical Sciences of FBIH. https://doi.org/10.17305/bjbms.2018.2756.

Snigdha, Kirti, Karishma Sanjay Gangwani, Gauri Vijay Lapalikar, Amit Singh, and Madhuri Kango-Singh. 2019. "Hippo Signaling in Cancer: Lessons from Drosophila Models." *Frontiers in Cell and Developmental Biology*. Frontiers Media S.A. https://doi.org/10.3389/fcell.2019.00085.

Soleimani, Atena, Mehran Pashirzad, Amir Avan, Gordon A. Ferns, Majid Khazaei, and Seyed Mahdi Hassanian. 2019. "Role of the Transforming Growth Factor-β Signaling Pathway in the Pathogenesis of Colorectal Cancer." *Journal of Cellular Biochemistry* 120 (6): 8899–907. https://doi.org/10.1002/jcb.28331.

Song, Kai, and Maryam Farzaneh. 2021. "Signaling Pathways Governing Breast Cancer Stem Cells Behavior." *Stem Cell Research and Therapy*. BioMed Central Ltd. https://doi.org/10.1186/s13287-021-02321-w.

Song, Yanlin, Zhenfei Bi, Yu Liu, Furong Qin, Yuquan Wei, and Xiawei Wei. 2023. "Targeting Ras–Raf–MEK–ERK Signaling Pathway in Human Cancer: Current Status in Clinical Trials." *Genes and Diseases*. KeAi Communications Co. https://doi.org/10.1016/j.gendis.2022.05.006.

Strumberg, Dirk. 2002. "The Ras-Raf-MEK-ERK Pathway in the Treatment of Cancer." www.karger.com/journals/onkFax+497614520714.

Syed, Viqar. 2016. "TGF-β Signaling in Cancer." *Journal of Cellular Biochemistry* 117 (6): 1279–87. https://doi.org/10.1002/jcb.25496.

Tennis, Meredith, Michelle Van Scoyk, and Robert A. Winn. 2007. "Role of the Wnt Signaling Pathway and Lung Cancer." *Journal of Thoracic Oncology* 2 (10): 889–92. https://doi.org/10.1097/JTO.0b013e318153fdb1.

Teoh, Seong Lin, and Srijit Das. 2016. "The Emerging Role of the Hippo Pathway in Lung Cancers: Clinical Implications." *Current Drug Targets* 18 (16). https://doi.org/10.2174/1389450117666160907153338.

Thomas, S. J., J. A. Snowden, M. P. Zeidler, and S. J. Danson. 2015. "The Role of JAK/STAT Signalling in the Pathogenesis, Prognosis and Treatment of Solid Tumours." *British Journal of Cancer* 113 (3): 365–71. https://doi.org/10.1038/bjc.2015.233.

Toss, Angela, Marta Venturelli, Chiara Peterle, Federico Piacentini, Stefano Cascinu, and Laura Cortesi. 2017. "Molecular Biomarkers for Prediction of Targeted Therapy Response in Metastatic Breast Cancer: Trick or Treat?" *International Journal of Molecular Sciences*. MDPI AG. https://doi.org/10.3390/ijms18010085.

Trocoli, Aurore, and Mojgan Djavaheri-Mergny. 2011. "The Complex Interplay between Autophagy and NF-κB Signaling Pathways in Cancer Cells." *American Journal of Cancer Research* 1, no. 5: 629.

Vaghari-Tabari, Mostafa, Gordon A. Ferns, Durdi Qujeq, Ali Nosrati Andevari, Zahra Sabahi, and Soheila Moein. 2021. "Signaling, Metabolism, and Cancer: An Important Relationship for Therapeutic Intervention." *Journal of Cellular Physiology*. John Wiley and Sons Inc. https://doi.org/10.1002/jcp.30276.

Villanueva, Augusto, Philippa Newell, Derek Y. Chiang, Scott L. Friedman, and Josep M. Llovet. 2007. "Genomics and Signaling Pathways in Hepatocellular Carcinoma." *Seminars in Liver Disease*. https://doi.org/10.1055/s-2006-960171.

Wang, Juan, Baiping Cui, Xiaojie Li, Xinyue Zhao, Taomin Huang, and Xiaolei Ding. 2023. "The Emerging Roles of Hedgehog Signaling in Tumor Immune Microenvironment." *Frontiers in Oncology*. Frontiers Media S.A. https://doi.org/10.3389/fonc.2023.1171418.

Wang, Y., W. Ding, C. Chen, Z. Niu, M. Pan, and H. Zhang. 2015. "Roles of Hippo Signaling in Lung Cancer." *Indian Journal of Cancer*. Medknow Publications. https://doi.org/10.4103/0019-509X.168949.

Wei, Changran, Ying Wang, and Xiangqi Li. 2018. "The Role of Hippo Signal Pathway in Breast Cancer Metastasis." *OncoTargets and Therapy*. Dove Medical Press Ltd. https://doi.org/10.2147/OTT.S157058.

White, Bryan D., Andy J. Chien, and David W. Dawson. 2012. "Dysregulation of Wnt/β-Catenin Signaling in Gastrointestinal Cancers." *Gastroenterology*. W.B. Saunders. https://doi.org/10.1053/j.gastro.2011.12.001.

Whittaker, S., R. Marais, and A. X. Zhu. 2010. "The Role of Signaling Pathways in the Development and Treatment of Hepatocellular Carcinoma." *Oncogene* 29 (36): 4989–5005. https://doi.org/10.1038/onc.2010.236.

Xia, Longzheng, Shiming Tan, Yujuan Zhou, Jingguan Lin, Heran Wang, Linda Oyang, Yutong Tian, et al. 2018. "Role of the NFκB-Signaling Pathway in Cancer." *OncoTargets and Therapy* 11 (April): 2063–73. https://doi.org/10.2147/OTT.S161109.

Xia, Yifeng, Shen Shen, and Inder M. Verma. 2014. "NF-KB, an Active Player in Human Cancers." *Cancer Immunology Research*. https://doi.org/10.1158/2326-6066.CIR-14-0112.

Xiao, Yi, and Jixin Dong. 2021. "The Hippo Signaling Pathway in Cancer: A Cell Cycle Perspective." *Cancers*. MDPI. https://doi.org/10.3390/cancers13246214.

Xu, Qijin, Xiangpeng Yuan, Gentao Liu, Keith L. Black, and John S. Yu. 2008. "Hedgehog Signaling Regulates Brain Tumor-Initiating Cell Proliferation and Portends Shorter Survival for Patients with PTEN-Coexpressing Glioblastomas." *Stem Cells* 26 (12): 3018–26. https://doi.org/10.1634/stemcells.2008-0459.

Yang, Sufang, and Guohua Liu. 2017. "Targeting the Ras/RAF/MEK/ERK Pathway in Hepatocellular Carcinoma." *Oncology Letters*. Spandidos Publications. https://doi.org/10.3892/ol.2017.5557.

Yang, Yue, Xiaoman Li, Ting Wang, Qianqian Guo, Tao Xi, and Lufeng Zheng. 2020. "Emerging Agents That Target Signaling Pathways in Cancer Stem Cells." *Journal of Hematology and Oncology*. BioMed Central Ltd. https://doi.org/10.1186/s13045-020-00901-6.

Yang, Yun, Wen Long Ye, Ruo Nan Zhang, Xiao Shun He, Jing Ru Wang, Yu Xuan Liu, Yi Wang, Xue Mei Yang, Yu Juan Zhang, and Wen Juan Gan. 2021. "The Role of TGF- β Signaling Pathways in Cancer and Its Potential as a Therapeutic Target." *Evidence-Based Complementary and Alternative Medicine*. Hindawi Limited. https://doi.org/10.1155/2021/6675208.

Yeung, Benjamin, Jihang Yu, and Xiaolong Yang. 2016. "Roles of the Hippo Pathway in Lung Development and Tumorigenesis." *International Journal of Cancer*. Wiley-Liss Inc. https://doi.org/10.1002/ijc.29457.

Yousefi, Hassan, Afshin Bahramy, Narges Zafari, Mahsa Rostamian Delavar, Khoa Nguyen, Atousa Haghi, Tahmineh Kandelouei, et al. 2022. "Notch Signaling Pathway: A Comprehensive Prognostic and Gene Expression Profile Analysis in Breast Cancer." *BMC Cancer* 22 (1). https://doi.org/10.1186/s12885-022-10383-z.

Yousefnia, Saghar, Farzad Seyed Forootan, Shiva Seyed Forootan, Mohammad Hossein Nasr Esfahani, Ali Osmay Gure, and Kamran Ghaedi. 2020. "Mechanistic Pathways of Malignancy in Breast Cancer Stem Cells." *Frontiers in Oncology*. Frontiers Media S.A. https://doi.org/10.3389/fonc.2020.00452.

Zaballos, Miguel A., and Pilar Santisteban. 2017. "Key Signaling Pathways in Thyroid Cancer." *Journal of Endocrinology* 235 (2): R43–61. https://doi.org/10.1530/joe-17-0266.

Zhan, T., N. Rindtorff, and M. Boutros. 2017. "Wnt Signaling in Cancer." *Oncogene*. Nature Publishing Group. https://doi.org/10.1038/onc.2016.304.

Zhang, Maoduo, Ying Yi Zhang, Yongze Chen, Jia Wang, Qiang Wang, and Hezhe Lu. 2021. "TGF-β Signaling and Resistance to Cancer Therapy." *Frontiers in Cell and Developmental Biology*. Frontiers Media S.A. https://doi.org/10.3389/fcell.2021.786728.

Zhang, Tao, Chao Ma, Zhiqiang Zhang, Huiyuan Zhang, and Hongbo Hu. 2021. "NF-KB Signaling in Inflammation and Cancer." *MedComm*. John Wiley and Sons Inc. https://doi.org/10.1002/mco2.104.

Zhang, Ya, and Xin Wang. 2020. "Targeting the Wnt/β-Catenin Signaling Pathway in Cancer." *Journal of Hematology and Oncology*. BioMed Central Ltd. https://doi.org/10.1186/s13045-020-00990-3.

Zhao, Hui, Tianqi Ming, Shun Tang, Shan Ren, Han Yang, Maolun Liu, Qiu Tao, and Haibo Xu. 2022. "Wnt Signaling in Colorectal Cancer: Pathogenic Role and Therapeutic Target." *Molecular Cancer*. BioMed Central Ltd. https://doi.org/10.1186/s12943-022-01616-7.

Zhou, Binghan, Wanling Lin, Yaling Long, Yunkai Yang, Huan Zhang, Kongming Wu, and Qian Chu. 2022. "Notch Signaling Pathway: Architecture, Disease, and Therapeutics." *Signal Transduction and Targeted Therapy*. Springer Nature. https://doi.org/10.1038/s41392-022-00934-y.

Zinatizadeh, Mohammad Reza, Bettina Schock, Ghanbar Mahmoodi Chalbatani, Peyman Kheirandish Zarandi, Seyed Amir Jalali, and Seyed Rouhollah Miri. 2021. "The Nuclear Factor Kappa B (NF-KB) Signaling in Cancer Development and Immune Diseases." *Genes and Diseases*. Chongqing University. https://doi.org/10.1016/j.gendis.2020.06.005.

14 Cell Signaling Pathways in Endocrine Diseases

Gauhar Rehman and Adil Sher
Department of Zoology, Abdul Wali khan University Mardan Pakistan
* **Correspondence:** Dr. Gauhar Rehman, Department of Zoology, Abdul Wali khan University Mardan Pakistan

14.1 INTRODUCTION

Drug development is strongly dependent on signaling pathways, which are critical for understanding the complicated molecular mechanisms that drive illness genesis and progression. Cellular signaling is a highly regulated communication mechanism that enables cells to respond to external stimuli while maintaining homeostasis [1]. Signaling pathways carry information from cell surface receptors to intracellular effectors, which influences a variety of cellular functions. Dysregulation or incorrect activation of these pathways is frequently associated with the development of illnesses such as cardiovascular disease, diabetes, cancer, and neurological disorders [2]. Understanding the molecular nuances of signaling networks aids in identifying therapeutic targets and designing innovative medications that alter specific signaling pathways. The concept of a "pathway," consisting of proteins or subcellular activities with precise functional connections, represents various aspects of health and illness [3]. An illustration of the fundamental processes controlling developmental, physiological, and neurological procedures is necessary for the understanding of health and disease as well as the development of biomedical therapies for complex disorders affecting the system. The idea of a pathway, a well-defined topological complex that specifies a variety of biological processes and the functional relationships among them has proven to be one of the most potent formalisms [4]. The idea that the connectome and pathway states provide the local specifications for the information required for prediction and control is implied in this metaphor. By pinpointing a deterministic, manageable collection of dynamics at a single level of association and downplaying the complication of the background, larger scale systems, and history that the network is immersed in, this causality perspective is ideally suited to mechanistic approaches in the biological sciences [5, 6]. It directs methods in data science concerning the types of data that are collected and utilized for modeling, as well as the models that examine these data to find correlations and causal linkages.

14.2 DIABETES AND CELL SIGNALING PATHWAYS

Diabetes is a multifaceted metabolic illness caused by imbalances in the capacity of the body to maintain insulin and glucose homeostasis. Diabetes has become an epidemic and remains a major public health issue. According to estimates from 2007, 23.6 million Americans (7.8% of the US population) suffered from diabetes, and by 2010, the disease would impact 210 million people globally [7]. Over the next 20 years, it is anticipated that these numbers will rise by 50%, placing a substantial financial burden on people and healthcare systems around the globe [8]. The total annual economic cost of diabetes in the US in 2007 was projected to be $174 billion [7]. Diabetes mellitus is a family of metabolic disorders categorized by consistently high blood sugar levels, which can be caused by deficiencies in the production of insulin by beta cells in the pancreas or by low insulin sensitivity in specific organs. In those with type 1 diabetes, the immune system destroys beta cells that produce insulin, requiring

 DOI: 10.1201/9781003594901-14

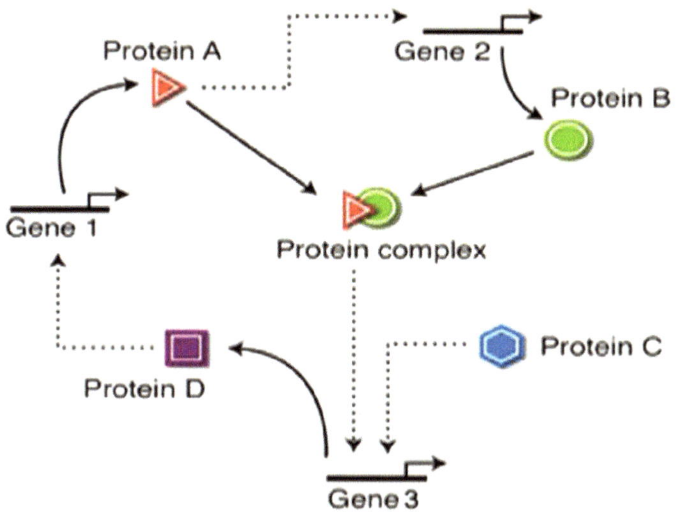

FIGURE 14.1 The generic GRN/pathway paradigm is a common conceptual framework for understanding biological consequences such as protein interactions and gene-regulatory networks. It depicts node relationships as well as the degree and direction of their interaction, which frequently includes loops and self-loops. The system's dynamics are adequately represented by the topology of connections and parameters, which draws a causal border around a group of molecular actors and focuses on interactions. This offers a mechanical, local, single-level organizational perspective. Image courtesy of Jeremy Guay.

the use of exogenous insulin. Peripheral insulin resistance and insufficient compensatory insulin production are the signs of type 2 diabetes, which is often associated with obesity. Diabetes affects a number of important signaling pathways that control insulin activity and glucose balance. When insulin connects to its receptor tyrosine kinase (RTK), cascades that promote the uptake and storage of glucose are activated, including PI3K/AKT and mitogen-activated protein kinase (MAPK). Adipose tissue and skeletal muscle disruption of these pathways are linked with insulin resistance in type 2 diabetes. Falling glucose levels trigger the glucagon pathway, which in turn causes the liver to increase glycogenolysis and gluconeogenesis through various effectors cAMP/PKA (A-cAMP–dependent protein kinase) that phosphorylates substrates involved in the metabolism of sugar, lipids, and glycogen. Diabetes patients' uncontrolled glucose production is a result of this pathway's dysregulation [9].

14.2.1 Insulin Signaling Pathway

The insulin signaling system in pancreatic β-cells consists of insulin receptors (IR) and insulin receptor substrate (IRS). These include IRS, IR tyrosine kinase, PI3K, and its downstream effectors, such as protein kinase B (PKB), MAPK, MAPK kinase (MAPKK/MEK), raf protein, and rat sarcomas (Ras) proteins [10, 11]. IRS, which exists in two main types in β-cells, IRS2 and IRS1, is stimulated by activated IR. Insulin production that is dependent on glucose is strictly regulated by IRS 1, which is mostly found in islet β-cells. In contrast, IRS2 is mostly found in the periphery of the pancreatic islets and is expressed at low levels in β-cells. Its function is to positively modulate the compensatory mechanism of β-cells [12]. The proliferation, growth, and mitosis of β-cells are ultimately modulated by the regulatory mechanisms that activate IRS1 (calcium channels and PI3K, among others) and IRS2 (Ras, MAPK, PI3K, and other pathways) [13, 14].

14.2.2 PI3K–AKT/PKB Signaling Pathway

The intricate network of mediators orchestrates the control of β-cell function, in which the PI3K signaling pathway plays a significant role [15]. AKT/PKB functions as a downstream target of

PI3K and controls the growth of β-cells by modifying its numerous downstream genes, including GSK3, mTOR (mammalian target of rapamycin), and Forkhead box protein O1 (FOXO1) [16, 17]. The phosphorylation of PI3K/AKT signaling proteins affects the transcription activity of FOXO1, a member of the phosphorylase Fox gene family [18]. Pancreatic and duodenal home box 1 (Pdx1), which helps maintain the similar structures of β-cells, such as the control of mark gene expression for genes like GLUT2 (glucose transporter 2) [18], preservation of proper cell function, and insulin secretion, has been found to be weakly inhibited by FOXO1 in β-cells [19]. This suggests that FOXO1 performs a crucial part in preserving the function of β-cells via altering Pdx1. It has been shown that GSK3 is a downstream target of several signaling pathways and is involved in the control of cell cycle and cell proliferation [20]. According to reports, β-cell function is negatively regulated by GSK3 stimulation, which keeps Pdx1 stable and at its subcellular location [21]. Moreover, FOXO1, which is triggered by phosphorylated PI3K AKT, suppresses GSK3 when mTOR is activated [22]. mTOR is an essential part of the two mTOR complexes, mTORC2 and mTORC1, as a member of the serine/threonine protein kinase family [23, 24]. PI3K AKT, AMP-activated growth factors, protein kinase, and other related signaling proteins of mTOR affect a variety of metabolic indices, insulin secretion, cellular survival or death, dedifferentiation, pancreatic β-cell function and resistance, stem cell differentiation, and cellular metabolism [25]. According to Maiese [26], PI3K AKT was activated by a feedback mechanism when mTOR and mTORC1 were downregulated, blocking all of the expected beneficial effects of therapy. After PI3K AKT inhibited the activation of TSC1/2, an upstream negative regulator of mTORC1, S6K1, and 4E BPs stimulated mRNA translation and ribosome manufacturing, ultimately leading to an increase in β-cell proliferation [27]. Moreover, P21cip1 and p27kip1 were cell cycle inhibitors. PI3K AKT raised P27kip1 by activating the FOXO1 gene, which inhibited the proliferation of β-cells, and decreased P21cip1 by activating the MDM2/P53 gene, which increased β-cell proliferation. The immortalized mouse β-cell line, which is only expressed in the IR A isoform, was shown to be susceptible to the mitogenic response generated by IGF1. This additionally boosted the proliferation of β-cells, mainly through mTORC1 [28]. Additionally, early on in the 60% partial pancreatectomy, mTORC1 primarily used the cyclin D2 pathway to have a compensatory effect on the mass of β-cells. Though few have looked into protein metabolism, recent investigations have concentrated on the process underlying inulin-mediated glucose and fat metabolism. For example, it has been documented that hyperglycemia stimulated the growth of compensatory β-cells to fulfill the insulin demand brought on by hyperglycemia [29]. In a recent study on lipid metabolism in pregnant female mice, animals were randomized to consume standard chow (MO SC) or a high-fat diet (MO HF) sans prior preparation. When compared to MO SC mice, it was found that MO HF caused β-cell proliferation and survival to fail by downregulating IRS1, PI3K, and GLUT2 protein and upregulating insulin protein and FOXO1 [30]. Furthermore, it has been presented that a high-fat diet has an adverse outcome on the proliferation of β-cells, which causes the development of insulin resistance and type 2 diabetes mellitus (T2DM) traits [31, 32]. Additionally, preadipocyte factor 1 (Pref-1) might individually promote the release of insulin via AKT Rab43, according to Rhee *et al.* [33]. The relationship between the associated signaling pathways of β-cell proliferation and protein metabolism has also been an issue of debate. According to Chen *et al.* [16], FOXO1 may suppress albumin expression, but other proteins, like phycocyanin, have been shown to increase β-cell proliferation by controlling PI3K AKT signaling and downregulating FOXO1, as well as improve diabetes in mice by inducing insulin signaling and glucokinase expression in the pancreas [34]. Exendin 4 has also been observed to stimulate PI3K AKT signaling, which may lead to an intermediary ligand-binding activation phase containing homologous receptors and increase β-cell proliferation in some cases.

14.2.3 Calcium-Mediated Signaling Pathway

Calcium is a crucial second messenger that controls the production of hormones and enzymes, including insulin, and plays a role in apoptosis. Calcium homeostasis, the preservation of the body's natural Ca2+ content, is essential for cell viability [35]. Prior research suggests that β-cell death might be accelerated and type 1 diabetes mellitus (T1DM) could originate from a breakdown of calcium homeostasis [35]. PI3K signaling promoted autocrine function and inhibited β-cell development by activating calcium-mediated signaling pathways, increasing intracellular Ca2+ concentration, and inhibiting endoplasmic reticulum Ca2+ ATPase [36, 37]. Increases in intracellular calcium in calcium-mediated signaling cause calmodulin (CaM) to become active. CaM has two substrates: transcription coactivator 2, which is regulated by carbohydrate response element binding (CREB); and K Ras, which stimulates the antiproliferative ras association domain family 1A (Rasf1a) pathway and reduces cell proliferation. Thus, K Ras could have a negative regulatory function in β-cell proliferation, and K Ras gene downregulation could promote proliferation by lowering Rassf1a activity. ERKs, often referred to as classical MAPKs, are cellular signaling molecules that are diffusely expressed and involved in the regulation of cell proliferation [38]. ERK1 (MAPK3) and ERK2 (MAPK1) have been shown to be the primary ERK forms expressed in pancreatic β-cells. Generally, a ligand stimulates the small GTPase Ras upon binding to its corresponding tyrosine kinase or G protein-coupled receptor. This produces activated Ras GTP, which phosphorylates cRaf, activates MEK, and finally activates ERK1/2 [38]. Furthermore, it has been documented that a variety of growth factors and hormones activate ERK1/2 to cause β-cell proliferation. For instance, it was shown that activation of ERK1/2 was necessary for the stimulation of the platelet-derived growth factor receptor. This resulted in the overexpression of the polycomb group protein (EZH2) enhancer of zeste homolog 2, the suppression of the tumor suppressor protein P16LINK4, and eventually the proliferation of adult β-cells. [39]. ERK communication was also revealed to play a pivotal role in 3,5,3′-triidothyronine (T3)-induced β-cell proliferation [40]. Furthermore, reactive oxygen species promoted the development of β-cells via controlling the (ERK1/2) pathway [41] and activation of the ERK1/2 signaling pathway caused visfatin's anti-apoptotic action [42]. A mice model of diabetes was created in research by Ozaki *et al.* [43] using leptin receptor-deficient (db/db) mice and high-fat diet KKAy mice. These mice showed that insulin resistance linked to obesity may be addressed with an inhibitor of MEK, suggesting that type 2 diabetes may benefit from ERK signaling modulation [43].

14.2.4 Growth Factor Signaling Pathway

Many growth factors have been associated with the control of β-cell proliferation, including IGF 1, members of the nerve growth factor (NGF) superfamily, members of the transforming growth factor beta (TGFβ) superfamily, placental growth factor (PLGF), and their corresponding receptors (PLGFR, IGF 1R, TGFβR, and NGFR). These mitogen signals have been linked to a number of downstream pathways that regulate β-cell survival and proliferation, such as PI3K AKT mTOR, ERK, and JAK–STAT [44].

14.2.5 EGF Family Signaling Pathway

B-cell proliferation has been linked to a number of downstream epidermal growth factor (EGF) family proteins, including TGFα, betacellulin (BTC), EGF, HB (heparin binding), and epiregulin [45]. In β-cells, for example, BTC-dependent mitogenic signals were responsible for IRS overexpression and the activation of both ErbB2 and EGF/ErbB1 receptors. Additionally, BTC has been connected to the IRS PI3K signaling pathway, which controls AKT and FOXO1 via signaling through PDK1 (3′ phosphoinositide-dependent protein kinase 1) [38]. EGF and TGFα, which belong to the EGF superfamily, have been shown to temporarily activate ERK1/2, however, the connection was not sufficient to cause β-cell proliferation [38]. Moreover, constitutively activated EGFR guarded against harm and possibly death to β-cells [39]. Furthermore, it has been shown that FOXM1-mediated cell proliferation, EGFR signaling, and mTOR activation may interact to enhance β-cell enlargement in response to an abundance of nutrients [46]. Additional investigation into the survival mechanism of β-cells has revealed that Raf-1/MEK/ERK and EGF signaling may regulate β-cell proliferation, mainly increasing the half-life of β-cells.

14.2.6 NGF Signaling Chain

NGF controls neuronal cell improvement and also affects non-neuronal cell function [47, 48]. NGF is one of the peptides that is held by β-cells in islet β-cells within secretory vesicles. When β-cells connect with their receptors, it can cause autocrine and paracrine actions when the right stimuli are applied [48]. P75 neurotrophin receptor (P75 NTR) and tropomyosin receptor kinase A (trkA) are two membrane receptors by which NGF principally works. trkA is activated upon phosphorylation and primarily regulates the stimulation of signals related to cell survival, proliferation, and differentiation [49], whereas P75 NTR, in its NGF-unbound and NGF-bound forms, recruits certain intracellular effectors [50]. According to recent research, islet NGF was critical in regulating insulin secretion to preserve glucose biphasic secretion as well as baseline insulin secretion [51].

14.2.7 TGFb Superfamily Signaling Pathway

It has been established that several TGFβ superfamily members, such as TGFβ, inhibins, follistatins, activins, bone morphogenetic proteins, and Mullerian-inhibiting substances, work in β-cells [52], yet it's still unclear what the precise underlying mechanisms are. The TGFβ superfamily was essential in controlling the development and maturity of pancreatic endocrine function. It has been linked to several developmental events, such as proliferation, differentiation, and growth arrest [53, 54]. This mechanism was specifically coordinated by interactions between intracellular TGFβ signaling mediators and the TGFβ signaling inhibitor Smad7 [53]. According to a new study on TGFβ signaling, macrophages produce TGFβ1 to activate Smad7 expression, which in turn promotes the growth of β-cells [54]. Following partial pancreatic duct ligation (PDL), Smad7 was likewise elevated in β-cells [55]. TGFβ superfamily members often influence downstream Smad proteins in β-cells, which usually inhibits proliferation [56]. As a result, β-cell proliferation may be encouraged by TGFβ signaling suppression [55, 57]. TGFβ1 bound to the downstream targets Ras, AKT, and ERK1/2 in refined human pancreatic duct cells, where it caused the epithelial–mesenchymal transition and prevented development into β-cells [58]. TGFβ signaling increase may be one of the fundamental processes enabling pancreatic islet cell growth [59]. The proliferation and differentiation of BCD (β-cell-derived cells) were inhibited when TGFβ signaling was blocked by short hairpin RNA (shRNA) against TGFβ1R, such as Alk5 shRNA [60]. This suppression may have been linked to the AKT FOXO1 pathway [61].

14.2.8 PLGF and Pref-1 Signaling Pathway

Lastly, gestational PLGF can promote the production of growth factors by pancreatic islet endothelial cells, which in turn activates the PI3K AKT pathway and boosts β-cell proliferation [62]. According to prior work, Pref-1 changed the expression of FOXO1 and Pdx1 and separately promoted the phosphorylation of ERK1/2 and AKT.

14.2.9 Hormone Signaling Pathway

The following hormones have proven to be crucial for β-cell proliferation, survival, insulin secretion growth, and differentiation: growth hormone (GH), thyroid hormone (TH), insulin, prolactin (PRL), and glucagon-like peptide (GLP 1), as well as their corresponding receptors (GHR, TR, PRLR, and GLP 1R, respectively). They may also be convoluted in JAK–STAT, PI3K AKT, and ERK crosstalk [56]. TRs in many cell lines showed a stimulatory impact of T3 on cellular proliferation. It was proposed that T3 might stimulate the proliferation of β-cells in the pancreas based on TR expression in the organ. The investigation showed that the ERK signaling pathway was responsible for T3-induced proliferation of pancreatic β-cells [63]. The incretin hormone (GLP 1), which is largely secreted by intestinal L cells in response to neurons, hormones, and nutrients has been the major focus of the current review. It may have a major role in the proliferation of β-cells and the production of insulin [64, 65]. GLP 1 and its receptor (GLP 1R) together have the potential to increase insulin production in general. Nevertheless, it has been shown that a variety of intricate protein interactions are necessary for GLP 1R activation [66]. Numerous new GLP 1R interactors have been discovered in recent research, such as progesterone receptor membrane component 1, the Ras superfamily proteins Rab5c and Rab5b, and ATPase H+ transporting lysosomal accessory protein 2 (ATP6Ap2). It performed dual functions by controlling insulin production and GLP 1R signaling via cAMP and EGF receptor signaling [67]. GLP 1 may activate many signaling pathways in β-cells, such as PI3K AKT, cAMP PKA, Wnt, and ERK1/2 [68, 69].

GLP1 bolstered glucose-stimulated insulin production in a hyperglycemic response assay, mostly via activating cAMP PKA signaling in the β-cell line INS 1E. It has been observed that GLP 1 binding to GLP 1R activates cAMP PKA. Furthermore, GLP 1 stimulated ErbB1/2 by secreting BTC via ADAM proteins [70]. Additionally, geniposide improves insulin secretion by activating GLP 1R in conjunction with the signaling of adenylyl cyclase (AC) and cAMP, according to a recent investigation [71]. Yang *et al.* [72] discovered that GW501516, a peroxisome proliferator-activated receptor β/δ agonist, can prevent lipotoxic apoptosis in β-cells via modulating GLP 1R transcription. The study demonstrated that β-catenin signaling is essential for the conditional chicken ovalbumin upstream promoter transcription factor II, a novel mediator of GLP 1 signaling needed for GLP 1 activation. TCF7L2 controlled the production of this transcription factor, which raised β-cell mass throughout the neonatal stage [73]. Several important hormone routes linked to β-cell proliferation have been discussed in this study; nevertheless, more investigation is needed to fully understand the role of hormones in β-cell proliferation.

14.2.10 Wnt Signaling Pathway

Extracellular wnt ligands, adenomatous polyposis coli protein, axin, GSK3β, disheveled protein, β-catenin, and Frizzled receptor are the main components of the wnt signaling pathway. The Wnt pathway is divided into three types (canonical Wnt pathway, noncanonical planar cell polarity pathway, and noncanonical Wnt/Ca^{2+} pathway) that is a planar the activated phospholipase C, a canonical wnt/β-catenin pathway, small GTPase, wnt/Ca2+, PKC, and the cell polarity (PCP) route, as well as a non-canonical pathway [74, 75]. Wnt signaling fragments have been shown to control the postnatal β-cell activities of insulin secretion, differentiation, proliferation, and survival as well as pancreatic endocrine development [76]. According to Bader *et al.* [77], wnt/PCP genes caused β-cell maturation and were in charge of functional heterogeneity. Being a firefly luciferase and wnt/PCP effector, Fltp is thought to be a marker gene that might help clarify the molecular processes underlying the plasticity and heterogeneity of islet cells. In the event of diabetes, it may be a target in endocrine cells for effective regeneration of functioning β-cells [78]. It has been shown that Wnt4, a regulator of β-cell proliferation, mainly controls canonical Wnt signaling [79, 80]. It was shown by Krutzfeldt *et al.* [78] that wnt4 expression and upregulation were typical of obese mice in β-cells. Moreover, Exendin-4 activation increased wnt4 expression in β-cells. Additionally, it has been revealed that wnt3 activates the wnt/β-catenin pathway, demonstrating an interaction with PI3K signaling, to modify β-cell proliferation and promote β-cell function [81]. Wnt3a was found to drive greater β-cell proliferation in vitro and activate the production of Cdk4 and cyclin D1 and D2 by Rulifson *et al.* [82]. In terms of stem cells, canonical WNT signaling promoted stem cell survival and proliferation by expressing downstream targets associated to β-catenin, namely, cMyc, which was advantageous for the transplantation of pluripotent stem cells in T1DM [79, 83]. Wnt7b was shown to be a regulator of pancreatic progenitor cell formation by Afelik *et al.* [80]. There is evidence that TCF7L2 (T cell factor 7-like 2) plays a major role in wnt/β-catenin signaling and is a powerful factor in the etiology and development of T2DM [84]. Specifically, the "at risk" or dominant negative version of TCF7L2 negatively impacted the growth of β-cells and the release of insulin [85]. Furthermore, in BCD (INS1) and human colorectal cancer (HCT116) cells, reduced high mobility group protein B1 led to decreased TCF7L2 mRNA expression, glucose-triggered insulin secretion, and TCF7L2 transcriptional regulatory activity. [86]. A study found that geniposide improves β-cell survival and proliferation via changing the β-catenin/TCF7L2 signaling pathway [87].

14.2.11 JAK–STAT Signaling Pathway

A comparatively new signaling pathway called JAK–STAT is linked to several cytokines, including interleukin (IL)-1, -2, and -6, as well as GH, leptin, placental lactogens, erythropoietin (EPO), and PRL hormones [88, 89]. Important biological processes, including proliferation, cell division, and death, are regulated in part by JAK–STAT [88]. Moreover, Janus kinases (JAKs) could function as a basic switch that has to be triggered by a cytokine in order to activate the signal transducer and activator of transcription proteins (STATs), which then permits downstream target activation and the consequent control of target gene transcription [90]. Active STATs have also been shown to stimulate the transcription of cytokine-inducible SH2 domain-containing proteins (SOCS), suppressors of cytokine signaling, and protein repressors of active STAT [91]. By blocking the upstream signals of JAK–STAT signaling, it is believed that SOCS proteins are crucial JAK–STAT signaling negative modulators. Of these, four have been identified as major players in the JAK–STAT pathway: SOC1, SOC2, SOC3, and CISH [92]. T1DM is instigated by a gradual damage of pancreatic β-cells and is intimately linked to an autoimmune attack [93]. β-cell dysfunction during the inflammatory progressive stage of pancreatic islets has been shown to be influenced by the overexpression of JAK–STAT molecules [94]. According to recent research, β-cell survival was facilitated by BRD0476, a novel activator of β-cell apoptosis, which blocked interferon γ-induced JAK2 and STAT1 activation. In contrast to generic JAK–STAT pathway suppressors, BRD0476 did not impair the kinase activity of any JAK proteins while blocking JAK–STAT signaling pathways through USP9X (ubiquitin-specific peptidase 9X) [95]. Regarding PRL, it has been shown that human β-cells, maybe because they lacked PRL receptors, were unable to proliferate in response to PRL. Notably, overexpression of murine STAT5a might overcome β-cell growth failure since it would cause cdk4 and cyclin D1 to be upregulated and nuclear translocated. There is a connection between β-cell cycle activity and STAT5 signaling [96]. Moreover, it has been documented that EPO stimulates β-cell proliferation via JAK2–STAT5 when acting through the EPO receptor [97]. According to Groef *et al.* [98], PDL induced the production of many cytokines in β-cells that functioned as positive ligands for STAT3 signaling. Additionally, using IL-6 to activate STAT3 improved β-cell cycling.

14.2.12 TLR4 Signaling Pathway

TLR4 (toll-like receptor 4), which is typically expressed in the inborn immune system, is connected to diabetes mellitus [99]. TLR4 expression in T2DM may rise as a result of the nuclear factor κB (NFκB) signaling pathway being activated. Although NFκB has not been linked to the inflammatory response in diabetes, it could have a significant impact on insulin response and glucose metabolism. The binding of NFκB to the promoters of innate immunity genes, including those pertaining to pro-inflammatory cytokines such as IL-1β and tumor necrosis factor α (TNFα), is thought to have a positive feedback effect on the expression of TLR4 [100, 101]. Sepehri *et al.* [101] found that, regardless of age, sex, blood sugar level, or body mass index, TLR4 was upregulated in the peripheral blood mononuclear cells of T2DM patients. Moreover, TLR4 is thought to be a key contender in the associated T2DM problems, such as atherosclerosis and nephropathy [102, 103].

14.3 SIGNALING PATHWAYS IN CANCER

Two of the most essential signaling cascades that are often altered in cancer are the pathways for (MAPK) Ras/Raf/MEK/ERK and (mTOR) PTEN/PI3K/AKT/mammalian target of rapamycin. In addition, there is an emerging indication that these pathways can collaborate to improve the endurance of altered cells [103]. Both solid tumors and hematologic malignancies commonly harbor activating mutations of Raf and Ras, which activate their downstream substrates ERK1/2 and MEK1/2 [104]. It led to the therapeutic development of small molecule inhibitors that target certain elements of the MAPK cascade. Examples of these include inhibitors of farnesyl transferase (like tipifarnib), inhibitors of Raf-1 (like sorafenib), and inhibitors of MEK1/2 (like AZD6244). Similarly, one of the most often altered pathways in solid tumor malignancies is the PI3K/AKT/mTOR pathway. For instance, PTEN mutation or inactivation frequently occurs in these cancers, activating PI3K, and its downstream targets AKT and mTOR [105]. Although there is now a lot of interest in the development of AKT and PI3K inhibitors, the majority of clinical experience has been gained with mTOR inhibitors like rapamycin and its variants. To control protein translation and survival, mTOR combines many signals controlled by growth hormones, cellular stress, and energy needs. It binds to two distinct signaling complexes in order to function: the regulatory-associated protein of (Mtor) mTORC1/RAPTOR and the mTOR complex, which is rapamycin-insensitive and retorts to nutrient availability and growth factors; and mTORC2/rapamycin-insensitive companion of mTOR (RICTOR), which is rapamycin-insensitive and responds primarily to growth factors [106]. It's interesting to note that rapamycin has been shown to produce a negative feedback loop in which mTORC1 suppression leads to AKT activation [107]. In contrast, in some systems, continuous exposure to rapamycin may prevent TORC2 assembly and interfere with AKT signaling [108]. On the other hand, it is unclear how mTOR inhibition affects the MAPK pathway.

14.3.1 mTOR and MAPK Interactions

There has been evidence of interaction between ERK and mTOR [109], which suggests that this network may have a feedback loop. In this case, this phenomenon might be used medicinally. These concepts are supported by the findings of two different investigations published in the *Journal of Clinical Investigation* (*JCI*). The authors of the study by Waugh Kinkade *et al.* [110] used a transgenic prostate cancer model lacking PTEN to examine the causes of concurrent therapy with an inhibitor of mTOR (rapamycin) and PD0325190 (MEK1/2/ERK1/2). According to their findings, these substances efficiently inhibited their targets and, when taken together, worked in concert to stop the development of prostate cancer cells both in vivo and in vitro. Remarkably, the significant downregulation of cell proliferation was functionally associated with the upregulation of Bim, a pro-apoptotic Bcl-2 member that only requires BH3, whose quantity is known to be elevated by MEK/ERK inactivation [111]. Prostate cancer development in patient specimens was linked to ERK activation and the PTEN/PI3K/AKT/mTOR pathway. Moreover, the investigators discovered that a combination of mTOR and MEK/ERK inhibition was useful in the adjuvant situation. The regimen's inhibitory effects on prostate cancer cell development were predominantly directed contrary to androgen-independent illness, rather than dependent disease, which was an intriguing and somewhat surprising discovery. The researchers concluded that treating advanced cancer or using it as an adjuvant may benefit from an approach that combines inhibition of mTOR and MEK/ERK, especially if the condition is androgen-liberated.

In a related investigation, Carracedo *et al.* [112] found that ERK activation was seen in tumor cell samples taken from patients receiving mTOR inhibitor RAD001 therapy in a neo-adjuvant context. In a model of breast cancer cells, the same occurrences were noted. Pharmacologic and genetic methods both demonstrated the participation of a negative regulatory S6K/PI3K/AKT/mTOR-dependent pathway in ERK activation. Surprisingly the inhibitory consequences of this method on prostate cancer cell proliferation in the in vitro condition was not directly connected to the triggering of death, but a discernible rise in apoptosis was seen in animal experiments. The authors concluded that pharmacologic agents that block ERK activation significantly increase the efficacy of clinically relevant mTOR inhibitors. They also found that when cancer cells are exposed to mTORC1 inhibitors, the MAPK pathway is activated in response to the discharge of the S6K1-related "brake" on PI3K and Ras.

14.4 INFLAMMATION AND CELL SIGNALING

One of the best physiological processes that shields our body from damage, illness, exposure to toxins or foreign particles, or stress is inflammation [113]. Various immune system cell types and signaling pathways are involved in the intricate process of inflammation. It is a dynamic process that is determined by its location (native or systemic), intensity (medium, modest, or severe), and duration (severe or chronic). The innate and adaptive immune systems employ inflammation as a potent mechanism to preserve tissue and cell homeostasis [114]. A significant pro-inflammatory cytokine, IL-6, is produced as a 26 kDa protein [115]. IL-6 was produced by a variety of cell types, consisting of T cells, B cells, monocytes, tumor cells, and macrophages [116]. IL-6 initially attaches to the mIL-6R (membrane-bound IL-6 receptor) in classical IL-6 signaling. The JAK–STAT signaling pathway can then be activated by the combination of IL-6 and mIL-6R, which can also cause the homodimerization of glycoprotein 130 (gp130). But in the body, mIL-6R is exclusively expressed on certain leukocytes and hepatocytes. Interestingly, it has been shown that bodily fluids including blood and urine include soluble forms of IL-6R, or sIL-6R [117, 118]. Two distinct methods might produce sIL-6R: ADAM17's proteolysis of the membrane-bound receptor and the translation of an mRNA with variable splicing [119, 120]. The affinity of this sIL-6R for IL-6 binding is comparable to that of the membrane-bound receptor. sIL-6R can activate IL-6 signaling on gp130-only cells without mIL-6R expression. Since every cell in the body expresses gp130, the IL-6/sIL-6R complex, also known as trans-signaling, has the potential to activate every single cell in the body [121]. After interaction with mIL-6R, IL-6 is quickly internalized in excited cells, and this internalization has the ability to stop the signaling. Nevertheless, the IL-6/sIL-6R complex only partially internalizes, resulting in an extended period of IL-6 signaling [122]. The four members of JAK are TYK2, JAK1, JAK3, and JAK2. It is a significant IL-6 signaling downstream effector [123]. Non-receptor cytoplasmic tyrosine kinases, or JAKs, catalyze the transfer of a phosphate group to a substrate protein. JAK brings gp130 closer to itself in reaction to IL-6, and gp130 phosphorylates the STATs [124]. The genes that encode the seven members of the STAT family are STAT1, STAT2, STAT4, STAT6, STAT5B, STAT5A, and STAT3. A wide range of signaling outcomes are produced by the unique coupling between JAK and STAT proteins, which is caused by the presence of distinct family members of these two proteins. The transcription factor STAT has four core domains— the DNA-binding domain, coiled coil domain, Src homology 2 (SH2) domain, and linker domain–together with an unspoiled Tyr residue at the C-terminus [125, 126]. Through reciprocal pTyr–SH2 domain contacts, STAT can phosphorylate and then undergo either heterodimerization (e.g., STAT1–STAT3) or homodimerization (e.g., STAT3–STAT3) [127]. After translocating to the nucleus of the cell, the dimer attaches with DNA to control the expression of many genes, including COX2, TWIST, VEGF, and Cyclin D1 [128, 129]. As one of the main JAK signal transducers, STAT3 is essential for controlling the variability of biological processes, with cell proliferation and cell division. Various inhibitory molecules, such as SOCS, PIAS, and SHP-1 and SHP-2, play a crucial role in controlling STAT3 signaling under typical conditions [130, 131]. WhereasPIAS is a nuclear factor that suppresses transcription and phosphorylation of STAT3, SOCS binds to JAKs to reduce STAT3 activity [132]. By dephosphorylating STAT3, protein tyrosine phosphatases SHP1 and SHP-2 controlled STAT3 activity adversely. Because STAT3 is involved in IL-6-signaling, among other physiological and pathological processes, it is closely associated with inflammation [123, 133].

14.5 OBESITY AND CELL SIGNALING PATHWAY

One of the main intracellular signal transduction pathways, the STAT–JAK pathway, is a crucial downstream mediator for a number of growth factors, hormones, and cytokines. Tyrosine phosphorylation can be used to activate the whole STAT protein family (STAT1, 2, 3, 4, 5a, 5b, and 6) in response to growth factor and cytokine stimulation. The STAT–JAK is an important intracellular signaling route that regulates cytokines, hormones, and growth factors. Tyrosine phosphorylation can activate the STAT protein family in response to growth factor and cytokine stimulation [134]. JAKs are activated when growth factors or cytokines bind to their corresponding receptors. This allows them to trans-phosphorylate the receptor's cytoplasmic tail on tyrosine residues as well as each other [135, 136]. Next, the receptor subunits provide STATs a place to bind, and these proteins are phosphorylated as well [137]. The nucleus is where the phosphorylated STATs relocate, attach to certain DNA sequences, and control the transcription of desired genes [138]. Obesity is caused either directly by the disruption of the JAK–STAT signaling system or indirectly through its interactions with additional signaling pathways such as PI3K and MAPK. Because JAK–STAT mediates the leptin-regulated energy balance, the JAK–STAT signaling pathway is associated with the melanocortin pathway [139]. Leptin receptor (LEPR), which is expressed as a dimer at the plasma membrane and is involved in leptin signaling, phosphorylates itself when receptor-associated JAK2 is activated. Subsequently, LEPR interacts with STAT3 and STAT5. JAK2 then phosphorylates them so that they can operate as transcription factors [140, 141]. Controlling food intake requires LEPR to activate STAT3/STAT5 [142, 143]. Furthermore, the suppressor of cytokine signaling 3 is expressed in response to phosphorylated STAT3, and it functions as a feedback inhibitor of the leptin signaling pathway [144]. To preserve energy balance, leptin binding to LEPR causes Rho-kinase 1 to be activated downstream. This phosphorylates and activates JAK2 [137]. Additionally, the binding causes JAK2 to engage with IRS2-mediated PI3K pathway activation and SH2-Bβ, which encourages IRS1 [145, 146]. Subsequently, it stimulates POMC transcription, raises the expression of carboxypeptidase, and processes POMC more thoroughly to produce α-melanocyte-stimulating hormone (αMSH) while suppressing appetite [147]. Conversely, reduced leptin sensitivity in POMC neurons is linked to central nervous system (CNS) inhibition of JAK–STAT activity. Hepatic steatosis, or the buildup of fat in the liver, is a characteristic of obesity [148]. Growth factors and cytokines affect this process through the JAK–STAT signaling pathway [149, 150]. Investigation has repeatedly indicated that insulin resistance and elevated expression of gluconeogenic genes are caused by hepatocyte-specific STAT3 deficiency [151, 152]. On the other hand, hepatocyte STAT3 activation may stop steatosis. Hepatic fat formation is reduced in obese mice treated with STAT3-inducing cytokines (IL-6 and IL-22) or STAT3 overexpression [153, 154]. It has been established that the hepatic growth factor–JAK2–STAT5–IGF1 axis plays a critical role in lipid metabolism. Growth factor is essential for the hepatic synthesis of IGF1 through the activation of JAK2 and STAT5. Although the exact mechanism by which low growth factor levels cause obesity is debatable, it may be related to increased hepatic steatosis and impaired lipolysis in adipose tissue [155]. Intracellular lipid buildup and concomitant activation of STAT1 and STAT3 are caused by loss of STAT5 signaling. Moreover, data indicates that mice with a hepatocyte-specific JAK2 deletion experience spontaneous steatosis as early as two weeks of life, but they also demonstrate protection against insulin resistance and glucose intolerance brought on by a high-fat diet [156]. Leptin can also stimulate the peripheral JAK–STAT signaling pathway [157, 158]. In adipose tissue, for example, HFD-induced leptin secretion raises the expression of the STAT3 target gene producing caveolin-1, which negatively feedback reduces leptin signaling [159]. Cernkovich *et al.* used an adipocyte-specific STAT3 mouse colony to investigate the function of STAT3 in adipocytes further. They found that adipocyte hypertrophy was associated with increased body weight and adipose tissue mass, indicating that STAT3 stimulates lipolysis and inhibits adipogenesis [160]. Additionally, animals missing Tyk2 exhibit progressively higher levels of obesity as a result of impaired BAT development, suggesting that Tyk's activation of STAT3 is necessary for BAT function [161]. By lowering insulin sensitivity and raising adipocyte inflammation, STAT4 also plays a part in the pathogenesis of obesity [162]. Adipocyte dysfunction and insulin resistance are also caused by increased levels of interferon-γ and JAK–STAT1 signaling in obese individuals [163, 164]. Since JAK proteins are the main upstream kinases needed for STAT activity, they have an effect on adipose function as well. In mice, adipocyte-specific JAK2 deletion causes adipolysis-related obesity [165], but in vitro browning of human adipocytes and increased expression of UCP1 are caused by pharmacological suppression of JAK–STAT [166].

14.6 CONCLUSION

To conclude, we have highlighted the importance and mechanisms of the following cell signaling pathways: insulin signaling pathway, PI3K–AKT/PKB signaling pathway, calcium-mediated signaling pathway, growth factor signaling pathway, EGF family signaling pathway, NGF signaling chain, TGFβ superfamily signaling pathway, PLGF and Pref-1 signaling pathway, hormone signaling pathway, Wnt signaling pathway, JAK–STAT signaling pathway, TLR4 signaling pathway, and mTOR and MAPK interactions. Multiple therapeutic strategies may be used for the treatment of diabetes, cancer, inflammation, and obesity. Identification of appropriate molecular targets within these pathways may aid to develop novel strategies in the treatment of diabetes and improve the outcome for patients.

REFERENCES

1. Ahmad HA, Seemab K, Wahab F and Khan MI: Signaling pathways in drug development. In: Rauf A, editor. *Drug Development and Safety*. London: IntechOpen; 2024.
2. El-Mahdy HA, Elsakka EG, El-Husseiny AA, Ismail A, Yehia AM, Abdelmaksoud NM, Elshimy RA, Noshy M and Doghish AS: miRNAs role in bladder cancer pathogenesis and targeted therapy: Signaling pathways interplay–A review. *Pathol Res Pract* 242: 154316, 2023.
3. Mathews J, Chang AJ, Devlin L and Levin M: Cellular signaling pathways as plastic, proto-cognitive systems: Implications for biomedicine. *Patterns* 4(5): 100737, 2023.
4. Emmert-Streib F, Dehmer M and Haibe-Kains: Gene regulatory networks and their applications: understanding biological and medical problems in terms of networks. *Front Cell Dev Biol* 2: 38, 2014. https://doi.org/10.3389/fcell.2014.00038.
5. Brash DE: Rethinking causation for data-intensive biology: constraints, cancellations, and quantized organisms: causality in complex organisms is sculpted by constraints rather than instigators, with outcomes perhaps better described by quantized patterns than rectilinear pathways. *Bio-essays* 42: e1900135, 2020. https://doi.org/10.1002/ bies.201900135.
6. Bizzarri M, Palombo A and Cucina A: Theoretical aspects of systems biology. *Prog Biophys Mol Biol* 112: 33–43, 2013. https://doi.org/ 10.1016/j.pbiomolbio.2013.03.019.
7. Centers for Disease Control and Prevention: *National diabetes fact sheet: general Information and national estimates on diabetes in the United States, 2007.* Atlanta: U.S. Department of Health and Human Services, Centers for Disease Control and Prevention; 2008.
8. International Diabetes Federation Diabetes e-Atlas; 2006. Available at: http://www. eatlas.idf.org Accessed 6 July 2010.
9. Rains JL and Jain SK: Oxidative stress, insulin signaling, and diabetes. *Free Radic Biol Med* 50(5): 567–575, 2011.
10. Saltiel AR and Kahn CR: Insulin signaling and the regulation of glucose and lipid metabolism. *Nature* 414: 799–806, 2001.
11. Pende M, Kozma SC, Jaquet M, Oorschot V, Burcelin R, Le Marchand–Brustel Y, Kluperman J, Thorens B and Thomas G: Hypoinsulinaemia, glucose intolerance and diminished beta–cell size in S6K1–deficient mice. *Nature* 408: 994–997, 2000.
12. Arajuo EP, Amaral ME, Souza CT, Bordin S, Ferreira F, Saad MJ, Boschero AC, Maqalhaes EC and Velloso LA: Blockade of IRS1 in isolated rat pancreatic islets improves glucose–induced insulin secretion. *FEBS Lett* 531: 437–442, 2002.
13. Bringhenti I, Ornellas F, Mandarim–de–Lacerda CA and Aguila MB: The insulin–signaling pathway of the pancreatic islet is impaired in adult mice offspring of mothers fed a high–fat diet. *Nutrition* 32: 1138–1143, 2016.
14. Ou Y, Ren Z, Wang J and Yang X: Phycocyanin ameliorates alloxan–induced diabetes mellitus in mice: involved in insulin signaling pathway and GK expression. *Chem Biol Interact* 247: 49–54, 2016.
15. Kaneko K, Ueki K, Takahashi N, Hashimoto S, Okamoto M, Awazawa M, Okazaki Y, Ohsugi M, Inabe K, Umehara T, et al: Class IA phosphatidylinositol 3–kinase in pancreatic β cells controls insulin secretion by multiple mechanisms. *Cell Metab* 12: 619–632, 2010.
16. Chen Q, Lu M, Monks BR and Birnabaum MJ: Insulin is required to maintain albumin expression by inhibiting forkhead box o1 protein. *J Biol Chem* 291: 2371–2378, 2016.
17. Wang C, Chen X, Ding X, He Y, Gu C and Zhou L: Exendin–4 promotes beta cell proliferation via PI3K/Akt signaling pathway. *Cell Physiol Biochem* 35: 2223-2232, 2015
18. Katoh M and Katoh H: Human FOX gene family (Review). *Int J Oncol* 25: 1495–1500, 2004.

19. Seyer P, Vallois D, Poitry–Yamate C, Schutz F, Metref S, Tarussio D, Maechler P, Staels B, Lanz B, Grueter R, et al: Hepatic glucose sensing is required to preserve β cell glucose competence. *J Clin Invest* 123: 1662–1676, 2013.
20. Huang L, Jiang X, Gong L and Xing D: Photoactivation of Akt1/GSK3β Isoform–specific signaling axis promotes pancreatic β–cell regeneration. *J Cell Biochem* 116: 1741–1754, 2015.
21. Boucher MJ, Selander L, Carlsson L and Edlund H: Phosphorylation marks IPF1/PDX1 protein for degradation by glycogen synthase kinase 3–dependent mechanisms. *J Biol Chem* 281: 6395–6403, 2006.
22. Morral N: Novel targets and therapeutic strategies for type 2 diabetes. *Trends Endocrinol Metab* 14: 169–175, 2003.
23. Foster KG and Fingar DC: Mammalian target of rapamycin (mTOR): Conducting the cellular signaling symphony. *J Biol Chem* 285: 14071–14077, 2010.
24. Duzgn Z, Eroglu Z and Biray–Avci C: Role of mTOR in glioblastoma. *Gene* 575: 187–190, 2016.
25. Wang J, Yang X and Zhang J: Bridges between mitochondrial oxidative stress, ER stress and mTOR signaling in pancreatic β cells. *Cell Signal* 28: 1099–1104, 2016.
26. Maiese K: Targeting molecules to medicine with mTOR, autophagy and neurodegenerative disorders. *Br J Clin Pharmacol* 82: 1245-1266, 2016.
27. Kulkarni RN, Mizrachi EB, Ocana AG and Stewart AF: Human β-cell proliferation and intracellular signaling: Driving in the dark without a road map. *Diabetes* 9: 2205–2213, 2012.
28. Escribano O, Gomez–Hernandez A, Diaz–Castroverde S, Nevado C, Garcia G, Otoro YF, Perdomo L, Beneit N and Benito M: Insulin receptor isoform A confers a higher proliferative capability to pancreatic β cells enabling glucose availability and IGF–1 signaling. *Mol Cell Endocrinol* 409: 82–91, 2015.
29. Velazquez-Garcia S, Valle S, Rosa TC, Takane KK, Demirci C, Alvarez-Perez JC, Mellado-Gil JM, Ernst S, Scott DK, Vasavada RC and Alonso LC: Activation of protein kinase C-ζ in pancreatic β-cells in vivo improves glucose tolerance and induces β-cell expansion via mTOR activation. *Diabetes* 60(10): 2546–2559, 2011.
30. McCurdy CE and Klemm DJ: Adipose tissue insulin sensitivity and macrophage recruitment: Does PI3K pick the pathway. *Adipocyte* 2: 135–142, 2013.
31. Bringhenti I, Moraes–Teixeira JA, Cunha MR, Ornellas F, Mandarim–de–lacerda CA and Aguila MB: Maternal obesity during the preconception and early life periods alters pancreatic development in early and adult life in male mouse offspring. *PLoS One* 8: e55711, 2013.
32. Cerf ME, Williams K, Chapman CS and Louw J: Compromised beta–cell development and beta–cell dysfunction in weanling offspring from dams maintained on a high–fat diet during gestation. *Pancreas* 34: 347–353, 2007.
33. Rhee M, Lee H, Kim JW, Ham DS, Park HS, Yang HK, Shin JY, Cho JH, Kim YB, Youn BS, et al: Preadipocyte factor 1 induces pancreatic ductal cell differentiation into insulin–producing cells. *Sci Rep* 6: 23960, 2016.
34. Gao Y, Liao G, Xiang C, Yang X, Cheng X and Ou Y: Effects of phycocyanin on INS–1 pancreatic β–cell mediated by PI3K/Akt/FoxO1 signaling pathway. *Int J Biol Macromol* 83: 185–194, 2016.
35. Ahn C, An BS and Jeung EB: Streptozotocin induces endoplasmic reticulum stress and apoptosis via disruption of calcium homeostasis in mouse pancreas. *Mol Cell Endocrinol* 412: 302–308, 2015.
36. Khan S, Yan–Do R, Duong E, Wu X, Bautista A, Cheley S, MacDonald PE and Braun M: Autocrine activation of P2Y1 receptors couples Ca (2+) release in human pancreatic beta cells. *Diabetologia* 57: 2235–2245, 2014.
37. Roper MG, Qian WJ, Zhang BB, Kulkarni RN, Kahn CR and Kennedy RT: Effect of the insulin mimetic L-783,281 on intracellular Ca2 and insulin secretion from pancreatic beta–cells. *Diabetes* 51 (Suppl): S43–S49, 2002.
38. Oh YS, Shin S, Lee YJ, Kim EH and Jun HS: Betacellulin–induced beta cell proliferation and regeneration is mediated by activation of ErbB–1 and ErbB–2 receptors. *PLoS One* 6: e23894, 2011.
39. Hakonen E, Ustinov J, Eizirik DL, Sariola H, Miettinen PJ and Otonkoski T: In vivo activation of the PI3K–Akt pathway in mouse beta cells by the EGFR mutation L858R protects against diabetes. *Diabetologia* 57: 970–979, 2014.
40. Kim TK, Lee JS, Jung HS, Ha TK, Kim SM, Han N, Lee EJ, Kim TN, Kwon MJ, Lee SH, et al: Triiodothyronine induces proliferation of pancreatic β–cells through the MAPK/ERK pathway. *Exp Clin Endocrinol Diabetes* 122: 240–245, 2014.
41. Hoarau E, Chandra V, Rustin P, Scharfmann R and Duvillie B: Pro–oxidant/antioxidant balance controls pancreatic β-cell differentiation through the ERK1/2 pathway. *Cell Death Dis* 5: e1487, 2014.
42. Xiang RL, Mei M, Su YC, Li L, Wang JY and Wu LL: Visfatin protects rat pancreatic β–cells against IFN–γ–induced apoptosis through AMPK and ERK1/2 signaling pathways. *Biomed Environ Sci* 28: 169–177, 2015.

43. Ozaki KI, Awazu M, Tamiya M, Iwasaki Y, Harada A, Kugisaki S, Tanimura S and Kohno M: Targeting the ERK signaling pathway as a potential treatment for insulin resistance and type 2 diabetes. *Am J Physiol Endocrino Metab* 310: E643–E651, 2016.
44. Wang H, Gambosova K, Cooper ZA, Holloway MP, Kassai A, Izquierdo D, Cleveland K, Boney CM and Altura RA: EGF regulates surviving stability through the Raf–1/ERK pathway in insulin–secreting pancreatic β–cells. *BMC Mol Biol* 11: 66, 2010.
45. Ernesto BM, Kulkarni RN, Scott DK, Mauvais–Jarvis F, Stewart AF and Garcia–Ocana A: Human β–cell proliferation and intracellular signaling part 2: Still driving in the dark without a road map. *Diabetes* 63: 819–831, 2014.
46. Zarrouki B, Benterki I, Fontes G, Peyot ML, Seda O, Prentki M and Poitout V: Epidermal growth factor receptor signaling promotes pancreatic β–cell proliferation in response to nutrient excess in rats through mTOR and FOXM1. *Diabetes* 63: 982–993, 2014.
47. Zhang CC, Yin X, Cao CY, Wei J, Zhang Q and Gao JM: Chemical constituents from hericium erinaceus and their ability to stimulate NGF–mediated neurite outgrowth on PC12 cells. *Bioorg Med Chem Lett* 25: 5078–5082, 2015.
48. Rosenbaum T, Vidaltamayo R, Sanchez–Soto MC, Zentella A and Hiriart M: Pancreatic beta cells synthesize and secrete nerve growth factor. *Proc Natl Acad Sci USA* 95: 7784–7788, 1998.
49. Freund V and Frossard N: Expression of nerve growth factor in the airways and its possible role in asthma. *Prog Brain Res* 146: 335–346, 2004.
50. Roux PP and Barker PA: Neurotrophin signaling through the P75 neurotrophin receptor. *Prog Neurobiol* 67: 203–233, 2002.
51. Pingitore A, Caroleo MC, Cione E, Castanera–Gonzalez R, Huang GC and Perasud SJ: Fine tuning of insulin secretion by release of nerve growth factor from mouse and human islet β–cells. *Mol Cell Endocrinol* 436: 23–32, 2016.
52. Ahmad FK, Zhiheng H and King GL: Molecular targets of diabetic cardiovascular complications. *Curr Drug Targets* 6:487–494; 2005
53. El–Gohary Y, Tulachan S, Guo P, Welsh C, Wiersch J, Prasadan K, Paredes J, Shiota C, Xiao X, Wada Y, et al: Smad signaling path- ways regulate pancreatic endocrine development. *Dev Biol* 378: 83–93, 2013.
54. Xiao X, Wiersch J, El–Gohary Y, Guo P, Prasadan K, Paredes J, Welsh C, Shiota C and Gittes GK: TGFβ receptor signaling is essential for inflammation–induced but not β–cell work- load–induced β–cell proliferation. *Diabetes* 62: 1217–1226, 2013.
55. Blum B, Roose AN, Barrrandon O, Maehr R, Arvanites AC, Davidow LS, Davis JC, Peterson QP, Rubin LL and Melton DA: Reversal of β cell de–differentiation by a small molecule inhibitor of the TGFβ pathway. *Elife* 3: e02809, 2014.
56. Huang Y and Chang Y: Regulation of pancreatic islet beta-cell mass by growth factor and hormone signaling. *Prog Mol Biol Transl Sci* 121: 321–349, 2014.
57. Xiao X, Gaffar I, Guo P, Wiersch J, Fischbach S, Peirish L, Song Z, El–Gohary Y, Prasadan K, Shiota G and Gittes GK: M2 macrophages promote beta–cell proliferation by up–regulation of SMAD7. *Prov Natl Acad Sci USA* 111: E1211–E1220, 2014.
58. Shin JA, Hong OK, Lee HJ, Jeon SY, Kim JW, Lee SH, Cho JH, Lee JM, Choi YH, Chang SA, et al: Transforming growth factor–β induces epithelial to mesenchymal transition and suppresses the proliferation and trans differentiation of cultured human pancreatic duct cells. *J Cell Biochem* 112: 179–188, 2011.
59. Lei C, Zhou X, Pang Y, Mao Y, Lu X, Li M and Zhang J: TGF–β signaling prevents pancreatic beta cell death after proliferation. *Cell Prolif* 48: 356–362, 2015.
60. Toren–Haritan G and Efrat S: TGFβ pathway inhibition re-differentiates human pancreatic islet β cells expanded in vitro. *PLoS One* 9: e0139168, 2015.
61. Feng Z, Zi Z and Liu X: Measuring TGF–β ligand dynamics in culture medium. *Methods Mol Biol* 1344: 379–389, 2016.
62. Li J, Ying H, Cai G, Guo Q and Chen L: Pre–Eclampsia–associated reduction in placental growth factor impaired beta cell proliferation through PI3K signaling. *Cell Physiol Biochem* 36: 34-43, 2015.
63. Talior I, Yarkoni M, Bashan N and Eldar-Finkelman H: Increased glucose uptake promotes oxidative stress and PKC-δ activation in adipocytes of obese, insulinresistant mice. *Am J Physiol Endocrinol Metab* 285: E295–E302, 2003.
64. Lee YS and Jun HS: Anti–diabetic actions of glucagon–like peptide–1 on pancreatic beta–cells. *Metabolism* 63: 9–19, 2014.
65. Tian L and Jin T: The incretin hormone GLP–1 and mechanisms underlying its secretion. *J Diabetes* 8: 753–765, 2016.

66. Dai FF, Bhattacharjee A, Liu Y, Batchuluun B, Zhang M, Wang XS, Huang X, Luu L, Zhu D, Gaisano H and Wheeler MB: A novel GLP1 receptor interacting protein ATP6ap2 regulates insulin secretion in pancreatic beta cells. *J Biol Chem* 290: 25045–25061, 2015.
67. Zhang M, Robitaille M, Showalter AD, Huang X, Liu Y, Bhattacharjee A, Willard FS, Han J, Froese S, Wei L, et al: Progesterone receptor membrane component 1 is a functional part of the glucagon–like peptide–1 (GLP–1) receptor complex in pancreatic β cells. *Mol Cell Proteomics* 13: 3049–3062, 2014.
68. Campbell JE and Drucker DJ: Pharmacology, physiology, and mechanisms of incretin hormone action. *Cell Metab* 17: 819–837, 2013.
69. Lavine JA and Attie AD: Gastrointerstinal hormones and the regulation of β–cell mass. *Ann NY Acad Sci* 1212: 41–58, 2010.
70. Talior I, Yarkoni M, Bashan N and Eldar-Finkelman H: Increased glucose uptake promotes oxidative stress and PKC-δ activation in adipocytes of obese, insulin resistant mice. *Am J Physiol Endocrinol Metab* 285: E295–E302, 2003.
71. Zhang Y, Ding Y, Zhong X, Guo Q, Wang H, Gao J, Bai T, Ren L, Guo Y, Jiao X and Liu Y: Geniposide acutely stimulates insulin secretion in pancreation β–cells by regulating GLP–1 receptor/cAMP signaling and ion channels. *Mol Cell Endocrinol* 430: 89–96, 2016.
72. Yang Y, Tong Y, Gong M, Lu Y, Wang C, Zhou M, Yang Q, Mao T and Tong N: Activation of PPARβ/δ protects pancreatic β cells from palmitate–induced apoptosis by upregulating the expression of GLP–1 receptor. *Cell Signal* 26: 268–278, 2014.
73. Boutant M, Ramos OH, Tourrel–Cuzin C, Movassat J, llias A, Vallois D, Planchais J, Pegorier JP, Schuit F, Petit PX, et al: COUP–TFII controls mouse pancreatic β–cell mass through GLP–1–β–catenin signaling pathways. *PLoS One* 7: e30847, 2012.
74. Heller C, Kuhn MC, Mulders–Opgenoorth B, Schott M, Willenberg HS, Scherbaum WA and Schinner S: Exendin–4 upregulates the expression of wnt–4, a novel regulator of pancreatic β–cell proliferation. *Am J Physiol Endocrinol Metab* 301: E864–E872, 2011.
75. Xue G, Romano E, Massi D and Mandala M: Wnt/β-catenin signaling in melanoma: Preclinical rationale and novel therapeutic insights. *Cancer Treat Rev* 49: 1–12, 2016.
76. Schinner S, Ulgen F, Papewalis C, Schott M, Woelk A, Vidal–Puig A and Scherbaum WA: Regulation of insulin secretion, glucokinase gene transcription and beta cell proliferation by adipicyte–derived wnt signaling molecules. *Diabetologia* 51: 147–154, 2008.
77. Bader E, Migliorini A, Gegg M, Moruzzi N, Gerdes J, Roscioni SS, Bakhti M, Brandl E, Irmler M, Beckers J, et al: Indentification of proliferative and mature β–cells in the islets of Langerhans. *Nature* 535: 430–434, 2016.
78. Krutzfeldt J and Stoffel M: Regulation of wingless–type MMTV integration site family (WNT) signaling in pancre- atic islets from wild–type and obese mice. *Diabetologia* 53: 123–127, 2010.
79. Zhang YQ, Morris JP, Yan W, Schofield HK, Gurney A, Simeone DM, Millar SE, Hoey T, Hebrok M and Pasca di Magliano M: Canonical wnt signaling is required for pancreatic carcinogenesis. *Cancer Res* 73: 4909–4922, 2013.
80. Afelik S, Pool B, Schmerr M, Penton C and Jensen J: Wnt7b is required for epithelial progenitor growth and operates during epithelial–to–mesenchymal signaling in pancreatic development. *Dev Biol* 399: 204–217, 2015.
81. Gui S, Yuan G, Wang L, Zhou L, Xue Y, Yu Y, Zhang J, Zhang M, Yang Y and Wang DW: Wnt3a regulates proliferation, apoptosis and function of pancreatic NIT–1 β cells via activation of IRS2/PI3K signaling. *J Cell Biochem* 114: 1488–1497, 2013.
82. Rulifson IC, Karnik SK, Heiser PW, Ten Berge D, Chen H, Gu X, Taketo MM, Nusse R, Hebro M and Kim SK: Wnt signaling regulates pancreatic beta cell proliferation. *Proc Nati Acad Sci USA* 104: 6247–6252, 2007.
83. He X, Han W, Hu SX, Zhang MZ, Hua JL and Peng S: Canonical wnt signaling pathway contributes to the proliferation and survival in porcine pancreatic stem cells (PSCs). *Cell Tissue Res* 362: 379–388, 2015.
84. Lyssenko V: The transcription factor 7–like 2 gene and increased risk of type 2 diabetes: An update. *Curr Opin Clin Nutr Metab Care* 11: 385–392, 2008.
85. Takamoto I, Kubota N, Nakaya K, Kumagai K, Hashimoto S, Kubota T, Inoue M, Kajiwara E, Katsuyama H, Obata A, et al: TCF7L2 in mouse pancreatic beta cells plays a crucial role in glucose homeostasis by regulating beta cell mass. *Diabetologia* 57: 542–553, 2014.
86. Zhou Y, Oskolkov N, Shcherbina L, Ratti J, Kock KH, Su J, Martin B, Oskolkova MZ, Goransson O, Bacon J, et al: HMGB1 binds to the rs7903146 locus in TCF7L2 in human pancreatic islets. *Mol Cell Endocrinol* 430: 138–145, 2016.

87. Yao DD, Yang L, Wang Y, Liu C, Wei YJ, Jia XB, Yin W and Shu L: Geniposide promotes beta–cell regeneration and survival through regulating β–catenin/TCF7L2 pathway. *Cell Death Dis* 6: e1746, 2015.
88. Kiu H and Nicholson SE: Biology and significance of the JAK/STAT signaling pathways. *Growth Factors* 30: 88-106, 2012.
89. Brooks AJ, Dai W, O'Mara ML, Abankwa D, Chhabra Y, Pelekanos RA, Gardon O, Tunny KA, Blucher KM, Morton CJ, et al: Mechanism of activation of protein kinase JAK2 by the growth hormone recetor. *Science* 344: 1249783, 2014.
90. Liongue C, Taznin T and Ward AC: Signaling via the CytoR/JAK/STAT/SOCS pathway: Emergence during evolu- tion. *Mol Immunol* 71: 166–175, 2016.
91. Linossi EM, Babon JJ, Hilton DJ and Nicholson SE: Suppression of cytokine signaling: The SOCS perspective. *Cytokine Growth Factor Rev* 24: 241–248, 2013.
92. Trengove MC and Ward AC: SOCS proteins in development and disease. *Am J Exp Clin Immunol* 2: 1–29, 2013.
93. Fleyel T, BrorssonC, Nielsen LB, Miani M, Bang–BerthelsenCH, Friedrichsen M, Overgaard AJ, Berchtold LA, Wiberg A, Poulsen P, et al: CTSH regulates β-cell function and disease progression in newly diagnosed type 1 diabetes patients. *Proc Natl Acad Sci USA* 111: 10305–10310, 2014.
94. Stanley WJ, Litwak SA, Quah HS, Tan SM, Kay TW, Tiganis T, de Haan JB, Thomas HE and Gurzov EN: Inactivation of protein tyrosine phosphatases enhances interferon signaling in pancre- atic islets. *Diabetes* 64: 2489–2496, 2015.
95. Chou DH, Vetere A, Choudhary A, Scully SS, Schenone M, Tang A, Gomez R, Burns SM, Lundh M, Vital T, et al: Kinase–independent small–molecule inhibition of JAK–STAT signaling. *J Am Chem Soc* 137: 7929–7934, 2015.
96. Chen H, KleinbergerJW, Takane KK, Salim F, Fiaschi–Taesch N, Pappas K, Parsons R, Jiang J, Zhang Y, Liu H, et al: Augmented stat5 signaling bypasses multiple impediments to lactogen–mediated proliferation in human β–cells. *Diabetes* 64: 3784–3797, 2015.
97. Choi D, Schroer SA, Lu SY, Wang L, Wu X, Liu Y, Zhang Y, Gaisano HY, Wagner KY, Wu H, et al: Erythropoietin protects against diabetes through direct effects on pancreatic beta cells. *J Exp Med* 207: 2831–2842, 2010.
98. De–Groef S, Renmans D, Cai Y, Leuckx G, Roels S, Staels W, Gradwohl G, Baeyens L, Heremans Y, Martens GA, et al: STAT3 modulates β–cell cycling in injured mouse pancreas and protects against DNA damage. *Cell Death Dis* 7: e2272, 2016.
99. Shao L, Zhang P, Zhang Y and Ma A: Inflammatory unbalance of TLR3 and TLR4 in PCI patients with or without type 2 diabetes mellitus. *Immunol Lett* 161: 81–88, 2014.
100. Kruger B, Yin N, Zhang N, Yadav A, Coward W, Lai G, Zang W, S Heeger P, Bromberg JS, Murphy B, et al: Islet–expressed TLR2 and TLR4 sense injury and mediate early graft failure after transplantation. *Eur J Immunol* 40: 2914–2924, 2010.
101. Sepehri Z, Kiani Z, Nasiri AA, Mashhadi MA, Javadian F, Haghighi A, Kohan F, Bahari A and Sargazi A: Human Toll like receptor 4 gene expression of PBMCs in diabetes mellitus type 2 patients. *Cell Mol Biol* 61: 92–95, 2015.
102. Verzila D, Cappuccino L, D' Amato E, Villaggio B, Gianiorio F, Mij M, Simonato A, Viazzi F, Salvidia G and Garibotto G: Enhanced glomerular Toll–like receptor 4 expression and signaling in patients with type 2 diabetic nephropathy and microalbuminuria. *Kidney Int* 86: 1229–1243, 2014.
103. Baldan A, Ferronato S, Olivato S, Malerba G, Scuro A, Veraldi GF, Gelati M, Ferrari S, Mariotto S, Pignati PF, et al: Cyclooxygenase 2, toll–like receptor 4 and interleukin 1β mRNA expression in atherosclerotic plaques of type 2 diabetic patients. *Inflamm Res* 63: 851–858, 2014.
104. Sebolt-Leopold JS and Herrera R: Targeting the mitogen-activated protein kinase cascade to treat cancer. *Nat Rev Cancer* 4: 937–947, 2004.
105. Manning BD and Cantley LC: AKT/ PKB signaling: Navigating downstream. *Cell* 129: 1261–1274, 2007.
106. Bjornsti MA and Houghton PJ: The TOR pathway: A targetfor cancertherapy. *Nat Rev Cancer* 4: 335–348, 2004.
107. O'Reilly KE, et al: mTOR inhibition induces upstream receptor tyrosine kinase signaling and activates *Akt Cancer Res* 66: 1500–1508, 2006.
108. Sarbassov DD, et al: Prolonged rapamycin treatment inhibits mTORC2 assembly and Akt/ *PKB Mol Cell* 22: 159–168, 2006.
109. Mitsiades N, et al: Biologic sequelae of nuclear factor-kappa B blockade in multiple myeloma: therapeutic applications. *Blood* 99: 4079–4086, 2002.

110. Kinkade CW, et al: Targeting AKT/mTOR and ERK MAPK signaling inhibits hormone-refractory prostate cancer in a preclinical mouse model. *J Clin Invest* 118: 3051–3064, 2008.
111. Ley R, et al: Extracellular signal-regulated kinases 1/2 are serum-stimulated "Bim(EL) kinases" that bind to the BH3-only protein Bim(EL) causing its phosphorylation and turnover. *J Biol Chem* 279: 8837–8847, 2004.
112. Carracedo A, et al: Inhibition of mTORC1leadsto MAPK pathway activation through a PI3Kdependent feedback loop in human cancer. *J Clin Invest* 118: 3065–3074, 2008.
113. Chovatiya R, Medzhitov R: Stress, inflammation, and defense of homeostasis. *Mol Cell* 54(2): 281–288, 2004.
114. Yeung YT, Aziz F, Guerrero-Castilla A and Arguelles S: Signaling pathways in inflammation and anti-inflammatory therapies. *Current Pharmaceutical Design* 24(14): 1449–1484, 2018.
115. Chiu CP, Moulds C, Coffman RL, Rennick D, Lee F: Multiple biological activities are expressed by a mouse interleukin 6 cDNA clone isolated from bone marrow stromal cells. *Proc Natl Acad Sci USA* 1988; 85(19): 7099–7103.
116. Kishimoto T: The biology of interleukin-6. *Blood* 74(1): 1–10, 1989.
117. Heaney ML, Golde DW: Soluble cytokine receptors. *Blood* 87(3): 847–857, 1996.
118. Sutherland AG, Deehan DJ: Urinary soluble IL-6 receptor levels after cardiopulmonary bypass. *Cytokine* 12(9): 1441–1443, 2020.
119. Horiuchi S, Koyanagi Y, Zhou Y, et al: Soluble interleukin-6 receptors released from T cell or granulocyte/macrophage cell lines and human peripheral blood mononuclear cells are generated through an alternative splicing mechanism. *Eur J Immunol* 24(8): 1945–1948, 1994.
120. Müllberg J, Schooltink H, Stoyan T, et al: The soluble interleukin-6 receptor is generated by shedding. *Eur J Immunol* 23(2): 473–480, 1993.
121. Scheller J, Ohnesorge N, Rose-John S: Interleukin-6 transsignalling in chronic inflammation and cancer. *Scand J Immunol* 63(5): 321–329, 2006.
122. Rose-John S: IL-6 trans-signaling via the soluble IL-6 receptor: importance for the pro-inflammatory activities of IL-6. *Int J Biol Sci* 8(9): 1237–1247, 2012.
123. Rane SG, Reddy EP: Janus kinases: components of multiple signaling pathways. *Oncogene* 19(49): 5662–5679, 2000.
124. Rawlings JS, Rosler KM, Harrison DA: The JAK/STAT signaling pathway. *J Cell Sci* 117(Pt 8): 1281–1283, 2004.
125. Bromberg J, Darnell JE Jr: The role of STATs in transcriptional control and their impact on cellular function. *Oncogene* 19(21): 2468–2473, 2000.
126. Darnell JE Jr: STATs and gene regulation. *Science* 277(5332): 1630–1635, 1997.
127. Heinrich PC, Behrmann I, Müller-Newen G, Schaper F, Graeve L: Interleukin-6-type cytokine signalling through the gp130/Jak/STAT pathway. *Biochem J* 334(Pt 2): 297–314, 1998.
128. Leslie K, Lang C, Devgan G, et al: Cyclin D1 is transcriptionally regulated by and required for transformation by activated signal transducer and activator of transcription 3. *Cancer Res* 66(5): 2544–2552, 2006.
129. Gong J, Xie J, Bedolla R, et al: Combined targeting of STAT3/NFκB/COX-2/EP4 for effective management of pancreatic cancer. *Clin Cancer Res* 20(5): 1259–1273, 2014.
130. Yamada S, Shiono S, Joo A, Yoshimura A: Control mechanism of JAK/STAT signal transduction pathway. *FEBS Lett* 534(1–3): 190–196, 2003.
131. Shuai K: Modulation of STAT signaling by STAT-interacting proteins. *Oncogene* 19(21): 2638–2644, 2000.
132. Starr R, Hilton DJ. SOCS: suppressors of cytokine signalling. *Int J Biochem Cell Biol* 30(10): 1081–1085, 1998.
133. Wang Y, van Boxel-Dezaire AHH, Cheon H, Yang J, Stark GR: STAT3 activation in response to IL-6 is prolonged by the binding of IL-6 receptor to EGF receptor. *Proc Natl Acad Sci USA* 110(42): 16975–16980, 2013.
134. Shuai K: Modulation of STAT signaling by STAT-interacting proteins. *Oncogene* 19: 2638–2644, 2000.
135. Shuai K: The STAT family of proteins in cytokine signaling. *Prog Biophys Mol Biol* 71: 405–422, 1999.
136. Corry J, et al: Activation of STAT transcription factors by the Rho-family *GTPases Biochem Soc Trans* 48: 2213–2227, 2020.
137. Huang H, et al: Rho-kinase regulates energy balance by targeting hypothalamic leptin receptor signaling. *Nat Neurosci* 15: 1391–1398, 2012.
138. Horvath CM: STAT proteins and transcriptional responses to extracellular signals. *Trends Biochem Sci* 25: 496–502, 2000.

139. Harris M, et al: Transcriptional regulation of the thyrotropin-releasing hormone gene by leptin and melanocortin signaling. *J Clin Invest* 107: 111–120, 2001.
140. Banks AS, et al: Activation of downstream signals by the long form of the leptin receptor. *J Biol Chem* 275: 14563–14572, 2000.
141. Li C, et al: Leptin receptor activation of SH2 domain containing protein tyrosine phosphatase 2 modulates Ob receptor signal transduction. *Proc Natl Acad Sci USA* 96: 9677–9682, 1999.
142. Bates SH, et al: STAT3 signalling is required for leptin regulation of energy balance but not reproduction. *Nature* 421: 856–859, 2003.
143. Zhang J, et al: The soluble leptin receptor neutralizes leptin-mediated STAT3 signalling and anorexic responses in vivo. *Br J Pharm* 158: 475–482, 2009.
144. Galic S, et al: Suppressor of cytokine signalling (SOCS) proteins as guardians of inflammatory responses critical for regulating insulin sensitivity. *Biochem J* 461: 177–188, 2014.
145. Anderwald C, et al: Short-term leptin-dependent inhibition of hepatic gluconeogenesis is mediated by insulin receptor substrate-2. *Mol Endocrinol* 16: 1612–1628, 2002.
146. Kim YB, et al: In vivo administration of leptin activates signal transduction directly in insulin-sensitive tissues: overlapping but distinct pathways from insulin. *Endocrinology* 141: 2328–2339, 2000.
147. Plum L, et al: The obesity susceptibility gene Cpe links FoxO1 signaling in hypothalamic pro-opiomelanocortin neurons with regulation of food intake. *Nat Med* 15: 1195–1201, 2009.
148. Perry RJ, et al: The role of hepatic lipids in hepatic insulin resistance and type 2 diabetes. *Nature* 510: 84–91, 2014.
149. Wei CC, et al: Magnesium reduces hepatic lipid accumulation in yellow catfish (Pelteobagrus fulvidraco) and modulates lipogenesis and lipolysis via PPARA, JAK-STAT, and AMPK pathways in hepatocytes. *J Nutr* 147: 1070–1078, 2017.
150. Dai Z, et al: Depletion of suppressor of cytokine signaling-1a causes hepatic steatosis and insulin resistance in zebrafish. *Am J Physiol Endocrinol Metab* 308: E849–E859, 2015.
151. Inoue H, et al: Role of STAT-3 in regulation of hepatic gluconeogenic genes and carbohydrate metabolism in vivo. *Nat Med* 10: 168–174, 2004.
152. Moh A et al: STAT3 sensitizes insulin signaling by negatively regulating glycogen synthase kinase-3 beta. *Diabetes* 57: 1227–1235, 2008.
153. Hong F: et al: Interleukin 6 alleviates hepatic steatosis and ischemia/reperfusion injury in mice with fatty liver disease. *Hepatology* 40: 933–941, 2004.
154. Ki SH, et al: Interleukin-22 treatment ameliorates alcoholic liver injury in a murine model of chronic-binge ethanol feeding: Role of signal transducer and activator of transcription 3. *Hepatology* 52: 1291–1300, 2010.
155. Lichanska AM, et al: How growth hormone controls growth, obesity and sexual dimorphism. *Trends Genet* 24: 41–47, 2008.
156. Shi SY, et al: Hepatocyte-specific deletion of Janus kinase 2 (JAK2) protects against diet-induced steatohepatitis and glucose intolerance. *J Biol Chem* 287: 10277–10288, 2012.
157. Li X, et al: Microfluidic systems for studying dynamic function of adipocytes and adipose tissue. *Anal Bioanal Chem* 410: 791–800, 2018.
158. Seth M, et al: Leptin and obesity. *Physiol Int* 107: 455–468, 2020.
159. Singh P, et al: Leptin signaling in adipose tissue: role in lipid accumulation and weight gain. *Circ Res* 111: 599–603, 2012.
160. Cernkovich ER, et al: Adipose-specific disruption of signal transducer and activator of transcription 3 increases body weight and adiposity. *Endocrinology* 149: 1581–1590, 2008.
161. Derecka M, et al: Tyk2 and Stat3 regulate brown adipose tissue differentiation and obesity. *Cell Metab* 16: 814–824, 2012.
162. Dobrian AD et al: STAT4 deficiency reduces obesity-induced insulin resistance and adipose tissue inflammation. *Diabetes* 62: 4109–4121, 2013.
163. Sun S et al: Mechanisms of inflammatory responses in obese adipose tissue. *Annu Rev Nutr* 32: 261–286, 2012.
164. Richard AJ, et al: The role of JAK-STAT signaling in adipose tissue function. *Biochim Biophys Acta* 1842: 431–439, 2014.
165. Shi SY, et al: Adipocyte-specific deficiency of Janus kinase (JAK) 2 in mice impairs lipolysis and increases body weight, and leads to insulin resistance with ageing. *Diabetologia* 57: 1016–1026, 2014.
166. Moisan A, et al: White-to-brown metabolic conversion of human adipocytes by JAK inhibition. *Nat Cell Biol* 17: 57–67, 2015.

15 Cell Signaling in Neurological Disorders

Muhammad Ibrar Khan[1,2,3], *Muhammad Junaid*[2,4], *Sana Zafar*[1,2], *and Salman Khan*[1,2*]
[1]Pharmacological Sciences Research Lab, Department of Pharmacy, Faculty of Biological Sciences, Quaid-i-Azam University, Islamabad, Pakistan
[2]Department of Pharmacy, Faculty of Biological Sciences, Quaid-i-Azam University, Islamabad, Pakistan
[3]Department of Pharmacology, Faculty of Pharmacy, Capital University of Science and Technology, Islamabad, Pakistan
[4]Dosage Forms Research and Development Lab, Department of Pharmacy, Faculty of Biological Sciences, Quaid-i-Azam University, Islamabad, Pakistan
*Associate Professor, Department of Pharmacy, Faculty of Biological Sciences, Quaid-i-Azam University, Islamabad, Pakistan
* **Corresponding Author:** Dr. Salman Khan, Associate Professor, Department of Pharmacy, Faculty of Biological Sciences, Quaid-i-Azam University, Islamabad, Pakistan

15.1 INTRODUCTION

Neurodegenerative (ND) diseases are defined by the steady degeneration of certain groupings of neurons. Although the basic cause of neuronal death remains unrevealed, the risk probability amplifies with age. Interestingly, the accumulation of protein clusters in the central nervous system (CNS) seems to be in the background of many NDs. Over time, these deposits of insoluble peptides or proteins may build, causing heightened toxicity in elderly neurons. The aggregation of neuronal deterioration-associated peptides or proteins, such as Aβ1–42 peptides, excessively phosphorylated tau protein, or α-synuclein, disturbs intracellular pathways in both neurons and glial cells. The implications of this altered intracellular signaling are apparent in reduced mitochondrial homeostasis, protein quality control, disordered autophagy and lysosomal regulation, the presence of maladaptive innate immune reactions, synaptic toxicity, neuroinflammation, and stress granules [1–3].

15.1.1 Importance of Understanding Cell Signaling in Neurological Disorders

The underlying mechanism involved in the pathogenesis of neurological disorders is extremely complex. Cell signaling plays a pivotal role in neurodegeneration and neuroinflammation. These pathways are integral in understanding normal nervous system homeostasis, as they are responsible for the functioning, thus providing molecular insight into the pathophysiology of neurological disorders. Recent findings revealed that disruption of cell signaling occurs during neurological disorders, suggesting their significance [4, 5].

Therefore, understanding the cell signaling involved in the pathophysiology of neurological disorders is of utmost importance. Moreover, insights into abnormal cell signaling pathways proposed potential targets for pharmacological intervention. By understanding the specific molecular processes involved in neurological disorders, researchers and clinicians can develop novel therapeutic

DOI: 10.1201/9781003594901-15

strategies by developing targeted drugs to modulate these signaling abnormalities. The current chapter aims to evaluate key signaling pathways involved in the pathophysiology of neurological disorders and to explore various therapeutic targets for the treatment of neurological disorders [5, 6].

15.1.2 Fundamentals of Cell Signaling in the Nervous System

A nerve cell, also known as a neuron, is the essential component and functional unit of the nervous system responsible for transmitting information via both electrical and chemical means. Inside the neuronal cell, transmission of information occurs through the movement of an electrical charge or an impulse. Anatomically, the neuron consists of three major parts: (1) the axon, a thin and lengthy fiber responsible for transmitting information via a nerve impulse to nearby neurons; (2) the cell body, also called a soma, is the core section of the neuron having a nucleus and is responsible for maintaining all the fundamental activities of a neuronal cell; and (3) dendrites, tree-like structures with slender fibers that receive the information from other neuronal cells [7, 8].

Nerve impulses often traverse large distances throughout the body. For example, when you step on a sharp item without shoes, the sensory information is conveyed from the foot to the brain, then nerve impulses return to the leg muscles, causing them to tighten in foot withdrawal. Numerous neurons are engaged in this process, needing a sophisticated communication mechanism to rapidly send messages between neuronal cells. Furthermore, as single neurons may stretch up to 3 feet in length, a quick conveyance mechanism inside the neurons themselves is necessary to transport each signal from the receipt point to the place where it is given to an adjacent cell. Two techniques have progressed for sending nerve signals: first, electrical impulses are transported over the cell membrane inside cells, and, second, for intercellular interactions, electrical signals are generally turned into chemical signals delivered by neurotransmitters (NTs) [7, 9, 10].

15.1.2.1 Signal Transmission Within Nerve Cells

The mechanism that stimulates the transmission of electrical impulses across neurons relies on a voltage gradient, simply referred to as potentials, that exists between the interior and exterior of the cell. An asymmetrical arrangement of charged ions such as K^+, Ca^{2+}, Na^+, and Cl^- is predominantly behind the generation of membrane potential. Specific protein channels govern the passage of ions across the cell membrane. Opening or shutting of these channels happens in response to the expulsion of NTs or fluctuations in the cell's membrane potential. Consequently, the reorganization of electric charge could impact the voltage differential across the membrane. Attenuation of the voltage gradient is referred to as depolarization. In the event that the depolarization surpasses a particular limit or threshold, an action potential or impulse that travels along the neuron is triggered. Multiple approaches demonstrate that the impulse expands unidirectionally, approaching the tip of the axon. The start of an action potential is frequently referred to as "firing" [7, 11–13].

15.1.2.2 Signal Transmission Between Cells

Neurons usually communicate through small gaps known as synaptic clefts, with each neuron having the capability to interact with hundreds of thousands of other neurons. The transmission of signal occurs via the release of NTs from a presynaptic neuron into the synaptic space. The NT then activates (excitatory NT) or inhibits (inhibitory NT) the postsynaptic neuron by binding to the surface receptors on the postsynaptic neuron. The discharge of NTs emanates from presynaptic neurons, some of which might branch out to link with multiple different postsynaptic neurons. While dendrites are important for receiving neuronal impulses, receptors may be present in numerous areas of the cell. There are around 100 distinct NTs, and each individual neuron makes and releases just one or a few kinds of NTs. However, a cell may contain receptors on its surface for numerous NTs [9, 13].

To get beyond the synaptic gap, the cell's electrical interaction must be changed into a chemical one. The transformation happens anytime an action potential ends up at the axon, culminating in the phenomenon of depolarization initiating Ca^{2+} influx. The elevation in intracellular Ca2+ concentration leads the neuron to commit its function that for the most part comprises the diffusion of the NTs through the neuronal synapse. Two basic kinds of receptors emerge that cause unique responses in the effector cell: (i) receptors that elicit acute and ephemeral responses, herein, referred to as ligand-gated ion channels, and (ii) receptors that operate via the axillary messenger, thereby providing chronic yet enduring responses [9, 14].

15.1.2.2.1 Ligand-Gated Ion Channel Receptors

When a NT molecule electrostatically adsorbs onto a receptor that operates by the mechanism of a ligand-gated ion channel, a passageway or channel opens, accelerating the ionic flux across the membrane (refer to Figure 15.1). High flux of the positive charges disturbs the normal gradient culminating in the cellular depolarization in a way where the area closest to the receptor will be affected accordingly in comparison to the distal areas. Still, as an intricate system, the neuronal potential is translated to the peripheral areas until the neuron performs its function. Because this condition is beneficial to the future propagation of the depolarization mechanism as a syncytium, these ligand-gated channel sensors that are responsive to positive ions are called excitatory. If the opposite situation occurs wherein the negative ions are channeled in a process of repolarization, hyperpolarization ensues. The development of such a situation renders it very difficult for a neuron to activate and perform its actions. NTs that operate via this mechanism or induce refractiveness in neurons are termed inhibitory [9, 15].

15.1.2.2.2 Second Messenger-Linked Receptors

Auxiliary messengers (e.g., G proteins) refer to the molecules responsible for traversing the signals from the interstitial spaces to the cytoplasm of neurons. Second messenger-linked receptors are receptive to NTs, such as 3,4 dihydroxy phenethyl amine or dopamine (DA), and cause a sophisticated cascade of chemical events that may either provide the activation energy for the physiological function or stop the conduction of the electrical impulse altogether (see Figure 15.1). The NTs may bind on the allosteric site and/or on the intrinsic presynaptic sites commencing a feedback system that may affect the neuronal conduction in the posterity [9, 16].

A neuron has so many distinct sensors on its cellular landscape that some of the signals the neuron receives will have inhibitory repercussions, whereas others will be of excitatory consequence. The integration of the cell of these constantly conflicting inputs selects whether the neuron will release NTs, an action potential, and therefore have an influence on the operation of neurons [7, 9].

15.2 CELL SIGNALING DYSFUNCTION IN NEUROLOGICAL DISORDERS

15.2.1 Role of TLRs and NF-κB Signaling in Neurological Disorders

Toll-like receptors (TLRs)/NF-κB signaling chains are implicated in the pathogenesis of various CNS disorders. There is accumulating evidence of the significance of TLR4/NF-κB in numerous ND illnesses, notably Alzheimer's disease (AD), Huntington's disease (HD), Parkinson's disease (PD), dementia, and amyotrophic lateral sclerosis (ALS). Furthermore, TLR4/NF-κB is involved in demyelinating illnesses, such as multiple sclerosis (MS) and optic neuritis (ON), and acute neurological disorders, including headache, migraine, and epilepsy, as well as in CNS illnesses and traumatic brain damage. Moreover, TLR4/NF-κB has a role in the regulation of cognitive processes, namely, memory and cognition, which are the culprits for the pathobiology of psychiatric illnesses, including depression, schizophrenia, and anxiety [17–21].

Neuronal dysfunction causes several intracellular signaling cascades to create different inflammation, causing chemokines and cytokines, further contributing to various reactive oxygen species (ROS) being generated. Notwithstanding neurons, the microglia and astrocytes are each influenced

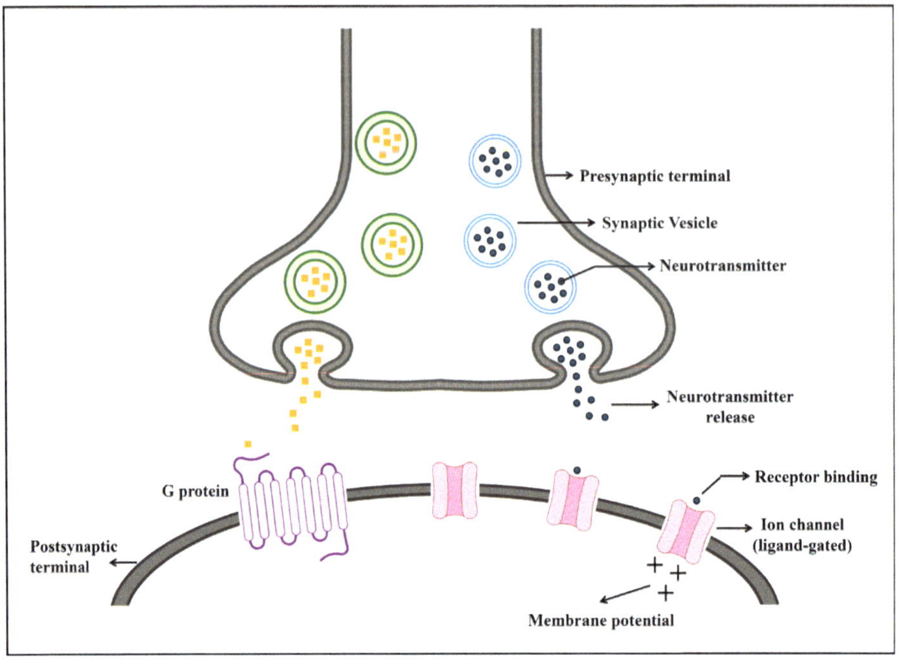

FIGURE 15.1 Signal propagation across the synaptic fissure. The transmission of signals across the neuronal synapse entails the tethering of neurotransmitters (NTs) (blue circles) to receptor proteins serving as ligand-catalyzed ion channels (ligand gated). This electrostatic interaction between the NTs and the receptor stimulates the opening of these channels, perhaps triggering depolarization in the membrane area proximal to the channel. Once the threshold is reached, the process of depolarization ensues, which consequently stimulates the opening of more ion channels, which may eventually give forth an action potential. NTs (marked as yellow circles) that connect to receptors related to axillary messengers trigger a complicated cascade of chemical reactions capable of inducing modifications in cell activity. The illustration clears up the confusion regarding the fact that it is G protein cascade that is responsible for the mechanics.

by this signaling cascade. The effect of this release can be deemed favorable, neutral, or harmful for cell survival. TLRs initiate and communicate their downstream pathway to promote NF-κB and pro-IL-1β, either of which are linked with neuroinflammation and contribute to the emergence of numerous age-related neurological illnesses [19, 22].

A variety of studies on inflammatory indicators have shown the relevance of TLR4/NF-κB in aging-related NDs, such as AD, MS, and ischemic strokes. As one ages, the brain's pro-inflammatory gene regulation and transcription enhance; consequently, TLR4/NF-κB transcription levels vary and participate in age-linked neurodegeneration. Moreover, they are also involved in brain damage following injury, when glial cells respond and express different cytokines and chemokines surrounding the lesion region. In a mouse model of brain injury, TLR2 was observed increased by microglia in the hippocampus zone. In contrast, TLR2 deficits reduce cytokine, microglial, and chemokine activation and expression [23].

TLR4 is also substantially involved in the glial cell production and activation of NF-κB, as well as the onset of inflammatory cytokines, such as TNF-α, IL-1β, and IL-6, manufactured in the nervous system across myriad of animal models. Both TLR2 and TLR4 connections contribute to the activation of glial cells along with additional inflammatory cytokines that are considered answerable for inflammation in the damaged brain. Yet, in glioma—a glial cell tumor—TLR9 has been transcribed substantially and was found to be favorable in a clinical examination [24].

Pharmacological substances such as polyphenols, flavonoids, and curcumins may suppress the activation of inflammatory mediators via TLR4/NF-κB/STAT communication intervention, whereas others might prevent neuronal death by changing the TLR4/MyD88/NF-κB pathway in microglia/macrophages. Indeed, dementia etiology is multifaceted and yet to be effectively explained because targeting TLR4/NF-κB could unveil a number of biochemical and pharmacological aspects connected to ND disorders. Thus, activating TLR4/NF-κB signaling modulation by natural resources can provide unique therapeutic possibilities for the treatment of ND illnesses [20, 24, 25].

15.2.2 Role of MAPK Signaling in Neurological Disorders

Among the cell-catalyzed cascade systems responsible for the consolidation of the input signals to the cell with those of gene expression is the MAPK (mitogen-activated protein kinase) superfamily. The consequence of the forgoing linkup ends up being the modulation in the cellular phenotypes. For optimal functioning, the MAPK pathway has to be constituted of various kinases of which three of them are of prime importance. Each of the subtypes of p48 kinases is divided into many further subtypes such as p38 kinases, extracellular signal-regulated kinases (ERKs), and the Jun amino-terminus kinases (JNKs), also termed stress-activated kinases (SAPKs) [26].

Recent advances proposed that the glial cells have a crucial role in the commencement of MAPK/ERK signaling. The aftermath of this prompts the development of diseases including PD, HD, AD, ALS, and prion disease. Furthermore, in addition to the glial cells, the inflammatory cytokines, indigenous immune cells, and astrocytes do exhibit their significance in the precipitation of ND illnesses via the MAPK/ERK cascade. This causes the development of the pro-inflammatory state in cerebral tissues via the activation of inflammatory pathways and the release of cytokines, chemokines, and nitric oxides (NOs). The stimulation of the glial cells, including but not limited to the astrocytes and glial cells, and the indigenous immune cells of the brain cause the pro-inflammatory state with the generation of NO, cytokines, and chemokines, and the participation of inflammatory-related pathways. The majority of the elements of these networks are cytosolic substrates of ERK, indicating a crucial role of the MAPK pathway in the formation or retaining of such an abnormal condition and, therefore, the deleterious occurrences responsible for the particular manifestation of the neurodegeneration [26, 27].

The characteristic features of PD include the progressive and selective death of dopaminergic neurons located in the ventral substantia nigra regions, reactive gliosis, and the presence of Lewy bodies, which are intracytoplasmic deposits of α-synuclein. This is because α-synuclein stimulates p38, ERK, and JNK cascades in human microglial cells, facilitating the generation of IL-1β and TNF-α, hence promoting inflammation. Repetitive functional lesions, symptoms, and clinical indications are produced as neurons in the substantia nigra die off, which causes a lack of dopamine at their intended destinations, the striatum, and other basal ganglia nuclei. Treatment methods that result in the dephosphorylation of p38 and ERK1/2 have been shown in many animal studies to help alleviate the ND changes associated with PD. Research also shows phosphorylated ERK-mediated fibrillar packages under coarse distinct cytoplasmic granular accumulations surrounding Lewy bodies in PD patients' substantia nigra, suggesting a possible link in the dopaminergic neurodegeneration between mitochondrial functionality and MAPK/ERK signaling [28].

MAPK pathways drastically activate with AD. Amyloid beta (Aβ), the primary feature that renders amyloid plaques, forms the critical connection with ERK cascade stimulation. In this regard, it has been proven in both in vivo and in vitro research that persistently high levels of Aβ enhance the dysregulation of hippocampus ERK signaling. Additionally, elevated p-ERK was detected in cerebral isolates of individuals with AD. Similarly, it is the overproduction of ROS, thought to be the consequence of Aβ, responsible for triggering p38 MAPK, thereby catalyzing the volcanic phosphorylation of tau. The generation of phosphorylated tau is crucial for neurological pathologies and is a characteristic feature of AD. Although modern investigations substantiate that p38 and JNK MAPK serve a key part in ALS, ERK route alteration is also entailed, since SOD1(G93A) transgenic mice furnish an impairment in axonal transport connected with a decline in the expression of ERK connecting with an elevation of JNK and caspase-8. Similarly, dysregulation of MAPK

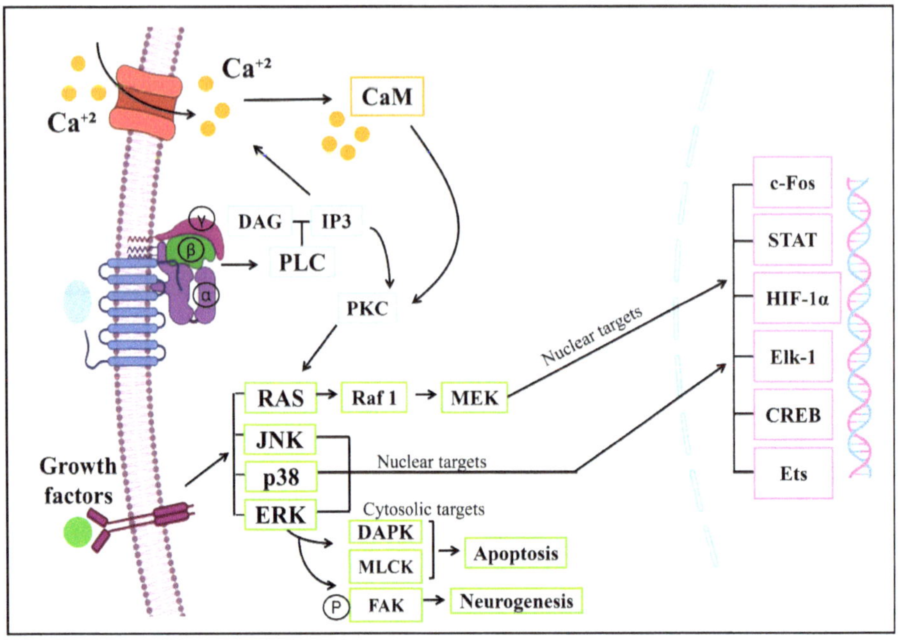

FIGURE 15.2 The main metabolic pathways associated with MAPK/ERK activation are illustrated. Two main receptors may activate the ERK pathway: (i) the receptor tyrosine kinase (RTK) and (ii) the G protein-coupled receptor (GPCR). (iii) Protein kinase C (PKC) and calcium-calmodulin (CaM) are two other cytosolic pathways that may start the route. Through their involvement in assembling the route's components, scaffolding proteins may direct the stimulation of the pathway. Phosphorylation of ERK may trigger nuclear or cytoplasmic alterations, activating gene sets involved in plastic modifications, migration, ultimate fate determination, and brain development. Cytoplasmic p-ERK has the potential to regulate processes such as apoptosis and neurogenesis.

signaling is a hallmark in the development of other neurological illnesses such as HD, MS, and prion disorders. Our findings strongly suggest identifying novel therapeutic medicines that target MAPK/ERK signaling versus neuropathological alterations [28, 29].

15.2.3 Role of VEGF/VEGFR Signaling in Neurological Disorders

Vascular endothelial growth factor (VEGF) is a vasoactive molecule with essential trophic effects for brain development. Previous reports indicate the effects of VEGF on the brain with cerebrovascular and ND diseases. Blood–brain barrier (BBB) dysfunction is the most common pathological characteristic of many neurological disorders such as ALS, AD, and PD. Neuronal injury or damage is accompanied by increased expression and secretion of VEGF, which exerts neuroprotective and neurorestorative effects [30]. However, recent studies revealed dysregulation of VEGF expression in multiple neurological disorders, suggesting the importance of VEGF in the pathogenesis of these disorders [31, 32].

In AD, an increased level of VEGF occurs in the cerebral cortex, accompanied by cerebral hypoperfusion, BBB dysfunction, and neuronal loss. Similarly, upregulation of VEGF signaling in cerebrospinal fluid occurs during PD associated with leaky vasculature and white matter lesions. Moreover, ALS is characterized by a decreased level of VEGF in the cerebral cortex, which further leads to insufficient neuroprotection, resulting in degeneration of motor neurons and BBB dysfunction. VEGF/VEGFR signaling plays a significant role in the pathogenesis of epileptic disorders [31]. Despite the neurotropic effects of VEGF/VEGFR signaling in the CNS, such as stimulation of neurite outgrowth and inhibition of neuronal death, past evidence claimed that VEGF/VEGFR signaling in epileptic brain exerts negative effects such as BBB dysfunction, excessive angiogenesis, and vascular malformations. These findings suggest the role of VEGF/VEGFR signaling in neurological disorders, therefore, potential therapeutic strategies should specifically target VEGF/VEGFR signaling that significantly attenuates BBB dysfunction and neuronal loss in neurological disorders [33, 34].

15.2.4 Role of Nrf2 Signaling in Disorders

Numerous ND illnesses such as AD or PD are characterized by enhanced generation of ROS in conjunction with oxidative stress. To minimize oxidative stress-produced cell damage, the production of cytoprotective enzymes that might involve GSH, GST, and SOD is elevated to start the body's natural (endogenous) antioxidant response pathway. The expression of these antioxidant enzymes is controlled by the transcription factor nuclear factor erythroid 2-related factor 2 (Nrf2). Moreover, Nrf2 signaling reduces mitochondrial dysfunction and has considerable anti-inflammatory potential. Past research suggested the significance of neuroinflammation alongside mitochondrial dysfunction in several ND disorders. Hence, Nrf2 has emerged as a possible therapeutic target to overcome neurological diseases [35–39].

Recent results demonstrated that mitochondrial failure, oxidative stress, and neuroinflammation are the cornerstones of most neurological illnesses as validated in many animal models of HD, AD, PD, and MS. Using these models, the research has proven a substantial importance for the Nrf2. Animal models of HD generated by 3-nitropropionic acid (3-NP) substantially lowered Nrf2 expression. Likewise, Nrf2 knock-out mice have been shown to be much more sensitive to both 3-NP and malonic acid-induced lesions in the striatum. Similarly, an increase in Nrf2 signaling has been related to neuroprotection in animal models of PD. Interestingly, Nrf2 downregulation led to inferior anti-oxidant response when the mice were subjected to 1-methyl-4-phenyl-1,2,3,6-tetrahydropyridine (MPTP)-treated mice. This unsurprisingly was culprit behind ROS accumulation culminating in cytotoxicity to the neuronal cells. However, activation of Nrf2 by siRNA-induced suppression of KEAP1, promisingly reduced neuroinflammation and oxidative stress by boosting the expression of antioxidant enzymes while lowering the generation of TNF-α and IL-1β. Furthermore, suppression of Nrf2 in numerous in vivo models of AD has been related to increased accumulation of Aβ and phosphorylated tau proteins. Additionally, loss of Nrf2 signaling dramatically accelerated neurodegeneration and worsened cognitive impairment via activation of glial cells, potentiating oxidative stress, and neuroinflammation. However, activation of Nrf2 signaling effectively reversed oxidative stress and neuroinflammation while boosting the antioxidant defense system. These effects greatly amplify neuroprotection showing the major involvement of Nrf2 signaling in neurological diseases [37, 40–43].

Numerous animal models of MS reiterate the role of neuroinflammation, oxidative stress, and mitochondrial dysfunction. The mitochondrial dysfunction is quite significant because it emerges acutely in the disease progression, even manifesting prior to the development of the symptoms concerning in vivo autoimmune encephalomyelitis (EAE). Overproduction of ROS, followed by damage of the peripheral nervous system (PNS), for the development of oxidative stress and the enzymes concerning the process has also been reported in EAE mice models. Additionally, EAE mice are connected with lower frequency of B cell MHCII, TH1, and, TH17 in respective manner. The finding concurred with the MS model induced by LPS (lipopolysaccharide), in which mice demonstrated augmented concentrations of cytokines and enhanced production of ROS leading to oxidative stress. Multiple studies propose that Nrf2 plays a vital role in the pathogenesis of MS. Recent studies suggest that Nrf2 is also significant in MS. EAE induction in animals that has a correlation to loss of Nrf2 have culminated in faster onset with low prognosis after EAE therapy that was accompanied by enhanced glial activation and worsened spinal cord injury and axonal degeneration as well as elevated the concentrations of pro-inflammatory cytokines. Conversely, Nrf2-activating drugs have shown favorable effects in MS model systems [37, 42, 44].

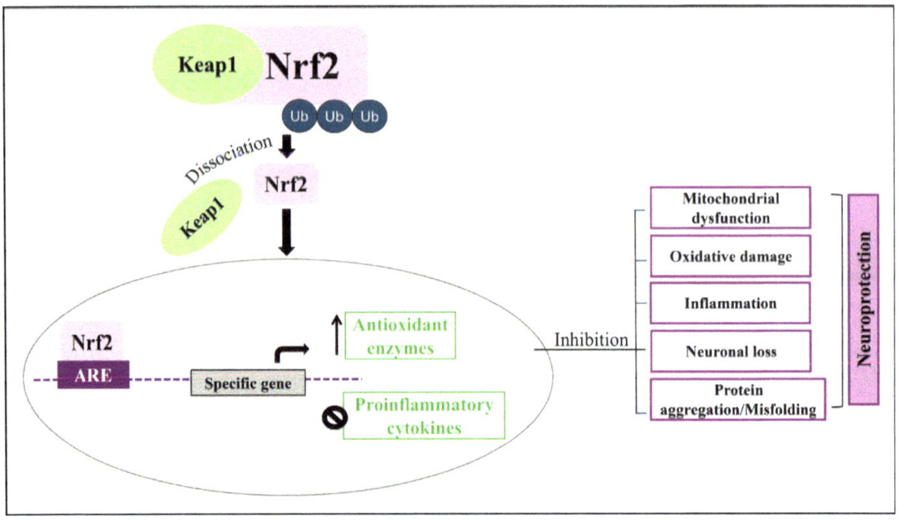

FIGURE 15.3 Role of Nrf2 signaling in neurodegenerative (ND) disorders.

It is becoming obvious that Nrf2 might represent an intriguing target for therapy in disorders of the brain. Its activation attenuates numerous processes related to ND conditions, notably mitochondrial failure, neuroinflammation, and oxidative stress (Figure 15.3). As evident from the significance of Nrf2, the drugs bearing the mechanism of Nrf2-activation have already been approved by the Food and Drug Administration for use in MS. Friedreich's ataxia have been studied on the same premises, whereby the research has been steadily progressing. However, the literature varies in the fact that it reveals the aforementioned activators might be necessary for many other NDs too [37, 45, 46].

15.2.5 Role of Notch Signaling in Neurological Disorders

A lipid bilayer membranous protein, Notch performs its function as a receptor for the ligands termed as, Delta and Jagged. This, in turn, helps the development of a relationship with membranes and the surface of surrounding cells. Whenever there is an incidence of interaction with the ligands, the notch receptor is enzymatically cleaved inside its transmembrane component by an enzymatic constituent of presenilin-1 triggering the dissociation of a notch intracellular region, which migrates to the nucleus and subsequently influences gene expression, as shown in Figure 15.5. As the mechanistic cascade concludes, the receptor is responsible for the growth and development of the CNS composed of controlling neural stem cell (NSC) proliferation, presence, regeneration, and maturity. Another example of the receptor exhibiting its significance is in the post-mitotic neurons in the adult CNS, whereby its activation enables structural and functional plasticity, including pathways linked with learning and memory. Furthermore, activation of notch signaling causes the release of free radicals, ROS, reactive nitrogen species, pro-inflammatory cytokines, pro-inflammatory mediators, and activation of glial cells, which is accompanied by oxidative stress and neuroinflammation, resulting in neurodegeneration (Figure 15.6). Interestingly, recent studies imply that the forgoing cascade is prevalent in neurons, glia, and NSCs, and is responsible for pathological abnormalities that arise in disorders such as stroke, AD, and CNS malignancies. Investigations of research animals reveal the promise of medications focused on such pathways as curative modalities for various distinct CNS illnesses [47–53].

The two age-related NDs, namely, AD and PD, are occurring with more frequency than usual, especially as the mean lifespan for the general population grows. And although ALS and HD are

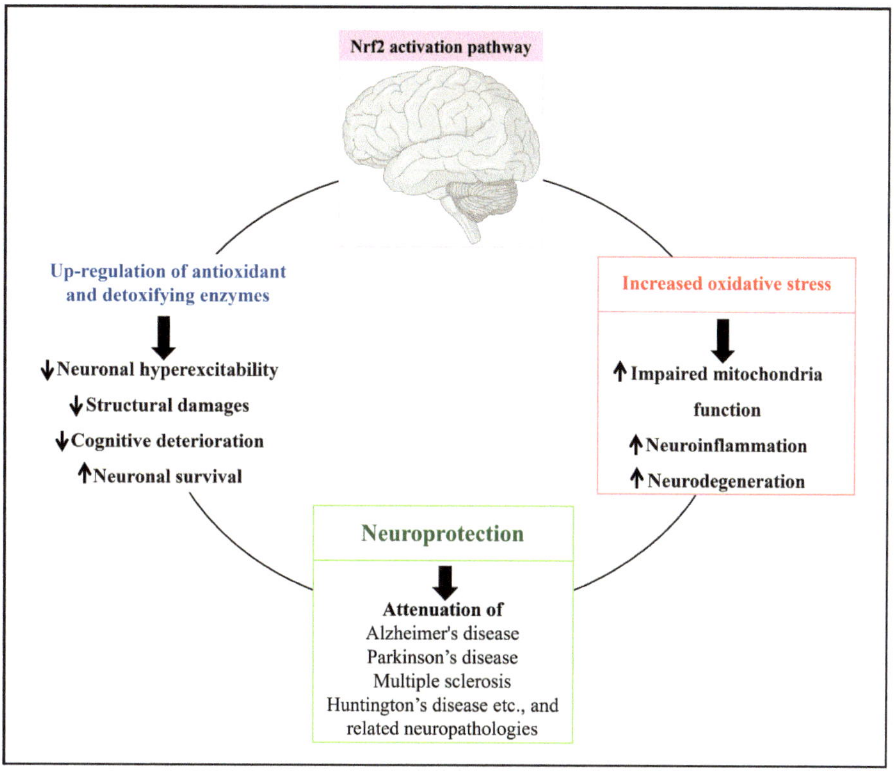

FIGURE 15.4 Neuroprotective potential of Nrf2 against neurological disorders.

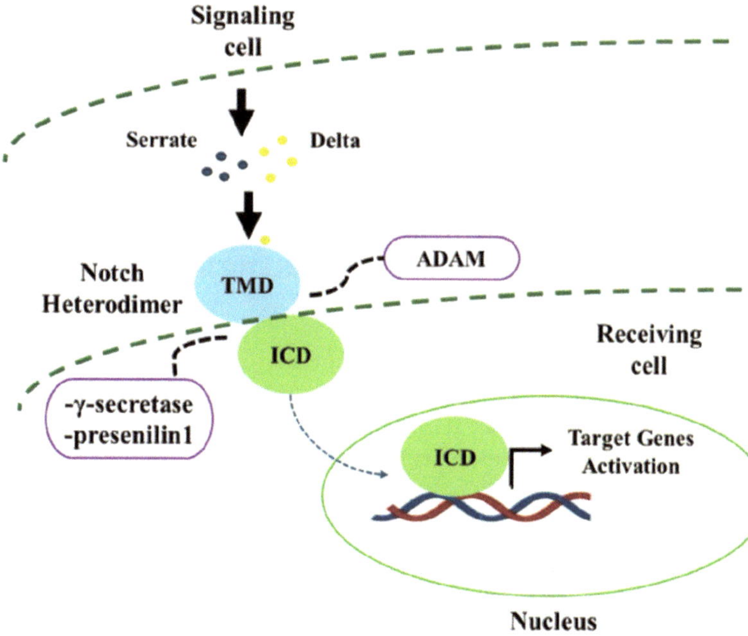

FIGURE 15.5 Notch pathway activation. The release of delta (yellow color circles) or serrate (blue color circles) ligands interact with the transmembrane domain (TMD) of the heterodimer notch receptor, resulting in their proteolytic cleavage by ADAM or γ-secretase/presenilin-1 which is accompanied by dissociation of the intracellular domain (ICD). Subsequently, the ICD translocates into the nucleus and causes activation of target genes.

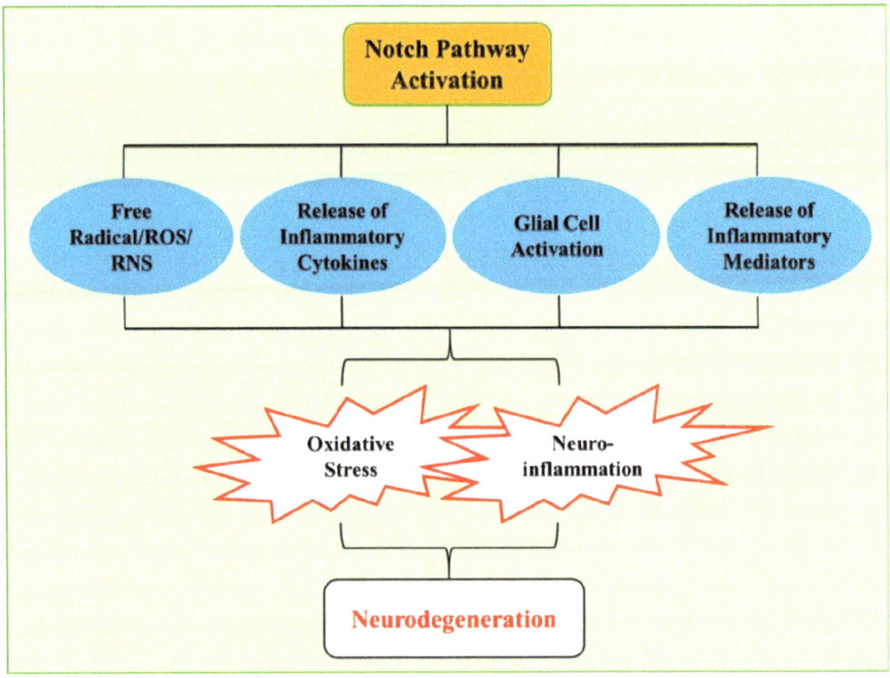

FIGURE 15.6 Role of notch receptor activation in neurodegeneration.

less common, they are arguably equally damaging. An overwhelming amount of evidence shows that AD may be strongly influenced by notch-1 signaling, but the literature is scarce with regard to the probable involvement of notch-1 in HD, PD, and ALS. AD is a condition marked by the placement of Aβ, defacement of synaptic function and degeneration, inflammation, and apoptosis of neurons in brain areas that are critical for memory and intellectual functioning. The recognition that presenilin-1/γ-secretase is responsible for breaking the amyloid precursor protein (APP) to create Aβ and also cleaves notch-1 to form the notch intracellular domain (NICD) has prompted tremendous interest in decoding the actions of notch-1 signaling in AD [47, 53, 54].

In AD and Pick's disease, immunohistochemistry analysis of postmortem hippocampal tissues shows elevated notch-1 expression levels in this brain area. Dementia and AD-like symptoms exist in people with Down syndrome. Analysis of notch-1 target gene transcription in postmortem brain tissue and fibroblasts from Down syndrome patients provides more evidence of excessive stimulation of the notch-1 signaling system. The results of a study showed that notch-1 signaling is associated with amyloidogenesis, since there was an increase in the quantity of notch-1 target gene transcription in the brain tissue of AD patients. New research has shown that APP is instantly linked to both the notch-1 and notch-2 cascades. It is not yet known how the APP-notch connection contributes to AD and related diseases. Just as Aβ is synthesized after APP fragmentation by γ-secretase, a notch-1 Aβ-like peptide is dissolved upon destruction of notch-1 by γ-secretase. The amount of notch-1 Aβ-like peptide produced is higher in cells that have presenilin-1 polymorphisms associated with AD [47, 53, 55].

15.2.6 Ubiquitin Signaling in Neurological Disorders

Proteinopathies, which cause a buildup of misfolded proteins within the brain, are often associated with neuronal death in neurological disorders. This is because these disorders are caused by either genetic abnormalities or an imbalance in protein homeostasis. Two main degradation pathways, the

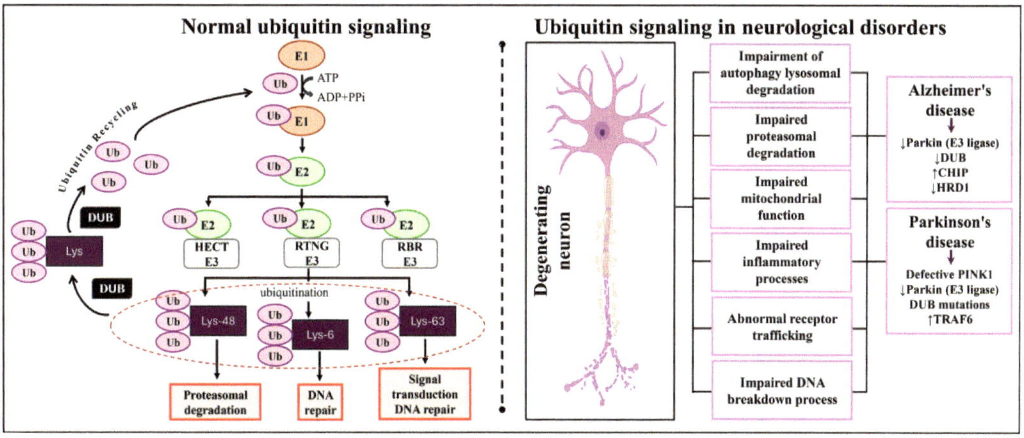

FIGURE 15.7 Ubiquitin signal transduction, whether its degradation is independent or not, plays a critical part in neuronal efficiency and lifespan. However, the disrupted transmission of both pathways may trigger neuronal death. Impaired clearance of misfolded proteins in confluence to the ubiquitinated protein agglomerates (such as htt in HD; α-synuclein in PD; Aβ and tau in AD; SOD1 and TDP-43 in ALS) owing to impeded impaired autophagic–lysosomal breakdown and/or proteasomal degradation is a feature of NDs. Impaired mitophagy: The adverse effect of dysfunctional mitochondria is amplified by diminished ubiquitin-mediated mitophagy due to defective PINK1 or the E3 ubiquitin ligase Parkin. Pathological mitochondria leave behind a trail of destruction: ROS (free radicals), DAMPS (such as mtDNA), and dysregulated inflammation. These factors exacerbate the inflammatory process. They trigger the formation of inflammasomes, which release pro-inflammatory cytokines such as IL-18 and IL-1β and other harmful substances through the reactive astrocytes and microglia. The immunoproteasome, which is overexpressed in microglia, further amplifies the inflammation, leading to disease progression. Abnormal receptor trafficking: The endolysosomal pathway, which is controlled by NEDD4, affects the availability of crucial neuronal receptors, such as EGF and AMPA, on the cell surface. Defective DNA damage response: The ubiquitin system is disrupted in the DNA breakdown process (for example, by altered ubiquitination of histones).

ubiquitin–proteasome and the autophagy–lysosomal networks, are vital. Both of them contribute to the removal of abnormal or misfolded proteins by metabolizing them and preventing their cellular accumulation, leading to toxicity that would eventually lead to better mental health. Both of these degradative processes depend on the modification of targets with ubiquitin. Aging is the major risk factor for most neurological illnesses, including ALS, AD, and PD. With aging there is a general loss in proteasomal breakdown and autophagy, and a parallel formation of potentially neurotoxic protein clusters of Aβ, tau, α-synuclein, SOD1, and TDP-43. An often overlooked but significant component of these aggregates is ubiquitin, suggesting that these protein aggregates might function as an adaptive response to dangerous misfolded three-dimensional peptides or as evidence of disordered ubiquitin-mediated breakdown producing detrimental aggregation. In addition, non-degradative ubiquitin signaling is crucial for homoeostatic mechanisms required for the lifetime and sentience of neurons, including DNA damage repair responses, mitochondrial physiology, and receptor transmissions, as well as playing a role in inflammatory processes [56, 57].

15.3 EMERGING SIGNALING TARGETS FOR NEUROLOGICAL DISORDERS

Signaling pathways are crucial modulators for multitudinous physiopathological processes, and the aberrant activation of certain signaling pathways may lead to disease development in various contexts. As a result, the signaling pathways have grown as a vital instrument by which the occurrence and advancement of diseases may be explored, culminating in the creation of novel medicaments [56, 58].

Multiple investigations demonstrate the involvement of several signaling pathways in the progression of neurological disease. The activation of TLR/NF-κB signaling greatly contributes to the pathophysiology of neurological illness by generating neurodegeneration and neuroinflammation. Any amount of injury to the neuronal tissue induces TLR4/NF-κB signaling leading to the production of pro-inflammatory cytokines such as IL-6, TNF-α, and IL-1β. Likewise, the activated TLR4/NF-κB signaling boosts the mRNA expression of iNOS and COX-2. These further lead to augmentation in the oxidative stress and consequent spike in the mediators such as NO and the release of pro-inflammatory mediators. As a consequence, neuronal damage is potentiated by oxidative damage and neuroinflammation followed by a reduction in neuronal signaling. Similarly, the deregulation of MAPK signaling is a characteristic of the pathophysiology of numerous neurological illnesses. Several studies indicated that upregulation of MAPK signaling occurs in several ND disorders such as PD. Moreover, Nrf2 signaling is responsible for neuroprotective benefits ascribed to its anti-inflammatory and antioxidant capability. Furthermore, numerous other pathways encompassing the notch/ubiquitin/VEGF signaling assume a key role in neuronal homeostasis and sentience. However, impaired modulation of these signals might induce neuronal damage followed by exacerbation of neurological illness [17, 30, 40, 56, 59, 60].

Neurological illnesses, such as AD, PD, MS, HD, ALS, and Batten disease, which essentially is a storage disease associated with lysosomes, are particularly upsetting, contingent upon their progressive deterioration of cognitive and physical abilities. Moreover, the available treatment techniques have limited efficiency in slowing the development of neurological problems and are connected to considerable negative responses. The limited number of successful therapies for this and other nervous system problems has led to a large exploration of different therapeutic approaches, such as the modulation of cell signaling. There is an urgent necessity for finding unique treatment ways to conquer neurological illness and enhance patient quality of life. Consequently, in-depth investigation and comprehension of all these signaling pathways (TLR4/NF-κB/MAPK/Nrf2/VEGF/Notch/Ubiquitin) show potential for the creation of innovative therapeutic techniques that specifically target and control particular signals to overcome neurological illness [17, 30, 59–62].

Astrocytes serve several tasks, including providing assistance and maintaining the optimum environment for the functioning and survival of neurons. The variety and complexity of these activities unambiguously suggest that the precise functioning of astrocytes is crucial for the physiological operation of the CNS, and any disturbance to it may contribute to the start and progression of numerous neurological disorders. In order to transpose the successful repair of astrocyte dysfunction in preclinical investigations to clinical practice, it is important to develop improved and reliable ways for delivering therapeutic drugs selectively to astrocytes and for replacing astrocytes completely [63, 64].

15.4 CONCLUSION

Neurological disorders are highly debilitating conditions and significantly diminish the quality of life. The pathophysiology of neurological disorders is extremely complex and needs to be further explored. Molecular signaling plays a vital role in the pathophysiology of these devastating conditions. Any pathological insult triggers the dysregulation of a widespread range of cellular signaling, such as TLR4/NF-κB/MAPK/Nrf2/VEGF/Notch/Ubiquitin signaling, which is accompanied by neuroinflammation, neurodegeneration, decreased neural plasticity, apoptosis, and attenuation of neurogenesis. Current treatments have limited therapeutic efficacy. Therefore, novel therapeutic agents are required to selectively target these signaling pathways and inhibit neuroinflammation and neurodegeneration in order to overcome neurological disorders (Figure 15.8).

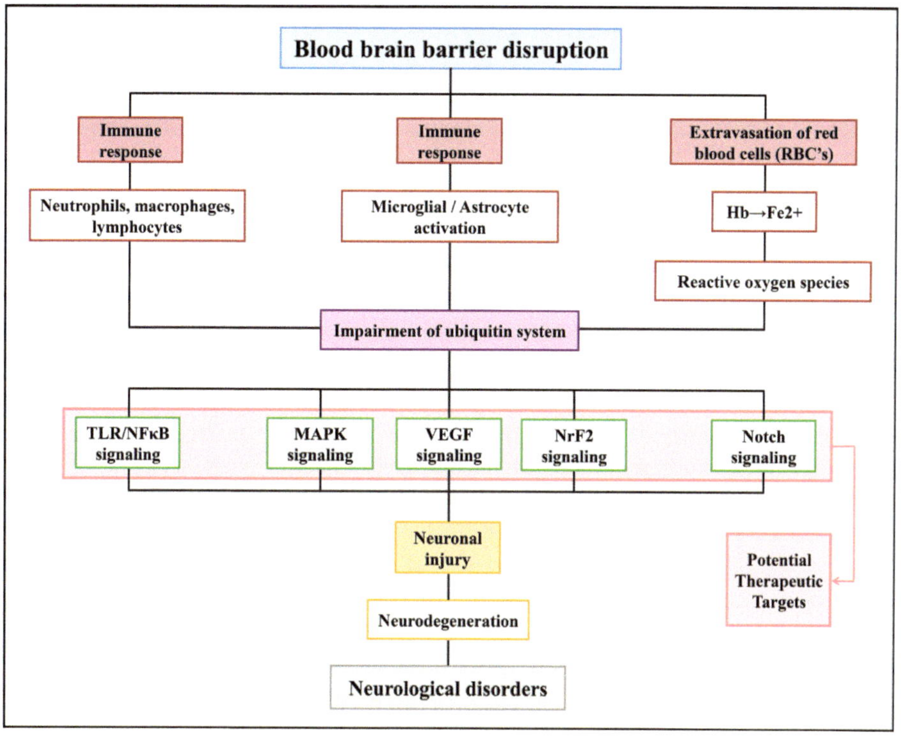

FIGURE 15.8 Role of cell signaling in neurological disorders and potential therapeutic targets.

REFERENCES

1. Zugaza, J.L., Cell signaling in neurodegeneration. *International Journal of Molecular Sciences*, 2021. **22**(16): p. 8978.
2. Feigin, V.L., et al., The global burden of neurological disorders: translating evidence into policy. *The Lancet Neurology*, 2020. **19**(3): p. 255–265.
3. Lorenzo, D.N., R.J. Edwards, and A.L. Slavutsky, Spectrins: molecular organizers and targets of neurological disorders. *Nature Reviews Neuroscience*, 2023. **24**(4): p. 195–212.
4. Collier, J.J., et al., Mitochondrial signalling and homeostasis: from cell biology to neurological disease. *Trends in Neurosciences*, 2023. **46**: p. 168–180.
5. Moujalled, D., A. Strasser, and J.R. Liddell, Molecular mechanisms of cell death in neurological diseases. *Cell Death & Differentiation*, 2021. **28**(7): p. 2029–2044.
6. Mathews, J., et al., Cellular signaling pathways as plastic, proto-cognitive systems: implications for biomedicine. *Patterns*, 2023. **4**(5).
7. The principles of nerve cell communication. *Alcohol Health Research World*, 1997. **21**(2): p. 107–178.
8. Liang, Y., et al., PEDOT: PSS-based bioelectronic devices for recording and modulation of electrophysiological and biochemical cell signals. *Advanced Healthcare Materials*, 2021. **10**(11): p. 2100061.
9. Neuronal communication. *Nature Structural & Molecular Biology*, 2019. **26**(7): p. 527.
10. Romero, G., et al., Modulating cell signalling in vivo with magnetic nanotransducers. *Nature Reviews Methods Primers*, 2022. **2**(1): p. 92.
11. Mousavi, A., CXCL12/CXCR4 signal transduction in diseases and its molecular approaches in targeted-therapy. *Immunology Letters*, 2020. **217**: p. 91–115.
12. Kennedy, R.H. and R. Silver, Neuroimmune signaling: cytokines and the central nervous system. In *Neuroscience in the 21st Century: From Basic to Clinical*. Springer, 2022. p. 883–922.
13. Calvo-Rodriguez, M. and B.J. Bacskai, Mitochondria and calcium in Alzheimer's disease: from cell signaling to neuronal cell death. *Trends in Neurosciences*, 2021. **44**(2): p. 136–151.
14. Fischer, I., J.N. Dulin, and M.A. Lane, Transplanting neural progenitor cells to restore connectivity after spinal cord injury. *Nature Reviews Neuroscience*, 2020. **21**(7): p. 366–383.

15. SenGupta, S., C.A. Parent, and J.E. Bear, The principles of directed cell migration. *Nature Reviews Molecular Cell Biology*, 2021. **22**(8): p. 529–547.
16. Zhang, T., et al., The core of maintaining neuropathic pain: crosstalk between glial cells and neurons (neural cell crosstalk at spinal cord). *Brain and Behavior*, 2023. **13**(2): p. e2868.
17. Ashayeri Ahmadabad, R., et al., Toll-like receptor signaling pathways: novel therapeutic targets for cerebrovascular disorders. *International Journal of Molecular Sciences*, 2021. **22**(11): p. 6153.
18. Dong, X., et al., Physcion protects rats against cerebral ischemia-reperfusion injury via inhibition of TLR4/NF-kB signaling pathway. *Drug Design, Development and Therapy*, 2021. **15**: p. 277–287.
19. Zhou, J., et al., Icariside II attenuates lipopolysaccharide-induced neuroinflammation through inhibiting TLR4/MyD88/NF-κB pathway in rats. *Biomedicine & Pharmacotherapy*, 2019. **111**: p. 315–324.
20. Guo, T.-T., et al., Neuroprotective effects of sodium butyrate by restoring gut microbiota and inhibiting TLR4 signaling in mice with MPTP-induced Parkinson's disease. *Nutrients*, 2023. **15**(4): p. 930.
21. Khan, M.I., et al., Anti-nociceptive effects of magnolol via inhibition of TRPV1/P2Y and TLR4/NF-κB signaling in a postoperative pain model. *Life Sciences*, 2023. **312**: p. 121202.
22. Dong, J., Y. Liao, and B. Wu, TAK-242 ameliorates epileptic symptoms in mice by inhibiting the TLR4/NF-κB signaling pathway. *Annals of Translational Medicine*, 2022. **10**(14): 769.
23. Yang, Y., et al., Luteolin alleviates neuroinflammation via downregulating the TLR4/TRAF6/NF-κB pathway after intracerebral hemorrhage. *Biomedicine & Pharmacotherapy*, 2020. **126**: p. 110044.
24. Azam, S., et al., Regulation of toll-like receptor (TLR) signaling pathway by polyphenols in the treatment of age-linked neurodegenerative diseases: focus on TLR4 signaling. *Frontiers in Immunology*, 2019. **10**: p. 1000.
25. Zusso, M., et al., Ciprofloxacin and levofloxacin attenuate microglia inflammatory response via TLR4/NF-kB pathway. *Journal of Neuroinflammation*, 2019. **16**(1): p. 1–12.
26. Albert-Gascó, H., et al., MAP/ERK signaling in developing cognitive and emotional function and its effect on pathological and neurodegenerative processes. *International Journal of Molecular Sciences*, 2020. **21**(12): p. 4471.
27. Naqvi, B.R., et al., Pharmacological effects of matrine in nitroglycerine-induced migraine pain by inhibiting cytokines and oxidative stress. *Natural Product Sciences*, 2024. **30**(3): p. 227–236.
28. Gravandi, M.M., et al., Therapeutic targeting of Ras/Raf/MAPK pathway by natural products: a systematic and mechanistic approach for neurodegeneration. *Phytomedicine*, 2023. **115**: p. 154821.
29. Park, J., et al., Mitochondrial ROS govern the LPS-induced pro-inflammatory response in microglia cells by regulating MAPK and NF-κB pathways. *Neuroscience Letters*, 2015. **584**: p. 191–196.
30. Ureña-Guerrero, M.E., et al., Neuroprotective and neurorestorative effects of Epo and VEGF: perspectives for new therapeutic approaches to neurological diseases. *Current Pharmaceutical Design*, 2020. **26**(12): p. 1263–1276.
31. Shim, J.W. and J.R. Madsen, VEGF signaling in neurological disorders. *International Journal of Molecular Sciences*, 2018. **19**(1): p. 275.
32. Matsuno, H., et al., Association between vascular endothelial growth factor-mediated blood–brain barrier dysfunction and stress-induced depression. *Molecular Psychiatry*, 2022. **27**(9): p. 3822–3832.
33. Ogaki, A., Y. Ikegaya, and R. Koyama, Vascular abnormalities and the role of vascular endothelial growth factor in the epileptic brain. *Frontiers in Pharmacology*, 2020. **11**: p. 20.
34. Ahmad, A. and M.I. Nawaz, Molecular mechanism of VEGF and its role in pathological angiogenesis. *Journal of Cellular Biochemistry*, 2022. **123**(12): p. 1938–1965.
35. Zamanian, M.Y., et al., Targeting Nrf2 signaling pathway by quercetin in the prevention and treatment of neurological disorders: an overview and update on new developments. *Fundamental & Clinical Pharmacology*, 2023. **37**(6): p. 1050–1064.
36. Dinkova-Kostova, A.T., R.V. Kostov, and A.G. Kazantsev, The role of Nrf2 signaling in counteracting neurodegenerative diseases. *The FEBS Journal*, 2018. **285**(19): p. 3576–3590.
37. Johnson, D.A. and J.A. Johnson, Nrf2—a therapeutic target for the treatment of neurodegenerative diseases. *Free Radical Biology and Medicine*, 2015. **88**: p. 253–267.
38. Petri, S., S. Körner, and M. Kiaei, Nrf2/ARE signaling pathway: key mediator in oxidative stress and potential therapeutic target in ALS. *Neurology Research International*, 2012. **2012**.
39. Zafar, S., et al., Daidzein attenuated paclitaxel-induced neuropathic pain via the down-regulation of TRPV1/P2Y and up-regulation of Nrf2/HO-1 signaling. *Inflammopharmacology*, 2023. **31**(4): p. 1977–1992.
40. Uddin, M.S., et al., Emerging promise of sulforaphane-mediated Nrf2 signaling cascade against neurological disorders. *Science of the Total Environment*, 2020. **707**: p. 135624.
41. Sivandzade, F., A. Bhalerao, and L. Cucullo, Cerebrovascular and neurological disorders: protective role of NRF2. *International Journal of Molecular Sciences*, 2019. **20**(14): p. 3433.

42. Calkins, M.J., et al., The Nrf2/ARE pathway as a potential therapeutic target in neurodegenerative disease. *Antioxidants & Redox Signaling*, 2009. **11**(3): p. 497–508.
43. Saha, S., et al., A perspective on Nrf2 signaling pathway for neuroinflammation: a potential therapeutic target in Alzheimer's and Parkinson's diseases. *Frontiers in Cellular Neuroscience*, 2022. **15**: p. 787258.
44. Yang, Y., et al., An overview of the molecular mechanisms and novel roles of Nrf2 in neurodegenerative disorders. *Cytokine & Growth Factor Reviews*, 2015. **26**(1): p. 47–57.
45. Baba, M.Z., S. Gomathy, and U. Wahedi, Role of Nrf2 pathway activation in neurological disorder: a brief review. *Journal of Pharmacology and Pharmacotherapeutics*, 2022. **13**(3): p. 229–238.
46. Upadhayay, S. and S. Mehan, Targeting Nrf2/HO-1 anti-oxidant signaling pathway in the progression of multiple sclerosis and influences on neurological dysfunctions. *Brain Disorders*, 2021. **3**: p. 100019.
47. Kapoor, A. and D.A. Nation. Role of Notch signaling in neurovascular aging and Alzheimer's disease. in *Seminars in Cell & Developmental Biology*. Elsevier, 2021.
48. Ho, D.M., S. Artavanis-Tsakonas, and A. Louvi, The Notch pathway in CNS homeostasis and neurodegeneration. *Wiley Interdisciplinary Reviews: Developmental Biology*, 2020. **9**(1): p. e358.
49. Kim, S., M. Lee, and Y.K. Choi, The role of a neurovascular signaling pathway involving hypoxia-inducible factor and notch in the function of the central nervous system. *Biomolecules & Therapeutics*, 2020. **28**(1): p. 45.
50. Zhou, B., et al., Notch signaling pathway: architecture, disease, and therapeutics. *Signal Transduction and Targeted Therapy*, 2022. **7**(1): p. 95.
51. Zhang, Y., et al., Notch signaling pathway: a new target for neuropathic pain therapy. *The Journal of Headache and Pain*, 2023. **24**(1): p. 87.
52. Liu, C., et al., Activation of the Notch signaling pathway and cellular localization of notch signaling molecules in the spinal cord of SOD1-G93A ALS model mice. *Neuroscience*, 2020. **432**: p. 84–93.
53. Lathia, J.D., M.P. Mattson, and A. Cheng, Notch: from neural development to neurological disorders. *Journal of Neurochemistry*, 2008. **107**(6): p. 1471–1481.
54. Rafique, M., et al., Development of dexibuprofen loaded nano transfersomal gel with enhanced biopharmaceutical performance in complete Freund's adjuvant induced arthritis model. *Journal of Drug Delivery Science and Technology*, 2024. **98**: p. 105928.
55. Li, K.-X., et al., Astrocyte-neuron communication mediated by the Notch signaling pathway: focusing on glutamate transport and synaptic plasticity. *Neural Regeneration Research*, 2023. **18**(10): p. 2285–2290.
56. Schmidt, M.F., et al., Ubiquitin signalling in neurodegeneration: mechanisms and therapeutic opportunities. *Cell Death & Differentiation*, 2021. **28**(2): p. 570–590.
57. Ciechanover, A. and Y.T. Kwon, Degradation of misfolded proteins in neurodegenerative diseases: therapeutic targets and strategies. *Experimental & Molecular Medicine*, 2015. **47**(3): p. e147–e147.
58. Sun, J. and G. Nan, The extracellular signal-regulated kinase 1/2 pathway in neurological diseases: a potential therapeutic target. *International Journal of Molecular Medicine*, 2017. **39**(6): p. 1338–1346.
59. Singh, S. and T.G. Singh, Role of nuclear factor kappa B (NF-κB) signalling in neurodegenerative diseases: an mechanistic approach. *Current Neuropharmacology*, 2020. **18**(10): p. 918–935.
60. Ahmed, T., et al., Map kinase signaling as therapeutic target for neurodegeneration. *Pharmacological Research*, 2020. **160**: p. 105090.
61. Panaro, M.A., et al., The emerging role of curcumin in the modulation of TLR-4 signaling pathway: focus on neuroprotective and anti-rheumatic properties. *International Journal of Molecular Sciences*, 2020. **21**(7): p. 2299.
62. Khan, A., et al., Matrine alleviates neurobehavioral alterations via modulation of JNK-mediated caspase-3 and BDNF/VEGF signaling in a mouse model of burn injury. *Psychopharmacology*, 2020. **237**: p. 2327–2343.
63. Valori, C.F., et al., Astrocytes: emerging therapeutic targets in neurological disorders. *Trends in Molecular Medicine*, 2019. **25**(9): p. 750–759.
64. Lee, H.-G., M.A. Wheeler, and F.J. Quintana, Function and therapeutic value of astrocytes in neurological diseases. *Nature Reviews Drug Discovery*, 2022. **21**(5): p. 339–358.

16 Cell Signaling and Diagnosis

Areej Fatima[1], Muhammad Ammad Jamil[1], Maleeha Azhar[1], Aroosa Younis Nadeem[1], and Adeeb Shehzad[1,2]
[1]Department of Biomedical Sciences, School of Mechanical and Manufacturing Engineering, National University of Sciences & Technology, H-12 Islamabad, Pakistan
[2]Biodiversity Unit, Research Center, Dhofar University, Salalah, Oman

16.1 INTRODUCTION

Cell signaling is the communication of cells with their external and internal environments by recognizing and responding to signaling molecules on their surfaces [1]. Cells require specific receptors to react to external signals, including neurotransmitters, hormones, and growth factors. A ligand is a molecule that binds to the receptor; distinct types of cells have different receptors that only bind to specific types of ligand molecules. Some cells share the same kind of receptors, even though these receptors respond differently to extracellular stimuli.

Ligands bind to the receptors at their extracellular site; once they move into the intracellular site, the signal is then transmitted in two ways: either by recruiting proteins or by activating the second messenger. The outcome is the same: a protein at the top of intracellular signaling pathways gets activated or deactivated, resulting in a specific cellular response. These signaling paths are like superhighways transmitting information within the cell, and each consists of a series of distinct proteins that work in a specific sequence [2].

Healthy tissue carefully regulates cell signaling, which controls the growth and division cycle, replicative senescence, angiogenesis, apoptosis, and gene expression signatures that determine cellular identity. However, in cases of disease, disruptions can occur in cell signaling, leading to uncontrolled cell proliferation, dysfunctional cell death processes, abnormal immune responses, metabolic irregulating, or disrupted neural signaling. As a result, these deregulated signaling pathways underlie the pathogenesis of several human diseases. While the extent to which these signaling pathways are affected in disease lineages is not fully understood, there is hope. Identifying signaling aberrations associated with specific types of disease can serve as disease biomarkers and therapeutic targets, paving the way for more effective disease diagnosis and treatment [3].

In recent years, several biomarkers have been identified for diagnosing various diseases. These biomarkers are molecules involved in signaling pathways, including those related to growth factors, cell cycle regulation, apoptosis, cell adhesion, and immune checkpoint control. For instance, the human epidermal growth factor receptor 2 (HER2) is a biomarker found in growth factor pathways, while the tumor protein p53 (encoded by the TP53 gene) is a biomarker involved in regulating the cell cycle and apoptosis. Similarly, E-cadherin is a cell adhesion molecule that can serve as a biomarker, and programmed death 1 and programmed death-ligand 1 (PD-1/PD-L1) are modulators of immune checkpoint control that can be used as biomarkers. Other molecules related to exosomes, DNA, RNA, or epigenetic modifications can also serve as useful biomarkers [4].

This chapter will explore cell signaling pathways and their corresponding biomarkers, which play a critical role in detecting diseases early. We will focus on three signaling pathways: nuclear-mediated, receptor-mediated, and extracellular. Each of these pathways has unique characteristics and a distinct set of biomarkers. By understanding these pathways and the biomarkers used in diagnosis, we can gain insight into disease progression, development, and treatment strategies.

DOI: 10.1201/9781003594901-16

16.2 MECHANISM OF CELL SIGNALING

Cell signaling is a crucial process for the proper functioning of cells, wherein a cell receives signals from its environment and sends messages to its interior. This process involves three distinct steps:

1. A signaling molecule (ligand) binds to the cell's surface receptor
2. The ligand-receptor complex triggers a series of events, such as conformational changes in ion channels, increased production of second messengers, and recruitment of cytoplasmic proteins

Enzymes (e.g., phosphatases and protein kinases) are activated, leading to a biological response (Figure 16.1)

This signal amplification sets off a downstream cascade of molecular events that ultimately result in the desired cellular response [5].

Cell signaling is a highly intricate process that occurs via multiple pathways. Among these, the Ras/Raf/MAPK pathway is one of the most well-defined signal transduction pathways, which is regulated through a negative feedback mechanism. The Ras superfamily comprises genes that encode small GTP-binding proteins, which play a critical role in this pathway [6].

This pathway is initiated when an external signal, like a growth factor, binds to a cell surface receptor, known as a receptor tyrosine kinase (RTK). This binding activates the RTK, leading to the replacement of guanosine diphosphate (GDP) with guanosine triphosphate (GTP) in the small G-proteins of Ras. This activates downstream signaling through Raf kinase activation via the binding of Raf kinases with active Ras [7]. Once Raf is activated, it phosphorylates MAPK (mitogen-activated protein kinase), a group of serine/threonine kinases known as MEK in mammals, which triggers a

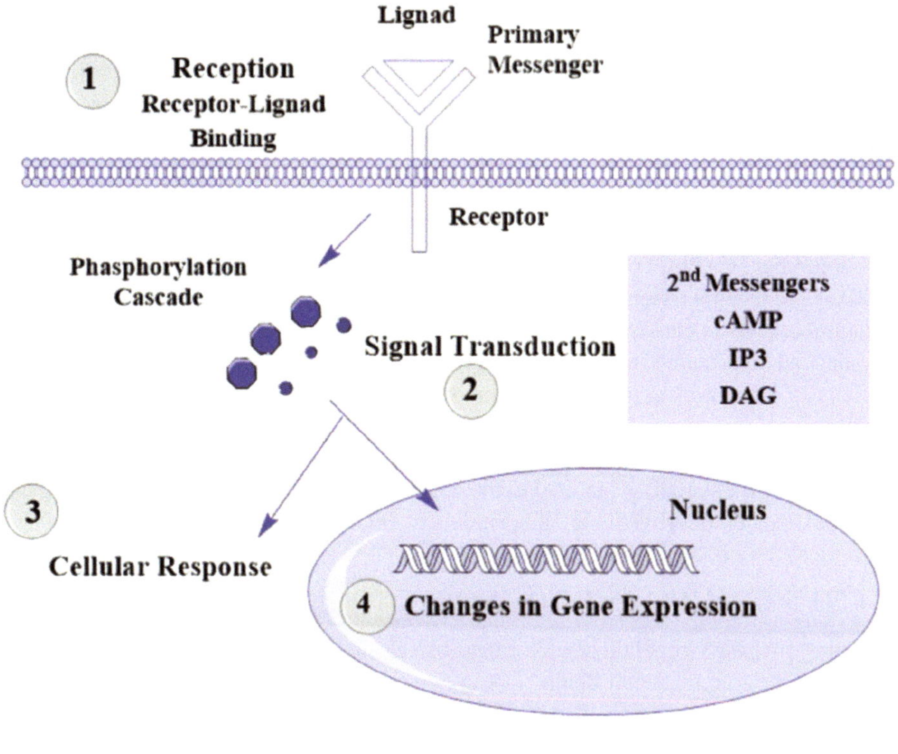

FIGURE 16.1 Mechanisms of cell signaling.

signaling cascade [8]. The active MAPK is transported to the nucleus, phosphorylating numerous transcription factors, including ERK 1 and 2 [9]. These transcription factors then control the expression of genes involved in the cell's differentiation, proliferation, and survival [10]. Dysregulation of the Ras/Raf/MAPK pathway has been linked to several diseases, including cancer [11].

The activation of the PI3K/AKT/mTOR pathway begins with the binding of extracellular ligands to G-protein-coupled receptors (GPCRs) or receptor tyrosine kinases (RKTs). This binding triggers a series of events, including the activation of phosphoinositide 3-kinase (PI3K). PI3Ks are lipid enzymes related to viral oncoprotein activity and are divided into three classes: class I (further divided into IA and IB), class II (PI3K-C2), and class III (Vps34) [12].

PI3K then activates phosphatidylinositol 3,4,5-triphosphate (PIP3) through phosphorylation of phosphatidylinositol 4,5-bisphosphate (PIP2). PIP3 activation, in turn, activates AKT at the plasma membrane. AKT is the central downstream mediator of PI3K and is recognized as an essential component of the insulin receptor in intracellular signaling. [13]. AKT activation phosphorylates many downstream targets involved in cell growth and metabolism, and it also phosphorylates mTOR (which forms two complexes, mTORC1 and mTORC2). mTORC1 phosphorylates downstream mediators such as 4E-BP1 and S6K, which promote cell growth, metabolism, and protein synthesis [14]. In cancer, dysregulation of the PI3K/AKT/mTOR pathway results in abnormal cell growth, survival, and proliferation [15].

The Wnt family of secreted glycolipoproteins significantly influences cell proliferation, cell polarity, and tissue homeostasis through signaling mechanisms. This signaling pathway is initiated when Wnt ligands bind to Frizzled (Fz) receptors and co-receptors, resulting in either canonical or non-canonical signaling pathways [16]. The activation of the canonical Wnt pathway occurs through the binding of Wnt to Fz and LRP5/6, disrupting the destruction complex that would otherwise degrade β-catenin. As a result of this disruption, β-catenin accumulates and localizes to the nucleus, where it serves as a transcriptional coactivator alongside transcription factors from the TCF/LEF family [17].

Two non-canonical Wnt pathways exist:

1. **The planar cell polarity (PCP) pathway:** The PCP pathway is triggered by the binding of Wnt to Fz and its co-receptor, which recruits Dishevelled (Dsh) and activates Rho. Rho, in turn, activates ROCK, which regulates the cytoskeleton [18].
2. **The Wnt/calcium pathway:** This pathway helps to modulate calcium release from the endoplasmic reticulum (ER) to regulate intracellular calcium levels. Following ligand binding, the Fz receptor directly interacts with a trimeric G-protein, activating either phospholipase C (PLC) or phosphodiesterase (PDE). PLC cleaves PIP2 into diacylglycerol (DAG) and inositol trisphosphate (IP3), prompting calcium release from the ER. Increased calcium and DAG can activate Cdc42 and calcineurin, while CaMKII (calcium/calmodulin-dependent protein kinase II) induces the activation of NFAT (nuclear factor of activated T cells). PDE, however, prevents calcium release from the ER by inhibiting protein kinase G (PKG).

Dysregulation in Wnt signaling can lead to the development of many diseases, including cardiovascular diseases, cancer, developmental abnormalities, and neurodegenerative disorders [19].

16.3 CELL SIGNALING AS A DIAGNOSTIC TOOL

Cell signaling pathways regulate a wide range of cellular functions, such as differentiation, proliferation, and programmed cell death. Disruptions in these complex networks are increasingly recognized as hallmarks of numerous diseases [20]. Analyzing or examining the molecular variations within these pathways provides an innovative diagnostic approach that holds the potential for enhancing disease detection and targeted therapy [21]. Current diagnostic strategies often rely on the presence or absence of specific disease markers. However, delving deeper into cell signaling analysis reveals the functional changes within these pathways, offering a more precise approach.

In oncology, this method is particularly promising. Mutations in oncogenes disrupt the growth and survival signaling pathways, providing a unique fingerprint for different tumor types [22]. Cell signaling analysis illuminates not only cancer but also the development of autoimmune and infectious diseases. Some pathogens exploit the host cell signaling pathways to enter or manipulate cellular functions for their replication [23]. By uncovering these aberrant signaling patterns, clinicians can identify new and novel therapeutic targets, tailored to the precise molecular characteristics of the tumor, providing a reassuring level of effectiveness [24]. Understanding these instances of hijacking paves the way for innovative therapeutic approaches that disrupt the pathogen's reliance on host cell signaling.

Moreover, autoimmune diseases, characterized by the immune system's attack on healthy tissues, can also be linked to disrupted signaling pathways controlling immune tolerance [25]. Detecting these signaling pathway malfunctions offers opportunities for enhanced diagnosis and the development of specific targeted therapies, empowering us with the hope of controlling abnormal immune responses [26].

16.3.1 Nuclear-Mediated Signaling

Signaling molecules are crucial in transferring regulatory signals within and outside the cells. In various signaling pathways, these molecules act as ligands, receptors, transcription factors, and protein kinases [27]. The signal transduction network consists of diverse pathways characterized by a detailed regulatory system through mutual interactions [28]. The nuclear factor kappa B (NF-κB) signaling pathway plays a complex role in regulating various physiological and pathological processes, and it has the potential to interact with other molecules directly or indirectly. This can trigger interactions with other signaling pathways, such as MAPK, TGF-β, JAK-STAT, PI3K/AKT, Wnt, Hedgehog, and Notch signaling [29]. These pathways work together in biological processes like cell proliferation, survival, development, differentiation, inflammation, immunity, tumorigenesis, and death [30]. Additionally, the toll-like receptor (TLR) family members are involved in NF-κB signaling by identifying the antigenic components of microorganisms [31] (Figure 16.2).

16.3.1.1 Nuclear-Cytoplasmic Transport

The transportation of molecules between the nuclear and cytoplasmic regions is facilitated by various proteins, including importins, exportins, and 14-3-3 family members. The exportin family, which consists of at least seven proteins, plays a key role in transferring a wide range of binding partners from the nucleus to the cytoplasm through nuclear pore complexes [32]. These exportins utilize the energy from RanGTP to transport their clients effectively. Exportin proteins are recognized for their ability to transport certain classes of molecules or proteins [33]. However, exportin 1 (XPO1), which is the most extensively studied exportin protein, stands out as it can bind to a wide range of proteins with distinct functions through the leucine-rich nuclear export signal (NES) recognition. Due to its diverse client interactions, XPO1 is likely to have an impact on almost every cellular process [34].

It is not surprising that exportin proteins are associated with various types of cancer, where they are frequently overexpressed and contribute to tumor progression and chemotherapy resistance. A recent study has revealed that exportin 1 can serve as a prognostic marker for different types of cancers, such as gastric, head and neck squamous cell, lung, ovarian, pancreatic, brain, osteosarcoma, and breast cancers. For instance, in breast cancer, Yu and colleagues have demonstrated that the cytoplasmic G-protein estrogen receptor (GPER) enhancement depends on XPO1, and can induce a switch to aerobic glycolysis in cancer-associated fibroblasts (CAFs) [35]. This metabolic modification of CAFs leads to the supply of energy from mitochondria in the form of pyruvate and lactate to surrounding breast cancer cells, thereby providing them with multidrug resistance [35]. In myeloma

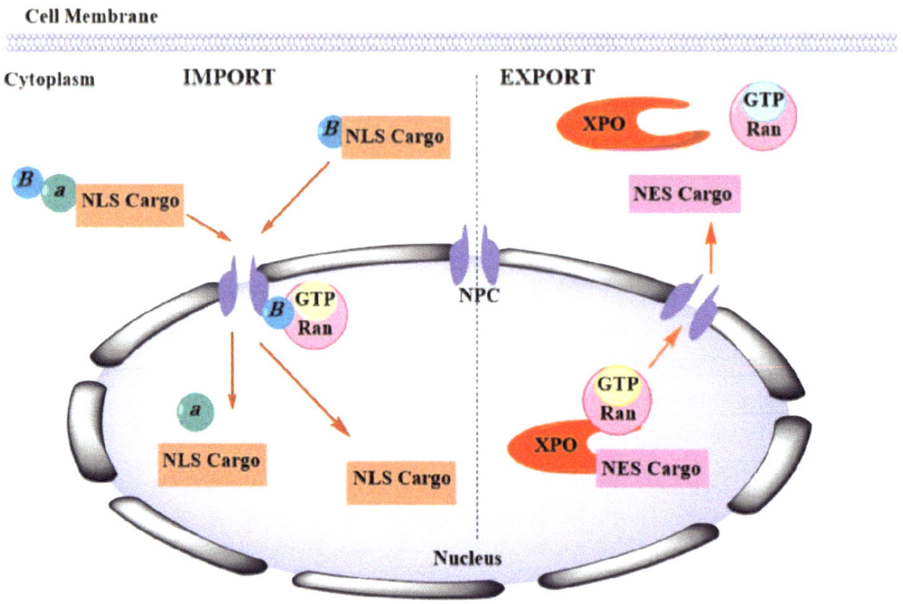

FIGURE 16.2 Nuclear-mediated signaling.

cells, DNA-bound topoisomerase IIα is targeted by doxorubicin inside the nucleus for cell death induction. However, myeloma cells can develop resistance to topoisomerase IIα by enhancing the controlled export of XPO1 [36].

In cancer research, XPO1 has long been a well-known focus, but exploring the role of other exportin family members is an emerging area of investigation. Among these, cellular apoptosis susceptibility protein (CAS, also known as XPO2) is the most well-studied. It acts as the transporter for importin-1α that helps regulate cell proliferation and apoptosis [37]. XPO2 is associated with various types of cancer, where it is believed to enhance cancer cell survival, invasion, and metastasis [38]. Exportin 6, responsible for transporting actin-profilin complexes out of the nucleus, is overexpressed in lung and prostate cancer, and its silencing with siRNA induces cell death in non-small-cell lung cancer [39]. Exportin 5, a nuclear transporter of pre-microRNAs (miRNAs), shows upregulation in lung, prostate, skin, and colorectal cancer. Previous studies have shown that in these cancers, exportin 5 either blocks the export of tumor suppressor miRNAs or enhances the nuclear export of oncogenic miRNAs [40]. Lastly, exportin 7, characterized by its broad substrate specificity, is upregulated in the cytoplasm of ovarian cancer cells, correlating with poor patient survival [41]. Additionally, there is evidence indicating that XPO7 acts as an oncogene in pancreatic cancer when mutated [42]. Recent studies also showed that XPO7 likely interacts with over 200 export clients. Other exportins may also have similarly numerous clients. XPO7 and XPO2 help promote the epithelial-to-mesenchymal transition when they errantly bind to CD47 and stabilize the E-cadherin/β-catenin complex. This transition is an essential factor in cancer metastasis. Considering the numerous clients, further research is required to understand the mechanisms through which exportin proteins contribute to cancer development [43].

16.3.1.2 Protein Subcellular Distribution

The relocation and accumulation of proteins in the cytoplasm from the nucleus can disrupt cellular homeostatic balance and provide cancerous cells with traits that allow uncontrolled expansion and growth. Dysregulated control of subcellular protein distribution can result in abnormalities in cell cycle regulation, resistance to programmed cell death, enhanced invasion, and metastasis. Among the frequently mislocated proteins are those crucial for cell division. For example, retinoblastoma

(Rb) protein is vital for controlling tumor progression and cell growth. Dysregulation of this protein can lead to not only the development of retinoblastoma tumors but also to the onset of most other cancers [44]. In normal mouse and human fibroblast cells, Rb is localized within the nucleus. However, in embryonic mouse cells with a mutation in Cdk4R24C, Rb moves to the cytoplasm; Jiao et al. revealed that this movement of Rb is controlled by exportin proteins and cyclin-dependent kinases (CDKs) [45]. Translocation of Rb to the cytoplasm removes the essential barrier that modulates cell division control, whose absence can facilitate oncogenic transformation.

CDK inhibitor 1B (p27) serves as a critical inhibitor of cell cycle progression and as a tumor suppressor. Nevertheless, in HER2-positive breast cancer cells, nuclear p27 levels are reduced while cytoplasmic levels increase, leading to a worsened prognosis and enhanced metastasis risk among patients [46]. Like p27 and Rb, survivin is also a protein linked with cell cycle regulation, but its localization regulates its functional activity. Specifically, nuclear survivin acts as a regulator of mitosis, while cytoplasmic survivin supports tumorigenesis and inhibits apoptosis [47].

As described with Rb, nuclear export of cell cycle regulatory proteins can result in the loss of their desired function inside the nucleus. This was further illustrated in primary tumor cells from colon cancer, hepatocellular carcinoma, neuroblastoma, and breast cancer, where enhanced cytoplasmic p53 results in loss of its tumor suppressor function [48]. This can expose cells to hallmarks of cancer progression and development and dangerous genomic instability.

Overexpression of typically nuclear CDC25 phosphatases, which help regulate the cell cycle, is associated with the progression of numerous human cancers. When associated with a poorer prognosis, they frequently undergo mislocalization to the cytoplasm. When CDC25A and CDC25B are overexpressed in esophageal cancer, they relocate to the cytoplasm, where they are associated with cell survival [49]. Additionally, CDC25A has been shown to enhance cell survival by inhibiting apoptosis in various cancer types. In ovarian cancer cells, CDC25A is located in the nucleus, where it can't inhibit apoptosis, but in rat 423 cells, CDC25A is located in the cytoplasm, where it can inhibit apoptosis [43].

16.3.2 Receptor-Mediated Signaling

Receptor-mediated signaling is widely recognized as a crucial mechanism for transferring extracellular information [50]. Ligands or signals bind to receptors present in the cell membrane to initiate the mechanism, and the binding triggers secondary signals (known as secondary messengers), such as inositol phosphates, cyclic adenosine monophosphate, and calcium. These secondary messengers then regulate the function and activity of intracellular proteins (including phosphatase, proteases, and kinases) [51]. These proteins can then change cellular behavior by modifying the expression of transcription factors, inducing changes in membrane permeability, and altering ion channel activity. Receptor-mediated signal transduction, triggered by ligand-receptor binding, regulates endocrine, autocrine, and paracrine interactions essential for organogenesis and cell differentiation. In particular, signaling peptides and their receptors, such as SHH, FGF, BMP, WNT, NOTCH, PDGF, and VEGF, are engaged in the organogenesis of various organs, including the lung [52].

16.3.2.1 VRAC and LRRC8A–E

Among the major pivotal chloride channels present in the cell membrane and characterized by LRRC8A and four other similar group members (LRRC8B–E), the volume-regulated anion channel (VRAC) has been implicated in all cancer cases [53]. VRAC has been studied for its contribution to the cell cycle regulation in nasopharyngeal carcinoma, human ovarian cancer, small-cell lung cancer, glioblastoma, and cervical cancer, using broad-spectrum pharmacological VRAC blockers such as DIDS, DCPIB, and NPPB [54]. Furthermore, some reports have indicated the involvement of VRAC in migration and multidrug resistance in multiple cancers has also been affected. Specifically, a VARC current was recorded in pancreatic duct adenocarcinoma cell lines [55]. Therefore, VRAC emerges as a potential therapeutic target for pancreatic adenocarcinoma

(PAAD). Research has shown that VRAC, as part of LRRC8A–E, has been linked to proliferation, metastasis, and drug resistance in cancer cells, indicating its potential as a molecular target in regulating tumor progression. Yet, LRRC8A–E prognostic significance and biological role in PAAD remain poorly understood [56].

A study explored the expression of LRRC8A–E across multiple cancers using independent datasets from TCGA data in GEPIA and ONCOMINE. In PAAD tissues, the expression of LRRC8A was found to be upregulated compared to normal tissues, and its levels increased as PAAD progressed [57]. It was reported that LRRC8 heteromers are vital components of VRAC, and LRRC8A serves as its key subunit. Therefore, further investigation is required to elucidate the vital role and molecular mechanisms of LRRC8A in cancer management [58].

As invasion, migration, proliferation, and drug resistance are key cancer cell hallmarks, further investigation is needed to determine whether LRRC8A's downregulation inhibits these hallmarks in cancer cells [59]. Some studies have concluded that LRRC8A inhibition may serve as a potential therapeutic target in pancreatic adenocarcinoma. Adjacent genes of LRRC8A were predominantly associated with focal adhesion and PI3K-AKT pathways. A previous study revealed that the PI3K-AKT pathway served as a downstream signaling process responsible for LRRC8A's regulatory impact on cancer cell proliferation [60]. Also, LRRC8A-Pi3K-AKT-mTOR pathways were found to control esophageal squamous cell carcinoma migration [61].

Significantly, EPHA2 exhibits a strong association with LRRC8A in the majority of PAAD studies. LRRC8A suppression results in reduced EPHA2 expression, which was shown to regulate prostaglandin-endoperoxide synthase involvement in anti-tumor immunity. Furthermore, LRRC8A is vital for the development of T cells, and recent findings have illuminated its integral role in tumor immunotherapy [62]. Heteromeric VRAC channels consisting of LRRC8A, along with LRRC8C or LRRC8E, mediate cGAMP transportation to adjacent cells, triggering activation of the innate immune sGAS-cGAMP-STING pathway in murine macrophages and vascular epithelial cells. The immune invasion results from TIMER suggest that LRRC8A impacts patient clinical outcomes through interaction with immune infiltration in PAAD. Moreover, immune infiltration analysis displayed a critical correlation between LRRC8A and the levels of infiltration by several immune cells. It was hypothesized that LRRC8A/C heteromeric channels promote cGAMP transportation and p53 signaling in T-cell activation [63].

16.3.3 Extravesicular Communication Pathways

Various approaches have been utilized for earlier cancer detection, including examination of patients' medical records and clinical histories, biochemical test results, functional imaging data analysis, and examination of molecular surgical findings. Identifying stable biomarkers for earlier cancer detection is vital for lowering patient mortality rates [64]. Using non-invasive methods, an ideal diagnostic approach should accurately identify specific tumor biomarkers during premetastatic stages. Despite recent progress in biomarker discovery, only a handful of potential biomarkers have been effectively utilized for diagnostic purposes in clinics [65]. There is a clear demand for developing accurate, novel, and less invasive methods for early cancer diagnosis. Conventional procedures, including biopsy and bodily fluid samples containing circulating tumor cells, tumor RNA, tumor DNA, and exosomes, present distinct biomarkers for cancer detection. Bodily fluids also assist with medical decision-making, providing clinical information throughout treatment and impacting therapeutic decisions.

Recently, exosomes have emerged as an accurate and unique source of biomarkers (miRNAs) for detecting different cancer types, including hepatocellular carcinoma, pancreatic cancer, breast cancer, melanoma, ovarian cancer, prostate cancer, lung cancer, and gastrointestinal cancer [66]. Exosomal cargo molecules and miRNAs have gained attention as biomarker candidates. Furthermore, exosome-based vaccines using dendritic cell therapy have emerged as a novel treatment method for tumor growth suppression [67]. In blood samples of pancreatic cancer patients, the presence of glypican 1 (GPC1) in exosomes has been identified. Glypican 1, a membrane-bound heparin sulfate proteoglycan, shows upregulation in various tumors, including breast, pancreatic, glioma, and colorectal cancer. This upregulation is measured or assessed by mass spectrometry and liquid chromatography. Circulating exosomal GPC1 and its regulatory miRNAs, miR-96-5p and miR-149, have proven valuable for diagnosing colorectal cancer and also act as targets for colorectal cancer treatment and other cancer types [68] (Table 16.1).

16.3.3.1 Exosomes

While solid biopsy remains the preferred method for pathological diagnosis and serves as the foundation for cancer treatment, it can be invasive, challenging to perform, and subject to tumor heterogeneity. Therefore, utilizing tumor-derived exosomes (TDEs) as a non-invasive liquid biopsy offers substantial benefits for precise and accurate diagnosis, as well as personalized medicine. This is due to the correlation between TDEs and the formation of metastatic niches, immune evasion, and tumor progression [64].

Numerous studies have explored exosomes as a means of early cancer detection and precise diagnosis. For example, research has shown that exosomal DNA in the initial phase of pancreatic cancer is highly likely to display KRAS mutations. Furthermore, elevated levels of glypican 1 (GPC1) were discovered in circulating exosomes from patients with colorectal cancer and pancreatic ductal carcinoma, suggesting that this observation could serve as an early detection tool for digestive system tumors [75]. A clinical trial (NCT03032913) was conducted to analyze diagnostic accuracy by detecting the presence of CTCs and GPC1-containing exosomes presence for assessment and comparison. For the early diagnosis of colorectal cancer, lncRNA (BCAR4) and the two mRNAs, KRTAP5-4 and MAGEA3, present potential options [76].

In addition to exosomal RNAs, various proteins found in exosomes have shown promise in distinguishing between cancerous and non-cancerous patients. Sandfeld-Paulsen et al. conducted a study utilizing exosome arrays containing 49 antibodies, revealing that CD171 and CD151 were key molecules for identifying lung cancer patients from those without cancer [77].

Meanwhile, Chen et al. utilized mass spectrometry (MS) to identify 144 phosphorylated proteins that were distinctly elevated in exosomes from breast cancer patients compared to non-cancer patients. Numerous other studies are investigating the potential use of exosomes in cancer diagnosis and prognosis, some of which have progressed to clinical trials [78].

16.3.3.2 CircRNAs

Animal cells possess relatively stable circular RNAs (circRNAs), which can modulate the expression of targeted genes through various mechanisms [79]. These circRNAs exhibit a covalently closed loop structure without any 5'-3' polarity or poly A tail. They also contain various miRNA

TABLE 16.1
Association of Biomarkers with Various Diseases

Disease	Biomarker	Signaling Pathways	References
Cancer	CA125, PSA, CEA C↑	MAPK, Ras/Raf, PI3K/AKT, JAK/STAT, Wnt	[69]
Neurological Disorders	α-Synuclein ↓, NFL ↑	Notch, MAPK, PI3K/AKT	[70]
Autoimmune Diseases	Anti-CCP, RF ↑	TGF-β, NF-κB, JAK/STAT	[71]
Cardiovascular	Troponin, BNP ↑	cAMP/PKA, MAPK	[72]
Infectious Diseases	HIV RNA, HPV, HBsAg ↑	NF-κB, JAK/STAT	[73]
Diabetes	HbA1c, Fasting Blood Glucose ↑	Raf/Ras/MEK/ MAPK, PI3K/AKT	[74]

binding sites called miRNA response elements (MREs), which enable them to function as miRNA decoys [80]. Additionally, by serving as RNA-binding protein decoys, they can regulate mRNA expression [81]. Furthermore, circRNAs can bind directly to specific mRNA targets to regulate their expression of mRNA. Through binding with miRNAs, circRNAs can play a vital role in cancer regulation and participate in cancer-related signaling pathways, including the Wnt/beta-catenin pathway, phosphatidylinositol 3-kinase (PI3K)/protein kinase B (AKT) intracellular signaling pathway, and MAPK [82].

Studies have shown that circRNA has specific expression patterns and stability in regulating gene expression; therefore, it has the potential to act as a biomarker for numerous diseases. Moreover, circRNAs can be valuable for the diagnosis and treatment of cancer, as they can either promote oncogenesis or suppress the tumors [82].

In clinical settings, circular RNAs have emerged as promising biomarkers for detecting different types of human cancer. These circular RNAs can be detected in bodily fluids, tissues, and blood samples. However, circRNAs must possess desirable traits such as stability, sensitivity, and specificity to be effective biomarkers. Some examples of circRNAs that have shown potential as effective biomarkers include hsa_circ_001569 in colorectal cancer, hsa_circ in squamous cell cancer (**circ_CDYL** (has_circ_0008285)), circ_CDYL, hsa_circ_0029 in gastric cancer, and hsa_circ_0005075 in hepatocellular carcinoma. Additionally, newly discovered circRNAs hold promise as potential biomarkers for cancer detection [82–85].

16.3.3.3 HPV

The integration of high-risk HPV (HR-HPV) viral DNA integration into the host chromosomal DNA is the primary cause of pre-neoplastic cell formation. This happens when clones of cells with deregulated viral oncogenes expression develop in basal and parabasal cell layers, resulting in invasive carcinoma formation in the cervix. As such, DNA-based assays can be formulated to identify the presence of DNA responsible for cervical cancer. The detection of cervical cancer relies primarily on the HPV DNA molecular marker, and clinical evaluation considers various molecular abnormalities such as DNA mutations, chromosomal anomalies, apoptotic markers, cell cycle checkpoints, oncogenes, tumor suppressor gene expression, and epigenetic regulation (hypermethylation) as markers [86].

Developing cervical cancer is an uncommon outcome of a common cervical infection with a high-risk HPV type. To develop, maintain, and progress to CIN 3, a persistent high-risk HPV infection is required. The HPV DNA test is utilized as the fundamental screening method and has demonstrated greater sensitivity than cytology in several clinical studies. Women who have had both Pap and HPV tests can be reassured that they are not at risk of undetected CIN 2, CIN 3, or cervical cancer. HPV testing is more sensitive in identifying these conditions. Therefore, a negative result in both tests indicates that the woman is not at risk of developing the aforementioned conditions [87, 88].

Widschwendter et al. have highlighted that serum HPV DNA can be a helpful tool in identifying early recurrence in cervical cancer. Meanwhile, as reported by Campitelli et al., HPV mutation (insertion) is a reliable molecular marker for circulating tumor DNA (ctDNA) in individuals with HPV-related cervical cancer. Detectable ctDNA was found in the majority of patients with cervical cancer at stage I or beyond, and the concentration of ctDNA was observed to correspond to the tumor burden in these individuals [89].

16.3.4 Immune Response Signaling

Cells have the remarkable ability to perceive, react to, and integrate various signals from their environment. In the immune system, these signals include interactions with neighboring cells through direct contact, recognizing soluble factors such as cytokines that can travel long distances through the bloodstream, and detecting pathogens and their products. Once cell surface receptors are activated, a cascade of internal signal transduction events is set in motion [90]. This ultimately leads to the activation of transcription factors and other proteins that regulate various biological processes, such as gene expression, phagocytosis, cell death, cell division, and secretion. The signal transduction machinery comprises distinct cell surface receptors, effector signaling components, and regulatory proteins that work together in a coordinated manner to ensure proper cellular responses to environmental cues [91]. The signaling cassettes function to recognize, amplify, and integrate external signals to elicit the appropriate cellular response [91].

16.3.4.1 T-Helper 17.1 Cells

Recently, the understanding of sarcoidosis as a disease caused by Th1 cells was based on the presence of specific cytokines (including IFN-γ, IL-12, and IL-18) associated with bronchoalveolar lavage fluid (BALF) and serum of type 1 sarcoidosis patients. However, it is now widely recognized that sarcoidosis is influenced by a combination of factors linked to both Th1 and Th17 cells [91]. In sarcoidosis, Th17 cells (which typically produce IL-17A and IL-17F) have also been found to produce IFN-γ, resulting in what are now known as Th17.1 cells. These cells are believed to arise from conventionally polarized Th17 cells. Th17 cells possess a high degree of plasticity and have demonstrated the ability to transform into a Th1-like phenotype, with the majority of cells producing IFN-γ. Researchers have identified Th17 and Th17.1 cells in patients with sarcoidosis. These T cells exhibit various chemokine receptors and cytokine effector functions. The quantity of Th17.1 cells was notably higher in the bronchoalveolar lavage fluid of sarcoidosis patients compared to healthy individuals [91, 92]. Th17.1 cells could potentially be used as immunological markers for sarcoidosis diagnosis, but further research with larger cohorts is needed to confirm their diagnostic significance. Moreover, a strong correlation was found between the number of Th17.1 cells in BALF and chronic sarcoidosis progression, making these cells a valuable prognostic marker for (chronic) pulmonary sarcoidosis [93].

16.3.4.2 Cancer Stem Cells

Cancer stem cells (CSCs) are believed to be responsible for the dissemination of tumors and the creation of metastases in the abdominal cavity. CSCs are characterized by their ability to proliferate indefinitely and provide a foundation for the most diverse cells that constitute the tumor mass. Typically, CSCs constitute only a small fraction of the tumor. Numerous investigations have uncovered distinct indicators of CSCs that are considered vital in metastasis formation. Moreover, the Wnt, Notch, and Hedgehog (Hh) signaling pathways are believed to be crucial in the epithelial-mesenchymal transition (EMT) process [94]. Consequently, genetic analysis offers several diagnostic options [95]. Tumor classification and the selection of appropriate treatment can be determined using CSC markers. Numerous techniques are available for identifying and extracting CSCs from tumors, such as fluorescent-activated cell sorting (FACS) and magnetic-activated cell sorting (MACS), which use surface and intracellular markers. MACS, a nonparametric isolation method, is the fastest, while FACS, a multiparametric separation method, offers more options. While some cancers, particularly blood cancers, have specific markers available, it is crucial to identify the markers for ovarian CSCs [94, 95] (Figure **16.3**).

Research studies have shown that CD133 is one of the commonly used cell surface markers for CSC isolation. The presence of CD133 in cancerous tissue has also been linked to higher tumor aggressiveness. This conclusion has also been drawn for glioma cancer stem cells (C6) [96]. However, the presence or absence of CD133 alone is not a definitive indicator of malignant tumor classification. Upon implantation of the CD133+ cell fraction, ovarian tumors were found to consist of both CD133+ and CD133− cells. Nonetheless, the CD133+ cells exhibited a more substantial capacity for tumorigenesis than their CD133− counterparts. This suggests that these positive cells may give rise to multiple phenotypic groups [94].

Furthermore, it has been shown that cells lacking CD133 expression can form tumors containing both positive and negative cell populations. However, it remains unclear whether these

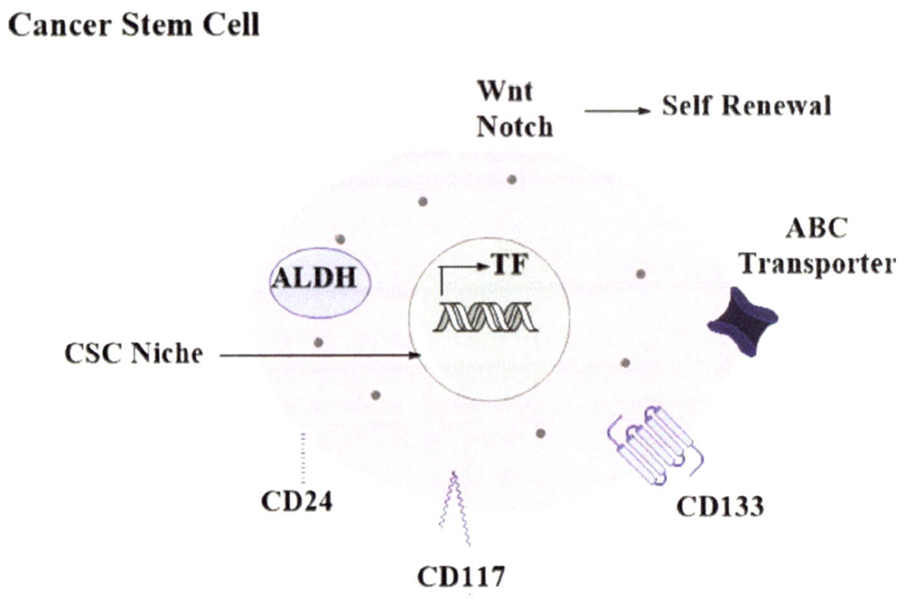

FIGURE 16.3 Cancer stem cell and biomarkers.

CD133- cells can generate CD133+ cells or if the presence of a small fraction of CD133+ cells in the purified CD133- population induces their appearance [97]. Alternatively, CD133+ cells may be derived from a distinct population of tumor CSCs. Several research teams have investigated the presence of CD133 antigens in ovarian cancer cells [98]. Studies have discovered that CD133+ cells are present in both benign and malignant tumor cells, as well as normal ovarian tissue. However, the percentage of these cells in normal tissue or metastases is lower than in primary ovarian tumors. Tissue microarray analysis has shown that CD133 expression is linked with high-grade serous ovarian cancer, ascites, advanced stages of the disease, and poor response to treatment [99].

The Aldefluor technique is a valuable tool for assessing aldehyde dehydrogenase activity (ALDH), a critical marker of CSCs. ALDH is involved in detoxification through the conversion of retinol to retinoic acid and the oxidation of aldehydes, and its presence is indicative of CSCs [100]. In ovarian tumors, the co-occurrence of both CD133+ and ALDH+ cells has been linked to chemotherapy resistance, tumor growth, and initiation in mice. Furthermore, the potential of the ALDH1 isoform as a CSC marker in ovarian cancer has been investigated, but the findings have been inconsistent [101]. While the expression of ALDH1 and CD44 was found to promote treatment resistance in ovarian cancer cells, ALDH1 was identified as a favorable prognostic factor in ovarian cancer therapy in research conducted by Chang et al. [102].

Identifying cancer stem cells can be accomplished through the use of the side population (SP) and ABC transporter expression. Recently, unique CD44+/MyD88+ epithelial ovarian cancer stem cells were discovered to be responsible for tumor initiation, even after surgical and chemotherapeutic intervention [102]. The TLR2-MyD88-NFκB pathway, which promotes tumor healing and self-renewal, is pro-inflammatory in nature. A recent study has underscored the significance of HMGA1, a chromatin remodeling protein, in the functionality of ovarian cancer stem cells. The study employed a 3D culture of ovarian cancer cells, such as SKOV3, PA1, and A2780, and revealed that increased HMGA1 expression correlated with the expression of stemness markers. The deletion of HMGA1 resulted in the reduction of proliferation, spheroid-forming capabilities, and expression of stemness markers. Furthermore, the deletion rendered the cells more sensitive to chemotherapy [103].

By utilizing the FACS technique, it becomes feasible to categorize ovarian cancer stem cells (OCSCs) based on specific indicators. The separated cells exhibited characteristic phenotypic qualities of CSCs, such as resistance to therapy, self-regeneration, and tumor proliferation. No observable tumor development was observed from the sorted CSCs, mirroring the typical ovarian cancer progression in humans. However, it is essential to consider that a variety of internal and external factors, as well as the method of subcutaneous transplantation of the isolated cells into mice, could contribute to this outcome [104]. The most frequent metastases in human patients affect the areas adjacent to the ovary, such as the peritoneum, mesentery, bladder, colon, and others. It is worth noting that the varying stimuli of human organisms and the weakened immune response in experimental mice should not be underestimated [105]. The potential for tumors to develop from OCSCs is dependent upon the unique profile of specific markers. It has yet to be established whether populations of CSCs with different marker profiles have genetic, epigenetic, or molecular differences. It has not been accurately determined if isolated OCSCs can replicate the original heterogeneity of the tumor after transfer into a mouse model. The presence of spheroid complexes in the peritoneal cavity of patients with advanced disease may contribute to the capability of OCSCs to metastasize and develop spheroids [106].

The presence of typical CSC markers in ovarian cancer indicates that the identification of a single marker is not enough to confirm the presence of OCSCs due to their complexity. Determining the populations of cells may depend on knowledge of their entire marker profiles and biochemical properties. Thus far, no specific CSC marker has been identified that occurs solely in the ovary, as the existing markers are also found in various other cancers [107]. Previous studies indicate that ovarian cancer cells containing CSC markers are typically linked to a higher degree of malignancy, as well as an increased risk of relapse and resistance to treatment, resulting in a less favorable prognosis for the patient. As a result, research in this field is ongoing, with a focus on identifying specific markers for ovarian CSCs and examining the associated signal pathways that could potentially lead to new diagnostic and therapeutic options [108].

16.4 CELL SIGNALING RECOGNITION METHODS

Living cells, such as neurons, bone cells, and somatic cells, are intricate microstructures containing various functional subunits at the "nano-size" level. These subunits are linked by biochemical and physical signaling pathways, requiring biosignal processing within the cell structure. When investigating the broad range of ultrastructural patterns found in more complex cells, their diversity may seem perplexing at first glance [109].

Analytical light microscopy is a valuable technique for monitoring the dynamics of cellular signal transduction. A notable example is the use of fluorescent calcium indicators, which have shed light on the significance of receptor-mediated oscillations of intracellular Ca^{2+} as a primary regulatory element in cell signaling [110]. In addition to Ca^{2+}, attention has shifted to other ion species, as transmembrane potentials coupled to such ion gradients can be informative in understanding cell mode pathophysiology [111].

From a biological standpoint, it seems that ion gradients played a major role in the development of complex cell structures and communication pathways both inside and outside the cell. This process is highly complex and redundant, with many factors influencing the transmission of signals through a dynamic network of pathways. Regulatory circuits, cross-talk between pathways, nonlinear biochemical reactions, and the division of cells into quasi-autonomous subsystems all contribute to the intricate nature of intracellular signaling [112]. Cells can respond in various ways to the chemical and physical signals they receive from their environment, such as growth and mitosis, metabolic changes, protein production and release, alterations in cell shape and movement, programmed cell death, and signal transmission to neighboring cells. This process involves complex signal evaluation, which poses a significant challenge to the field of theoretical biology. To address this challenge, systems biology studies all the elements in a biological system, such as a cell, and their interactions in response to external changes. Developing a mathematical model that describes the dynamic behavior of biological systems at the system level is the ultimate aim of systems biology [113].

Improved understanding of pathological cellular states is one of the many benefits of in silico modeling, providing therapeutic targets for treating diseases and enabling drug testing. However, the relevance of such models to real cells is worth exploring, as the accuracy of computer models depends on the completeness and quality of data collected from laboratories worldwide [110]. These data include details on the number and activity of receptors and enzymes, the concentration and binding kinetics of enzyme substrates and regulatory effector substances, as well as the corresponding functional relationships of these effectors. Additionally, the nonlinear kinetics of biochemical systems pose a challenge to predicting system dynamics over a longer time scale, and models can overlook the memory effects that occur in natural cells [110].

16.4.1 Biological Systems Modeling

Various techniques have been developed to numerically describe cellular signaling pathways, including structured biological modeling (SBM). SBM addresses the challenge of translating vast amounts of experimental data on a biological system into a coherent model that captures its key regulatory processes [114]. The ultimate goal of SBM is to understand how a biological system, such as a cell, transforms incoming signals, like hormone stimulation of a cell surface receptor, into a specific output, such as oscillatory waves in the intracellular concentration of calcium ions that convey vital messenger information for numerous intracellular processes [115].

By incorporating pathway data from databases into data flow diagrams and interpreting them mathematically, stochastic simulations can be specified for processes such as induced Ca2+ oscillation. Generally, stochastic methods are better suited for highly nonlinear and multidimensional systems compared to models using deterministic differential equations. SBM has successfully provided mathematical descriptions of specific cellular subsystems. It also serves as a valuable tool for constructing data flow diagrams that can be used as the foundation for computer simulations of even more complex processes, such as the malignant transformation of tumor cells. However, the predictive ability of these simulations is limited by the incompleteness and inconsistency of existing databases [116].

16.4.2 Biohybrid Structures for Cellular Response Monitoring

Cells' intricate signal processing abilities make it clear that a reductionist approach (such as manipulating a small number of genetic elements to achieve changes in biological functions) will not lead to favorable outcomes. While advancements in gene expression and protein pattern research have shed light on certain aspects of systems biology complexity, there is no direct correlation between this data and cellular systems' dynamic metabolic and physiological characteristics [117].

A successful approach to studying cell system biology involves the development of (primary) human cell models that integrate various cell types. When combined with automated assays, these models can assess disease-related cellular responses that result from pharmaceutical or environmental factors [118]. This approach eliminates the need for complex in silico models, which may be a long way off in the future. It offers valuable insights, even without in-depth knowledge of the specific biochemical targets involved, as multiple pathways are activated simultaneously, allowing for analysis of the resulting response pattern [119].

The response profile created provides a unique representation of the input signals from the analyzed cell system. This measurement can include a combination of external factors or a group of proteins that can be continuously monitored using "biohybrid devices." These devices consist of microelectronic sensor chips with cell cultures grown directly on them, placed in a fluidic life-support system that mimics the cellular environment [120]. In contrast to advanced analytical methods such as microscopy, image analysis, and NMR spectroscopy, biohybrid devices are user-friendly and compact tools that can effectively track and analyze real-time changes in materials [121].

The direct and extracellular measurement method does not require labeling procedures, and cellular function remains undisturbed. The sensor chips, with compact dimensions and a cell growth area of around 50 mm2, only require a cell count between 104 and 105. This allows for the growth of complex primary cell systems, such as human tumor samples, which include a collection of tumor cells and various stroma cells grown on the chip surface [122].

The intricacies of the distinct microsensor components—potentiometric, amperometric, and impedimetric structures—have been previously documented. In summary, cells' metabolic function is gauged by utilizing extracellular sensors for both pH and pO2. The pH levels outside the cell indicate acidic metabolic byproducts being expelled across the cell membrane [123].

In most tumor cells, the level of dissolved oxygen (pO2) experiences a significant increase. This increase can serve as an indicator of the activity of the mitochondrial respiratory chain. Adherent cell growth and subtle changes in cell morphology can be detected by impedimetric electrode structures [124]. Any pharmacological or environmental signal inputs are highly likely to affect metabolic rates and the structure of the cytoskeleton. The addition of agonists to different types of cellular receptors is a known factor that affects cell metabolism and morphology in a specific and dynamic manner, depending on the activated pathways.

For virtually any biological experiment, statistical analysis of the results is essential. When working with biohybrid structures, there is a specific need for reliable and automated assays that can model intricate cellular microenvironments. To meet this demand, medium-density plates with multiparameter glass sensor chips are being developed. These plates are used in climate-controlled boxes and are linked to pipetting robots for liquid handling, as well as microscope platforms for enhanced cell imaging and optical readout [125].

It is essential to develop a practical solution for electric contacts and to appropriately preprocess and compress the collected data. One of the significant benefits of this technology is the ability to incorporate on-chip circuitry for signal amplification, data analysis, and sensor self-testing. On-chip sensor multiplexing is crucial for constructing 96- or 384-multiwell arrays, as managing the number of electric connections required would otherwise be difficult [126]. Additionally, individual chips can form arrays of sensors, such as an array of (ion-sensitive field-effect transistors (ISFETs)) for measuring pH gradients directly beneath growing cells with high local resolution. The information obtained from such multisensor arrays relies on the whole signal pattern rather than a single sensor's characteristics, necessitating a different type of biosignal processing—pattern recognition. Pattern recognition is a common challenge in biological and technical applications such as image analysis, automatic speech recognition, and optical character recognition [127].

16.5 CONCLUSION

Cell signaling pathways are intricate processes crucial for cells to function and survive. When these pathways are disrupted, they can contribute to the development of various diseases, such as cancer, infectious diseases, and autoimmune disorders. By analyzing the molecular abnormalities that occur within these pathways, researchers can develop innovative diagnostic techniques that offer the potential for improved disease classification and targeted therapy. This can lead to the creation of new therapeutic approaches that can help regulate aberrant immune responses. Furthermore, a deeper understanding of the complex structure of living cells, including their biochemical and physical signaling pathways, can help researchers gain insights into the pathogenesis of various diseases. Systems biology, which aims to develop mathematical models of the system-level dynamics of biological systems, can provide a comprehensive understanding of how cellular signal processing works and how cellular patterns relate to functional outcomes. All of this can contribute to developing new and better ways to diagnose and treat various diseases.

REFERENCES

1. Maloy, S. and K. Hughes, *Brenner's encyclopedia of genetics*. 2013: Academic Press.
2. Loewenstein, W.R., *The touchstone of life: molecular information, cell communication, and the foundations of life*. 1999: Oxford University
3. Janmey, P.A. and C.A.J.A.R.B.E. McCulloch, Cell mechanics: integrating cell responses to mechanical stimuli. 2007. **9**: p. 1–34.
4. Alberts, B., et al., *Essential cell biology*. 2015: Garland Science.
5. Horih, S.I.J.N., Basic neurochemistry: molecular, cellular, and medical aspects. 1989. **39**(3): p. 460–460-b.
6. Seto, E.S., et al., When cell biology meets development: endocytic regulation of signaling pathways. 2002. **16**(11): p. 1314–1336.
7. Valencia, A., et al., The ras protein family: evolutionary tree and role of conserved amino acids. 1991. **30**(19): p. 4637–4648.
8. Crespo, P., J.J.C. Leon, and M.L.S. CMLS, Ras proteins in the control of the cell cycle and cell differentiation. 2000. **57**: p. 1613–1636.
9. Shehzad, A. and Lee, Y.S. Molecular mechanisms of curcumin action: signal transduction. 2013. **39**(1): p. 27–36.
10. Molina, J.R. and A.A. Adjei, The Ras/Raf/MAPK pathway. *J Thorac Oncol*, 2006. **1**(1): p. 7–9.
11. Butnariu, L.I., et al., The genetic architecture of vascular anomalies: current data and future therapeutic perspectives correlated with molecular mechanisms. 2022. **23**(20): p. 12199.
12. Katso, R., et al., Cellular function of phosphoinositide 3-kinases: implications for development, immunity, homeostasis, and cancer. 2001. **17**(1): p. 615–675.
13. Kriplani, N., et al., Class I PI 3-kinases: function and evolution. 2015. **59**: p. 53–64.
14. Whiteman, E.L., et al., Role of Akt/protein kinase B in metabolism. 2002. **13**(10): p. 444–451.
15. Vahidnezhad, H., L. Youssefian, and J.J.J.O.I.D. Uitto, Molecular genetics of the PI3K-AKT-mTOR pathway in genodermatoses: diagnostic implications and treatment opportunities. 2016. **136**(1): p. 15–23.
16. Praveena, R., Regulation of Wnt/Wg signaling pathway by chromatin organizer SATB1. 2016, Dept. of Biology.
17. Hayat, R., M. Manzoor, and A.J.C.B.I. Hussain, Wnt signaling pathway: a comprehensive review. 2022. **46**(6): p. 863–877.
18. Valkenburg, K.C., et al., Wnt/β-catenin signaling in normal and cancer stem cells. 2011. **3**(2): p. 2050–2079.
19. Johnson, M.L., N.J.R.I.E. Rajamannan, and M. Disorders, Diseases of Wnt signaling. 2006. **7**: p. 41–49.
20. Jin, Z., W.S.J.C.B. El-Deiry, and therapy, Overview of cell death signaling pathways. 2005. **4**(2): p. 147–171.
21. Chae, Y.K., et al., Path toward precision oncology: review of targeted therapy studies and tools to aid in defining "actionability" of a molecular lesion and patient management support. 2017. **16**(12): p. 2645–2655.
22. Wareham, L.K., et al., Solving neurodegeneration: common mechanisms and strategies for new treatments. 2022. **17**(1): p. 23.
23. Brambilla, E. and A.J.E.R.J. Gazdar, Pathogenesis of lung cancer signalling pathways: roadmap for therapies. 2009. **33**(6): p. 1485–1497.
24. Gambardella, V., et al., Personalized medicine: recent progress in cancer therapy. 2020. **12**(4): p. 1009.
25. De Monerri, N.C.S. and K.J.T.A.j.o.p. Kim, Pathogens hijack the epigenome: a new twist on host-pathogen interactions. 2014. **184**(4): p. 897–911.
26. Muñoz-Carrillo, J.L., et al., Immune system disorders: hypersensitivity and autoimmunity. 2018. **1**(30): p. 10.5772.
27. Nair, A., et al., Conceptual evolution of cell signaling. 2019. **20**(13): p. 3292.
28. Kestler, H.A., et al., Network modeling of signal transduction: establishing the global view. 2008. **30**(11-12): p. 1110–1125.
29. Hayden, M.S., S.J.G. Ghosh, and development. *Signaling to NF-κB*, 2004. **18**(18): p. 2195–2224.
30. Espinosa-Sánchez, A., et al., Therapeutic targeting of signaling pathways related to cancer stemness. 2020. **10**: p. 1533.
31. Tata, J.R.J.N.r.M.c.b., Signalling through nuclear receptors. 2002. **3**(9): p. 702–710.
32. Weis, K.J.T.i.b.s., Importins and exportins: how to get in and out of the nucleus. 1998. **23**(5): p. 185–189.

33. Fornerod, M., et al., CRM1 is an export receptor for leucine-rich nuclear export signals. 1997. **90**(6): p. 1051–1060.
34. Mahipal, A., M.J.P. Malafa, and therapeutics, Importins and exportins as therapeutic targets in cancer. 2016. **164**: p. 135–143.
35. Yu, T., et al., Cytoplasmic GPER translocation in cancer-associated fibroblasts mediates cAMP/PKA/CREB/glycolytic axis to confer tumor cells with multidrug resistance. 2017. **36**(15): p. 2131–2145.
36. Turner, J.G., et al., Human multiple myeloma cells are sensitized to topoisomerase II inhibitors by CRM1 inhibition. 2009. **69**(17): p. 6899–6905.
37. Monian, P. and X.J.J.O.B.C. Jiang, The cellular apoptosis susceptibility protein (CAS) promotes tumor necrosis factor-related apoptosis-inducing ligand (TRAIL)-induced apoptosis and cell proliferation. 2016. **291**(5): p. 2379–2388.
38. Böni, R., et al., Expression of the proliferation and apoptosis-associated CAS protein in benign and malignant cutaneous melanocytic lesions. 1999. **21**(2): p. 125–128.
39. Hao, J., et al., Elevated XPO6 expression as a potential prognostic biomarker for prostate cancer recurrence. 2016. **8**(1): p. 44–55.
40. Wu, K., et al., The role of exportin-5 in microRNA biogenesis and cancer. 2018. **16**(2): p. 120–126.
41. Cáceres-Gorriti, K.Y., et al., RAN nucleo-cytoplasmic transport and mitotic spindle assembly partners XPO7 and TPX2 are new prognostic biomarkers in serous epithelial ovarian cancer. 2014. **9**(3): p. e91000.
42. Aksu, M., et al., Xpo7 is a broad-spectrum exportin and a nuclear import receptor. 2018. **217**(7): p. 2329–2340.
43. Holmes, T.R., S. Dindu, and L.A.J.M.C. Hansen, Aberrant localization of signaling proteins in skin cancer: implications for treatment. 2019. **58**(9): p. 1631–1639.
44. Giacinti, C. and A.J.O. Giordano, RB and cell cycle progression. 2006. **25**(38): p. 5220–5227.
45. Jiao, W., et al., Nucleocytoplasmic shuttling of the retinoblastoma tumor suppressor protein via Cdk phosphorylation-dependent nuclear export. 2006. **281**(49): p. 38098–38108.
46. Denicourt, C., et al., Relocalized p27Kip1 tumor suppressor functions as a cytoplasmic metastatic oncogene in melanoma. 2007. **67**(19): p. 9238–9243.
47. Dohi, T., et al., Mitochondrial survivin inhibits apoptosis and promotes tumorigenesis. 2004. **114**(8): p. 1117–1127.
48. Ozaki, T. and A.J.C. Nakagawara, Role of p53 in cell death and human cancers. 2011. **3**(1): p. 994–1013.
49. Nishioka, K., et al., Clinical significance of CDC25A and CDC25B expression in squamous cell carcinomas of the oesophagus. 2001. **85**(3): p. 412–421.
50. Gonda, A., et al., Internalization of exosomes through receptor-mediated endocytosis. 2019. **17**(2): p. 337–347.
51. Chinoy, N.J.C. and B. Science, Second messengers in signal transduction and cell signaling. 2010. **151**.
52. Basson, M.A.J.C.S.H.p.i.b., Signaling in cell differentiation and morphogenesis. 2012. **4**(6): p. a008151.
53. Stauber, T.J.B.C., The volume-regulated anion channel is formed by LRRC8 heteromers–molecular identification and roles in membrane transport and physiology. 2015. **396**(9–10): p. 975–990.
54. Caramia, M., et al., The volume-regulated anion channel in glioblastoma. 2019. **11**(3): p. 307.
55. Hoffmann, E.K. and I.H.J.P.T.O.T.R.S.B.B.S. Lambert, Ion channels and transporters in the development of drug resistance in cancer cells. 2014. **369**(1638): p. 20130109.
56. Xu, R., et al., LRRC8A is a promising prognostic biomarker and therapeutic target for pancreatic adenocarcinoma. 2022. **14**(22): p. 5526.
57. Zuccolini, P., et al. The VRAC blocker DCPIB opens Ca2+-activated BK channels and increases intracellular Ca2+ in melanoma and pancreatic duct carcinoma cell lines. 20th IUPAB Congress. 2021.
58. Qiu, Z., et al., SWELL1, a plasma membrane protein, is an essential component of volume-regulated anion channel. 2014. **157**(2): p. 447–458.
59. Liu, T. and T.J.I.J.O.M.S. Stauber, The volume-regulated anion channel LRRC8/VRAC is dispensable for cell proliferation and migration. 2019. **20**(11): p. 2663.
60. Konishi, T., et al., LRRC8A expression influences growth of esophageal squamous cell carcinoma. 2019. **189**(10): p. 1973–1985.
61. Lu, P., et al., SWELL1 promotes cell growth and metastasis of hepatocellular carcinoma in vitro and in vivo. 2019. **48**: p. 100–116.
62. Kumar, L., et al., Leucine-rich repeat containing 8A (LRRC8A) is essential for T lymphocyte development and function. 2014. **211**(5): p. 929–942.
63. Concepcion, A.R., et al., The volume-regulated anion channel LRRC8C suppresses T cell function by regulating cyclic dinucleotide transport and STING–p53 signaling. 2022. **23**(2): p. 287–302.

64. Shehzad, A., et al., Extracellular vesicles in cancer diagnostics and therapeutics. Pharmacology & Therapeutics, 2021. **223**: p. 107806.
65. Drabovich, A.P., et al., Toward an integrated pipeline for protein biomarker development. 2015. **1854**(6): p. 677–686.
66. Kosaka, N., et al., Exploiting the message from cancer: the diagnostic value of extracellular vesicles for clinical applications. 2019. **51**(3): p. 1–9.
67. André, F., et al., Exosomes as potent cell-free peptide-based vaccine. I. Dendritic cell-derived exosomes transfer functional MHC class I/peptide complexes to dendritic cells. 2004. **172**(4): p. 2126–2136.
68. Li, J., et al., GPC 1 exosome and its regulatory mi RNA s are specific markers for the detection and target therapy of colorectal cancer. 2017. **21**(5): p. 838–847.
69. Desai, S. and A.K. Guddati, Carcinoembryonic antigen, carbohydrate antigen 19-9, cancer antigen 125, prostate-specific antigen and other cancer markers: a primer on commonly used cancer markers. *World Journal of Oncology*, 2023. **14**(1): p. 4.
70. Yuan, A. and R.A. Nixon, Neurofilament proteins as biomarkers to monitor neurological diseases and the efficacy of therapies. *Frontiers in Neuroscience*, 2021. **15**: p. 689938.
71. Taylor, P., et al., A systematic review of serum biomarkers anti-cyclic citrullinated peptide and rheumatoid factor as tests for rheumatoid arthritis. *Autoimmune Diseases*, 2011. **2011**.
72. Bosello, S., et al., Cardiac troponin T and NT-proBNP as diagnostic and prognostic biomarkers of primary cardiac involvement and disease severity in systemic sclerosis: a prospective study. *European Journal of Internal Medicine*, 2019. **60**: p. 46–53.
73. Yousefpouran, S., et al., The assessment of selected MiRNAs profile in HIV, HBV, HCV, HIV/HCV, HIV/HBV Co-infection and elite controllers for determination of biomarker. *Microbial Pathogenesis*, 2020. **147**: p. 104355.
74. Sherwani, S.I., et al., Significance of HbA1c test in diagnosis and prognosis of diabetic patients. *Biomarker Insights*, 2016. **11**: p. BMI. S38440.
75. Janjua, K.A., R. Shahzad, and A. Shehzad, Development of novel cancer biomarkers for diagnosis and prognosis. In *Cancer biomarkers in diagnosis and therapeutics*. 2022: Springer. p. 277–343.
76. Lafitte, M., C. Lecointre, and S. Roche, Roles of exosomes in metastatic colorectal cancer. *American Journal of Physiology-Cell Physiology*, 2019. **317**(5): p. C869–C880.
77. Sandfeld-Paulsen, B., et al., Exosomal proteins as prognostic biomarkers in non-small cell lung cancer. *Molecular Oncology*, 2016. **10**(10): p. 1595–1602.
78. Tsai, C.-F., et al., Large-scale determination of absolute phosphorylation stoichiometries in human cells by motif-targeting quantitative proteomics. *Nature Communications*, 2015. **6**(1): p. 6622.
79. Huang, J., et al., Mechanism and function of circular RNA in regulating solid tumor radiosensitivity. *International Journal of Molecular Sciences*, 2022. **23**(18): p. 10444.
80. Xia, S., et al., CSCD: a database for cancer-specific circular RNAs. *Nucleic Acids Research*, 2018. **46**(D1): p. D925–D929.
81. Zhu, Y.-J., et al., Circular RNAs negatively regulate cancer stem cells by physically binding FMRP against CCAR1 complex in hepatocellular carcinoma. *Theranostics*, 2019. **9**(12): p. 3526.
82. Salzman, J., Circular RNA expression: its potential regulation and function. *Trends in Genetics*, 2016. **32**(5): p. 309–316.
83. Mai, S., Z. Zhang, and W. Mi, Upregulation of circ_PVT1 and circ_001569 indicate unfavorable prognosis in colorectal cancer. *Annals of Clinical & Laboratory Science*, 2021. **51**(1): p. 55–60.
84. Li, P.-F., et al., Non-coding RNAs and gastric cancer. *World Journal of Gastroenterology: WJG*, 2014. **20**(18): p. 5411.
85. Yang, Y., et al., Effects of MiR-194-5p on the proliferation and invasion of laryngeal cancer cells by regulating the expression of smurf1 and activating the mTOR signaling pathway. *Cellular and Molecular Biology*, 2022. **68**(6): p. 79–83.
86. Chen, H.-P., et al., Serological and viraemic status of human cytomegalovirus infection in patients with colorectal cancer is not correlated with viral replication and transcription in tumours. *Journal of General Virology*, 2016. **97**(1): p. 152–159.
87. Meng, Z., The theoretical study of DNA-small molecule interaction: a microscopic perspective. 2017.
88. Petry, K.U., HPV and cervical cancer. *Scandinavian Journal of Clinical and Laboratory Investigation*, 2014. **74**(sup244): p. 59–62.
89. Galati, L., et al., Detection of circulating HPV16 DNA as a biomarker for cervical cancer by a bead-based HPV genotyping assay. *Microbiology Spectrum*, 2022. **10**(2): p. e01480–1421.
90. Rodriguez-Ruiz, M.E., et al., Immunological impact of cell death signaling driven by radiation on the tumor microenvironment. *Nature Immunology*, 2020. **21**(2): p. 120–134.

91. Huang, H., et al., mTOR signaling at the crossroads of environmental signals and T-cell fate decisions. *Immunological Reviews*, 2020. **295**(1): p. 15–38.
92. Arger, N.K., S. Machiraju, and P.G. Woodruff, T-Bet expression in peripheral Th17. 0 cells is associated with pulmonary function changes in sarcoidosis. *Frontiers in Immunology*, 2020. **11**: p. 537289.
93. Zhang, H., et al., Imbalanced distribution of regulatory T cells and Th17. 1 cells in the peripheral blood and BALF of sarcoidosis patients: relationship to disease activity and the fibrotic radiographic phenotype. *Frontiers in Immunology*, 2023. **14**: p. 1185443.
94. Nowicki, A., et al., Ovarian cancer and cancer stem cells—cellular and molecular characteristics, signaling pathways, and usefulness as a diagnostic tool in medicine and oncology. *Cancers*, 2021. **13**(16): p. 4178.
95. Wilczyński, J.R., M. Wilczyński, and E. Paradowska, Cancer stem cells in ovarian cancer—a source of tumor success and a challenging target for novel therapies. *International Journal of Molecular Sciences*, 2022. **23**(5): p. 2496.
96. Jafarian, A.H., et al., Evaluation of CD133 and CD56/NCAM expression in Wilms tumor and their association with prognostic factors. *Iranian Journal of Basic Medical Sciences*, 2020. **23**(7): p. 853.
97. Lv, S., et al., Emerging role of interactions between tumor angiogenesis and cancer stem cells. *Journal of Controlled Release*, 2023. **360**: p. 468–481.
98. Ikram, D., et al., CD133 act as an essential marker in ovarian carcinogenesis. *Asian Pacific Journal of Cancer Prevention: APJCP*, 2021. **22**(11): p. 3525..
99. Liu, C.L., et al., Characteristics of CD133-sustained chemoresistant cancer stem-like cells in human ovarian carcinoma. *International Journal of Molecular Sciences*, 2020. **21**(18): p. 6467.
100. da Trindade, G.A., et al., Expression of a tumor stem cell marker (Aldehyde dehydrogenase 1-ALDH1) in benign epithelial odontogenic lesions. *Head and Neck Pathology*, 2022. **16**(3): p. 785–791.
101. Khalifa, S., et al., Aldehyde dehydrogenase 1A1 expression in ovarian epithelial tumors. *Journal of Obstetrics, Gynecology and Cancer Research*, 2022. **3**(1): p. 13–18.
102. Chang, B., et al., ALDH1 expression correlates with favorable prognosis in ovarian cancers. *Modern Pathology*, 2009. **22**(6): p. 817–823.
103. Huldani, H., et al., Mechanisms of cancer stem cells drug resistance and the pivotal role of HMGA2. *Pathology-Research and Practice*, 2022. **234**: p. 153906.
104. Ahmed, N., et al., Cancerous ovarian stem cells: obscure targets for therapy but relevant to chemoresistance. *Journal of Cellular Biochemistry*, 2013. **114**(1): p. 21–34.
105. Aguilar-Gallardo, C., et al., Overcoming challenges of ovarian cancer stem cells: novel therapeutic approaches. *Stem Cell Reviews and Reports*, 2012. **8**: p. 994–1010.
106. Ahmed, N. and K.L. Stenvers, Getting to know ovarian cancer ascites: opportunities for targeted therapy-based translational research. *Frontiers in Oncology*, 2013. **3**: p. 256.
107. Ahmed, N., K. Abubaker, and J.K. Findlay, Ovarian cancer stem cells: molecular concepts and relevance as therapeutic targets. *Molecular Aspects of Medicine*, 2014. **39**: p. 110–125.
108. Zeimet, A., et al., Ovarian cancer stem cells. *Neoplasma*, 2012. **59**(6): p. 747.
109. Martin, G.S., Cell signaling and cancer. *Cancer Cell*, 2003. **4**(3): p. 167–174.
110. Wolf, B., et al., Cellular signaling: aspects for tumor diagnosis and therapy. 2007: 164–168.
111. Bals, S., et al., Spontaneous and experimentally evoked [Ca2+] i-transients in cardiac myocytes measured by means of a fast Fura-2 technique. *Cell Calcium*, 1990. **11**(6): p. 385–396.
112. Hill, S.M., Receptor crosstalk: communication through cell signaling pathways. *The Anatomical Record: An Official Publication of the American Association of Anatomists*, 1998. **253**(2): p. 42–48.
113. Shajib, M.S., et al., Characterization of serotonin signaling components in patients with inflammatory bowel disease. *Journal of the Canadian Association of Gastroenterology*, 2019. **2**(3): p. 132–140.
114. Feig, M., J. Karanicolas, and C.L. Brooks III, MMTSB Tool Set: enhanced sampling and multiscale modeling methods for applications in structural biology. *Journal of Molecular Graphics and Modelling*, 2004. **22**(5): p. 377–395.
115. Cavada, B.S., et al., ConA-like lectins: high similarity proteins as models to study structure/biological activities relationships. *International Journal of Molecular Sciences*, 2018. **20**(1): p. 30.
116. Cheng, X. and J.C. Smith, Biological membrane organization and cellular signaling. *Chemical Reviews*, 2019. **119**(9): p. 5849–5880.
117. Tsien, R.Y., Fluorescent probes of cell signaling. *Annual Review of Neuroscience*, 1989. **12**(1): p. 227–253.
118. Chowdhury, S. and R.R. Sarkar, Comparison of human cell signaling pathway databases—evolution, drawbacks and challenges. *Database*, 2015. **2015**: p. 16.

119. Welf, E.S. and J.M. Haugh, Signaling pathways that control cell migration: models and analysis. *Wiley Interdisciplinary Reviews: Systems Biology and Medicine*, 2011. **3**(2): p. 231–240.
120. Smith, R.S.H., et al., Hybrid living materials: digital design and fabrication of 3D multimaterial structures with programmable biohybrid surfaces. *Advanced Functional Materials*, 2020. **30**(7): p. 1907401.
121. Sievers, J., et al., Customizing biohybrid cryogels to serve as ready-to-use delivery systems of signaling proteins. *Biomaterials*, 2021. **278**: p. 121170.
122. Li, C., et al., Highly robust and soft biohybrid mechanoluminescence for optical signaling and illumination. *Nature Communications*, 2022. **13**(1): p. 3914.
123. Fang, Y., Non-invasive optical biosensor for probing cell signaling. *Sensors*, 2007. **7**(10): p. 2316–2329.
124. Bouquerel, C., et al., Precise and fast control of the dissolved oxygen level for tumor-on-chip. *Lab on a Chip*, 2022. **22**(22): p. 4443–4455.
125. Johnson, G.T., et al., Building the next generation of virtual cells to understand cellular biology. *Biophysical Journal*, 2023. **122**(18): p. 3560–3569.
126. Arya, S.K., et al., Advances in complementary-metal–oxide–semiconductor-based integrated biosensor arrays. *Chemical Reviews*, 2015. **115**(11): p. 5116–5158.
127. Cho, Y., et al., Ion-selective carbon nanotube field-effect transistors for monitoring drug effects on nicotinic acetylcholine receptor activation in live cells. *Sensors*, 2020. **20**(13): p. 3680.

17 Signal Transduction and Targeted Therapies

Ayesha Khan, Ayesha Riaz, Asima Saeed, and Muhammad Imran Khan
Department of Biomedical Sciences, Pak Austria Fachhochschule: Institute of Applied Sciences and Technology, Haripur, Khyber Pakhtunkhwa, Pakistan

17.1 INTRODUCTION TO SIGNAL TRANSDUCTION

Signal transduction encompasses the myriad processes by which cells respond to external signals and translate them into appropriate cellular behaviors and responses. This sophisticated communication network is crucial for regulating various cellular activities, including proliferation, differentiation, migration, and apoptosis (1). Signal transduction involves the binding of ligands (such as hormones, growth factors, and neurotransmitters) to their respective receptors, triggering a cascade of biochemical events that ultimately lead to a specific cellular response (2). These signaling pathways are fundamental to an organism's normal physiological function and play a crucial role in disease pathology when dysregulated (3). Understanding the intricate mechanisms of signal transduction is crucial for developing targeted therapeutic interventions to correct the molecular imbalances underlying various disorders.

17.1.1 Historical Perspectives

The study of signal transduction has evolved significantly over the past few decades, from the initial discovery of hormone receptors to today's complex, multidimensional understanding of cellular signaling networks. In the early 20th century, the first insights into how cells communicate with their environment were gained, primarily through work on hormones and their effects on target tissues. This period saw the identification of neurotransmitters and the concept of chemical messengers. Only with the advent of molecular biology techniques in the latter half of the century could researchers delve deeper into the molecular basis of signal transduction. Key milestones include the discovery of G protein-coupled receptors (GPCRs) in the 1970s, the elucidation of tyrosine kinase receptors, and the mapping of intracellular signaling cascades such as the mitogen-activated protein kinase/extracellular signal-regulated kinase (MAPK/ERK) signaling pathway. These discoveries laid the foundation for the current era of targeted therapy, which aims to specifically modulate signaling pathways dysregulated in disease. The historical development of signal transduction research highlights the complexity of the field and the collaborative effort required to decipher the complex web of cellular communication.

17.2 KEY SIGNAL TRANSDUCTION PATHWAYS

17.2.1 Receptor Tyrosine Kinases (RTKs)

RTKs are an important class of cell surface receptors that play critical roles in various cellular processes, including growth, differentiation, migration, metabolism, and apoptosis (4). These enzymes are activated when their specific ligands bind, resulting in the phosphorylation of tyrosine residues

DOI: 10.1201/9781003594901-17

within the receptor itself and in downstream signaling molecules (5, 6). This activation triggers a cascade of signaling pathways that ultimately determine cell fate decisions (7).

17.2.1.1 Structure and Function of RTKs

RTKs are characterized by a ligand-binding extracellular domain, a single transmembrane helix, and an intracellular domain with tyrosine kinase activity (8). Ligand binding to the extracellular domain triggers dimerization or oligomerization of the receptors, activating the kinase domain. This leads to autophosphorylation and the subsequent recruitment of signaling proteins to the intracellular domain, initiating various signaling cascades (9).

17.2.1.2 Signaling Pathways and Cellular Responses

RTKs regulate several critical pathways, including the PI3K/Akt signaling pathway, which is involved in cell survival and growth, and the Ras/MAPK signaling pathway, which influences cell proliferation and differentiation. Dysregulation of RTK signaling is associated with numerous pathological conditions, including cancer (10). Aberrant activation of RTKs, whether through overexpression, mutation, or autocrine stimulation, can lead to uncontrolled cell proliferation, evasion of apoptosis, angiogenesis, and metastasis (11).

17.2.1.3 RTKs in Disease and Therapeutic Targeting

The significance of RTKs in oncogenesis has made them attractive targets for cancer therapy. Several therapeutic agents targeting RTKs have been developed and approved for clinical use. These include monoclonal antibodies that inhibit ligand binding and small molecule tyrosine kinase inhibitors (TKIs) that block the receptors' catalytic activity (12). The efficacy of these agents in treating various types of cancer highlights the importance of RTKs in disease pathology and the potential of targeted therapies. RTK's role in disease pathogenesis, particularly in cancer, underscores the importance of understanding RTK signaling pathways (13). Ongoing research and development of targeted therapeutic agents hold promise for improving treatment outcomes in diseases caused by RTK dysregulation. Future efforts will likely focus on overcoming therapeutic resistance, minimizing toxicity, and further refining the clinical application of RTK-targeted therapies (14). Despite their success, RTK-targeted therapies still face challenges such as drug resistance and toxicity (10). Resistance mechanisms include mutations in the kinase domain that prevent drug binding, activation of alternative signaling pathways, and adaptive responses that bypass the inhibited pathway (15). Addressing these challenges requires a better understanding of RTK signaling networks and the development of novel therapeutic strategies, such as combination therapies and targeting downstream effectors (16).

17.2.2 G Protein-Coupled Receptors

GPCRs form the largest and most diverse family of cell surface receptors, mediating cellular responses to a diverse array of external signals. These receptors are involved in numerous physiological processes, including sensory perception, immune responses, and neurotransmission, making them critical targets for pharmacological intervention (17). This section discusses current knowledge of GPCRs, focusing on their classification, structural characteristics, and physiological roles.

17.2.2.1 Classification and Types of GPCRs

GPCRs are classified into several major families based on their structural and functional properties: Class A (rhodopsin-like), Class B (secretin receptor family), Class C (metabotropic glutamate/pheromone), Class D (fungal mating pheromone receptors), Class E (cyclic AMP receptors), and Class F (Frizzled/Smoothened) (18). Class A is the largest and most studied, containing receptors for light-sensitive compounds, odors, pheromones, hormones, and neurotransmitters.

Class A (rhodopsin-like) receptors interact with a wide variety of ligands, from ions and small molecules to peptides and proteins. This versatility underlies their involvement in numerous signaling pathways (19).

Class B (secretin receptor family) receptors primarily bind to peptide hormones. They regulate metabolic processes, including insulin secretion and glucose homeostasis (20).

Class C (metabotropic glutamate/pheromone) receptors are distinguished by a large extracellular domain that binds ligands such as glutamate, facilitating neurotransmission and calcium signaling (21).

Classes D, E, and F include less studied receptors involved in developmental processes and signal transduction pathways specific to fungi (Class D), cAMP signaling pathways (Class E), and Wnt signaling (Class F) (22).

17.2.2.2 Structural Characteristics

GPCRs share a common structural motif of seven transmembrane α-helices connected by intra- and extracellular loops. The extracellular loops and N-terminus region typically recognize and bind ligands, while the intracellular loops interact with G proteins to trigger intracellular signaling cascades. This framework's structural diversity enables the high specificity and selectivity of ligand binding and signal transduction observed among the different GPCR families (23).

17.2.2.3 Physiological Roles and Significance

GPCRs' physiological importance cannot be overemphasized, given their involvement in almost every biological process. They are critical for the cardiovascular system, the central nervous system, and the immune system, among others. GPCRs are implicated in the pathophysiology of numerous diseases, including neurological disorders, cancer, diabetes, and heart disease, making them a focus of medical research and drug development (24). In fact, a significant proportion of drugs currently on the market target GPCRs, highlighting their value as therapeutic targets (25).

17.2.3 Notch Signaling Pathway

The Notch signaling pathway is a highly conserved cell communication system that plays a crucial role in regulating cell fate determination, differentiation, proliferation, and apoptosis (26). It is instrumental in the development and maintenance of various tissues in multicellular organisms and contributes to numerous diseases, including cancer when dysregulated (Artavanis-Tsakonas, Rand, & Lake, 1999). This section delves into the mechanism of the Notch signaling pathway, highlighting key studies that have contributed to our current understanding (27).

17.2.3.1 Mechanism of Notch Signaling

The Notch signaling pathway is initiated when a Notch ligand, on the surface of a signaling cell, binds to the Notch receptor on a neighboring receiving cell. This interaction leads to two proteolytic cleavages of the Notch receptor. An ADAM family metalloprotease mediates the first cleavage, and the γ-secretase complex performs the second, releasing the Notch intracellular domain (NICD) (27). The NICD then translocates to the nucleus, where it associates with the DNA-binding protein RBP-J (also known as CSL or Suppressor of Hairless). This association converts RBP-J from a repressor to an activator of transcription, leading to the expression of Notch target genes such as Hes and Hey family genes (28).

17.2.3.2 Notch Ligands, Receptors, and Signaling Regulation

There are multiple Notch receptors (NOTCH1-4) and ligands, classified into two families: Delta-like (DLL1, DLL3, DLL4) and Jagged (JAG1, JAG2), each exhibiting distinct but overlapping patterns of expression and function (29). The specificity and outcome of Notch signaling depend on the specific ligand-receptor interaction, which is further modulated by cellular context and fringe glycosyltransferases that modify the receptors (30). Notch signaling is finely regulated at multiple levels, including ligand-receptor interactions, post-translational modifications of the Notch receptor, and nuclear events affecting target gene transcription (31). Endocytosis plays a vital role in regulating Notch signaling, as the ubiquitination of ligands and receptors dictates their internalization and degradation, thus modulating signal intensity (32). The dynamics of NICD in the nucleus, its interaction with co-activators and co-repressors, and chromatin modifiers are also crucial for the pathway's transcriptional output (33).

17.2.3.3 Functional Implications and Clinical Significance

The Notch signaling pathway influences a wide array of developmental processes and diseases. Its role in cell fate decisions makes it critical for the development of tissues and organs (34). Dysregulation of Notch signaling contributes to various pathologies, including T-cell acute lymphoblastic leukemia (T-ALL), where activating mutations of NOTCH1 are prevalent (35). Notch signaling also maintains stem cells and promotes cancer stem cell development, presenting potential targets for therapeutic intervention (36).

17.2.4 Wnt/β-Catenin Signaling

The Wnt/β-Catenin signaling pathway is highly conserved across species and plays a critical role in embryonic development and adult tissue homeostasis (37). It is initiated when Wnt proteins bind to the Frizzled (Fz) family of receptors and the low-density lipoprotein receptor-related protein 5/6 (LRP5/6) co-receptors on the cell surface, activating Dishevelled (Dvl) proteins (38). Without Wnt signaling, destruction complex Axin, adenomatous polyposis coli (APC), glycogen synthase kinase 3 (GSK-3β), and casein kinase 1 (CK1) continuously mark β-catenin for degradation. GSK-3β and CK1 phosphorylate β-catenin, triggering its ubiquitination and subsequent proteasomal degradation (39). Wnt binding inhibits the destruction complex, stabilizing and accumulating β-catenin in the cytoplasm before it translocates into the nucleus (40). There, β-catenin acts as a co-activator for T-cell factor/lymphoid enhancer factor (TCF/LEF) transcription factors, leading to the transcription of Wnt target genes that regulate cell fate, proliferation, and migration (41).

17.2.4.1 Regulation and Pathological Implications

The regulation of Wnt/β-Catenin signaling is multifaceted, involving several feedback mechanisms and crosstalk with other signaling pathways. For example, the Wnt pathway can be negatively regulated by the action of secreted Frizzled-related proteins (sFRPs), which bind to Wnt proteins and prevent their interaction with Frizzled receptors (42). The pathway also interacts significantly with other signaling cascades, such as the Notch and Hedgehog pathways, allowing for intricate control of cellular processes. Dysregulation of Wnt/β-Catenin signaling contributes to a variety of diseases, most notably cancer, where mutations in pathway components, such as APC or β-catenin, aberrantly activate the pathway and promote tumorigenesis (43). In colorectal cancer, mutations in the APC gene commonly result in constitutive activation of Wnt signaling and uncontrolled cell proliferation (44).

17.2.5 Hedgehog Signaling Pathway

The Hedgehog (Hh) signaling pathway is a crucial regulator of cell growth, differentiation, and tissue patterning during embryonic development and maintains stem cells in adults. Dysregulation of this pathway occurs in various cancers and developmental disorders, making it a focal point of research in developmental biology and oncology (43).

17.2.5.1 Signaling Mechanism

The Hh pathway is initiated when the Hh ligand binds to its receptor Patched (PTCH1), relieving PTCH1's repression of Smoothened (SMO), a G protein-coupled receptor-like protein. In the absence of Hh, PTCH1 inhibits SMO, preventing the activation of the downstream signaling cascade (45). Hh binding to PTCH1 allows SMO to accumulate and transduce the signal to the cytoplasm, activating the GLI family of transcription factors (45). In the cytoplasm, multiple kinases, including PKA, GSK3β, and CK1, regulate GLI proteins through complex processes, promoting their phosphorylation, which leads to partial degradation and the formation of GLI repressor forms (46). However, an active Hh signal counteracts these modifications, resulting in full-length GLI activators that translocate to the nucleus and induce the expression of Hh target genes (47). The primary cilium, a microtubule-based organelle projecting from the cell surface, plays a critical role in Hh pathway regulation (48). Components of the Hh signaling machinery, including PTCH1, SMO, and GLI proteins, dynamically localize to the primary cilium, and their correct localization is essential for the proper Hh signal transmission and interpretation (49).

17.2.5.2 Significance and Implications

The Hh pathway's role in regulating developmental processes and maintaining tissue homeostasis underscores its importance in biology and medicine. Its involvement in basal cell carcinoma, medulloblastoma, and other cancers has sparked interest in developing inhibitors targeting various components of the pathway, particularly SMO and GLI proteins (Yauch et al., 2008). These therapeutic strategies aim to modulate aberrant Hh signaling in tumors while minimizing the impact on the pathway's normal function in adult tissues.

17.2.6 MAPK/ERK Pathway

The MAPK/ERK pathway is a crucial signaling cascade that regulates various cellular processes, including proliferation, differentiation, and survival. A range of extracellular stimuli, such as growth factors, cytokines, and environmental stressors, activate this pathway, initiating a phosphorylation cascade involving several key proteins (50).

17.2.6.1 Mechanism of the MAPK/ERK Pathway

The MAPK/ERK pathway begins with the activation of the Ras protein, a small GTPase that acts as a molecular switch. Activated Ras facilitates the recruitment and activation of the RAF kinase family members (ARAF, BRAF, and CRAF), which in turn phosphorylate and activate MEK1 and MEK2 (51). These MEK kinases specifically phosphorylate ERK1 and ERK2 on both tyrosine and threonine residues, fully activating them. Activated ERKs translocate to the nucleus, where they phosphorylate various transcription factors, thus modulating gene expression patterns that affect cell fate decisions (52). Scaffold proteins, which bind to specific pathway components, guide the spatial and temporal aspects of signaling, achieving MAPK/ERK signaling specificity (53). This specificity is crucial for differential cellular responses to various stimuli.

17.2.6.2 Biological Functions and Clinical Implications

The MAPK/ERK pathway plays a pivotal role in the control of cell cycle progression and contributes to the pathogenesis of numerous cancers (54). Mutations in components of the MAPK/ERK pathway, especially in the BRAF and Ras genes, occur in a significant proportion of human tumors, causing constitutive activation of the pathway and uncontrolled cell proliferation (55). Given its central role in cell proliferation and survival, researchers have targeted the MAPK/ERK pathway for cancer therapy. Inhibitors targeting MEK and BRAF have shown promise in clinical trials, particularly for melanoma patients with specific BRAF mutations (55). However, resistance to these targeted therapies often develops, necessitating further research into combination therapies and novel targets within the pathway.

17.2.7 PI3K/Akt/mTOR Pathway

The PI3K/Akt/mTOR signaling cascade begins when growth factors bind to RTKs, activating PI3K. Activated PI3K catalyzes the conversion of phosphatidylinositol 4,5-bisphosphate (PIP2) to phosphatidylinositol 3,4,5-trisphosphate (PIP3) (56). The tumor suppressor phosphatase and tensin homolog (PTEN) counteracts this reaction by dephosphorylating PIP3 back to PIP2, serving as a critical negative regulator of the pathway (57). PIP3 serves as a docking site for Akt, also known as protein kinase B (PKB), facilitating its translocation to the plasma membrane where phosphoinositide-dependent kinase-1 (PDK1) and the mammalian target of rapamycin complex 2 (mTORC2) phosphorylate and activate it (58). Activated Akt phosphorylates various downstream targets, including the TSC1/2 (tuberous sclerosis complex) protein complex, leading to activating mTOR complex 1 (mTORC1) (59). mTORC1 regulates several key cellular processes by phosphorylating targets such as S6 kinase (S6K1) and the eukaryotic translation initiation factor 4E-binding protein 1 (4E-BP1), which participate in protein synthesis, ribosome biogenesis, and nutrient transport. mTORC1 activation enhances cellular growth and proliferation, while its inhibition leads to autophagy, a process of cellular catabolism (60).

17.2.7.1 Therapeutic Implications

PI3K/Akt/mTOR pathway dysregulation frequently occurs in cancer, where mutations in PI3K, PTEN loss, and constitutive activation of Akt and mTOR contribute to malignant transformation, tumor growth, angiogenesis, and resistance to apoptosis (61). Given its central role in tumorigenesis, researchers have focused on the PI3K/Akt/mTOR to develop targeted cancer therapies. Several inhibitors targeting different pathway components are currently under investigation or in clinical use, offering new avenues for cancer treatment (62).

17.2.8 JAK/STAT Pathway

The Janus kinase (JAK)/signal transducer and activator of transcription (STAT) pathway is a cornerstone of cellular signaling, pivotal in mediating responses to cytokines and growth factors across a wide array of biological processes, including immune regulation, hematopoiesis, and cell growth (63). This section delves into the JAK/STAT pathway mechanism and its biological implications, drawing upon a wealth of research studies and reviews in this field.

17.2.8.1 Mechanism of the JAK/STAT Pathway

The JAK/STAT signaling cascade initiates when extracellular ligands, such as cytokines or growth factors, bind to their cognate cell surface receptors, typically type I and type II cytokine receptors, which lack intrinsic kinase activity (64). This binding induces a conformational change in the receptor, bringing together two JAK family kinases associated with the receptor's intracellular domain. The JAK kinases, including JAK1, JAK2, JAK3, and Tyk2, then phosphorylate each other on specific tyrosine residues, a process known as trans-phosphorylation, thereby becoming fully activated (65). Upon activation, JAKs phosphorylate tyrosine residues on the receptor tails, creating docking sites for STAT proteins. JAKS then phosphorylate STATs at the receptor on a critical tyrosine residue. Phosphorylated STATs dimerize through reciprocal Src homology 2 (SH2) domain-phosphotyrosine interactions, enabling their nuclear translocation (66). In the nucleus, STAT dimers bind to specific DNA response elements, activating the transcription of target genes involved in cell proliferation, differentiation, apoptosis, and immune responses (67). The termination and modulation of JAK/STAT signaling are equally critical for maintaining cellular homeostasis. Suppressors of cytokine signaling (SOCS) proteins, protein inhibitors of activated STATs (PIAS), and tyrosine phosphatases such as SHP1/2 provide feedback inhibition, preventing prolonged or inappropriate activation of the pathway (68).

17.2.8.2 Biological Implications and Clinical Relevance

Dysregulation of the JAK/STAT pathway occurs in a wide range of diseases, including cancers, autoimmune conditions, and inflammatory disorders, highlighting its importance in maintaining normal cellular function and its potential as a target for therapeutic interventions (69). For instance, constitutive activation of the JAK/STAT pathway characterizes myeloproliferative neoplasms and certain leukemias, often driven by mutations in JAK2 (70). This has led to the development and FDA approval of JAK inhibitors, such as ruxolitinib, for treating myelofibrosis and polycythemia vera (71). Moreover, the JAK/STAT pathway's role in immune regulation has spurred interest in targeting this pathway for autoimmune and inflammatory diseases. JAK inhibition offers a means to modulate the immune response, providing therapeutic benefits in conditions such as rheumatoid arthritis and psoriasis (72).

17.3 MECHANISMS OF DYSREGULATION IN DISEASE

17.3.1 Genetic Alterations Leading to Dysregulation

Genetic alterations play a pivotal role in the dysregulation of cellular processes, contributing significantly to the pathogenesis of many diseases. These alterations can occur at various levels, from mutations in individual genes to chromosomal abnormalities, and can affect the function of critical proteins involved in signaling pathways, cellular growth, and apoptosis (73). This section explores how genetic alterations lead to cellular dysregulation, focusing on mutations, gene amplifications, deletions, and epigenetic modifications.

17.3.1.1 Mutations

Mutations, which are changes in a gene's DNA, are among the most direct ways that genetic alterations can lead to disease. These mutations can be inherited or acquired and can result in malfunctioning protein production or loss of protein expression (74). For instance, point mutations in the proto-oncogene BRAF occur in various cancers, including melanoma, where they lead to constitutive activation of the MAPK signaling pathway, promoting uncontrolled cell proliferation (75).

17.3.1.2 Gene Amplifications and Deletions

Gene amplifications and deletions alter gene dosage within the genome, leading to the overexpression or complete absence of certain proteins. Amplification of the HER2 gene, which encodes a receptor tyrosine kinase, occurs in approximately 20% of breast cancers and results in increased signaling through pathways that drive cell division and survival, contributing to tumor growth and progression (76).

17.3.1.3 Chromosomal Abnormalities

Chromosomal abnormalities, including translocations and inversions, can disrupt gene function by altering gene structure or expression. The Philadelphia chromosome, a reciprocal translocation between chromosomes 9 and 22, exemplifies this, resulting in the BCR-ABL fusion gene. The BCR-ABL protein has constitutive tyrosine kinase activity, leading to the activation of multiple signaling pathways and the development of chronic myeloid leukemia (CML) (77).

17.3.1.4 Epigenetic Modifications

Epigenetic modifications, such as DNA methylation and histone modification, can also lead to the dysregulation of gene expression without altering the DNA sequence. Aberrant methylation of promoter regions can silence tumor suppressor genes, as seen in the hypermethylation of the CDKN2A promoter in various cancers, which reduces expression of the p16INK4a tumor suppressor and contributes to cell cycle dysregulation (78).

17.3.1.5 Implications for Therapy and Research

Understanding disease-causing genetic alterations is crucial for developing targeted therapies. By identifying specific mutations or alterations that drive disease progression, researchers can design drugs that specifically target these aberrant proteins or pathways. For example, identifying BCR-ABL in CML led to the development of imatinib, a tyrosine kinase inhibitor that specifically targets the BCR-ABL protein, revolutionizing CML treatment (79).

In conclusion, genetic alterations are a fundamental mechanism leading to the dysregulation of cellular processes and the development of disease. Studying these alterations not only illuminates disease pathogenesis but also enables the development of targeted therapeutic strategies, offering hope for more effective and personalized treatments (Figure 17.1; Table 17.1).

17.3.2 Signal Transduction in Cancer

Cancer is characterized by uncontrolled cell growth and proliferation resulting from dysregulated signal transduction pathways (80). These pathways, under normal physiological conditions, govern cell growth, differentiation, survival, and apoptosis. However, in cancer, mutations and aberrant expression of signaling components lead to continuous activation of these pathways, driving tumorigenesis and cancer progression. This section explores how signal transduction contributes to cancer, focusing on key pathways commonly implicated in the disease (81).

17.3.2.1 Oncogenes and Tumor Suppressors

Cancer often develops when mutations occur in genes encoding components of signaling pathways (82). These genes can be broadly classified into two categories: oncogenes and tumor suppressor genes. Oncogenes, when mutated or overexpressed, lead to the activation of signaling pathways that promote cell proliferation and survival. Examples include the Ras family of GTPases and the MYC transcription factor, which are involved in numerous signaling pathways, including the MAPK and PI3K/Akt pathways, respectively (83). Conversely, tumor suppressor genes, when inactivated by mutations, result in the loss of negative regulation of these pathways, further contributing to cancer development. The TP53 gene, encoding the p53 protein, is a well-known tumor suppressor that regulates the cell cycle and apoptosis in response to DNA damage and other stress signals (84).

17.3.2.2 Key Signaling Pathways in Cancer

17.3.2.2.1 MAPK/ERK Pathway

The MAPK/ERK pathway plays a critical role in cell proliferation and differentiation (80). In many cancers, mutations in components of this pathway, such as BRAF, KRAS, and NRAS, lead to its

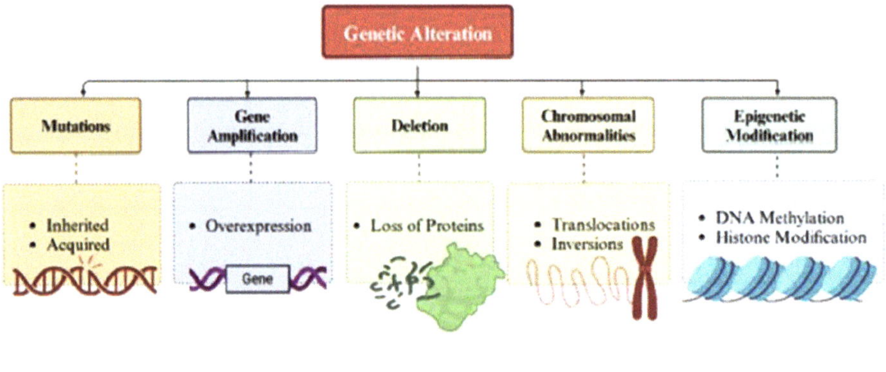

FIGURE 17.1 Genetic alterations in disease.

TABLE 17.1.
Types of Genetic Alterations Leading to Dysregulation in Disease

Type of Genetic Alteration	Description	Example Diseases	Example Gene(s)	Implications
Mutations	Changes in a gene's DNA sequence, leading to malfunctioning or loss of protein expression	Various cancers, including melanoma	BRAF	Constitutive activation of signaling pathways promoting uncontrolled cell proliferation
Gene Amplifications	Increase in the number of copies of a gene, leading to overexpression of the protein	Breast cancer	HER2	Increased signaling through pathways driving cell division and survival, contributing to tumor growth
Gene Deletions	Loss of gene segments, leading to the absence of certain proteins	–	–	Can result in the loss of tumor suppressor genes, contributing to tumorigenesis
Chromosomal Abnormalities	Structural changes in chromosomes, such as translocations and inversions, altering gene structure or expression	Chronic myeloid leukemia (CML)	BCR-ABL (Philadelphia chromosome)	Production of fusion proteins with constitutive tyrosine kinase activity, activating multiple signaling pathways
Epigenetic Modifications	Changes in gene expression without altering the DNA sequence, through mechanisms like DNA methylation and histone modification	Various cancers	CDKN2A (p16^INK4a)	Silencing of tumor suppressor genes, contributing to cell cycle dysregulation

constitutive activation, promoting oncogenic transformation and tumor growth. Researchers have developed targeted therapies, such as BRAF and MEK inhibitors, to block this pathway in cancers with specific mutations (85).

17.3.2.2.2 PI3K/AKT/mTOR Pathway

The phosphatidylinositol 3-kinase (PI3K)/AKT/mammalian target of rapamycin (mTOR) pathway is pivotal in controlling cell survival and growth (86). Activating mutations in PIK3CA, which encodes the p110α catalytic subunit of PI3K, or inactivating mutations and deletions of PTEN, a tumor suppressor that inhibits PI3K, are common in various cancers. This leads to the activation of Akt and subsequent stimulation of mTOR, promoting protein synthesis and cell growth. Inhibitors targeting PI3K, Akt, and mTOR are currently being used or evaluated in clinical trials (87).

17.3.2.2.3 JAK/STAT Pathway

The JAK/STAT pathway is essential for the response to cytokines and growth factors (88). Its dysregulation, through activating mutations in JAK or constitutive activation of STAT proteins, contributes to cancer cell proliferation and survival, particularly in hematological malignancies (89). JAK inhibitors, such as ruxolitinib, have shown efficacy in treating myeloproliferative neoplasms (90).

17.3.2.3 Crosstalk and Redundancy

A hallmark of cancer signal transduction is the extensive crosstalk and redundancy among pathways. This interconnectivity allows cancer cells to bypass targeted therapies aimed at a single pathway component, leading to drug resistance (91). For example, inhibition of the MAPK pathway can lead to compensatory activation of the PI3K pathway, sustaining cancer cell growth despite treatment. Understanding these complex interactions among signaling pathways is crucial for developing more effective combination therapies (83).

17.3.3 Signal Transduction in Neurodegenerative Diseases

Neurodegenerative diseases, such as Alzheimer's disease, Parkinson's disease (PD), amyotrophic lateral sclerosis (ALS), and Huntington's disease (HD), are characterized by the progressive loss of neuronal structure or function, including neuronal death (92). Various signal transduction pathways contribute to the pathogenesis of these diseases, highlighting the complexity of cellular mechanisms underlying neurodegeneration (93). This section explores the current understanding of signal transduction in neurodegenerative diseases, focusing on key pathways, their dysregulation, and the implications for therapeutic strategies (94).

17.3.3.1 Alzheimer's Disease and the Role of Amyloid Precursor Protein Signaling

In Alzheimer's disease, the amyloid precursor protein (APP) and its cleavage products, including amyloid-beta (Aβ) peptides, play a crucial role in disease pathogenesis (95). Aβ accumulation triggers a cascade of events leading to neuronal dysfunction and death, partly through dysregulation of signal transduction pathways. Studies have shown that Aβ interacts with various cell surface receptors, such as the cellular prion protein (PrP^C), to activate intracellular kinases like Fyn, subsequently impairing synaptic function (96). Additionally, Aβ oligomers can dysregulate calcium homeostasis, further affecting signaling pathways involved in cell survival and apoptosis (97).

17.3.3.2 Parkinson's Disease and Mitogen-Activated Protein Kinase Pathways

Parkinson's disease involves the loss of dopaminergic neurons in the substantia nigra pars compacta and the accumulation of Lewy bodies, primarily composed of α-synuclein (98). The MAPK pathways, including ERK, c-Jun N-terminal kinase (JNK), and p38 MAPK, contribute to PD pathogenesis (99). Oxidative stress and mitochondrial dysfunction, common in PD, can activate these pathways, leading to neuronal apoptosis. Specifically, α-synuclein aggregates can activate MAPK pathways, contributing to neuroinflammation and neuronal death (100).

17.3.3.3 Amyotrophic Lateral Sclerosis and the PI3K/Akt Pathway

The PI3K/Akt signaling pathway is critical for cell survival, and its dysregulation has been associated with the pathogenesis of amyotrophic lateral sclerosis (101). ALS is characterized by the progressive degeneration of motor neurons, leading to muscle weakness and atrophy. Mutations in superoxide dismutase 1 (SOD1), TAR DNA-binding protein (TARDBP or TDP-43), and C9orf72 are among the genetic alterations linked to ALS. These genetic changes can disrupt PI3K/Akt signaling, reducing neuronal survival and increasing susceptibility to apoptosis (102).

17.3.3.4 Huntington's Disease and Neurotrophic Factor Signaling

Huntington's disease results from the expansion of Cytosine-Adenine-Guanine (CAG) repeats in the Huntingtin (HTT) gene, producing mutant huntingtin protein (mHTT) (103). mHTT interferes with brain-derived neurotrophic factor (BDNF) signaling, a critical growth factor for neuronal survival and function. The disruption of BDNF signaling pathways, particularly those involving the receptor TrkB and downstream effectors like ERK and Akt, contributes to neuronal dysfunction and death in HD (104).

17.3.4 Signal Transduction in Cardiovascular Diseases

Cardiovascular diseases (CVDs) are the leading cause of mortality worldwide, underpinned by diverse etiologies yet often sharing common pathological pathways involving aberrant signal transduction. This section focuses on the critical role of signal transduction mechanisms in the pathogenesis of CVDs, including hypertension, atherosclerosis, heart failure, and myocardial infarction, highlighting advances in understanding that have paved the way for novel therapeutic targets.

17.3.4.1 Signal Transduction Pathways in Cardiovascular Health and Disease

Signal transduction maintains cardiovascular homeostasis, mediating the effects of neurohormonal signaling on vascular tone, cardiac contractility, and cell survival. Dysregulation of these signaling pathways can lead to maladaptive responses, contributing to CVD development and progression.

a) **G protein-coupled receptors in hypertension and heart failure**: GPCRs, including β-adrenergic receptors (β-ARs), angiotensin II type 1 receptors (AT1Rs), and endothelin receptors, are extensively involved in cardiovascular regulation (105). Overstimulation of β-ARs and AT1Rs has been linked to hypertension and heart failure, promoting vasoconstriction, sodium retention, and adverse cardiac remodeling (106). Conversely, stimulation of β-ARs can also activate cardioprotective signaling pathways, highlighting the complexity of GPCR-mediated signal transduction in the cardiovascular system (107).
b) **The role of RTKs in atherosclerosis**: RTKs, including vascular endothelial growth factor (VEGF) receptors and platelet-derived growth factor (PDGF) receptors, play critical roles in endothelial function and vascular remodeling (108). Dysregulation of RTK signaling can contribute to atherosclerosis by promoting endothelial dysfunction, smooth muscle cell proliferation, and migration, as well as inflammation (109).
c) **Nuclear factor kappa b (NF-κB) in inflammation and CVD**: NF-κB is a key regulator of inflammatory responses, and its activation contributes to the pathogenesis of atherosclerosis and myocardial infarction (110). Activation of NF-κB in endothelial cells, smooth muscle cells, and macrophages leads to the expression of pro-inflammatory cytokines and adhesion molecules, contributing to the inflammatory milieu of atherosclerotic plaques (111).
d) **MAPKs in cardiac hypertrophy and heart failure**: The MAPK family, including ERK1/2, JNK, and p38 MAPK, is involved in cellular stress responses. Activation of these kinases can lead to cardiac hypertrophy and heart failure, with ERK1/2 activation associated with compensatory hypertrophy, while JNK and p38 MAPK are linked to maladaptive remodeling and apoptosis (112).

17.3.4.2 Therapeutic Implications

Understanding CVD signal transduction pathways has not only enhanced our understanding of disease mechanisms but also identified potential therapeutic targets. Angiotensin receptor blockers (ARBs) inhibit AT1Rs to treat hypertension, while beta-blockers blockade β-ARs for heart failure, both modulating deleterious signaling pathways. Researchers are investigating emerging therapies targeting RTKs and MAPKs, offering hope for more precise interventions in atherosclerosis and heart failure (113). However, challenges remain, including achieving specificity in targeting signaling molecules to minimize off-target effects and the complexity of signaling networks that may compensate for the inhibited pathway. Future research should focus on unraveling the intricate web of signal transduction in CVD, emphasizing biomarker identification for personalized therapy and exploring combination treatments to target multiple signaling pathways simultaneously.

17.4 TARGETED THERAPIES: PRINCIPLES AND PRACTICE

17.4.1 Rationale for Targeted Therapies

Targeted therapies have transformed disease management and treatment, particularly for cancer. Unlike traditional cytotoxic agents that indiscriminately affect rapidly dividing cells, targeted therapies aim to selectively interfere with specific molecular targets that are involved in the growth, progression, and spread of diseases (114). This section explores the rationale behind targeted therapies, focusing on their development, mechanism of action, and impact on disease treatment.

17.4.1.1 Development and Evolution

Molecular biology and genomics spurred the development of targeted therapies, revealing the complex signaling networks driving disease pathogenesis. Identifying oncogenes, tumor suppressor genes, and other cellular pathways dysregulated in cancer and other diseases has provided a plethora of molecular targets for therapeutic intervention. The rationale for targeted therapy lies in its precision; specifically targeting these molecular aberrations potentially offers higher efficacy and lower toxicity than conventional therapies (115).

17.4.1.2 Mechanism of Action

Targeted therapies work by interfering with specific molecules or pathways essential for tumor growth and survival. These can be broadly classified into two categories: small molecule inhibitors and monoclonal antibodies. Small molecule inhibitors easily enter cells and disrupt target protein function, such as tyrosine kinases, within the cell. Monoclonal antibodies, being larger, cannot easily enter cells but can target cell surface antigens or receptors, blocking growth-promoting signals from the external environment or marking cancer cells for destruction by the immune system (Goldberg, 2015).

17.4.1.3 Clinical Impact

The clinical implementation of targeted therapies has significantly improved patient outcomes in various diseases. In cancer, targeted therapies have revolutionized treatment for several malignancies. For example, discovering the BCR-ABL fusion protein in chronic myeloid leukemia led to the development of imatinib, a tyrosine kinase inhibitor, transforming CML from a fatal disease to a manageable condition (116). Similarly, trastuzumab, a monoclonal antibody targeting HER2, has significantly improved survival rates for HER2-positive breast cancer patients (117).

Despite their success, targeted therapies face several challenges, including the development of resistance, identifying appropriate biomarkers for patient selection, and the high cost of therapy. Resistance can arise through various mechanisms, such as mutations in the target protein, activation of alternative signaling pathways, or phenotypic changes in tumor cells (118). Addressing these challenges requires ongoing research into understanding the complex biology of diseases, developing combination therapies, and using comprehensive genomic profiling to guide treatment decisions. The rationale for targeted therapies is rooted in the desire to achieve more effective and less toxic treatments by exploiting specific molecular characteristics of diseases. The success of targeted therapies in improving patient outcomes underscores precision medicine's importance. However, realizing the full potential of these therapies requires continued research, addressing the challenges of resistance, patient selection, and cost, and exploring innovative strategies to further enhance their efficacy and accessibility.

17.4.2 Design and Development of Targeted Therapeutic Agents

The development of targeted therapeutic agents represents a significant advance in the treatment of various diseases, particularly cancer, chronic inflammatory conditions, and certain genetic disorders. This shift from traditional, non-specific therapies to targeted approaches has been underpinned by a profound understanding of the molecular and cellular mechanisms of diseases. This section explores the methodologies, challenges, and progress in designing and developing these innovative treatments.

17.4.2.1 Methodologies in Target Identification and Validation

The inception of a targeted therapeutic strategy begins with identifying and validating a target that plays a critical role in disease pathogenesis. This process involves comprehensively exploring the disease's molecular landscape through genomic, proteomic, and bioinformatics analyses. High-throughput screening (HTS) and genome-wide association studies (GWAS) have been instrumental in identifying potential targets by revealing genetic mutations, aberrant signaling pathways, and dysregulated proteins associated with diseases (119).

Following identification, the validation of these targets is crucial to ascertain their relevance and therapeutic potential. Techniques such as RNA interference (RNAi), CRISPR-Cas9 gene editing, and conditional knockout models elucidate the target's function and its impact on disease pathology (120).

17.4.2.2 Drug Design and Optimization

Once validated, the next step involves designing agents that can modulate the target's activity. This phase often employs structure-based drug design (SBDD) and computer-aided drug design (CADD) techniques, leveraging the target's three-dimensional structure to identify and optimize small molecules or biologics that can interact with it effectively (121). Medicinal chemistry and iterative cycles of synthesis and testing refine the efficacy, specificity, stability, and bioavailability of these agents (Anderson, 2003). Biologics, including monoclonal antibodies, peptide-based drugs, and gene therapies, are targeted agents designed to interact with targets with high specificity. Advances in recombinant DNA technology and antibody engineering have facilitated the development of these complex molecules, which can inhibit or modulate their targets' function with precision (122).

17.4.2.3 Preclinical and Clinical Development

Preclinical testing in cell-based assays and animal models evaluates the therapeutic agent's safety, efficacy, and pharmacokinetic profile. These studies provide critical data for designing clinical trials and obtaining regulatory approval for human studies. Clinical development proceeds through phases I to III, assessing the drug's safety, dosage, efficacy, and side effects in progressively larger patient cohorts (123). This stage demonstrates the therapeutic benefit over existing treatments and reveals the drug's risk-benefit profile in the targeted patient population. Despite successes, the development of targeted therapies faces challenges, including complex disease pathogenesis, interpatient variability, drug resistance, and the high cost of drug development. Overcoming these hurdles requires innovative drug design approaches, such as multi-target drugs, drug combinations to prevent resistance, and personalized medicine strategies that tailor therapies based on the individuals' genetic makeup (124).

17.4.3 Monoclonal Antibodies

Monoclonal antibodies (mAbs) are revolutionary therapeutic agents that have transformed the treatment landscape for a myriad of diseases, including cancers, autoimmune disorders, and infectious diseases. This section examines the development, mechanisms of action, and therapeutic applications of monoclonal antibodies, underscoring their significance in modern medicine.

17.4.3.1 Development of Monoclonal Antibodies

The advent of monoclonal antibodies can be traced back to Köhler and Milstein's seminal work in 1975 developing hybridoma technology, a method enabling the production of monoclonal antibodies (125). This groundbreaking innovation earned the Nobel Prize in Physiology or Medicine in 1984 and laid the foundation for the therapeutic application of monoclonal antibodies (125).

17.4.3.2 Mechanisms of Action

Monoclonal antibodies (mAbs) are a cornerstone of targeted therapy, leveraging various mechanisms to exert their therapeutic effects. They bind with unparalleled specificity to target antigens, which include proteins on cancer cell surfaces or molecules entangled in disease pathways. This precision targeting directly attacks diseased cells while minimizing collateral damage to healthy tissues. Additionally, mAbs can neutralize disease-promoting ligands or receptors, such as tumor necrosis factor-alpha (TNF-α) in autoimmune diseases, effectively halting the disorder's progression (126). Some mAbs use antibody-dependent cellular cytotoxicity (ADCC), enlisting the help of immune cells to annihilate targeted cells, a strategy that has proven pivotal in cancer treatment (127). Another pathway through which mAbs deliver their therapeutic payload is complement-dependent cytotoxicity (CDC), wherein the binding of mAbs to target cells triggers the complement system, resulting in the lysis of these cells (127). Moreover, mAbs can modulate immune responses, exemplified by checkpoint inhibitors that enhance the immune system's capacity to combat cancer. These diverse mechanisms make monoclonal antibodies powerful tools against various diseases, demonstrating their indispensable role in modern medicine.

17.4.3.3 Therapeutic Applications

The therapeutic applications of mAbs show their remarkable versatility and growing impact across a range of medical conditions. In oncology, mAbs such as trastuzumab and rituximab have been pivotal, targeting HER2 in breast cancer and CD20 in B-cell malignancies, respectively (128). These antibodies interfere with specific molecules crucial for cancer cell growth and survival, offering a targeted treatment approach that has significantly improved outcomes for many patients. Beyond cancer, mAbs have transformed autoimmune disease management. For example, infliximab, an anti-TNF-α antibody, has markedly changed the treatment landscape for conditions like rheumatoid arthritis and Crohn's disease, providing relief for patients who were previously unresponsive to conventional therapies. For cardiovascular diseases, innovative mAbs such as alirocumab and evolocumab target PCSK9 to effectively lower is low-density lipoprotein (LDL) cholesterol levels, offering a novel therapeutic avenue for patients at high risk of cardiovascular events (129). Moreover, mAbs' utility extends to infectious diseases, with agents like palivizumab against respiratory syncytial virus (RSV) and mAbs developed for COVID-19, highlighting their role in both prevention and treatment (130). This broad application range underscores monoclonal antibodies' significant potential to address unmet medical needs, heralding a new era of precision therapy.

Despite their success, the development and use of mAbs face challenges, including high production costs, potential immune reactions, and the development of resistance. Future research aims to overcome these challenges, optimize therapeutic efficacy, and expand the range of diseases treatable with mAb therapy. Ongoing monoclonal antibody technology advances, including the development of bispecific antibodies and antibody-drug conjugates, promise to further enhance their therapeutic potential, offering hope for more effective and personalized treatment options across a broad spectrum of diseases (Figures 17.2 and 17.3).

17.4.4 Small Molecule Inhibitors

Small molecule inhibitors are crucial therapeutic agents for targeted cancer therapy, offering a means to interfere with specific molecular pathways that are dysregulated in cancer cells (131). Unlike biologics such as monoclonal antibodies, small molecules can often penetrate the cell membrane to target intracellular signaling pathways directly, providing a versatile approach to inhibiting oncogene activity and halting cancer progression. This section explores the mechanisms, applications, and challenges associated with small molecule inhibitors, underpinned by key studies and clinical trials in the field.

17.4.4.1 Mechanism of Action and Clinical Applications

Small molecule inhibitors typically bind to the active site of enzymes or receptor ligand-binding domains, thereby blocking their activity. In cancer treatment, they often target kinases—enzymes

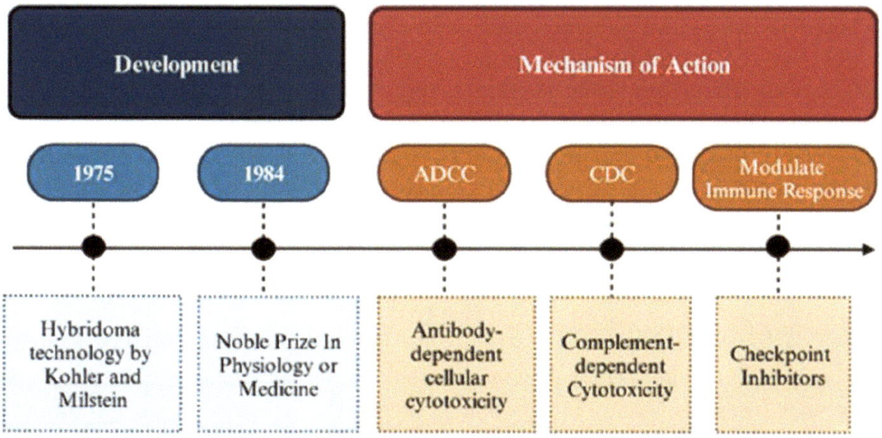

FIGURE 17.2 Development of monoclonal antibody therapy.

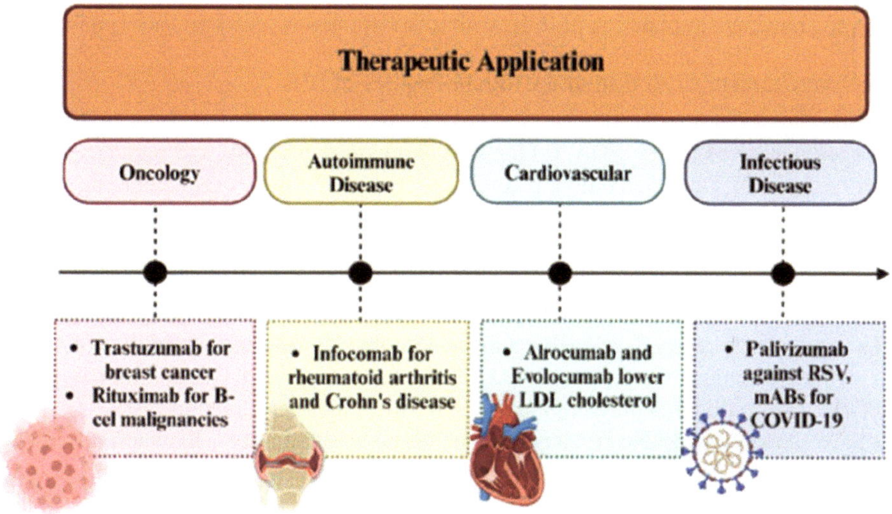

FIGURE 17.3 Therapeutic applications of monoclonal antibodies.

that phosphorylate other proteins, a critical regulatory mechanism in many signaling pathways. By inhibiting specific kinases involved in cancer cell proliferation and survival, such as the BCR-ABL tyrosine kinase in CML or the epidermal growth factor receptor (EGFR) in non-small cell lung cancer (NSCLC), these agents can exert potent anti-tumor effects (132).

Imatinib (Gleevec), which targets the BCR-ABL fusion protein in CML, was one of the first successful small molecule inhibitors. Its FDA approval in 2001 marked a significant milestone, demonstrating the efficacy of targeting molecular aberrations in cancer (133). Since then, a slew of kinase inhibitors have been developed and approved for various cancers, including gefitinib and erlotinib for EGFR-mutated NSCLC, vemurafenib for BRAF-mutated melanoma, and sorafenib for hepatocellular carcinoma (134). The development of these inhibitors is often guided by the identification of genetic mutations or alterations in tumor cells through molecular diagnostics, exemplifying the move towards personalized medicine in oncology. For example, EGFR inhibitors are most effective in NSCLC patients with specific EGFR mutations, underscoring the importance of genetic profiling in treatment selection (135).

The clinical utility of small molecule inhibitors is often hampered by the development of resistance, a common issue that limits their long-term efficacy. Resistance can arise through various mechanisms, such as the mutation of the target enzyme, activation of compensatory signaling pathways, or drug efflux. To combat resistance, researchers are exploring combination therapies, next-generation inhibitors with broader activity spectra, and strategies targeting the tumor microenvironment (136). Furthermore, achieving small molecule inhibitor specificity remains challenging, as many kinases have similar active sites, leading to off-target effects and toxicity. Ongoing advances in drug design and more sophisticated screening techniques aim to enhance these agents' selectivity and safety profiles.

17.4.5 Immune Checkpoint Inhibitors

Immune checkpoint inhibitors (ICIs) have revolutionized the landscape of cancer therapy by harnessing the body's immune system to fight malignancies. These therapeutics target immune checkpoints—molecular brakes that normally maintain immune homeostasis and prevent autoimmunity but can be co-opted by tumors to evade immune surveillance (137). The most clinically significant checkpoints to date include the cytotoxic T-lymphocyte-associated protein 4 (CTLA-4) and the programmed cell death protein 1 (PD-1) pathways, along with its ligand PD-L1 (138). This section examines the mechanism of action of ICIs, their clinical impact, and associated challenges.

17.4.5.1 Mechanism of Action and Clinical Impact of ICIs

CTLA-4 checkpoint inhibitors were the first to be clinically developed, with ipilimumab as the pioneering agent approved by the FDA. CTLA-4 is expressed on T cells and regulates the amplitude of the early stages of T-cell activation, primarily in lymph nodes. By outcompeting the costimulatory receptor CD28 for its ligands, CTLA-4 inhibits T-cell activation. Ipilimumab blocks CTLA-4, enhancing T-cell activation and proliferation (139).

The PD-1 pathway operates later in the immune response and in peripheral tissues. PD-1, expressed on T cells, binds to PD-L1, often overexpressed on tumor cells and tumor-infiltrating immune cells (140). This interaction causes T-cell exhaustion and inhibits the anti-tumor immune response. Agents such as nivolumab and pembrolizumab block PD-1, while atezolizumab, durvalumab, and avelumab target PD-L1, preventing this inhibitory signal and restoring T-cell function (141).

ICIs have shown remarkable efficacy across a broad spectrum of malignancies, including melanoma, NSCLC, renal cell carcinoma (RCC), urothelial carcinoma, and Hodgkin lymphoma, among others. These agents have led to durable responses and improved survival rates in some patients. For instance, nivolumab has significantly improved survival in patients with advanced melanoma, NSCLC, and RCC (142).

There are several challenges associated with the use of ICIs. Not all patients respond to these therapies, with response rates varying widely across different tumor types and individual patients. ICIs can also lead to immune-related adverse events (irAEs) affecting any organ system and ranging from mild to life-threatening conditions. Managing irAEs requires prompt recognition and treatment, often using immunosuppressive agents, which can complicate the care of patients undergoing treatment with ICIs (143). The future of ICIs lies in improving patient selection through identifying biomarkers predictive of response, understanding and overcoming mechanisms of resistance, and developing novel combination strategies to enhance efficacy while minimizing toxicity. Researchers are also studying additional immune checkpoints as potential therapeutic targets (Table).

Recent studies have demonstrated that combining nivolumab (PD-1 inhibitor) and ipilimumab (CTLA-4 inhibitor) improves survival rates in patients with advanced melanoma compared to either agent alone, suggesting synergistic effects of blocking both pathways (144).

Current research is investigating other immune checkpoints like LAG-3 and TIGIT (T-cell immunoglobulin and ITIM domain) as potential therapeutic targets to expand the ICI repertoire (145).

TABLE 17.2.
Summary of Immune Checkpoint Inhibitors: Agents, Mechanisms, Clinical Impacts, and Challenges

Checkpoint	Agent	Mechanism of Action	Clinical Impact	References	Key Challenges
CTLA-4	Ipilimumab	Blocks CTLA-4, enhancing T-cell activation and proliferation in early stages, primarily in lymph nodes	Used in melanoma, improves immune response with durable responses	Hodi et al., 2010	Immune-related adverse events (irAEs), variable patient responses, need for biomarkers for predicting response
PD-1	Nivolumab	Blocks PD-1, preventing T-cell exhaustion by inhibiting binding to PD-L1 on tumor cells	Improved survival in advanced melanoma, NSCLC, and RCC	Motzer et al., 2015	irAEs, management of irAEs, resistance mechanisms
PD-1	Pembrolizumab	Similar to Nivolumab, it blocks PD-1, enhancing T-cell function and anti-tumor activity	Effective in melanoma, NSCLC, shows long-lasting responses in a subset of patients	Topalian et al., 2012	Managing irAEs, optimizing patient selection and treatment strategies
PD-L1	Atezolizumab	Targets PD-L1, preventing it from binding to PD-1 and activating the PD-1 pathway, thus restoring T-cell function	Used in urothelial carcinoma and NSCLC, shows significant efficacy in PD-L1-expressing tumors		irAEs, identification of responsive patient populations
PD-L1	Durvalumab	Inhibits PD-L1, enhancing immune response by restoring T-cell activity	Approved for urothelial carcinoma and NSCLC, beneficial in tumors with high PD-L1 expression		Similar to other ICIs: irAEs, patient response variability
PD-L1	Avelumab	Blocks PD-L1, enhancing T-cell-mediated immune responses against tumors	Shows promise in Merkel cell carcinoma and urothelial carcinoma, improved survival rates in these cancers		Managing irAEs, enhancing efficacy, reducing toxicity

17.4.6 Gene Therapy

Gene therapy offers a revolutionary approach to treating and potentially curing a wide range of diseases by correcting or modifying underlying genetic abnormalities. This therapeutic strategy involves introducing, altering, or removing genes within an individual's cells and biological tissues to treat disease. Gene therapy now encompasses not only monogenic diseases but also complex conditions such as cancer, cardiovascular diseases, and infectious diseases (146). This section explores the mechanisms, advances, challenges, and future directions of gene therapy, based on current research and clinical findings.

17.4.6.1 Mechanisms and Vectors of Gene Therapy

Gene therapy operates through various mechanisms, including replacing a mutated gene with a healthy copy, inactivating a mutated gene that is functioning improperly, and introducing a new or modified gene to help treat a disease. The delivery of therapeutic genes into patient cells is facilitated by vectors, often viruses that have been genetically engineered to carry human DNA. Adenoviruses, adeno-associated viruses (AAV), lentiviruses, and retroviruses are commonly used vectors due to their efficiency in gene delivery and ability to target a wide range of cell types (147).

17.4.6.2 Advances in Gene Therapy

Gene therapy has achieved several significant milestones. One of the earliest successes was treating severe combined immunodeficiency (SCID), also known as "bubble boy disease," through retroviral-mediated gene transfer (147). Recently, gene therapies have been approved for a variety of conditions. For example, voretigene neparvovec (Luxturna) became the first directly administered gene therapy approved in the United States to treat patients with a rare form of inherited vision loss (148) (Figure 17.4).

The development of clustered regularly interspaced short palindromic repeats (CRISPR)-Cas9 technology for genome editing has further revolutionized gene therapy, offering precise, efficient, and versatile gene modification. CRISPR-Cas9 can correct gene mutations at their genomic locations, offering potential cures for a myriad of genetic disorders (149).

17.4.6.3 Ethical, Regulatory Considerations, and Challenges in Gene Therapy

Despite its promise, gene therapy faces several challenges. Vector-related immunogenicity, potential insertional mutagenesis, and difficulty targeting the correct cells or tissues are significant hurdles. Moreover, high development and delivery costs pose accessibility and affordability issues (150). Gene therapy advances have also sparked ethical and regulatory debates, particularly regarding germline modifications, which could have heritable effects. The international community has called for a cautious approach, emphasizing the need for comprehensive ethical reviews and stringent regulatory oversight to prevent misuse and ensure patient safety (151).

17.5 CHALLENGES AND OPPORTUNITIES

17.5.1 Drug Resistance Mechanisms

The development of drug resistance remains a formidable challenge in disease treatment, particularly for cancer and infectious diseases. It not only undermines the efficacy of existing therapeutic agents but also highlights the necessity for continuous research and development of new strategies to combat resistance mechanisms (152). This section examines the underlying mechanisms of drug resistance, based on a range of studies and reviews that shed light on this complex phenomenon.

Drug resistance can manifest through various biological mechanisms, broadly categorized into intrinsic (pre-existing) resistance and acquired resistance. Intrinsic resistance stems from

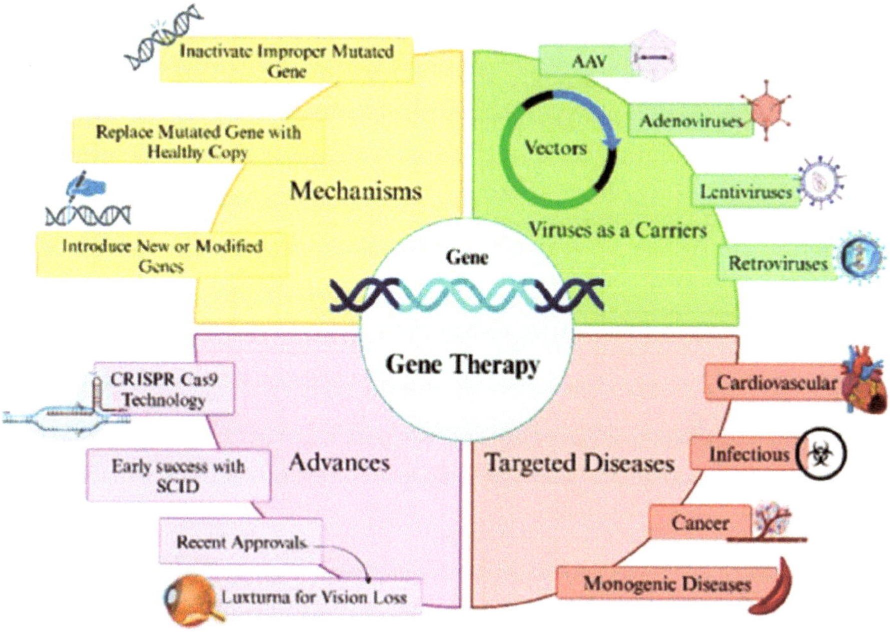

FIGURE 17.4 Insights on gene therapy.

inherent characteristics of the cells or pathogens that render them naturally resistant to certain drugs. Acquired resistance develops through genetic mutations, epigenetic alterations, or changes in cellular processes in response to drug exposure (151).

17.5.1.1 Genetic Mutations and Alterations

A primary mechanism of drug resistance involves genetic mutations that alter the drug target, thereby reducing the drug's efficacy. In cancer, mutations in drug target genes can change the target's structure or expression level, rendering the drug ineffective. For example, mutations in the BCR-ABL gene confer resistance to imatinib in chronic myeloid leukemia (153). Similarly, in infectious diseases, mutations in the bacterial ribosomal RNA can confer resistance to antibiotics like macrolides (154).

17.5.1.2 Drug Efflux Pumps

Another significant mechanism is the upregulation of drug efflux pumps, which actively transport drugs out of the cells, reducing their intracellular concentration and effectiveness. The overexpression of ATP-binding cassette (ABC) transporters, such as P-glycoprotein (P-gp), often causes multidrug resistance in cancer cells (155).

17.5.1.3 Alterations in Drug Metabolism

Cells can also develop resistance by altering the metabolism of drugs, either by increasing the metabolism of the drug into inactive forms or by decreasing its activation into active forms. Cytochrome P450 enzymes, responsible for metabolizing many drugs, can be differentially expressed in resistant cells, affecting drug efficacy (156).

17.5.1.4 Epigenetic Modifications

Epigenetic changes, such as DNA methylation and histone modification, can also contribute to drug resistance by altering the expression of genes involved in drug response (157). For instance, hypermethylation of the O^6-methylguanine-DNA methyltransferase (MGMT) gene promoter has been associated with resistance to alkylating agents in gliomas (158).

17.5.1.5 Microenvironmental Factors

The tumor microenvironment can also play a crucial role in drug resistance, creating a physical and biochemical barrier to drug delivery. Hypoxia, acidic pH, and the presence of stromal cells can modulate drug sensitivity and resistance through various mechanisms, including altering drug penetration, modifying target sensitivity, and inducing a protective response in cancer cells (159) (Figure 17.5).

17.5.2 Biomarkers and Personalized Medicine

The evolution of biosignaling research has significantly contributed to the burgeoning field of personalized medicine, with biomarkers playing a pivotal role in this paradigm shift. Personalized medicine, which tailors medical treatment to each patient's individual characteristics, has become increasingly feasible due an improved understanding of the molecular basis of diseases, particularly through the lens of biosignaling pathways. Biomarkers, which are biological molecules found in blood, other body fluids, or tissues, provide critical information about normal or pathological processes and pharmacologic responses to therapeutic intervention (160). This section explores the current state and future prospects of biomarkers in personalized medicine, focusing on their discovery, validation, and clinical implementation.

17.5.2.1 The Role of Biomarkers in Personalized Medicine

Biomarkers serve as the cornerstone of personalized medicine, offering a molecular snapshot of an individual's health status. They have diverse applications, including disease risk assessment, diagnosis, prognosis, and monitoring response to therapy, all of which are integral to customizing patient care (161). Regarding biosignaling, biomarkers can indicate the activity of disease-related signaling pathways, thus identifying potential therapeutic targets.

For example, in cancer, identifying specific biomarkers related to signal transduction pathways has led to the development of targeted therapies. HER2, a receptor tyrosine kinase involved in the MAPK and PI3K pathways, serves as a biomarker for aggressive breast cancer, guiding the use of HER2-targeted therapies such as trastuzumab (162). Similarly, the presence of BCR-ABL fusion protein, resulting from a specific chromosomal translocation in CML, is targeted by imatinib, a tyrosine kinase inhibitor (163).

17.5.2.2 Challenges in Biomarker Development

Despite their potential, biomarker clinical implementation faces many challenges. These include needing robust validation studies to confirm the clinical relevance of biomarkers, developing

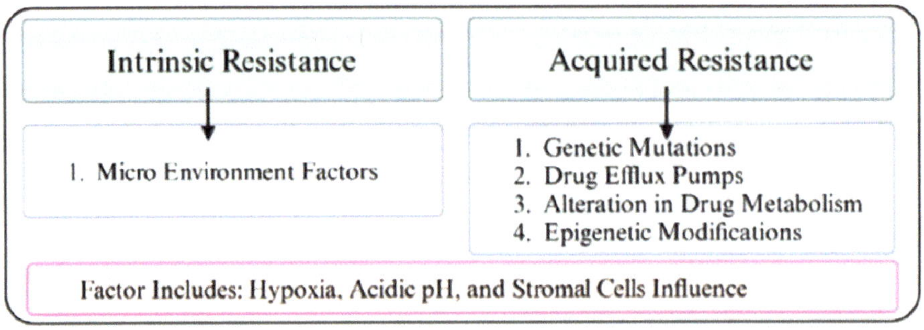

FIGURE 17.5 Types of resistances.

standardized assays for their measurement, and integrating biomarker testing into clinical workflows (164). Furthermore, disease heterogeneity, especially in cancer, means that a single biomarker may not provide a complete picture, necessitating biomarker panels or combinatorial approaches for accurate disease characterization (165).

17.5.3 Overcoming Resistance to Targeted Therapies

Targeted therapies have revolutionized the treatment of various diseases, particularly cancer, by exploiting specific molecular abnormalities in diseased cells. However, the development of resistance to these therapies remains a formidable challenge, necessitating a deeper understanding of the underlying biosignaling mechanisms to devise effective counterstrategies. This section explores the biosignaling perspectives on resistance to targeted therapies and identifies potential ways to overcome it.

17.5.3.1 Mechanisms of Resistance in Biosignaling Pathways

Resistance to targeted therapies can emerge through various biosignaling mechanisms, including genetic mutations, activation of alternative signaling pathways, and adaptive cellular responses. Genetic mutations in target molecules can directly confer resistance by altering the drug-binding site, rendering the therapy ineffective. For example, mutations in the BCR-ABL gene confer resistance to imatinib in chronic myeloid leukemia patients (166). Alternatively, cancer cells can activate compensatory signaling pathways that bypass the blocked pathway, sustaining cell proliferation and survival. The activation of the PI3K/Akt/mTOR pathway in response to targeted therapy inhibition of the MAPK/ERK pathway in melanoma exemplifies this (167).

17.5.3.2 Strategies to Overcome Resistance

17.5.3.2.1 Combination Therapy

One strategy to overcome resistance is the use of combination therapy, targeting multiple signaling pathways simultaneously. This approach can prevent the activation of compensatory pathways and mitigate the development of resistance. A study by Long et al. (2014) demonstrated that combining BRAF inhibitors with MEK inhibitors significantly delayed the onset of resistance in melanoma patients by inhibiting both primary and compensatory signaling pathways (168).

17.5.3.3 Adaptive Therapy

Adaptive therapy, based on the principle of evolutionary dynamics, aims to maintain a population of therapy-sensitive cells to suppress the growth of resistant cells. This approach involves the intermittent application of targeted therapy, allowing sensitive cells to outcompete resistant ones during therapy breaks, thereby controlling tumor growth over time (169).

17.5.3.4 Targeting the Tumor Microenvironment

The tumor microenvironment plays a crucial role in the development of resistance to targeted therapies. Factors secreted by the stromal cells can activate survival pathways in cancer cells. Targeting these microenvironmental interactions, such as inhibiting the CAF-derived factors or modulating the immune microenvironment, can enhance the efficacy of targeted therapies (170).

17.5.3.5 Exploiting Synthetic Lethality

Synthetic lethality arises when the simultaneous impairment of two genes leads to cell death, whereas impairment of only one does not. Identifying synthetic lethal partners of resistance genes can provide new targets for therapy. For instance, poly (ADP-ribose) polymerase inhibitors (PARP) exploit the synthetic lethality between PARP enzymatic activity and BRCA1/2 mutations in breast and ovarian cancers (171)

17.5.4 The Future of Signal Transduction Research and Targeted Therapy

Signal transduction research and targeted therapy are on the verge of transformative breakthroughs. Advances in understanding cellular signaling mechanisms have built a robust foundation, but the journey forward is both exhilarating and challenging. Advances in genomic and proteomic technologies, combined with the integration of artificial intelligence and computational modeling, promise to accelerate the discovery and development of novel therapeutic targets.

Future signal transduction research will likely focus on unraveling the complexity of signaling networks. This involves deciphering the spatial and temporal dynamics of signal transduction within the cellular context, understanding the crosstalk between different signaling pathways, and elucidating the role of the microenvironment in modulating these processes. Precision medicine will play a pivotal role, as individual genetic variations significantly impact signaling pathways and their response to therapies. Tailoring treatments to the molecular profile of diseases, particularly in oncology, will enhance therapeutic efficacy and reduce adverse effects (172).

Furthermore, the development of targeted therapies will benefit from innovative drug delivery systems that ensure the specificity and stability of therapeutic agents, minimizing off-target effects. Exploring non-traditional therapeutic modalities, such as RNA-based therapies and CRISPR-Cas9 mediated gene editing, offers exciting possibilities for modulating signal transduction pathways at the genetic level (173). Challenges remain, particularly in overcoming drug resistance, a common setback in targeted therapy. Understanding the mechanisms behind resistance and developing combination therapies or novel agents to circumvent these obstacles will be crucial. Additionally, fostering collaboration across disciplines and between academia and industry will be essential to effectively translate research findings into clinical applications.

17.6 FUTURE PERSPECTIVES

Several promising directions and emerging trends are poised to further revolutionize our understanding of biosignaling pathways and their implications in health and disease.

17.6.1 Precision Medicine and Personalized Biosignaling Therapies

Integrating biosignaling with precision medicine promises tailor-made therapies designed according to an individual's genetic makeup. Identifying unique biosignaling signatures in diseases, particularly in cancer and autoimmune disorders, enables more effective targeted treatments with fewer side effects than traditional therapies. Advances in genomic sequencing and bioinformatics are crucial for this endeavor, enabling the detailed mapping of signaling pathways and their alterations in individual patients (173). The challenge lies in translating this vast array of molecular data into clinically actionable insights, necessitating interdisciplinary collaboration among molecular biologists, clinicians, and computational scientists.

17.6.1.1 The Role of Non-coding RNAs in Biosignaling

Non-coding RNAs (ncRNAs), including microRNAs (miRNAs) and long non-coding RNAs (lncRNAs), are critical regulators of gene expression intricately involved in biosignaling pathways (174). These molecules influence signaling cascades by modulating the stability and translation of mRNAs encoding signaling proteins or directly interacting with proteins to affect their function (175). Future research will likely unravel more about the complex roles of ncRNAs in biosignaling, offering new avenues for therapeutic intervention and diagnostic markers.

17.6.1.2 Exploiting Synthetic Biology in Biosignaling

Synthetic biology, which combines engineering principles with biological sciences, offers innovative strategies to redesign or create new biosignaling pathways. Scientists can create bespoke, precisely controlled signaling systems by engineering cells with synthetic receptors (176). This approach has significant implications for regenerative medicine, drug development, and biosensors. For instance, engineered T cells with synthetic receptors have been developed for cancer immunotherapy, demonstrating synthetic biology's potential to provide new solutions to longstanding challenges (177).

17.6.1.3 Understanding the Interplay Between Biosignaling Pathways

Biosignaling is not a collection of isolated pathways but a complex network of interactions. The crosstalk between different signaling pathways allows cells to integrate multiple signals and respond appropriately to their environment. Future biosignaling research will increasingly focus on understanding these interactions at a systems level, using tools such as proteomics and systems biology (178). This holistic approach will uncover new regulatory mechanisms and potential therapeutic targets that are not apparent when pathways are studied in isolation.

17.6.1.4 Technological Advancements and Biosignaling

The continued development of high-throughput technologies and single-cell analysis tools will deepen our understanding of biosignaling pathways. These technologies allow for the analysis of signaling dynamics at unprecedented resolution and scale, shedding light on the heterogeneity within cell populations and the spatial organization of signaling components within cells (179). The integration of artificial intelligence and machine learning with these data-rich technologies holds the potential to decode complex biosignaling networks and predict their behavior under different conditions.

REFERENCES

1. Asl ER, Amini M, Najafi S, Mansoori B, Mokhtarzadeh A, Mohammadi A, et al. Interplay between MAPK/ERK signaling pathway and MicroRNAs: a crucial mechanism regulating cancer cell metabolism and tumor progression. *Life Sciences*. 2021;278:119499.
2. Girault J-A, Greengard P. Principles of signal transduction. In: *Neurobiology of Mental Illness*. 5th ed. Oxford University Press; 2018
3. Mehta M, Dhanjal DS, Paudel KR, Singh B, Gupta G, Rajeshkumar S, et al. Cellular signalling pathways mediating the pathogenesis of chronic inflammatory respiratory diseases: an update. *Inflammopharmacology*. 2020;28:795–817.
4. Huang XL, Khan MI, Wang J, Ali R, Ali SW, Kazmi A, et al. Role of receptor tyrosine kinases mediated signal transduction pathways in tumor growth and angiogenesis—New insight and futuristic vision. *International Journal of Biological Macromolecules*. 2021;180:739–52.
5. Critchley WR, Pellet-Many C, Ringham-Terry B, Harrison MA, Zachary IC, Ponnambalam S. Receptor tyrosine kinase ubiquitination and de-ubiquitination in signal transduction and receptor trafficking. *Cells*. 2018;7(3):22.
6. Nair A, Chauhan P, Saha B, Kubatzky KF. Conceptual evolution of cell signaling. *International Journal of Molecular Sciences*. 2019;20(13):3292.
7. Neben CL, Lo M, Jura N, Klein OD. Feedback regulation of RTK signaling in development. *Developmental Biology*. 2019;447(1):71–89.
8. Wintheiser GA, Silberstein P. Physiology, tyrosine kinase receptors. StatPearls Publishing, Treasure Island (FL); 2019. PMID: 30860767
9. Trenker R, Jura N. Receptor tyrosine kinase activation: from the ligand perspective. *Current Opinion in Cell Biology*. 2020;63:174–85.
10. Butti R, Das S, Gunasekaran VP, Yadav AS, Kumar D, Kundu GC. Receptor tyrosine kinases (RTKs) in breast cancer: signaling, therapeutic implications and challenges. *Molecular Cancer*. 2018;17:1–18.
11. Zhang N, Li Y. Receptor tyrosine kinases: biological functions and anticancer targeted therapy. *MedComm*. 2023;4(6):e446.
12. Roskoski Jr R. Small molecule inhibitors targeting the EGFR/ErbB family of protein-tyrosine kinases in human cancers. *Pharmacological Research*. 2019;139:395–411.

13. Aschner Y, Downey GP. The importance of tyrosine phosphorylation control of cellular signaling pathways in respiratory disease: pY and pY not. *American Journal of Respiratory Cell and Molecular Biology*. 2018;59(5):535–47.
14. Schwark K, Messinger D, Cummings JR, Bradin J, Kawakibi A, Babila CM, et al. Receptor tyrosine kinase (RTK) targeting in pediatric high-grade glioma and diffuse midline glioma: pre-clinical models and precision medicine. *Frontiers in Oncology*. 2022;12:922928.
15. Ward RA, Fawell S, Floc'h N, Flemington V, McKerrecher D, Smith PD. Challenges and opportunities in cancer drug resistance. *Chemical Reviews*. 2020;121(6):3297–351.
16. Torres-Ayuso P, Brognard J. Combing the cancer genome for novel kinase drivers and new therapeutic targets. *Cancers*. 2019;11(12):1972.
17. Azam S, Haque ME, Jakaria M, Jo S-H, Kim I-S, Choi D-K. G-protein-coupled receptors in CNS: a potential therapeutic target for intervention in neurodegenerative disorders and associated cognitive deficits. *Cells*. 2020;9(2):506.
18. Gao H, Li Y, Wang M, Song X, Tang J, Feng F, et al. Identification and expression analysis of G protein-coupled receptors in the Miridae insect Apolygus lucorum. *Frontiers in Endocrinology*. 2021;12:773669.
19. Vu O, Bender BJ, Pankewitz L, Huster D, Beck-Sickinger AG, Meiler J. The structural basis of peptide binding at class AG protein-coupled receptors. *Molecules*. 2021;27(1):210.
20. Mann E, Sunni M, Bellin MD. *Secretion of insulin in response to diet and hormones*. Pancreapedia: The Exocrine Pancreas Knowledge Base; 2020.
21. Goudet C, Rovira X, Rondard P, Pin J-P, Llebaria A, Acher F. Modulation of metabotropic glutamate receptors by orthosteric, allosteric, and light-operated ligands. *Structure and Function of GPCRs*. 2019:253–84.
22. Bassilana F, Nash M, Ludwig M-G. Adhesion G protein-coupled receptors: opportunities for drug discovery. *Nature Reviews Drug Discovery*. 2019;18(11):869–84.
23. Seyedabadi M, Gharghabi M, Gurevich EV, Gurevich VV. Receptor-arrestin interactions: the GPCR perspective. *Biomolecules*. 2021;11(2):218.
24. Lymperopoulos A, Cora N, Maning J, Brill AR, Sizova A. Signaling and function of cardiac autonomic nervous system receptors: insights from the GPCR signalling universe. *The FEBS Journal*. 2021;288(8):2645–59.
25. Hauser AS, Chavali S, Masuho I, Jahn LJ, Martemyanov KA, Gloriam DE, et al. Pharmacogenomics of GPCR drug targets. *Cell*. 2018;172(1):41–54. e19.
26. Fasoulakis Z, Daskalakis G, Theodora M, Antsaklis P, Sindos M, Diakosavvas M, et al. The relevance of notch signaling in cancer progression. *Notch Signaling in Embryology and Cancer: Notch Signaling in Cancer*. 2021:169–81.
27. Shahi MH. *Notch cell signaling pathway and brain tumors: Role of signaling pathways in brain tumorigenesis*. Springer; 2023. p. 29–37.
28. Mutvei AP. *The role of notch signaling in cancer and metastasis*. Karolinska Institutet (Sweden); 2014.
29. Anusewicz D, Orzechowska M, Bednarek AK. Notch signaling pathway in cancer—review with bioinformatic analysis. *Cancers*. 2021;13(4):768.
30. Pandey A, Niknejad N, Jafar-Nejad H. Multifaceted regulation of Notch signaling by glycosylation. *Glycobiology*. 2021;31(1):8–28.
31. Goruganthu MU, Shanker A, Dikov MM, Carbone DP. Specific targeting of notch ligand-receptor interactions to modulate immune responses: a review of clinical and preclinical findings. *Frontiers in Immunology*. 2020;11:556819.
32. Antfolk D. *Regulation of notch signaling by intracellular trafficking*; Åbo Akademi University, 2020.
33. Pillidge Z. *Transcription and chromatin dynamics in the notch signalling response*; University of Cambridge, 2019.
34. Sachan N, Sharma V, Mutsuddi M, Mukherjee A. Notch signalling: multifaceted role in development and disease. *The FEBS Journal*. 2023;291: 4.
35. Toribio ML, González-García S. Notch partners in the long journey of T-ALL pathogenesis. *International Journal of Molecular Sciences*. 2023;24(2):1383.
36. Yadav AK, Desai NS. Cancer stem cells: acquisition, characteristics, therapeutic implications, targeting strategies and future prospects. *Stem Cell Reviews and Reports*. 2019;15:331–55.
37. Hayat R, Manzoor M, Hussain A. Wnt signaling pathway: a comprehensive review. *Cell Biology International*. 2022;46(6):863–77.
38. Xin Y. *Cholesterol binding is a determinant of canonical Wnt signaling activity*. University of Illinois at Chicago; 2018.

39. Anand AA, Khan M, Kar D. The molecular basis of Wnt/β-catenin signaling pathways in neurodegenerative diseases. *International Journal of Cell Biology.* 2023;2023:8.
40. Shah K, Kazi JU. Phosphorylation-dependent regulation of WNT/Beta-catenin signaling. *Frontiers in Oncology.* 2022;12:858782.
41. Sharma A, Mir R, Galande S. Epigenetic regulation of the Wnt/β-catenin signaling pathway in cancer. *Frontiers in Genetics.* 2021;12:681053.
42. Zmijanac Partl J, Karin V, Skrtic A, Nikuseva-Martic T, Serman A, Mlinarec J, et al. Negative regulators of Wnt signaling pathway SFRP1 and SFRP3 expression in preterm and term pathologic placentas. *The Journal of Maternal-Fetal & Neonatal Medicine.* 2018;31(22):2971–9.
43. Xia R, Xu M, Yang J, Ma X. The role of Hedgehog and Notch signaling pathway in cancer. *Molecular Biomedicine.* 2022;3(1):44.
44. Nguyen LH, Goel A, Chung DC. Pathways of colorectal carcinogenesis. *Gastroenterology.* 2020;158(2):291–302.
45. Zhang Q, Jiang J. Regulation of hedgehog signal transduction by ubiquitination and deubiquitination. *International Journal of Molecular Sciences.* 2021;22(24):13338.
46. Huang D, Wang Y, Tang J, Luo S. Molecular mechanisms of suppressor of fused in regulating the hedgehog signalling pathway. *Oncology Letters.* 2018;15(5):6077–86.
47. Pietrobono S, Gagliardi S, Stecca B. Non-canonical hedgehog signaling pathway in cancer: activation of GLI transcription factors beyond smoothened. *Frontiers in Genetics.* 2019;10:465464.
48. Ho EK, Stearns T. Hedgehog signaling and the primary cilium: implications for spatial and temporal constraints on signaling. *Development.* 2021;148(9):dev195552.
49. Liu H, Kiseleva AA, Golemis EA. Ciliary signalling in cancer. *Nature Reviews Cancer.* 2018;18(8):511–24.
50. Zlobin A, Bloodworth JC, Osipo C. Mitogen-activated protein kinase (MAPK) signaling. *Predictive Biomarkers in Oncology: Applications in Precision Medicine.* 2019:213–21.
51. Bahar ME, Kim HJ, Kim DR. Targeting the RAS/RAF/MAPK pathway for cancer therapy: from mechanism to clinical studies. *Signal Transduction and Targeted Therapy.* 2023;8(1):455.
52. Bohn S, Hexemer L, Huang Z, Strohmaier L, Lenhardt S, Legewie S, et al. State-and stimulus-specific dynamics of SMAD signaling determine fate decisions in individual cells. *Proceedings of the National Academy of Sciences.* 2023;120(10):e2210891120.
53. Brown MD, Sacks DB. Protein scaffolds in MAP kinase signalling. *Cellular Signalling.* 2009;21(4):462–9.
54. Guo YJ, Pan WW, Liu SB, Shen ZF, Xu Y, Hu LL. ERK/MAPK signalling pathway and tumorigenesis. *Experimental and Therapeutic Medicine.* 2020;19(3):1997–2007.
55. Dillon M, Lopez A, Lin E, Sales D, Perets R, Jain P. Progress on Ras/MAPK signaling research and targeting in blood and solid cancers. *Cancers.* 2021;13(20):5059.
56. He Y, Sun MM, Zhang GG, Yang J, Chen KS, Xu WW, et al. Targeting PI3K/Akt signal transduction for cancer therapy. *Signal Transduction and Targeted Therapy.* 2021;6(1):425.
57. Fusco N, Sajjadi E, Venetis K, Gaudioso G, Lopez G, Corti C, et al. PTEN alterations and their role in cancer management: are we making headway on precision medicine? *Genes.* 2020;11(7):719.
58. Fayard E, Xue G, Parcellier A, Bozulic L, Hemmings BA. Protein kinase B (PKB/Akt), a key mediator of the PI3K signaling pathway. *Phosphoinositide 3-kinase in Health and Disease.* 2011;1:31–56.
59. Meikle L, Pollizzi K, Egnor A, Kramvis I, Lane H, Sahin M, et al. Response of a neuronal model of tuberous sclerosis to mammalian target of rapamycin (mTOR) inhibitors: effects on mTORC1 and Akt signaling lead to improved survival and function. *Journal of Neuroscience.* 2008;28(21):5422–32.
60. Laplante M, Sabatini DM. mTOR signaling in growth control and disease. *Cell.* 2012;149(2):274–93.
61. Rascio F, Spadaccino F, Rocchetti MT, Castellano G, Stallone G, Netti GS, et al. The pathogenic role of PI3K/AKT pathway in cancer onset and drug resistance: an updated review. *Cancers.* 2021;13(16):3949.
62. Janku F, Yap TA, Meric-Bernstam F. Targeting the PI3K pathway in cancer: are we making headway? *Nature Reviews Clinical Oncology.* 2018;15(5):273–91.
63. Bharadwaj U, Kasembeli MM, Robinson P, Tweardy DJ. Targeting janus kinases and signal transducer and activator of transcription 3 to treat inflammation, fibrosis, and cancer: rationale, progress, and caution. *Pharmacological Reviews.* 2020;72(2):486–526.
64. Lin CM, Cooles FA, Isaacs JD. Basic mechanisms of JAK inhibition. *Mediterranean Journal of Rheumatology.* 2020;31(Suppl 1):100.
65. Yang L. *Antagonizing JAK-STAT signaling by porcine reproductive and respiratory syndrome virus.* University of Maryland, College Park; 2018.
66. Puigdevall L, Michiels C, Stewardson C, Dumoutier L. JAK/STAT: why choose a classical or an alternative pathway when you can have both? *Journal of Cellular and Molecular Medicine.* 2022;26(7):1865–75.

67. Laudisi F, Cherubini F, Monteleone G, Stolfi C. STAT3 interactors as potential therapeutic targets for cancer treatment. *International Journal of Molecular Sciences.* 2018;19(6):1787.
68. Lee C, Cheung ST. STAT3: an emerging therapeutic target for hepatocellular carcinoma. *Cancers.* 2019;11(11):1646.
69. Xue C, Yao Q, Gu X, Shi Q, Yuan X, Chu Q, et al. Evolving cognition of the JAK-STAT signaling pathway: autoimmune disorders and cancer. *Signal Transduction and Targeted Therapy.* 2023;8(1):204.
70. Greenfield G, McMullin MF, Mills K. Molecular pathogenesis of the myeloproliferative neoplasms. *Journal of Hematology & Oncology.* 2021;14:1–18.
71. Bose P, Verstovsek S. JAK inhibition for the treatment of myelofibrosis: limitations and future perspectives. *Hemasphere.* 2020;4(4):e424.
72. Jamilloux Y, El Jammal T, Vuitton L, Gerfaud-Valentin M, Kerever S, Seve P. JAK inhibitors for the treatment of autoimmune and inflammatory diseases. *Autoimmunity Reviews.* 2019;18(11):102390.
73. Zhu S, Jin J, Gokhale S, Lu AM, Shan H, Feng J, et al. Genetic alterations of TRAF proteins in human cancers. *Frontiers in Immunology.* 2018;9:2111.
74. Fabris L, Fiorotto R, Spirli C, Cadamuro M, Mariotti V, Perugorria MJ, et al. Pathobiology of inherited biliary diseases: a roadmap to understand acquired liver diseases. *Nature Reviews Gastroenterology & Hepatology.* 2019;16(8):497–511.
75. Tian Y, Guo W. A review of the molecular pathways involved in resistance to BRAF inhibitors in patients with advanced-stage melanoma. *Medical Science Monitor: International Medical Journal of Experimental and Clinical Research.* 2020;26:e920957.
76. Joshi H, Press MF. *Molecular oncology of breast cancer: the breast.* Elsevier; 2018. p. 282–307. e5.
77. Boni C, Sorio C. Current views on the interplay between tyrosine kinases and phosphatases in Chronic Myeloid Leukemia. *Cancers.* 2021;13(10):2311.
78. Inoue K, Fry EA. Aberrant expression of p16INK4a in human cancers–a new biomarker? *Cancer Reports and Reviews.* 2018;2(2).
79. Osman AE, Deininger MW. Chronic Myeloid Leukemia: modern therapies, current challenges and future directions. *Blood Reviews.* 2021;49:100825.
80. !!! INVALID CITATION!!!.
81. Yang L, Shi P, Zhao G, Xu J, Peng W, Zhang J, et al. Targeting cancer stem cell pathways for cancer therapy. *Signal Transduction and Targeted Therapy.* 2020;5(1):8.
82. Marei HE, Althani A, Afifi N, Hasan A, Caceci T, Pozzoli G, et al. p53 signaling in cancer progression and therapy. *Cancer Cell International.* 2021;21(1):703.
83. Shorning BY, Dass MS, Smalley MJ, Pearson HB. The PI3K-AKT-mTOR pathway and prostate cancer: at the crossroads of AR, MAPK, and WNT signaling. *International Journal of Molecular Sciences.* 2020;21(12):4507.
84. Alvarado-Ortiz E, de la Cruz-López KG, Becerril-Rico J, Sarabia-Sánchez MA, Ortiz-Sánchez E, García-Carrancá A. Mutant p53 gain-of-function: role in cancer development, progression, and therapeutic approaches. *Frontiers in Cell and Developmental Biology.* 2021;8:607670.
85. Huang L, Guo Z, Wang F, Fu L. KRAS mutation: from undruggable to druggable in cancer. *Signal Transduction and Targeted Therapy.* 2021;6(1):386.
86. Xu Z, Han X, Ou D, Liu T, Li Z, Jiang G, et al. Targeting PI3K/AKT/mTOR-mediated autophagy for tumor therapy. *Applied Microbiology and Biotechnology.* 2020;104:575–87.
87. Starzyńska A, Sejda A, Adamska P, Marvaso G, Sakowicz-Burkiewicz M, Adamski Ł, et al. Prognostic value of the PIK3CA, AKT, and PTEN mutations in oral squamous cell carcinoma: literature review. *Archives of Medical Science: AMS.* 2021;17(1):207.
88. Xin P, Xu X, Deng C, Liu S, Wang Y, Zhou X, et al. The role of JAK/STAT signaling pathway and its inhibitors in diseases. *International Immunopharmacology.* 2020;80:106210.
89. Purohit M, Gupta G, Afzal O, Altamimi ASA, Alzarea SI, Kazmi I, et al. Janus kinase/signal transducers and activator of transcription (JAK/STAT) and its role in Lung inflammatory disease. *Chemico-Biological Interactions.* 2023;371:110334.
90. Luo Q, Xiao Z, Peng L. Effects of ruxolitinib on infection in patients with myeloproliferative neoplasm: a meta-analysis. *Hematology.* 2021;26(1):663–9.
91. Torrisi F, Alberghina C, D'Aprile S, Pavone AM, Longhitano L, Giallongo S, et al. The hallmarks of glioblastoma: heterogeneity, intercellular crosstalk and molecular signature of invasiveness and progression. *Biomedicines.* 2022;10(4):806.
92. Pathak N, Vimal SK, Tandon I, Agrawal L, Hongyi C, Bhattacharyya S. Neurodegenerative disorders of alzheimer, parkinsonism, amyotrophic lateral sclerosis and multiple sclerosis: an early diagnostic approach for precision treatment. *Metabolic Brain Disease.* 2021:1–38.

93. Schmidt MF, Gan ZY, Komander D, Dewson G. Ubiquitin signalling in neurodegeneration: mechanisms and therapeutic opportunities. *Cell Death & Differentiation.* 2021;28(2):570–90.
94. Serafino A, Giovannini D, Rossi S, Cozzolino M. Targeting the Wnt/β-catenin pathway in neurodegenerative diseases: recent approaches and current challenges. *Expert Opinion on Drug Discovery.* 2020;15(7):803–22.
95. Xia Q, Yang X, Shi J, Liu Z, Peng Y, Wang W, et al. The protective A673T mutation of amyloid precursor protein (APP) in Alzheimer's disease. *Molecular Neurobiology.* 2021;58(8):4038–50.
96. Crestini A, Santilli F, Martellucci S, Carbone E, Sorice M, Piscopo P, et al. Prions and neurodegenerative diseases: a focus on Alzheimer's disease. *Journal of Alzheimer's Disease.* 2022;85(2):503–18.
97. Calvo-Rodriguez M, Kharitonova EK, Bacskai BJ. Therapeutic strategies to target calcium dysregulation in Alzheimer's disease. *Cells.* 2020;9(11):2513.
98. Choong CJ, Mochizuki H. Neuropathology of α -synuclein in Parkinson's disease. *Neuropathology.* 2022;42(2):93–103.
99. Dinter E, Saridaki T, Diederichs L, Reichmann H, Falkenburger BH. Parkinson's disease and translational research. *Translational Neurodegeneration.* 2020;9:1–11.
100. Picca A, Calvani R, Coelho-Junior HJ, Landi F, Bernabei R, Marzetti E. Mitochondrial dysfunction, oxidative stress, and neuroinflammation: intertwined roads to neurodegeneration. *Antioxidants.* 2020;9(8):647.
101. Choi H-J, Cha SJ, Lee J-W, Kim H-J, Kim K. Recent advances on the role of gsk3β in the pathogenesis of amyotrophic lateral sclerosis. *Brain Sciences.* 2020;10(10):675.
102. Kirola L, Mukherjee A, Mutsuddi M. Recent updates on the genetics of amyotrophic lateral sclerosis and frontotemporal dementia. *Molecular Neurobiology.* 2022;59(9):5673–94.
103. Tabrizi SJ, Flower MD, Ross CA, Wild EJ. Huntington disease: new insights into molecular pathogenesis and therapeutic opportunities. *Nature Reviews Neurology.* 2020;16(10):529–46.
104. Albert-Gascó H, Ros-Bernal F, Castillo-Gómez E, Olucha-Bordonau FE. MAP/ERK signaling in developing cognitive and emotional function and its effect on pathological and neurodegenerative processes. *International Journal of Molecular Sciences.* 2020;21(12):4471.
105. Song Y, Woo AY-H, Zhang Y, Xiao R. Cardiac β-adrenoceptor signaling: the new insight on an old target in the therapy of cardiovascular disease. *International Journal of Drug Discovery and Pharmacology.* 2022;3:1–10.
106. Elia A, Fossati S. Autonomic nervous system and cardiac neuro-signaling pathway modulation in cardiovascular disorders and Alzheimer's disease. *Frontiers in Physiology.* 2023;14:1060666.
107. Westhoff M, Dixon RE. Mechanisms and regulation of cardiac CaV1. 2 trafficking. *International Journal of Molecular Sciences.* 2021;22(11):5927.
108. Grismaldo A, Sobrevia L, Morales L. Role of platelet-derived growth factor c on endothelial dysfunction in cardiovascular diseases. *Biochimica et Biophysica Acta (BBA)-General Subjects.* 2022;1866(10):130188.
109. Sawma T, Shaito A, Najm N, Sidani M, Orekhov A, El-Yazbi AF, et al. Role of RhoA and Rho-associated kinase in phenotypic switching of vascular smooth muscle cells: implications for vascular function. *Atherosclerosis.* 2022;358:12–28.
110. Cheng W, Cui C, Liu G, Ye C, Shao F, Bagchi AK, et al. NF-κB, A potential therapeutic target in cardiovascular diseases. *Cardiovascular Drugs and Therapy.* 2023;37(3):571–84.
111. Bonetti J, Corti A, Lerouge L, Pompella A, Gaucher C. Phenotypic modulation of macrophages and vascular smooth muscle cells in atherosclerosis—nitro-redox interconnections. *Antioxidants.* 2021;10(4):516.
112. Yue J, López JM. Understanding MAPK signaling pathways in apoptosis. *International Journal of Molecular Sciences.* 2020;21(7):2346.
113. Zhang Z, Yang Z, Wang S, Wang X, Mao J. Targeting MAPK-ERK/JNK pathway: a potential intervention mechanism of myocardial fibrosis in heart failure. *Biomedicine & Pharmacotherapy.* 2024;173:116413.
114. Alam A, Farooq U, Singh R, Dubey V, Kumar S, Kumari R, et al. Chemotherapy treatment and strategy schemes: a review. *Open Access Journal of Toxicology.* 2018;2(5):21.
115. Gambardella V, Tarazona N, Cejalvo JM, Lombardi P, Huerta M, Roselló S, et al. Personalized medicine: recent progress in cancer therapy. *Cancers.* 2020;12(4):1009.
116. Manley PW, Stiefl NJ. Progress in the discovery of BCR-ABL Kinase inhibitors for the treatment of leukemia. *Cancer II.* 2018:1–37.
117. Patel A, Unni N, Peng Y. The changing paradigm for the treatment of HER2-positive breast cancer. *Cancers.* 2020;12(8):2081.

118. Haider T, Pandey V, Banjare N, Gupta PN, Soni V. Drug resistance in cancer: mechanisms and tackling strategies. *Pharmacological Reports*. 2020;72(5):1125–51.
119. Kaushik AC, Mehmood A, Wei D-Q, Nawab S, Sahi S, Kumar A. *Cheminformatics and bioinformatics at the interface with systems biology: bridging chemistry and medicine*. Royal Society of Chemistry; 2023.
120. Lamprecht Tratar U, Horvat S, Cemazar M. Transgenic mouse models in cancer research. *Frontiers in Oncology*. 2018;8:268.
121. Kesh M, Goel S. *Target-based screening for lead discovery: CADD and informatics in drug discovery*. Springer; 2023. p. 141–73.
122. Lee AC-L, Harris JL, Khanna KK, Hong J-H. A comprehensive review on current advances in peptide drug development and design. *International Journal of Molecular Sciences*. 2019;20(10):2383.
123. Alsultan A, Alghamdi WA, Alghamdi J, Alharbi AF, Aljutayli A, Albassam A, et al. Clinical pharmacology applications in clinical drug development and clinical care: a focus on Saudi Arabia. *Saudi Pharmaceutical Journal*. 2020;28(10):1217–27.
124. Zhou Y, Peng S, Wang H, Cai X, Wang Q. Review of personalized medicine and pharmacogenomics of anti-cancer compounds and natural products. *Genes*. 2024;15(4):468.
125. Ortiz Hidalgo C. Immunohistochemistry in historical perspective: knowing the past to understand the present. Immunohistochemistry and Immunocytochemistry: *Methods and Protocols*. 2022:17–31.
126. Jang D-i Lee A-H, Shin H-Y, Song H-R, Park J-H, Kang T-B, et al. The role of tumor necrosis factor alpha (TNF-α) in autoimmune disease and current TNF-α inhibitors in therapeutics. *International Journal of Molecular Sciences*. 2021;22(5):2719.
127. Rahman MA, Islam F, Hasan M, Joya IS, Mithila MM, Islam MF, et al. Monoclonal antibody: a cell specific immunotherapy to treat cancer. *International Journal of Basic & Clinical Pharmacology*. 2023;12(2):290.
128. Kardile V, Kulkarni A, Nadar B, Saldanha T. Monoclonal antibodies in oncology: a decade of novel options. *Cell Biochemistry and Biophysics*. 2023;81(3):395–408.
129. Nicholls SJ. PCSK9 inhibitors and reduction in cardiovascular events: current evidence and future perspectives. *Polish Heart Journal (Kardiologia Polska)*. 2023;81(2):115–22.
130. Taylor PC, Adams AC, Hufford MM, De La Torre I, Winthrop K, Gottlieb RL. Neutralizing monoclonal antibodies for treatment of COVID-19. *Nature Reviews Immunology*. 2021;21(6):382–93.
131. Zheng S, Liu J, Wu Y, Huang TL, Wang G. Small-molecule inhibitors of Wnt signaling pathway: towards novel anticancer therapeutics. *Future Medicinal Chemistry*. 2015;7(18):2485–505.
132. Metibemu DS, Akinloye OA, Akamo AJ, Ojo DA, Okeowo OT, Omotuyi IO. Exploring receptor tyrosine kinases-inhibitors in Cancer treatments. *Egyptian Journal of Medical Human Genetics*. 2019;20(1):1–16.
133. Bhavani N, Reddy MS. Imatinib mesylate: recent drug used in oncology. *GSC Biological and Pharmaceutical Sciences*. 2023;22(2):173–86.
134. Pottier C, Fresnais M, Gilon M, Jérusalem G, Longuespée R, Sounni NE. Tyrosine kinase inhibitors in cancer: breakthrough and challenges of targeted therapy. *Cancers*. 2020;12(3):731.
135. Koopman B, Cajiao Garcia BN, Kuijpers CC, Damhuis RA, van der Wekken AJ, Groen HJ, et al. A nationwide study on the impact of routine testing for EGFR mutations in advanced NSCLC reveals distinct survival patterns based on EGFR mutation subclasses. *Cancers*. 2021;13(14):3641.
136. Marin-Acevedo JA, Dholaria B, Soyano AE, Knutson KL, Chumsri S, Lou Y. Next generation of immune checkpoint therapy in cancer: new developments and challenges. *Journal of Hematology & Oncology*. 2018;11:1–20.
137. Hargadon KM, Johnson CE, Williams CJ. Immune checkpoint blockade therapy for cancer: an overview of FDA-approved immune checkpoint inhibitors. *International Immunopharmacology*. 2018;62:29–39.
138. Qin S, Xu L, Yi M, Yu S, Wu K, Luo S. Novel immune checkpoint targets: moving beyond PD-1 and CTLA-4. *Molecular Cancer*. 2019;18:1–14.
139. Graziani G, Lisi L, Tentori L, Navarra P. *Monoclonal antibodies to CTLA-4 with focus on ipilimumab*. Interaction of Immune and Cancer Cells: Springer; 2022. p. 295–350.
140. Ai L, Xu A, Xu J. Roles of PD-1/PD-L1 pathway: signaling, cancer, and beyond. *Regulation of Cancer Immune Checkpoints: Molecular and Cellular Mechanisms and Therapy*. 2020; 1248:33–59.
141. Liu J, Chen Z, Li Y, Zhao W, Wu J, Zhang Z. PD-1/PD-L1 checkpoint inhibitors in tumor immunotherapy. *Frontiers in Pharmacology*. 2021;12:731798.
142. Topalian SL, Hodi FS, Brahmer JR, Gettinger SN, Smith DC, McDermott DF, et al. Five-year survival and correlates among patients with advanced melanoma, renal cell carcinoma, or non–small cell lung cancer treated with nivolumab. *JAMA Oncology*. 2019;5(10):1411–20.

143. Postow MA, Sidlow R, Hellmann MD. Immune-related adverse events associated with immune checkpoint blockade. *New England Journal of Medicine*. 2018;378(2):158–68.
144. Friedman CF, Spencer C, Cabanski CR, Panageas KS, Wells DK, Ribas A, et al. Ipilimumab alone or in combination with nivolumab in patients with advanced melanoma who have progressed or relapsed on PD-1 blockade: clinical outcomes and translational biomarker analyses. *Journal for Immunotherapy of Cancer*. 2022;10(1).
145. Ziogas DC, Theocharopoulos C, Lialios P-P, Foteinou D, Koumprentziotis I-A, Xynos G, et al. Beyond CTLA-4 and PD-1 inhibition: novel immune checkpoint molecules for melanoma treatment. *Cancers*. 2023;15(10):2718.
146. Haritha PN, Uma S, Nagaratna DP, Chaitanya S, Gunasekharan V. Gene therapy–A review. *International Journal of Biopharmaceutics*. 2012;3(1):55–64.
147. Li C, Samulski RJ. Engineering adeno-associated virus vectors for gene therapy. *Nature Reviews Genetics*. 2020;21(4):255–72.
148. Ferraz Sallum JM, Godoy J, Kondo A, Kutner JM, Vasconcelos H, Maia A. The first gene therapy for RPE65 biallelic dystrophy with voretigene neparvovec-rzyl in Brazil. *Ophthalmic Genetics*. 2022;43(4):550–4.
149. Liu N, Olson EN. CRISPR modeling and correction of cardiovascular disease. *Circulation Research*. 2022;130(12):1827–50.
150. Dunbar CE, High KA, Joung JK, Kohn DB, Ozawa K, Sadelain M. Gene therapy comes of age. *Science*. 2018;359(6372):eaan4672.
151. Wang X, Zhang H, Chen X. Drug resistance and combating drug resistance in cancer. *Cancer Drug Resistance*. 2019;2(2):141.
152. Coller BS. Ethics of human genome editing. *Annual Review of Medicine*. 2019;70:289–305.
153. Chandrasekhar C, Kumar PS, Sarma PVGK. Novel mutations in the kinase domain of BCR-ABL gene causing imatinib resistance in chronic myeloid leukemia patients. *Scientific Reports*. 2019;9(1):2412.
154. Ero R, Yan X-F, Gao Y-G. Ribosome protection proteins—"New" players in the global arms race with antibiotic-resistant pathogens. *International Journal of Molecular Sciences*. 2021;22(10):5356.
155. Amawi H, Sim H-M, Tiwari AK, Ambudkar SV, Shukla S. ABC transporter-mediated multidrug-resistant cancer. *Drug Transporters in Drug Disposition, Effects and Toxicity*. 2019:549–80.
156. Kaur G, Gupta S, Singh P, Ali V, Kumar V, Verma M. Drug-metabolizing enzymes: role in drug resistance in cancer. *Clinical and Translational Oncology*. 2020;22:1667–80.
157. Karami Fath M, Azargoonjahromi A, Kiani A, Jalalifar F, Osati P, Akbari Oryani M, et al. The role of epigenetic modifications in drug resistance and treatment of breast cancer. *Cellular & Molecular Biology Letters*. 2022;27(1):52.
158. Kinslow CJ, Mercurio A, Kumar P, Rae AI, Siegelin MD, Grinband J, et al. Association of MGMT promoter methylation with survival in low-grade and anaplastic gliomas after alkylating chemotherapy. *JAMA Oncology*. 2023;9(7):919–27.
159. Seebacher NA, Krchniakova M, Stacy AE, Skoda J, Jansson PJ. Tumour microenvironment stress promotes the development of drug resistance. *Antioxidants*. 2021;10(11):1801.
160. Califf RM. Biomarker definitions and their applications. *Experimental Biology and Medicine*. 2018;243(3):213–21.
161. Iovanna J. Implementing biological markers as a tool to guide clinical care of patients with pancreatic cancer. *Translational Oncology*. 2021;14(1):100965.
162. Asgari-Karchekani S, Aryannejad A, Mousavi SA, Shahsavarhaghighi S, Tavangar SM. The role of HER2 alterations in clinicopathological and molecular characteristics of breast cancer and HER2-targeted therapies: a comprehensive review. *Medical Oncology*. 2022;39(12):210.
163. Amarante-Mendes GP, Rana A, Datoguia TS, Hamerschlak N, Brumatti G. BCR-ABL1 tyrosine kinase complex signaling transduction: challenges to overcome resistance in chronic myeloid leukemia. *Pharmaceutics*. 2022;14(1):215.
164. Chen J, Zheng N. Accelerating protein biomarker discovery and translation from proteomics research for clinical utility. *Bioanalysis*. 2020;12(20):1469–81.
165. Walk EE, Yohe SL, Beckman A, Schade A, Zutter MM, Pfeifer J, et al. The cancer immunotherapy biomarker testing landscape. *Archives of Pathology & Laboratory Medicine*. 2020;144(6):706–24.
166. Martínez-Castillo M, Gómez-Romero L, Tovar H, Olarte-Carrillo I, García-Laguna A, Barranco-Lampón G, et al. Genetic alterations in the BCR-ABL1 fusion gene related to imatinib resistance in chronic myeloid leukemia. *Leukemia Research*. 2023;131:107325.
167. Czarnecka AM, Bartnik E, Fiedorowicz M, Rutkowski P. Targeted therapy in melanoma and mechanisms of resistance. *International Journal of Molecular Sciences*. 2020;21(13):4576.

168. Saei A, Eichhorn PJA. Adaptive responses as mechanisms of resistance to BRAF inhibitors in melanoma. *Cancers.* 2019;11(8):1176.
169. Hockings HA. *Developing adaptive therapy to suppress the evolution of treatment resistance in high-grade serous ovarian cancer.* Queen Mary University of London; 2021.
170. Zhang C, Fei Y, Wang H, Hu S, Liu C, Hu R, et al. CAFs orchestrates tumor immune microenvironment—A new target in cancer therapy? *Frontiers in Pharmacology.* 2023;14:1113378.
171. Ashworth A, Lord CJ. Synthetic lethal therapies for cancer: what's next after PARP inhibitors? *Nature Reviews Clinical Oncology.* 2018;15(9):564–76.
172. Manzari MT, Shamay Y, Kiguchi H, Rosen N, Scaltriti M, Heller DA. Targeted drug delivery strategies for precision medicines. *Nature Reviews Materials.* 2021;6(4):351–70.
173. Tu W, Tu Y, Tan C, Zhong H, Xu S, Wang J, et al. Elucidating the role of T-cell exhaustion-related genes in colorectal cancer: a single-cell bioinformatics perspective. *Functional & Integrative Genomics.* 2023;23(3):259.
174. Sha G, Zhang W, Jiang Z, Wang D, Tang D, Zhao Q. Exosomal non-coding RNA: a new frontier in diagnosing and treating pancreatic cancer: a review. *International Journal of Biological Macromolecules.* 2024:130149.
175. Roux PP, Topisirovic I. Signaling pathways involved in the regulation of mRNA translation. *Molecular and Cellular Biology.* 2018;38:37.
176. Manhas J, Edelstein HI, Leonard JN, Morsut L. The evolution of synthetic receptor systems. *Nature Chemical Biology.* 2022;18(3):244–55.
177. Su FY, Mac QD, Sivakumar A, Kwong GA. Interfacing biomaterials with synthetic T cell immunity. *Advanced Healthcare Materials.* 2021;10(15):2100157.
178. Monti C, Zilocchi M, Colugnat I, Alberio T. Proteomics turns functional. *Journal of Proteomics.* 2019;198:36–44.
179. Yang BA, Westerhof TM, Sabin K, Merajver SD, Aguilar CA. Engineered tools to study intercellular communication. *Advanced Science.* 2021;8(3):2002825.

18 Role of Cell Signaling in Drug Development

Taimoor Musharaf, Anoosha Ahmed, Muhammad Asim Siddiqui, Wajahat Sultan Abbasi, and Saadia Andleeb
Atta ur Rahman School of Applied Biosciences, National University of Sciences & Technology, Pakistan

18.1 INTRODUCTION

Cell signaling plays a critical role in drug development by influencing the safety, efficacy, and mechanism of action of drugs. Dysregulation of signaling pathways can cause disease, influenced by many factors including infections, hereditary mutations, and environment (Aggarwal et al., 2007). This dysregulation disrupts protein expression in many signaling pathways, resulting in altered processes such as apoptosis, growth, and differentiation.

Conventionally, drug discovery has followed physiologically based methods, traditionally targeting systemic aspects. This traditional approach has aimed to establish drugs with a broad effect on the entire physiological system. The complexity and diversity of diseases, which result from a variety of underlying causes, have raised awareness of the shortcomings of this broad systemic approach.

In response to this dilemma, there is a shift towards a more targeted approach to drug development. In 1908, Paul Ehrlich (Strebhardt & Ullrich, 2008) launched a revolutionary search for "magic bullets" that would specifically target the components of infectious organisms rather than those of the host. Ehrlich used his expertise in treating infectious disorders with medications sourced from the German dye industry to develop his receptor theory. His strategy involved focusing on the invaders' receptors, which were different from the host's receptors. This novel approach aimed to greatly reduce the risk of negative consequences in patients. Ehrlich's groundbreaking work established a more selective and less risky approach to drug therapy by focusing on receptors specific to infectious agents, paving the way for the creation of safer and more potent drugs.

Cellular signaling can be targeted at three basic levels to aid drug development: signal reception, intracellular components, and intercellular responses.

Many signaling molecules bind to receptors in the cells. The majority of drugs (Rajasethupathy et al., 2005) currently in development target these receptors. This highlights the important role that receptor-targeted interventions play in drug development. By manipulating and targeting receptor activity, drugs can impact cellular signaling pathways, providing an approach that can lead to effective treatment of various diseases.

Intracellular components of signaling pathways can also be targeted by drugs. Kinase inhibitors (J. Zhang et al., 2009), for instance, are a family of drugs that inhibit signaling cascades by blocking specific enzymes involved in these pathways. This level of intervention is essential for illnesses where dysregulated intracellular signaling leads to pathogenesis.

Drugs can be designed to influence intercellular communication. Immunomodulatory drugs (Yao et al., 2013), for example, target immune response-related signaling pathways in an effort to either strengthen or weaken particular immune system functions.

DOI: 10.1201/9781003594901-18

This chapter provides a detailed exploration of cell signaling's molecular intricacies, encompassing target identification, diverse drug discovery approaches, and the impact of signaling pathways on human diseases, and concludes with insights into personalized medicine. It offers a comprehensive study, including case studies, clinical considerations, and an examination of emerging challenges and trends in cell signaling-driven drug development.

18.2 TARGET IDENTIFICATION AND VALIDATION

Signaling diseases can arise from alteration of signaling cascades in the body. Treating these signaling issues has been a major point of concern. A plethora of research has been carried out in this regard and is still ongoing (Behar et al., 2013).

Cell signaling plays an integral role in drug development. Its importance stems from the fact that cell signaling is how cells respond to the stimuli taking place in their surroundings and maintain homeostasis. Intracellular signaling provides a major breakthrough for drug therapy by targeting various components and molecules in the signaling cascade (Sygnature Discovery, n.d.). These can be modulated and altered to treat certain diseases. Determining key molecules in the signaling cascade is important for drug discovery and development. Various molecules can be inactivated, and receptors targeted to enable drug development and disease treatment.

Cell signaling pathways provide an efficacious means to test different drug targets. This method has lowered the drug failure rate compared to other methods of drug development (Abhyankar et al., 2018a).

Identifying key signaling molecules in cell signaling pathways can be a major breakthrough in drug development. These signaling pathways are responsible for connecting important cellular receptors with cellular machinery. It is crucial to identify specific targets actually involved in cellular functions. Multiple drugs can be developed to address the over-activation of these signaling cascades and their concomitant implications (Mavers et al., 2009).

18.2.1 Development of Target-Specific Drugs

Target-specific drugs provide a more focused approach to treating diseases. They target and destroy the central molecule involved in disease progression and spread. The preliminary step is to determine whether that target molecule is involved in causing the disease or not. Once confirmed, targeted therapies using specific drugs are employed. Specific inhibitors, including chemicals and antibodies, can be used in this process.

18.2.2 Target Identification

This is the process of identifying potential drug targets within cell signaling pathways (Schenone et al., 2013).

18.2.3 Properties of an Ideal Target

An ideal target possesses these traits:

- It plays a role in the pathophysiology of the disease.
- It is localized to specific areas in the body.
- It can be assayed easily by high-throughput screening.
- Target regulation by a small molecule drug is favorable, and potential side effects can be predicted using phenotypic data (UCL, n.d.).

Several advancements in cell signaling have led to breakthroughs in target identification and drug development. Some well-known examples of targeted drugs include trastuzumab for breast cancer treatment, gefitinib and erlotinib for lung cancer treatment, and imatinib and rituximab for treating hematopoietic tumors (Kudo, 2012).

Targeted drugs have been developed to act on various receptors such as epidermal growth factor receptor (EGFR) and vascular endothelial growth factor (VEGFR). Kinase inhibitors have also been effectively developed to treat various cancers. While systemic chemotherapy has been the major treatment for multiple cancer types, the advent of targeted drug therapies has led to proposals to use systemic chemotherapy in conjunction with targeted approaches for better and specific treatment. This combined approach may provide more long-lasting and efficacious treatment for various diseases, especially cancers. Researchers are comparing the efficacy of existing drugs with the latest studies to develop the most viable and applicable treatment options.

18.2.4 Techniques for Target Identification

Multiple methods exist for target identification in signaling cascades:

18.2.4.1 Direct Biochemical Methods

Biochemical affinity purification determines drug binding to target molecules, providing information on efficacy and toxicity due to direct interactions with human/mammalian proteins. This method involves analyzing chromatographic fractions after exposing extracts to immobilized compounds to observe enzyme activity, followed by elution. This technique has identified both natural and synthetic targets, including thalidomide's toxic effects.

However, this method is best suited for abundant drug targets that bind to high-affinity ligands. Control beads are coated with inactive analogs.

Recent affinity-based methods use ultraviolet light or chemical-induced cross-linking to capture low-abundance and low-affinity proteins for the small molecule. This approach employs a biased study using specific enzyme activity to reduce nonspecific background noise.

Stable isotope labeling by amino acids in cell culture has also been used. This method provides an unbiased assessment of direct and indirect targets.

18.2.4.2 Genetic Interaction and Genomic Methods

These methods use DNA and RNA for target identification, employing genetic modifiers (suppressors or enhancers) and gene interaction principles. Target identification can combine small molecule and RNA interference (RNAi) disruption results. Genome-scale RNAi experiments can determine phenotypes similar to those caused by small molecules.

18.2.4.3 Computational Inference Methods

Computational methods provide information on drug targets of small molecules and offer analytical knowledge from proteomic and genomic studies. They can identify new drug targets for existing drugs, analyze ligand structures to predict drug targets, and use structure-based methods to predict protein and small molecule interactions.

18.2.5 Target Validation

Target validation is a crucial step in the drug development process. These drugs target certain molecules and act as inhibitors in signaling cascades. However, it is essential to determine whether the target is actually instrumental in the disease process, which is termed target validation. The role of a specific molecule in disease prognosis must be confirmed before designing drugs for it (Smith, 2003).

Target validation techniques ensure that the drug does not lead to toxicity or inefficacy. They involve employing pre-requisite techniques to confirm that the drug that has been developed to target a specific part of the pathway is indeed effective and does not have harmful implications (Rousseaux et al., 2023).

Target validation involves an extensive study of the disease mechanism, how a particular molecule or substance can be an essential target in disease prognosis, and validating its role in that pathway. A combination of techniques and assays is used to ensure that the molecule or target in question is an important precursor of disease initiation and sustenance. Animal models are an important source of confirmation to identify and validate the role of targets in certain diseases, enabling effective treatment initiation. However, it is crucial to identify whether or not similarities exist in the course of a disease in humans and animals to determine the efficacy of target identification and validation.

Validating a specific target for treating a disease is an extremely difficult and tedious task. It can be done through genetic observations and analysis, targeting either a gene, its product, or a molecule in the signaling cascade.

18.2.6 Techniques for Target Validation

Some important techniques for target validation are as follows:

18.2.6.1 Model Interactions

Computer model interactions provide a comparatively fast way to determine target and drug interactions. By knowing the structure of the target protein's active site, potential drug-binding sites can be predicted, and drug candidates can be determined.

This technique determines the three-dimensional structure and function of the target protein and compares it with pre-existing literature. Several techniques, like NMR and X-ray crystallography, can be employed to determine protein structure.

18.2.6.2 Sense Reversal

Another important method for target validation is gene alteration and changing the production rate of a specific target protein. Once the expression of the target protein is reduced, its physiological effects can be determined and its role as the target can be confirmed or rejected. Examples of this technique include RNAi, gene knockouts, and antisense technology (Smith, 2003).

Antisense technology is one of the most efficacious techniques in this regard. It alters gene expression by using oligonucleotide sequences that are to destroy specific mRNA.

18.2.6.3 The Proteomics Approach

One drawback of manipulating gene expression is that several isoforms of a protein can be altered, so the specific protein might not be targeted. To address this issue, proteomics analysis can be used to identify specific protein targets. Proteomics alters the cellular makeup of a specific protein, instead of changing multiple proteins.

18.2.6.4 Validation In Vivo

This technique uses animal models for target validation. However, a major challenge is identifying and developing the correct animal models for relevant human diseases. Specific genes are knocked out in the animal model. Mouse models have been effective in vivo targets in this technique.

Zebrafish is another significant animal model alternative for in vivo validation. Due to the similarity in disease-causing genes between humans and zebrafish, it has proved to be an important animal model in target validation.

18.3 DRUG DISCOVERY APPROACHES IN CELL SIGNALING

After target validation, the next important step is developing a suitable, highly efficacious drug for treating the disease. Drug discovery and development in cell signaling involves creating small molecule drugs through various techniques. These small molecules are organic compounds that influence key points in disease-related signaling cascades, targeting proteins and entering cells. Their small size makes them suitable for a wide array of applications. They can easily enter and penetrate cells, perform their function efficiently, be consumed orally, and target intracellular machinery. These small molecule drugs can be developed through rational drug design, high-throughput screening, or obtained from natural resources (Ma et al., 2021).

18.3.1 Rational Drug Design

Rational drug design creates drug molecules that bind to specific targets. This approach prevents random testing of thousands of molecules, allowing for efficient drug design. It focuses on existing knowledge regarding the structure and function of molecules (Bioexcel, n.d.).

In rational drug design, the structure of the target and other important aspects are determined along with molecular and biological characteristics.

Rational drug design utilizes computational methods to develop suitable drugs (Mahapatra & Karuppasamy, 2022). Two common approaches in rational drug design are structure-based and ligand-based drug design.

As the name indicates, structure-based drug design focuses on the three-dimensional structure of the target, using computational analysis. Specific structures of binding sites and targets are determined, and this information is used to develop drugs.

However, when three-dimensional structures are unavailable, ligand-based drug design is used. This technique determines potential interactions between drug-like molecules and targets, using this information to develop suitable drugs.

18.3.2 High-Throughput Screening

High-throughput screening (HTS) is an important technique for conducting many tests in a short time. These tests can be chemical or biological. This technique is significant in drug discovery for identifying and classifying relevant biological compounds (Pusterla, 2019).

This technique has gained immense popularity in drug discovery. It screens a large number of substances against their targets, capable of screening thousands of compounds and expediting the drug discovery process. It classifies the drugs based on toxicological, pharmacokinetic, and metabolic parameters, reducing overall drug development costs by avoiding repetitive processes (Szymański et al., 2011).

HTS involves several steps: target identification, reagent preparation, assay development, and high-throughput library screening.

HTS focuses on the alterations in chemical synthesis, allowing simultaneous chemical testing of many. By combining techniques like homogeneous time resolved-fluorescence (HTRF) and fluorescence resonance energy transfer (FRET), compounds can be identified easily.

High-throughput screening involves primary and secondary screening. Primary screening is less quantitative and specific than secondary screening. Compounds with positive primary screening results proceed to secondary screening. Both heterogeneous and homogeneous assays can be performed, with heterogeneous assays being more sensitive.

This technique uses cell-based assays to screen molecules involved in different biological activities. Microtiter plates with two-dimensional monolayer cell cultures are used for microarrays. Living cells and the cellular responses can be analyzed as biomolecules are displayed on solid-support microarrays. These tests determine both the efficacy and toxicity of compounds.

After these processes, clinical testing of the discovered drug begins for marketing approval. The next crucial step is ensuring drug safety, confirming no harmful impacts. The rate of drug development and approval varies widely, with many drugs rejected in clinical trials due to toxic effects (Field MJ, 2010).

18.3.3 Biologics of Drug Development

Protein therapeutics for disease treatment have gained significant attention in recent years. A major advantage of biologics is their high specificity and larger surface area for binding, resulting in fewer off-target effects compared to small molecule drugs.

Different types of biologic therapies include growth factors, cytokines, and monoclonal antibodies. Monoclonal antibodies have revolutionized cancer and inflammatory disease treatment.

18.3.3.1 Monoclonal Antibodies (MABs)

Monoclonal antibodies are laboratory-made proteins that mimic natural antibodies in our blood. Each monoclonal antibody consists of copies of a single antibody type. They are an effective treatment option for cancer, recognizing specific proteins on cell surfaces. Different monoclonal antibodies target signaling cascades in various cancer types. Their mode of action depends upon the targeted protein (Cancer Research UK, n.d.).

Monoclonal antibodies have proven highly effective against cancer, offering advantages over small molecule drugs (Shepard et al., 2017). Their high selectivity and specificity reduce off-target binding and toxicity. Recent advancements include using monoclonal antibodies to target multiple substances simultaneously, such as both IL-17 and tumor necrosis factor (TNF)-α, increasing their therapeutic potential (Aggarwal et al., 2007).

18.3.4 Examples of Targeted Drug Development in Cell Signaling

In cancer, signaling pathways with dysregulated protein kinases can be identified.

In one study, monoclonal antibodies targeting the Her-2 receptor and EGFR were used to formulate drugs for colon cancer treatment, namely cetuximab and trastuzumab.

18.3.4.1 Treatment of Chronic Myelogenous Leukemia

Chronic myelogenous leukemia can be treated by inhibiting BCR/ABL kinase through the drug imatinib mesylate.

Gleevec, an inhibitor binding to the inactive form of BCR-ABL, was initially effective. However, tumor cells eventually developed resistance, reducing its efficacy.

18.3.4.2 Treatment of Lung, Ovarian, and Breast Cancer

EFGR overexpression is a leading cause of cancer in the lungs, ovaries, and breasts, often due to EGFR mutation. Inhibitors can treat cancer with EGFR upregulation. One approach uses drugs like gefitinib and erlotinib, which are competitive adenosine triphosphate (ATP) inhibitors at EGFR's tyrosine kinase domain. Cetuximab, a monoclonal antibody, can also target EGFR.

In breast cancer patients, ErbB2 is upregulated in 25–30% of cases. Trastuzumab, a humanized monoclonal antibody, has been approved for breast cancer treatment. This protein kinase inhibitor acts on the HER2 receptor.

18.3.4.3 Treatment of Metastatic Colorectal Cancer

Bevacizumab, a recombinant humanized monoclonal antibody, inhibits human vascular endothelial growth factor (VEGF) by binding to it. It is used for treating metastatic colorectal cancer by inhibiting the formation of new blood vessels.

18.3.4.4 Treatment of Autoimmune Diseases

Infliximab, a chimeric monoclonal antibody, has been approved to reduce the *TNF-α* levels in the body. Another such inhibitor is etanercept, a recombinant human soluble *TNF-α* receptor fusion protein. These drugs treat conditions caused by TNF overexpression, including rheumatoid arthritis and autoimmune diseases. Adalimumab, a human monoclonal antibody, is used to treat Crohn's disease.

18.3.4.5 Treatment of Melanomas

Bortezomib, a proteasome inhibitor, inhibits the proteasome enzyme involved in cell growth regulation and function. By inhibiting proteasome function, it can prevent several types of melanomas.

18.3.4.6 Treatment of Arthritis

Cyclooxygenase enzymes, which convert arachidonic acid into prostaglandins, are implicated in cancer and inflammation. Elevated levels of these enzymes can indicate inflammation. Celecoxib, an FDA-approved cyclooxygenase-2 inhibitor, is used to treat rheumatoid arthritis in adults, osteoarthritis, and familial adenomatous polyposis (FAP), in which polyps cover the rectum and colon.

18.3.4.7 Treatment of B Cell Lymphomas

Rituximab, a chimeric anti-CD20 IgG1 monoclonal antibody, is used to treat B cell lymphomas. It shows greater efficacy when combined with cyclophosphamide, doxorubicin, vincristine, and prednisone.

18.3.4.8 Preventing High Cholesterol Levels and Coronary Diseases

High cholesterol levels can lead to coronary diseases. Statins, derived from *Aspergillus terreus*, inhibit HMG-CoA reductase, the rate-limiting enzyme in cholesterol biosynthesis. However, high doses of statins can be extremely harmful.

18.4 SIGNALING PATHWAYS INVOLVED IN VARIOUS DISEASES AND POTENTIAL DRUG TARGETS

Cell signaling pathways play critical roles in various diseases, including cancer and neurodegenerative diseases. This section discusses these pathways and their potential therapeutic targets.

18.4.1 Cancer

Cancer, characterized by uncontrolled cell proliferation, is caused by multiple factors and alterations in basic cell division mechanisms (Matthews et al., 2022). In 2021, over 19.3 million new cancer cases were reported worldwide, with about 10 million cancer-related deaths (Ferlay et al., 2021).

Cancer-related inherited lesions include oncogenes (e.g., MYC, KIT, BRAF, and Ras) and tumor suppressor genes (e.g., PTEN, TP53, and BRCA1) (P. J. Campbell et al., 2020). Signaling pathways controlling crucial cellular processes are associated with potential treatment for cancer onset. The most commonly mutated pathways in cancers are the PI3K/AKT/mTOR (Peng et al., 2022) and the Ras/MAPK pathways (B. B. Campbell et al., 2021). Drugs targeting these pathways provide therapeutic options for cancer treatment.

18.4.1.1 PI3K/AKT/mTOR Signal Transduction Pathway in Cancer Progression

The PI3K pathway is activated when growth factors bind to receptor tyrosine kinases (RTKs) or G protein-coupled receptors (GPCRs), leading to protein kinase B (AKT) activation (Xie & Weiskirchen, 2020). AKT activation results in the phosphorylation of mTOR, p70S6 kinase 1 (p70S6K), and eukaryotic translation factor 4E-binding protein 1 (4EBP1), controlling gene transcription associated with cancer aggressiveness. AKT also promotes cell death resistance by downregulating forkhead box O3 (FOXO3) and upregulating mouse double minute 2 homolog (MDM2) and pro-survival proteins such as Bcl2 and X-linked inhibitor of apoptosis protein (XIAP). mTORC2 fully activates AKT and induces actin reorganization, resulting in cell motility (Iksen et al., 2021). AKT regulates mTOR complex 1 (mTORC1) activation, which phosphorylates and recruits eIF4F, inhibiting autophagy. This activates S6K1, triggering S6 and resulting in protein synthesis and cell growth (Nepstad et al., 2020). Dysregulation of this pathway can lead to cancer (see Figure 18.1).

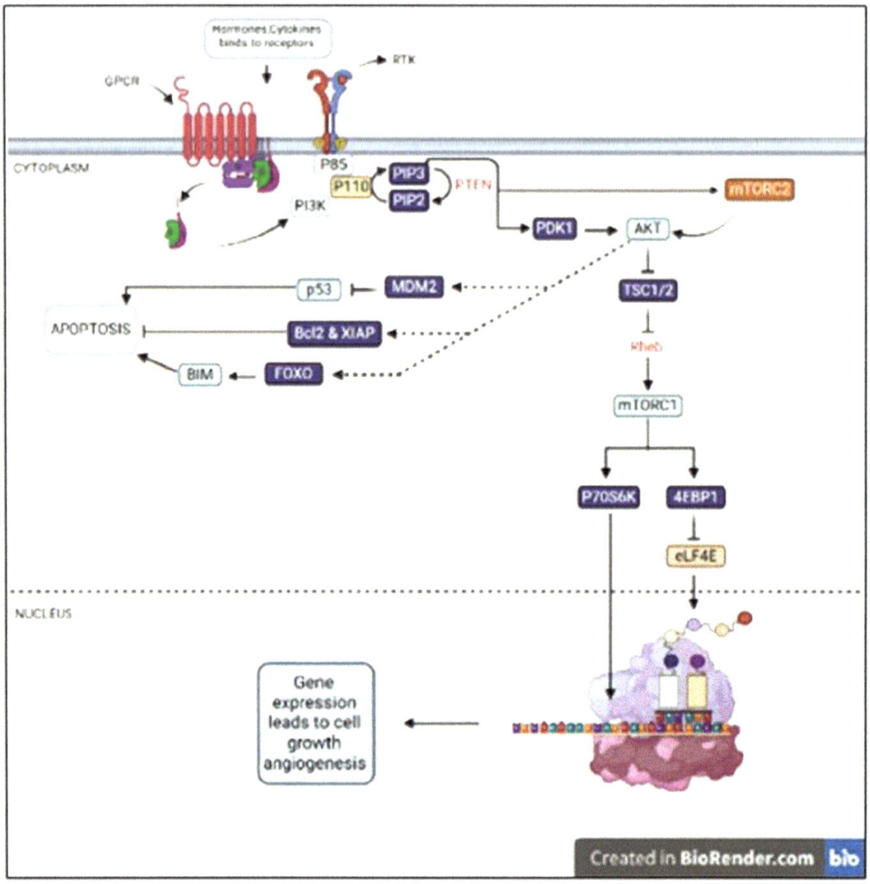

FIGURE 18.1 Overview of PI3K/AKT/mTOR signal transduction pathway in cancer progression. The PI3K pathway is initiated by cytokines and hormones attaching to receptors (RTKs or GPCRs), activating AKT, and resulting in mTOR, p70S6K, and 4EBP1 activation. This pathway regulates gene transcription linked to cancer progression, triggers cell survival, and stimulates mTORC1. mTORC1 hinders autophagy and activates S6, leading to protein synthesis and cell growth. AKT, controlled by mTORC2, also stimulates cell motility *(created in BioRender.com).*

18.4.1.2 Drugs Targeting PI3K/AKT/mTOR Signal Transduction Pathway

18.4.1.2.1 PI3K-Specific Inhibitors

Scientists have developed three categories of PI3K-specific inhibitors based on their pharmacokinetic properties and ability to bind and interact with ATP-binding clefts: pan-PI3K inhibitors, isoform-selective PI3K inhibitors, and dual PI3K/mTOR inhibitors (Peng et al., 2022).

Pan-class I inhibitors block p110 isoform catalytic properties. Buparlisib (NVP-BKM120), a pan-class I inhibitor, showed limited efficacy in PI3K-mutated recurrent glioblastoma, as it resulted in incomplete inhibition of the PI3K pathway (Wen et al., 2019), highlighting the need for more effective inhibitors. However, buparlisib plus paclitaxel proved effective for metastatic head and neck squamous cell carcinoma (Zhan et al., 2021). Other pan-PI3K inhibitors include pilaralisib, sonolisib, ZSTK474, B591, TG-100-115, CH5132799, and copanlisib (Peng et al., 2022).

In 2021, the FDA approved umbralisib (a PI3Kδ/CK1ϵ inhibitor) for treating marginal zone lymphoma and follicular lymphoma (Fowler et al., 2021). Other isoform-selective PI3K inhibitors include alpelisib (a PI3kα inhibitor), serabelisib (PI3Kα), duvelisib (PI3kδ/γ), GSK2636771 (PI3Kβ), AZD8835 (PI3Kδ/α), and parsaclisib (PI3Kδ).

Dual PI3K/mTOR inhibitors interact with the ATP-binding cleft, minimizing the kinase activity of both PI3K and mTOR (Hu et al., 2020). These inhibitors, including dactolisib (BEZ235), paxalisib (GDC-0084), apitolisib (GDC-0980), gedatolisib (PF-05212384), and voxtalisib (SAR245409, XL765), have shown effectiveness in early trials for cancer treatment.

18.4.1.2.2 AKT Inhibitors

AKT is also an effective drug target for cancer treatment. AKT inhibitors (NITULESCU et al., 2016) are categorized based on their inhibition mechanisms and chemical frameworks. One type is ATP-competitive inhibitors, which include isoquinoline-5-sulfonamides such as H-89. At high concentrations, H-89 induces apoptosis in ovarian carcinoma cells (Kilanowska et al., 2022). However, ATP-competitive inhibitors have limitations, including non-selectivity towards AKT isozymes and low selectivity towards closely related kinases. These limitations prompted scientists to look for more selective inhibitors, leading to the development of allosteric inhibitors (Wu et al., 2010). Allosteric inhibitors are more specific and less toxic. Examples include 2,3-diphenylquinoxaline and analogs, alkyl phospholipids (ALPs), and indole-3-carbinol and analogs.

18.4.1.2.3 mTOR Inhibitors

These inhibitors are effective anticancer agents that inhibit cancer cell proliferation. One example is rapamycin and its derivatives, which have shown anti-tumor capabilities in clinical trials (Faivre et al., 2006) (Figure 18.2).

18.4.1.3 Ras/MAPK Pathway Signal Transduction Pathway in Cancer Progression

This pathway is commonly mutated in cancer. K-Raw oncogene mutations are found in codons 12, 13, 59, and 61, with codons 12 and 61 having the highest frequency of mutations. K-Raw mutations cause 15–50% of lung cancers and 72–90% of pancreatic cancers (Mascaux et al., 2005).

The MAPK pathway includes Ras and kinase cascades, including ERK, MEK, and MAPK. KSR (Kinase suppressor of Ras) triggers MAPK activation. Mutated Ras genes hyperactivate MAPK signaling, resulting in cell proliferation. DUSP6, a component of MAPK phosphatases, dephosphorylates ERK through negative feedback. In lung adenocarcinoma, NKW2-1 triggers DUSP6 transcription, which is then transported to the ribosome for translation. DUSP6 deactivates cytosolic ERK, while DUSP5 inhibits nuclear ERK (Nussinov et al., 2022).

KRAS helps tumors escape from the immune system in several ways. It decreases MHC-I expression and upregulates PD-L1 expression. It also recruits immunosuppressive immune cells by increasing cytokine secretion (Huang et al., 2021) (Figure 18.3).

18.4.1.4 Drugs Targeting Ras/MAPK Pathway

Recent studies on Ras structure, function, and signaling have facilitated the development of Ras targeting therapies. Direct suppression of Ras oncogenic signals involve RNA interference, targeting upstream and downstream proteins, and directly targeting Ras (Downward, 2003).

18.4.1.4.1 EGFR Inhibitors

When activated, the EGFR family triggers GDP to GTP conversion, activating Ras. This activation can be inhibited by EGFR inhibitors such as cetuximab (Muhammad et al., 2013).

18.4.1.4.2 RAF inhibitors

Rapidly Accelerated Fibrosarcoma (RAF) kinases are a family of serine/threonine kinases that play a crucial role in the RAS-RAF-MEK-ERK signaling pathway. These RAF inhibitors are a class of drugs designed to target and inhibit the activity of RAF kinases. RAF inhibitors can completely suppress the MAPK pathway. RAF dimer inhibitors are particularly effective and include

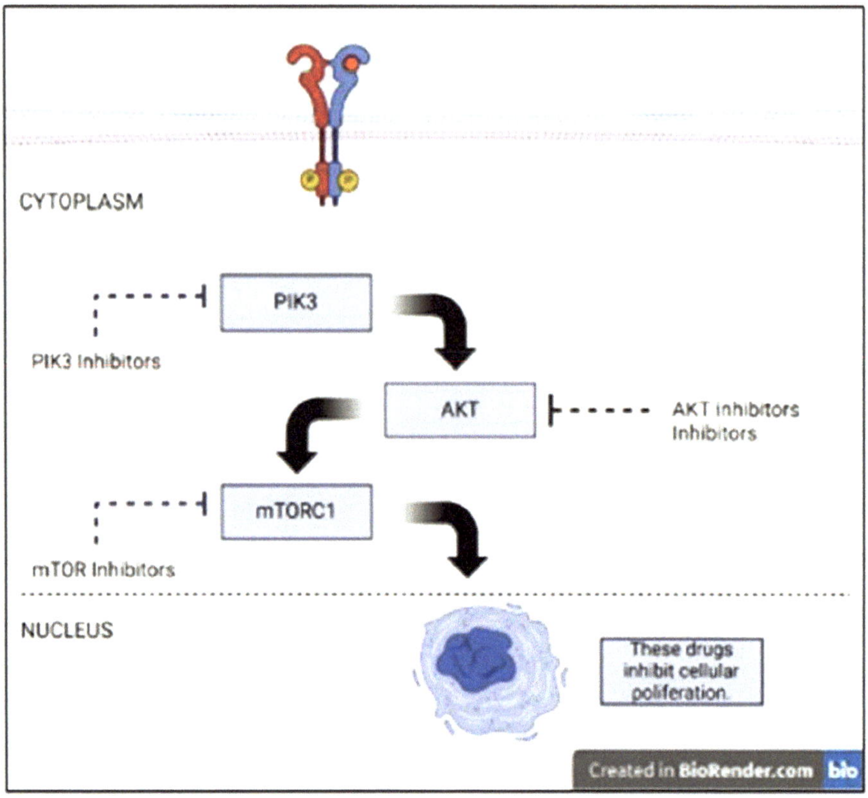

FIGURE 18.2 Different drug targets in the PI3K/AKT/mTOR signal transduction pathway *(created in BioRender.com)*.

belvarafenib, LY3009120, and LXH-254. These drugs attach to RAF in a DFG-motif 'out,' αC helix 'in' active position (Vasta et al., 2023). They are effective against both RAF monomers and dimers. Belvarafenib and LXH-254 are currently in phase I clinical trials for tumor treatment.

4.1.4.3 MEK Inhibitors

Some MEK1/2 inhibitors have shown promising therapeutic effects in uveal melanoma and melanoma with NRAS mutations, which have not responded well to RAF-targeted therapy (Zhao & Adjei, 2014). Three clinically approved MEK inhibitors (Han et al., 2021) are trametinib, cobimetinib, and binimetinib.

18.4.1.4.4 ERK Inhibitors

ERK-targeting inhibitors have been developed based on the understanding of ERK components and signaling. In the past 20 years, many selective ERK inhibitors have been developed, each with a different mechanism of action. Several ERK inhibitors have shown good anticancer properties but have been approved for market use (Pan et al., 2022). Examples include GDC-0994, KO-947, and ulixertinib (Figure 18.4).

18.4.2 Neurodegenerative Diseases

Neurodegeneration is characterized by structural and functional loss of neurons that can result in cognitive disabilities such as dementia. Age and genetic mutations are some of the causes of

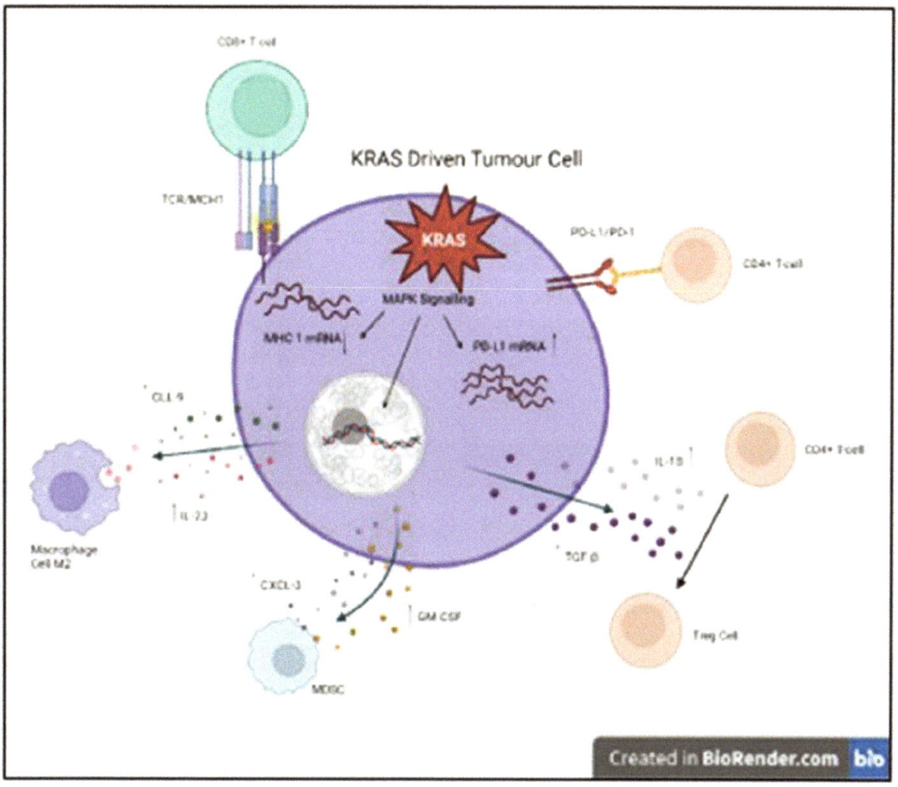

FIGURE 18.3 KRAS facilitates tumor escape from the immune system *(created in BioRender.com).*

neurodegenerative diseases. Age may be a causative factor in Alzheimer's and Parkinson's disease, whereas Huntington's disease may be caused by genetic mutations (Cell Signaling Technology, n.d.).

This section discusses signaling pathways involved in Alzheimer's and Parkinson's disease, along with potential drug targets for their treatment.

18.4.2.1 Alzheimer's Disease

Alzheimer's disease (AD) is the most widespread form of dementia, resulting in high mortality rates worldwide. Some of the main causes of this disease are extracellular amyloid-β (Aβ) aggregation, inflammation, and intracellular hyperphosphorylation (Kanti Das et al., 2014).

Here, we will discuss Aβ-induced neurotoxicity and microglial response (Guo et al., 2020) in AD.

Amyloid precursor protein (APP is a transmembrane protein that is primarily involved in cell adhesion and synaptic plasticity) cleavage generates Aβ in acidified regions such as late endosomes, which is then discharged from neurons. Extracellular Aβ assembly sequentially forms Aβ oligomer aggregates (oAβ), fibrils, and finally amyloid plaques. Aβ disrupts synaptic function through long-term potentiation (LTP is a long-lasting increase in synaptic strength that is primarily concerned with learning and memory) inhibition and LTD enhancement. oAβ binds to a range of neuronal Aβ receptors, such as EphA4, PrPc, LiLRB2, EphB2, and N-methyl-D-aspartate receptors (NMDARs are a type of ionic glutamate receptors that are involved in long term potentiation and long term depression, ultimately affecting synaptic plasticity and memory formation), causing synaptotoxicity. SORLA can block EphA4-mediated synaptic and cognitive dysfunction caused by oAβ exposure. Fyn kinase is a significant regulator of NMDAR-mediated oAβ neurotoxicity. oAβ also activates caspase-3, leading to altered mitochondrial function and aggravated synaptic dysfunction.

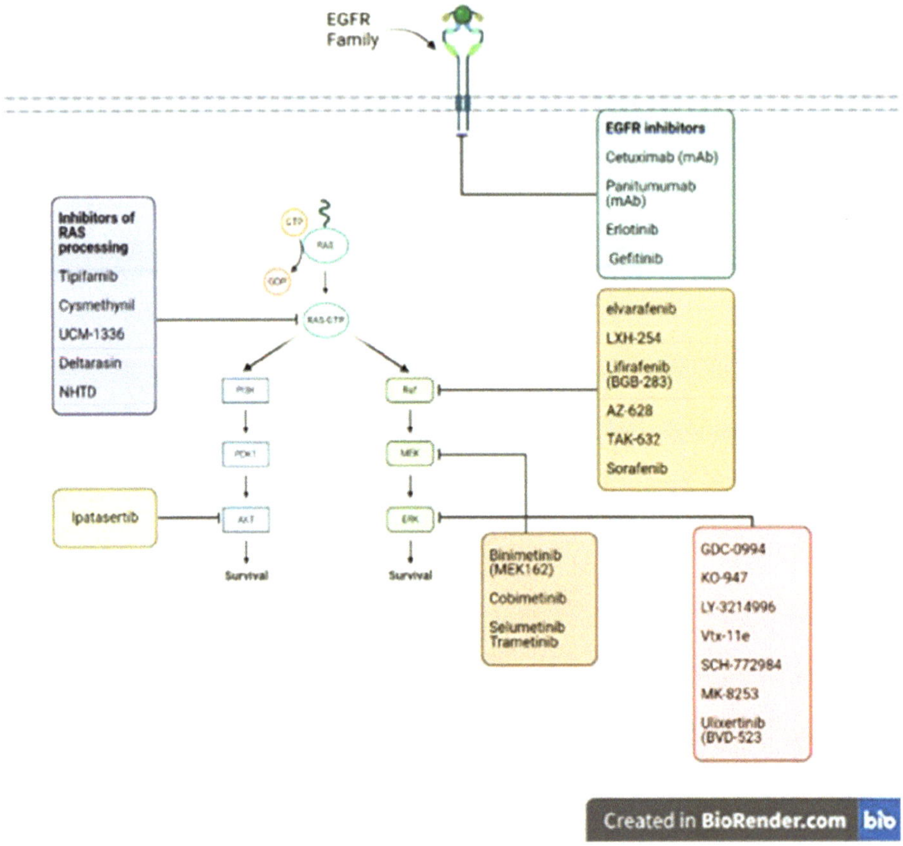

FIGURE 18.4 Therapeutic strategies targeting the Ras/MAPK pathway *(created with BioRender.com).*

oAβ may activate microglia by binding to putative Aβ receptors TREM2, LRP1, RAGE, TLR4, and CD36. DAP12, an adaptor protein for TREM2, activates the spleen tyrosine kinase (SYK is a type of cytoplasmic tyrosine kinase that plays a crucial role in immune cell activation, phagocytosis, and cell adhesion) pathway when Aβ binds to TREM2, causing causes Aβ disintegration. Astrocyte-produced Apolipoprotein E (APOE is a protein that is involved in the transportation of lipid molecules in the body) attaches to Aβ, increasing Aβ/APOE interactions with LRP1. Astrocytes can be activated when activated microglia release proinflammatory substances like TNF-α, IL-1, IL-6, and IL-8. Moreover, oAβ may initiate astrocyte activation through α7-nAchR, CaSR, CD36, CD47, and AQP4, potentially leading to neural damage through extracellular glutamate dyshomeostasis.

18.4.2.1.1 Targeting Amyloid-β (Aβ) Peptide for Treatment of AD

AD results in amyloid-β (Aβ) peptide accumulation in the brain, so targeting these peptides can be an effective treatment.

Five medications have been approved for AD treatment to date. Four are categorized as cholinesterase inhibitors (CIs) (Alhazmi & Albratty, 2022): galantamine, donepezil, rivastigmine, and tacrine. Except for donepezil, which treats severe or late-stage Alzheimer's disease, most are approved for mild-to-moderate stages. Tacrine is banned in the US as it causes liver toxicity. The fifth medication is memantine, an N-methyl-D-aspartate (NMDA) receptor antagonist that blocks glutamate activity, protecting neurons. However, these medications only ease symptoms rather than modify the disease. Therefore, disease-modifying therapies for AD are urgently needed.

18.4.2.2 Parkinson's Disease

Parkinson's disease (PD) affects 2–3% of the population and is the second-most common neurodegenerative disorder. Its pathological causes are neuronal loss in the substantia nigra, leading to striatal dopamine deficiency and intracellular inclusions containing α-synuclein aggregates (Poewe et al., 2017).

The ubiquitin-proteasome and lysosomal autophagy systems regulate intracellular α-synuclein homeostasis. α-synuclein accumulation (Yi et al., 2022) can occur when these pathways are dysregulated. Mitochondrial dysfunction, reactive oxygen species (ROS), and neuroinflammation may lead to α-synuclein accumulation. Furthermore, mutations in genes such as LRRK2, DJ-1, Parkin, and Pink1 can cause mitochondrial dysfunction, increasing cell death.

18.4.2.2.1 Drugs Targeting Pathways Involved in Parkinson's Disease

Current novel targets for PD include molecular chaperones, neuromelanin, neuroinflammation, mitochondrial dysfunction, the ubiquitin-proteasome system, cocaine-amphetamine-regulated transcripts, Abelson protein, and synaptic vesicle glycoprotein 2 (Pingale & Gupta, 2021). Larger studies are needed to confirm these drug targets' efficacy in PD. Gene therapy and various metal chelators have also shown potent therapeutic effects. Multi-target approaches show better efficiency for PD treatment (Rai et al., 2021).

18.5 PRECLINICAL TESTING AND MODEL SYSTEMS

18.5.1 Introduction to Preclinical Testing

Preclinical testing comes before the clinical trials and is critical for measuring the efficacy and safety of potential drug candidates. Before human evaluation, pharmaceutical products undergo rigorous safety assessments using in vitro models and preclinical species (Holt & Nuttall, 2013). Preclinical safety research aims to evaluate drug-associated risks, as mandated by regulatory bodies, to minimize risks in human clinical trials (ITR Laboratories Canada Inc., n.d.).

Preclinical efficacy research focuses on predicting drug benefits and establishing the drug's functionality in treating targeted diseases under specific usage conditions (ITR Laboratories Canada Inc., n.d.). These studies identify potential drug toxicities, unwanted side effects, and pharmacokinetics and pharmacodynamics. Preclinical studies can improve overall drug development success rates and reduce failure risks in clinical trials by identifying potential issues early on (Sun et al., 2022). While regulatory agencies provide extensive guidance for preclinical safety studies, there is little guidance for preclinical efficacy study design and objectives (Langhof et al., 2018). Therefore, successful preclinical testing is paramount for developing safe and effective drugs and ultimately for successful clinical trials (ITR Laboratories Canada Inc., n.d.).

18.5.2 Risks of Not Conducting Preclinical Testing for Drug Development

Several risks are associated with drug development if no preclinical testing is conducted:

- Efficacy failure: There is a higher risk of clinical trial failure without sufficient preclinical efficacy research, due to a lack of evidence of the drug's effectiveness in treating the target disease (Kimmelman & Federico, 2017).
- Safety concerns: Without proper preclinical safety studies, it is difficult to predict potential risks associated with a drug candidate. This can lead to safety issues during clinical trials or after market release (ITR Laboratories Canada Inc., n.d.).

- Wasted resources and time: Over half of the drugs that reach later stages of clinical trials fail due to lack of efficacy. Inadequate preclinical testing can result in significant financial losses and wasted research time (Kimmelman & Federico, 2017).
- Regulatory challenges: Regulatory bodies require studies such as genotoxicity tests for candidate drug safety assessment. Lack of preclinical testing can make it difficult to meet these regulatory requirements (Andrade et al., 2016).
- Reduced confidence in drug development: A well-designed preclinical plan is crucial for clinical success, providing information on drug safety, tolerability, and pharmacology related to the desired therapeutic outcome (Grzemski, n.d.).

18.5.3 In Vitro and In Vivo Model Systems for Preclinical Testing in Drug Discovery

In vitro and in vivo model systems play a crucial role in preclinical testing during drug discovery, helping researchers better understand animal cell behavior in controlled environments. These models also provide significant insights into drug efficacy and safety in humans (Arango MT, 2013). Mimicking physiological conditions is essential for the relevance and predictive value of these models (Langhans, 2018).

In vitro models offer several advantages, including better variable control, enhanced reproducibility, and the ability to study cellular and molecular mechanisms (Bédard et al., 2020). However, these models have limitations in accurately reflecting human physiological responses (Marei et al., 2022). Three-dimensional (3D) cell culture techniques such as hydrogels, spheroids, and organoids are being developed to better mimic the in vivo environment and provide more reliable and predictable results (Badr-Eldin et al., 2022).

In vivo models like zebrafish and *Caenorhabditis elegans* can be used at various stages of drug development, including target discovery and preclinical trials (Williams & Hong, 2011). These organisms offer a starting point for rational drug design and can help researchers avoid difficulties associated with conventional target-based drug design (Williams & Hong, 2011).

Microfabricated physiological models, such as microfluidics-based models, are being developed to enhance drug effectiveness by imitating various physiological conditions for in vitro drug screening. These models can provide a more accurate depiction of drug behavior (Marei et al., 2022; Ugolini et al., 2016).

18.5.4 Role of Cell Signaling in Drug Development

Understanding cell signaling is essential for predicting drug responses and improving drug efficacy. Cell signaling pathways facilitate intracellular immune cell communication, protection from foreign invaders, and regulation of transcriptional processes governing immune cell development (Abhyankar et al., 2018b).

The drug development industry has shown increased interest in systems biology approaches due to their potential to reduce failure risk and increase drug development success and efficiency. This approach studies cell signaling networks as graphs with nodes and edges (Abhyankar et al., 2018b). Signaling pathway activation status can serve as mechanism-based biomarkers for predicting drug sensitivity and providing insight into molecular modes of drug action (Amadoz et al., 2015).

For example, dysregulated cellular signaling pathways and genetic alterations in cancers like medullary thyroid carcinoma have prompted the development of targeted molecular therapies that inhibit specific signaling pathways, showing promise in treatment (Giunti et al., 2013). Intracellular signaling is a prime target for drug design, and assays measuring intracellular messengers and phosphoproteins can provide a comprehensive assessment of a candidate drug's influence on cellular signaling pathways (Sygnature Discovery, n.d.).

18.6 CLINICAL TRIALS AND REGULATORY CONSIDERATIONS IN CELL SIGNALING FOR DRUG DEVELOPMENT

18.6.1 Clinical Trials

Clinical trials (phases I–IV) evaluate the effectiveness and safety of new treatments in drug development. Each phase has a distinct function, helping researchers answer different questions.

Phase I trials study drug treatment safety and identify potential side effects in a small group of 20 to 80 people. The main goal is to ascertain the maximum dosage of a novel therapy or drug that can be administered without severe side effects (NIH, n.d.-b).

Phase II trials involve a larger group of 100 to 300 people to assess drug effectiveness and study safety more thoroughly. These trials evaluate safety, pharmacokinetics, and pharmacodynamics, and may provide information critical for planning phase III trials, such as determining optimal doses, frequency, routes of administration, and endpoints (Umscheid et al., 2011).

Phase III trials include a larger group of around 1000 to 3000 participants to confirm the effectiveness of the novel drug or treatment. They are conducted to demonstrate and/or confirm the efficacy of the candidate treatment or drug. They also seek to determine the prevalence of common adverse reactions and estimate their incidence. Phase III trials assess the novel drug's safety and efficacy in comparison to the current standard treatment (NIH, n.d.-b).

In the last phase of clinical trials, Phase IV, after FDA approval and public availability, researchers track the drug's safety in the population. This enables them to look for more details regarding benefits, appropriate use, and potential long-term or rare side effects (NIH, n.d.-b).

These trial phases are crucial for ensuring that the new drugs or treatments developed are safe and effective for use by the public. They are conducted in a sequence, with each phase building upon the information collected from the previous one to provide a comprehensive understanding of the treatment's safety and efficacy.

18.6.2 Regulatory Aspects

Regulatory considerations have a significant impact on the drug development process, ensuring that emerging treatments meet stringent safety and efficacy standards for public use. For drugs that target cell signaling pathways, specific regulatory aspects need to be addressed to thoroughly assess their safety and effectiveness.

These regulatory considerations involve coordination between drug sponsors, clinical researchers, and regulatory authorities such as the FDA. All drugs must move through a pipeline of discovery, investigation, clinical trials, and FDA approval before becoming available to consumers (P. Zhang & Winston, 2023). The development process includes compliance with regulatory requirements, including adherence to good practices, to ensure that regulatory bodies can appropriately assess and evaluate the drug development program in a standardized manner (Mitchel, 2000).

For cell signaling-based drugs to treat diseases with novel concepts, it is essential to examine signal transduction pathways within the body (L. Wang & Umezawa, 2021a). Drug development should consider the fundamental signal transduction involved in diseases, and inhibitors of cellular signal transduction that are over-activated in diseased cells or tissues, should be useful as rational drugs for each disease (L. Wang & Umezawa, 2021a). Additionally, as systems biology integrates the concepts of combination assays and phenotypic drug discovery, it plays a critical role in the discovery and development of new drugs, providing a more representative understanding of in vivo conditions (Abhyankar et al., 2018a).

In summary, regulatory considerations in drug development are significant for ensuring the safety and efficacy of new treatments. Understanding specific signal transductions in diseases and their inhibitors is critical for the development of rational, cell-signaling-based drugs for each disease. Moreover, systems biology plays a crucial role in drug discovery and development by providing a more representative understanding of in vivo settings, contributing to the precise evaluation of cell signaling-based drugs (Abhyankar et al., 2018a; Mitchel, 2000; L. Wang & Umezawa, 2021a; P. Zhang & Winston, 2023).

18.7 PERSONALIZED MEDICINE AND TARGETED THERAPIES

18.7.1 Role of Cell Signaling in Personalized Medicine

Cell Signaling contributes significantly to the development of personalized medicine by providing information about an individual's unique characteristics at the molecular, physiological, environmental exposure, and behavioral levels (Goetz & Schork, 2018). This information can be instrumental in adjusting interventions for specific diseases based on an individual's detailed characteristics. DNA sequencing, imaging protocols, proteomics, wireless health monitoring devices, and other emerging technologies have revealed greater inter-individual variation in disease processes, leading to the development of personalized medicine (Goetz & Schork, 2018).

Cell signaling information contributes to personalized medicine and treatment through the development of targeted therapies. These therapies utilize drugs to target specific pathogenic molecules or cells, allowing for precise and personalized treatment (R. C. Wang & Wang, 2023). For example, several medications, including antibodies like trastuzumab (Herceptin), gefitinib (Iressa), and imatinib (Gleevec), have been developed to block growth signals specific to cancer (R. C. Wang & Wang, 2023). Moreover, using patient-driven cells has shown success in cancer settings in personalized drug screening initiatives, as tumor biopsies can provide appropriate materials for drug screening (Goetz & Schork, 2018).

Successful personalized medicine requires effective therapeutic treatment tailored to each disease subtype (R. C. Wang & Wang, 2023). To achieve this, researchers should use pharmaco-omics to find personalized drugs and therapies, develop targeted therapies that specifically target known molecular subtypes, or examine how various drugs affect patient-derived cancer cells in vitro (functional precision medicine) should (R. C. Wang & Wang, 2023).

Molecular imaging, a recently emerged dynamic and exciting field of study, also plays an important role in personalized medicine by providing valuable information about disease state and treatment response through in vivo visualization of molecular and cellular processes (Salih et al., 2023).

18.7.2 Targeted Therapies

Cell signaling information contributes to personalized medicine by providing deeper insights into an individual's unique characteristics across molecular, physiological, environmental exposure, and behavioral levels. This information can be used to develop targeted therapies and tailor interventions for specific diseases. Targeted therapies focus on specific genetic changes or proteins that help cancerous cells grow, divide, and spread, using drugs based on patients' signaling profiles. These therapies can be non-selective, targeting multiple kinases in a signaling pathway, or highly selective, targeting only one kinase (Montoya et al., 2021; NIH, n.d.-a).

Examples of successful disease-specific targeted treatments include small molecule kinase inhibitors, hormone receptor targeting, PI3K-AKT signaling inhibition, immunotherapies, and monoclonal antibodies (Montoya et al., 2021; NIH, n.d.-a). Small molecule kinase inhibitors have been successful in treating chronic myelogenous leukemia (CML) with imatinib and ER-positive breast cancer with tamoxifen. Monoclonal antibodies like trastuzumab have shown success in treating HER2-positive breast cancer (Montoya et al., 2021; NIH, n.d.-a). Additionally, targeting the PI3K-AKT pathway has shown potential in treating various cancer types, including lung squamous cell carcinoma (Niu et al., 2022). These examples demonstrate the effectiveness of targeted therapies based on patients' signaling profiles in treating specific diseases.

18.8 DRUGS TARGETING CELL SIGNALING PATHWAYS

Cell signaling pathways are complex communication networks within cells that regulate various biological processes such as cell growth, proliferation, and response to external stimuli. These pathways are pivotal for maintaining cellular balance and regulating appropriate responses to changes in the cellular environment. Targeting cell signaling pathways has emerged as a promising approach in drug development, especially for the treatment of cancer and other ailments characterized by dysregulated signaling, offering a personalized and targeted therapeutic strategy. By interfering with key proteins involved in cell signaling, these drugs aim to disrupt the uncontrolled growth and survival signals in diseased cells. The development of cell signaling inhibitors marks a significant advancement in precision medicine, providing effective and tailored treatments for a range of conditions (L. Wang & Umezawa, 2021b).

Many drugs targeting cell signaling pathways have been developed and are used as common treatments for various diseases. Some examples include:

18.8.1 Tyrosine Kinase Inhibitors (TKIs)

TKIs are pharmacologic agents that inhibit the signal transduction pathways of protein kinases by blocking the stimulation of tyrosine kinases. TKIs block the active site of the kinase, preventing the phosphorylation of proteins often involved in cell proliferation or angiogenesis. TKIs are effective treatments for various forms of cancer and can induce long-term remission or increase the life span of some cancer patients . FDA-approved TKIs include imatinib, masitinib, and sunitinib. TKI combinations are considered a popular treatment option because they can prevent primary and secondary therapeutic resistance (Thomson et al., 2023) (Figure 18.5).

TKIs exert their pharmacological effects by competitively inhibiting the catalytic binding site of tyrosine kinases, disrupting signal transduction pathways. These pathways regulate various cellular

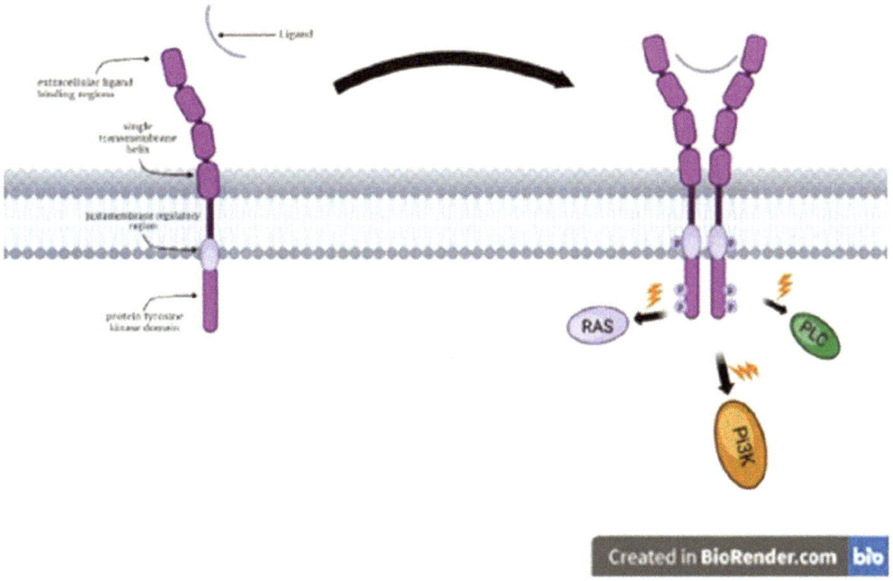

FIGURE 18.5 Activation of the tyrosine kinase receptor. Ligand binding strengthens bonds between monomeric receptors, forming an active dimer and activating the intracellular kinase. The three main effectors triggered are phospholipase C (PLC), phosphoinositide 3-kinase (PI3K), and rat sarcoma (RAS) *(created in BioRender.com).*

processes, such as cell growth, migration, differentiation, and apoptosis. Abnormal activation of tyrosine kinases may lead to dysregulated signal cascades, potentially resulting in malignancy and other pathologies. By blocking these initial signals, TKIs can be effective against mutated or overexpressed tyrosine kinases, making them valuable cancer treatments (Thomson et al., 2023).

18.8.2 Tumor Necrosis Factor Inhibitors

TNF inhibitors are drugs that stop inflammation and are used to treat various inflammatory conditions, including rheumatoid arthritis, psoriatic arthritis, juvenile idiopathic arthritis, inflammatory bowel disease (Crohn's and ulcerative colitis), ankylosing spondylitis, psoriasis, hidradenitis suppurativa, and refractory asthma (Bernstein, n.d.).

These inhibitors work by impeding the activity of TNF, a protein that causes inflammation in the body. TNF inhibitors can be administered by self-injection or intravenous infusion (Chan, 2023). FDA-approved examples include infliximab, adalimumab, etanercept, golimumab, and certolizumab (Eustice, 2022). The most significant side effects of TNF inhibitors are an increased risk of all types of infections, including tuberculosis (TB) and fungal infections, and an increased risk of certain types of skin cancer (Chan, 2023).

18.8.3 Cyclooxygenase-2 Inhibitors

Cyclooxygenase-2 (COX-2) inhibitors, also known as COX-2 selective nonsteroidal anti-inflammatory drugs (NSAIDs), produce prostaglandins that cause pain and inflammation. Unlike traditional NSAIDs, COX-2 inhibitors do not significantly affect the COX-1 enzyme, which maintains the stomach lining and platelet function. This selective inhibition leads to a reduced risk of gastrointestinal side effects compared to traditional NSAIDs. However, COX-2 inhibitors may increase the risk of cardiovascular events, leading to the withdrawal of some drugs and black box warnings on others. Examples of COX-2 inhibitors include celecoxib, rofecoxib, and valdecoxib. These drugs are used as painkillers and to manage inflammation in various conditions like osteoarthritis, rheumatoid arthritis, and acute pain.

18.8.4 Metaproterenol

Metaproterenol is a commonly used drug in the class of drugs known as adrenergic bronchodilators, which are inhaled to open the bronchial tubes in the lungs, relieving symptoms such as cough, wheezing, shortness of breath, and difficulty breathing by increasing airflow.

Metaproterenol functions as a relatively selective beta-2-adrenergic agonist, stimulating receptors in the smooth muscle of the lungs and uterus and affecting the vasculature of skeletal muscle, with little to no effect on alpha-adrenergic receptors. It is believed to affect the cAMP pathway by activating adenylate cyclase, an enzyme that produces cAMP, a cellular mediator (Figure 18.6).

Side effects of metaproterenol may include dizziness, drowsiness, nausea, gastrointestinal issues, fatigue, and tremors (Drugs.com, n.d.).

18.8.5 Anticholinergic Bronchodilators

Anticholinergic bronchodilators, also called antimuscarinics, are medications primarily used to treat chronic obstructive pulmonary disease (COPD) and, to a lesser extent, asthma. These medications block the action of cholinergic nerves, causing bronchodilation. They are usually administered through inhalation and sometimes via nebulization for sudden and severe symptoms (Scullion, 2007; Westby et al., 2004).

Anticholinergic bronchodilators work by blocking the action of acetylcholine at the neuromuscular junction, relaxing airway smooth muscle and dilating airways, helping alleviate symptoms of

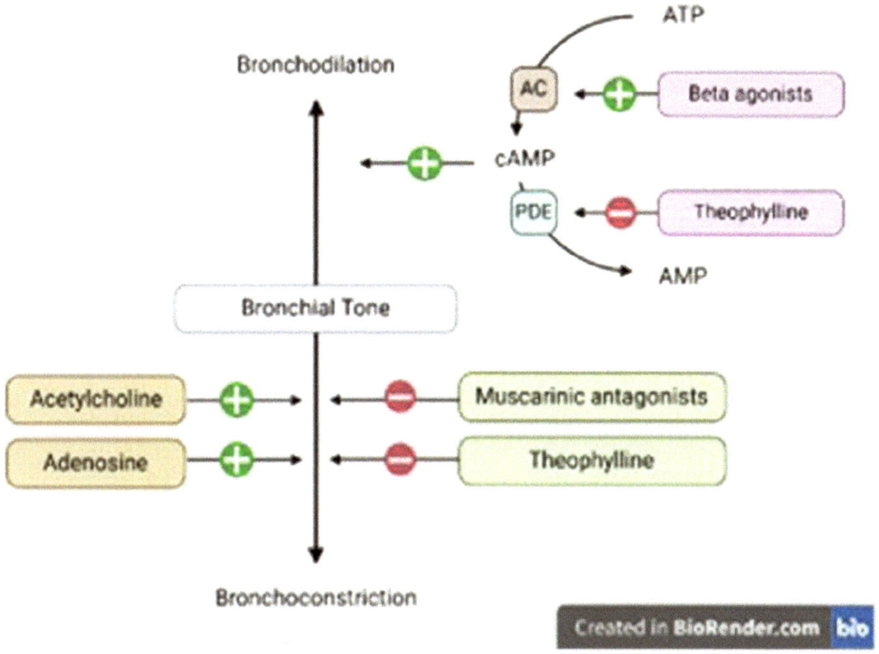

FIGURE 18.6 cAMP promotes bronchodilation. β-adrenoceptor agonists can increase intracellular cAMP levels *(created in BioRender.com).*

COPD and asthma. Examples such as glycopyrronium bromide, tiotropium, and aclidinium bromide competitively inhibit the binding of acetylcholine to its receptors, relaxing smooth muscle in the airway. These drugs antagonize muscarinic acetylcholine receptors, which are widely expressed in airway smooth muscle cells, preventing contraction and promoting airway dilation (NHS, 2022).

Anticholinergic bronchodilators are generally unsuitable as controller medications due to their short duration of action. However, they may offer clinical benefits as add-on therapy for patients with uncontrolled asthma or severe COPD.

18.8.6 Other Examples

Various other drugs targeting cell signaling pathways have been developed. VEGF and bcr-abl inhibitors have been developed as cellular signal transduction inhibitors for diseases including cancer (L. Wang & Umezawa, 2021b). For respiratory diseases, drugs targeting cell signaling pathways include glucocorticosteroids, β2-adrenoceptor agonists, anticholinergic bronchodilators, antiallergic drugs, and leukotriene-targeting drugs (Nature America Inc., 2000). In cancer treatment, drugs targeting signaling pathways in cancer stem cells have been developed, including those targeting the Wnt, Hedgehog, Notch, Hippo, and autophagy signaling pathways (Yang et al., 2020). These examples demonstrate the diverse application of drugs that utilize cell signaling pathways for therapeutic intervention.

18.8.7 Problems

The production and use of signal-transduction-targeted drugs face various challenges. Knowledge about these pathways is often limited, and key pathway components suspected of participating in disease must be validated as potential drug targets. These drugs require extensive preclinical and clinical validation, which are not trivial feats, as testing is required in a cause-effect format. Several groups are addressing these challenges by deciphering signal transduction pathways, improving our knowledge, and better equipping us for the discovery of new drug forms beneficial for humanity (Nahorski, 2006).

Drugs that target cell signaling pathways often have potential side effects. Targeted therapies may lead to liver conditions such as hepatitis and elevated liver enzymes, and in some cases skin problems (Vlasceanu et al., 2017). Severe side effects can limit the use of these drugs and therapies, decreasing their value and effectiveness, which ultimately negatively affects patients (Zhu et al., 2016a). In cancer treatment, the inhibition of critical signaling pathways can impact normal cell function and lead to side effects (Lim et al., 2018).

These drugs may affect other cellular pathways, leading to unprecedented side effects. Furthermore, the inhibition of one signaling component may be compensated for by increased signaling through alternative pathways, also leading to side effects. This compensation can occur through different mechanisms, such as the activation of compensatory opposing signaling pathways, increased resistance to targeted therapy by reactivation of specific signaling pathways, or the development of drug tolerance due to compensatory mechanisms. Off-target effects of these drugs could also lead to the activation of alternative pathways, resulting in side effects. The complexity of cellular signaling networks and the potential for compensatory mechanisms highlight the importance of understanding and addressing these issues in the development and administration of targeted therapies (Mavers et al., 2009; Nussinov et al., 2017). Some anticancer drugs that act as cell signaling inhibitors can attack not just cancer cells but also normal cells, leading to further complications (L. Wang & Umezawa, 2021b).

18.9 CHALLENGES IN CELL SIGNALING-DRIVEN DRUG DEVELOPMENT

Targeting specific components of the cellular signaling cascade has given rise to multiple highly effective drugs. However, there are challenges in employing this method of drug development, and more research and analysis is needed. This section discusses some of these challenges.

18.9.1 Functional Pleiotropy

Developing drugs targeting signaling pathways presents various challenges. One issue stems from cross-signaling in signaling cascades and interconnected signaling loops. As a result, if one part of the pathway is targeted, other parts might also be influenced and inhibited.

Many signaling mediators play roles in multiple physiological functions, a phenomenon known as functional pleiotropy. Different stimuli generate different signals that share networks. As these signals travel towards their specific effectors, they may interconnect. Inhibiting a particular molecule in this network can have undesirable and widespread effects, hindering the effectiveness of drugs targeting major signaling pathways like p53, MAPK, and IKK. This explains why drugs with high molecular specificity, such as inhibitors of MAPK, JNK, and IKK, have not shown promising results.

18.9.2 High Cost

Drug development is a rigorous, costly, and time-consuming process that can take years. Single-target drugs have proven expensive and uneconomical. For example, two doses of etanercept can cost $1,600 each, while infliximab administration of 10mg/kg up to eight doses can cost approximately $9,000 (Sun et al., 2022).

Despite these costs, 90% of these drugs fail during clinical trials. Drug failure can be due to toxicity, lack of efficacy, side effects, poor drug-like properties, and other factors (Sun et al., 2022).

18.9.3 Inefficacy of Single-Target Drugs

Targeting a single molecule should theoretically be effective in achieving desired therapeutic effects. However, this is not always the case. Single-target drugs are often ineffective in treating diseases involving multiple cell types, such as autoinflammatory diseases and diabetes.

18.9.4 Side Effects and Toxicity

Severe side effects occur with drugs targeting cellular signaling pathways. These can include infections, blood disorders, lymphomas, central nervous system disorders, and liver injuries.

Erlotinib, known as a 'wonder drug' due to its high efficacy, is associated with a wide variety of side effects. Likewise, inhibitors of VEGF, bcr-abl, proteosomes, COX-2, TNF, and RTKs have limited effectiveness against inflammatory diseases and cancers due to adverse side effects.

Bortezomib, which targets signaling pathways, has shown success in treating myeloma by inhibiting proteosomes. However, it reportedly regulates approximately 400 genes, potentially causing detrimental health conditions like gastrointestinal issues, peripheral neuropathy, and cardiovascular toxicity.

The drug celecoxib has caused toxicity in the stomach, kidneys, liver, and heart while showing a very low efficacy.

18.9.5 Need for Other Augmented Therapies to Increase Efficacy

Regarding kinase inhibitors, only a small number of breast cancer patients benefit from herceptin, and they generally develop resistance within one year. Another drug, trastuzumab, can be given to breast cancer patients but must be augmented with other therapies such as chemotherapeutic agents or alternative treatment modalities.

18.9.6 High Number of Mutations in Cancer Genes

Targeted therapies may have limited effectiveness in treating cancer due to the approximately 300 gene mutations present in a given cancer. Current targeted therapies need significant improvement in this regard.

Another challenge is determining validated drug interactions and their effects in both diseased and healthy models. This requires studying complex protein-protein interactions under different conditions and linking different variables and outcomes through information systems.

Unfortunately, target-specific drug therapy has not increased the number of drugs entering the market. Successfully developing a drug requires a major emphasis on efficacy and safety, rather than focusing solely on target selection (Abhyankar et al., 2018a).

18.10 FUTURE DEVELOPMENT

The development of drugs targeting cell signaling pathways holds great promise for treating various diseases, including cancer. Aberrant signaling through communication pathways can result in diseases, and signal transduction pathways are increasingly targeted for drug development (Costa et al., 2010). The prospects for drugs targeting cell signaling pathways are promising, with ongoing research aimed at identifying new drug targets and developing more effective therapies. Identifying vulnerable components in signaling pathways can contribute to developing drug therapies addressing aberrations in cancer signaling pathways (Zhu et al., 2016b). The specificity of response profiles to ligands generated by these drugs will improve precision medicine and help eliminate traditional medicine's negative effects. Much research is still needed in this area. Drug developers must create drugs targeting cell signaling pathways effectively and in a controlled manner that minimizes adverse effects.

18.11 CONCLUSION

Cell signaling plays an important role in drug development, helping identify potential drug targets. Signaling pathways are involved in various diseases such as cancer, Alzheimer's, and Parkinson's disease, and drugs are developed to act at various stages of signaling pathways, thus treating diseases or reducing their severity. Preclinical and clinical testing are crucial in drug development. Emerging trends in disease treatment include personalized medicine and targeted therapies. While drug development targeting signaling pathways and molecules has many advantages, it also has some disadvantages. These therapies often have high failure rates, and some are nonspecific. Despite these challenges, cell signaling will remain the most important aspect of success in drug development. Ethical considerations should also be addressed when using cell signaling for disease treatment.

REFERENCES

Abhyankar, V., Bland, P., & Fernandes, G. (2018). The role of systems biologic approach in cell signaling and drug development responses—A mini review. *Medical Sciences*, *6*(2), 43. https://doi.org/10.3390/medsci6020043

Aggarwal, B. B., Sethi, G., Baladandayuthapani, V., Krishnan, S., & Shishodia, S. (2007). Targeting cell signaling pathways for drug discovery: An old lock needs a new key. *Journal of Cellular Biochemistry*, *102*(3), 580–592. https://doi.org/10.1002/jcb.21500

Alhazmi, H. A., & Albratty, M. (2022). An update on the novel and approved drugs for Alzheimer disease. *Saudi Pharmaceutical Journal*, *30*(12), 1755–1764. https://doi.org/10.1016/j.jsps.2022.10.004

Amadoz, A., Sebastian-Leon, P., Vidal, E., Salavert, F., & Dopazo, J. (2015). Using activation status of signaling pathways as mechanism-based biomarkers to predict drug sensitivity. *Scientific Reports*, *5*(1), 18494. https://doi.org/10.1038/srep18494

Andrade, E. L., Bento, A. F., Cavalli, J., Oliveira, S. K., Schwanke, R. C., Siqueira, J. M., Freitas, C. S., Marcon, R., & Calixto, J. B. (2016). Non-clinical studies in the process of new drug development - Part II: Good laboratory practice, metabolism, pharmacokinetics, safety and dose translation to clinical studies. *Brazilian Journal of Medical and Biological Research*, *49*(12). https://doi.org/10.1590/1414-431x20165646

Arango MT, Q.-R. P. C. J. et al. (2013). Cell culture and cell analysis. In S. Y. R.-V. A. et al. Anaya JM (Ed.), *Autoimmunity: From Bench to Bedside*. El Rosario University Press.

Badr-Eldin, S. M., Aldawsari, H. M., Kotta, S., Deb, P. K., & Venugopala, K. N. (2022). Three-dimensional in vitro cell culture models for efficient drug discovery: Progress so far and future prospects. *Pharmaceuticals*, *15*(8), 926. https://doi.org/10.3390/ph15080926

Bédard, P., Gauvin, S., Ferland, K., Caneparo, C., Pellerin, È., Chabaud, S., & Bolduc, S. (2020). Innovative human three-dimensional tissue-engineered models as an alternative to animal testing. *Bioengineering*, *7*(3), 115. https://doi.org/10.3390/bioengineering7030115

Behar, M., Barken, D., Werner, S. L., & Hoffmann, A. (2013). The dynamics of signaling as a pharmacological target. *Cell*, *155*(2), 448–461. https://doi.org/10.1016/j.cell.2013.09.018

Bernstein, S. (n.d.). *TNF inhibitor drugs: Autoimmune disease treatments | WebMD*. Retrieved 26 December 2023, from https://www.webmd.com/rheumatoid-arthritis/tnf-inhibitor-inflammation

Bioexcel. (n.d.). *Rational drug design – BioExcel – centre of excellence for computation biomolecular research*. Retrieved 26 December 2023, from https://bioexcel.eu/research/projects/rational-drug-design/

Campbell, B. B., Galati, M. A., Stone, S. C., Riemenschneider, A. N., Edwards, M., Sudhaman, S., Siddaway, R., Komosa, M., Nunes, N. M., Nobre, L., Morrissy, A. S., Zatzman, M., Zapotocky, M., Joksimovic, L., Kalimuthu, S. N., Samuel, D., Mason, G., Bouffet, E., Morgenstern, D. A., … Tabori, U. (2021). Mutations in the RAS/MAPK pathway drive replication repair–deficient hypermutated tumors and confer sensitivity to MEK inhibition. *Cancer Discovery*, *11*(6), 1454–1467. https://doi.org/10.1158/2159-8290.CD-20-1050

Campbell, P. J., Getz, G., Korbel, J. O., Stuart, J. M., Jennings, J. L., Stein, L. D., Perry, M. D., Nahal-Bose, H. K., Ouellette, B. F. F., Li, C. H., Rheinbay, E., Nielsen, G. P., Sgroi, D. C., Wu, C. L., Faquin, W. C., Deshpande, V., Boutros, P. C., Lazar, A. J., Hoadley, K. A., … Zhang, J. (2020). Pan-cancer analysis of whole genomes. *Nature*, *578*(7793), 82–93. https://doi.org/10.1038/s41586-020-1969-6

Cancer Research UK. (n.d.). *Monoclonal antibodies | targeted cancer drugs | cancer research UK*. Retrieved 26 December 2023, from https://www.cancerresearchuk.org/about-cancer/treatment/targeted-cancer-drugs/types/monoclonal-antibodies

Cell Signaling Technology. (n.d.). *Overview of neurodegenerative diseases | cell signaling technology*. Retrieved 28 December 2023, from https://www.cellsignal.com/science-resources/overview-of-neurodegeneration

Chan, K. K. (2023, February). *Tumor necrosis factor (TNF) inhibitors | American college of rheumatology*. https://rheumatology.org/patients/tumor-necrosis-factor-tnf-inhibitors

Costa, L. G., Giordano, G., & Guizzetti, M. (2010). Cell signaling and neurotoxicity. *Comprehensive Toxicology, Second Edition*, *13*, 89–100. https://doi.org/10.1016/B978-0-08-046884-6.01307-5

Downward, J. (2003). Targeting RAS signalling pathways in cancer therapy. *Nature Reviews Cancer*, *3*(1), 11–22. https://doi.org/10.1038/nrc969

Drugs.com. (n.d.). *Metaproterenol side effects: Common, severe, long term*. Retrieved 26 December 2023, from https://www.drugs.com/sfx/metaproterenol-side-effects.html

Eustice, C. (2022, June 2). *TNF inhibitors: Uses, side effects, and more*. https://www.verywellhealth.com/tnf-blockers-what-you-need-to-know-189481

Faivre, S., Kroemer, G., & Raymond, E. (2006). Current development of mTOR inhibitors as anticancer agents. *Nature Reviews Drug Discovery*, *5*(8), 671–688. https://doi.org/10.1038/nrd2062

Ferlay, J., Colombet, M., Soerjomataram, I., Parkin, D. M., Piñeros, M., Znaor, A., & Bray, F. (2021). Cancer statistics for the year 2020: An overview. *International Journal of Cancer*, *149*(4), 778–789. https://doi.org/10.1002/ijc.33588

Field MJ, B. T. editors. (2010). Development of new therapeutic drugs and biologics for rare diseases. In *Rare diseases and Orphan products: Accelerating research and development*. National Academies Press. https://www.ncbi.nlm.nih.gov/books/NBK56179/

Fowler, N. H., Samaniego, F., Jurczak, W., Ghosh, N., Derenzini, E., Reeves, J. A., Knopińska-Posłuszny, W., Cheah, C. Y., Phillips, T., Lech-Maranda, E., Cheson, B. D., Caimi, P. F., Grosicki, S., Leslie, L. A., Chavez, J. C., Fonseca, G., Babu, S., Hodson, D. J., Shao, S. H., … Zinzani, P. L. (2021). Umbralisib, a dual PI3Kδ/CK1ε inhibitor in patients with relapsed or refractory indolent lymphoma. *Journal of Clinical Oncology*, *39*(15), 1609–1618. https://doi.org/10.1200/JCO.20.03433

Giunti, S., Antonelli, A., Amorosi, A., & Santarpia, L. (2013). Cellular signaling pathway alterations and potential targeted therapies for medullary thyroid carcinoma. *International Journal of Endocrinology*, *2013*, 1–16. https://doi.org/10.1155/2013/803171

Goetz, L. H., & Schork, N. J. (2018). Personalized medicine: Motivation, challenges, and progress. *Fertility and Sterility*, *109*(6), 952–963. https://doi.org/10.1016/j.fertnstert.2018.05.006

Grzemski, F. (n.d.). *What are the benefits of a well-designed, robust preclinical development plan? | Novotech CRO*. Retrieved 26 November 2023, from https://novotech-cro.com/faq/what-are-benefits-well-designed-robust-preclinical-development-plan

Guo, T., Zhang, D., Zeng, Y., Huang, T. Y., Xu, H., & Zhao, Y. (2020). Molecular and cellular mechanisms underlying the pathogenesis of Alzheimer's disease. *Molecular Neurodegeneration*, *15*(1), 40. https://doi.org/10.1186/s13024-020-00391-7

Han, J., Liu, Y., Yang, S., Wu, X., Li, H., & Wang, Q. (2021). MEK inhibitors for the treatment of non-small cell lung cancer. *Journal of Hematology & Oncology*, *14*(1), 1. https://doi.org/10.1186/s13045-020-01025-7

Holt, J. D. S., & Nuttall, J. P. (2013). *Preclinical safety evaluation* (pp. 55–78). https://doi.org/10.1007/82_2013_327

Hu, X., Xia, M., Wang, J., Yu, H., Chai, J., Zhang, Z., Sun, Y., Su, J., & Sun, L. (2020). Dual PI3K/mTOR inhibitor PKI-402 suppresses the growth of ovarian cancer cells by degradation of Mcl-1 through autophagy. *Biomedicine & Pharmacotherapy*, *129*, 110397. https://doi.org/10.1016/j.biopha.2020.110397

Huang, L., Guo, Z., Wang, F., & Fu, L. (2021). KRAS mutation: From undruggable to druggable in cancer. *Signal Transduction and Targeted Therapy*, *6*(1), 386. https://doi.org/10.1038/s41392-021-00780-4

Iksen, Pothongsrisit, S., & Pongrakhananon, V. (2021). Targeting the PI3K/AKT/mTOR signaling pathway in lung cancer: An update regarding potential drugs and natural products. *Molecules*, *26*(13), 4100. https://doi.org/10.3390/molecules26134100

ITR Laboratories Canada Inc. (n.d.). *Preclinical safety vs efficacy research - ITR Laboratories Canada Inc.* Retrieved 26 November 2023, from https://www.itrlab.com/preclinical-safety-vs-efficacy-research/

Kanti Das, T., Wati, M. R., & Fatima-Shad, K. (2014). Oxidative stress gated by fenton and haber weiss reactions and its association with Alzheimer's disease. *Archives of Neuroscience*, *2*(3). https://doi.org/10.5812/archneurosci.20078

Kilanowska, A., Ziółkowska, A., Stasiak, P., & Gibas-Dorna, M. (2022). cAMP-Dependent signaling and ovarian cancer. *Cells*, *11*(23), 3835. https://doi.org/10.3390/cells11233835

Kimmelman, J., & Federico, C. (2017). Consider drug efficacy before first-in-human trials. *Nature*, *542*(7639), 25–27. https://doi.org/10.1038/542025a

Kudo, M. (2012). Signaling pathway/molecular targets and new targeted agents under development in hepatocellular carcinoma. *World Journal of Gastroenterology*, *18*(42), 6005. https://doi.org/10.3748/wjg.v18.i42.6005

Langhans, S. A. (2018). Three-dimensional in vitro cell culture models in drug discovery and drug repositioning. *Frontiers in Pharmacology*, *9*. https://doi.org/10.3389/fphar.2018.00006

Langhof, H., Chin, W. W. L., Wieschowski, S., Federico, C., Kimmelman, J., & Strech, D. (2018). Preclinical efficacy in therapeutic area guidelines from the U.S. Food and drug administration and the European medicines agency: A cross-sectional study. *British Journal of Pharmacology*, *175*(22), 4229–4238. https://doi.org/10.1111/bph.14485

Lim, H., Poleksic, A., & Xie, L. (2018). Exploring landscape of drug-target-pathway-side effect associations. *AMIA Summits on Translational Science Proceedings*, *2018*, 132. /pmc/articles/PMC5961812/

Ma, Y.-S., Xin, R., Yang, X.-L., Shi, Y., Zhang, D.-D., Wang, H.-M., Wang, P.-Y., Liu, J.-B., Chu, K.-J., & Fu, D. (2021). Paving the way for small-molecule drug discovery. *American Journal of Translational Research*, *13*(3), 853–870.

Mahapatra, M. K., & Karuppasamy, M. (2022). Fundamental considerations in drug design. In *Computer aided drug design (CADD): From ligand-based methods to structure-based approaches* (pp. 17–55). Elsevier. https://doi.org/10.1016/B978-0-323-90608-1.00005-8

Marei, I., Abu Samaan, T., Al-Quradaghi, M. A., Farah, A. A., Mahmud, S. H., Ding, H., & Triggle, C. R. (2022). 3D Tissue-engineered vascular drug screening platforms: Promise and considerations. *Frontiers in Cardiovascular Medicine*, *9*. https://doi.org/10.3389/fcvm.2022.847554

Mascaux, C., Iannino, N., Martin, B., Paesmans, M., Berghmans, T., Dusart, M., Haller, A., Lothaire, P., Meert, A.-P., Noel, S., Lafitte, J.-J., & Sculier, J.-P. (2005). The role of RAS oncogene in survival of patients with lung cancer: A systematic review of the literature with meta-analysis. *British Journal of Cancer*, *92*(1), 131–139. https://doi.org/10.1038/sj.bjc.6602258

Matthews, H. K., Bertoli, C., & de Bruin, R. A. M. (2022). Cell cycle control in cancer. *Nature Reviews Molecular Cell Biology*, *23*(1), 74–88. https://doi.org/10.1038/s41580-021-00404-3

Mavers, M., Ruderman, E. M., & Perlman, H. (2009). Intracellular signal pathways: Potential for therapies. *Current Rheumatology Reports*, *11*(5), 378–385. https://doi.org/10.1007/s11926-009-0054-9

Mitchel, J. T. (2000). FDA relations during drug development. *Dialogues in Clinical Neuroscience*, *2*(3), 213–217. https://doi.org/10.31887/DCNS.2000.2.3/jtmitchel

Montoya, S., Soong, D., Nguyen, N., Affer, M., Munamarty, S. P., & Taylor, J. (2021). Targeted therapies in cancer: To be or not to be, selective. *Biomedicines*, *9*(11), 1591. https://doi.org/10.3390/biomedicines9111591

Muhammad, S., Jiang, Z., Liu, Z., Kaur, K., & Wang, X. (2013). The role of EGFR monoclonal antibodies (MoABs) cetuximab/panitumab, and BRAF inhibitors in BRAF mutated colorectal cancer. *Journal of Gastrointestinal Oncology*, *4*(1), 72–81. https://doi.org/10.3978/j.issn.2078-6891.2012.044

Nahorski, S. R. (2006). Pharmacology of intracellular signalling pathways. *British Journal of Pharmacology*, *147*(Suppl 1), S38. https://doi.org/10.1038/SJ.BJP.0706468

Nature America Inc. (2000). Signal transduction as a drug-discovery platform. *Nature Biotechnology*, *18*(S10), IT37–IT39. https://doi.org/10.1038/80076

Nepstad, I., Hatfield, K. J., Grønningsæter, I. S., & Reikvam, H. (2020). The PI3K-Akt-mTOR signaling pathway in human acute myeloid leukemia (AML) cells. *International Journal of Molecular Sciences*, *21*(8), 2907. https://doi.org/10.3390/ijms21082907

NHS. (2022, October 5). *Bronchodilators - NHS*. https://www.nhs.uk/conditions/bronchodilators/

NIH. (n.d.-a). *Targeted therapy for cancer - NCI*. Retrieved 28 November 2023, from https://www.cancer.gov/about-cancer/treatment/types/targeted-therapies

NIH. (n.d.-b). *The basics | National institutes of health (NIH)*. Retrieved 27 November 2023, from https://www.nih.gov/health-information/nih-clinical-research-trials-you/basics

Nitulescu, G. M., Margina, D., Juzenas, P., Peng, Q., Olaru, O. T., Saloustros, E., Fenga, C., Spandidos, D. A., Libra, M., & Tsatsakis, A. M. (2016). Akt inhibitors in cancer treatment: The long journey from drug discovery to clinical use (Review). *International Journal of Oncology*, *48*(3), 869–885. https://doi.org/10.3892/ijo.2015.3306

Niu, Z., Jin, R., Zhang, Y., & Li, H. (2022). Signaling pathways and targeted therapies in lung squamous cell carcinoma: mechanisms and clinical trials. *Signal Transduction and Targeted Therapy*, *7*(1), 353. https://doi.org/10.1038/s41392-022-01200-x

Nussinov, R., Tsai, C. J., & Jang, H. (2017). A new view of pathway-driven drug resistance in tumor proliferation. *Trends in Pharmacological Sciences*, *38*(5), 427. https://doi.org/10.1016/J.TIPS.2017.02.001

Nussinov, R., Tsai, C. J., & Jang, H. (2022). A new view of activating mutations in cancer. *Cancer Research, 82*(22), 4114. https://doi.org/10.1158/0008-5472.CAN-22-2125

Pan, X., Pei, J., Wang, A., Shuai, W., Feng, L., Bu, F., Zhu, Y., Zhang, L., Wang, G., & Ouyang, L. (2022). Development of small molecule extracellular signal-regulated kinases (ERKs) inhibitors for cancer therapy. *Acta Pharmaceutica Sinica B, 12*(5), 2171–2192. https://doi.org/10.1016/j.apsb.2021.12.022

Peng, Y., Wang, Y., Zhou, C., Mei, W., & Zeng, C. (2022). PI3K/Akt/mTOR pathway and its role in cancer therapeutics: Are we making headway? *Frontiers in Oncology, 12*. https://doi.org/10.3389/fonc.2022.819128

Pingale, T., & Gupta, G. L. (2021). Current and emerging therapeutic targets for Parkinson's disease. *Metabolic Brain Disease, 36*(1), 13–27. https://doi.org/10.1007/s11011-020-00636-w

Poewe, W., Seppi, K., Tanner, C. M., Halliday, G. M., Brundin, P., Volkmann, J., Schrag, A.-E., & Lang, A. E. (2017). Parkinson disease. *Nature Reviews Disease Primers, 3*(1), 17013. https://doi.org/10.1038/nrdp.2017.13

Pusterla, T. (2019, April 10). *High-throughput screening (HTS) | BMG LABTECH*. https://www.bmglabtech.com/en/blog/high-throughput-screening/

Rai, S., Singh, P., Varshney, R., Chaturvedi, V., Vamanu, E., Singh, M., & Singh, B. (2021). Promising drug targets and associated therapeutic interventions in Parkinson's disease. *Neural Regeneration Research, 16*(9), 1730. https://doi.org/10.4103/1673-5374.306066

Rajasethupathy, P., Vayttaden, S. J., & Bhalla, U. S. (2005). Systems modeling: A pathway to drug discovery. *Current Opinion in Chemical Biology, 9*(4), 400–406. https://doi.org/10.1016/j.cbpa.2005.06.008

Rousseaux, C. G., Bracken, W. M., & Guionaud, S. (2023). Overview of drug development. In *Haschek and Rousseaux's handbook of toxicologic pathology, volume 2: safety assessment environmental toxicologic pathology* (pp. 3–48). Elsevier. https://doi.org/10.1016/B978-0-12-821047-5.00016-6

Salih, S., Elliyanti, A., Alkatheeri, A., AlYafei, F., Almarri, B., & Khan, H. (2023). The role of molecular imaging in personalized medicine. *Journal of Personalized Medicine, 13*(2), 369. https://doi.org/10.3390/jpm13020369

Schenone, M., Dančík, V., Wagner, B. K., & Clemons, P. A. (2013). Target identification and mechanism of action in chemical biology and drug discovery. *Nature Chemical Biology, 9*(4), 232–240. https://doi.org/10.1038/nchembio.1199

Scullion, J. E. (2007). The development of anticholinergics in the management of COPD. *International Journal of Chronic Obstructive Pulmonary Disease, 2*(1), 33. https://doi.org/10.2147/COPD.2007.2.1.33

Shepard, H. M., Phillips, G. L., Thanos, C. D., & Feldmann, M. (2017). Developments in therapy with monoclonal antibodies and related proteins. *Clinical Medicine, 17*(3), 220–232. https://doi.org/10.7861/clinmedicine.17-3-220

Smith, C. (2003). Drug target validation: Hitting the target. *Nature, 422*(6929), 342–345. https://doi.org/10.1038/422341a

Strebhardt, K., & Ullrich, A. (2008). Paul Ehrlich's magic bullet concept: 100 years of progress. *Nature Reviews Cancer, 8*(6), 473–480. https://doi.org/10.1038/nrc2394

Sun, D., Gao, W., Hu, H., & Zhou, S. (2022). Why 90% of clinical drug development fails and how to improve it? *Acta Pharmaceutica Sinica B, 12*(7), 3049–3062. https://doi.org/10.1016/j.apsb.2022.02.002

Sygnature Discovery. (n.d.). *Intracellular signalling | drug discovery | sygnature discovery*. Retrieved 27 November 2023, from https://www.sygnaturediscovery.com/drug-discovery/bioscience/cellular-assays/intracellular-signalling/

Szymański, P., Markowicz, M., & Mikiciuk-Olasik, E. (2011). Adaptation of high-throughput screening in drug discovery—toxicological screening tests. *International Journal of Molecular Sciences, 13*(1), 427–452. https://doi.org/10.3390/ijms13010427

Thomson, R. J., Moshirfar, M., & Ronquillo, Y. (2023). Tyrosine kinase inhibitors. *StatPearls*. https://www.ncbi.nlm.nih.gov/books/NBK563322/

UCL. (n.d.). *Target identification and validation (small molecules) | UCL therapeutic innovation networks - UCL – University College London*. Retrieved 26 December 2023, from https://www.ucl.ac.uk/therapeutic-innovation-networks/target-identification-and-validation-small-molecules

Ugolini, G., Cruz-Moreira, D., Visone, R., Redaelli, A., & Rasponi, M. (2016). Microfabricated physiological models for in vitro drug screening applications. *Micromachines, 7*(12), 233. https://doi.org/10.3390/mi7120233

Umscheid, C. A., Margolis, D. J., & Grossman, C. E. (2011). Key concepts of clinical trials: A narrative review. *Postgraduate Medicine, 123*(5), 194–204. https://doi.org/10.3810/pgm.2011.09.2475

Vasta, J. D., Michaud, A., Zimprich, C. A., Beck, M. T., Swiatnicki, M. R., Zegzouti, H., Thomas, M. R., Wilkinson, J., Crapster, J. A., & Robers, M. B. (2023). Protomer selectivity of type II RAF inhibitors within the RAS/RAF complex. *Cell Chemical Biology*, *30*(11), 1354-1365.e6. https://doi.org/10.1016/j.chembiol.2023.07.019

Vlasceanu, G. M., Victor, L., Maricica, H., Raluca, T., Vlad, O., Gheorghe, I., Bolocan, A., Grumezescu, A. M., & Holban, A. M. (2017). Nanostructures for cancer therapy: From targeting to selective toxicology. *Nanostructures for Cancer Therapy*, 831–847. https://doi.org/10.1016/B978-0-323-46144-3.00030-1

Wang, L., & Umezawa, K. (2021a). Cellular signal transductions and their inhibitors derived from deep-sea organisms. *Marine Drugs*, *19*(4), 205. https://doi.org/10.3390/md19040205

Wang, L., & Umezawa, K. (2021b). Cellular signal transductions and their inhibitors derived from deep-sea organisms. *Marine Drugs*, *19*(4). https://doi.org/10.3390/MD19040205

Wang, R. C., & Wang, Z. (2023). Precision medicine: Disease subtyping and tailored treatment. *Cancers*, *15*(15), 3837. https://doi.org/10.3390/cancers15153837

Wen, P. Y., Touat, M., Alexander, B. M., Mellinghoff, I. K., Ramkissoon, S., McCluskey, C. S., Pelton, K., Haidar, S., Basu, S. S., Gaffey, S. C., Brown, L. E., Martinez-Ledesma, J. E., Wu, S., Kim, J., Wei, W., Park, M.-A., Huse, J. T., Kuhn, J. G., Rinne, M. L., … Ligon, K. L. (2019). Buparlisib in patients with recurrent glioblastoma harboring phosphatidylinositol 3-kinase pathway activation: an open-label, multicenter, multi-arm, phase II trial. *Journal of Clinical Oncology*, *37*(9), 741–750. https://doi.org/10.1200/JCO.18.01207

Westby, M. J., Benson, M. K., & Gibson, P. G. (2004). Anticholinergic agents for chronic asthma in adults. *The Cochrane Database of Systematic Reviews*, *2004*(3). https://doi.org/10.1002/14651858.CD003269.PUB2

Williams, C. H., & Hong, C. C. (2011). Multi-step usage of in vivo models during rational drug design and discovery. *International Journal of Molecular Sciences*, *12*(4), 2262–2274. https://doi.org/10.3390/ijms12042262

Wu, W.-I., Voegtli, W. C., Sturgis, H. L., Dizon, F. P., Vigers, G. P. A., & Brandhuber, B. J. (2010). Crystal structure of human AKT1 with an allosteric inhibitor reveals a new mode of kinase inhibition. *PLoS One*, *5*(9), e12913. https://doi.org/10.1371/journal.pone.0012913

Xie, J., & Weiskirchen, R. (2020). What does the “AKT” stand for in the name “AKT kinase”? Some historical comments. *Frontiers in Oncology*, *10*. https://doi.org/10.3389/fonc.2020.01329

Yang, Y., Li, X., Wang, T., Guo, Q., Xi, T., & Zheng, L. (2020). Emerging agents that target signaling pathways in cancer stem cells. *Journal of Hematology & Oncology*, *13*(1), 1–18. https://doi.org/10.1186/S13045-020-00901-6

Yao, S., Zhu, Y., & Chen, L. (2013). Advances in targeting cell surface signalling molecules for immune modulation. *Nature Reviews Drug Discovery*, *12*(2), 130–146. https://doi.org/10.1038/nrd3877

Yi, S., Wang, L., Wang, H., Ho, M. S., & Zhang, S. (2022). Pathogenesis of α-Synuclein in Parkinson’s disease: From a neuron-glia crosstalk perspective. *International Journal of Molecular Sciences*, *23*(23), 14753. https://doi.org/10.3390/ijms232314753

Zhan, Z.-J., Yao, W.-Y., Zhang, F., Qiu, W.-Z., Liao, K.-, Feng, J.-H., Tan, J.-Y., Liu, H., Yuan, T.-Z., Zheng, R.-H., & Yuan, Y.-W. (2021). The optimal second-line systemic treatment model for recurrent and/or metastatic head and neck squamous cell carcinoma: A Bayesian network meta-analysis. *Frontiers in Immunology*, *12*. https://doi.org/10.3389/fimmu.2021.719650

Zhang, J., Yang, P. L., & Gray, N. S. (2009). Targeting cancer with small molecule kinase inhibitors. *Nature Reviews Cancer*, *9*(1), 28–39. https://doi.org/10.1038/nrc2559

Zhang, P., Patel, P., & Winston, N. R. (2023). *Federal medication development regulation*. In: StatPearls [Internet]. Treasure Island (FL): StatPearls Publishing; 2024 Jan.

Zhao, Y., & Adjei, A. A. (2014). The clinical development of MEK inhibitors. *Nature Reviews Clinical Oncology*, *11*(7), 385–400. https://doi.org/10.1038/nrclinonc.2014.83

Zhu, P., Aliabadi, H. M., Uludag, H., & Han, J. (2016). Identification of potential drug targets in cancer signaling pathways using stochastic logical models. *Scientific Reports*, *6*(1), 1–11. https://doi.org/10.1038/srep23078

Index

Aberrant signaling, vii, 19, 40–42, 57, 58, 127, 141, 337, 365, 403
Abl kinase, 388
Actin, 134, 241, 338, 389
Acute lymphoblastic leukemia, 288, 356
Adenomatous polyposis coli (APC), 205, 279, 356
Afferent, 137, 234
Akt–mTOR signaling, 340
Alzheimer's disease, 20, 41, 51, 59, 134, 143, 165, 180, 321, 362, 393–394, 404
AMPK, 20, 218, 223–225
Amyloid precursor protein, 328, 362, 393
Anaplerosis, 221
Apolipoprotein E, 394
Apoptosis, 7, 17, 27, 28, 30, 31, 38, 40, 48, 55–57, 74, 77, 80, 92, 104, 115, 128, 129, 139, 164, 174, 181, 208–210, 218, 247, 250, 254, 261, 268–275, 278, 282, 290–293, 304, 307–309, 324, 328, 330, 334, 338, 339, 353–355, 358–360, 362, 363, 383, 389, 391, 400
Appendage outgrowth, 209
APP fragmentation, 328
Arginine, 174, 246
Autocrine signaling, 1, 4, 7, 46, 51

B-cell lymphomas, 20
Bcr-Abl inhibitors, 401
Bevacizumab, 20, 21, 388
Biomolecules, 100, 219, 387
Biosensors, 18, 19, 42, 375
B lymphocytes, 101, 118, 142, 143, 244
Bowel disease, 19, 48, 245, 249, 254, 287, 400
BRAF mutations, 357
Breast cancer, 20, 59, 140, 257, 260, 272, 274, 277, 280–282, 287–289, 291–294, 309, 337, 339–341, 359, 361, 364, 366, 372, 385, 388, 398, 403

Cadherin, 15, 53, 138, 139, 147, 149, 150, 210, 334, 338
Calcium/calmodulin-dependent kinase-2 (CaMK2), 135
Calcium channels, 54, 133, 143, 172, 180, 181, 224, 303
Calmodulin, 39, 74, 75, 135, 174–176, 305, 324, 336
cAMP response element (CRE), 55
Canonical WNT signaling, 205, 213, 279, 280, 307
Catecholamines, 133, 136, 138, 224
Cellular responses, 5, 12–14, 16–18, 20, 22, 28, 30, 31, 33, 36–38, 42, 47, 53–55, 74, 76, 106, 107, 159, 173, 212, 227, 343, 346, 354, 357, 373, 387
Complementarity, 95
Co-stimulatory, 109, 110, 113
Cross-presentation, 128
CSC population, 288, 294
cyclic adenosine monophosphate (cAMP), 1, 11, 48, 55, 218
Cytokines, 4, 5, 7, 9, 10, 16, 17, 27, 29–33, 50, 51, 55, 57–59, 77, 79, 80, 82, 92, 101, 104, 108–111, 113, 114, 116, 119, 127–130, 135–140, 143, 232, 234–240, 242, 244–255, 261, 271–274, 277, 287, 292, 308, 311, 321, 322, 325, 326, 330, 342, 343, 357, 358, 361, 363, 388

Developmental disorders, 20, 202, 356
Diabetes, 12, 14, 20, 21, 41, 59, 119, 120, 191, 198, 255, 302–305, 307, 308, 312, 341, 355, 403
Diacylglycerol, 12, 54, 74, 108, 110, 130, 131, 197, 224, 243, 336
Dichotomy, 114, 150
Dimerization, 10, 32, 33, 53, 54, 76, 77, 80, 190, 195, 208, 246, 248, 251, 272, 275, 286, 310, 354
DNA damage, 28, 259, 277, 329, 330, 360, 395
Docking proteins, 77, 192, 193, 203
Dopamine, 7, 27, 75, 136, 139, 143, 321, 323
Downstream signaling pathways, 10, 11, 17, 48, 60, 203
Drug development, 21, 181, 302, 355, 365, 375, 383–388, 390, 395–398, 402–404
Dsh-dependent, 206

Dynamics, 12, 15, 18, 19, 22, 33, 38, 42, 50, 56, 84, 85, 148, 155, 212, 302, 303, 345–347, 356, 373–375
Dynamin, 150, 151
Dynein, 131, 241
Dyskinesias, 165

Embryogenesis, 76, 202–204, 209
Endocrine signaling, 4, 8, 51, 52, 187
Endocrine cells, 52, 307
Endocytosis, 33, 103, 117, 126, 128, 148–152, 159, 166, 179, 236, 356
Epidermal growth factor receptor (EGFR), 10, 274, 367, 385
Epigenetic, 41, 42, 209, 214, 260, 268, 269, 274, 277, 293, 294, 334, 342, 345, 359, 361, 371
Epithelial-to-mesenchymal-transition, 210, 287, 338
Epitope, 94–98, 100, 107, 119
Exosomal cargo, 340
Extracellular matrix (ECM), 16, 268, 292

Farnesyl transferase, 57, 59, 309
Fas-associated death domain (FADD), 57
Fatty acids, 143, 219, 222, 224, 225, 235
Feedback mechanisms, 5, 6, 13, 14, 33, 113, 179, 211, 212, 285

Glucogenolysis, 58
Gluconeogenesis, 193, 224, 303
Glycosylation, 150, 159, 160, 166, 175
G-protein-coupled receptors, 47, 48, 52, 55, 60, 69, 70, 106, 219, 336
Greek key, 150

Helicobacter pylori, 249
Hematopoietic stem cells (HSC), 280
Heparin, 77, 236, 305, 341
Hepatitis, 249, 402
Herceptin, 260, 398, 403
Heterodimerization, 207, 310
Heterotrimeric G proteins, 9, 48, 206
Human epidermal growth factor receptor–2 (HER2), 274, 334
Hydrophobic, 11, 47, 49, 50, 53, 64, 97, 98, 148, 195, 242, 252
Hypertrophy, 56, 58, 59, 164, 311, 363

Immune senescence, 231, 233
Immunotherapy, 110, 112, 128, 129, 234, 257–260, 340, 375
Impaired lipolysis, 311
Indigenous immune cells, 323
Inflammation, 7, 27, 31–33, 41, 52, 58, 80, 92, 106, 113–119, 136–143, 218, 231, 232, 234–239, 244–251, 255–261, 271–273, 275, 285, 287, 292, 310–312, 319, 321–323, 325, 328–331, 334–337, 362, 363, 389, 393, 395, 400
Integrin signaling, 16, 17
Intracellular loop, 70, 170, 174, 175, 177, 355
Intricacies, 22, 83, 86, 107, 108, 110, 261, 270, 347, 384

K channels, 50, 162, 164, 169, 224
Keratinocytes, 50, 234, 245
Kinase activity, 76, 192, 203, 292, 308, 354, 358, 359, 361, 391

Lck, 110, 129, 130, 139, 196, 240, 241
Leptin, 20, 138, 224, 305, 308, 311
Low Density Lipoprotein Receptor (LDLR), 147, 149, 205, 278, 356

Matrix, 15–17, 20, 46, 50, 68, 115, 225, 234, 236, 250, 268, 269, 292
Maturation, 74, 106, 116, 117, 131, 232, 243, 274, 307
Membrane potential, 10, 49, 50, 53, 84, 133, 154, 157, 160–165, 168–174, 176, 177, 320, 345
MHC complexes, 103, 110, 127, 131, 132, 139
Modular, 35, 46, 47, 269
Multimolecular signaling complexes, 203

Neoproteins, 95
Neurotransmitter, 4–9, 12, 15, 47–51, 59, 67–70, 74, 126, 127, 132–136, 139, 140, 154, 155, 158, 162, 163, 166, 169, 171–176, 180, 320–322, 334, 353, 354
Notch signaling, 15, 53, 202, 207, 208, 218, 280, 288–290, 326, 337, 355, 356

Osmoregulation, 164, 177

Platelet-derived growth factor, 7, 59, 76, 203, 274, 305, 363
Pleiotropic, 42, 69, 115

Ras protein, 35, 248, 274, 303, 357
Retinal, 166, 179, 181
ROS formation, 220

Scaffolding proteins, 12, 40, 41, 324
Sedimentation, 99, 150, 245
Segments, 68, 95, 98, 102, 134, 161, 168, 170–172, 177, 178, 361
Sensitization, 119
Specificity, 13, 21, 29, 33, 35, 40, 48, 53, 72, 78, 79, 94, 96, 97, 107, 109, 113, 126–128, 150, 154, 187, 203, 234, 246, 247, 274, 338, 342, 355–357, 363, 365, 368, 374, 388, 402
Subcellular, 18, 19, 22, 33, 34, 40, 302, 304, 338, 403
Superfamily, 52, 68, 71, 76, 128, 129, 139, 148, 155, 156, 158, 190, 238, 273, 292, 305–307, 312, 323, 335
Synaptotagmins, 67, 132, 133

Targeted therapies, 14, 21, 43, 171, 337, 353, 354, 356–358, 360, 361, 363–365, 372–374, 384, 398, 402–404
T-cell pool, 233
Termination, 6, 8, 15, 29, 33–36, 53, 99, 202, 208, 210, 358
Tetrameric, 49, 68, 133, 161
Therapeutic approach, 41, 57, 58, 159, 181, 280, 330, 337, 347
Trafficking, 33, 34, 69, 158, 159, 170, 173, 179, 194, 227, 233, 247, 329
Transcription factors, 5, 6, 11, 14, 20, 28–32, 34, 37–41, 48, 52, 55, 56, 73, 79–81, 106, 120, 129, 135, 139, 174, 190, 194, 196, 205, 212, 231, 242, 243, 246, 247, 250–252, 255, 269, 271, 274, 282, 285, 286, 290, 311, 336–339, 343, 356, 357
Transmission, 11, 12, 48, 49, 53, 54, 71, 72, 106, 126, 134, 135, 154, 155, 158, 162, 163, 169, 173, 175, 176, 181, 187, 188, 190, 198, 205, 209, 249, 282, 320, 322, 329, 345, 354, 355, 357
Tryptophan, 195
TSC1/2, 304, 358
Tubulin, 282
Tumorigenesis, 274, 275, 278, 282, 285, 290, 293, 337, 339, 343, 356, 358, 360, 361

U.S. Food and Drug Administration, 326
Ubiquitination, 17, 33–35, 39, 41, 129, 159, 212, 329, 356
Ubiquitin E3 ligase, 221
Ubiquitin ligase activity, 212
Ubiquitin ligases, 35, 137, 212
Ulcerative colitis, 120, 249, 400

Variable, 95, 96, 98, 110, 119, 156, 160, 242, 249, 310, 369, 396, 403
Vascular endothelial growth factor (VEGF), 7, 52, 203, 218, 290, 324, 363, 388
Vascular smooth muscle cells, 179, 250
Vertebrates, 3, 70, 206, 231, 238
Vesicle, 67, 131–135, 141, 166, 169, 173, 176, 179, 194, 224, 306, 395
VHL, 221
Viral infection, 102, 114, 116, 130, 236
Viral oncogene, 204, 342
Viral proteins, 102, 152
Vitamin D, 50, 224
Voltage-gated channels, 50, 69

Wavelength, 85, 86, 99
Waves, 83, 259, 346
Wnt signaling, 3, 14, 39, 203, 205, 206, 212, 213, 218, 278–282, 307, 312, 336, 355, 356
Wound, 16, 17, 31, 115, 119, 204

X-linked inhibitor of apoptosis protein (XIAP), 389
X-ray crystallography, 83, 84, 161, 386

Yeast, 17, 69

ZAP70, 240
Zinc 49, 68, 245